Cordilleran Section
of the
Geological Society of America

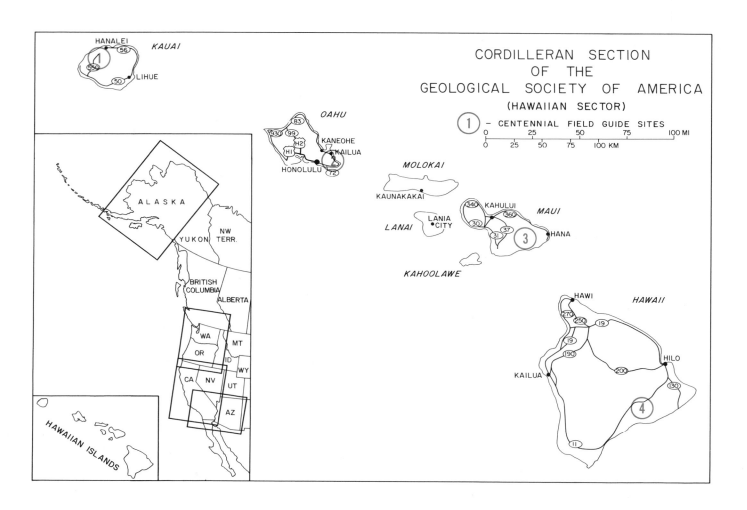

CORDILLERAN SECTION
OF THE
GEOLOGICAL SOCIETY OF AMERICA
(HAWAIIAN SECTOR)

(1) – CENTENNIAL FIELD GUIDE SITES

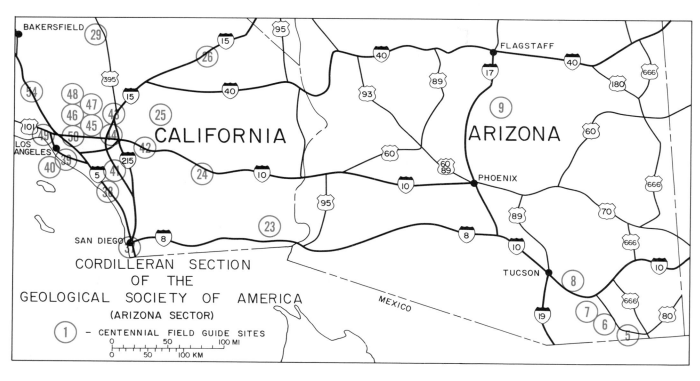

CORDILLERAN SECTION
OF THE
GEOLOGICAL SOCIETY OF AMERICA
(ARIZONA SECTOR)

(1) – CENTENNIAL FIELD GUIDE SITES

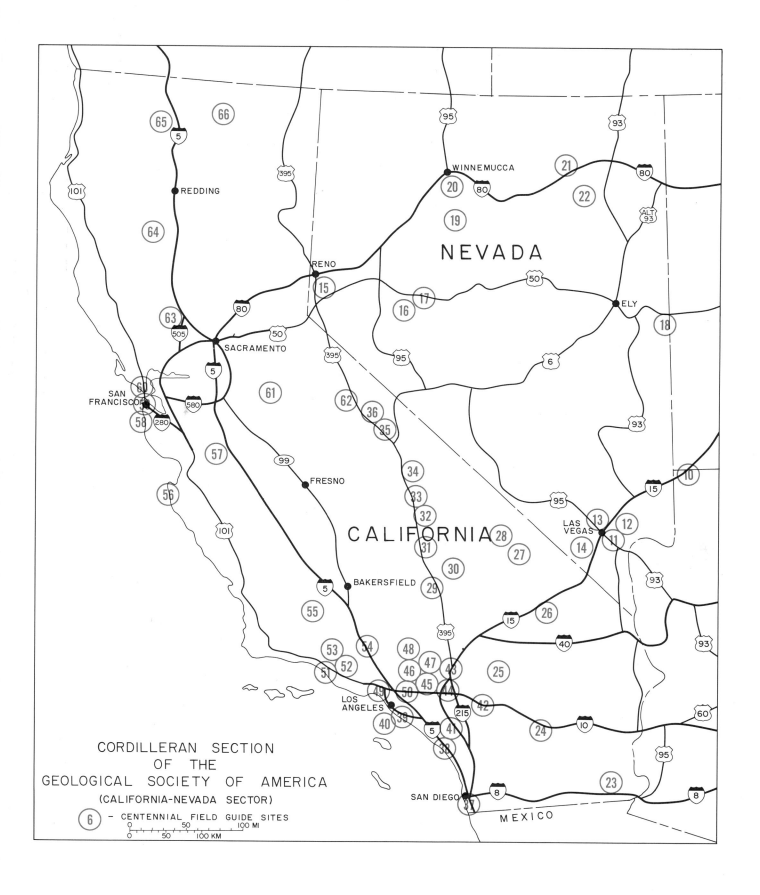

CORDILLERAN SECTION
OF THE
GEOLOGICAL SOCIETY OF AMERICA
(CALIFORNIA-NEVADA SECTOR)

⑥ - CENTENNIAL FIELD GUIDE SITES

0   50   100 MI
0   50   100 KM

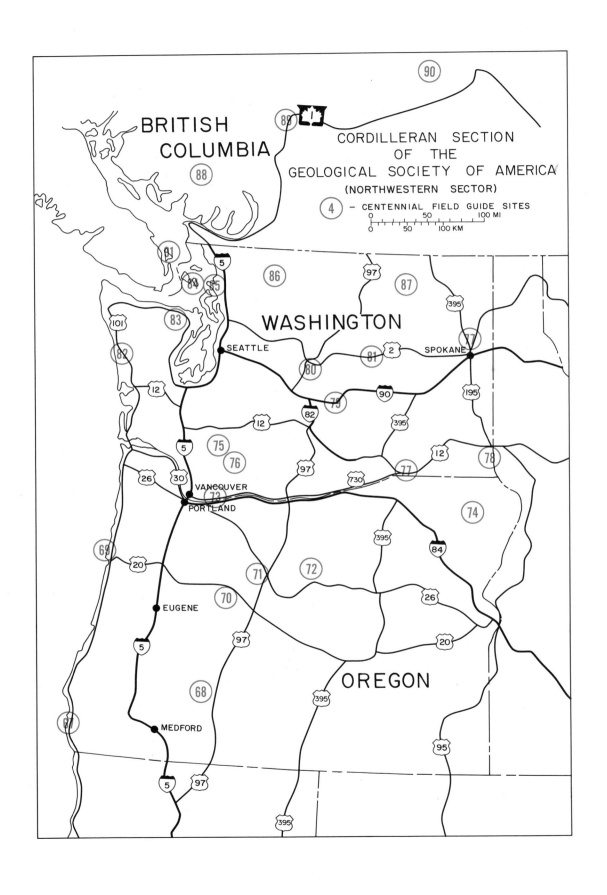

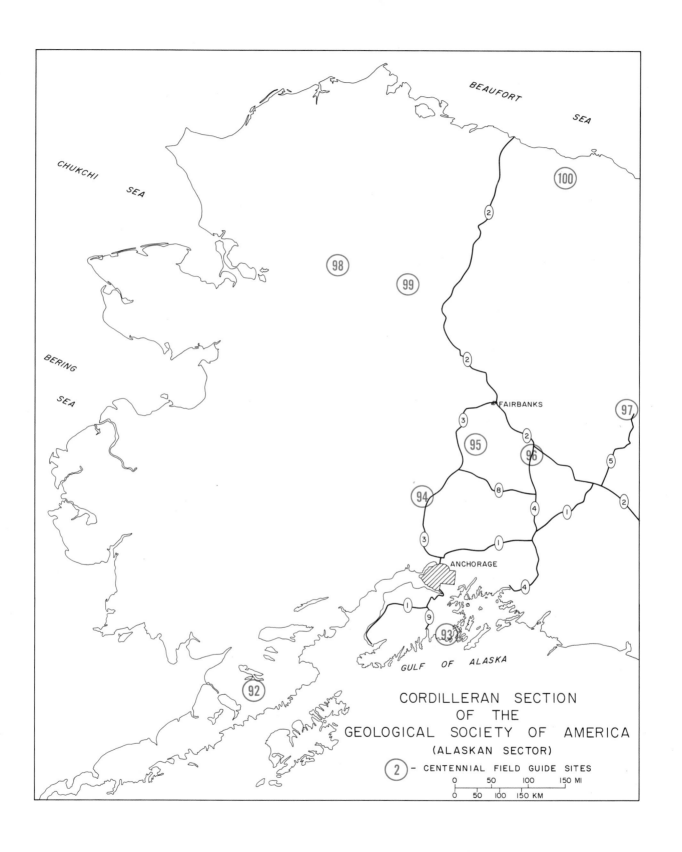

CHUKCHI SEA

BEAUFORT SEA

BERING SEA

100

98

99

2

2

FAIRBANKS

3

95

2

96

97

5

94

8

4

1

2

3

1

ANCHORAGE

4

1

9

93

4

92

GULF OF ALASKA

CORDILLERAN SECTION
OF THE
GEOLOGICAL SOCIETY OF AMERICA
(ALASKAN SECTOR)

2 - CENTENNIAL FIELD GUIDE SITES

0   50   100   150 MI
0   50   100 150 KM

Front Cover: View west from over Owens Valley, with the town of Lone Pine, California, in the foreground. Scarps produced during the 1872 Owens Valley earthquake are the Owens Valley fault scarp (A), and the Lone Pine fault scarp (B), site 33 in this volume. The Los Angeles aqueduct (C) crosses the area. The Alabama Hills in the middle ground (D) contain some of the youngest plutonic rocks in the Sierra Nevada batholith. Here, they intrude Jurassic metavolcanic rocks (E). The Sierra Nevada, with Mt. Whitney (F) (elevation 14,494 ft; 4,419 m), rise majestically as a fault-line scarp above the valley and contain plutonic rocks similar in age to the Alabama Hills in this area. Photo courtesy of Pierre Saint-Amand.

Centennial Field Guide Volume 1

Cordilleran Section
of the
Geological Society of America

Edited by

Mason L. Hill
14067 East Summit Drive
Whittier, California 90602

1987

Acknowledgment

Publication of this volume, one of the Centennial Field Guide Volumes of *The Decade of North American Geology Project* series, has been made possible by members and friends of the Geological Society of America, corporations, and government agencies through contributions to the Decade of North American Geology fund of the Geological Society of America Foundation.

Following is a list of individuals, corporations, and government agencies giving and/or pledging more than $50,000 in support of the DNAG Project:

ARCO Exploration Company
Chevron Corporation
Cities Service Company
Conoco, Inc.
Diamond Shamrock Exploration
    Corporation
Exxon Production Research Company
Getty Oil Company
Gulf Oil Exploration and Production
    Company
Paul V. Hoovler
Kennecott Minerals Company
Kerr McGee Corporation
Marathon Oil Company
McMoRan Oil and Gas Company
Mobil Oil Corporation
Pennzoil Exploration and Production
    Company

Phillips Petroleum Company
Shell Oil Company
Caswell Silver
Sohio Petroleum Corporation
Standard Oil Company of Indiana
Sun Exploration and Production Company
Superior Oil Company
Tenneco Oil Company
Texaco, Inc.
Union Oil Company of California
Union Pacific Corporation and
    its operating companies:
        Champlin Petroleum Company
        Missouri Pacific Railroad Companies
        Rocky Mountain Energy Company
        Union Pacific Railroad Companies
        Upland Industries Corporation
U.S. Department of Energy

Published by the Geological Society of America, Inc.
3300 Penrose Place, P.O. Box 9140, Boulder, Colorado 80301

Printed in U.S.A.

Library of Congress Cataloging-in-Publication Data
(Revised for vol. 1)

Centennial field guide.

"Prepared under the auspices of the regional
Sections of the Geological Society of America as
a part of the Decade of North American Geology
(DNAG) Project"—V. 6, pref.
    Vol.   : maps on lining papers.
    Includes bibliographies and index.
    Contents: v. 1, Cordilleran Section of the
Geological Society of America / edited by
Mason L. Hill—   —v. 6. Southeastern Section
of the Geological Society of America / edited by
Thorton L. Neathery.
    1. Geology—United States—Guide-books. 2. Geology—
Canada—Guide-books. 3. United States—Description
and travel—1981–   —Guide-books. 4. Canada—
Description and travel—1981– Guide-books.
5. Decade of North American Geology Project.
I. Geological Society of America
QE77.C46   557.3   86-11986
ISBN 0-8137-5406-2 (v. 6)

Contents

California Desert

Southern California

Contents

Alaska

Site Distribution

Topical cross-references for Field Guide Sites

Accreted Terranes
  Alaska: 93, 94
  British Columbia: 89
  California, northern: 57, 60, 64, 65
  Oregon: 69
  Washington: 80, 83, 84, 85

Active faults
  Alaska: 93
  California, desert: 24, 29, 33, 36
  California, southern, 42, 43, 44, 48, 49, 50, 55
  Nevada: 17, 19

Archaeological Geology
  Arizona, southern: 6

Cenozoic volcanism
  Alaska: 92
  British Columbia: 88
  California, desert: 31, 35, 36
  California, northern, 61, 62
  Hawaii: 1, 2, 3, 4
  Oregon: 68, 69, 70, 71, 72, 73
  Washington: 73, 75, 76, 78, 79, 81, 83

Economic Geology
  Alaska: 94, 95, 98
  Arizona, southern: 5
  California, desert: 30, 31, 32
  California, southern: 45, 46, 49, 51
  Nevada: 15

Engineer Geology
  Alaska: 93, 96
  California, desert: 24, 25, 29, 33
  California, northern: 58
  California, southern: 40, 42, 43, 44, 45, 48, 49, 50, 55
  Nevada: 17, 19
  Washington: 75

Geomorphology
  Alaska: 92, 93, 95, 96
  British Columbia: 88
  California, desert: 23, 25, 28, 30, 31, 32, 33, 34, 36
  California, northern: 58, 61, 66
  California, southern: 37, 38, 39, 40, 43, 44, 45, 51, 53, 55
  Hawaii: 1
  Nevada: 17, 19
  Oregon: 68, 71, 73
  Washington: 73, 75, 76, 77, 78, 79, 81, 85

Landslides:
  California, desert: 25
  California, northern: 58
  California, southern: 40, 45, 49
  Nevada: 11

Metamorphic and igneous rocks
  Alaska: 92, 93, 94, 98
  Arizona, southern: 5, 8
  British Columbia: 85, 89, 90
  California, desert: 26, 34, 35, 36
  California, northern: 57, 60, 61, 62, 64, 65
  California, southern: 40, 41, 46
  Oregon: 68, 69, 70, 71, 72, 74
  Washington: 76, 78, 79, 80, 83, 84, 85, 86, 87

Stratigraphy and sedimentology
  Alaska: 92, 93, 94, 95, 96, 97, 98, 99, 100
  Arizona, southern: 6, 7, 9
  Arizona, northwestern: 10
  British Columbia: 88, 91
  California, desert: 23, 24, 26, 27, 30, 32, 35
  California, northern: 56, 58, 59, 60, 61, 63, 64
  California, southern: 37, 38, 39, 40, 43, 44, 45, 47, 48, 49, 51, 52, 53, 54, 55
  Nevada: 11, 12, 13, 14, 16, 20
  Oregon: 67, 69, 72
  Washington: 76, 77, 78, 80, 81, 82, 83 84

Structural Geology
  Alaska: 93, 94, 98, 99, 100
  Arizona, southern: 7, 8
  British Columbia: 89, 90
  California, desert: 24, 26, 27, 28, 29, 33
  California, northern: 60, 62, 64
  California, southern: 38, 39, 42, 43, 44, 47, 48, 49, 50, 51, 53, 54, 55
  Nevada: 11, 13, 16, 17, 18, 19, 20, 21, 22
  Oregon: 73
  Washington: 73, 79, 80, 82, 83, 84, 85, 86, 87

Preface

The Cordilleran Section of the Geological Society of America, which includes southern Arizona, California, Nevada, Oregon, Washington, Alaska, Hawaii, and the Province of British Columbia in Canada, encompasses some of the most complex geology on the continent.

This volume is one of a six-volume set of Centennial Field Guides prepared under the auspices of the regional Sections of the Society as a part of the Decade of North American Geology (DNAG) Project. The intent of this volume is to highlight, for the geologic traveler and for students and professional geologists interested in major geologic features of regional significance, 100 of the best and most accessible geologic localities in the area of the Section. The leadership provided by the editor, Mason L. Hill, and the support provided to him by the Cordilleran Section of the Geological Society of America are greatly appreciated.

Drafting services were offered by the DNAG Project to those authors of field guide texts who did not have access to drafting facilities. Particular thanks are given here to Ms. Karen Canfield of Louisville, Colorado, who prepared final drafting of many figures from copy provided by the authors.

In addition to Centennial Field Guides, the DNAG Project includes a 29 volume set of syntheses that constitute *The Geology of North America,* and 8 wall maps at a scale of 1:5,000,000 that summarize the geology, tectonics, magnetic and gravity anomaly patterns, regional stress fields, thermal aspects, seismicity, and neotectonics of North America and its surroundings. Together, the synthesis volumes and maps are the first coordinated effort to integrate all available knowledge about the geology and geophysics of a crustal plate on a regional scale. They are supplemented, as a part of the DNAG project, by 23 Continent–Ocean Transects providing strip maps and both geologic and tectonic cross-sections strategically sited around the margins of the continent, and by several related topical volumes.

The products of the DNAG Project have been prepared as a part of the celebration of the Centennial of the Geological Society of America. They present the state of knowledge of the geology and geophysics of North America in the 1980s, and they point the way toward work to be done in the decades ahead.

Allison R. Palmer
Centennial Science Program Coordinator

Foreword

This Field Guide project began by invitation to the Fellows and Members of the Cordilleran Section to identify field sites of special geologic interest and significance in the area of the Section from which 100 of the best would be selected. Some 350 sites had been proposed by 1984 from which 114 were selected by a committee of geologists with a wide range of geologic expertise and areal experience, and authors were invited to submit texts with target lengths of two and four book pages, and not to exceed six book pages. Not all those invited agreed to participate or followed through on preparation of manuscripts; therefore other sites and authors continued to be obtained until, by January 1987, all of the required 100 manuscripts were in hand.

An effort was made to have an appropriate geographic spread and geologic diversity for the Field Guide sites. Obviously, the distribution of sites and balance of topics is irregular; the preponderance of sites in California may be justified by the larger number of geologists in, and the greater number of visiting geologists to, this State.

The papers are arranged geographically by States (although California is divided into three regions), and a topical cross reference tabulation is included. The intent is that this Field Guide will be useful to those visiting geologists who are unfamiliar with the areas described.

As coordinator for this DNAG project, I accept responsibility for any shortcomings of this Field Guide. I gratefully acknowledge the volunteer or invited help of J. E. Allen, M. C. Blake, Jr., J. C. Crowell, G. D. Davis, H. D. Drewes, R. W. Hazlett, W. H. Mathews, R. Moberly, C. G. Mull, P. D. Snavely, A. G. Sylvester, B. W. Troxel, and R. E. Wallace for the sites: M. L. Stout for special help at the beginning and end of the project; Jean MacKay of Pomona College Geology Department; my wife, Marie, and most especially for the patience and expertise of A. R. (Pete) Palmer for making it all come together.

Mason L. Hill
Editor, January 1987

Waimea Canyon, Kauai, Hawaii

David A. Clague and Wendy Bohrson, U.S. Geological Survey, 345 Middlefield Road, Menlo Park, California 94025

LOCATION

Kauai is the westernmost of the principal Hawaiian Islands. Figure 1 shows the road guide to Waimea Canyon starting in the town of Kekaha and ending in the town of Waimea. The entire trip can be made in a passenger car, with the exception of the optional hike down the Kukui trail. The road stops are at major viewpoints that have parking areas and facilities. Do not reverse the order of the starting and stopping points, since the road between the town of Waimea and the intersection of Waimea Canyon Road and Hawaii 55 is so steep that small passenger cars may overheat going uphill. The entire roadlog is just over 40 mi (64 km) long and can be completed in about 3 hours, including relatively brief stops; allow more time for longer stops. The starting point in the town of Kekaha is roughly 45 mi (72 km) and 1 hour from Lihue. The optional hike down the Kukui trail to the floor of Waimea Canyon can be completed in about 3 to 4 hours; the trail is steep, strenuous, and slippery when wet. Those interested in the hike may wish to spend a full day to allow time to explore the canyon floor.

SIGNIFICANCE OF SITE

Hawaiian volcanoes commonly evolve through a sequence of eruptive stages apparently beginning with the submarine eruption of small volumes of alkalic lava and then large volumes of tholeiitic lava. As the volcano emerges above sea level, the eruption of voluminous tholeiitic lava forms a shield volcano. During or at the end of the shield stage, a caldera forms, which is then filled with tholeiitic and/or alkalic lava. The alkalic postshield stage consists of the eruption of alkalic basalt and associated differentiated lava, which forms a thin veneer capping the shield volcano. A long period of erosion and volcanic quiescence is sometimes interrupted by small-volume eruptions of strongly alkalic lava during the alkalic rejuvenated stage. On Kauai, the lava that formed the shield volcano, ponded within the caldera, overflowed the caldera and ponded in a structural depression, and erupted during the rejuvenated stage is well exposed. The excellent exposures occur in Waimea Canyon where deep erosion has revealed the structural relations of these rock units. This field guide to Waimea Canyon will trace the geologic evolution of Kauai through these eruptive stages.

SITE INFORMATION

The island of Kauai consists of a single large shield volcano with a (now filled) summit caldera 10 to 12 mi (16 to 19 km) across. The Waimea Canyon Basalt has been subdivided into four members, which consist of tholeiitic basalt, olivine tholeiitic basalt, and picritic tholeiitic basalt. The Napali Member of the Waimea Canyon Basalt consists of the shield stage lava, which erupted about 5.1 m.y. ago. These flows dip outward 6 to 12°

from the center of the island and are generally between 3 and 15 ft (1 and 4.5 m) thick. In Waimea Canyon, about 80 percent of the exposed section is pahoehoe, whereas the remaining 20 percent is aa.

Kauai, unlike most Hawaiian volcanoes, has no well-defined rift zones; dikes radiate from the summit caldera in all directions but are more concentrated trending northeastward and west-southwestward from the summit. The Olokele and Makaweli Members of the Waimea Canyon Basalt consist of tholeiitic lava of the caldera collapse phase that filled the summit caldera and a younger 4-mile- (6.4-km)-wide graben on the south flank, respectively. Two other calderas formed on the flanks of the Kauai shield volcano: the Lihue Depression (7 to 10 mi; 11.2 to 16 km across), which was apparently not filled by tholeiitic lava, and the Haupu caldera in the southeast corner of the island (roughly 2 mi; 3.2 km across), which was filled with thick, ponded lava of the Haupu Member of the Waimea Canyon Basalt. These are the only flank calderas known in the Hawaiian Islands. The graben containing the Makaweli Member of the Waimea Canyon Basalt is also the only such graben known in the Hawaiian Islands.

The alkalic postshield stage is only poorly developed on Kauai, but several of the youngest lavas in the Olokele and Makaweli Members of the Waimea Canyon Basalt are hawaiite. Eruption of this differentiated postshield stage alkalic lava was followed by about 3 m.y. of volcanic quiescence and erosion. Finally, the alkalic rejuvenated stage Koloa Volcanics erupted from at least 40 vents concentrated on the south and east flanks of the shield. The widely scattered vents erupted between 2.6 and 0.5 m.y. ago. The lava, which contains xenoliths of spinel lherzolite and dunite, ranges from alkalic basalt, basanite, and nephelinite, to melilitite. The abundant vents located along the southeast coast erupted mainly alkalic basalt.

ROADLOG

Mi (km)

0 (0) Intersection of Hawaii 50 and 550 in Kekaha. Turn inland at sign to Waimea Canyon. The coastal plain beneath Kekaha consists of uplifted lagoonal deposits.

1.1 (1.8) Thin bedded flows of tholeiitic basalt and olivine tholeiitic basalt of the Napali Member of Waimea Canyon Basalt. These thin-bedded flows dip 6 to 12° radially outward from the center of the island and erupted during the shield stage about 5.1 m.y. ago. The cliff separating the coastal plain from the lava exposures is an ancient seacliff eroded at the time the lagoonal deposits formed.

3.7 (5.9) The roadway climbs up onto the deeply weathered original flow surface. Sugar cane is grown in the thick soils developed on this dip-slope.

5.1 (8.2) Good example of tropical spheroidal weathering in a

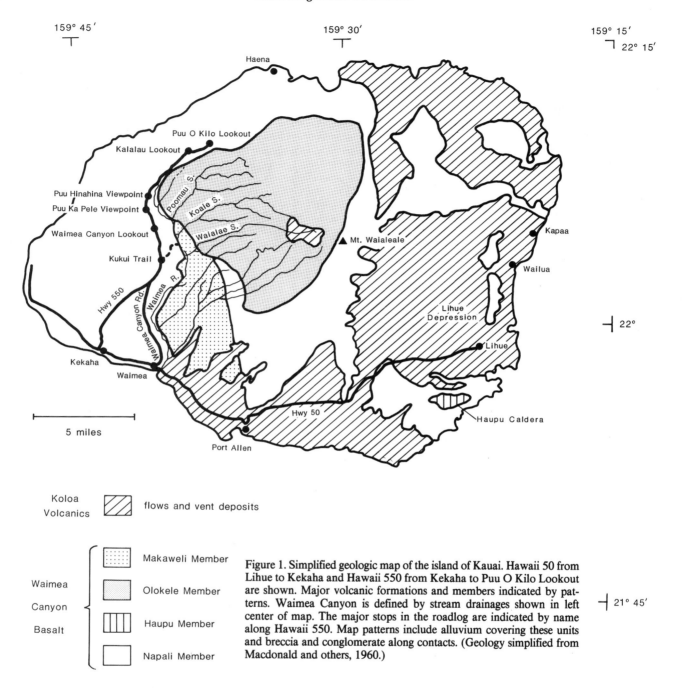

Figure 1. Simplified geologic map of the island of Kauai. Hawaii 50 from Lihue to Kekaha and Hawaii 550 from Kekaha to Puu O Kilo Lookout are shown. Major volcanic formations and members indicated by patterns. Waimea Canyon is defined by stream drainages shown in left center of map. The major stops in the roadlog are indicated by name along Hawaii 550. Map patterns include alluvium covering these units and breccia and conglomerate along contacts. (Geology simplified from Macdonald and others, 1960.)

thick aa flow of the Napali Member of the Waimea Canyon Basalt. East side of road.

7.6 (12.2) Intersection of Hawaii 550 and Waimea Canyon Road. The return trip is down Waimea Canyon Road.

9.6 (15.4) Kukui trailhead. Stop on return trip.

10.15 (16.2) Dike on east side of road. Other examples occur at 10.25, 10.5, and 11.15 mi (16.4, 16.8, and 17.8 km), all on the east side of the road. These dikes are part of the poorly defined dike swarm that radiates from the central caldera of Kauai. Most

other Hawaiian volcanoes have well-defined dike swarms aligned along the prominent rift zones. The dikes were feeders for overlying fissure eruptions.

11.2 (17.9) Turnout to Waimea Canyon Lookout. Short walk from parking area to lookout. The scenic view from this lookout may be photographed most effectively in the late afternoon or early morning when the shadows are deep, although early morning is better because of late afternoon rains. The photomosaic in Figure 2 outlines the geology of the Waimea Canyon Basalt. The

Waimea Canyon Basalt

M Makaweli Member
O Olokele Member
N Napali Member

Figure 2. Photomosaic taken from Waimea Canyon Lookout showing major volcanic members of Waimea Canyon Basalt. Right edge is looking southeast and left edge is looking north-northeast.

Napali Member of the Waimea Canyon Basalt is located to the west of the canyon. Numerous dikes cross-cut these outward-dipping flows. To the east and northeast across the canyon are thick, ponded flows of the Olokele Member of the Waimea Canyon Basalt. These flows ponded within the summit caldera and are commonly 50 ft (15 m) thick and flat-lying. To the east and southeast across the canyon are thick, ponded flows of the Makaweli Member of the Waimea Canyon Basalt. These flows postdate those filling the caldera and spilled out of the caldera and ponded within a structural graben on the south flank of the shield. The lava of the Makaweli Member of the Waimea Canyon Basalt dips gently to the south. Looking east-southeast, one can see the contact between the higher elevation Olokele Member of the Waimea Canyon Basalt to the north and the lower elevation Makaweli Member of the Waimea Canyon Basalt to the south.

11.9 (19) Additional scenic stop with small parking pullout on east side of road.

12.2 (19.5) Additional scenic stop with small parking pullout on east side of road. Puu Ka Pele is the small hill immediately to the south-southeast. It is a small pit crater filled with thick, ponded lava flows that are more resistant to weathering than the thin-bedded surrounding flows.

13.0 (20.8) Scenic stop with parking along east side of road. Thick, ponded lava of the Olokele Member of the Waimea Canyon Basalt is located east of the canyon, whereas thin, outward-dipping flows of the Napali Member of the Waimea Canyon Basalt are located west of the canyon.

14.1 (22.6) Puu Ka Pele viewpoint. Similar view to that at previous stop.

14.9 (23.8) Turnout to Puu Hinahina Viewpoint. There are two lookout points here. Lookout no. 1 is a view of Waimea Canyon looking southeast down the canyon. Olokele and Napali Members of Waimea Canyon Basalt occur on east and west sides of the canyon, respectively. The base of the canyon roughly fol-

lows the contact, which defines the caldera-bounding fault. In the distance to the southeast, the lower elevation block is the Makaweli Member of the Waimea Canyon Basalt. Lookout no. 2 provides a view of Niihau Island where the eroded tholeiitic shield (Paniau Basalt) and the low-lying plains of rejuvenated stage alkalic basalt (Kiekie Basalt) can be seen on a clear day. The cone north of Niihau is Lehua Island, which is a rejuvenated stage tuff cone. A similar tuff cone occurs at the southern point of Niihau Island.

16.7 (26.7) Kokee Lodge turnout on left. Continue on main road. The grassy valley to the northwest of the road marks the boundary fault of the ancient caldera and is the contact between the Napali and Olokele Members of the Waimea Canyon Basalt. Caldera-filling lava of the Olokele Member of the Waimea Canyon Basalt is located to the southeast, whereas the thin, dipping tholeiitic shield flows of the Napali Member of the Waimea Canyon Basalt are located to the northwest.

19.35 (31) Kalalau Lookout. Bear left into parking area. Thin-bedded flows of the Napali Member of the Waimea Canyon Basalt have been eroded by heavy runoff, forming "knife-edge" ridges. Some of these ridges are aligned along more resistant dikes. The amphitheatre shape of the valley is similar to the heads of all the larger canyons. The steep slopes plunging to the sea are wave-eroded cliffs. The valley is accessible by trail from the northeast end of Hawaii 56 near Haena. Make left turn coming out of parking area and continue toward northeast.

20.55 (32.9) Puu O Kilo Lookout. Another scenic view into Kalalau Valley. End of the road. Return along same road.

31.0 (49.6) Kukui Trailhead. A short walk of perhaps 660 ft (200 m) along a dirt trail takes one to a viewpoint into the canyon. The thin, outward-dipping flows directly beneath the viewpoint are shield-stage tholeiitic lava of the Napali Member of the Waimea Canyon Basalt. Across the canyon to the east, the lower block consists of lava ponded in the Makaweli graben, whereas the more distant higher elevation area consists of lava ponded inside the caldera. The Makaweli graben formed after the caldera had been filled to overflowing. Lava ponded in the graben generally erupted within the caldera and flowed southward into the graben. The brown ridge just west of the canyon floor is a conglomerate unit deposited by the ancestral Waimea River. The conglomerate is capped by an alkalic basalt flow of the rejuvenated stage Koloa Volcanics. This flow apparently originated within the boundaries of the caldera and flowed down the canyon, covering the conglomerate. It was probably at this time that the Waimea River changed course and began eroding the present river gorge. The flow of rejuvenated stage alkalic basalt is at an elevation of 1,840 ft (560 m), the present bed of the Waimea River below this area is at 640 ft (195 m), and the rim of the canyon is at about 2,700 ft (823 m). Given roughly 5 m.y. to erode the canyon, we estimate rates of erosion of about 410 ft/m.y. (125 m/m.y.) (5 in or 12.5 cm/1,000 yr). Using this erosion rate, we estimate that the age of the perched stream conglomerate and capping alkalic basalt flow of the Koloa Volcanics is about 2.9 m.y. This estimated age is slightly older than the oldest dated Koloa Volcanics flow, a nephelinite erupted 2.6 m.y. ago within the caldera.

The optional hike down the Kukui trail allows the opportunity to examine the flows of the Napali and Makaweli Members of the Waimea Canyon Basalt, as well as the conglomerate and overlying alkalic basalt flow of the Koloa Volcanics. Cobbles in the conglomerate include lava from the Koloa Volcanics, as well as cobbles of tholeiitic basalt presumably from the Napali, Olokele, and Makaweli Members of the Waimea Canyon Basalt. Figure 3 shows the trail and a simplified geologic map.

32.95 (52.7) Intersection of Hawaii 550 and Waimea Canyon Road. Take left fork onto Waimea Canyon Road. The road is steep, curved, and rough, so drive accordingly.

36.1 (57.8) Scenic stop with parking on east side of road.

37.4 (59.8) Scenic stop with view of Olokele Canyon to northeast. Looking southeast, the areas covered with sugar cane are rejuvenated stage flows of the Koloa Volcanics.

38.9 (62.2) Steep grade down the ancient seacliff.

39.7 (63.5) Intersection of Waimea Canyon Road and Hawaii 50. Turn left into town of Waimea.

40.3 (64.5) A typical flow of alkalic basalt of the Koloa Volcanics is exposed in the roadcut just past the bridge across the Waimea river and on the north side of the road.

EPILOGUE

The volcanic evolution of Kauai appears to be complete, although its geologic evolution continues with erosion and construction of fringing reefs. Farther west, along the Hawaiian-Emperor volcanic chain, we can see what Kauai will become in the future. Twelve-m.y.-old Gardner Pinnacles is the oldest rocky island exposed above sea level. Volcanoes older than and to the west of Gardner Pinnacles are coral atolls, such as Midway Island, and finally drowned atolls or guyots. Kauai, like the Hawaiian volcanoes before it, will eventually be reduced to a tiny rocky islet, then an atoll, and finally a guyot.

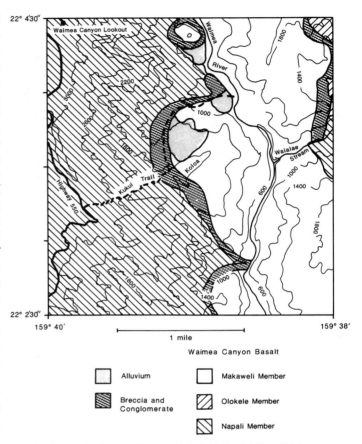

Waimea Canyon Basalt		

▦	Alluvium	▢	Makaweli Member
▨	Breccia and Conglomerate	▨	Olokele Member
		▨	Napali Member

Figure 3. Detailed map of optional Kukui Trail hike; 400-ft contours shown. Unit along lower half of trail is mapped as fault breccia bounding Makaweli graben. More recent work indicates this unit is a conglomerate capped by a rejuvenated stage flow of alkalic basalt of Koloa Volcanics (indicated on map by "Koloa"). The extent of this flow is uncertain; it is also unknown if other breccia units shown are conglomerate deposited by an earlier Waimea River. (Geology from Macdonald and others, 1960; discussed in the text.)

REFERENCES CITED

Clague, D. A., and Dalrymple, G. B., 1986, Age and petrology of the Koloa Volcanics, Kauai, Hawaii: Contributions to Mineralogy and Petrology (in press).

—— , 1986, The geology of the Hawaiian-Emperor Volcanic Chain: U.S. Geological Survey Professional Paper 1350 (in press).

Feigenson, M. D., 1984, Geochemistry of Kauai volcanics and a mixing model for the origin of Hawaii alkalic basalts: Contributions to Mineralogy and Petrology, v. 87, p. 109–119.

Macdonald, G. A., Davis, D. A., and Cox, D. C., 1960, Geology and groundwater resources of the island of Kauai, Hawaii: Hawaii Division of Hydrography Bulletin 13, 212 p. (includes geologic map of the island).

Macdonald, G. A., Abbott, A. A., and Peterson, F. L., 1983, Volcanoes in the Sea: Honolulu, University of Hawaii Press, 517 p.

McDougall, I., 1979, Age of shield-building volcanism on Kauai and linear migration of volcanism in the Hawaiian Islands: Earth and Planetary Science Letters, v. 46, p. 31–42.

Stearns, H. T., 1966, Road Guide to Points of Geologic Interest in the Hawaiian Islands: Palo Alto, California, Pacific Books.

Coastal and volcanic geology of the Hanauma Bay area, Oahu, Hawaii

Ralph Moberly and George P. L. Walker, Hawaii Institute of Geophysics, University of Hawaii, Honolulu, Hawaii 96822

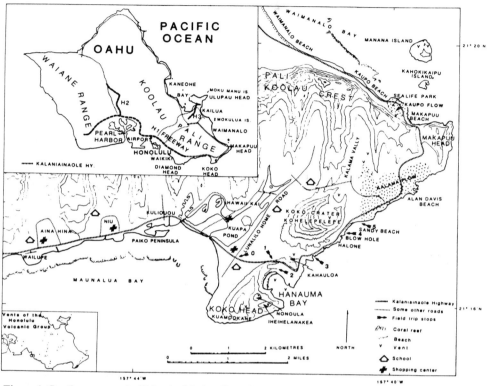

Figure 1. Southeastern end of island of Oahu, Hawaii, showing volcanic and coastal features and field guide stops in the Hanauma Bay-Koko Fissure area. Topographic contours at 200 ft (about 60 m). Upper inset: Kalanianaole Highway and other main roads of Oahu. Lower inset: Vents of flows, tuff cones, and cinder cones of Honolulu Volcanic Group (Quaternary).

LOCATION AND ACCESSIBILITY

The Hanauma Bay–Koko Fissure area occupies the southeastern tip of the island of Oahu, Hawaii (Fig. 1), and is shown on the Koko Head 7½-minute Quadrangle. The Geologic Map of Oahu (1:62,500; Stearns, 1939) is out of print. A 1:500,000 colored geologic map of the state and generalized chart shows the Honolulu and Koolau volcanic groups relative to other rock units (Bennison, 1974).

Access is easy by automobile. Buses of No. 1 route, marked "Lunalilo Home Road," give frequent service from Honolulu to the Lunalilo Home Road point described below as mileage point zero. Buses of No. 58 route travel from Waikiki to that point. Traffic on the winding, narrow highway between stops 2 and 4 demands caution as does high surf on the seacliff. The area is within a park offering generally free access.

SIGNIFICANCE

Southeastern Oahu was the site of sporadic volcanism in the past 0.5 m.y. Nowhere else in the United States are Surtseyan eruptive products and structures and a reef of the Indo-Pacific biogeographic realm as accessible and clearly exposed as in the area between Koko Head and Makapuu Head. This field guide is based on five closely adjacent stops, each at a parking area off Kalanianaole Highway.

Approach. From the end of Freeway H-1, travel 5.3 mi (8.5 km) southeast on Hawaii 72 to Lunalilo Home Road. The intersection can be identified by a stop light and the only pedestrian overpass spanning the highway. Ahead are the gullied crater slopes. Mileage is given along the highway and does not include side trips for parking.

Mile 0 (0 km). Intersection Hawaii 72 and Lunalilo Home Road. Koko Crater ahead on left. Continue southeast on Hawaii 72, climbing the flanks of Koko Head.

Mile 0.6 (1.0 km) Stop 1. Scenic Point parking area on left. General geologic setting of southeast Oahu. (Stearns and Vaksvik, 1935; Macdonald and others, 1983; Stearns, 1985).

Oahu is built of two large basaltic shield volcanoes, now

deeply eroded and partly covered by soil, younger volcanic rocks, and marine and non-marine sedimentary rocks. Remnants of the western and older shield make the Waianae Range. Koolau volcano was active at 2.3 Ma; its lavas lap against Waianae. After much erosion and isostatic sinking, volcanism was renewed across southeastern Oahu as the Honolulu Volcanic Group, which erupted sporadically from about 3 dozen vents over the past 0.5 m.y. (Fig. 1 inset).

Clockwise, the western skyline shows Diamond Head, Kaimuki Hill, and the Koolau Mountains (Fig. 2). Diamond Head is a tuff cone or ring of the Honolulu Group. Eruptions near sea level are usually explosive due to stream generated from the contact of magma with water, as described for stop 2. Presumably, strong tradewinds were blowing when Diamond Head erupted, building the down-wind southwest rim highest. The broad hill of Kaimuki is a small shield volcano formed of numerous thin flows from a Honolulu Group vent. Magmas of the Honolulu Volcanic Group are characteristically poorer in silica and richer in alkalis, and usually also calcium and magnesium, than most basalts. Nephelene (and melilite) may be present, giving basanites and nephelinites as typical lavas.

On the right hand is the leeward slope of the Koolau shield, built of countless flows of tholeiitic basalt, silica-saturated and rich in iron and magnesium but poor in alkalies. Unlike many Hawaiian shield volcanoes, Koolau does not have an alkali basalt cap. The smaller valleys are V-shaped, but the larger ones have steep sides, flat floors, and rounded, amphitheater-shaped heads. The flat, seaward-sloping upland ridges between valleys are planezes, triangular-shaped remnants parallel to the original volcanic slope. The steep valley walls show outcrops of the Koolau flows, dipping about 15° to the south (Hinds, 1931).

The farthest shoreline of Maunalua Bay is a small peninsula of basanitoid lava near Diamond Head. Nearer, the coastal plain of raised reef limestone interfingering with alluvium and ash is fringed by a reef that is 1,650 ft (500 m) wide. Ancient Hawaiian fishponds, stone-walled enclosures built on the reef flats at coastal springs or stream mouths, were filled in and subdivided at Wailupe and Niu (Fig. 1). Paiko Peninsula is a spit.

Reefs and beaches are strongly affected by the wave climate and recency of volcanism (Moberly, 1983). Tidal range is about 3.3 ft (1 m). The largest Hawaiian beaches are on the west sides of the islands, shaped mainly by waves refracted from north Pacific swell. The sand is almost entirely calcareous and well sorted, and these western beaches show marked seasonal changes. Benthic marine life is abundant, but reefs are either absent or are deep and without distinct reef flats and algal ridges. Northeast-facing coasts, dominated by tradewind-driven waves, have fringing reefs of moderate width and shallowness, interrupted by stream-mouth bays. Beaches are narrow; their sand is dominantly calcareous, fine grained, and moderately sorted. Southern coasts receive low, long-period swell generated in the southern hemisphere winter in the southern Pacific ocean. Reefs are generally shallowest and widest off the south shores. Some detrital sand and gravel is mixed in the poorly sorted beach and reef-veneering

Figure 2. View to west from stop 1, showing Diamond Head tuff cone (DH) and Kaimuki lava shield (K), the dissected slopes of Koolau shield volcano at right, a narrow coastal plain underlain by alluvium and reef limestone, and the active fringing reef (R). Lagoon (L) behind barrier beach is now the Hawaii Kai marina.

calcareous sediment. The reef and narrow beaches in sight are of this third type. The modal sediment is coarse sand or fine gravel. Fragments of red algae predominate, followed by moderate amounts of mollusk, foraminifer, and *Halimeda* components, with lesser amounts of basalt, coral, and echinoid components (Moberly and others, 1965; Moberly, 1968).

Hawaii Kai, a marina community planned by the late Henry J. Kaiser, surrounds the remnants of Kuapa Pond. After the Koko eruptions, a lagoon formed behind a bay-head barrier built across the sharp angle between the east-trending coast of the Koolau range and the southwest-trending coasts of Koko Crater and Koko Head.

Mile 0.7 (1.1 km) Stop 2. Side road to right to Hanauma Bay. Interior of tuff cones; reef and reef organisms.

Hanauma Bay is in a state park. No collecting of rocks, sea life, plants, or other natural things is permitted. Since 1967, when fishing and coral collection were prohibited, the marine life has flourished. The bay is in overlapping craters breached by the sea (Wentworth, 1926). The view from the rim shows a pocket beach protected by a fringing reef about 330 ft (100 m) wide. Sandy patches and channels are light blue. A low bench or terrace lines the shore. To the south lie other craters, also sources of the ash that built the Koko Head tuff cone.

The cratered tuff cones of Hanauma Bay result from a hydromagmatic (phreatomagmatic) eruption of Surtseyan type; the vents were in the sea, and copious amounts of water entered them during the eruption. The uprising magma was quenched and finely fragmented in contact with the water. The resulting tuff is predominantly composed of sub-millimetre sized ash. Lapilli, bombs, and blocks occasionally exceed 4 in (10 cm) in size (Fig. 3). Bombs represent fragments of the newly-erupted basanitic magma. They are gray to black and are identified by their ragged shape. Inconspicuous vesicles and small olivine crystals occur. Blocks are pieces of pre-existing rock broken off the vent walls and flung out by the explosions. They include Koolau basalts, many of which are highly vesicular, and reef limestone, and occasional pieces of tuff from an earlier hydrovolcanic tuff cone in this vicinity, no in situ outcrops of which have been identified.

Eruptions of the kind which generated Hanauma Bay have often been witnessed. One gave birth to the island of Surtsey in

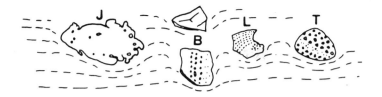

Figure 3. Typical shapes of the larger ejecta in the tuffs of Hanauma Bay: bomb (J), and lithic blocks of Koolau basalt (B), reef limestone (L), and an older hydromagmatic tuff (T).

the sea off Iceland in 1963 (Thorarinsson and others, 1964; Thorarinsson, 1967). During its most vigorous activity, explosions every few seconds to few minutes sent heavily ash-laden jets 1,000 to 3,000 ft (300 to 1000 m) into the air. Many of these jets generated small base surges when they collapsed. During the four years of the Surtsey eruption, the active vent shifted several times on Surtsey itself to produce overlapping craters, and also shifted from Surtsey to other points along about 1.9 mi (3 km) of fissure. The centers of activity in Hanauma Bay along the 7.4-mi (12-km) rift from a submarine cone 1.9 mi (3 km) southwest of Koko Head, northeast to Manana Island, could similarly have developed during several years of activity.

The tuff contains palagonite, a hydration product of basaltic glass that forms a hydrogel which crystallizes to a microfibrous texture and orange-brown color; its first traces were found on Surtsey 6 years after the eruption began (Jakobsson, 1978). Besides the gain of water, palagonitization involves a considerable loss of calcium, potassium, sodium, magnesium, and silicon; some of these removed constituents crystallize as zeolites and opal in the pore spaces, effectively lithifying the tuff (Hay and Iijima, 1968; Furnes, 1974).

Angular discordances at many places in the Hanauma Bay crater walls dip steeply inward and truncate bedded tuffs that dip inward at some places or outward at others. Slumping of loose ashes into the crater is common in hydrovolcanic eruptions, particularly when the active vent shifts position and the explosions then undermine part of the crater wall. The discordance planes are locally gullied (Fig. 4). Late in the eruption, mild fire-fountaining took place from two fissures on the north side of the bay (Fig. 5). Loose ash slumped into and widened the fissures, which then filled with spatter that locally coalesced and flowed. The Hanauma Bay craters are nested inside a somewhat earlier crater (Fig. 6).

The former Hanauma crater walls are now sea cliffs, forming as high waves and differential weathering cut notches under the friable jointed tuff, allowing blocks to fall into the sea or onto the bench bordering the shoreline, and there to be ground by the waves. The bench is planed across dipping tuff beds (Fig. 4). It rises from less than 3 ft (1 m) above mean sea level behind the reef, to 6 or 10 ft (2 or 3 m) seaward of the reef, and to more than 16 ft (5 m) along the headland points. The bench and notch do not represent a higher stand of the sea. Except behind the beach

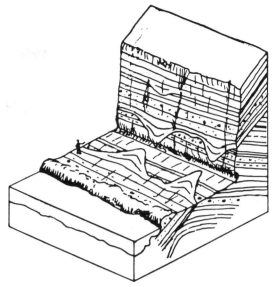

Figure 4. Block diagram to illustrate gullied discordances found at many points in the walls around Hanauma Bay. These discordances are due to slumping during the eruption, and the gullies were eroded by running water, possibly thrown out by the explosions. Figure on shore bench gives scale. The shore bench is forming under present-day conditions.

where the old notch is protected by the beach itself, today the notch is receding rapidly and the bench lowering slowly, according to surveys initiated by William Easton. Apparently the tuff that is saturated by saline groundwater is lithified firmly, but the zone above it that is alternately dry and wet from spray weathers quickly (Wentworth, 1938).

Abundant detrital olivine, eroded from the tuff, is concentrated as green beach placers. Marine habitats are numerous, ranging from tide pools to reef to sublittoral sand. Hawaiian shallow-water organisms are an impoverished part of the Indo-Pacific biogeographic realm with strongly endemic elements. A line of bore holes across the reef provided samples for paleoecologic and radiocarbon analysis. The sea breached the craters about 7,000 B.P. and grew upward at about 3 ft (1 m) per 300 years mainly in the interval 5,800 to 3,500 B.P. Since then, reef growth has been chiefly seaward, advancing about 3 ft (1 m) per 45 years (Easton and Olson, 1976).

The greatest mass contributor to the reef is calcareous red algae that secrete magnesian calcite. Living *Porolithon* and *Goniolithon* are the magenta-pink crusts at the reef edge as well as along the rocky shoreline beyond the reef. The principal corals (aragonite skeletons) are species of *Porites* and *Pocillopora*. In Hawaii the main reef coral genus of the Indo-Pacific realm, *Acropora*, is absent. Hawaiian corals were first described by the geologist James D. Dana, naturalist of the Wilkes U.S. Exploration Expedition of 1838-1842. Other carbonate-producing organisms include foraminifers and microgastropods, larger mollusks, and echinoids. The aragonitic green alga *Halimeda* is not common here, but the brown alga *Padina*, with its external film

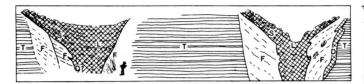

Figure 5. Fissures in the north wall of Hanauma Bay that opened late in the eruption and were widened by the slumping of loose ash (S), and from which brief fire-fountaining produced spatter (F) and small lava flows (L). A dike (D) was intruded along one boundary between undisturbed (T) and slumped ash.

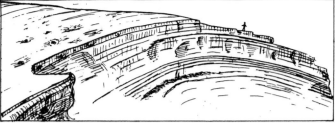

Figure 6. Antiform rim of the outer crater of Hanauma Bay 990 ft (300 m) northeast of the Toilet Bowl. The antiform grew by differential accumulation of ash, not by folding. An exhumed bedding plane is pocked with impact craters where the ash was struck by bombs and blocks.

of fine aragonite needles, is common on the reef flat and reef front. Boring filamentous blue-green algae and sponges break down the carbonate rubble, and parrotfishes eat live coral. Other brilliantly colored reef fish include surgeonfishes, wrasses, butterflyfishes, goatfishes, tangs, and damselfishes. Nearly 100 species of fish are known from Hanauma Bay (Gosline and Brock, 1976; Hobson and Chave, 1972; Doty, 1967).

The visiting geologist should walk seaward along the north shore-bench, observing the tuff, bench, reef flat, and bay shore, but avoiding overhanging or fissured blocks of the cliff. If high surf is crashing on the bench, do not go beyond the basalt exposures about 600 ft (200 m) from the beach (Fig. 5). The lava flow is a natural arch over the path now that some tuff has been eroded from under it. On days of calm seas, the next 1,300 ft (400 m) can be traversed to the gulch where the surging water is aptly but inelegantly named the Toilet Bowl, the site of several drownings. A basanitoid flow trickled down the gulch. Impact structures are common (Fig. 7).

Near the basalt arch, a swimmer equipped with face plate or goggles, snorkle, and sneakers or fins may jump into the water outside the reef, and swim and skin-dive over the fore reef, taking photographs with an underwater camera if desired. It is easy to see the three components of the reef structure: the wave-resistant framework of coral and coralline algae, the rippled fill of carbonate sand and rubble, and the binding of the fill by corals, worms, and other encrusting organisms. Floating with limbs outstretched so as to avoid banging hands and knees into coral, the swimmer can let the surge of the waves carry him across the algal ridge at the reef front. This is easiest on a rising tide in moderately calm seas, but possible in all but the lowest tides and roughest water, owing to the counterclockwise circulation in the bay. The algal ridge, with groove and buttress shape, rises 0.7 to 1.6 ft (0.2 to 0.5 m) above the general upper surfaces of the corals of the reef flat.

Pairs of properly trained scuba divers can examine the bay, out to the 100-ft (30-m) depth of the bay mouth. Strong currents, eddies and unpredictable large waves of the south side of the bay must be avoided. Even non-swimmers with a face plate, however, can enjoy the shallow sandy pockets on the reef flat, to observe the corals and abundant fish.

Continue along highway. Some tuff layers are especially rich in coral fragments, in pisolites from ash accreted in the air by raindrops, or in pyroxene crystals that weather out as coarse

black sand. Kahauloa Crater is a rifle and pistol range. Immediately beyond it the road crosses a 3-ft (1-m) thick basanitoid lava.

Mile 1.5 (2.4 km) Stop 3. Lanai Lookout Scenic Point parking area (on right). Emplacement and alteration of ash.

If the clouds allow, look east at the shields of Lanai (right), West Molokai (about 30 mi; 50 km at 105°T) and East Molokai (behind and slightly left).

The limestone-rich, yellow-brown to gray tephra around the site and in the adjacent seacliff (dangerous when waves are high!) was deposited mainly by volcanic base surges, which are short-lived ash hurricanes, generated by explosive eruptions in which the eruptive column is so heavily overloaded with ash that it collapses. They form instead of pyroclastic flows when the ash is more or less cold or damp. Though highly turbulent, base surges very rapidly deposit their load of entrained and saltating lapilli and ash. These deposits show draped antidune bedding, large wave-length but low amplitude cross bedding of well-laminated ash and lapilli, and bedding-sag features. Deposition is predominantly on the lee side of the dunes: this, together with the tendency for the lee side to be steeper than the stoss side, is an indicator of the flow-direction. These particular base surges travelled from west to east; they originated from nearby Kahauloa Crater. Their wavelength of 16 to 20 ft (5 to 6 m) may be a function of surge velocity: in the 1965 surge deposit of Taal volcano (Philippines) the wavelength decreased from 60 ft (20 m) near the crater rim to 16 ft (5 m) in distal areas (Moore, 1967).

Tuffs between the parking lot and the shore were erupted on land, evidenced by their mantle bedding, abundant well-preserved pisolites (accretionary lapilli) in several beds, and the pitting of several bedding plane surfaces by probable raindrop depressions. On the shore platform, small steep-sided inliers of reef limestone are exposed, below the tuff (Fig. 8). About 500 ft (150 m) northeast, erosional channels are clearly exposed within the lower tuff (Fig. 9).

These exposed slopes and road cuts show the nature of tephra alteration. Relatively fresh lithic-vitric ash and lapilli are nepheline-normative sideromelane glass, with some augite and olivine phenocrysts. Fresh ash may be 16 to 40 ft (5 to 12 m)

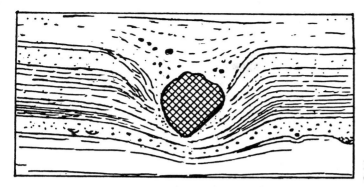

Figure 7. Large impact structure ("bomb sag") in the narrow gulch about 230 ft (70 m) inland from the Toilet Bowl. A 3.3 ft (1 m) block of limestone has penetrated or disturbed more than 5 ft (1.5 m) of ash.

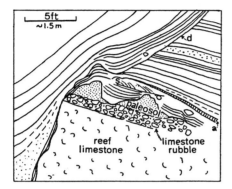

Figure 8. Small inlier of reef limestone and soil under tuff, at the seacoast below stop 3. The ash mantling must have been damp to be able to coat the steep limestone surface. One bed (a) contains accretionary lapilli. Slumping during the eruption produced slump deposits (s) and a discordance (D) in the tuff.

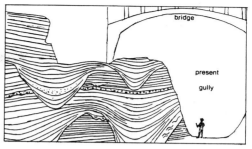

Figure 9. Erosion gullies, cut through and filled by tuffs from Koko Crater, 492 ft (150 m) northeast of stop 3. The gullies were eroded, presumably by water, on the slope of Koko Crater while the eruption was still in progress. Fisher (1977) suggested that additional erosion was caused by the passage of base surges.

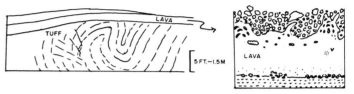

Figure 10. Synsedimentary folding seen in road cut near stop 3. The mass slumped down the steep flank of Koko Crater and before the overlying lava was erupted. The lava is aa, with a layer of spinose rubble along its top into which "fingers" of coherent basalt project. Vesicles are strongly deformed; surface tension operating against viscosity failed to reinstate a spherical form before the lava solidified. A patch of vesicles (V) marks a piece of rubble incorporated into the flow.

thick, depending on the present topography. Opal or montmorillonite or both may cement the glass. The sharp contact of gray glassy tuff over brown palagonite tuff cuts across bedding, roughly parallel to the gullied land surface. Undistorted textures in part-glass, part-palagonite pumice fragments at the diagenetic contact suggest there is little volume change during alteration of the glass. Much of that which is leached by cold percolating ground waters is precipitated as zeolite along with calcite, opal, and montmorillonite. Paragenesis of the zeolite is phillipsite first, then chabazite, thomsonite, gonnardite, natrolite, and the zeolite-like mineral analcime (Hay and Iijima, 1968).

The uppermost part of the deposit seen in a road cut 394 ft (120 m) northeast of the parking lot is non-lithified, more or less as it was at the end of the eruption, apart from some caliche deposited on the surface and along plant rootlets. The color boundary between the gray non-lithified ash and the dominantly yellow palagonite tuff locally cuts across the bedding. This same cut shows fine cross sections of dune-bedded base surge deposits. Excessive deposition that has prevented erosion on their stoss side led to the formation of climbing dunes.

A road cut 200 ft (60 m) northeast of the parking lot shows a thin aa lava flow resting on strongly folded ash (Fig. 10). This slump sheet is one of many along the seaward base of Koko Crater cone. During the eruption its foot was continually being eroded by the sea, so that slumping took place repeatedly during (and also no doubt after) the eruption.

A U-shaped channel is in the roadcut about 0.2 mi (0.3 km) beyond exit. Most rock is yellow to gray, fresh to opal-cemented tuff. Orange to red-brown tuff exposed in ravines and in a roadcut about 0.1 mi (160 m) farther is phillipsite- and chabazite-cemented palagonite.

Mile 2.2 (3.5 km) Stop 4. Blow Hole parking area. Coastal and volcanic features on flank of cone.

The principal scenic attraction here is a spouting horn. A seacave eroded into soft tuff layers has a fissure with two L-shaped steps from the cave up through the rocks. High swell rushing into the cave compresses the entrapped air and water, which ejects geyser-fashion.

Tuffs exposed on the benches to the south are strongly disturbed having slumped from Koko Crater cone during its eruption, because in a road cut 330 to 660 ft (100 to 200 m) southwest of the parking lot, they are mantled by undisturbed tuffs. Some of the disturbed tuffs contain ash-coated lapilli. These range in size up to about 2 in (5 cm) and are generally cored by a scoria fragment. It is thought that they grew by the snowball-like accretion of damp ash on lapilli that roll down the side of a steep cone. They lack the fine outer sheath possessed by accretionary lapilli of raindrop type. The beds in which they are concentrated have a discontinuous, lenticular form that indicates the lapilli moved into place by rolling or grain-flow.

Opposite the exit is a layer of tuff breccia. The 4 to 12 in (10 to 30 cm) blocks are the coarsest pyroclastic deposits of the Honolulu Volcanic Group.

Mile 2.5 (4.1 km) Stop 5. Sandy Beach parking lot. Pocket beach; aa flow.

Sandy Beach lies in the shallow coastal indentation between Koko Crater and Kalama lava flow and so owes its origin to those two eruptions. It takes its extreme popularity among Oahu's body surfers from being the nearest beach to Honolulu unobstructed by a reef. Heavy surf is dangerous; the foreshore steepens and a strong rip current sets to the south.

The beach is calcareous well-sorted medium sand. Abraded fragments of mollusk shells and foraminifers are followed in abundance by fragments of coralline algae, echinoids, and some coral. Deeper than 13 ft (4 m), the sloping bottom is rocky, except for a band of sand following a channel under the rip into deeper water (Clark, 1977).

The northeast end of Sandy Beach is Kalama basalt, the largest flow from the Koko fissure (Fig. 1). It filled the low area between the Koolau Range and the Koko tuff cones. The gray aa flow is nepheline basanite, with sparse olivine phenocrysts in a fine-grained groundmass of plagioclase, pyroxene, nepheline, and analcime (Winchell, 1947).

The highway had a different alignment before the 1 April 1946 tsunami from the Aleutian Islands destroyed much of it. A ranch northeast of Sandy Beach was destroyed except for part of the concrete wall of its swimming pool. Debris from the ranch was washed more than 1 mi (1.6 km) up into Kalama Valley. Tsunami run-up heights were about 31 ft (9.4 m) here, 11 ft (3.4 m) inside Hanauma Bay, and 37 ft (11.3 m) at Makapuu, (Fig. 1), the highest recorded on Oahu for the 1946 tsunami, although there were heights above 52 ft (16 m) on the north coasts of Kauai, Molokai, and Hawaii (Macdonald and others, 1947). Tsunamis in 1952, 1957, 1960, 1964, 1975, and 1986 did little or no damage to the Koko area. In the 1946 tsunami, 156 persons lost their lives in the Hawaiian islands, mainly in and near Hilo. A warning system was subsequently established and property damage was slight in 1952 and 1957. Many persons ignored the warnings in 1960, leading to 61 deaths in Hilo. Tsunami sirens are tested monthly, but Civil Defense officials fear that the decades since the last major destruction may have lulled citizens near the shore into ignoring the next large tsunami.

REFERENCES CITED

Bennison, A. P., 1974, Geological highway map of the State of Hawaii and the State of Alaska: American Association of Petroleum Geologists, United States Geological Highway Map Series, map 8.

Clark, J.R.K., 1977, The beaches of Oahu: Honolulu: University Press of Hawaii.

Doty, M. S., 1967, Pioneer intertidal population and the related general vertical distribution of marine algae in Hawaii: Blumea, v. 15, p. 95–105.

Easton, W. H., and Olson, E. A., 1976, Radiocarbon profile of Hanauma Bay, Oahu, Hawaii: Geological Society of America Bulletin, v. 87, p. 711–719.

Fisher, R. V., 1977, Erosion by volcanic base-surge density currents; U-shaped channels: Geological Society of America Bulletin, v. 88, p. 1287–1297.

Furnes, H., 1974, Volume relations between palagonite and authigenic minerals in hyaloclastites and its bearing on the rate of palagonitization: Bulletin Volcanologique, v. 38, p. 173–186.

Gosline, W. A., and Brock, V. E., 1976, Handbook of Hawaiian fishes: Honolulu, University Press of Hawaii.

Hay, R. L., and Iijima, A., 1968, Nature and origin of palagonite tuffs of the Honolulu Group on Oahu, Hawaii, in Coats, R. R., Hay, R. L., and Anderson, C. A., eds., Studies in volcanology, a memoir in honor of Howel Williams: Geological Society of America Memoir 116, p. 331–376.

Hinds, E. N., 1931, The relative ages of the Hawaiian landscapes: University of California, Bulletin of the Department of Geological Sciences, v. 20, p. 143–260.

Hobson, E. S., and Chave, E. H., 1972, Hawaiian reef animals: Honolulu, University Press of Hawaii.

Jakobsson, S. P., 1978, Environmental factors controlling the palagonitization of the Surtsey tephra, Iceland: Bulletin of the Geological Society of Denmark, Special Issue 27, p. 91–105.

Macdonald, G. A., Abbott, A. T., and Peterson, F. L., 1983, Volcanoes in the sea: The geology of Hawaii (2nd ed.): Honolulu, University of Hawaii Press, 517 p.

Macdonald, G. A., Shepard, F. P., and Cox, D. C., 1947, The tsunami of April 1, 1946, in the Hawaiian Islands: Pacific Science, v. 1, p. 21–37.

Moberly, R., 1968, Loss of Hawaiian littoral sand: Journal of Sedimentary Petrology, v. 38, p. 17–34.

Moberly, R., 1983, The ocean, in Armstrong, R. W., ed., Atlas of Hawaii (2nd edition): Honolulu, University of Hawaii Press, p. 53–58.

Moberly, R., Baver, L. D., and Morrison, A., 1965, Source and variation of Hawaiian littoral sand: Journal of Sedimentary Petrology, v. 35, p. 589–598.

Moore, J. G., 1967, Base surge in recent volcanic eruptions: Bulletin Volcanologique, v. 30, p. 337–363.

Stearns, H. T., 1939, Geologic map and guide of the island of Oahu, Hawaii: Hawaii Division of Hydrography, Bulletin 2.

—— , 1985, Geology of the State of Hawaii (2nd edition): Palo Alto, Pacific Books.

Stearns, H. T., and Vaksvik, K. N., 1935, Geology and ground-water resources of the island of Oahu, Hawaii: Hawaii Division of Hydrography, Bulletin 1, 479 p.

Thorarinsson, S., 1967, The Surtsey eruption and related scientific work: Polar Record, v. 13, p. 579–581.

Thorarinsson, S., Einarsson, T., Sigvaldason, G., and Elisson, G., 1964, The submarine eruption off the Vestmann Islands, 1963–64: Bulletin Volcanologique, v. 27, p. 435–445.

Wentworth, C. K., 1926, Pyroclastic geology of Oahu: B. P. Bishop Museum Bulletin 30, 121 p.

—— , 1938, Marine bench-forming processes: Water-level weathering: Journal of Geomorphology, v. 6, p. 6–32.

Winchell, H., 1947, Honolulu series, Oahu, Hawaii: Bulletin of the Geological Society of America, v. 58, p. 1–48.

Haleakala Crater, Maui, Hawaii

William R. Hackett, *Department of Geology, Idaho State University, Pocatello, Idaho 83209-0009*

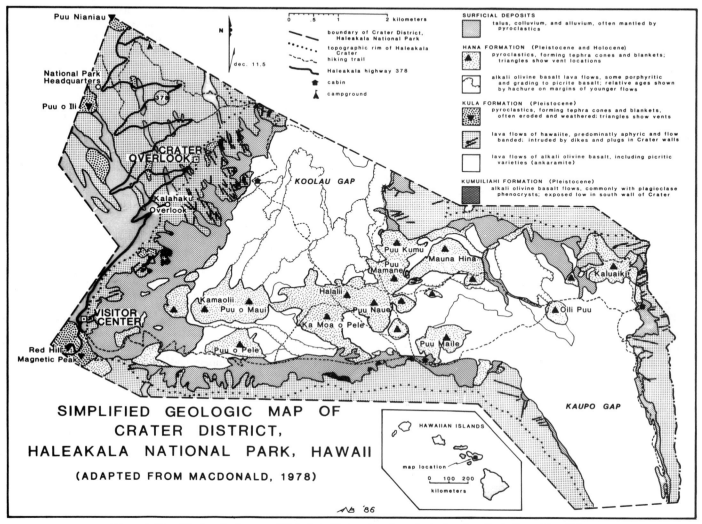

Figure 1. Simplified geologic map of Haleakala Crater, showing principal rock units, trails, and localities mentioned in text.

LOCATION AND ACCESS

Haleakala Crater occupies the summit region of east Maui, Hawaii (Fig. 1). The crater is the centerpiece of Haleakala National Park and access is via Hawaii 378, a well-marked paved road. The highway parallels the crater rim for 3 mi (5 km), and has several overlooks with parking areas. The crater itself is a Wilderness Area; access is restricted to travel on foot. An extensive system of hiking trails traverses the crater, with access at two points: Sliding Sands Trail leaves Hawaii 378 about 660 ft (200 m) south of the Visitor Center, and Halemauu Trail leaves Hawaii 378 about 3.2 mi (5.1 km) south of Haleakala National Park Headquarters. As no brief excursions to the crater floor are possible, two options are recommended: (1) a half-day trip, which

involves a stop at the Visitor Center for orientation, short walks along trails near the Visitor Center or at Red Hill, and then stops at several crater overlooks on the return drive; and (2) a full-day excursion (8 to 10 hours), which enters the crater via Sliding Sands Trail, travels across the crater floor, and departs via the Halemauu Trail. No drinking water is available in the crater. Park Headquarters should be contacted to obtain trail maps and up-to-date weather information.

Accommodation in Haleakala National Park is geared to camping only; no restaurants or food concessions are available. Hosmer Grove Campground is accessible from Hawaii 378, in the northwest corner of the Park. Holua Campground is within the crater and accessible only on foot; users must have a camping permit from the park. Cabins are available within the crater, but these must be reserved by written request (several months in advance, since cabins are allocated by a monthly lottery). It is recommended that persons, especially those planning on overnight trip, write well in advance for more detailed information and maps to the Superintendent, Haleakala National Park, P.O. Box 537, Makawao, Maui, Hawaii 96768.

GEOLOGICAL SIGNIFICANCE

Haleakala "Crater" (7 mi long, 2 mi wide, 0.5 mi deep; 11.5 × 3.6 km × 300 m) is not actually a volcanic crater, but rather one of the largest erosional depressions on earth. Development of the crater atop Haleakala, one of Hawaii's largest shield volcanoes, involved several hundred thousand years of stream erosion. The crater walls contain isolated exposures of early shield-forming lavas and extensive exposures of the late-stage alkalic capping, which overlies the tholeiitic shield of Haleakala. The crater is partially infilled by young posterosional lava flows and pyroclastics. Thus, the record of volcanism and erosion at Haleakala Crater illustrates a nearly complete sequence of Hawaiian volcanic evolution from early shield building, to late stage alkalic capping, to protracted dormancy and erosion, and finally to rejuvenation of volcanism that has continued into the past few centuries. Haleakala Highway 378 may be unique, since it is a paved road that travels from sea level to 10,000 ft (3,000 m) elevation in a distance of only about 50 mi (75 km).

REGIONAL GEOLOGY

The island of Maui is part of a huge volcanic massif comprising at least six major shield volcanoes. During the Pleistocene, the islands of Molokai, Lanai, Kahoolawe, and Maui were joined as a single large island, but are now separated by shallow seas. Maui itself consists of two major volcanoes. The older one is West Maui, which has gone through the principal stages of Hawaiian volcanism and is probably extinct. Haleakala is younger, and its lavas overlap those of West Maui volcano to form the isthmus connecting the two mountains.

Hawaiian volcanoes progress through a consistent pattern of life stages during their several-million-year history. Activity begins with quiet submarine eruptions. Shield volcanoes of broad, low profile are built on the sea floor, eventually breaching the surface. Thin, fluid lava flows of tholeiitic (low in alkalies with respect to silica) basalt comprise most of the shield, with very little pyroclastic material. Individual lava flows differ primarily in the amount of olivine phenocrysts present; some are picritic types called oceanites. Later in the shield-building stage, collapse craters or calderas often form on the summit. Commonly, two or three rift zones, marked by lines of vents and fissures, radiate from the summit regions of Hawaiian shields.

The primitive shield of Haleakala volcano is composed of thin pahoehoe and aa lava flows. Early shield lavas exposed near the seacoast are tholeiitic basalt, with alkali basalt flows in the upper part. The tholeiitic shield lavas are not exposed in Haleakala Crater, but isolated outcrops of thin alkali basalt lava flows at the base of the southern crater wall (Kumuiliahi Formation, Fig. 1) unconformably underlie later lavas and may represent the upper portion of the Haleakala shield. It is not known whether a caldera formed on Haleakala, since all evidence has been buried by later volcanic rocks.

Toward the end of the shield-building stage, changes are noted both in the compositions of lavas and in their styles of eruption. The frequency of eruptions decreases, gas content and proportion of felsic magma increase, and eruptions become more explosive, building large tephra cones, extensive ash blankets, and thick, viscous lava flows of intermediate composition. At Haleakala, passage into the later phase of more explosive volcanism was gradual, but at other Hawaiian centers it was interrupted by a long period of dormancy and erosion. Basalts of the alkalic capping contain relatively less silica and more alkalies than tholeiites, and are known as alkali basalt. Most contain olivine, and some are picritic varieties with many phenocrysts of green olivine and black augite. A common type of picritic alkali basalt is ankaramite, with augite phenocrysts more abundant than olivine. Many lavas are richer in feldspar and poorer in mafic minerals than the basalts. At Haleakala, an important lava type is hawaiite, a variety of alkaline andesite in which the feldspar is andesine. Far less abundant are mugearite, with oligoclase feldspar, and trachyte, in which the dominant feldspar is albite.

The Kula Formation (Fig. 1) represents the capping of alkalic lavas on the early tholeiitic shield of Haleakala. The formation is Pleistocene in age (0.4 to 0.8 Ma; McDougall, 1964), and is dominated by thick, light-colored hawaiite lava flows, with lesser alkali basalt and picrite. Near Haleakala summit, the thickness of Kula Formation exposed in the walls of Haleakala Crater is at least 2,500 ft (760 m), but near the seacoast the formation thins to 30 ft (10 m) or less. Kula eruptions took place from three rift zones, the two most prominent ones extending southwestward and east-northeastward from Haleakala summit. The third rift zone, extending north-northwestward, is less prominent but is still clearly marked by a line of tephra cones extending almost to the coast. These cones are clearly seen from Haleakala Highway 378.

Toward the end of the Kula period, the intervals between eruptions increased and local unconformities, soils, and alluvial deposits formed among the lava flows. Eventually, Kula volcanism ended and large stream valleys extended headward into the heart of the mountain, merging into a single great depression. The erosional depression is known today as Haleakala Crater; its origin is discussed in detail by Stearns (1942). Two stream valleys lead out of the depression through Koolau and Kaupo Gaps (Fig. 1). Projection of normal Hawaiian valley slopes suggests that the canyons may have been about 1.3 mi (2,100 m) deep at one time. The cutting of such huge valleys to form the S-shaped depression probably required several hundred thousand years.

Renewed eruptions occurred along the southwest and northeast rift zones, across the crater and down the flanks of

Haleakala to the sea. The north-trending rift zone was not reactivated. Eruptions within Haleakala Crater partially filled it with huge cinder cones, and lava flows spilled down the Koolau and Kaupo Gaps to the coast. Most of the ridge that formerly divided the depression was buried. Lava flows and pyroclastics of this renewed period of volcanism are designated as the Hana Formation (Fig. 1). The rocks are predominantly alkali basalts similar to those in the Kula Formation, but typically containing slightly more alkalies and less silica. Picrite basalt and ankaramite are common, but no hawaiite has been found in the Hana Formation within Haleakala Crater. The time of inception of Hana volcanism is unknown, but the formation is younger than the 0.45-m.y. age determined by McDougall (1964) for a lava flow in the upper Kula Formation. Allowing several hundred thousand years of erosion to form Haleakala Crater, the oldest Hana Formation lavas probably have ages of 200,000 years or less. No historical eruptions have been reported from within Haleakala Crater, but it must be recalled that historic time in Hawaii includes only about the last two centuries. The last-dated eruption in the Hana Formation occurred around 1790 A.D. (Oostdam, 1965), when lava flows from the southwest rift zone entered the sea at La Perouse Bay.

For more complete accounts of the geology of Haleakala, see Stearns and Macdonald (1942, p. 53-115, 277-311), Stearns (1985, p. 203–221), Macdonald and Abbott (1970, p. 50-51, 326-336), and Macdonald (1978). Discussions of petrography, geochemistry and petrogenesis of Haleakala lavas are found in Macdonald (1949), Macdonald and Katsura (1964), Macdonald (1968), Chen and Frey (1983, 1985), and Hegner and others (1986).

LOCALITY DESCRIPTIONS

Visitor Center

The Visitor Center at the rim of Haleakala Crater (Fig. 1) contains geologic displays, including a three-dimensional model of the mountaintop at a scale of about 1:8,000 (eight inches to the mile). The model illustrates the unusual S-shape of the erosional depression and is a good place to review Haleakala's geologic history. Lava flows exposed near the Visitor Center are representative of the Kula Formation, erupted prior to the crater-forming erosion. Ankaramite dikes are exposed below and to the south of the Visitor Center. The dark lava flow outside the building is also ankaramite, and contains abundant euhedral phenocrysts of black augite and iridescent green olivine set in a black matrix. Tephra blanketing the area are picritic alkali basalt lapilli of the posterosional Hana Formation; these deposits also contain euhedral augite crystals.

Hawaiite is the dominant lava type in the upper Kula Formation. South of the Visitor Center, a short trail spirals to the top of White Hill. This rock is dense, aphyric, pale-gray hawaiite with prominent flow structure expressed as platey jointing. Such jointing is common in andesitic lavas worldwide and reflects internal shearing of the viscous flows, in analogous fashion to shear planes

developed near the bases of glaciers. The lava beneath the White Hill hawaiite is petrographically similar, but contains scattered phenocrysts of olivine and augite. In the hill 0.2 mi (300 m) north of the Visitor Center, on the western rim of the crater, hawaiite lava contains phenocrysts of hornblende—a mineral rare in Hawaiian lavas.

Red Hill

This is the highest point on the island of Maui, 10,023 ft (3,056 m), and the third highest peak in the Hawaiian Chain. The panorama encompasses several of the other islands. On the island of Hawaii to the south, still-active Mauna Loa represents the shield-building stage of Hawaiian volcanoes. To the left of Mauna Loa, Mauna Kea also represents Haleakala's former appearance: a volcano whose summit rose to nearly 13,000 ft (4 km) and had an irregular profile due to the eruption of alkaline lavas as tephra cones and thick, viscous lava flows on its upper slopes. The West Maui mountains, Lanai and Molokai, are visible from Haleakala summit on clear days. Once much higher, they are now reduced to elevations of less than about 5,000 ft (1,500 m).

Haleakala Crater, perhaps 1.2 mi (2 km) deep at one time, is now only 0.5 mi (300 m) deep and partly infilled by young volcanics of the Hana Formation. Eruptions of alkali basalt and picrite occurred within the crater and along several rift zones. From Red Hill, a southwest-trending line of tephra cones, spatter mounds, lava flows, and fissures continues to the seacoast. Tephra cones within Haleakala Crater were formed in similar manner as Red Hill and are closely related in age and composition. These vents define an east-trending rift zone that traverses the crater floor, continues through the rain forest on the eastern slopes of Haleakala, and eventually reaches the coast near the town of Hana. The influence of moist, northeasterly trade winds on vegetation patterns is well displayed at Haleakala: the desert-like conditions of the western crater are a strong contrast to the verdant slopes of the eastern crater. A sense of scale and distance is lost in this vast landscape—the far cliffs to the east are 7 mi (11.5 km) away. Puu O Maui, the largest and best-formed tephra cone in Haleakala Crater, is about 2.5 mi (4 km) from Red Hill, and rises nearly 650 ft (200 m) from the crater floor.

The Red Hill parking area has been constructed within the summit crater of a small tephra cone, and short trails around the rim allow examination of the alkali basalt eruptive products. Most of the ejecta are oxidized scoriaceous lapilli with euhedral augite phenocrysts. During later, less-energetic stages of the eruption, large fluid bombs were ejected, forming an armor of welded spatter (agglutinate) over much of Red Hill. Many bombs were pasty enough to retain their airborne shapes, and can be found resembling baseballs, almonds, and ribbons. Frequently, the outer crusts are breadcrusted as a result of internal gas expansion.

Crater Overlook (Leleiwi Overlook)

A small parking area is located on the south side of Hawaii 378, 3.2 mi (4.8 km) from the Visitor Center and 6.1 mi (9.2 km) from Park Headquarters (Fig. 1). About 230 ft (70 m) west (downhill) of the parking area, a composite alkali basalt lava flow is exposed in a roadcut (Fig. 2). The aphyric base of the flow is alkali basalt, and shear planes are well developed. However, the upper 10 ft (3 m) are rich in phenocrysts of augite and olivine, the transition to porphyritic basalt occurring over a narrow zone 1 to 2 cm (inset, Fig. 2). Macdonald (1972) has interpreted the flow as representing the reversed stratigraphy of a magma chamber that had undergone crystal settling. The aphyric base represents the first-erupted magma from the upper portion of the chamber, which was depleted in phenocrysts. The remainder of the flow is porphyritic, and contains rare dunite inclusions 2 to 4 cm in diameter, which are probably fragments of olivine cumulate.

Use the crosswalk to get onto the path to the Crater Overlook. From the overlook, the form of stream-carved Keanae Valley is well displayed to the northeast (left). Here on the rim of the Koolau Gap, the Keanae Valley is 2.5 mi (4 km) wide, but narrows to less than 1.2 mi (2 km) as it reaches the sea at Keanae. Headward widening is characteristic of Hawaiian stream erosion, and results in broad amphitheater-headed valleys. The valleys are fed predominantly by ground water; as more surface and ground water are captured by the widening valley, tributary valleys become established. These are bowl-like in morphology and account for the scalloped appearance of the northwest rim of the crater.

Lava flows and pyroclastics of the Kula Formation are exposed in the cliffs below the overlook. The deposits are largely hawaiite lava flows, with less common ankaramite flows. Several hawaiite dikes are exposed below the overlook. To the northeast, the switchbacks of the Halemauu Trail traverse Kula pyroclastics and block-lava hawaiite flows. Several palagonite tuff horizons and transected tephra cones are visible in the Halemauu section. At the overlook enclosure, hawaiite lava flows occur with shear surfaces developed at their bases, and just below the enclosure, ankaramite lava flows occur. Both of these lava types are typical of the upper Kula Formation.

Several young lava flows of the Hana Formation are seen on

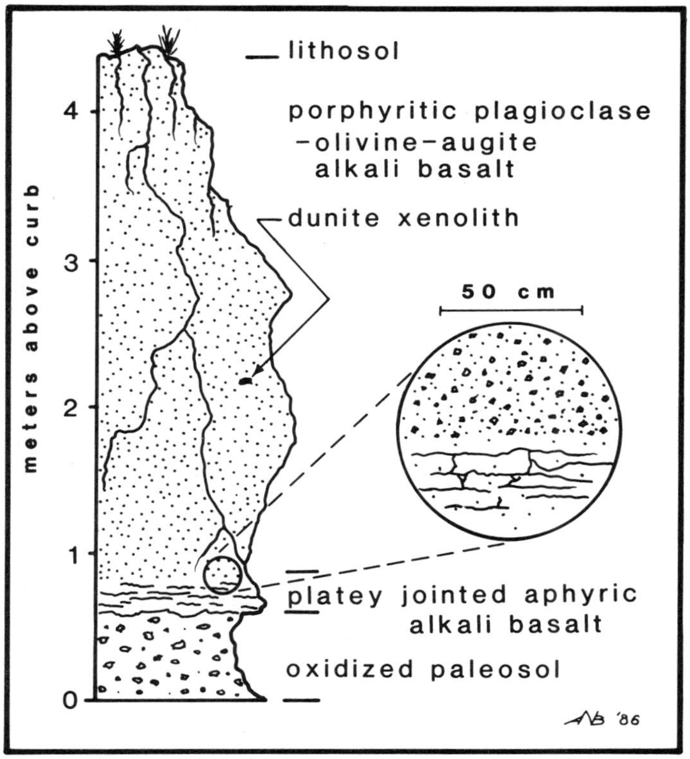

Figure 2. Measured section of a composite alkali basalt lava flow exposed in a cutting on Hawaii 378, 6.1-mi (9.1-km) road distance southeast of Haleakala National Park Headquarters. Inset shows rapid transition between porphyritic and aphyric alkali basalt.

the crater floor in Koolau Gap. These alkali basalt flows came from the bases of Kaluu O Ka Oo and Puu o Maui (Fig. 1). The porphyritic alkali basalt flow from Puu o Maui is thickly covered with tephra from this vent. Segments of the cinder cone have been rafted northward on the flow more than 1.2 mi (2 km) from source, forming angular monoliths that project above the top of the flow. Several of these can be seen from the overlook.

REFERENCES CITED

Chen, C., and Frey, F. A., 1983, Origin of Hawaiian tholeiites and alkalic basalt: Nature, v. 302, p. 785–789.

——— , 1985, Trace element and isotopic geochemistry of lavas from Haleakala Volcano, East Maui, Hawaii; Implications for the origin of Hawaiian basalts: Journal of Geophysical Research, v. 90, p. 8743–8768.

Hegner, E., Unruh, D., and Tatsumoto, M., 1986, Nd-Sr-Pb isotope constraints on the sources of West Maui volcano, Hawaii: Nature, v. 319, p. 478–480.

Macdonald, G. A., 1949, Hawaiian petrographic province: Geological Society of America Bulletin, v. 60, p. 1541–1596.

——— , 1968, Composition and origin of Hawaiian lavas, *in* Coats, R. R., Hay, R. L., and Anderson, C. A., eds., Studies in Volcanology: Geological Society of America Memoir 116, p. 477–522.

——— , 1972, Composite lava flows on Haleakala volcano, Hawaii: Geological Society of America Bulletin, v. 83, p. 2971–2974.

——— , 1978, Geologic Map of the Crater Section of Haleakala National Park, Maui, Hawaii: U.S. Geological Survey Miscellaneous Investigations Series

Map I-1088, scale 1:24,000, with 8 p. text.

Macdonald, G. A., and Abbott, A. T., 1970, Volcanoes in the Sea; The Geology of the Hawaiian Islands: Honolulu, University Press of Hawaii, 441 p.

Macdonald, G. A., and Katsura, T., 1964, Chemical composition of Hawaiian lavas: Journal of Petrology, v. 5, pt. 1, p. 82–133.

McDougall, I., 1964, Potassium-argon ages from lavas of the Hawaiian Islands: Geological Society of America Bulletin, v. 75, p. 107–128.

Oostdam, B. L., 1965, Age of lava flows on Haleakala, Maui, Hawaii: Geological Society of America Bulletin, v. 76, p. 393–394.

Stearns, H. T., 1942, Origin of Haleakala Crater, island of Maui, Hawaii: Geological Society of America Bulletin, v. 53, p. 1–14.

——— , 1985, Geology of the State of Hawaii: Palo Alto, California, Pacific Books, 335 p.

Stearns, H. T., and Macdonald, G. A., 1942, Geology and Ground-Water Resources of the Island of Maui, Hawaii: Hawaii Division of Hydrography Bulletin 7, 344 p.

Kilauea caldera and adjoining volcanic rift zones

Richard W. Hazlett, *Department of Geology, University of Hawaii at Hilo, Hilo, Hawaii 96720-4091*

LOCATION

Kilauea volcano is located at the southeastern corner of the Island of Hawaii (Fig. 1). Access to the site is easy by ordinary passenger vehicle. Major mainland and interisland airlines serve island airports, where numerous car rental agencies are located. Because Hawaii is a popular tourist destination, it is recommended that car reservations be placed well in advance of a visit. Hawaii 11 links Hilo, the largest town on the island, directly with the Kilauea summit area, a drive of about 31 mi (50 km). A map and information concerning the status of roads and trails within Hawaii Volcanoes National Park may be obtained at the Kilauea Visitor Center, a short side trip from Hawaii 11 (Fig. 2). The Visitor Center is a good place for one to begin exploring the caldera area.

SIGNIFICANCE

Few places on earth allow closer or more dramatic views of dynamic geological forces than the active volcanoes of Hawaii. Located at the approximate center of the Pacific Plate, the Hawaiian hot spot continues a process of island building that stretches back at least 73 Ma. Today, much of this constructional process is focused within the boundaries of Hawaii Volcanoes National Park, which includes the summits of two frequently erupting shield volcanoes: Kilauea and Mauna Loa. The Kilauea caldera area exemplifies the summit of a Hawaiian shield volcano in a youthful, subaerial stage of development (Macdonald and others, 1983; Clague and Dalrymple, 1987).

Because this area is protected for its natural beauty and scientific value, a set of regulations must be heeded by all visitors:

1) Collection of plant and rock material in Hawaii Volcanoes National Park is prohibited by federal law. If samples are required for teaching, empirical work, or related purposes, a collecting permit must be obtained at the Office of the Park Naturalist, Kilauea Visitor Center. Samples then may be collected from specific areas, out of public sight.

2) Permission must be secured at Kilauea Visitor Center to explore beyond established roads and trails. Certain areas without public access are extremely hazardous, or are closed for scientific research. Such sites may not be obvious to the untrained or unwary eye.

SITE INFORMATION

Kilauea caldera. Translated from Hawaiian, Kilauea means "the source of great spreading, or spewing" (Pukui and others, 1974). The summit area of Kilauea, one of the world's most active volcanoes, is marked by a set of single and composite pit craters, intersecting rift zones, and a small caldera measuring 2 by 3 mi (3.25 by 4.75 km) (Fig. 2). The highest walls of Kilauea caldera rise to 400 ft (120 m) on the northeastern side near Volcano House Hotel. At the opposite end of the caldera, on the southwestern side, the caldera walls drop to a shallow channel named The Outlet. Several recent lava flows originating within the caldera have flowed through The Outlet onto the nearby Kau Desert.

The floor of Kilauea caldera is occupied by a gently sloping lava shield culminating at Halemaumau Crater, a collapse pit that is a frequently active summit vent. The Halemaumau shield is an historic feature, built up during a period of frequent lava lake activity in the late 19th and early 20th centuries (Holcomb, 1987). To best appreciate that the floor of the caldera slopes upward to the rim of Halemaumau, one may take the Halemaumau Trail, which begins at Volcano House, leading 3.2 mi (5.2 km) across the floor to the Halemaumau parking area on Crater Rim Drive (Fig. 2).

From the Halemaumau Trail, one may examine at close range sites of recent fissure-vent eruptions on the caldera floor (Fig. 2). These are marked by spatter ramparts, some iridescent with hydrated glasses, or brownish-yellow with palagonite. Vents themselves are typically reddened from the high-temperature oxidation which accompanies fountaining of lava. Overhangs bordering some vents feature numerous lava stalactites.

Young vents near the southern wall of the caldera have become part of a solfatara field which has been active throughout historic time. Because fumaroles in this area emit toxic gases which may be lethal to some persons under certain atmospheric conditions, they should be avoided by visitors (Fig. 2). Recent studies of the fumarolic gases of Kilauea, and of related incrusting minerals, include Greenland (1987) and Casadevall and others (1987).

Also well displayed along the Halemaumau Trail is the shallow structure beneath the summit of Kilauea, shown in the enclosing caldera walls. Hundreds of thin lava flows and interflow breccias, punctured by fractures and intrusive bodies, lie exposed in the cliffsides. The most spectacular intrusive body is the Uwekahuna laccolith, visible as two light-colored lenticular masses near the foot of the wall beneath, and slightly to the east of the Volcano Observatory. The laccolith measures 80 by 960 ft (25 by 295 m). During the earthquake of 16 November 1983, it was partly buried by talus. Murata and Richter (1961, p. 424) examined a suite of rocks from this body, describing evidence to show "that the original mafic tholeiitic magma differentiated into tholeiitic picrite, tholeiitic olivine gabbro, and an aphanitic rock approaching quartz-basalt in composition. Mechanisms involved were an initial gravity settling of olivine and a final filter pressing of the residual liquid." The compositional range of the Uwekahuna rocks is greater than that of any other comagmatic suite yet studied at Kilauea.

R. W. Hazlett

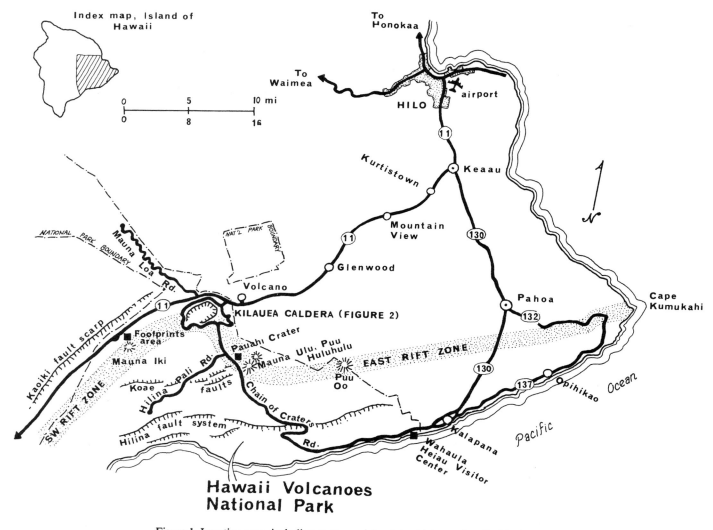

Figure 1. Location map, including some special points of interest described in the text.

Near Halemaumau Crater. A short walk from Crater Rim Road provides access to the rim of Halemaumau Crater. The walls of Halemaumau record a history of multi-stage collapse. The prominent circumferential ledge is all that remains of a former, shallower crater floor which subsided 150 ft (45 m) in September 1971. Since this collapse, lava has partly refilled the crater.

Along the path to the rim of Halemaumau is a landscape littered with blocks thrown out during a series of unusual phreatic explosions in May 1924. Among the rock types represented in this debris are intrusive lithologies analogous to those of the Uwekahuna laccolith, and sintered, thermally metamorphosed lavas. Detailed petrography of these rocks was done by Macdonald (1944) and Chapman (1947). Petrologists interested in learning more about the genesis and composition of Kilauea's lavas may refer to Wright and Helz (1987), and Tilling and others (1987), among many papers.

Degasing in Halemaumau occurs via fractures in the weakened crater walls instead of through the floor, which is plugged by relatively impervious, unfractured lavas.

Southwest rift zone. A fine view of the southwest rift zone of Kilauea may be seen from the Uwekahuna Overlook, beside the Jaggar Museum (Fig. 2). Another site, which provides direct access to large fissures along the rift zone, may be reached by taking Crater Rim Road to a pullout near the southwestern rim of Kilauea caldera (Fig. 2).

The rift zones of Kilauea are loci of extensional failure, magmatic intrusion, and eruption, linked by dikes to the central magma reservoir beneath the volcano's summit. Most of the growth of Kilauea has been accomplished by magmatic processes in its rift zones. Excellent studies of the volcanic plumbing system include Duffield and others (1982), Epp and others (1983), Ryan and others (1983), Dzurisin and others (1984), Decker (1987), and Klein and others (1987).

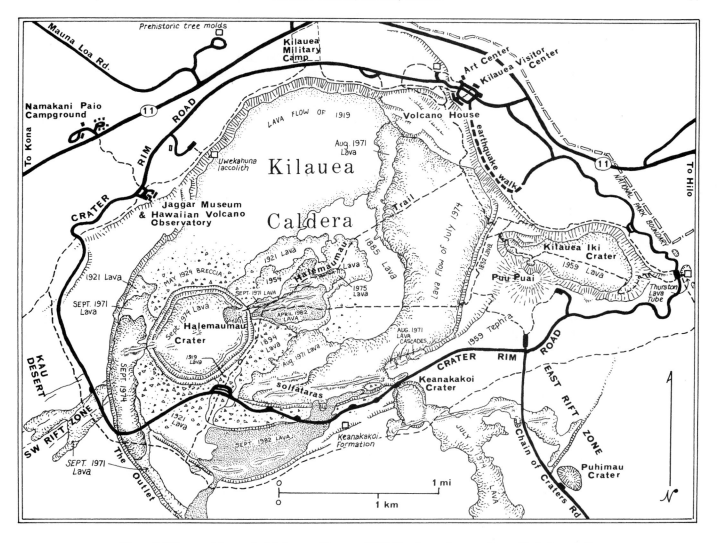

Figure 2. Kilauea caldera and vicinity, as of January 1987. Roads are shown as heavy black lines, trails as thin dashed lines. Historical lava flows are outlined, patterned, and dated. Ridge-like patterns in lava flows represent eruptive fissures with prominent spatter ramparts. Breccia symbol shows distribution of May 1924 explosion breccia from Halemaumau Crater.

A flank eruption typically begins with a period of intrusion as magma leaves the summit reservoir by injection into one of the two rift zones. This episode is marked by swarms of small earthquakes, accompanied by deflation of the summit to accomodate drainage of its reservoir. As magma proceeds "downrift," earthquake epicenters may shift correspondingly, allowing observers at the Hawaiian Volcano Observatory to discern the progress of the intrusion. Within a matter of hours, the molten dike may breach the surface to feed an eruption. Perhaps two-thirds to one-half of all intrusive events at Kilauea do not lead to eruptions. Instead, the intrusive bodies slowly cool, differentiate, and harden to form diabasic or gabbroic masses beneath the rift zones. Alternatively, the magma may remain in storage for years, even decades, only to be driven out by injection of new magma into the rift zone at a

later time. Petrologic evidence of this is documented in studies by Wright and Fiske (1971), and Wolfe and others (1987).

The southwest rift zone is delineated by lengthy fissure systems and scattered, small parasitic cones. Unlike the east rift zone, there are few large lava shields or pit craters in this area. The southwest rift zone is shorter, and has been less historically active than the east rift zone, suggesting that the southwestern flank of the volcano is less unstable than its eastern flank, perhaps due to the buttressing effect from Kilauea's older neighbor, Mauna Loa. Differential stresses set up between inflation in Kilauea's summit and southwest rift zone and inflation in Mauna Loa, has resulted in continuing seismicity along a fault system between the two volcanoes (Koyanagi and others, 1984; Endo, 1985). This is the Kaoiki fault zone, an escarpment of which may be seen at the foot of the slope of Mauna Loa, north of Hawaii 11, as one drives seaward down the western flank of Kilauea (Fig. 1).

An excellent stop for photographing the Kaoiki escarpment is also a site of historical interest in the southwest rift zone. This is the Mauna Iki Trail, the start of which may be reached by following Hawaii 11 to the Kau Desert pullout, 9.2 mi (14.2 km) west of the turnoff for Crater Rim Road. A 0.9 mi (1.5 km) walk along the trail leads one to a gallery of poorly preserved footprints. These footprints were left by Hawaiian victims of a powerful pyroclastic eruption originating at the summit of Kilauea about 1790 (Swanson and Christiansen, 1973). Occasionally, better preserved footprints are exposed in this area by wind erosion. If found, they should be left untouched, as the weathered ash layer in which they are preserved is quite fragile.

A bit more than a kilometer beyond the footprints exhibit, the Mauna Iki Trail leads one to the summit of Mauna Iki, a small lava shield formed during an eight-month long flank eruption in 1919–1920, the longest eruption on record for the southwest rift zone (Jaggar, 1947). From the summit of Mauna Iki, one may obtain an excellent panoramic view of the rift zone, and southern flank of neighboring Mauna Loa, all the way to Ka Lae, the south point of the island.

Deposits of the 1790 pyroclastic eruption. In addition to the Mauna Iki Trail, excellent exposures of 1790 ash and lapilli may be seen along Crater Rim Road in the southern portion of Kilauea caldera. Much of the modern caldera may have formed by collapse immediately precursory to this catastrophe (Holcomb, 1987). Pyroclastic eruptions are relatively rare occurrences at Kilauea (Decker and Christiansen, 1984). Within recorded history, less than one percent of several hundred eruptions observed here have been explosive in character. The most recent explosive event, a series of phreatic blasts at Halemaumau in 1924, was much less powerful, perhaps by an order of magnitude, than the 1790 eruption.

The finest easily accessible exposures of 1790 ash and lapilli, termed the Keanakakoi Formation, lie in a fault scarp along the edge of the caldera a few hundred meters west of Keanakakoi Crater (Fig. 2). These exposures may be easily reached by a short walk from Crater Rim Road. They consist of interlayered pyroclastic surge, flow, and airfall beds, locally deformed by bomb and block sags.

An older pyroclastic unit, comparable in lithology to the Keanakakoi, is exposed elsewhere near the foot of the caldera wall. This is the Uwekahuna Ash, which has been dated at 1000–2200 yr (Holcomb, 1987). The immediate cause of explosive volcanism at Kilauea is believed to be large-volume drainage of the summit magma reservoir, accompanied by rift intrusion, which allows entry of groundwater through fractured wallrock into drained, superheated reservoir conduits (Decker and Christiansen, 1984). Large pyroclastic eruptions such as those which formed the Keanakakoi and Uwekahuna deposits may occur with crude regularity at Kilauea, perhaps related to the level of the water table around the magma reservoir. As the water table rises, the probability of a pyroclastic eruption increases.

Large pyroclastic eruptions deplete the water table significantly, and the recharge process, lasting many centuries, is forced to begin anew (Robert W. Decker, pers. commun., 1986).

The Kilauea Iki area. Before the formation of Kilauea caldera, several large, coalescing lava shields similar to the present-day Halemaumau shield comprised the summit of Kilauea. The two-stage collapse of the Kilauea Iki (Trans.: Little Kilauea) shield occurred 350–500 bp, producing a depression in the shape of a figure-8 which, like Kilauea caldera itself, has been largely filled by subsequent lava flows, producing a relatively smooth, flat floor (Holcomb, 1987). The most recent eruption at Kilauea Iki took place in November 1959, forming the oxidized Puu Puai cinder cone on the southwestern rim, and filling the crater with lava to a depth of 400 ft (120 m) (Richter and others, 1970). Backdrain of lava into the vent at the base of Puu Puai during this eruption, together with contraction and cooling, have lowered the lava lake surface relative to its chilled margins along the base of the crater wall. The result is a conspicuous lava-subsidence terrace, or "bathtub ring," about 50 ft (15 m) high.

Since its formation in 1959, the lava lake in Kilauea Iki has been drilled repeatedly as part of a research program by the U.S. Geological Survey in conjunction with Sandia Laboratories to study the cooling and crystallization of small magma bodies. By 1981 it was concluded that no very fluid lava remained in the lava lake (Helz and Wright, 1983). Accounts of drilling operations on Hawaiian lava lakes are given in Peck and others (1979), Helz (1987), Wright and Helz (1987), and Wright and Swanson (1987).

Thurston Lava Tube lies near the eastern edge of Kilauea Iki, along Crater Rim Road (Fig. 2). This cave formed in a lava flow streaming from the top of the Kilauea Iki shield prior to its collapse. Perhaps during this collapse, several much smaller satellite craters were formed, one of which truncated Thurston Lava Tube. Hence, the approach to the tube is made by descending to the floor of a shallow pit crater. The path through the tube extends about 400 ft (120 m). Sinkholes downslope indicate that the tube system may extend for several kilometers.

During historical eruptions, lava tubes as long as 6.8 mi (11 km) have developed at Kilauea. The relative importance of lava tubes in the evolution of the Hawaiian landscape is described in Greeley (1987).

East rift zone (Chain of Craters Road). Near the Devastation parking area, Crater Rim Road intersects Chain of Craters Road. The route along the Chain of Craters provides easy access to the most spectacular features of the upper east rift zone of Kilauea, and to the steeply faulted southern flank of the volcano.

A set of aligned pit craters distinguished structure in the upper east rift zone from that of the middle and lower east rift zone, east of the national park boundary (Fig. 1). These pit craters were formed hundreds of years ago (Holcomb, 1987) by catastrophic collapse in the upper part of the rift zone as magma withdrew to intrude or erupt further downrift.

Historical eruptions in the east rift zone typically originate, as elsewhere on Kilauea, by the surface breaching of a dike to feed a curtain of lava fountains, which later dies out as the eruption becomes concentrated at one or a few points along the eruptive fissure. Pre-existing pit craters play no role in localizing eruptions, most of which occur outside the craters, on the surrounding gently sloping flank of the volcano. However eruptive fissures may coincidentally open across the floor, wall, or even along the rim of a pit crater. An excellent place to observe such relationships is Pauahi Crater, near the Mauna Ulu Turnoff (Fig. 1). Pauahi Crater is a coalescing pit crater, formed during at least three periods of collapse. During historical times, three small eruptions have broken out in the vicinity of Pauahi, two in 1973, and one in 1979. Eruptive fissures, in places lined with low spatter ramparts, lie along the crater rim just behind the visitor overlook, down the eastern wall of Pauahi Crater opposite the overlook, and at the base of the crater wall, out of sight, beneath the overlook. A very large bathtub ring in the southeastern pit of Pauahi indicates drainage of a large volume of lava through the fractured floor of the crater at the close of the November 1973 eruption.

In addition to pit craters and fissure systems, the east rift zone is marked by lava shields and other large, conical vent structures. One of the youngest lava shields in the rift zone is Mauna Ulu (Trans.: Growing Mountain), which dominates the southern skyline viewed from the Pauahi Crater overlook. Mauna Ulu developed during a nearly continuous eruption lasting from late 1969 to mid-1974 (Tilling and others, 1987b). The small, vegetated hill to the north of Mauna Ulu is Puu Huluhulu, a large spatter-flow cone formed about 500 years ago (Holcomb, 1987). Both features may be examined at close range by taking a 2 mi (3 km) roundtrip, moderately rugged trail from the Mauna Ulu parking area to the summit of Puu Huluhulu (Fig. 1).

From the summit of Puu Huluhulu, a spectacular panorama may be enjoyed, weather permitting. Between Mauna Ulu and the base of Puu Huluhulu is a well-developed perched lava pond, the feeding channel of which may be seen extending from the surface of the pond to the rim of Mauna Ulu Crater. Toward the summit of Kilauea may be seen several of the pit craters and fissure systems along Chain of Craters Road, portions of the caldera area, Kilauea Iki shield, and the summits of Hawaii's two largest volcanoes, Mauna Loa and Mauna Kea. Looking east from Puu Huluhulu one may see Alae lava shield–a low ridge extending from the left side of Mauna Ulu, and Makaopuhi Crater–largest and deepest of the pits in the Chain of Craters. Kane nui o Hamo is the large, prehistoric lava shield partially engulfed by Makaopuhi. Beyond Kane nui o Hamo are numerous other conical vent structures which delineate the middle and lower east rift zone as far as Cape Kumukahi (Fig. 1). Foremost of these vent structures is Puu Oo, a steep, 835-ft-high (255 m) cinder, spatter, and lava cone formed in a series of eruptions between June 1983 and July 1986.

South flank fault systems. The southern flank of Kilauea Volcano is cut by numerous normal faults which belong to two principal fault systems.

The Koae fault system intersects the east rift zone near Mauna Ulu and extends westward to the southwest rift zone. The Koae faults drop crust down-to-the-north. Though most scarps are at least several centuries old (Holcomb, 1987), they appear quite fresh as they cut only slightly weathered lavas. The best views of these faults may be seen looking south from Hilina Pali Road within a few kilometers of its intersection with Chain of Craters Road (Fig. 1). Fault development at various stages, including folding at the ends of faults, incipient fracture along fold axes, and full scarp formation are well exposed.

From the end of Hilina Pali Road, and from several overlooks along Chain of Craters Road south of Mauna Ulu, one may see outstanding views of the Hilina fault system. The Hilina fault scarps are much larger than those of the Koae fault system; several are hundreds of meters high. In addition, displacement is principally down-to-the-south, i.e., toward the sea. These faults give the southern flank of Kilauea a staircase topographic profile. They are believed to have originated from gravitational oversteepening of the structurally unsupported south flank due to magmatic intrusion and inflation in the adjoining rift zones (Swanson and others, 1976). At depth the Hilina faults flatten seaward, encompassing large rotational slump blocks.

The Hawaiian Volcano Observatory and Thomas A. Jaggar Museum (Fig. 2). The Hawaiian Volcano Observatory is operated by the U.S. Geological Survey. It is closed to the public, although group tours of the facility can be arranged through the National Park Service (Kilauea Visitor Center). Adjoining the Observatory is the Thomas A. Jaggar Museum, which contains exhibits explaining the research work of the Observatory, as well as current interpretations of the eruptive behavior of Kilauea. A popular publication summarizing Observatory history and operations was written by Heliker and others (1986).

REFERENCES CITED

Casadevall, T. J., Stokes, J. B., Greenland, L. P., Malinconico, L. L., Casadevall, J. R., and Furukawa, 1987, SO_2 and CO_2 emission rates at Kilauea Volcano, 1979–1984, *in* Decker, R. W., Wright, T. L., and Stauffer, P. H., eds., Volcanism in Hawaii: U.S. Geological Survey Professional Paper 1350, p. 771–780.

Chapman, R. W., 1947, Crystallization phenomena in volcanic ejecta from Kilauea, Hawaii: American Mineralogist, v. 32, p. 105–110.

Clague, D. A., and Dalrymple, G. B., 1987, The Hawaiian–Emperor volcanic chain; Part I. Geologic evolution, *in* Decker, R. W., Wright, T. L., and Stauffer, P. H., eds., Volcanism in Hawaii: U.S. Geological Survey Professional Paper 1350, p. 5–54.

Decker, R. W., 1987, Dynamics of Hawaiian volcanoes; An overview, *in* Decker, R. W., Wright, T. L., and Stauffer, P. H., eds., Volcanism in Hawaii: U.S. Geological Survey Professional Paper 1350, p. 997–1018.

Decker, R. W., and Christiansen, R. L., 1984, Explosive eruptions of Kilauea Volcano, Hawaii, *in* National Research Council and others, eds., Explosive volcanism; Inception, evolution, and hazards (Studies in Geophysics): Washington, D.C., National Academy Press, 122–132.

Duffield, W. A., Christiansen, R. L., Koyanagi, R. Y., and Peterson, D. W., 1982, Storage, migration, and eruption of magma at Kilauea Volcano, Hawaii, 1971–1972: Journal of Volcanology and Geothermal Research, v. 13, p. 237–307.

Dzurisin, D., Koyanagi, R. Y., and English, T. T., 1984, Magma supply and storage at Kilauea Volcano, Hawaii, 1956–1983: Journal of Volcanology and Geothermal Research, v. 21, p. 177–206.

Endo, E. T., 1985, Seismotectonic framework for the southeast flank of Mauna Loa Volcano, Hawaii [Ph.D. thesis]: Seattle, University of Washington, 349 p.

Epp, D., Decker, R. W., and Okamura, A. T., 1983, Relation of summit deformation to east rift zone eruptions of Kilauea Volcano, Hawaii: Geophysical Research Letters, v. 10, p. 493–496.

Greeley, R., 1987, The role of lava tubes in Hawaiian volcanoes, in Decker, R. W., Wright, T. L., and Stauffer, P. H., eds., Volcanism in Hawaii: U.S. Geological Survey Professional Paper 1350, p. 1589–1606.

Greenland, L. P., 1987, Hawaiian eruptive gases, in Decker, R. W., Wright, T. L., and Stauffer, P. H., eds., Volcanism in Hawaii: U.S. Geological Survey Professional Paper 1350, p. 759–770.

Heliker, C., Griggs, J. D., Takahashi, T. J., and Wright, T. L., 1986, Volcano monitoring at the U.S. Geological Survey's Hawaiian Volcano Observatory: Earthquakes and Volcanoes, v. 18, no. 1, 69 p.

Helz, R. T., 1987, Diverse olivine types in lava of the 1959 eruption of Kilauea Volcano, and their bearing on eruption dynamics, in Decker, R. W., Wright, T. L., and Stauffer, P. H., eds., Volcanism in Hawaii: U.S. Geological Survey Professional Paper 1350, p. 691–722.

Helz, R. T., and Wright, T. L., 1983, Drilling report and core logs for the 1981 drilling of Kilauea Iki lava lake (Kilauea Volcano, Hawaii): U.S. Geological Survey Open–File Report 83–326, 66 p.

Holcomb, R. T., 1987, Eruptive history and long-term behavior of Kilauea Volcano, in Decker, R. W., Wright, T. L., and Stauffer, P. H., eds., Volcanism in Hawaii: U.S. Geological Survey Professional Paper 1350, p. 261–350.

Jaggar, T. A., 1947, Origin and development of craters: Geological Society of America Memoir 21, 508 p.

Klein, F. W., Koyanagi, R. Y., Nakata, J. S., and Tanigawa, W. R., 1987, The seismicity of Kilauea's magma system, in Decker, R. W., Wright, T. L., and Stauffer, P. H., eds., Volcanism in Hawaii: U.S. Geological Survey Professional Paper 1350, p. 1019–1186.

Koyanagi, R. Y., Endo, E. T., Tanigawa, W. R., Nakata, J. S., Tomori, A. H., and Tamura, P. N., 1984, Kaoiki, Hawaii, earthquake of November 16, 1983; A preliminary compilation of seismographic data at the Hawaiian Volcano Observatory: U.S. Geological Survey Open–File Report 84–798, 35 p.

Macdonald, G. A., 1944, Unusual features in ejected blocks at Kilauea Volcano: American Journal of Science, v. 242, p. 322–326.

Macdonald, G. A., Abbott, A. T., and Peterson, F. L., 1983, Volcanoes in the sea; The geology of Hawaii (2nd ed.): Honolulu, The University of Hawaii Press, 517 p.

Murata, K. J., and Richter, D. H., 1961, Magmatic differentiation in the Uwekahuna laccolith, Kilauea caldera, Hawaii: Journal of Petrology, v. 2, pt. 3, p. 424–437.

Peck, D. L., Wright, T. L., and Decker, R. W., 1979, The lava lakes of Hawaii: Scientific American, v. 241, p. 114–129.

Pukui, M. K., Ebert, S. H., and Mookini, E. T., 1974, Placenames of Hawaii: Honolulu, University of Hawaii Press, 289 p.

Richter, D. H., Eaton, J. P., Murata, K. J., Ault, W. U., and Krivoy, H. L., 1970, Chronological narrative of the 1959–60 eruption of Kilauea Volcano, Hawaii: U.S. Geological Survey Professional Paper 537–E, p. E1–E73.

Ryan, M. P., Blevins, J.Y.K., Okamura, A. T., and Koyanagi, R. Y., 1983, Magma reservoir subsidence mechanics; Theoretical summary and application to Kilauea Volcano, Hawaii: Journal of Geophysical Research, v. 88, p. 4147–4181.

Swanson, D. A., and Christiansen, R. L., 1973, Tragic base surge in 1790 at Kilauea Volcano: Geology, v. 1, 83–86.

Swanson, D. A., Duffield, W. A., and Fiske, R. S., 1976, Displacement of the south flank of Kilauea Volcano; The result of forceful intrusion of magma into the rift zones: U.S. Geological Survey Professional Paper 963, 39 p.

Tilling, R. I., Wright, T. L., and Millard, H. T., Jr., 1987a, Trace-element chemistry of Kilauea and Mauna Loa lava in space and time; A reconnaissance, in Decker, R. W., Wright, T. L., and Stauffer, P. H., eds., Volcanism in Hawaii: U.S. Geological Survey Professional Paper 1350, p. 641–690.

Tilling, R. I., Christiansen, R. L., Duffield, W. A., Endo, E. T., Holcomb, R. T., Koyanagi, R. Y., Peterson, D. W., and Unger, J. D., 1987b, The 1972–1974 Mauna Ulu eruption, Kilauea Volcano; An example of quasi-steady-state magma transfer, in Decker, R. W., Wright, T. L., and Stauffer, P. H., eds., Volcanism in Hawaii: U.S. Geological Survey Professional Paper 1350, p. 405–470.

Wright, T. L., and Fiske, R. S., 1971, Origin of the differentiated and hybrid lavas of Kilauea Volcano, Hawaii: Journal of Petrology, v. 12, p. 1–65.

Wright, T. L., and Helz, R. T., 1987, Recent advances in Hawaiian petrology and geochemistry, in Decker, R. W., Wright, T. L., and Stauffer, P. H., eds., Volcanism in Hawaii: U.S. Geological Survey Professional Paper 1350, p. 625–640.

Wright, T. L., and Swanson, D. A., 1987, The significance of observations at active volcanoes, in Mysen, B. O., ed., Magmatic processes; Physiochemical principles: The Geochemical Society Special Publication No. 1 (in press).

Wolfe, E. W., Garcia, M. O., Jackson, D. B., Koyanagi, R. Y., Neal, C. A., and Okamura, A. T., 1987, The Puu Oo eruption of Kilauea Volcano, episodes 1–20, January 3, 1983, to June 8, 1984, in Decker, R. W., Wright, T. L., and Stauffer, P. H., eds., Volcanism in Hawaii: U.S. Geological Survey Professional Paper 1350, p. 471–508.

Supergene enriched fluidized breccia ore, Lavender open pit copper mine, Warren (Bisbee) Mining District, Cochise County, Arizona

Donald G. Bryant, 1516 E. Oxford Lane, Englewood, Colorado 80110

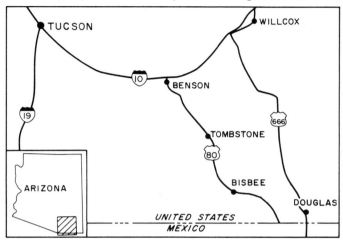

Figure 1. Road map of southern Arizona showing the location of Bisbee.

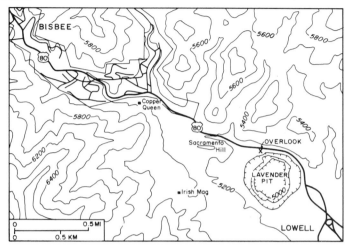

Figure 2. Topographic map of the Lavender open pit copper mine showing overlook site and various landmarks of the district.

LOCATION

The Lavender open pit is approximately 80 mi (130 km) southeast of Tucson, Arizona, and can easily be reached by car from either Tombstone or Douglas on paved roads (Fig. 1). The overlook or observation site for the Lavender open pit copper mine is on U.S. 80 between Bisbee and Lowell. The topography and a few of the more famous landmarks—such as the Copper Queen mine, which was the first discovery, the Irish Mag mine, and Sacramento Hill (Ransome, 1904)—are shown on Figure 2.

SIGNIFICANCE

Unlike many of the porphyry copper deposits of Arizona, the Lavender open pit mineralization is in a fluidized [intrusive] breccia pipe, not porphyry, and is Nevadan (130 Ma), not Laramide, in age. The fluidized breccia is composed of exotic fragments and boulders that have been transported from below in a fine-grained or rock flour matrix of pulverized rock and transported sulfides. The transporting media was a circulating hot water slurry flowing along preexisting fractures. The ore minerals actually mined in the Lavender pit were not in the porphyry rocks but were transported fragmental sulfides in the matrix and mineralized fragments and boulders with later supergene chalcocite enrichment. The mineralization in the porphyries was limited to pyrite with a thin coat of chalcocite, and was very low grade.

SITE INFORMATION

The general distribution of the different rock types exposed in the Lavender pit is shown in Figure 3. The approximate position of the overlook is in the upper central portion of the block diagram. The history of the intrusive complex is as follows: (1) intrusion of quartz porphyry into pre-Cretaceous rocks accompanied by massive pyrite and pervasive silica; (2) formation of the Contact intrusion breccia on the periphery of the quartz porphyry (protoclastic?); (3) intrusion of the quartz-feldspar porphry; (4) emplacement of the fluidized breccia; (5) oxidation and chalcocite enrichment; and (6) deposition of the Cretaceous Glance conglomerate. Originally the base-metal mineralization was thought to be penecontemporaneous with, or after, the fluidized breccia (Bryant and Metz, 1966; Bryant, 1968). Recent studies of polished thin sections of representative mineralization (Bryant, 1974) indicate that the sulfides are pre-breccia and have been transported as fragments, not in solution. The original fragmentation occurred during faulting and was not the result of explosive processes. Fluidized breccia dikes, sills, and irregular elongated masses along the intersection of faults and brittle rocks occur underground (Bryant, 1968). Some features of Lavender fluidized breccias are illustrated in Figure 4. The underground mine is flooding at this time, but visits into the pit for science can be arranged by contacting the Copper Queen Branch of Phelps Dodge at 602-434-3621. The mailing address is Copper Queen Branch, Phelps Dodge Corporation, Highway 92, Bisbee, Arizona 85603.

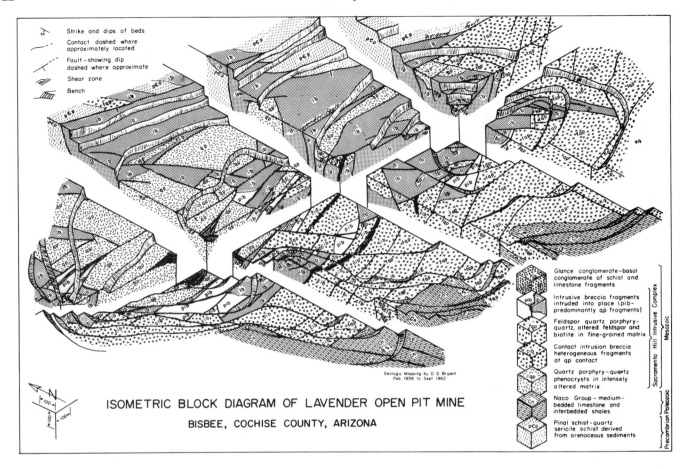

Figure 3. Geology of the Lavender open pit copper mine.

Figure 4. Left—fragment of Pinal Schist (Precambrian) transported several thousand feet (meters) and smoothed by abrasion during the fluidization process. Middle—large rounded boulder of Bolsa Quartzite (Middle Cambrian) transported several thousand feet (meters). Range rod below boulder is extended to 7 ft (2.1 m). Right—polished slab of mineralized Lavender fluidized breccia showing the distribution of fragments, character of the matrix, fragment rounding, and fragmental sulfides including chalcopyrite, sphalerite, bornite, and pyrite.

REFERENCES CITED

Bryant, D. G., 1968, Intrusive breccias and associated ore of the Warren (Bisbee) mining district, Cochise County, Arizona: Economic Geology, v. 63, p. 1–12.

—— , 1974, Intrusive breccias, fluidization, and ore magmas: Colorado Mining Association Mining Yearbook, p. 54–58.

Bryant, D. G., and Metz, H. E., 1966, Geology and ore deposits of the Warren mining district, *in* Titley, S. R., and Hicks, C. L., eds., Geology of the porphyry copper deposits, southwestern North America: University of Arizona Press, p. 189–203.

Ransome, F. L., 1904, Geology and ore deposits of the Bisbee Quadrangle, Arizona: U.S. Geological Survey Professional Paper 21, 168 p.

6

Curry Draw, Cochise County, Arizona: A late Quaternary stratigraphic record of Pleistocene extinction and paleo-Indian activities

C. Vance Haynes, Jr., Departments of Anthropology and Geosciences, University of Arizona, Tucson, Arizona 85721

LOCATION

Curry Draw, a typical discontinuous gully, is an easterly flowing tributary of the upper San Pedro River in Cochise County, Arizona (Fig. 1). The arroyo, which appears as an unnamed drainage on the Lewis Springs 7½-minute Quadrangle, exposes the Murray Springs Clovis site in the SW¼SE¼-Sec.26,T21S,R21E, (31°34′13″N, 110°10′40″W). The site where the stratigraphy is best exposed is reached on foot by following the draw downstream 0.65 mi (1.1 km) from where it crosses Moson Road or by following the abandoned Southern Pacific railroad grade that parallels the draw on the north side. The site is on property of the Bureau of Land Management, and permission to visit it must be obtained from the Safford District Office, 425 East 45th Street, Safford, Arizona 85546.

SIGNIFICANCE OF SITE

The stratigraphy and geomorphology of Curry Draw reveal an unusually complete record of late Quaternary depositional, pedological, and erosional events controlled by changing climate. If water table levels with respect to channel configuration and sediments can be taken as approximate indicators of climatic change, some gross indications of late Quaternary change can be read from the alluvial stratigraphy.

The Murray Springs Clovis site is unique in that it contains three distinct activity areas where a band of Clovis hunters killed mammoth and several bison and occupied a small camp site during two or three brief visits 11,000 years ago (Haynes, 1968, 1973, 1974, 1976, 1978, 1979, and 1980). The buried occupation surface is clearly displayed in the arroyo walls as an erosional contact at the base of a distinctive black organic mat that preserved artifacts and extinct animal bones in their original position, and mammoth tracks (Fig. 2), just as they were left 11,000 years ago (Haynes and Hemmings, 1968). The late Quaternary stratigraphic framework (Fig. 3), more complete than at any of the dozen known Clovis sites in stratigraphic context, is impressive to see because of the marked color contrasts of the sedimentary units and the excellent exposures along 1.6 mi (2.57 km) of arroyo walls between the site and the San Pedro floodplain to which the stratigraphy is directly traceable. Correlations throughout the upper San Pedro Valley are augmented by over 100 radiocarbon dates (Fig. 4).

DESCRIPTION AND DISCUSSION

The valley of Curry Draw is cut into early Pleistocene alluvial and lacustrine deposits of the Saint David Formation (Gray,

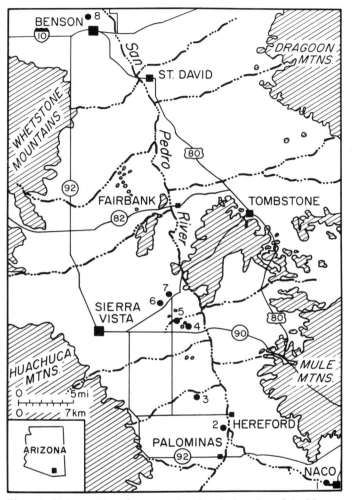

Figure 1. Index and location maps showing the location of the Murray Springs Clovis site (5) in Curry Draw, Cochise County, Arizona. Other Clovis sites are Naco (1), Lehner (2), and Escapule (4). Probable Clovis sites are the Leikum and Navarrete mammoth sites near Naco (1), and the Hargis bison site (3). Possible Clovis sites include the Schaldack mammoth site (6), the Donnet mammoth site (7), and the Grey-Seff faunal locality (8). Open circles indicate "grass circle" depressions discussed in the text.

1967; Johnson and others, 1975; Haynes, 1981). Gravel straths of the Nexpa Formation, inset into the Saint David Formation, are probably middle Pleistocene and appear to represent a long period of net degradation following the Brunhes/Matuyama paleomagnetic boundary. The Millville Formation, which is beyond

Figure 2. Mammoth tracks exposed by the archaeological excavations at the Murray Springs Clovis site extend along the right bank of a buried spring-fed creek, across the creek bed (lower right), along the left bank (middle ground), and end at the skeleton of an adult female *mammuthus columbi* on an erosion remnant of Coro Marl and surrounded by Clovis artifacts.

the limit of radiocarbon dating, contains the remains of Rancholabrean fauna scattered among poorly sorted muddy sand and gravel alluvium reflecting episodes of high energy stream discharge (Haynes, 1980). In cross section, the eroded Millville Formation is no thicker than the Holocene alluvium but occupies a channel three to five times wider than the Holocene channels, suggesting the erosion of a much broader valley floor.

The Murray Springs Formation, unconformably overlying the Millville, consists of a basal olive green, laminated, lacustrine mudstone, the Sobaipuri Mudstone Member, conformably overlain by the white, lacustrine Coro Marl Member. A basal sand, the Moson Member, is mainly confined to spring feeders as facies of the Sobaipuri Member and extends downward into older units. The springs indicate groundwater support of a late Pleistocene lake in Curry Draw. The mudstone, up to 3 ft (1 m) thick, contains a dark gray to black organic band that radiocarbon dates

the upper 4 in (10 cm) to 29,000 ± 2,000 B.P. (A-896). Squeeze-ups into the overlying marl occur at several places, and 2 to 5 mm thick mudstone partings, commonly accompanied by concentrations of gastropod and pelecypod shells, occur in the lower half of the marl. In some areas the Moson Sand Member extends horizontally within the top of the Millville Formation, along the Millville-Sobaipuri contact, and cuts across the contact in a few places indicating formation after deposition of the Millville and Sobaipuri units. It is believed to be due to the subsurface flow of groundwater leaving a lag accumulation as finer-grained clasts were removed by subsurface discharge.

The Murray Springs Formation, with Rancholabrean faunal remains, has yielded radiocarbon dates on both carbonate and organic fractions indicating Coro Marl deposition between 27,500 and 12,000 B.P. The late Wisconsinan along Curry Draw was a time of emergent groundwater and very low gradients. Ecological conditions were more mesic than at any time since (Martin, 1963). Mammoths, camels, horses, and bison were abundant in the area, but evidence for early man is lacking.

The Coro Marl is unconformably overlain by the Lehner Ranch Formation representing the early Holocene. A small channel, saucer-shaped in cross section and truncating the Murray Springs Formation, contains medium to coarse channel sand, the Graveyard Gulch Member, with abundant charcoal producing radiocarbon dates between 13,000 and 10,900 B.P. A black organic algal mat, the Clanton Ranch Member, conformably overlies the channel sand, but is unconformably in contact with the Murray Springs and older units (Fig. 3). The black mat, as much as 1 ft (0.3 m) thick in low areas, pinches out up slope and interfingers with a white marl pond facies in a few low areas. Both organic and carbonate radiocarbon dates indicate deposition between 10,800 and 9,700 B.P. in local ponds and groundwater seeps.

Clovis artifacts, bones of their prey, and charcoal from their fires are concentrated in the upper 4 in (10 cm) of the Graveyard Gulch channel sand and along the erosional contact at the base of the Clanton Ranch mudstone, which perfectly preserves and demarks the Clovis occupation surface. Artifacts and thousands of waste flakes were found exactly as they had been left by the Clovis people about 11,000 years ago. Eight radiocarbon dates on the Clovis-age charcoal, identified as ash *(Fraxinus)*, average 10,900 ± 50 B.P.

In the mammoth kill area numerous large, shallow depressions, presumably mammoth footprints, occurred in a swath along the right bank, across the channel sand, and up the left bank to where the partially articulated skeleton of an adult female mammoth lay surrounded by Clovis artifacts (Fig. 2). The near perfect preservation of these mammoth tracks in the soft channel sand indicates a lack of discharge sufficient to erase the tracks. The discharge must have been nil, yet subsequent deposition of the black algal mat and white marl facies indicates seepy ground and shallow ponds. This microstratigraphy suggests that a brief drop in the water table 11,000 years ago led to a dry or nearly dry stream bed that was followed by a gradual rise in the water table

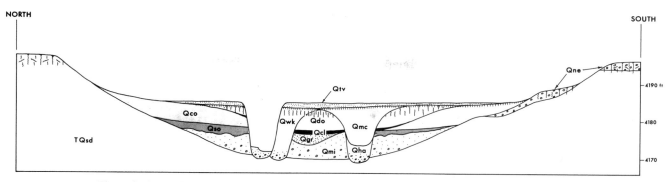

Figure 3. Generalized geologic cross section of Curry Draw showing the stratigraphic relationship of units.

soon after the mammoth crossing. Fossil pollen from the same level at the Lehner Clovis site, 10 mi (16 km) to the south (Fig. 1, loc. 2), indicates that the water table rise was a regional event (Mehringer and Haynes, 1965). Growth of the black mat was such that the mammoth tracks were preserved essentially intact. The edges and squeezed-up ridges were only slightly rounded before burial. The indicated drop and rise in the local water table is probably the result of changes in local base level and/or climatic change rather than tectonic activity.

The Clanton Ranch Member of the Lehner Ranch Formation is conformably overlain, except for localized disconformities, by the Donnet Ranch Member, a massive, silty, fine sandy loam that is believed to have been deposited by slope washing of aeolian silt and sand from adjacent valley slopes. Deposition took place between 9,500 and 8,000 B.P. with no distinct channel in evidence. This lack of a channel facies indicates that aggradation occurred gradually over a grassy swale topography. Alluvial pollen studies of the same unit at the Lehner site indicate that floodplain vegetation became increasingly more xerophytic as aggradation progressed (Mehringer and Haynes, 1965; Mehringer, 1967). This, plus the stratigraphic-sedimentologic evidence, suggests that the less effective moisture conditions were the result of a falling regional water table.

The Lehner Ranch Formation is unconformably overlain by the Escapule Ranch Formation consisting of three middle to late Holocene alluvial fill members each separated by paleosols and arroyo type erosional contacts. Each member is further subdivisible into two or three subunits separated by similar erosional contacts. The lower Weik Ranch Member fills the first true arroyo channel of the late Quaternary stratigraphic record. Its deposition between 6,500 and 4,300 B.P. began a cyclical regime of arroyo cutting and filling that has continued up to the present time.

The lower part of the Weik Member consists of channel sands, gravels, and chunks of older bank materials overlain by interbedded pond and slope wash deposits with pollen, indicating more mesic floodplain conditions between 6,500 and 4,500 B.P. than at any other time in the Holocene record (Martin, 1963; Mehringer and Haynes, 1965). The paucity of alluvial fills between 8,000 and 6,500 B.P. has prevented the recovery of alluvial

pollen for this erosional hiatus, but pollen from dune facies in New Mexico suggests drier conditions with less effective moisture at about 7,000 B.P. (Schoenwetter, personal communication; Mehringer, personal communication). The trend toward more xeric conditions appears to have started during the latter phases of deposition of the Donnet Ranch Member and culminated with arroyo cutting soon after 8,000 B.P. The filling that began 6,500 B.P. appears to have coincided with a net rise of the water table and may have been in response to it, promoting the growth of vegetation that trapped sediment in the channel. The 1,500 to 1,000 year hiatus between 8,000 and 6,500 B.P. was a period of erosion and channel widening during which much of the Murray Springs and Lehner formations were removed.

Each Holocene alluvial fill is made up mostly of slope-wash alluvium locally interbedded with pond deposits in part derived from slope wash but representing the winnowed fines. These fluvio-lacustrine deposits give way in the upper parts of the fills to loamy slope wash showing various degrees of pedogenesis and bioturbation, and pollen that indicates increasingly more xeric floodplain vegetation (Mehringer, personal communication). It is clear that down-valley transport of sediment, predominant in the early history of arroyo aggradation, gave way to predominantly slope-wash alluvium derived from adjacent or nearby slopes in the later history of each alluvial fill, the upper half foot (few decimeters) of which reflect the onset of the driest or most xeric part of the preserved alluvium. The subsequent hiatus may represent an even drier part of the alluvial cycle, but pollen-bearing sediments have not been preserved.

During the final half of each aggradational part of the cycle, fluvial discharge down Curry Draw was inadequate for removing slope wash faster than it was accumulating. As net aggradation ensued, the valley cross section became progressively less steep sided (less arroyo-like) as the shallow swale configuration developed. Thus, discharge became less confined as aggradation progressed and enhanced further aggradation. During the same time, a falling water table may have caused a reduction of grass cover. The combination of reduced vegetative cover and lower water table set the stage for the next episode of arroyo cutting by lowering the threshold for erosion.

The modern episode of arroyo cutting began in 1916 when

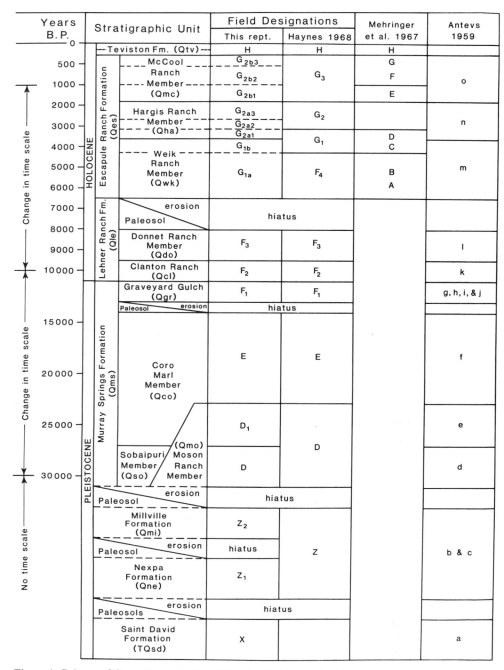

Years B.P.	Stratigraphic Unit		Field Designations		Mehringer et al. 1967	Antevs 1959
			This rept.	Haynes 1968		
0	—Teviston Fm. (Qtv)—		H	H	H	
500	Escapule Ranch Formation (Qes)	McCool Ranch Member (Qmc)	G_{2b3}	G_3	G	
1000			G_{2b2}		F	o
			G_{2b1}		E	
2000		Hargis Ranch Member (Qha)	G_{2a3}	G_2		
3000			G_{2a2}			n
			G_{2a1}	G_1	D	
4000		Weik Ranch Member (Qwk)	G_{1b}		C	
5000			G_{1a}	F_4	B	m
6000					A	
7000	Lehner Ranch Fm. (Qle)	erosion		hiatus		
8000		Paleosol				
		Donnet Ranch Member (Qdo)	F_3	F_3		l
9000						
10000		Clanton Ranch (Qcl)	F_2	F_2		k
		Graveyard Gulch (Qgr)	F_1	F_1		g, h, i, & j
		Paleosol erosion		hiatus		
15000	Murray Springs Formation (Qms)	Coro Marl Member (Qco)	E	E		f
20000						
25000			D_1			e
		(Qmo) Sobaipuri / Moson Member Ranch (Qso) / Member		D		
30000			D			d
		erosion		hiatus		
		Paleosol				
		Millville Formation (Qmi)	Z_2			
		Paleosol erosion	hiatus	Z		b & c
		Nexpa Formation (Qne)	Z_1			
		Paleosols erosion		hiatus		
		Saint David Formation (TQsd)	X			a

Figure 4. Column of Curry Draw stratigraphic units showing radiocarbon chronology and correlation with other unit designations.

excessive runoff deepened the ruts of a wagon road that led down a grassy swale where the Murray family had pastured dairy cows before 1911. Up to 1.6 ft (0.5 m) of sand and gravel alluvium, the Teviston Formation, covering the historic swale (floodplain) was flushed from a headcut about 0.6 mi (1 km) upstream of that at the Clovis site. Thus Curry Draw is a typical discontinuous gully whose floodplain was an undissected grassy swale at the beginning of the twentieth century. The remains of a protohistoric

bison cow and fetal calf were found between the Teviston and Escapule Ranch formations (Agenbroad and Haynes, 1975). Two main tributaries of Curry Draw, the east and west swales, remain essentially undissected, but aerial photography reveals that headcuts started up the east swale as soon as the main head-cut reached it shortly before 1955.

Curry Draw today does not extend to the Huachuca Mountains. Instead, the channel becomes broader and shallower in the headward reach until it becomes imperceptible about 3 mi (5 km) from the base of the mountains. Uplands along the draw are relatively flat interfluvial remnants of the middle Pleistocene pediment surface extending from the Huachuca Mountain front to the San Pedro River. Some broader interfluves have a thin (up to 8 in; 20 cm) veneer of late Pleistocene to Holocene silty loam believed to be an accumulation of wind-blown silt. Grass "circles," very shallow circular depressions up to 985 ft (300 m) diameter, on some of the broader flats (Fig. 1) are filled with up to 3.3 ft (1 m) of this silty loam bearing a dark gray to black paleosol and overlying a much stronger red calcic paleosol of late Pleistocene age. The Holocene paleosol has produced bulk-sample radiocarbon dates indicating minimum ages of pedogenesis of between 3,000 and 6,000 B.P. The paleosol is overlain by up to 2 in (5 cm) of gray powdery silt representing the latest Holocene increment of aeolian deposition. This thin unit is observed on most of the grassy swales of the area and on many uneroded terrace flats along the San Pedro River. It is buried by the Teviston Formation in some places, and a similar unit is buried by the McCool Member of the Escapule Formation in others.

From these observations it is apparent that there have been significant contributions of aeolian silt to the landscape at various times throughout the Holocene and probably the late Pleistocene. This material has been washed from valley slopes at various times to become a significant component of the valley fills. It appears to have been a major component of the Donnet and Clanton Members, indicating that a thicker aeolian blanket was deposited during late Wisconsinan time than at any time since. This veneer of aeolian silt mixed with slope wash, and its loss by erosion, has exerted a pronounced edaphic effect on vegetational changes over the past 10,000 years, and perhaps before. The thicker silty loam accumulations today, as exemplified by the grass circles, preserve the mesquite-grassland community as "islands" surrounded by the Chihuahuan desert scrub community where the silty loam has been removed, leaving a substrate of caliche or older units impregnated with calcium carbonate. The cycles of upland deposition and erosion implied by the alluvial cycles must, at times, have had a profound influence on the changes of vegetation as the substrate changed. Vegetational changes would, in turn, strongly influence infiltration and slope erosion. Valley sides covered by desert grassland would have a dramatic effect on geohydrologic processes along Curry Draw compared to today's desert scrub on the highly calcareous substrate. This edaphic influence was undoubtedly a major factor in affecting the modern episode of arroyo cutting and the changing landscape (Hastings and Turner, 1965).

The sand and gravel of the Millville Formation occupies a channel several times larger than the modern arroyo. Therefore, much larger ephemeral discharges appear to be represented for the period immediately preceding 30,000 B.P. The Murray Springs Formation, consisting of springlaid sand, paludal mudstone, and lacustrine marl, indicates a rising piezometric surface from about 30,000 B.P. to about 27,000 B.P. and quasi stability from 27,000 to 16,000 B.P. This was followed by a falling water table and

degradation from perhaps 14,000 to 13,000 B.P., when the Pleistocene lake dried up and the Murray Springs Formation was entrenched by a small seep-fed stream.

The Lehner Ranch Formation represents a transition during the early Holocene from the high effluent (perennial) flow conditions of the late Pleistocene to the influent (ephemeral) flows of the middle to late Holocene. A brief but significant fluctuation of the water table indicating a relatively wet-dry-wet sequence is indicated by the detailed stratigraphy at the Murray Springs Clovis site where the Clovis occupation coincides with the relatively dry episode. The subsequent wet episode was accompanied by aggradation on a grassy swale kept moist by a shallow water table until about 9,000 B.P. when a trend toward less effective moisture and a falling water table ensued. Concommitant loss of erosion, inhibiting vegetation and the development of soil cracks along grassy swales, led to instability. This was culminated by arroyo cutting about 7,500 B.P. Cycling between arroyo filling, quasi stability, and arroyo cutting continued with increasing frequency over the next 7,000 years in response to climatically controlled water table levels fluctuating between influent and effluent flow conditions.

Misuse of floodplains, beginning during the nineteenth century, may have augmented erosional instability and triggered arroyo cutting sooner than it would have occurred otherwise, but the frequency of the preceding two cycles indicates that entrenchment was within a few decades of occurring anyway. Several 3.3- to 6.6-ft (1- to 2-m) terraces inset within Curry Draw today could be the result of complex response to base level lowering (Schumm and Parker, 1973), but aggradation of each major Holocene fill to or near the level of the previous fill is more likely related to Holocene climatic fluctuations.

Since the beginning of excavations at the Murray Springs site in 1966, the south headcut has cut headward 115 ft (35 m) in 19 years at an average rate of 6 ft/yr (1.8 m/yr). This is roughly half the rate of the previous 47 years as measured from aerial photography. The reduction in the rate of cutting is due to three factors: (1) An earthen dam for a stock pond was completed 1.7 mi (2.8 km) upstream in 1966, (2) a dam for a wastewater evaporation pond 0.7 mi (1.2 km) upstream was completed in 1978, and (3) a backhoe stratigraphic trench excavated and filled in 1970, having eroded since, has diverted all but the highest discharges to the north headcut. Whereas this has reduced the rate of headward erosion at the site, it will eventually lead to bank erosion of the western end by the creation of a sharp bend.

The stratigraphy of the Lehner Clovis site (Fig. 1, loc. 2) 11 mi (18 km) SSE of Murray Springs (Haury and others, 1959), while nearly identical to that of Curry Draw, is not as well exposed. On the other hand, fluvial sand radiocarbon dated at between 13,000 and 14,000 B.P. in Lehner Arroyo (Haynes, 1982, unit E$_1$) is not represented in Curry Draw. Other Clovis sites and potential Clovis sites in the upper San Pedro Valley are shown on Figure 1 and indexed as follows: (1) Naco Clovis site (Haury and others, 1953), Leikum mammoth site (Haynes, 1968), and Navarrete mammoth site, (3) Hargis bison site

(Haynes, 1968), (4) Escapule Clovis site (Hemmings and Haynes, 1969), (6) Schaldack mammoth site, (7) Donnet mammoth site, and (8) Grey-Seff site (Gray, 1967; Haynes, 1968).

REFERENCES CITED

Agenbroad, L. D., and Haynes, C. V., Jr., 1975, *Bison bison* remains at Murray Springs, Arizona: The Kiva, v. 40, p. 309–313.

Antevs, E., 1959, Geological age of the Lehner mammoth site: American Antiquity, v. 25, p. 31–34.

Gray, R. S., 1967, Petrology of the upper Cenozoic non-marine sediments in the San Pedro Valley, Arizona: Journal of Sedimentary Petrology, v. 37, p. 774–789.

Hastings, J. R., and Turner, R. M., 1965, The changing mile; An ecological study of vegetation change with time in the lower mile of an arid and semiarid region: University of Arizona Press, 317 p.

Haury, E. W., Antevs, E., and Lance, J. R., 1953, Artifacts with mammoth remains, Naco, Arizona: American Antiquity, v. 19, p. 1–24.

Haury, E. W., Sayles, E. B., and Wasley, W. W., 1959, The Lehner mammoth site: American Antiquity: v. 25, p. 2–30.

Haynes, C. V., Jr., 1968, Preliminary report on the late Quaternary geology of the San Pedro Valley, Arizona: Southern Arizona Guidebook, Arizona Geological Society III, p. 79–96.

——, 1973, Exploration of a mammoth-kill site in Arizona: National Geographic Society Research Reports, 1966 Projects, p. 125–126.

——, 1974, Archaeological investigations at the Clovis Site at Murray Springs, Arizona, 1967: National Geographic Society Research Reports, 1967 Projects, p. 145–147.

——, 1976, Archaeological investigations at the Murray Springs Site, Arizona, 1968: National Geographic Society Research Reports, 1968 Projects, p. 165–171.

——, 1978, Archaeological investigations at the Murray Springs Site, Arizona, 1969: National Geographic Society Research Reports, 1969 Projects, p. 239–242.

——, 1979, Archaeological investigations at the Murray Springs Site, Arizona, 1970: National Geographic Society Research Reports, 1970 Projects, p. 261–267.

——, 1980, Archaeological investigations at the Murray Springs Site, Arizona, 1971: National Geographic Society Research Reports, 1971 Projects, v. 12, p. 347–353.

——, 1981, Geochronology and paleonenvironments of the Murray Springs Clovis Site, Arizona: National Geographic Society Research Reports, v. 13, p. 243–251.

——, 1982, Archaeological investigations at the Lehner Site, Arizona, 1974–75: National Geographic Society Research Reports, 1973 Project, p. 325–334.

Haynes, C. V., Jr., and Hemmings, E. T., 1968, Mammoth-bone shaft wrench from Murray Springs, Arizona: Science, v. 159, p. 186–187.

Hemmings, E. T., 1970, Early man in the San Pedro Valley, Arizona [Ph.D. thesis]: Tucson, University of Arizona, 236 p.

Hemmings, E. T., and Haynes, C. V., Jr., 1969, The Escapule mammoth and associated projectile points, San Pedro Valley, Arizona: Journal of Arizona Academic Science, v. 5, p. 184–188.

Johnson, N. M., Opdyke, N. D., and Lindsay, E. H., 1975, Magnetic polarity stratigraphy of Pliocene-Pleistocene terrestrial deposits and vertebrate faunas, San Pedro Valley, Arizona: Geological Society of America Bulletin, v. 86, p. 5–12.

Martin, P. S., 1963, The last 10,000 years, a fossil pollen record of the American Southwest: Tucson, University of Arizona Press, 87 p.

Mehringer, P. J., Jr., 1967, Pollen analysis and the alluvial chronology: The Kiva, v. 32, p. 96–101.

Mehringer, P. J., Jr., and Haynes, C. V., Jr., 1965, The pollen evidence for the environment of early man and extinct mammals of the Lehner mammoth site, southeastern Arizona: American Antiquity, v. 31, p. 17–23.

Mehringer, P. J., Jr., Martin, P. S., and Haynes, C. V., Jr., 1967, Murray Springs, A mid-postglacial pollen record from southern Arizona: American Journal of Science, v. 265, p. 786–797.

Schumm, S. A., and Parker, R., 1973, Implications of complex response of drainage systems for Quaternary alluvial stratigraphy: Nature, v. 243, p. 99–100.

Paleozoic stratigraphic section in Dry Canyon, Whetstone Mountains, Cochise County, Arizona

Chester T. Wrucke and Augustus K. Armstrong, U.S. Geological Survey, 345 Middlefield Road, Menlo Park, California 94025

LOCATION AND ACCESSIBILITY

The Whetstone Mountains are in southeastern Arizona along the boundary between Cochise and Pima Counties, about 42 mi (70 km) southeast of Tucson (Fig. 1). Benson, 15 mi (25 km) to the northeast, and Sierra Vista, about the same distance to the south, are the nearest towns. Most of the Whetstone Mountains, including the Dry Canyon area, are in Coronado National Forest. Dry Canyon, on the lower southeast flank of the mountains, is in the Benson 15-minute Quadrangle and the Apache Peak and McGrew Spring 7½-minute Quadrangles. Along the canyon, and particularly on the ridge on its south side, a thick sequence of paleozoic rocks is exposed, which is the focus of this report.

The Dry Canyon area (Fig. 2) is reached via Arizona 90, which trends south toward Fort Huachuca and Sierra Vista from its junction with I-10, about 2.7 mi (4.5 km) west of Benson. At a point 13 mi (21.5 km) south of I-10, a dirt road, marked by a simple ranch gate, leads westward a few mi (km) across public land into Dry Canyon. This road is best traveled using a four-wheel-drive vehicle, but it was passable with difficulty for passenger cars in 1982.

SIGNIFICANCE OF SITE

The Whetstone Mountains, together with the Mustang Mountains to the southwest, constitute a southwest-tilted fault block in the Basin and Range Province. The mountain block is structurally simple and has remarkably continuous sequences of unmetamorphosed Paleozoic sedimentary rocks, broken only by a few minor faults. All Paleozoic formations, Cambrian to Permian in age, that are widely distributed in southeastern Arizona are present, but it is the unusually complete succession of Pennsylvanian and Permian rocks that particularly distinguishes the Whetstone stratigraphic section. Formations of these ages were defined by Gilluly and others (1954) from exposures at numerous localities in central Cochise County, but in the Whetstone Mountains they are well displayed in a single, nearly unbroken sequence. The Dry Canyon area offers a complete and superbly exposed succession of Paleozoic rocks (Figs. 3, 4), which is so conveniently accessible that it provides an ideal reference section for the Paleozoic strata of southeastern Arizona.

SITE INFORMATION

Paleozoic rocks of the Dry Canyon area range in age from Middle Cambrian to Late Permian. Basal Cambrian strata rest on a surface of gentle relief developed on the Early Proterozoic Pinal

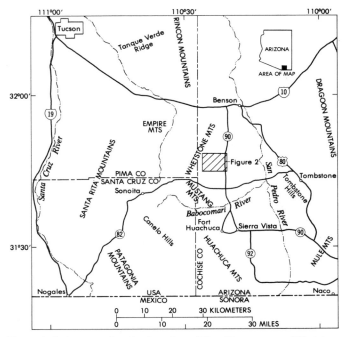

Figure 1. Index map showing location of Dry Canyon area, Whetstone Mountains.

Schist and crop out on the south side of French Joe Canyon in the northeastern part of the area. Rocks higher in the Cambrian section are exposed on the north side of Dry Canyon. The overlying west-dipping section of Devonian, Mississippian, Pennsylvanian, and Permian rocks extends to the crest of the range and is cut by a few discontinuous thrust faults that probably resulted from minor structural adjustments during Laramide tectonism and by a few normal faults resulting from Laramide and late Cenozoic movements. Paleozoic strata at the range crest and on the west flank are overlain unconformably by Cretaceous sedimentary rocks of the Bisbee Group, which was deposited on a locally rugged surface carved deeply into rocks of Early and Late Permian age. The distribution of the stratigraphic units and the structural relationships reported here are taken mainly from the geologic map of the Benson Quadrangle (Creasey, 1967). Additional information on the general geology of the Whetstone Mountains and a petrographic and facies analysis of the Paleozoic rocks were reported by Wrucke and Armstrong (1984).

Descriptions of the stratigraphic units reported here and interpretation of the depositional environments are based on field examinations, systematically collected samples, and the literature. Thicknesses of most of the units are those reported for strati-

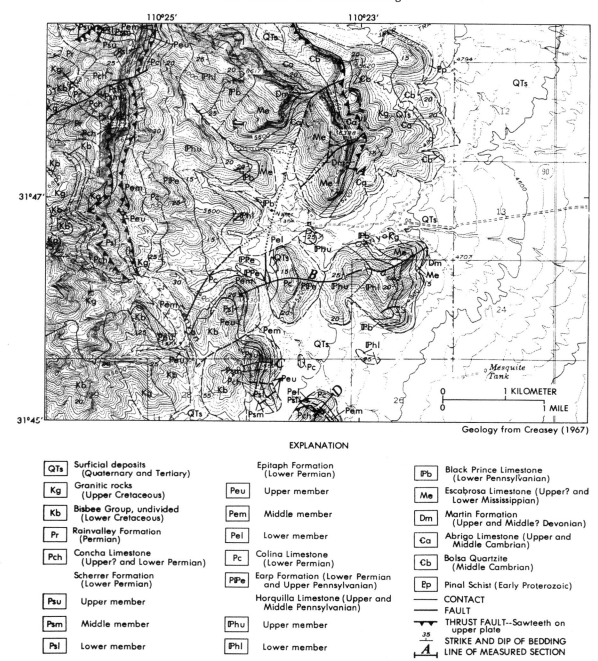

EXPLANATION

QTs	Surficial deposits (Quaternary and Tertiary)		Epitaph Formation (Lower Permian)	
Kg	Granitic rocks (Upper Cretaceous)	Peu	Upper member	
Kb	Bisbee Group, undivided (Lower Cretaceous)	Pem	Middle member	
Pr	Rainvalley Formation (Permian)	Pel	Lower member	
Pch	Concha Limestone (Upper? and Lower Permian)	Pc	Colina Limestone (Lower Permian)	

Scherrer Formation (Lower Permian)

Psu	Upper member
Psm	Middle member
Psl	Lower member

| PPe | Earp Formation (Lower Permian and Upper Pennsylvanian) |

Horquilla Limestone (Upper and Middle Pennsylvanian)

| Phu | Upper member |
| Phl | Lower member |

Pb	Black Prince Limestone (Lower Pennsylvanian)
Me	Escabrosa Limestone (Upper? and Lower Mississippian)
Dm	Martin Formation (Upper and Middle? Devonian)
Ca	Abrigo Limestone (Upper and Middle Cambrian)
Cb	Bolsa Quartzite (Middle Cambrian)
Ep	Pinal Schist (Early Proterozoic)

――― CONTACT
――― FAULT
▼▼▼ THRUST FAULT--Sawteeth on upper plate
╱35 STRIKE AND DIP OF BEDDING
A LINE OF MEASURED SECTION

Geology from Creasey (1967)

Figure 2. Geologic map of Dry Canyon area showing location of measured stratigraphic sections.

graphic sections measured in the Dry Canyon area at localities A–D, which are shown in Figure 2 and referenced in the descriptions that follow.

STRATIGRAPHIC SECTION

BISBEE GROUP, UNDIVIDED (LOWER CRETACEOUS)— Mostly gray, brown, and pale-red to dark-reddish-brown marine and nonmarine siltstone, sandstone, and shale, and locally abundant pelecypod- and gastropod-bearing limestone. The Glance Conglomerate

consisting of closely packed limestone pebbles and cobbles in fanglomerate deposits 0 to 300 ft (0 to 91.4 m) thick, but mostly 0 to 20 ft (0 to 6.1 m) thick, locally exists at the base (Tyrrell, 1957).

Angular Unconformity

RAINVALLEY FORMATION (PERMIAN)—Light-gray, medium- to thin-bedded limestone, olive-gray, thin-bedded dolomite, and very light-gray sandstone. Carbonate beds are algal laminates, locally containing abundant fragments of gastropods, cephalopods, and pelecypods. Basal beds are calcareous, well-sorted quartz sandstones that

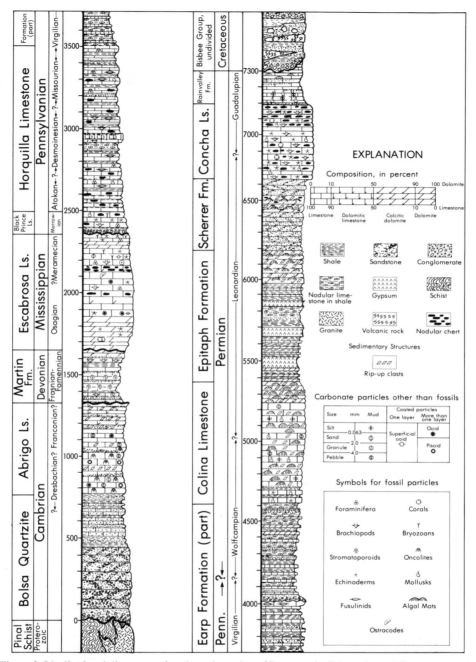

Figure 3. Idealized and diagrammatic columnar section of Proterozoic, Paleozoic, and Cretaceous rocks in the Whetstone Mountains.

total 3 to 5 ft (0.9 to 1.5 m) in thickness. Deposition of the formation took place in shallow subtidal to intertidal environments. Its stratigraphic position suggests a Guadalupian (Early and Late Permian) age for the unit in this area (Ross and Tyrrell, 1965). Thickness, as measured by Tyrrell (1957) about 4 mi (6.7 km) northwest of the Dry Canyon area, is 182 ft (55.5 m).

CONCHA LIMESTONE (UPPER? AND LOWER PERMIAN) —Divided by Tyrrell (1957) and Ross and Tyrrell (1965) into three informal members:

Upper member—Dark-gray, cliff-forming, medium- to thick-bedded limestone containing abundant brown chert and a rich fauna of cephalopods and gastropods. Tan chert, persistent throughout southeastern Arizona (Bryant and McClymonds, 1961), marks the base of the member. Only the lower part of the member is exposed in the Dry Canyon area. Thickness, as measured about 4 mi (6.7 km) northwest of the Dry Canyon area, is 49 ft (14.9 m), as interpreted here from Creasey (1967, his units 1 and 2), and 272 ft (82.9 m) according to Tyrrell (1957).

Middle member—Dark-gray, thick-bedded packstone containing small amounts of chert and abundant crinoids, bryozoans, and brachiopods.

Thickness, as measured by Tyrrell (1957) at section C, is 148 ft (45.1 m).

Lower member—Light- to dark-gray, cliff-forming, chert-rich wacke-stones and packstones in beds 1 to 6 ft (0.3 to 1.8 m) thick, containing abundant crinoids, echinoderms, and mollusks. Has bioclastic material that is poorly sorted and cross-bedded. The chert is nodular to lenticular. Thickness, as measured by Tyrrell (1957) at section C, is 137 ft (41.8 m). The abundant and diversified invertebrate fauna indicates deposition in well-circulated marine waters above wave base. Fusulinids collected in the upper member are considered Leonardian (Early Permian) by Tyrrell (1957), and early Guadalupian (Early and Late Permian) by Ross (Ross and Tyrrell, 1965). Thus the age for the Concha in this area is considered to be Early and Late(?) Permian. Thickness of the composite section of the formation reported here, as measured by Tyrrell (1957), is 557 ft (169.8 m).

SCHERRER FORMATION (LOWER PERMIAN)—Sandstone, limestone, dolomite, and marl. Divided into:

Upper member—White to pink and brown, fine- to medium-grained, cross-bedded sandstone. Thickness at section C, as measured by Creasey (1967), is 81 ft (24.7 m) and 118 ft (36.0 m), as measured by Tyrrell (1957).

Middle member—Light- to dark-gray, thick-bedded cherty limestone and olive-gray dolomite in beds as much as 6 ft (1.8 m) thick. Carbonate beds contain algal laminations, weakly developed stromatolites, litho-clasts, mud cracks, and sparry dolomite. The sparry dolomite is in cavities formerly occupied by gypsum and anhydrite. Thickness at section C, as measured by Creasey (1967), is 139 ft (42.4 m) and 154 ft (46.9 m), as measured by Tyrrell (1957).

Lower member—White to red sandstone similar to that of upper member. Thickness at section C, as measured by Creasey (1967), is 340 ft (103.6 m) and 198 ft (60.4 m), as measured by Tyrrell (1957). Sedimentary structures present in carbonate rocks indicate that deposition took place in subtidal to high intertidal environments. The sandstones may have accumulated as shallow marine deposits or as beach sands. In this area, the Scherrer is probably Leonardian (Early Permian) in age on the basis of its fossils and stratigraphic position (Ross and Tyrrell, 1965). Thickness of the formation at section C, as measured by Creasey (1967), is 560 ft (170.7 m) and 470 ft (143.3 m), as measured by Tyrrell (1957).

EPITAPH FORMATION (LOWER PERMIAN)—Dolomite, limestone, sandstone, marl, and gypsum. Divided into:

Upper member—Light- to medium-gray and bluish-gray limestone and dolomite, locally cherty, in beds 1 to 4 ft (0.3 to 1.2 m) thick. Thickness, as measured by Creasey (1967) at section B, is 200 ft (61.0 m).

Middle member—Interlayered yellowish-gray to medium-gray and deep-red, commonly thin-bedded limestone, dolomite, marl, and sandstone, locally containing rarely exposed beds of gypsum (Graybeal, 1962). The thickest known accumulation of gypsum in the Whetstone Mountain area is 33 ft (10.1 m) in SE¼Sec.4,T.20S.,R.19E., 2 mi (3.6 km) south of the Dry Canyon area. Thickness, as measured by Creasey (1967) at section B, is 858 ft (261.5 m).

Lower member—Light- to dark-gray, thin-bedded dolomite and minor marl and sandstone. Thickness, as measured by Creasey (1967) at section D, is 140 ft (42.7 m). Carbonate rocks in the formation range from lime mudstones to pure dolomites. They contain abundant mud cracks, mud chips, algal lamina-tions, gypsum pseudomorphs, and well-developed algal stromatolites. Commonly forms debris-covered slopes. Probably deposited on an en-closed shallow shelf in which evaporation exceeded influx of marine waters, resulting in saline intertidal and supratidal depositional environments similar to the sabkhas of the modern Persian Gulf. Contains few fossils. Age of the Epitaph fauna has been reported by Gilluly and others (1954) as Leonardian (Early Permian). Thickness of the formation, as measured by Creasey (1967), is 1,198 ft (365.2 m) and 973 ft (296.6 m), as measured by Tyrrell (1957).

Figure 4. View toward the south of Devonian, Mississippian, Pennsyl-vanian, and Permian strata on south side of Dry Canyon.

COLINA LIMESTONE (LOWER PERMIAN)—Medium- to dark-gray lime mudstone and subordinate grayish olive-green dolomite. A few limestone beds are black on fresh surfaces. Beds are 1 to 4 ft (0.3 to 1.2 m) thick. Algal mats, well-developed domal stromatolites, and mud chips and cracks are abundant. Many limestone beds have sparry dolomite in vugs formed by the dissolution of gypsum. Thin beds and lenses of molluscan bioclastic material present locally. Gastropods and echinoid spines locally abundant. The sedimentary structures present indicate deposition in a subtidal to supratidal environment. Gilluly and others (1954) reported the age of the Colina as Wolfcampian and Leo-nardian (Early Permian). Thickness, as measured by Creasey (1967) at section D, is 180 ft (54.9 m) and 493 ft (150.3 m) at section B, as measured by Tyrrell (1957).

EARP FORMATION (LOWER PERMIAN AND UPPER PENNSYLVANIAN)—Pale-red to reddish-brown, thin-bedded, nodu-lar, calcareous sandstone, siltstone, and shale interlayered with subordi-nate light- to medium gray to pinkish-gray limestone and sparse, gray to tan dolomite. Ross and Tyrrell (1965), Creasey (1967), and Hayes and Raup (1968) reported a conspicuous chert-pebble conglomerate 2 to 6 ft (0.6 to 1.8 m) thick a little below the middle of the formation. Limestone beds commonly are a foot or more thick and in some areas occur throughout the formation, but dolomite beds occur only near the top. These dolomites weather bright orange and contain algal mats, oncolites, stromatolites, abundant dolomite pseudomorphs after gypsum and an-hydrite, mud cracks, rip-up clasts, worm burrows, and fragments of gastropods and other mollusks. The lower part of the formation suggests a shallow-water, near-shore to shoaling-upward depositional environ-ment and influxes of terrigenous sediments. The dolomites near the top of the formation indicate subtidal, intertidal, and supratidal conditions. This mixture of environments resulted from transgressive and regressive seas during the rise of the Defiance-Zuni highlands in northern Arizona (Butler, 1971; Peirce, 1976; Ross, 1978). The Earp Formation is Virgil-ian to Wolfcampian in age (Late Pennsylvanian and Early Permian) (Gilluly and others, 1954; Ross and Tyrrell, 1965). Thickness at section B, as measured by Tyrrell (1957), is 578 ft (176.2 m) and 801 ft (244.1 m), as measured by Creasey (1967).

HORQUILLA LIMESTONE (UPPER AND MIDDLE PENN-SYLVANIAN)—Divided into:

Upper member—Interlayered sequences, commonly 12 to 30 ft (3.7 to 9.1 m) thick, of light- to medium-gray and pinkish lime mudstone, bioclastic wackestone, ooid packstone, and orange-red to dark-red, poorly exposed, calcareous siltstone and sandstone. About one-third of the member is limestone, generally fossiliferous. Some limestone beds in the lower half of the member contain sparse, brownish-gray to brownish-black chert nodules. Thickness at section B, as measured by Tyrrell (1957), is 619 ft (188.7 m) and 907 ft (276.5 m), as measured by Creasey (1967).

Lower member—Light- to dark-gray lime mudstone and packstone in beds 1 to 3 ft (0.3 to 0.9 m) thick and a few sequences of thin-bedded, light-gray to dark-red, calcareous shale and siltstone 12 to 36 ft (3.7 to 11.0 m) thick. Gray, brown, and black chert nodules and beds are common in the limestone, which also contains abundant remains of algae, foraminifers, echinoderms, mollusks, brachiopods, and solitary corals. The uppermost part of the lower member consists of a 30-ft (9-m)-thick biostrome of *Chaetetes* sp., *Syringopora* sp., and solitary corals. Most of the member forms a slope and ledge topography. The base of the formation is placed at the bottom of the lowest limestone ledge above the slope-forming beds at the top of the Black Prince Limestone (Fig. 4). Thickness, as measured by Creasey (1967) at section B and reinterpreted here to exclude his basal unit of the Horquilla Limestone (reassigned here to the Black Prince Limestone), is 319 ft (97.2 m). The limestone in the formation is interpreted as having been deposited in less than 100 ft (about 30 m) of water as lime mud and calcareous sand made of bioclastic material and abundant micropellets and peloids. Local hardground is characterized by colonies of bryozoans and crinoids and a rich fauna of echinoderms, brachiopods, and mollusks. Silty and sandy beds resulted from flooding of terrigenous material derived from the rising Defiance-Zuni highland in northern Arizona. Ross and Tyrrell (1965) and B. L. Mamet (written communication, 1978) have shown from studies of foraminifers that the lowest 112 ft (34.1 m) of the formation is of Atokan age (Middle Pennsylvanian), the overlying 350 ft (106.7 m) is Desmoinian (Middle Pennsylvanian), and the higher parts are Missourian and Virgilian (Late Pennsylvanian). Thickness of the formation, as measured by Tyrrell (1957), is 1,012 ft (308.5 m) and 1,226 ft (373.7 m), as reinterpreted here from Creasey (1967).

BLACK PRINCE LIMESTONE (LOWER PENNSYLVANIAN) —Mostly light- to medium-gray, in places pinkish-gray, bioclastic limestone in beds commonly 1 to 4 ft (0.3 to 1.2 m) thick, containing sparse nodules of brown to gray chert. Bioclastic debris consists of fragmented calcareous algae, foraminifers, ostracods, brachiopods, bryozoans, corals, and mollusks. The basal 16 to 40 ft (4.9 to 12.2 m) is composed of chert-pebble conglomerate, thin-bedded to massive chert, and maroon shaly siltstone. The chert locally contains cavities lined with quartz crystals. The formation crops out as ledges, except for the shale and conglomerate at the base, which forms slopes. Accumulated in a warm, shallow, nutrient-rich sea that transgressed, an uneven, weathered erosional surface developed on the Escabrosa Limestone. Conglomerate and siltstone at the base of the formation are weathered residue from the Escabrosa. The Black Prince was considered Mississippian by Gilluly and others (1954), but Nations (1963) reported a Pennsylvanian age for the formation, based on specimens of *Millerella* from the Johnny Lyon Hills, 12 mi (20 km) north-northeast of Benson, and *Chaetetes* from the Gunnison Hills, 18 mi (30 km) northeast of Benson. A Pennsylvanian age for the formation in the Whetstone Mountains was suggested by Ross and Tyrrell (1965), based on fusulinid correlations. B. L. Mamet (written communication, 1976) stated that the Black Prince Limestone in the Whetstone Mountains contains a large assemblage of calcareous algae and foraminifers of Morrowan age (zone 20 of Mamet [Armstrong and Mamet, 1977]; Early Pennsylvanian). On these bases, an Early Pennsylvanian age is accepted for the unit in this area. Goldhammer and Elmore (1984) described the Black Prince Limestone as a shelf, tidal-flat and paleosol facies in a shoaling upward sequence. Paleosol occurs as calcrete horizons and as nodular limestone beds representing terra rossa soils. Thickness at section B as measured by Tyrrell (1957) is 116 to 131 ft (35.4 to 39.9 m), as measured by Mamet (written communication, 1976) is 165 ft (50.3 m), and as measured by Creasey (1967) is 128 ft

(39.0 m) (here reinterpreted to exclude his lower unit of the Black Prince Limestone and to include his basal unit of the Horquilla Limestone).

Disconformity

ESCABROSA LIMESTONE (UPPER? AND LOWER MISSISSIPPIAN)—Light- to medium-gray, massive-appearing, generally thick-bedded limestone and dolomite. The upper 220 ft (67.1 m) of the formation is medium light-gray to medium-gray, thick-bedded limestone and dolomite. Limestone beds commonly are crinoidal packstones. Dark-gray chert nodules are abundant in the basal 33 ft (10.1 m) of this upper part. The middle 262 ft (79.9 m) of the formation is light-gray, bioclastic wackestone and packstone in beds about 1 to 4 ft (0.3 to 1.2 m) thick, composed of fragments of crinoids, bryozoans, and brachiopods. Some beds are rich in ooids, calcareous algae, and foraminifers. A few beds of dolomite, sandy limestone, and limestone containing chert nodules are present near the middle. The lowermost 230 ft (70.1 m) of the formation consists of dolomite in beds commonly 1 to 6 ft (0.3 to 1.8 m) thick and composed of dolomite rhombs (100 to 200 μm) in size and dolomite pseudomorphs of crinoids and syringoporids. The formation typically weathers to steep, ledgey slopes and prominent cliffs. It was deposited in a shallow, open marine environment rich in echinoderms, corals, bryozoans, and brachiopods. Armstrong and Mamet (1978) reported that conodonts, foraminifers, and calcareous algae indicate the base of the formation in this area to be of Osagean age (zone 7; Early Mississippian) and the top to be probably of Meramecian age (zone 10 of Mamet [Armstrong and Mamet, 1977]; Late Mississippian). Thus, the age of the formation in this area is regarded as Early and Late(?) Mississippian. Thickness at section B as measured by Tyrrell (1957) is 555 ft (169.2 m), as measured by Armstrong and Mamet (1978) is 712 ft (217.0 m), and as measured by Creasey (1967) is 742 ft (226.2 m) (here reinterpreted to include his basal unit of the Black Prince Limestone and to exclude the basal 10 ft (3.0 m) of strata that he included in the Escabrosa Limestone).

Disconformity

MARTIN FORMATION (UPPER AND MIDDLE? DEVONIAN)—Light-purplish-gray to dark-blue-gray, thin- to medium-bedded limestone, light-gray to dark-blue-gray and tan, thin- to thick-bedded dolomite, and subordinate tan to yellowish-tan sandstone and dolomitic sandstone. Minor shale occurs near the top of the formation. The limestone is peletoidal lime mudstone that contains small but variable amounts of dispersed fine- to medium-sized quartz grains. The dolomite appears to have been a peletoidal lime mudstone that has been largely replaced by low-porosity dolomite in subhedral rhombs 100 to 200 μm in size. Sparse intracrystalline voids are filled with calcite. The sandstone forms a few sequences 1 to 12 ft (0.3 to 3.7 m) thick in the lower third of the unit. Sandstone near the top of the formation is bimodal, consisting of 5 percent of rounded grains 0.3 to 0.7 mm in size, concentrated in laminae among angular grains commonly about 0.1 mm across. Atrypid brachiopods and stromatoporoids occur in the carbonate rocks. The formation was deposited in shallow marine to subtidal environments. Mostly Frasnian (Late Devonian), with some probable Givetian-age (Middle Devonian) rocks at the base. Beds of probable Famennian age (Late Devonian), about 30 ft (about 9 m) thick, at the top of the formation were reassigned to the Percha Formation by Schumacher (1978), but the name "Percha" has not been used in Arizona by other authors. Basal 10 to 15 ft (3.0 to 4.6 m) of the Escabrosa Limestone, as described by Creasey (1967), is here interpreted as belonging to the Famennian (Late Devonian) part of the Martin. The Martin Formation in this area is considered to be Middle(?) and Late Devonian, based on invertebrate macrofossils and conodonts (Schumacher, 1978). Thickness, as measured by Creasey (1967) at section A, is 301 ft (91.7

m), to which we add, as discussed above, an additional 10 ft (3.0 m) of strata that he assigned to the Escabrosa Limestone, for a total thickness of 311 ft (94.8 m).

Disconformity

ABRIGO LIMESTONE (UPPER AND MIDDLE CAMBRIAN) —In this area, consists of three members. Descriptive information taken in part from Hayes (1978):

Upper member—Dark-purple to greenish-black sandy limestone and subordinate sandy dolomite and dolomitic sandstone. Most beds are thin and laminated. Cross-laminations in thin, lenticular sets are common. The sand is mostly medium-grained quartz, but glauconite is abundant in many beds and accounts for the dark color of the rocks. Some dolomite beds consist of flat-pebble, intraformational conglomerate composed of rip-up clasts of dolomite. The formation weathers light-brown and probably is of intertidal and supratidal origin. Thickness, as measured by Creasey (1967, his units 1 and 2) on the north side of French Joe Canyon, 1 mi (1.6 km) north of the Dry Canyon area, is 145 ft (44.2 m).

Middle member—Dark-gray to dark-blue-gray lime mudstone that weathers light- to medium-gray and forms beds commonly 2 to 10 cm thick, although a few beds are as thick as 90 cm. Many beds are bioturbated; some beds contain oncolites. A few beds are limestone conglomerates composed of rip-up clasts of lime mudstone. Subordinate shale and siltstone form planar and distinctive wavy laminations in the limestone, but laminated olive-gray shale and siltstone are the dominant rock types in a few stratigraphic intervals as thick as 50 ft (15.2 m). The middle member is interpreted to have formed in subtidal and intertidal depositional environments. Thickness, as interpreted from measurements made by Creasey (1967, his units 3 through 18) on the north side of French Joe Canyon, 1 mi (1.6 km) north of the Dry Canyon area, is 406 ft (123.7 m).

Lower member—Gray, tan, and olive-tan shale, medium-gray limestone, and tan siltstone mostly in thin, laminated beds. These rocks form a section 4 to 33 ft (1.2 to 10.1 m) thick, in which either limestone or siltstone and shale are dominant. The basal 4 ft (1.2 m) consists of

medium-grained, thin-bedded orthoquartzite containing a few shale partings. Probably formed in an intertidal environment. Thickness, as interpreted from measurements made by Creasey (1967, his units 19 through 42) on the north side of French Joe Canyon, 1 mi (1.6 km) north of the Dry Canyon area, is 315 ft (96.0 m).

Thickness of the formation, as measured by Creasey (1967), is 866 ft (264.0 m).

BOLSA QUARTZITE (MIDDLE CAMBRIAN)—Grayish-red-purple to white quartzite and subordinate sandstone. The formation is mostly medium- to coarse-grained orthoquartzite, composed of subangular to subrounded grains that have prominent quartz overgrowths. Basal beds are poorly sorted, feldspathic gritstone to fine-pebble conglomerate, in which the larger grains are milky vein quartz. The formation becomes finer grained upward. Higher beds contain abundant fine- to medium-size grains and interbeds of brown to green shale. The quartzite forms resistant layers 1 to 4 ft (0.3 to 1.2 m) thick that are composed of sets of cross-strata commonly 10 to 20 cm thick, thought to have been formed by wave ripples. The formation is considered by Hayes (1978) to have accumulated in an offshore bar or beach environment. Thickness, as measured by Tyrrell (1957) on the north side of French Joe Canyon, about 0.7 mi (1.2 km) north of the Dry Canyon area, is 451 ft (137.5 m).

Unconformity

PINAL SCHIST (EARLY PROTEROZOIC)—Yellowish-gray to medium-gray well-foliated, finely layered, fine-grained schist, including muscovite-quartz, cordierite-andalusite-muscovite, and cordierite-muscovite-chlorite-quartz varieties in exposures mostly northeast of the Dry Canyon area. Most rocks contain small amounts of biotite; some have trace amounts of tourmaline. The rocks are strongly foliated. A few outcrops exhibit a mineral lineation in the plane of the foliation. Locally, the rocks are warped into small folds that have a weak fracture cleavage parallel to the fold plane. The schist is interpreted as consisting of regionally metamorphosed shales and silty or sandy shales. Intruded by Proterozoic alaskite and quartz monzonite in northern parts of the Whetstone Mountains.

REFERENCES CITED

Armstrong, A. K., and Mamet, B. L., 1977, Carboniferous Microfacies, Microfossils, and Corals, Lisburne Group, Arctic Alaska: U.S. Geological Survey Professional Paper 849, 144 p.
—— , 1978, The Mississippian System of southwestern New Mexico and southeastern Arizona, in Callender, J. F., Wilt, J. C., and Clemons, R. E., eds., Land of Cochise, Southeastern Arizona: New Mexico Geological Society Guidebook, Twenty-ninth Field Conference, p. 183–192.
Bryant, D. L., and McClymonds, N. E., 1961, Permian Concha Limestone and Rainvalley Formation, southeastern Arizona: American Association of Petroleum Geologists Bulletin, v. 45, p. 1324–1333.
Butler, W. C., 1971, Permian sedimentary environments in southeastern Arizona: Arizona Geological Society Digest, v. 9, p. 71–94.
Creasey, S. C., 1967, Geologic map of the Benson Quadrangle, Cochise and Pima Counties, Arizona: U.S. Geological Survey Miscellaneous Geologic Investigation Map I-470, scale 1:48,000.
Gilluly, J., Cooper, J. R., and Williams, J. S., 1954, Late Paleozoic Stratigraphy of Central Cochise County, Arizona: U.S. Geological Survey Professional Paper 266, 49 p.
Goldhammer, R. K., and Elmore, R. D., 1984, Paleosols capping regressive carbonate cycles in the Pennsylvanian Black Prince Limestone, Arizona: Journal of Sedimentary Petrology, v. 54, no. 4, p. 1124–1137.
Graybeal, F. T., 1962, The Geology and Gypsum Deposits of the Southern Whetstone Mountains, Cochise County, Arizona [M.S. thesis]: Tucson, University of Arizona, 80 p.
Hayes, P. T., 1978, Cambrian and Ordovician rocks of southeastern Arizona and

southwestern New Mexico, in Callender, J. F., Wilt, J. C., and Clemons, R. E., eds., Land of Cochise, Southeastern Arizona: New Mexico Geological Society Guidebook, Twenty-ninth Field Conference, p. 165–173.
Hayes, P. T., and Raup, R. B., 1968, Geologic Map of the Huachuca and Mustang Mountains, southeastern Arizona: U.S. Geological Survey Miscellaneous Geologic Investigations Map I-509, scale 1:48,000.
Nations, J. D., 1963, Evidence for a Morrowan age for the Black Prince Limestone of southeast Arizona: Journal of Paleontology, v. 37, p. 1252–1264.
Peirce, H. W., 1976, Elements of Paleozoic tectonics in Arizona: Arizona Geological Society Digest, v. 10, p. 37–57.
Ross, C. A., 1978, Pennsylvanian and Early Permian depositional framework, southeastern Arizona, in Callender, J. F., Wilt, J. C., and Clemons, R. E., eds., Land of Cochise, southeastern Arizona: New Mexico Geological Society Guidebook, Twenty-ninth Field Conference, p. 193–200.
Ross, C. A., and Tyrrell, W. W., Jr., 1965, Pennsylvanian and Permian fusilinids from Whetstone Mountains, Arizona: Journal of Paleontology, v. 39, p. 615–635.
Schumacher, D., 1978, Devonian stratigraphy and correlations in southeastern Arizona, in Callender, J. F., Wilt, J. C., and Clemons, R. E., eds., Land of Cochise, Southeastern Arizona: New Mexico Geological Society Guidebook, Twenty-ninth Field Conference, p. 175–181.
Tyrrell, W. W., Jr., 1957, Geology of the Whetstone Mountain Area, Cochise and Pima Counties, Arizona [Ph.D. thesis]: New Haven, Yale University, 171 p.
Wrucke, C. T., and Armstrong, A. K., 1984, Geologic Map of the Whetstone Roadless Area, Cochise and Pima Counties, Arizona: U.S. Geological Survey Miscellaneous Field Studies Map MF 1614-B, scale 1:48,000.

Saguaro National Monument, Arizona: Outstanding display of the structural characteristics of metamorphic core complexes

George H. Davis, Department of Geosciences, University of Arizona, Tucson, Arizona 85721

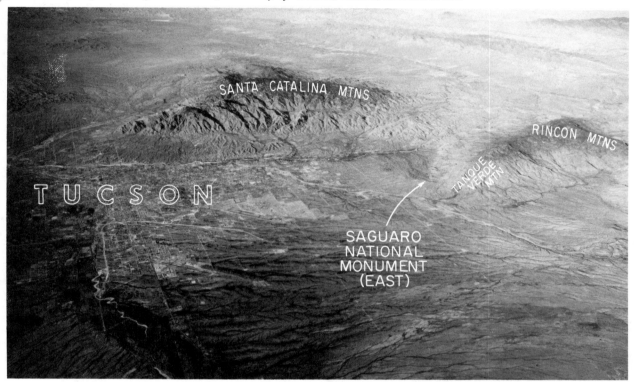

Figure 1. U-2 photograph showing the location of the field site with respect to the city of Tucson and the Santa Catalina, Tanque Verde, and Rincon mountains.

LOCATION AND ACCESSIBILITY

The Saguaro National Monument Centennial Field Site lies on the eastern outskirts of Tucson, Arizona (Fig. 1), along the flank of the Tanque Verde Mountains. Access to the site is via the Loop Drive, a paved road which serves visitors to Saguaro National Monument (east) (Fig. 2). The stops, which provide a clear view of the structural characteristics of metamorphic core complexes, are located near the half-way point of Loop Drive. Stops 1 through 4 need take only 1 hour. Allow yourself a second hour to see the geology at Stops 5 and 6, and a third hour to examine the structures at Stop 7. The walking is easy within the area of Stops 1 through 4. There are some slopes to descend and ascend in reaching Stops 5 and 6. Those who visit Stop 7 should be prepared for hillslopes and some sharp, steep-sided vertical ribs of limestone bedrock. In general, however, hiking within the site is straightforward and scenic.

Saguaro National Monument has beautiful desert vegetation, including saguaro, ocotillo, palo verde, prickly pear, and barrel cacti. Deer and javalina commonly may be seen in the vicinity of Stops 5 through 7, especially late in the afternoon. A canteen of cold water will be a comfort on warm days and a necessity on hot days. Be on the lookout for the hazards of cactus spines, thorny shrubs, scorpions, and rattlesnakes. Snakes are seldom seen during the months of November through mid-March. The rattlesnakes generally do not venture out during the heat of summer days.

Be advised that you are not permitted to collect anything within Saguaro National Monument.

SIGNIFICANCE

More than any other locality, the geology of this site served as the basis for distinguishing and classifying the fundamental structural characteristics of metamorphic core complexes as presented in Davis (1977, 1980) and Davis and Coney (1979). This site was the showpiece used in presenting the diagnostic features of metamorphic core complexes to participants in the Geological Society of America Penrose Conference on Cordilleran Metamorphic Core Complexes, held in Tucson in May, 1977 (Crit-

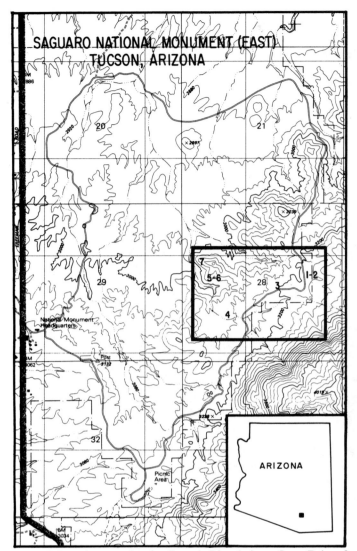

Figure 2. Location map showing Stops 1–7 in relation to Loop Drive of Saguaro National Monument (east).

tenden and others, 1980). It is now recognized that the mylonites, ultramylonites, and cataclasites exposed at the Saguaro National Monument site are among the very finest. The fault rocks and structures, taken together, provide an illuminating, stop-action glimpse of the progressive, ductile through brittle shearing which is responsible for fashioning the characteristics of metamorphic core complexes.

SITE INFORMATION

Beginning at Stop 1, and proceeding through Stop 7, an opportunity is presented to traverse the fundamental components of a metamorphic core complex (Fig. 3): mylonite and ultramylonite derived from Precambrian and Tertiary quartz monzonites;

chloritically altered microbreccias derived from the mylonite and ultramylonite; fine-grained microbreccias (cataclasite and ultra-cataclasite) derived from the chloritically altered breccias; the detachment fault (or décollement) separating cataclastic and mylonitic rocks below from nonmylonitic, noncataclastic cover rocks above; and some folded Paleozoic cover rocks.

The distribution and character of each of these components of the Rincon Mountains metamorphic core complex are presented in Figure 4, a structural geologic map of the site. Note that the locations of the seven stops highlighted in this guide are posted on the structure map. The relationship of the geology of this small area to that of the entire Tanque Verde and Rincon mountains can be appreciated by studying the geologic map of the Rincon Valley Quadrangle, prepared by Harald Drewes (1978). This useful map generally has been available for purchase at the Visitors Center at Saguaro National Monument. Drewes' mapping shows quite clearly that the site of interest lies on the northwestern flank of Tanque Verde arch, a gently southwest-plunging antiform of lineated quartzo-feldspathic gneiss.

Stop 1 reveals the character of one of the two major phases of mylonite gneiss exposed in the Tanque Verde and Rincon mountains. It is identical texturally and compositionally to a mylonitic gneissic phase interpreted to be mylonitized 1400 Ma quartz monzonite. Shakel and others (1977) extracted zircons from this phase of gneiss in the nearby Santa Catalina Mountains, and using Ur-Pb isotopic analysis, found them to be Proterozoic. The results of progressive mylonitization of unfoliated Precambrian porphyritic quartz monzonite to coarse-grained mylonitic gneiss can be observed in metamorphic core complexes at a number of locations in southern Arizona.

The mylonite gneiss is beautifully expressed in the physiography of the Santa Catalina and Rincon Mountains. The rock is so strongly layered that it resembles a thick stack of gently dipping, massively bedded sedimentary and/or volcanic rocks. However, in outcrop view the bedrock is seen to be mylonitic quartzo-feldspathic gneiss, and the layering proves to be just one of many expressions of the penetrative foliation within the gneiss.

The mylonitic gneiss at Stop 1 is coarse-grained augen gneiss with abundant large feldspar porphyroclasts (0.5 cm to 6 cm). (Stops 1A and 1B are additional sites where the mylonitic gneisses are especially interesting.) The mylonitic nature of the fabric is especially evident in microscopic view. Ribbon-quartz

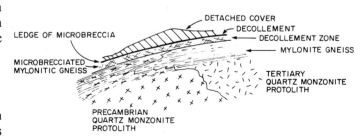

Figure 3. Schematic diagram showing the fundamental structural components of metamorphic core complexes.

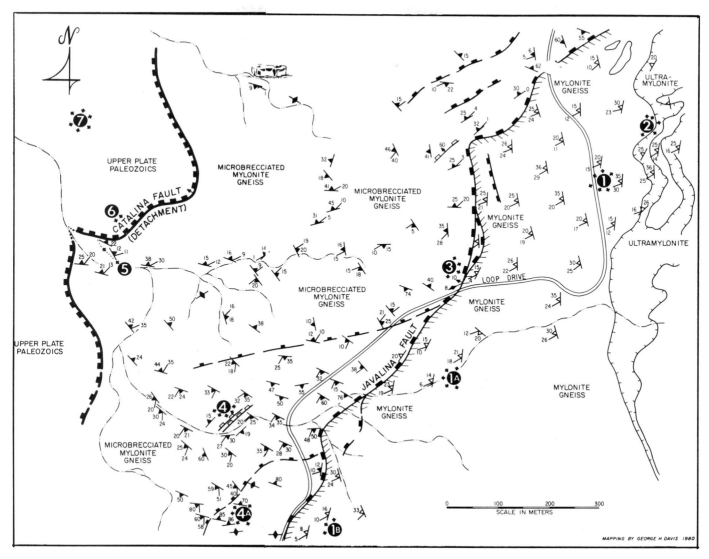

Figure 4. Structural geologic map of the field site, showing locations of stops.

layers and laminae wrap around the feldspars in a gently undulating habit. Alignment of nonequidimensional feldspar porphyroclasts also contributes to foliation expression. The strike of foliation at this stop is north-south. The foliation dips gently, typically less than 30°, to the west.

Mineral lineation is exquisitely developed. It lies within the plane of foliation and thus is everywhere gently plunging. The lineation shows an extraordinary degree of preferred orientation; N60°E-S60°W. Lineation is defined mainly by the structural response of quartz and quartz laminae to crystal-plastic deformation.

The strain significance of the mylonite gneiss has become understood through detailed structural analysis. The dominant foliation (S) is the XY plane of flattening, and the lineation direction is the X direction of greatest elongation. Locally, C

surfaces cut the mylonite as planes of slip, spaced at 1 cm or less. These dip more steeply than S and reveal a normal-slip sense of simple shear along a S60°W line. This same movement plan is revealed by asymmetric folds and sheath folds whose low-dipping axial surfaces are axial planar to the S surfaces.

Although the mylonites observed at this stop were derived from Precambrian rocks, the mylonitization event was actually Tertiary in age (Davis and Coney, 1979). This is clear because the same fabrics are seen in a phase of mylonitic gneiss that is derived from Tertiary quartz monzonite. This second dominant phase of mylonitic gneiss is abundantly exposed in the Santa Catalina and Rincon Mountains. Its protolith is interpreted to be approximately 50 Ma in age, based on radiometric dating (Shakel and others, 1977; Keith and others, 1980). This rock expresses itself as a light-colored, fine- to medium-grained garnet- and mica-bearing mylonitic gneiss. It crops out within short hiking distance of Stop 1 but is not featured in this chapter.

The mylonites at Stop 1 probably formed initially at depths of 6 to 7.2 mi (10 to 12 km). Fabrics and structures are thought to have originated during simple shear within a 0.6- to 1.2-m- (1- to 2-km-) thick regional shear zone, which accommodated normal-slip simple shear during regional extension (Davis, 1981a, 1983, 1986). The line of movement is parallel to mineral lineations: N60°E, S60°W.

Stop 2 provides a glimpse of steel gray to black ultramylonite, which comprises a shear zone separating the coarse-grained mylonite gneisses derived from Precambrian quartz monzonite and the finer grained mylonite gneisses derived from Tertiary garnet-mica quartz monzonite. The ultramylonite is seen first at Stop 2 as a float, but can be traced to modest bedrock exposures in small washes and along low ridges. The ultramylonite is very fine grained and contains tiny yet visible white broken feldspar chips, irregular in size and shape, floating in the ultramylonitic quartz-rich matrix. Mylonitic aplite and pegmatite layers and laminae accentuate the foliation. Penetrative folding within the ultramylonite is evident from place to place. The folds are intrafolial, tight to isoclinal, overturned to recumbent structures whose hinges lie in the plane of foliation. The ultramylonite is an exceedingly deformed rock.

Sense of shear can be determined for the ultramylonitic shear zone. The muscovite micas within the ultramylonite are thought to have formed in the late stages of shearing and are preferentially oriented in an imbricate, shingled manner. The nature of imbrication relative to the foliation surfaces discloses sense of shear, which is normal-slip shear along a S60°W line. Using the "fish flash" technique developed by Stephen J. Reynolds (State of Arizona Bureau of Geology), the preferred alignment of the micas can be easily documented in the field. Simply hold a piece of the ultramylonite in the flat of your hand, and view the foliation surface in the direction of lineation and at a low angle. If the mica "fish" reflect light into your eyes like an array of tiny mirrors, you can be sure that you are looking in the direction of shearing of the upper part of the sample with respect to the lower. If the surface simply looks dull when so viewed, rotate the sample by 180° in the palm of your hand (about a vertical axis) and view the foliation surface once again in the direction of lineation. You should see a difference.

The ultramylonite shear zone can be traced as a discrete unit for many kilometers within the Rincon and Tanque Verde Mountains. On Drewes' (1978) map of the Rincon Valley Quadrangle it is shown as schist that can be traced throughout much of the Rincon Mountains. The mylonite almost always occurs at the interface of the two main phases of mylonite, which were brought into contact by the shearing displacements.

Stop 3 is located on the trace of a gently west-dipping fault, here named the Javalina fault, which separates mylonite and ultramylonite on the footwall (to the east) from chloritically altered brecciated mylonites on the hanging wall (see Fig. 4). The trace of the fault trends north-northeast and can be readily

mapped because of the profound contrast in the physical condition and geometric character of the rocks on either side. The specific location of Stop 3 is one of the few places where the Javalina fault has some physical expression, in the form of a gently dipping surface underlain by relatively highly fractured, hematite-stained mylonitic gneiss.

It is useful to hike a short distance to the north along the fault trace (as shown in Fig. 4) and compare and contrast the rocks on either side. To the east the rocks are normal mylonite gneiss, with large feldspar porphyroclasts of rounded to elliptical shapes, set in a matrix of mainly quartz and feldspar. Fracturing is ordinary, and alteration of feldspar is insignificant, in the typical nonbrecciated mylonite gneiss. In contrast, the brecciated mylonite gneisses to the west of the Javalina fault are marked by extreme fracturing, microfaulting, and alteration. Grain size is greatly reduced by cataclastic deformation. Alteration assemblages of chlorite, epidote, hematite, and manganese transform fresh, light-colored mylonite gneiss to blue-green brecciated and microbrecciated counterparts. On weathered surfaces the foliation and lineation, so conspicuous in the typical mylonite gneiss, is masked and sometimes nearly obliterated by fracturing, faulting, brecciation, and alteration. Where orientations of relict foliation and lineation are measured within the breccias, they are commonly found to be disoriented with respect to foliation and lineation in the footwall mylonites (see Fig. 4).

Stop 4 provides a look at the most graphic outcrop expressions of the brecciated and microbrecciated mylonite gneisses (see Fig. 4). Excellent exposures are found all along the wash leading to Stop 5. Different intensities of brecciation are evident. Where brecciation is weak, feldspar porphyroclasts remain very large (1 to 2 cm), but they are quite fractured. Some show significant pull-apart structure and chlorite crystal-fiber veining. Within strongly microbrecciated mylonites there is a decrease in size of feldspar porphyroclasts to an average of 0.15 to 0.25 mm. The feldspar porphyroclasts that survive are noticeably more fractured than those in the weakly brecciated mylonites. Chlorite content shows an increase and is commonly associated with opaques and epidote. Veinlets of chlorite and epidote plus calcite are moderately abundant. Where microbrecciation has been most intense, the average size of feldspar chips is about 0.1 mm, the largest feldspar porphyroclasts being only 0.3 mm or so. The fine-grained quartzo-feldspathic groundmass has an average grain size of only 0.0005 mm. Abundant veinlets of chlorite and epidote follow innumerable microfaults which pervade the rock.

Mapping the internal structural fabric of the brecciated and microbrecciated mylonitic rocks reveals the presence of discrete fault-bounded blocks which have rotated with respect to one another (Davis and others, 1981; DiTullio, 1983) (see Fig. 4). Overall, the relict foliation in the zone of breccia departs radically from the normal north-northeast strike to an average which is northwest. The typical east-northeast lineation which characterizes nonbrecciated mylonites shifts to other orientations, notably more southerly. Locally the foliation is vertical, and it is possible that relict foliation in some fault-bounded blocks has been overturned, although this is difficult to prove. From place to place, faults are well exposed at the outcrop scale. In fact, one of the diagnostic structural characteristics of large areas of brecciated mylonite gneiss is the presence of long, continuous, very gently dipping planar fracture partings. These are expressions of faults which may have accommodated meters or tens of meters of fault translation.

The zone of microbrecciation appears to have been produced by prolonged superposed faulting and fracturing under conditions of elevated fluid pressure (Davis and Coney, 1979). Breccia fragments at all scales are themselves seen to be composed of breccia, and this suggests that microbrecciation was carried out by a progressive deformation through time. The nature of fracturing and the pervasive alteration records the role of fluids and fluid pressure in the fault-induced conversion of mylonites to microbreccias. Although the microbrecciation and rotation of fabrics obviously postdated the formation of the original mylonites, the structural movements which produced microbrecciation and rotational faulting seem to have been coordinated with the geometry of shearing which fashioned the mylonites. This is shown in part by the tendency of the strike of rotated relict foliation in the microbreccias to be aligned roughly at right angles to the trend of lineation in the footwall mylonites.

Stop 5 is the ledge of cataclasite and ultracataclasite which marks the very top of the zone of microbrecciated mylonite gneiss (see Fig. 4). The top surface of the ledge of cataclasite is the Santa Catalina fault, the structure to be examined at Stop 6. The cataclasite has no directional fabric whatsoever, and the entire gently dipping, 13-ft- (4-m-) thick, tabular ledge of cataclasite is discordant to the moderately steeply dipping relict foliation in underlying microbrecciated mylonitic gneiss. The cataclasite formed at the expense of microbreccia, presumably in the waning stages of the episode(s) of cataclastic deformation. The ledge of cataclasite and ultracataclasite may be an expression of the crushing and grinding of microbreccia within a zone of coalescing fault surfaces. Cataclasis was carried out at a time and at a structural level such that no upper-plate cover rocks became interleaved structurally with either the cataclasites or the microbreccias.

Stop 6 is the Santa Catalina fault, the detachment or décollement which separates nonmylonitic, noncataclastic cover rocks above from cataclasites, microbreccias, and mylonites below (see Fig. 4). It marks the very top of the ledge of cataclasite.

The Santa Catalina fault has a sinuous trace that can be followed for 42 mi (70 km) along the front of the Santa Catalina, Tanque Verde, and Rincon mountains (Banks, 1976; Drewes, 1978; Pashley, 1966). Its form is that of a cylindrically folded surface plunging 15° to 30° to the southwest, parallel to the dominant direction of mineral lineation in the mylonite gneiss. It wraps around the nose of Tanque Verde arch. Although foldlike at the scale of the entire Santa Catalina-Rincon mountain front, the fault in outcrop view is typically planar and gently dipping, as at Stop 6. Striations can be seen and measured from place to place along the surface, and locally (although not at this site) the

fault surface is polished and deeply grooved. The dominant direction of movement on the fault is S60°W, the same as the orientation of mineral lineation in the mylonites.

The Santa Catalina fault appears to be a low-angle normal-slip fault which helped accommodate crustal extension in mid-Tertiary time. Its hanging wall contains faulted strata as young as 20 Ma. The sense of movement along the fault is the same as that of the simple shear which created the microbreccias and the mylonites. This coordination suggests that the fundamental components of the Rincon Mountains metamorphic core complex may have been fashioned through a progressive ductile through brittle normal-slip shearing and faulting (Davis, 1981a, 1983, 1986). As a result of sustained normal-slip shearing, mylonites forming within the approximately 1-mi- (2-km-) thick shear zone (at a depth of approximately 6 to 7 mi; 10 to 12 km) were steadily raised to higher levels characterized by lower temperatures and pressures. The continued shearing and faulting became more brittle. Mylonite was converted to microbreccia, and microbreccia was progressively transformed to cataclasite. Final movements were achieved not within a zone of deformation, but along a discrete fault surface, the Santa Catalina fault.

Shearing and faulting resulted in more than 6 mi (10 km) of normal-slip translation of the cover rocks. Cover rocks exposed at the site were translated from locations to the east-northeast. The extensive normal-slip shearing and faulting quite naturally brought rocks of strikingly different original depth levels into juxtaposition: unmetamorphosed, nonmylonitic hanging-wall cover rocks rest on chlorite-stable microbreccias, which in turn rest on biotite-stable quartzo-feldspathic mylonites.

When the ductile-brittle shear zone formed, it probably dipped at an angle of 45° or more, not at its presently observed inclination of 30° or less. The zone may have steadily rotated to progressively shallower inclinations as the crust which it occupied was progressively stretched (Davis, 1983, 1986).

Many workers have emphasized that the Santa Catalina fault formed originally as a Laramide thrust fault, which accommodated tens of kilometers of northeasterly transport (for example, Moore and others, 1941; Drewes, 1976, 1978, 1981). As of the time of writing of this guide, the question of the ancestry of the Santa Catalina fault continues to be hotly pursued through investigations of gross structural relations, microstructures, and geochronology. Bykerk-Kauffman (1983, 1986), for example, has recognized penetrative fabrics within the Santa Catalina Mountains that may in part reflect east-directed thrust movements during the Laramide. Davis (1986) has portrayed the uncanny structural relationships that might result from normal-slip simple shearing of a geologic column that had already been subjected to the kind of superposed faulting that distinguishes southeastern Arizona as a whole (Davis, 1981b).

Stop 7 brings us to cover rocks which lie above the Santa Catalina fault. Rocks ranging from Precambrian to Miocene in age can be expected to be found in the hanging wall of the Santa Catalina fault. All tend to be brittlely deformed and commonly the cover rocks are back-rotated along bounding faults which are intercepted by the Santa Catalina fault at depth. In some places, including this stop, the cover rocks display fold structures. Some of the fold structures may be Laramide compressional features which, in the mid-Tertiary, were passively transported to their present locations by movements associated with the Santa Catalina fault. However, the geometry of most of the folds appears to be compatible with their formation during southwest-directed normal-slip simple shearing under moderately ductile to moderately brittle conditions.

The deformed cover rocks above the Santa Catalina fault at Stop 7 are Pennsylvanian–Permian limestones and interbedded fine-grained clastics of the Earp Formation. Fold structures and faults within what appear to be five discrete plates or sheets were mapped in fine detail by Davis and Frost (1976). The folded and faulted limestones are heavily veined by thin, locally penetrative, calcite veins. Abrupt truncations and offsets of veins suggest the

effects of pressure solution. The structures of Stop 7, taken together, constitute a small museum of well-exposed, provocative structures.

POSTSCRIPT

As you leave the site, continue clockwise along Loop Drive; you will come to a location known as "Javalina Rocks" within less than 1,300 ft (400 m). Javalina Rocks is a distinctive outcrop of moderately microbrecciated mylonite gneiss, which forms a beautifully exposed west-dipping homocline. The rocks invite a short stop to climb about and closely examine the fault rocks. Continuing further around Loop Drive, you will spot the turn-off to the Monument's picnic area. The picnic area contains ramadas, water, and restrooms, and is positioned within close walking distance of excellent ledge exposures of mylonite gneisses that crop out in steep drainages on the flank of Tanque Verde Mountain.

REFERENCES CITED

Banks, N. G., 1976, Reconnaissance geologic map of the Mount Lemmon Quadrangle, Arizona: U.S. Geological Survey Miscellaneous Field Studies Map MF-747, scale 1:62,500.

Bykerk-Kauffman, A., 1983, Structural investigation of a transition zone between metamorphic tectonites and their unmetamorphosed equivalents, Buehman Canyon, Arizona [M.S. thesis]: University of Arizona, 79 p.

——— , 1986, Multiple episodes of ductile deformation within the lower plate of the Santa Catalina metamorphic core complex: Arizona Geological Society Digest, v. 15, p. 460–463.

Crittenden, M. D., Jr., Coney, P. J., and Davis, G. H., eds., 1980, Cordilleran metamorphic core complexes: Geological Society of America Memoir 153, 490 p.

Davis, G. H., 1977, Characteristics of metamorphic core complexes, southern Arizona: Geological Society of America Abstracts with Programs, v. 9, no. 7, p. 944.

——— , 1980, Structural characteristics of metamorphic core complexes, southern Arizona, *in* Crittenden, M. D., Jr., Coney, P. J., and Davis, G. H., eds., Cordilleran metamorphic core complexes: Geological Society of America Memoir 153, p. 35–77.

——— , 1981a, Metamorphic core complexes; Expressions of regional ductile stretching and rotational, listric (?) faulting: Geological Society of America Abstracts with Programs, v. 13, no. 2, p. 51.

——— , 1981b, Regional strain analysis of the superposed deformations in southeastern Arizona and the eastern Great Basin, *in* Dickinson, W. R., and Payne, W. D., eds., Relation of tectonics to ore deposits in the southern Cordillera: Arizona Geological Society Digest, v. 14, p. 155–172.

——— , 1983, Shear-zone model for the origin of metamorphic core complexes: Geology, v. 11, no. 6, p. 342–347.

——— , 1986, A shear zone model for the structural evolution of metamorphic core complexes in southeastern Arizona: Geological Society of London Symposium Volume on Crustal Extension, p. 247–266.

Davis, G. H., and Coney, P. J., 1979, Geological development of metamorphic core complexes: Geology, v. 7, no. 3, p. 120–124.

Davis, G. H., Gardulski, A. F., and Anderson, T. H., 1981, Structural and structural petrological characteristics of some metamorphic core complex terranes

in southern Arizona and northern Sonora, *in* Ortlieb, L., and Roldan, Q. J., eds., Geology of northwestern Mexico and southern Arizona: Geological Society of America Cordilleran Meeting Field Guides and Papers, Hermosilla, Sonora, Mexico, p. 323–368.

Davis, G. H., and Frost, E. G., 1976, Internal structure and mechanism of emplacement of a small gravity-glide sheet, Saguaro National Monument (east), Tucson, Arizona, *in* Wilt, J. C., and others, eds.: Arizona Geological Society Digest, v. 10, p. 287–304.

DiTullio, L. D., 1983, Fault rocks of the Tanque Verde Mountain decollement zone, Santa Catalina metamorphic core complex, Tucson, Arizona [M.S. thesis]: University of Arizona, Tucson, 85 p.

Drewes, H., 1976, Laramide tectonics from Paradise to Hells Gate, southeastern Arizona, *in* Wilt, J. C., and others, eds.: Arizona Geological Society Digest, v. 10, p. 151–167.

——— , Geologic map and sections of the Rincon Valley quadrangle, Pima County, Arizona: U.S. Geological Survey Miscellaneous Investigations Map No. I-997, scale 1:48,000.

——— , 1981, Tectonics of southern Arizona: U.S. Geological Survey Professional Paper 1144, 96 p.

Keith, S. B., Reynolds, S. J., Damon, P. E., Shafiqullah, M., Livingston, D. E., and Pushkar, P. D., 1980, Evidence for multiple intrusion and deformation within the Santa Catalina-Rincon-Tortolita crystalline complex, southeastern Arizona, *in* Crittenden, M. D., Jr., Coney, P. J., and Davis, G. H., eds., Cordilleran metamorphic core complexes: Geological Society of America Memoir 153, p. 217–267.

Moore, B. N., Tolman, C. F., Butler, B. S., and Kernon, R. M., 1941, Geology of the Tucson quadrangle, Arizona: U.S. Geological Survey Open-File Report, 20 p.

Pashley, E. F., 1966, Structure and stratigraphy of the central, northern, and eastern parts of the Tucson basin, Pima County, Arizona [Ph.D. thesis]: University of Arizona, 273 p.

Shakel, D. W., Silver, L., and Damon, P. E., 1977, Observations on the history of the gneiss core complex, Santa Catalina Mountains, southern Arizona: Geological Society of America Abstracts with Programs, v. 9, p. 1169.

An ancestral Colorado plateau edge: Fossil Creek Canyon, Arizona

H. Wesley Peirce, Bureau of Geology and Mineral Technology, 845 N. Park Avenue, Tucson, Arizona 85719

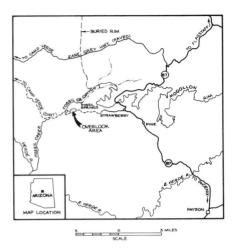

Figure 1. Index map showing road access to the Fossil Creek Canyon Site. Inset map shows general location in central Arizona.

LOCATION

The geologic features of principal interest at this site are revealed in the rugged walls of Fossil Creek Canyon in central Arizona. Access in most weather conditions is by a partially paved gravel road that intersects Arizona 87 at Strawberry (Pine 7½-minute Quadrangle, and Fig. 1). The most accessible panoramic overlook is where the gravel road crosses the lip of the south wall. The ideal viewing location, however, is about 0.5 mi (0.8 km) to the northeast near a gravel pit that is clearly marked on the Strawberry 7½-minute Quadrangle. Access to this overlook site is either by foot or by vehicle (other than sedans), and is a short walk from the borrow pit to a nearby vantage point. Excellent general access by foot to the canyon itself is provided by an old Jeep trail that branches from the pit road. It, too, is indicated on the Strawberry map. The heart of the canyon is now designated as Wilderness; therefore, vehicles are prohibited.

Fossil Creek flows southwest and joins the Verde River. The creek flows intermittently above Fossil Springs and perennially below. The steady spring flow serves two small hydroelectric plants downstream. Canyon headwaters are in several branch canyons that are headcutting the cliffy southern edge of the Colorado plateau province (the Mogollon Rim or Escarpment). The canyon-floor elevation near the overlook area is about 4,200 ft (1,260 m) above sea level, and the immediate rims are near 5,700 ft (1,710 m), a general elevation difference of 1,500 ft (450 m). Locally, higher elevations reach 6,500 ft (1,950 m), so that maximum relief exceeds 2,000 ft (600 m).

Fossil Creek Canyon is near the junction of two distinct segments of the Mogollon Rim: the Tonto segment to the east and the Mormon segment to the north (Peirce, 1984, p. 9). The canyon itself is at the extreme southern end of the Mormon

segment, in which the ancestral rim zone is buried by volcanics believed to be largely late Miocene in age. The Tonto segment, on the other hand, is relatively free of volcanics.

SITE

The principal significance of the Fossil Creek Canyon site is that it reveals two distinct episodes of large-scale erosional attrition along the southern margin of the Colorado plateau. The older (relief-causing) event predates the accumulation of a sequence of 1,500–2,000 ft (450–600 m) of nearly flat-lying, late Miocene volcanic rocks. The younger event, which postdates the volcanics and is part of the modern canyon-cutting cycle, has cut through the entire volcanic sequence. Recognition of an older, major episode of erosion along the plateau margin complicates traditional geologic understanding and compromises the view that only the modern canyon-cutting cycle requires explanation.

The postvolcanic incision of Fossil Creek reveals contrasting wall rocks on either side of a zone of active springs (Fossil Springs). To the southwest the canyon walls are dominated by an exposed thickness of 1,800 ft (540 m) of volcanic rocks of late Miocene age that are only slightly deformed. To the northeast, however, somewhat more deformed Paleozoic strata are dominant. The exposed Paleozoic section, spread over a lateral distance of six miles, ranges from the Mississippian Redwall Limestone to the Permian Kaibab Formation, a stratigraphic interval of about 2,500 ft (750 m) (Fig. 2). The overlying section of Tertiary rocks to the northeast is thin to absent.

The contact between the Tertiary and Paleozoic rocks is an irregular surface of erosion having an exposed relief of about 2,200 ft (330 m) (Figs. 3, 4). Approximately 1,350 ft (405 m) of this relief is acquired within a lateral distance of less than 0.75 mi (1.2 km). Fossil Springs and its extensive ancestral travertine buildup occur within this zone of rapid change in relief.

It is clear that the nearly flat-lying sequence of layered volcanic rocks was deposited above and against an irregular topography developed on Paleozoic strata (Twenter, 1962, p. 107). The lowermost unit of Tertiary age is an undated gravel of un-

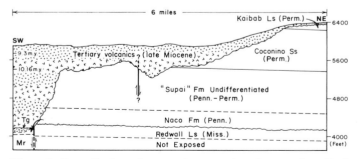

Figure 2. Generalized northeast-southwest section along north wall of Fossil Creek Canyon illustrating the buried ancestral Colorado Plateau edge.

Figure 3. Looking northward across Fossil Creek Canyon at a relatively undisturbed sequence of late Miocene volcanic rocks. These rocks overlie Paleozoic strata on a largely buried, irregular erosion surface that rapidly gains elevation eastward in the vicinity of ancestral travertine deposits of Fossil Springs. Dashed lines are projections into subsurface of Paleozoic stratigraphic units locally exposed within the canyon but largely buried beneath volcanic and younger materials in this scene.

Figure 4. Looking northward across Fossil Creek Canyon at a relatively undisturbed 1,800-ft (540-m) sequence of late Miocene volcanics. Light-colored tuff bed indicates minor postvolcanic faulting. Underlying older rocks in this view are buried just west (downstream) of Figure 3. Flume serves a small hydroelectric plant.

known total thickness that underlies the volcanics (Weir and Beard, 1983). This gravel occurrence appears analogous to one exposed in West Clear Creek Canyon to the north (Peirce and others, 1979, p. 14; Ulrich and Bielski, 1983) and is interpreted to represent fluvial deposits that accumulated in a topographic low at the base of an ancestral Mogollon Rim (plateau edge). Subsequently, volcanic flows and pyroclastics infilled and subdued much of the remaining irregularity. In the case of Fossil Creek Canyon, as elsewhere along the plateau margin, erosional relief generated by the modern canyon-cutting cycle tends to equal that produced earlier, although it differs in its geometry. This is important because the emphasis, which is due to its conspicuousness, tends to be placed only on the work of the modern erosional cycle.

The preponderance of the volcanics in the walls of Fossil Creek Canyon is believed to be late Miocene in age. A basalt flow along the road on the south side near the top of the sequence gave a K/Ar age of 9.30 ± 0.21 Ma. Another basalt flow, about 30 ft (90 m) lower in the sequence, was dated at 10.16 ±0.22 Ma (Peirce and others, 1979, p. 7). The latter locality is about 1,000 ft (300 m) above the canyon floor. Datable rocks have not been found in this lower interval; therefore, the maximum volcanic age is not known with certainty. Whatever their age, the underlying gravels are older, and the pre-gravel erosion surface is the oldest of all of the observed Tertiary phenomena. Peirce and others (1979) have suggested an Oligocene(?) age for an ancestral erosional Colorado Plateau edge in central Arizona.

Regionally, faulting is an important aspect of the landscape development story. It should be emphasized, however, that significant faulting is notably lacking at the Fossil Creek Canyon site. The focus is thus on erosion as the dominant mechanism for creating the two clear-cut manifestations of topographic relief. In both cases, downcutting is distinctive and leads to regional tectonic considerations. Relative uplift is implied in these relationships. A late Miocene "plateau uplift" concept has been commonly invoked to induce the modern canyon-cutting cycle, which includes Grand Canyon. McKee and McKee (1972) suggested that such an uplift took place between 10 Ma and 5 Ma. These workers, however, did not reckon with the older erosional manifestations that have as much relief along the southern plateau margin as do many of the modern canyons. Recognition of an earlier uplift history seems unavoidable. Until this earlier tectonism is acknowledged, a proper understanding of the geologic history of the southern margin of the Colorado Plateau province in Arizona cannot be realized.

Fossil Creek Canyon is a convenient locality to use in illustrating this problem. There are also many other localities along the plateau margin that clearly indicate that there is much to be learned, and relearned, about Cenozoic geologic history in central Arizona.

REFERENCES CITED

McKee, E. H., and McKee, E. D., 1972, Pliocene uplift of the Grand Canyon region—time of drainage adjustment: Geological Society of America Bulletin, v. 83, n. 7, p. 1923–1932.

Peirce, H. W., 1984, The Mogollon Escarpment: Arizona Bureau of Geology and Mineral Technology Fieldnotes, v. 14, no. 2, p. 8–11.

Peirce, H. W., Damon, P. D., and Shafiqullah, M., 1979, An Oligocene(?) plateau edge in Arizona: Tectonophysics, v. 61, p. 1–24.

Twenter, F. R., 1962, The significance of volcanic rocks in the Fossil Creek area, in Weber, R. H., and Peirce, H. W., eds., Mogollon Rim Region: New Mexico Geological Society, 13th Field Conference Guidebook, p. 107–108.

Ulrich, G. E., and Bielski, A. M., 1983, Mineral resource potential of the West Clear Creek Roadless Area, Yavapai and Coconino Counties, Arizona: U.S. Geological Survey, Map MF-1555-A, pamphlet p. 1–9.

Weir, G. W., and Beard, L. S., 1983, Geologic map of the Fossil Springs Roadless Area, Yavapai, Gila, and Coconico Counties, Arizona: U.S. Geological Survey, Map MF-1568-C.

Virgin River Gorge; Boundary between the Colorado Plateau and the Great Basin in northwestern Arizona

R. L. Langenheim, Jr., Department of Geology, 245 NHB, 1301 West Green Street, University of Illinois, Urbana, Illinois 61801
M. K. Schulmeister, Illinois State Water Survey, 101 Island Avenue, Batavia, Illinois 60510

LOCATION

The Virgin River Gorge and one of its tributaries are traversed by I-15 between milepost 12, about 13 mi (21 km) east of Mesquite, Nevada, and Milepost 28, about 12 mi (19 km) west of St. George, Utah (Fig. 1). Geologic features are readily observable from the road and from a scenic viewpoint in the Virgin River Canyon Recreation Area, which is entered from an interchange between mileposts 18 and 19. The site is in the northern half of the Purgatory Canyon, Arizona, 7½-minute Quadrangle and in the northeastern quarter of the Littlefield, Arizona, 15-Minute Quadrangle.

SIGNIFICANCE

The Virgin River crosses the boundary between the Colorado Plateau and the Basin and Range physiographic province in a deep gorge separating the Beaverdam Mountains and Virgin Mountains (Fig. 1). Little-deformed, nearly flat-lying rocks of the westernmost plateau are well exposed east of the "Narrows." The gorge cuts across a broad anticline broken by numerous basin-and-range–type faults. The Grand Wash fault, the western boundary of the Colorado Plateau, is not obvious from the road, but is well exposed. The Beaver Dam–Virgin Mountain block is separated from the intermountain valley containing Beaverdam Wash and the Virgin River by a major range-front fault at the lower end of the Narrows. Late Cambrian through Middle Permian rocks exposed in the gorge (Fig. 1) are characteristic of the transition from miogeosynclinal deposition, dominant to the northwest, and platform deposits on the plateau to the east. In addition, the Narrows is a fine example of the results of accelerated erosion that accompanied elevation of the Colorado Plateau and extensional deformation in the Basin and Range during the late Tertiary and Quaternary. Ancient channel and terrace deposits are prominent in the gorge. Classic alluvial fans, basin-filling deposits, and caliche crusts border the west face of the Beaver Dam-Virgin Mountains and the intermontane valley to the west.

SITE DESCRIPTION

Late and Middle Cambrian Bonanza King Dolomite, Dunderberg Shale, and Nopah Dolomite, Late Devonian Muddy Peak Limestone, Mississippian Monte Cristo Group, and Mississippian-Pennsylvanian Callville Limestone crop out along the road in the Narrows between milepost 12.8 at the western end of the gorge and milepost 17.3, where the Sullivans Canyon and Cedar Wash faults cross the highway (Fig. 1).

The Cambrian Bonanza King Dolomite and Nopah Dolomite crop out in steep rounded cliffs and benches and the intervening Dunderberg Shale forms the most prominent of the benches between mileposts 13.5 and 16.8 (Figs. 1, 2). The Bonanza King Dolomite in the vicinity of the gorge is brown-weathering, fine- to medium-grained dolomite with 33 ft (9 m) of interbedded limestone and dolomite resting on a notably glauconitic layer near the top (Steed, 1980). The glauconitic layer forms a notable bench and is exposed in a road cut at milepost 15.8. The Dunderberg Shale, thinly interbedded green shale and dolomite, is cleanly exposed in roadcuts near mileposts 13.9 and 16.8. The Nopah Dolomite is mostly very fine-grained, light-gray dolomite, and weathers somewhat ligher brown than the underlying Bonanza King Dolomite.

Cambrian rocks of the Virgin Gorge are at the southeastern limit of characteristic Middle and Late Cambrian miogeosynclinal or carbonate platform deposits. The same formations crop out in the Mormon Mountains, the first range to the northwest, where the mostly Middle Cambrian Bonanza King Dolomite is 1,914 ft (580 m) thick and the Late Cambrian Nopah Dolomite is 627 ft (190 m) thick (Wernicke and others, 1984). Farther northwest, these units are two or three times as thick. Cambrian rocks to the east on the platform differ greatly; they consist of the basal Tapeats Sandstone, a middle Bright Angel Shale, and an upper argillaceous Muav Limestone along the Grand Canyon (McKee and Resser, 1945). The Grand Canyon rocks, however, are all older than those in the Virgin Gorge; late-Middle Cambrian and Late Cambrian rocks have been removed by erosion at the canyon.

The hiatus at the post-Nopah Dolomite unconformity in the Virgin Gorge is represented by some of the uppermost Nopah Dolomite, the Ordovician Pogonip Dolomite, Eureka Quartzite, Ely Springs Dolomite, and, probably, part of the basal Devonian sequence in the Mormon Mountains—more than 1,221 ft (370 m) of rock (Wernicke and others, 1984). Farther west, in the Arrow Canyon Range, 649.5 ft (197 m) of Silurian Laketown Dolomite and Early and Middle Devonian Piute Formation intervene between the Ely Springs Dolomite and Late Devonian rocks. All of these rocks are absent along the Grand Canyon to the southeast.

Late Devonian Muddy Peak Limestone crops out in the slope of ledges and cliffs just below the high, sheer cliffs of the Monte Cristo Group on the north wall of the Narrows between mileposts 13.5 and 16.8 (Figs. 1, 2, 3). On the south wall of the canyon the Muddy Peak Limestone outcrop, as well as that of the Monte Cristo Group, is visible from the road between milepost 13.5 and 14.3. The lower part of the Muddy Peak Limestone is

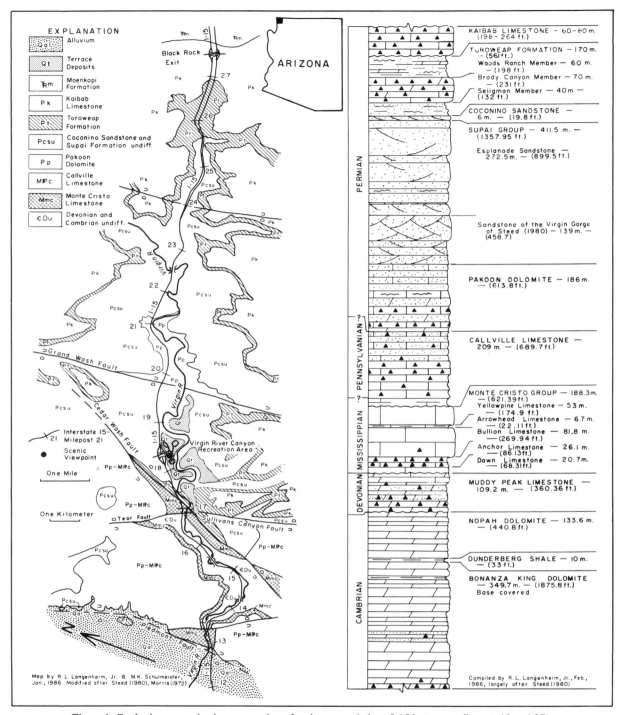

Figure 1. Geologic map and columnar section of rocks exposed along I-15 between mileposts 12 and 27, Arizona.

almost entirely medium-grained dolomite, but includes fairly prominent nodular layers of flint and rusty-weathering quartzite (Steed, 1980). The flint and quartzite distinguish the formation from the underlying Nopah Dolomite. The upper part of the Muddy Peak Limestone is lighter colored, includes substantial amounts of limestone, and is, in part, fine grained (Steed, 1980).

Devonian rocks thin on the platform to the southeast and only scattered remnants of Temple Butte Limestone occur in the eastern part of the Grand Canyon. To the northwest, Devonian rocks thicken substantially and Early and Middle Devonian sed-

Figure 2. Late Cambrian through Early Pennsylvanian rocks, view down the canyon from milepost 17. 1 = bench formed by Dunderberg Shale, 2 = base of Muddy Peak Limestone, 3 = base of Dawn Limestone, 4 = base of Anchor Limestone, 5 = base of Bullion Limestone, 6 = Arrowhead Limestone, 7 = base of Callville Limestone, and 8 = top of shaley basal part of Callville Limestone.

Figure 3. View west-southwest across the Cedar Wash fault from "Scenic Viewpoint" in the Virgin River Recreation Area. The Esplanade Sandstone, Coconino Sandstone, and the Brushy Canyon and Seligman Members of the Toroweap Formation are exposed in the foreground and in the prominent butte in the middle of the picture. Cliffs behind are on the upthrown block of the Cedar Wash fault and expose Cambrian through Mississippian rocks. The Callville Limestone and Permian rocks crop out along the skyline. 1 = base of Toroweap Formation. The Coconino Sandstone is the thin, white, cliffy ledge just below the contact. 2 = base of Monte Cristo Group, 3 = base of shaley bench in lowermost Callville Limestone, and 4 = base of ledge and cliff sequence in the Callville Limestone. Smooth slopes rising to the top of the ridge on the right include Pakoon Dolomite and Supai Group outcrops.

iments are also present. Wernicke and others (1984) reported 726 ft (220 m) of Sultan Limestone in the Mormon Mountains, and Langenheim and others (1962) cited 1970 ft (597 m) of Devonian rocks in the Arrow Canyon Range.

The Early and Middle Mississippian Monte Cristo Group crops out in a high, sheer cliff that rises precipitously above the ledge and slope exposures of the Muddy Peak Limestone (Figs. 1, 2, 3). The Yellowpine, Arrowhead, and Bullion Limestones are cleanly exposed in a roadcut from milepost 12.8 through 13.4. Prominent coral biostromes crop out in the vicinity of milepost 13. The lowermost Dawn Limestone is dominantly medium- to coarse-grained limestone with minor interbedded dolomite (Steed, 1980). The formation is chert free, and rests unconformably on the underlying Muddy Peak Limestone. The Anchor Limestone is sharply delineated by abundant nodular chert interbedded with fine- to coarse-grained bioclastic limestone. The chert weathers dark brown or black, and the unit forms either an indentation in the Monte Cristo cliff or a minor bench. The overlying Bullion Limestone is predominantly very thick-bedded, coarse-grained, bioclastic limestone and comprises most of the sheer cliff. The Arrowhead Limestone is composed of thin nodular bioclastic limestone beds separated by shale breaks. It is readily identified from a distance because it crops out as a light-colored, narrow stripe two-fifths of the way down the brown, sheer cliff. The uppermost formation of the Monte Cristo Group, the Yellowpine Limestone, is thick-bedded, medium- to coarse-grained limestone in the lower half and very fine-grained, somewhat thinner bedded limestone in the upper portion. The formation makes up the upper two-fifths of the sheer cliff and is abruptly succeeded by bench-forming basal Callville Limestone.

The Dawn Limestone correlates with the Whitmore Wash Member of the Redwall Limestone on the platform to the southeast. The Anchor Limestone correlates with the Thunder Springs Member, the Bullion Limestone with the Mooney Falls Member, and the Arrowhead and Yellowpipe Limestones with the Horseshoe Mesa Member. These units are strikingly uniform and continuous across the miogeosynclinal-platform boundary, in marked contrast to patterns in the pre-Mississippian sequence and in the Carboniferous and pre-Toroweap Limestone rocks above the Monte Cristo or Redwall Limestone. The Redwall Limestone thins from about 490.5 ft (148.6 m) in the eastern Grand Canyon (McKee and Gutschick, 1969) and the Monte Cristo Group thickens to 924 ft (280 m) in the Mormon Mountains (Wernicke and others, 1984).

The Late Mississippian and Pennsylvanian Callville Limestone is not exposed at the roadside, but crops out above the Monte Cristo Group north of the highway from the lower entrance of the Narrows to the Cedar Wash fault (Fig. 1), and south of the highway from the lower entrance of the Narrows to milepost 14.4. The basal 62.7 ft (19 m) of the Callville Limestone crops out in a well-defined bench separating the cliff of the Monte Cristo Limestone from the markedly cyclic low cliff and bench succession above. This covered slope is underlain by interbedded thin limestone and argillaceous beds (Steed, 1980). Very fine-grained limestone, with chert in nodules and nodular layers, interbedded with thin-bedded argillaceous limestone dominates the lower third of the Callville Limestone above the basal shaly member. Above this, the cliffy limestone beds in the middle third of the formation are more coarse grained and, in the upper third, sandy limestone and calcareous sandstone is dominant. Along with the change in lithology, dominantly gray weathering tones are replaced by gray brown at the top of the Callville Formation. Steed (1980) placed the formation boundary within a gradational change to more dolomitic rocks, which he also considers the Pennsylvanian-Permian boundary, on the basis of lithologic correlations. The abrupt basal contact of the Callville Limestone is disconformable.

The Callville Limestone is replaced by 1,980 ft (600 m) of Bird Springs Formation in the Mormon Mountains and becomes more calcareous. The Mississippian and Pennsylvanian lower part of the Bird Spring Group, farther northwest in the Arrow Canyon Range, is 2,112 ft (640 m) thick (Langenheim and others, 1985). Much of the thickening results from progressive loss of hiatus at the basal unconformity and in unconformities within the limestone. Steed (1980) correlated the Callville Limestone of the Virgin Gorge with the Wahatomigi, Manakacha, and Wescogome Formations of the Supai Group in the Grand Canyon area. Detrital rocks, mostly brown to red, progressively replace carbonates in these formations eastward on the platform.

Permian rocks of the Pakoon Dolomite, Supai Group, Coconino Sandstone, Toroweap Limestone, and Kaibab Limestone are exposed along the road, on the valley wall east of the Cedar Wash fault (milepost 17.3) and in continuous sequence between mileposts 19.8 and 26.9 east of the Grand Wash fault.

The Pakoon Dolomite crops out in brownish slopes above the Callville Limestone north of the highway between the west front of the Beaverdam Mountains and the Cedar Wash fault. Exposures of the upper part of the Pakoon Dolomite are truncated on the west by the Grand Wash fault at milepost 19.8 but are continuous along the highway to milepost 21.3 (Figs. 1, 3, 4). The lower half of the Pakoon Dolomite is made up of cyclically alternating beds of limestone and dolomite, including sandy and cherty layers (Steed, 1980). The upper half is interbedded sandstone, sandy limestone, and very fine-grained limestone (Steed, 1980).

Permian limestone of the Bird Spring Formation replaces the Pakoon Dolomite in the Mormon Mountains where Wernicke and others (1984) reported 3,960 ft (1,200 m) of Bird Spring Formation and Permian redbeds in a section superficially resembling that of the Virgin Gorge. In the Arrow Canyon Range, however, there are no redbeds above the Permian part of the Bird Spring Formation. Here the total thickness is unknown, but it appears to be much thicker than that of the Pakoon Dolomite in the Virgin Gorge. McKee (1982) reported that the Pakoon Dolomite is abruptly replaced by a much thinner red detrital sequence in the lower part of the Esplanade Sandstone. Steed (1980) cited an equally abrupt disappearance of the Pakoon Dolomite to the east, but suggested that it may have been removed by erosion.

Figure 4. View northwest up the canyon from milepost 20.6, showing the Pakoon Dolomite through the Kaibab Limestone in characteristic Colorado Plateau terrain. 1 = approximate top of the Pakoon Dolomite, 2 = approximate top of Steed's (1980) "Sandstone of the Virgin Gorge," 3 = top of the Esplanade Sandstone. Thin white stripe at the top of the Esplanade Sandstone is the outcrop of the Coconino Sandstone. 4 = base of the Kaibab Limestone.

Steed's (1980) "Sandstone of the Virgin River" forms steep, ledge-broken slopes above the Pakoon Dolomite between the Grand Wash fault and the crossing of the Virgin River at milepost 22.5 (Figs. 1, 4). These rocks are visible on both sides of the highway east of the Grand Wash fault; they crop out adjacent to the road between the Virgin River Recreation Area and the Grand Wash fault and cover much of the west slope of the Beaverdam Mountains north of the Virgin Gorge. The lower half of the "Sandstone of the Virgin River" is dominated by thick beds of yellowish, fine-grained sandstone resting on an abrupt contact above the softer beds of the Pakoon Dolomite. The upper half is lighter colored, more massive, and causes a steeper, more cliff-like slope. The contact between the two halves of the "Sandstone of the Virgin River" is well-marked on the west slope of the ridge trending north of the highway from the deep cut at milepost 20; as seen from the viewpoint in the Virgin River Recreation Area. The basal contact of the "Sandstone of the Virgin Gorge" also is visible on this slope, but is much less obvious.

Steed (1980) includes this unit in the lower part of the Supai Group; he stated that it closely resembles the Queantoweap Sandstone of McNair (1951). Steed (1980), however, did not recognize the Queantoweap Sandstone and considered "the precise stratigraphic relationships between the Queantoweap Sandstone, the Esplanade Sandstone of the Supai Group, and the Hermit Formation . . . beyond the scope of [his] study" (Steed, 1980, p. 104). McKee (1982) recognized widespread unconformities at the top and bottom of each formation in the Supai Group, and thus provided a basis for correlating these units. As yet, however, the detailed work required to recognize these unconformities in the Virgin Gorge has not yet been accomplished, and subdivision and correlation within the Supai Group remains ambiguous. The Queantoweap Sandstone or "Sandstone of the Virgin Gorge" loses its identity and thins substantially eastward on the platform (McKee, 1982); it is abruptly replaced by carbonate rocks in the Permian part of the Bird Spring Group to the northwest. The sandstone appears to be a well-sorted accumulation along the margin between more marine miogeosynclinal conditions to the northwest and a detrital coastal platform to the southeast. The "Sandstone of the Virgin Gorge" may or may not be part of the 1,980 ft (600 m) of Permian redbeds reported by Wernicke and others (1984) in the Mormon Mountains.

The Esplanade Sandstone crops out along the highway from milepost 22 plus to milepost 25.7 (Figs. 1, 3, 4). It forms steep slopes and cliffs below the Toroweap Formation and above the "Sandstone of the Virgin Gorge" north and south of the highway between milepost 17.3 and milepost 25.7. The lower part of the Esplanade Sandstone is very faintly cross-bedded cream and red sandstone with widely spaced thin interbeds of bright-red argillaceous siltstone. Massive cliff-forming beds of white sandstone dominate above. The base of the white sandstone crosses the road at milepost 21.3. The base of the Esplanade Sandstone is arbitrarily placed at the point where cross bedding becomes inconspicuous, and flat-bedded, massive sandstone becomes dominant.

In the eastern part of the Grand Canyon, the Esplanade Sandstone thins to about 330 ft (100 m). To the northwest, the sequence thickens and sandstone is replaced by marine carbonate rocks of the Bird Spring Group.

The Coconino Sandstone crops out as a thin, discontinuous, angular cliff of strongly cross-bedded white sandstone above the Esplanade Sandstone and below the bench underlying the Seligman Member of the Toroweap Formation. It is exposed in the roadcut at milepost 25.7 and crops out as a narrow, white band at the very top of the Esplanade Sandstone (Figs. 1, 3, 4). The sandstone is fine to coarse grained and is bimodally sorted. The base of the Coconino Sandstone is an abrupt change from very faintly cross-bedded to strongly cross-bedded sandstone.

The Coconino Sandstone is absent northwest of the Virgin Gorge, but thickens to the southeast. It is 400 ft (121 m) thick in the eastern part of the Grand Canyon. Its maximum thickness of 1,000 ft (330 m) is at Pine, Arizona, on the Mogollon Rim south of Flagstaff (Baars, 1962). The sandstone coarsens to the southeast and the cross bedding is inclined to the northwest, indicating a southeastern source. Steed (1980) suggests that the 'feather edge' of the Coconino in the Virgin Gorge may be alluvial rather than eolian.

The Toroweap Formation crops out along the highway between mileposts 25.7 and 26.9 (Figs. 1, 3, 4). It includes the lower or two gray cliffs and the slopes above and below the skyline on both sides of the highway between the Grand Wash fault and milepost 25.7. It also crops out extensively in cliffs north and south of the highway between the Cedar Wash–Sullivans Canyon faults and the Grand Wash fault and in an outlier west of the Piedmont fault about 2.5 mi (4 km) north of the Virgin River. The Brady Canyon Member, beta member of McKee (1938),

forms a prominent limestone cliff. The bench below is underlain by the Seligman Member, gamma member of McKee (1938), and the bench above is formed by the uppermost Woods Canyon Member, alpha member of McKee (1938).

The Seligman Member is interbedded gypsiferous siltstone, gypsum, and poorly consolidated sandstone. It rests conformably on the Coconino Sandstone. It thickens to 305 to 330 ft (92 to 100 m) in the Mormon Mountains to the northwest (Bissell, 1969) and thins, becoming more sandy, to the southeast (Rawson and Turner-Peterson, 1980). Rawson and Turner-Peterson (1980) interpreted facies changes in the Seligman Member as a transition from coastal sabkha environments on the northwest to a continental sabhka on the southeast.

The Brady Canyon Member is interbedded skeletal packstone and wackestone in the Virgin Gorge, with the more openmarine packstone progressively replaced by wackestone, pelletal packstone, and quartzose dolomite to the southeast. The Brady Canyon Member thickens to about 315 ft (95 m) in the Mormon Mountains (Bissell, 1969) but progressively thins to the southeast to a pinchout along a roughly north-south line from just west of Flagstaff to the junction of the Little Colorado and Colorado rivers (Rawson and Turner-Peterson, 1980). The contact between the Brady Canyon and Seligman Members is conformable and, on a regional basis, gradational.

The Woods Ranch Member is gypsiferous siltstone, dolomite, and red and white sandstone in the Virgin Gorge. It is about 150 ft (45 m) thick in the Mormon Mountains to the northwest, where Bissell (1969) reported it dominated by gypsum and dolomite. The member thins to the southeast and passes into a continental sabkha deposit more or less along the Grand Canyon (Rawson and Turner-Peterson, 1980). The basal contact of the Woods Ranch Member is conformable and gradational.

The Kaibab Limestone crops out along the highway between milepost 26.9 and 27.8 (Figs. 1, 3, 4). It also crops out above the Toroweap Formation on the skyline north and south of the highway between Black Rock Interchange and the Grand Wash fault, and in cliffs above the Toroweap Formation between the Grand Wash fault and the Cedar Wash–Sullivans Canyon faults. Only the Fossil Mountain Member, beta member of McKee (1938), is visible, because the overlying Harrisburg Member (Reeside and Bassler, 1922) cannot be seen from the road and is poorly exposed east of Black Rock Interchange. The Fossil Ridge Member is cliff-forming, fossiliferous limestone.

In the Mormon Mountains, the Kaibab Limestone thickens to about 430 ft (130 m). To the southeast along the Grand Canyon, thicknesses increase, but at the eastern end of the canyon the dolomite is replaced in facies transition by detrital rocks (McKee, 1938).

Widespread Pleistocene and Holocene channel and terrace deposits between mileposts 17.3 and 23 record successive stages in downcutting by the Virgin River. Extensive terraces formed along the river in the relatively depressed block between the Grand Wash fault and the Sullivans Canyon fault (Fig. 1). Channel fills exposed in road cuts between mileposts 20 and 23.5 are composed of sedimentary rock fragments and prominent boulders and cobbles of basalt from the St. George and Hurricane areas upstream (Fig. 5). Alluvial fans abut the west face of the Beaverdam and Virgin mountains, and the entrenched Virgin River west of the mountains exposes valley-fill deposits capped by a prominent caliche layer.

There are four structural blocks in the Virgin Gorge area. East of the Grand Wash fault, a block of gently dipping, little-deformed rocks belongs to the Colorado Plateau. The rocks between the Grand Wash and Cedar Wash–Sullivan Canyon faults are more intensively deformed and have been referred to as a transitional block between the Colorado Plateau and Basin and Range provinces (Moore, 1972). Rocks in the Narrows between the Sullivans Canyon and Piedmont faults belong to the Beaver Dam–Virgin Mountain block, part of the Basin and Range province. Many of the northwest-trending faults of minor displacement mapped by Steed (1950) in the Narrows are not shown in Figure 1. In addition, Nielson (1982) briefly cited specific exposures of low-angle faults north of the Virgin Gorge and on the west side of the Beaverdam Mountains. These are variously referred to as thrust faults of the Sevier orogeny or as gravity slides associated with Tertiary block faulting. The wide valley west of the Piedmont fault is a good example of a Basin and Range type intermontane basin. According to Wernicke and others (1982, 1984), the Grand Wash and Piedmont faults are breakaway faults, here overlapping en echelon, separating the Colorado Plateau and Basin and Range provinces. Bohannon (1986) cited seismic evidence for 2.4 to 3 mi (4 to 5 km) of structural relief between the western edge of the Colorado Plateau and Beaver-

Figure 5. Filled channel in Supai Group at milepost 21.7. Channel and terrace deposits include abundant basaltic boulders and cobbles brought down the ancestral Virgin River from the St. George and Hurricane areas.

dam Wash. Uplift of the Beaverdam–Virgin Mountain block along the Sullivans Canyon–Cedar Wash faults, however, appears anomalous in this respect. Wernicke and others (1982, 1984) asserted that the region between the plateau and the Sierra Nevada has undergone about 85 mi (140 km) of post-Miocene crustal extension. The plateau has been greatly uplifted without extensional faulting since the Pliocene, thus initiating downcutting of the Virgin River and the Grand Canyon.

REFERENCES CITED

Baars, D. L., 1962, Permian System of Colorado Plateau: American Association of Petroleum Geologists Bulletin, v. 46, p. 149–218.

Bissell, H. J., 1969, Permian and Lower Triassic transition from the shelf to basin (Grand Canyon, Arizona, to Spring Mountain Nevada, *in* Baars, D. L., Bowman, F. G., Jr., and Spencer, C. W., eds., Geology and natural history of the Grand Canyon region: Four Corners Geological Society, 5th Annual Field Conference, Guidebook, p. 135–169.

Bohannon, R. E., 1986, Transition from stable to extended crust in the Virgin Mountains region of southeastern Nevada and northwestern Arizona: Geological Society of America Abstracts with Programs, v. 18, no. 1, p. 342–343.

Langenheim, R. L., Jr., and others, 1962, Paleozoic section in the Arrow Canyon Range, Clark County, Nevada: American Association Petroleum Geologists Bulletin, v. 46, p. 592–609.

—— , and others, 1985, Preliminary report of the brachiopod fauna, Arrow Canyon section, southern Nevada U.S.A.: Congres International Stratigraphie Geologie Carbonifere, 10eme, Madrid, Compte Rendu, v. 2, p. 425–432.

McKee, E. D., 1938, The environment and history of the Toroweap and Kaibab Formations of northern Arizona and southern Utah: Carnegie Institution of Washington Publication 492, 268 p., 48 pl.

—— , 1982, The Supai Group of the Grand Canyon: U.S. Geological Survey Professional Paper 1173, 504 p.

McKee, E. D., and Gutschick, R. C., eds., 1969, History of the Redwall Limestone of northern Arizona: Geological Society of America Memoir 114, 726 p.

McKee, E. D., and Resser, C. E., 1945, Cambrian stratigraphy of the Grand Canyon region: Carnegie Institution of Washington Publication 563, 232 p.

McNair, A. H., 1951, Paleozoic stratigraphy of part of northwestern Arizona: American Association of Petroleum Geologists Bulletin, v. 35, p. 503–541.

Moore, R. T., 1972, Geology of Virgin and Beaverdam Mountains, Arizona: Arizona Bureau of Mines Bulletin 186, 65 p., geological map and sections.

Nielson, R. L., 1982, Post-Pennsylvanian structural evolution of southwestern Utah, *in* Nielson, D. L., ed., Overthrust belt of Utah, 1982 Symposium and Conference: Utah Geological Association Publication 10, p. 101–115.

Rawson, E. R., and Turner-Peterson, C. E., 1980, Paleogeography of northern Arizona during the deposition of the Permian Toroweap Formation, *in* Fouch, T. D., and Magathen, E. R., eds., Paleozoic paleogeography of the west-central United States, Rocky Mountain Paleogeography Symposium 1: Rocky Mountain Section, Society of Economic Paleontologists and Mineralogists, p. 341–352.

Reeside, J. B., Jr., and Bassler, H., 1922, Stratigraphic sections in southwestern Utah and northwestern Arizona: U.S. Geological Survey Professional Paper 129, p. 53–77.

Steed, D. A., 1980, Geology of the Virgin River Gorge, northwest Arizona: Brigham Young University Geology Studies, v. 27, p. 96–115.

Wernicke, B., Spencer, J. E., Burchfiel, B. C., and Guth, P. L., 1982, Magnitude of crustal extension in southern Great Basin: Geology, v. 10, p. 499–502.

Wernicke, B., Guth, P. E., and Axen, G. J., 1984, Tertiary extensional tectonics in the Sevier thrust belt of southern Nevada, *in* Lintz, J., Jr., ed., Western geological excursions (Geological Society of America 1984 annual meeting guidebook): Geological Society of America and Mackay School of Mines, University of Nevada, Reno, v. 4, p. 473–510.

Wilson, E. D., and Moore, R. T., 1959, Geologic map of Mohave County, Arizona: Arizona Bureau of Mines.

Landslide and debris-flow deposits in the Thumb Member of the Miocene Horse Spring Formation on the east side of Frenchman Mountain, Nevada: A measure of basin-range extension

Stephen L. Salyards, Seismological Lab, California Institute of Technology, Pasadena, California 91125
Eugene M. Shoemaker, U.S. Geological Survey, Flagstaff, Arizona 86001

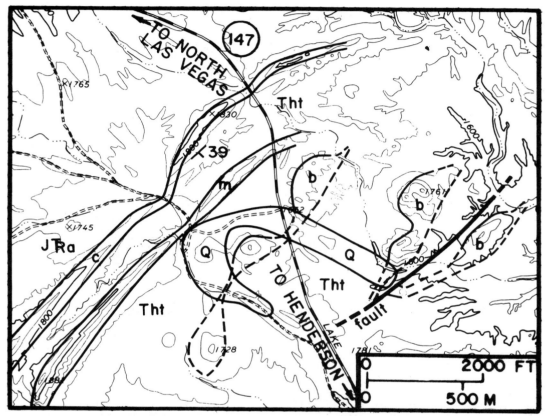

Figure 1. Index map of the landslide locality. Tht, Thumb Member of Miocene Horse Spring Formation; specific units of the Thumb are discussed in text: b, breccia lens; m, marl layer; c, basal pebble-conglomerate layer; JRa, Jurassic and Triassic Aztec Sandstone; Q, Quaternary alluvium and terrace deposits. Contacts and fault dashed where approximately located, dotted where concealed.

LOCATION

From the south: From Henderson, Nevada, take Lake Mead Drive (Nevada 147). At 7.1 mi (11.4 km) east of U.S. 95, turn north onto Northshore Road and go 3.2 mi (5.1 km) to Lake Mead Boulevard. The site is 3.7 mi (6 km) north on Lake Mead Boulevard.

From the north: From North Las Vegas, take Lake Mead Boulevard (also Nevada 147) east from I-15. At 11.4 mi (18.3 km) from I-15 a paved road from the east (left) intersects Nevada 147. Continue on 147 for 1.8 mi (2.9 km) to the site.

The site is marked by the high-tension power line that crosses the road. Park along the side of the road or on the dirt roads that intersect the main road in this area.

SIGNIFICANCE

The features of interest at this locality are masses of breccia that cap low hills on both sides of the road. The breccia occurs as lenses in the Thumb Member of the Miocene Horse Spring Formation, which overlies the Aztec Sandstone of Triassic(?) and Jurassic age (Fig. 1). These lenses are enclosed within pink, fine-grained, argillaceous and gypsiferous sandstone. The age of the Thumb here is about 17 Ma, as determined by fission-track dating of tuffaceous units (Bohannon, 1984, p. 5). The breccia lenses in the Thumb are composed of Precambrian igneous and metamorphic rock debris. The lenses include both coarse, clast-supported, typically monolithologic, "megabreccia" landslide

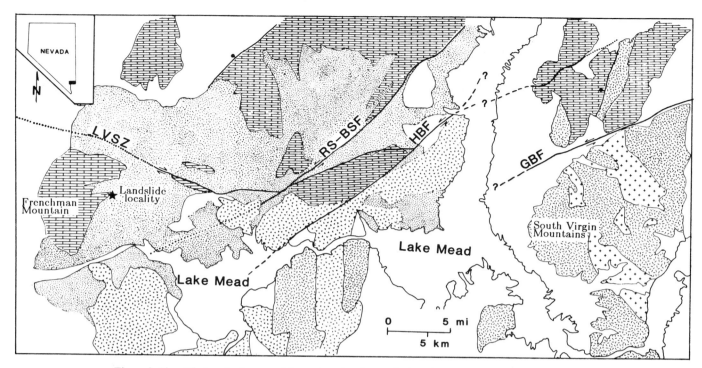

Figure 2. Simplified geologic map of the region showing geologic units and major faults. Random dash pattern, undifferentiated Precambrian igneous and metamorphic rocks; crosses, Precambrian Gold Butte Granite; block pattern, Paleozoic and Mesozoic strata; V pattern, Tertiary volcanics; dot pattern, Miocene and Pliocene sedimentary rocks; no pattern, Quaternary alluvium. LVSZ, Las Vegas Valley shear zone. Faults of Lake Mead fault system: GBF, Gold Butte fault; RS-BSF, Rogers Spring–Bitter Spring Valley fault; HBF, Hamblin Bay fault. Shear zone and faults dashed where approximately located, dotted where concealed, queried where questionable; arrows indicate relative movement; bar and ball on downthrown side.

masses and matrix-supported, heterogeneous, debris-flow depos-its (conglomerates of subangular to subrounded clasts). The probable source of the Precambrian rock debris is the South Virgin Mountains 37 mi (60 km) to the east (Fig. 2), the nearest location where Precambrian bedrock of the same lithology is now exposed. If the South Virgin Mountains are the source, crustal extension in this region may be as much as 37 mi (60 km). Extension of the crust south of the site occurred earlier than extension to the north. Differential extension was accommodated on two major fault systems: the northwest-trending, right-lateral Las Vegas Valley shear zone, and the northeast-trending, left-lateral Lake Mead fault system (Fig. 2).

DISCUSSION

The distinctive features of this site were described by Long-well (1974), who recognized the breccia lenses of exotic rocks in the Thumb Member and suggested that they indicate the amount of offset on the Las Vegas Valley shear zone. Concurrent work by Anderson (1973), however, demonstrated large left-lateral offset on the Hamblin Bay fault, a strand of the Lake Mead fault system

that intersects the Las Vegas Valley shear zone. Displacement on the Lake Mead fault system has also contributed to tectonic sepa-ration of the breccia lenses from their source region.

Low hills east of Nevada 147 are capped by a coarse breccia (eastern exposures of b in Fig. 1) composed of clasts of por-phyritic rapakivi granite. All of the granite breccia exposed on these hills was probably once connected as one large landslide body. The clasts are identical to the Precambrian Gold Butte Granite, which is exposed in the South Virgin Mountains (Vol-borth, 1962). The clasts are unsorted, millimeter to meter size. Many of the larger ones are fractured, and they break apart when removed from the breccia. These fractures probably were formed during emplacement of the breccia lens. Large clast size suggests proximity to the source.

Material underlying the breccia shows two scales of defor-mation. In places immediately below the breccia is a sandstone layer, generally less than 30 but locally as much as 70 cm thick, that lacks bedding. Angular granite clasts 1–2 cm across are found a centimeter or two down in the unbedded layer. This sandstone layer has the same composition as the underlying sand-stone; its lack of bedding suggests that it was formed by shearing of the sandstone when the landslide was emplaced. Below this "mixed" layer, sandstone beds are locally deformed to depths as much as 10 ft (3 m). Small folds and rip-up structures are present in places along the interface between the breccia and the sandstone. These structures generally are only centimeters across, and the deformation generally affects the lower-bedded material to depths of only a few centimeters. The deformation consistently shows that the direction of transport of the overlying breccia was to the southwest. The breccia is generally concordant with undeformed underlying beds, but in some places it is discordant. The shallow depth of deformation below the breccia in most places suggests that the lower contact generally represents the ground surface on which the breccia was emplaced.

A hill topped by a power transmission tower just east of Nevada 147 (Fig. 1) is capped by a lower layer of breccia and conglomerate lenses composed of both clast-supported breccia and matrix-supported debris-flow material. The clast-supported breccia is similar in texture to the breccia lenses farther east but is different in composition. Matrix-supported clasts are poorly sorted, subangular to subrounded, and contain a mixture of igneous and metamorphic rock types, the most common being biotite-quartz gneiss, amphibolite, and nonporphyritic granite. The base of this material is inversely graded and appears to be gradational with the underlying sandstone. A hill on the west side of the road is also capped by debris-flow deposits below which the sandstone is not deformed.

A conglomerate (c in Fig. 1) at the base of the Thumb Member contains well-rounded pebbles from the Paleozoic and Mesozoic formations found on the Colorado Plateau as well as in the northern part of the South Virgin Mountains and nearby Frenchman Mountain (Fig. 2). Most of the pebbles are limestone, but sandstone pebbles, probably derived from the underlying Aztec Sandstone, are common. A ridge-forming unit above the pebble conglomerate consists of white marl with a variable sand content (m in Fig. 1).

The Tertiary geologic history recorded by the local stratigraphy begins with deposition of a basal conglomerate in a fluvial environment. The base of the section here is not well dated but probably is no more than a few million years older than tuffs higher in the section determined by fission-track methods to be 17 Ma (Bohannon, 1984, p. 5). Lacustrine beds onlap the fluvial deposits. At times the depositional basin was closed, as shown by the presence of evaporites. Initial normal faulting of the Frenchman Mountain block may have begun about 20 Ma. Soon after this faulting, Precambrian basement was exposed somewhere nearby, and Precambrian rocks were deposited in the basin as landslides and debris flows. The debris flows could have traveled a long distance, but the landslide debris indicates proximity to the source, probably 6 mi (10 km) or less. The discrete layers of breccia may have been deposited during periods of more active

normal faulting that increased the uplift rate and provided favorable conditions for large landslides. The occurrence of Phanerozoic clasts at the base of the Thumb and of Precambrian clasts higher in the section suggests that episodes of faulting exposed increasingly older units from which these clasts were derived.

Westward tectonic transport of the Frenchman Mountain block away from the South Virgin Mountains and the Colorado Plateau was related to crustal extension of this part of the Basin and Range Province. Extension was more active first in the southern part of the Basin and Range Province and then in the northern part of the province. The structural boundary between the southern and northern parts is approximately at the latitude of this site. Differential extension was accommodated initially by the right-lateral Las Vegas Valley shear zone that bounds the north side of Frenchman Mountain (Fig. 2). A total of 42 mi (67 km) of offset (27 mi; 43 km of slip and another 15 mi; 24 km of bending across the shear zone) is obtained by correlating thrusts across Las Vegas Valley (see Longwell, 1974). If Longwell's estimate of the offset on the Lake Mead fault system is removed, the landslide breccias at this site are separated from their probable source by about 12 mi (20 km). The slip on the Las Vegas Valley shear zone probably decreases progressively to the east, however, and may die out near the South Virgin Mountains. Later differential extension was accommodated southeast of the site by the northeast-trending, left-lateral Lake Mead fault system. This fault system shows 28 to 34 mi (45 to 55 km) of offset summed across at least three major faults (Fig. 2). The timing of these fault systems is unclear, but field relations show at least some simultaneous activity; Bohannon (1984, p. 59) suggested that motion on both the Las Vegas Valley shear zone and the Lake Mead fault system began about 17 Ma. Detailed mapping by Shoemaker (unpublished) in the region of intersection of the Las Vegas Valley shear zone and the Lake Mead fault system shows that the right-lateral Las Vegas Valley shear zone is consistently offset by the intersecting left-lateral faults.

REFERENCES CITED

Anderson, R. E., 1973, Large-magnitude late Tertiary strike-slip faulting north of Lake Mead, Nevada: U.S. Geological Survey Professional Paper 794, 18 p.

Bohannon, R. G., 1984, Nonmarine sedimentary rocks of Tertiary age in the Lake Mead region, southeastern Nevada and northwestern Arizona: U.S. Geological Survey Professional Paper 1259, 72 p.

Longwell, C. R., 1974, Measure and date of movement on Las Vegas Valley shear zone, Clark County, Nevada: Geological Society of America Bulletin, v. 85, p. 985–990.

Volborth, A., 1962, Rapakivi-type granites in the Precambrian complex of Gold Butte, Clarke County, Nevada: Geological Society of America Bulletin, v. 73, p. 813–831.

Contribution number 4216, Division of Geological and Planetary Sciences, California Institute of Technology, Pasadena, California 91125.

12

Paleozoic stratigraphy of Frenchman Mountain, Clark County, Nevada

Stephen M. Rowland, Department of Geoscience, University of Nevada, Las Vegas, Nevada 89154

LOCATION AND ACCESSIBILITY

Frenchman Mountain lies on the eastern edge of Las Vegas, Nevada. Access is via Lake Mead Boulevard from I-15 in North Las Vegas (Fig. 1). The entire route of this field trip is on paved road and is easily passable by passenger car or bus. Allow approximately three hours.

A small-scale geologic map of Clark County, including the area described in this field guide, is available as Plate 1 of Longwell and others (1965). Stops 1 and 2 are on the Las Vegas NE 7½-minute Quadrangle. The rest of the stops are on the Frenchman Mountain 7½-minute Quadrangle. A road log of this same area, with less detailed stratigraphy but also including the Keystone Thrust and Valley of Fire, is available in Bohannon and Bachhuber (1979).

SIGNIFICANCE

Frenchman Mountain is a fault block of eastward-dipping Paleozoic, Mesozoic, and Cenozoic sedimentary rocks lying on Precambrian crystalline basement. It is the westernmost cratonal section in the Basin and Range Province and therefore marks the western edge of the craton and eastern margin of the Cordilleran miogeocline in this region. The Paleozoic section is essentially a Grand Canyon section lying on its side, with 100-percent exposure and unrestricted, easy access. The Mesozoic section, which is representative of the southwestern Colorado Plateau, is less resistant and consequently is less convenient to examine. Exposures of the Miocene nonmarine sediments of the Horse Spring Formation are excellent and convenient for study. Frenchman Mountain is probably the easiest place to examine a representative Phanerozoic stratigraphic section of the southwestern portion of the North American craton. Due to space restrictions, only the Paleozoic section is described in this guide. Mesozoic and Cenozoic units are shown in Figure 2 and may be observed from the numbered stops circled in Figure 1.

SITE DESCRIPTIONS FOR STOPS 1–4

Structure section A-A′-A″ (Fig. 2a,b) parallels the route of the field trip, as indicated in Figure 1. The units visible from field trip Stops 2 through 5 are indicated in Figure 2b, whereas the stratigraphic locations of Stops 6 through 7 are indicated in Figure 2a.

Take the Lake Mead Boulevard East off-ramp from I-15. Drive east about 6.5 mi (10.4 km) to Hollywood Boulevard at the foot of Frenchman Mountain. Although the road log begins at the corner of Hollywood and Lake Mead Boulevards (Stop 1), it is preferable to read the text for Stop 1 while traveling east on Lake Mead Boulevard well before arriving at the Stop. If you are able to do this, proceed directly to Stop 2.

Mile 0 (0 km) (Lake Mead and Hollywood Boulevards: Stop 1: West face of Frenchman Mountain. As shown in

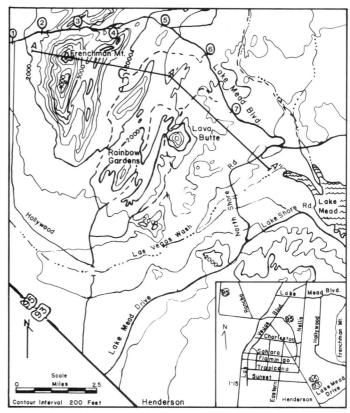

Figure 1. Index map. Circled numbers are field trip stops.

Figure 2b, most of the west face of Frenchman Mountain is composed of dolomite. The pink and black hills at the base of the mountain are Precambrian granite and schist. Cambrian sediments comprise more than half the west face, with Devonian and Mississippian beds making up the remainder (Fig. 2b). The letter "E" (for Eldorado High School) lies about in the middle of the Cambrian Bright Angel Formation. The orange layer is the lower part of the Banded Mountain Member of the Cambrian Bonanza King Formation. When viewed from several miles away, the most conspicuous unit is the Devonian Sultan Formation, which forms a prominent band of dark-gray dolomite about two-thirds of the way up. The uppermost strata on the left-hand peak are light-gray dolomites of the Mississippian Monte Cristo Formation. A little bit of the less-resistant Pennsylvanian Callville Formation is visible on the right-hand peak.

The field trip route follows Lake Mead Boulevard eastward between Frenchman Mountain on the south and Sunrise Mountain on the north (Fig. 1). The Sunrise Mountain block contains the same stratigraphy as the Frenchman Mountain block, but the

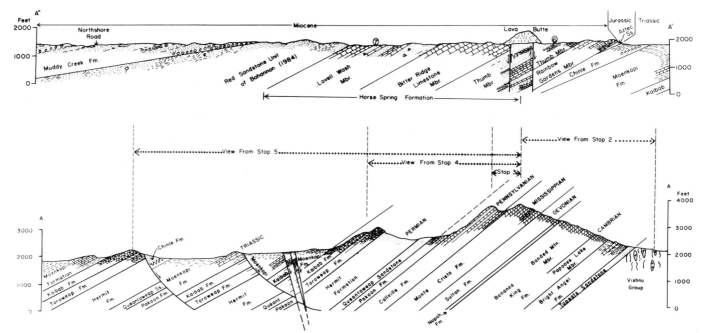

Figure 2a: Structure section A″–A′; stops 6 and 7 indicated. b: Structure section A′–A.

two blocks are separated by the Frenchman Fault. This fault displays a left-lateral sense of offset and has brought Permian beds in the Sunrise block into contact with Cambrian beds in the Frenchman block.

Proceed eastward on Lake Mead Boulevard.

Mile 1.0 (1.6 km) (Lake Mead Boulevard adjacent to Tapeats-Vichnu contact): Stop 2: Great Unconformity—Cambrian through Mississippian Units. The "great unconformity" is the white exposure across the wash on the south side of the road. Here, coarse, cross-bedded quartz sandstones of the Tapeats Sandstone (132 ft; 40 m) can be observed to lie nonconformably on the granites and garnet-biotite schists of the Vishnu Group. As in some places within the Grand Canyon (Babcock and others, 1976), the presence of large porphyroblasts of potassium feldspar in the granitic rocks and the diffuse pattern of granitic masses suggest that the plutonic rocks are largely the result of partial melting of portions of schist.

The Vishnu Group at Frenchman Mountain has not been dated. In the Grand Canyon, radiometric ages range from about 1.2 to 1.7 Ga (Babcock and othes, 1976), suggesting that a major Proterozoic orogenic event occurred around 1.7 Ga. Thus the "great unconformity" represents approximately 1 b.y. of geologic time.

Figure 3, which shows the strata visible from this point, was drawn from the top of the ridge on the north side of the road. The short climb up this ridge is rewarded with an excellent overview of the Cambrian through Mississippian strata of Frenchman Mountain.

The Bright Angel Formation (584 ft; 177 m) displays characteristics of both the craton and miogeocline. The prominent burrow-mottled limestone in the upper half of the Bright Angel at

Frenchman Mountain is equivalent to the Jangle Limestone Member of the miogeoclinal Carrara Formation in the Death Valley region (Palmer and Halley, 1979); it can also be traced southeastward onto the Colorado Plateau where it becomes the Meriwitica and Tincanebits dolomite tongues of the Bright Angel (compare McKee, 1976, Fig. 5). Northward, in the Pioche District of Lincoln County, Nevada, this limestone is called the Lyndon Limestone, and the underlying and overlying shales are called the Pioche and Chisholm Shales, respectively (Longwell and others, 1965).

Although the Frenchman Mountain Paleozoic section is quite similar to the section in the Grand Canyon in gross stratigraphy, there are some important differences. Most significantly, the Paleozoic of Frenchman Mountain is about twice as thick as the Paleozoic section of the central Grand Canyon. This doubling of thickness reflects a general westward thickening of all systems, but the Cambrian accounts for a great deal of it. As in the Grand Canyon, the Ordovician and Silurian Systems are absent here, but the Middle and Upper Cambrian dolomites of the Bonanza King (2,000 ft; 610 m) and Nopah Formations (110 ft; 33 m) are several hundred meters thicker at Frenchman Mountain than the partially age-equivalent Muav Limestone in the central Grand Canyon.

The conspicuous dark-gray dolomites of the Devonian Sultan Formation (574 ft; 174 m) are correlative with the less conspicuous Temple Butte Limestone in the Grand Canyon. The Paleozoic stratigraphic nomenclature evolved more or less independently in southern Nevada and in the Grand Canyon region. Consequently, some units that are very similar in both areas are known by different names. Much of the southern Nevada Paleozoic nomenclature originated with Hewitt (1931). The Mississip-

Figure 3. View from stop 2, drawn from top of ridge on north side of Lake Mead Boulevard.

Figure 4. View from stop 3, drawn from top of small hill on north side of Lake Mead Boulevard.

pian Monte Cristo Formation at Frenchman Mountain, for example, is very similar to the Redwall Limestone of the Colorado Plateau but is named for the Monte Cristo Mine in the Goodsprings District.

In addition to being thicker than correlative strata in northern Arizona, the southern Nevada Paleozoic carbonate units have, in many places, been altered to coarse crystalline dolomite. As a result, sedimentary structures and fossils are often obscured or obliterated. The Monte Cristo is a good example of this. At Frenchman Mountain it is mostly dolomite and is only sparsely fossiliferous.

Proceed eastward on Lake Mead Boulevard. Stop 3 is at the crest of the hill where the road curves to the right.

Mile 2.2 (3.5 km) (Crest of Hill): Stop 3: Mississippian and Pennsylvanian Units (Fig. 4). Park at the crest of the hill on the right-hand side of the road. Walk up onto the small hill adjacent to the road.

The Monte Cristo Formation (800 ft; 242 m) forms the right-hand peak of Frenchman Mountain, whereas the Callville Formation forms the left peak. The nonresistant redbeds in the notch are sometimes included within the Callville, as in Figure 2b, and sometimes they are separated out as the Indian Springs Formation. The Mississippian-Pennsylvanian boundary lies

within the Indian Springs interval (Langenheim and Webster, 1979).

The Callville Formation (980 ft; 297 m) is predominantly a fossiliferous marine limestone with dark-weathering beds of cross-bedded sandstone in the upper part. This carbonate unit is correlative with the Pennsylvanian portion of the siliciclastic Supai Group of the Grand Canyon and marks the most significant facies change between the Colorado Plateau and southern Nevada. Westward across Las Vegas Valley, the Pennsylvanian thickens dramatically into the rhythmically bedded Bird Spring Formation.

The hill you are standing on belongs to the Virgin Limestone Member of the Triassic Moenkopi Formation. Across Lake Mead Boulevard are good exposures of the Permian Toroweap and Kaibab Formations. These units will be viewed from a distance at Stop 4 but are more accessible for examination here.

Continue eastward on Lake Mead Boulevard.

Mile 2.9 (4.6 km) (Junction with Dirt Road on Right). Do *not* turn here, but slow down and prepare to turn right in 0.3 miles. This turn will seem to come up quickly, so watch your odometer.

Mile 3.2 (5.1 km) (Junction with Old Paved Road). Turn right onto the old paved road.

Figure 5. View from stop 4, drawn from top of small hill next to old paved road.

Mile 3.3 (5.3 km) (Small Hill on Right Side of Old Paved Road): Stop 4: Permian Units (Fig. 5). Overlying the Callville Limestone are dolomites and gypsum of the Pakoon Formation (884 ft; 268 m). The Pennsylvanian-Permian boundary lies within the Pakoon, but the absence of fossils makes the exact position uncertain.

The Queantoweap Sandstone (376 ft; 114 m) is a distinctive, friable, fine-grained, white to pinkish quartz sandstone that is correlative with the Esplanade Sandstone of the Supai Group on the Colorado Plateau (Bissell, 1969). Both the Pakoon and Queantoweap were named and defined by McNair (1951).

Overlying the Queantoweap is a thick sequence (900 ft; 273 m) of poorly exposed red shale and fine-grained sandstone of the Hermit Formation. It is everywhere a valley former. The Las Vegas city dump ("Sunrise Sanitary Landfill") is located within this same Hermit strike valley 5 mi (8 km) south of this point.

Overlying the Hermit is the Coconino Sandstone. Although the Coconino is a prominent cliff-forming unit in the central Grand Canyon, it thins westward (in contrast to the westward thickening of most of the Paleozoic units). At Frenchman Mountain it is only 106 ft (32 m) thick and rarely is well exposed, usually being covered by talus from the overlying units. From this point the only visible outcrop of Coconino is in a steep gulley just beyond the fault (see the central portion of Fig. 5).

From the Hermit Formation through the Triassic units, the stratigraphic terminology in common use in southern Nevada is identical with that of the southwestern Colorado Plateau.

Overlying the Coconino are the Toroweap (436 ft; 132 m) and Kaibab (644 ft; 195 m) Formations. Together these two formations consist of two cliff-forming cherty limestone layers sandwiched between three gypsiferous, slope-forming horizons. The lower three layers of this five-part sequence are, from bottom to top, the Seligman, Brady Canyon, and Woods Ranch Members of the Toroweap Formation. The upper two layers are the Fossil Mountain and Harrisburg Members of the Kaibab Formation (Fig. 5) (Bissell, 1969). The two cherty limestone layers (Brady Canyon Member of the Toroweap and Fossil Mountain Member of the Kaibab) are everywhere prominent ridge formers. The abundant nodules of black-weathering chert are always present in southern Nevada, but are not present in these units on the Colorado Plateau. The Harrisburg Member of the Kaibab is sporadically exposed in the Frenchman Mountain area, but it is not visible from the route of this field guide.

Proceed eastward on the old paved road. At the junction with Lake Mead Boulevard you may turn left and return to Las Vegas or turn right and proceed eastward upsection into the Mesozoic and Tertiary. Stops 5, 6, and 7, indicated in Figures 1 and 2, are convenient points from which to view the Mesozoic and Miocene units. For a description of the Miocene units, see Bohannon (1984).

REFERENCES CITED

Babcock, R. S., Brown, E. H., and Clark, M. D., 1976, Geology of the older Precambrian rocks of the Upper Granite Gorge of the Grand Canyon, *in* Breed, W. J., and Roat, E., eds., Geology of the Grand Canyon, 2nd ed.: Flagstaff, Arizona, Museum of Northern Arizona, p. 2–19.

Bissell, H. J., 1969, Permian and Lower Triassic transition from shelf to basin (Grand Canyon, Arizona to Spring Mountains, Nevada), *in* Baars, D. L., ed., Geology and Natural History of the Grand Canyon Region: Four Corners Geological Society Guidebook, Fifth Field Conference, p. 135–169.

Bohannon, R. G., 1984, Nonmarine Sedimentary Rocks of Tertiary Age in the Lake Mead Region, Southeastern Nevada and northwestern Arizona: U.S. Geological Survey Professional Paper 1259, 72 p.

Bohannon, R. G., and Bachhuber, F., 1979, Road log from Las Vegas to Keystone thrust area and Valley of Fire via Frenchman Mountain, *in* Newman, G. W., and Goode, H. D., eds., Basin and Range Symposium and Great Basin Field Conference: Denver, Colorado, Rocky Mountain Association of Geologists Guidebook, 1979, p. 579–596.

Hewitt, D. F., 1931, Geology and Ore Deposits of the Goodsprings Quadrangle, Nevada: U.S. Geological Survey Professional Paper 162.

Langenheim, R. L., Jr., and Webster, G. D., 1979, Stop descriptions–sixth day, *in* Beus, S. S., and Rawson, R. R., eds., Carboniferous Stratigraphy in the Grand Canyon Country, Northern Arizona and Southern Nevada, Field Trip No. 13, Ninth International Congress of Carboniferous Stratigraphy and

Geology: Falls Church, Virginia, American Geological Institute, p. 53–60.

Longwell, C. R., Pampeyan, E. H., Bower, B., and Roberts, R. J., 1965, Geology and Mineral Deposits of Clark County, Nevada: Nevada Bureau of Mines Bulletin 62, 218 p.

McKee, E. D., 1976, Paleozoic rocks of Grand Canyon, *in* Breed, W. J., and Roat, E., eds., Geology of the Grand Canyon, 2nd ed.: Flagstaff, Arizona, Museum of Northern Arizona, p. 42–64.

McNair, A. H., 1951, Paleozoic stratigraphy of part of northwestern Arizona: Bulletin of the American Association of Petroleum Geologists, v. 35, p. 503–541.

Palmer, A. R., and Halley, R. B., 1979, Physical stratigraphy and trilobite biostratigraphy of the Carrara Formation (Lower and Middle Cambrian) in the southern Great Basin: U.S. Geological Survey Professional Paper 1047, 131 p.

ACKNOWLEDGMENTS

Many UNLV students contributed to this compilation. Figures 3, 4, and 5 were drawn by Roberta Goodwill, Figure 2 was drafted by Jerry Baughman, and Figure 1 was drafted by Tammy Howard. The following students measured portions of stratigraphic sections: W. M. Alsup, E. L. Basham, K. M. Burrows, R. J. Gegenheimer, M. S. Hug, J. M. Lombardo, R. A. Mocabee, C. S. Rodell, U. O. Rupprecht, and B. A. Wright.

The Keystone and Red Spring thrust faults in the La Madre Mountain area, eastern Spring Mountains, Nevada

Gary J. Axen, Geology Department, Northern Arizona University, Flagstaff, Arizona 86011

LOCATION OF SITES

The La Madre Mountain area lies about 9 mi (15 km) west of Las Vegas, Nevada (Fig. 1). Access is provided to sites 1 through 5 by side roads off of West Charleston Boulevard, and site 6 is reached from U.S. 95 northwest of town. The sites are meant to be visited in consecutive order, and the mileages given begin at the I-15 and West Charleston exit. Although the first five sites are accessible by any vehicle, site 6 may necessitate a vehicle with a high center. Four-wheel drive is not necessary.

U.S. Geological Survey topographic maps for the area are the La Madre Mountain and Blue Diamond NE 7½-minute Quadrangles. Site 4 is on the easternmost edge of the Mountain Springs 15-minute Quadrangle.

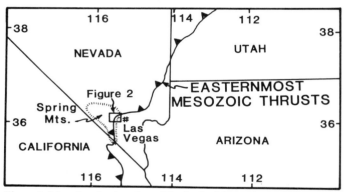

Figure 1. Map showing location of the Spring Mountains and Figure 2 in relation to the eastern edge of the thrust belt.

SIGNIFICANCE OF THE LA MADRE MOUNTAIN AREA

The Spring Mountains are one of the Nevada ranges that have been least affected by Tertiary extensional tectonics, and therefore provide one of the best areas in the Great Basin to study the Mesozoic thrust belt (Burchfiel and others, 1974). The La Madre Mountain area (Fig. 2) contains spectacular exposures of the two easternmost major thrusts at this latitude, the Keystone and Red Spring thrusts (Longwell, 1926; Hewett, 1931). Part or possibly all of the area is underlain by the lesser Bird Spring thrust of Hewett (1931). Also exposed are high-angle faults with complex histories of multiple movements. Cross-cutting fault relations and the synorogenic conglomerate of Brownstone Basin provide important timing constraints on structural events in the Spring Mountains; in this area, thrusting did not progress from west to east through time, as in the Idaho-Wyoming belt. Quartz-rich fault gouge exposed locally

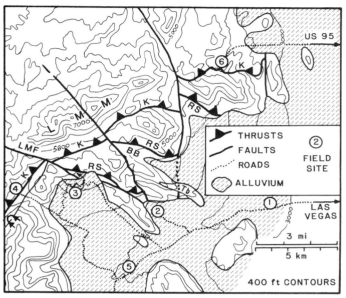

Figure 2. Map of the La Madre Mountain (LMM) area, showing field sites. K, Keystone thrust; RS, Red Spring thrust; LMF, La Madre fault; BB, Brownstone Basin; Tb, Tertiary(?) landslide breccia.

along the Keystone thrust provides information about the subsurface extent of the Red Spring allochthon, and allows inferences to be made about mechanics of movement on the Keystone thrust.

DIRECTIONS AND DISTANCES

(mi) *(km)* Proceed west on West Charleston Boulevard from junction with I-15.

4.6 *7.4* Rainbow Boulevard and West Charleston (continue west).

8.6 *13.8* *Site 1.* Pull off road on right.

13.6 *21.9* West Charleston and Calico Basin Road. Turn right onto Calico Basin Road.

14.7 *23.6* Pavement ends at "T" junction. Red Spring to left, private lands (no trespassing) to right. Walk straight ahead onto low hill for *Site 2.*

15.8 *25.4* Return to West Charleston, turn right (west).

17.5 *28.2* West Charleston and Red Rock Scenic Loop Drive. Turn right onto loop.

23.3 *37.5* Loop Drive and White Rock Spring Road. Turn right.

23.9 *38.4* End of White Rock Spring Road; park here (Fig. 3). Cross the wash to the right (northeast) and walk cross-country toward the black hill for about 660 ft (200 m) until you meet an abandoned Jeep trail. Walk north on the Jeep trail, around the northwest end of the black hill to the Aztec Sandstone outcrop on the northeast side, to *Site 3*. Allow 15–20 minutes for the walk, 1–1½ hours from the time you leave the parking area until you return.

24.5   39.4   Return to the Loop Drive. Turn right.

26.0   41.8   Loop Drive and road to Willow Spring picnic area. Turn right.

26.6   42.8   Pavement ends. Parking. Walk or drive on the gravel road past the parking area for about 0.3 mile (0.5 km), cross the wash, and continue on the (washed out) road for about 660 ft (200 m) to the red roadcut in Triassic sediments. Walk up the toe of the ridge that the roadcut is in for about 330 ft (100 m) to *Site 4*.

27.2   43.8   Return to Loop Drive. Turn right.

32.2   51.8   Loop Drive and West Charleston. Turn left (east).

32.7   52.6   Turn left into paved pull-out, for *Site 5*. Return to Las Vegas via West Charleston.

44.7   71.9   West Charleston and Rainbow Boulevard. Turn left (north) on Rainbow Boulevard.

45.9   73.9   Rainbow Boulevard and U.S. 95. Take north entrance onto U.S. 95.

50.7   81.6   U.S. 95 and Lone Mountain Road. Turn left (west).

55.0   88.5   Road bends to right. Continue straight ahead on one lane, unimproved dirt road, through south side of gravel pit, and ahead into Box Canyon.

56.7   91.2   Cross dirt road, gravel test pit to left.

57.9   93.2   Road to right, continue ahead.

58.3   93.8   Road forks, bear left.

58.5   94.1   Road ends at loop, *Site 6* (Fig. 4).

SITE DESCRIPTIONS

Site 1 (Fig. 2) provides an overview of the major structural relations at La Madre Mountain. In the foreground are three northeast-tilted blocks of Paleozoic carbonates. On their southwest sides they are bounded by the Red Spring thrust, which dips about 30 to 35° to the northeast and placed the Paleozoic strata above the red Jurassic Aztec Sandstone exposed in the valleys. The thrust plate and underlying parautochthon are cut by high-angle normal(?) faults trending north to northwest, which bound the Paleozoic blocks on their eastern sides (Fig. 2; Davis, 1973; Axen, 1985). Therefore, bounding the prominent valley on the east (underlain by Aztec Sandstone) is the Red Spring thrust on the right, and a high-angle fault on the left.

The high skyline cliff in the rear is La Madre Mountain, which is composed of Paleozoic strata of the Keystone thrust sheet (Fig. 2). The high-angle faults continue into the Keystone plate, but their offset there is much less than in the Red Spring allochthon and parautochthon. Davis (1973) interpreted this to mean that one episode of movement on the high-angle faults

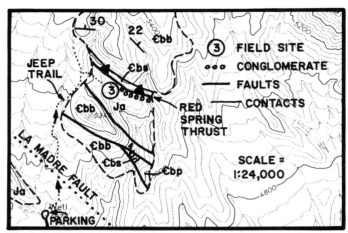

Figure 3. Detailed geologic map at site 3. Ja, Aztec Sandstone. Units in the Bonanza King Formation: €bb, Banded Mountain Member; exclusive of the "silty unit" (€bs); €bp, Papoose Lake Member (From Axen, 1985.)

tilted the Red Spring plate and predated emplacement of the Keystone thrust, and that later movement on the same faults cut the Keystone plate after its emplacement.

Site 2 has a view of the Red Spring thrust and one of the high-angle faults. On the dark peak in the foreground to the northwest, Middle Cambrian dolostone of the Bonanza King Formation overlies tan and red Jurassic Aztec Sandstone and thin lenses of the synorogenic conglomerate of Brownstone Basin (not visible from a distance). Bounding the block on the east and west are high-angle faults that cut the Red Spring plate and parautochthon. From here, only the eastern fault is visible.

To the northeast, beyond the cliffy Aztec Sandstone ridge in the foreground, is a ridge capped by dark, cemented landslide breccia, mainly composed of upper and middle Paleozoic carbonate fragments. This breccia overlaps one of the high-angle faults, which can be seen juxtaposing pink Aztec Sandstone on the left against the rubble-covered Virgin Limestone Member of the Moenkopi Formation near the east end of the ridge (Fig. 2). Note that the base of the breccia is quite irregular, in contrast to the planar thrust contact. The landslide has been mapped as both a thrust sheet (Longwell, 1926) and Mesozoic(?) landslide (Davis, 1973), and is currently considered Tertiary(?) in age (Axen, 1984), because it overlaps the high-angle fault that cuts the late Cretaceous Keystone thrust farther north.

Site 3 is at an exposure of the Red Spring thrust. As one walks from the parking area to the site, the buried trace of the La Madre fault is crossed. Here the La Madre fault has about 3.6 mi (6 km) of right-lateral separation, or about 4,290 ft (1,300 m) of northeast-side-down separation, and bounds the northeast side of a large horst(?) formed by the Aztec Sandstone cliffs to the southwest (Fig. 2).

At the site, the tan-weathering "silty unit" at the base of

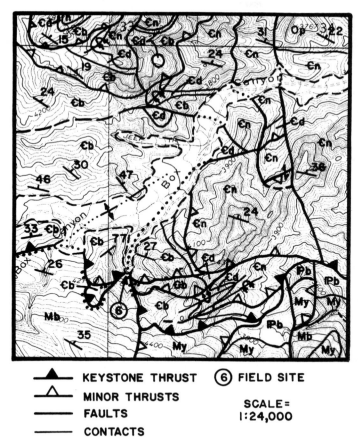

▲——— KEYSTONE THRUST ⑥ FIELD SITE

△——— MINOR THRUSTS

———— FAULTS   SCALE=

———— CONTACTS   1:24,000

Figure 4. Detailed geologic map at site 6. Pb, Bird Spring Formation; My and Mb, Yellow Pine and Arrowhead, and Bullion Limestone Members of the Monte Cristo Formation, respectively; Op, Pogonip(?) Group; €d and €n = Dunderberg Shale Member and upper main part of the Nopah Formation, respectively; €b, Banded Mountain Member of the Bonanza King Formation. Dots along the Keystone thrust show the location of quartz-rich fault gouge. (Simplified from Axen, 1985.)

the Banded Mountain Member of the Bonanza King Formation lies in thrust contact above the Jurassic Aztec Sandstone and the conglomerate of Brownstone Basin (Fig. 3). Throughout most of the Spring, Muddy, and Mormon Mountains, the frontal thrusts detached at or near the "silty unit," in the middle of a sequence of strong dolostone, rather than forming a décollement zone in the weaker shales below the Bonanza King Formation—a mechanically puzzling situation (Burchfiel and others, 1982).

At this site, the conglomerate of Brownstone Basin is composed of: (1) cobbles and reworked pebbles of jasper-rich Triassic Shinarump Conglomerate, (2) clasts of late Precambrian or early Cambrian quartzites, and (3) matrix and clasts of Aztec Sandstone (Davis, 1973). Paleocurrent analysis indicates transport to the east-northeast (Jones and others, 1984). The quartzites must have been derived from the Wheeler Pass allochthon to the west, the nearest thrust that carried those rocks, indicating that that thrust is older than the Red Spring

thrust (Axen, 1984; Jones and others, 1984). The Triassic detritus is believed to have originated in the overturned fold in parautochthonous rocks to the west (Fig. 2, site 4), and the reworked Aztec Sandstone is probably locally derived and/or from the same fold. In its thickest section, exposed in Brownstone Basin (Fig. 2), the conglomerate has a stratigraphically higher lithofacies composed of Paleozoic carbonate clasts (Davis, 1973; Axen, 1984; Jones and others, 1984). The abrupt change in clast type is interpreted to be due to the approach of the Red Spring thrust sheet as it overrode the land surface and its own detritus (Davis, 1973).

Due to its structural and stratigraphic setting, the conglomerate of Brownstone Basin is correlated with the Lavinia Wash sequence found by Carr (1980) below the correlative Contact thrust to the south. He obtained a K-Ar date of 150 ± 10 Ma on a tuff in the Lavinia Wash sequence. Thus the conglomerate of Brownstone Basin is thought to be Late Jurassic or Early Cretaceous(?) in age. This suggests that the Wheeler Pass thrust is pre–Late Jurassic in age, and that the Red Spring thrust is Late Jurassic in age.

At site 4, carbonate rocks of the Keystone sheet (to the north) are in thrust contact with steeply overturned parautochthonous Mesozoic strata. The thrust is exposed due north of the site, where a black carbonate layer lies on tan and red Triassic sediments. Although this fold appears to be a footwall syncline to the Keystone thrust, it is now thought that the fold formed early in the Red Spring event, and was overridden by the Red Spring allochthon (Axen, 1985; Jones and others, 1984). Uplift of the southwest side of the La Madre fault followed, and erosion entirely removed rocks of the Red Spring plate from this site (Davis, 1973), although remnants remain to the south, below the Keystone thrust (Wilson Cliffs thrust plate of Burchfiel and Royden, 1984).

Site 5 allows an overview of the relations between the Keystone and Red Spring thrusts and the La Madre fault (Fig. 2). The dip slopes to the southeast are composed of Permian Kaibab and Toroweap Limestones, which lie in stratigraphic continuity below the strike valley of Triassic rocks and the Aztec Sandstone cliffs to the west. The dark limestone in thrust contact above the cliffs is Middle Cambrian in age. It is part of the Wilson Cliffs allochthon of Burchfiel and Royden (1984), the equivalent of the Red Spring thrust, which can be seen to the right of the cliff and (at site 3), across the La Madre fault. The Keystone plate underlies the skyline ridge (La Madre Mountain) behind site 3. To the northeast of the La Madre fault, the Keystone plate overrode tilted rocks of the Red Spring plate, while to the southwest it overrode the uplifted horst(?) of parautochthonous rocks and remnants of the Wilson Cliffs/Red Spring plate.

The Keystone plate is cut by the La Madre fault, but its vertical separation is only 330 to 660 ft (100-200 m), as opposed to the 4,290 ft (1300 m) of separation of the Red Spring plate and parautochthon. In addition to the two movements necessary to explain this geometry (see the site 1 description), the La Madre fault acted as a tear fault during emplacement of the Keystone plate (Secor, 1962) and may have had Tertiary activity (Axen, 1985) related to the Las Vegas Valley shear zone, which bounds the northeast side of the Spring Mountains.

The Keystone thrust is well exposed at site 6. There the Banded Mountain Member of the Bonanza King Formation (Middle Cambrian) forms the canyon walls. It overlies the lighter gray dip slope of the Mississippian Bullion Member of the Monte Cristo Limestone in the Red Spring plate (Axen, 1984, 1985). The thrust is well exposed along the eastern side of the canyon. The Keystone thrust cuts upsection in its upper plate toward the east, by truncating the steep limb of an asymmetric anticline (Fig. 4). To the east is one of the few areas in the region where the décollement horizon is not near the "silty unit" of the Bonanza King Formation.

In this area the upper plate of the Keystone thrust is riddled with minor imbricate thrusts (Fig. 4), in contrast to the relatively undeformed upper plate rocks farther west. The imbricate nature of the upper plate can be seen on the skyline-forming mountain to the north. There the slope-forming, thin-bedded Dunderberg Shale is repeated by thrusts (Fig. 4). The change is believed to be due to a variation in the ramp geometry (Axen, 1984).

Scattered along the exposed trace of the Keystone thrust at site 6 are lenses and injections of light tan quartz-rich fault gouge, derived mainly from the Aztec Sandstone. The largest lens is exposed on the east side of the canyon, just above the alluvium, where it is slivered into the upper plate. The largest injection is roughly circular in plan view, and is exposed about 33 ft (10 m) above the thrust plane, high in the west wall of the canyon. This gouge indicates that the Keystone plate rode on parautochthonous strata to the northwest or west, and that the Red Spring plate must be absent at depth in that direction (Axen, 1984, 1985).

The injections of gouge appear to be finer grained than the lenses, and are not planar. These relations are interpreted to mean that the thrust moved by a stable sliding mechanism, and that high pore-fluid pressure may not have been important along this section of the thrust (Axen, 1984).

REFERENCES CITED

Axen, G. J., 1984, Thrusts in the eastern Spring Mountains, Nevada: Geometry and mechanical implications: Geological Society of America Bulletin, v. 95, no. 10, p. 1202–1207.

—— , 1985, Geologic map and description of structure and stratigraphy, La Madre Mountain, Spring Mountains, Nevada: Geological Society of America Map and Chart Series MC-51.

Burchfiel, B. C., Royden, L. H., 1984, The Keystone thrust fault at Wilson Cliffs, Nevada, is not the Keystone thrust: Implications: Geological Society of America Abstracts with Programs, v. 16, no. 6, p. 458.

Burchfiel, B. C., Fleck, R. J., Secor, D. T., Vincelette, R. R., and Davis, G. A., 1974, Geology of the Spring Mountains, Nevada: Geological Society of America Bulletin, v. 85, p. 1013–1022.

Burchfiel, B. C., Wernicke, B. P., Willemin, J. H., Axen, G. J., and Cameron, C. S., 1982, A new type of décollement thrusting: Nature, v. 300, no. 5892, p. 513–515.

Carr, M. D., 1980, Upper Jurassic to Lower Cretaceous(?) synorogenic sedimentary rocks in the southern Spring Mountains, Nevada: Geology, v. 8, no. 8, p. 385–389.

Davis, G. A., 1973, Relations between the Keystone and Red Spring thrust faults, eastern Spring Mountains, Nevada: Geological Society of America Bulletin, v. 84, p. 3709–3716.

Hewett, D. F., 1931, Geology and ore deposits of the Goodsprings Quadrangle, Nevada: U.S. Geological Survey Professional Paper 162, 172 p.

Jones, D. A., Schmitt, J. G., and Axen, G. J., 1984, Sedimentology, provinance, and tectonic significance of the Brownstone Basin conglomerate, Upper Jurassic–Lower Cretaceous(?), Spring Mountains, Nevada: Geological Society of America Abstracts with Programs, v. 16, no. 6, p. 552.

Longwell, C. R., 1926, Structural studies in southern Nevada and western Arizona: Geological Society of America Bulletin, v. 37, p. 551–584.

Secor, D. T., 1962, Geology of the central Spring Mountains, Nevada [Ph.D. thesis]: Stanford, California, Stanford University, 152 p.

14

Lower Paleozoic craton-margin section, northern Potosi Valley, southern Spring Mountains, Clark County, Nevada

John D. Cooper, Department of Geological Sciences, California State University, Fullerton, California 92634

INTRODUCTION

Potosi Valley, located in the southern Spring Mountains approximately 30 mi (48 km) southwest of Las Vegas, Nevada (Fig. 1), exhibits a well-exposed section of the lower Paleozoic Goodsprings Dolomite of Hewett (1931). The best section is exposed on the west side of northern Potosi Valley in the east face of a 4-mi- (6.4-km-) long strike ridge (Fig. 2) containing a 3,300 ft (1,000 m +) thick succession of Middle Cambrian through Middle to Upper Devonian carbonate rocks (Fig. 3). These strata provide an instructive look at the Cordilleran craton-margin to miogeoclinal transition sequence comprising part of an allochthonous assemblage in the upper plate of the Keystone thrust. This accessible area has much to offer the student interested in Great Basin stratigraphy, sedimentology, and structure.

From a stratigraphic perspective, the Goodsprings Dolomite includes regionally correlatable subdivisions, which are, in ascending order: the Bonanza King, Nopah, and Mountain Springs formations. In the middle of the sequence is the Dunderberg Shale, which is the basal member of the Nopah Formation (Figs. 3, 4) and is one of the most widespread and reliable stratigraphic markers in the southern Great Basin. The Mountain Springs Formation (Figs. 3, 4) contains the easternmost exposures of Ordovician strata south of the Las Vegas Valley shear zone and is punctuated by two significant disconformities (Fig. 4).

Good examples of primary depositional textures and fabrics representing a diverse spectrum of shallow-marine depositional environments are in great abundance. These, together with a variety of diagenetic features including secondary dolomite and

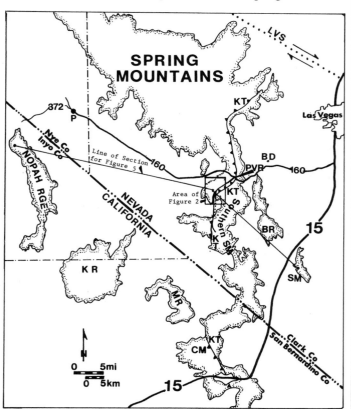

Figure 1. Regional location map showing southern Spring Mountains and northern Potosi Valley, south of Nevada State Highway 160 (note outline of area for Figure 2). KT, Keystone thrust; LVS, Las Vegas Valley shear zone; BD, Blue Diamond; PVR, Potosi Valley road; P, Pahrump; BR, Bird Spring Range; SM, Sheep Mountain; KR, Kingston Range; CM, Clark Mountain; MR, Mesquite Range.

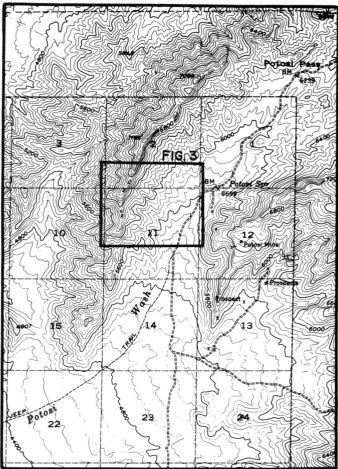

Figure 2. Northeastern part of U.S.G.S. Shenandoah Peak Nevada-California 15-minute Quadrangle showing northern Potosi Valley and strike ridge on western side of valley exposing described stratigraphic section. Section line is 1 mi (1.6 km).

chert, should be of interest to the carbonate petrologist. Geologic relationships along the Keystone thrust also can be observed.

LOCATION AND ACCESSIBILITY

Northern Potosi Valley is located in sections 1, 2, 11, 12, 14, and 15, R57E, T23S, Shenandoah Peak, Nevada-California Quadrangle, U.S.G.S. 15-minute topographic series (Fig. 2). The area is easily reached from the north, off Nevada State Highway 160, by a south-trending unpaved road (the Potosi Camp road) that can be traveled by standard passenger car in good weather. The junction of this Potosi Camp road with SR 160 is 19.5 mi (31.2 km) west of Interstate 15 (and 6.5 mi (10.4 km) west of Blue Diamond) and 34 mi (54.4 km) east of Pahrump, Nevada, where SR 160 joins SR 372 (to Shoshone, California). Traveling east from Pahrump, the turnoff to Potosi Valley is 1.85 mi (2.96 km) east of Mountain Springs summit; from both directions, it is well marked with a sign for Potosi Mountain. The main area of interest is between Potosi Pass and Potosi Wash (Fig. 2). The summit of Potosi Pass is 2.55 mi (4.08 km) south of the junction with SR 160. The best place to traverse the section (Fig. 3) is just south of Potosi Spring, approximately 2 mi (3.2 km) south of Potosi Pass summit. A vigorous, but not strenuous, hike up the ridge face between 5,500 and 6,500 ft (1,650 and 1,950 m) elevation provides a good view of the section. Passenger cars are not recommended south of the junction of Potosi Wash jeep trail and the Potosi Valley road (Fig. 2); however, four-wheel-drive or sturdy, high-clearance, two-wheel-drive vehicles can negotiate both roads southward to Sandy Valley.

STRATIGRAPHY

Introduction

The Goodsprings Dolomite was named and described by Hewett (1931), who assigned a Cambrian-Devonian age based on sparse fossils and stratigraphic position between known Cambrian and Devonian strata. In its type section at Sheep Mountain, the Goodsprings is 2,475 ft (750 m) thick and lies between the Middle Cambrian Bright Angel Shale (= Carrara Formation) and the Middle Devonian Sultan Formation. Gans (1974) subdivided the Goodsprings Dolomite of the Goodsprings, Nevada, district into one previously unnamed and two regionally known units. Along the western side of northern Potosi Valley, the lower half of the exposed thickness of the outcropping Goodsprings interval was recognized by Gans (1974) as the Banded Mountain Member of the Bonanza King Formation (Middle-Upper Cambrian). The upper half includes the regionally widespread Nopah Formation (Upper Cambrian) and the Mountain Springs Formation (Lower Ordovician-lower Middle Devonian; named by Gans, 1974). Overlying the Goodsprings Dolomite is the cliff-forming Middle Devonian Sultan Formation.

Bonanza King Formation

Banded Mountain Member. The lowest unit in the

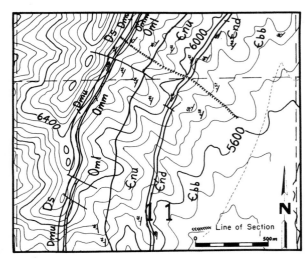

Figure 3. Geologic map of parts of sections 11 and 2 (see Fig. 2) showing subdivisions of Goodsprings Dolomite and suggested traverse (>>>>) through stratigraphic section. €bb Banded Mountain member of Bonanza King Formation; €nd-Dunderberg Shale member of Nopah Formation; €nu-upper Member of Nopah Formation; Oml, Omn, Dmu - lower, middle, and upper members of Mountain Springs Formation; Ds - Sultan Formation. Mapping by R. W. Goodman.

Goodsprings section along the western side of northern Potosi Valley is the Banded Mountain Member of the Bonanza King Formation. Hazzard and Mason (1936) named and described the Middle to Upper Cambrian Bonanza King Formation from its type section in the Providence Mountains. Palmer and Hazzard (1956) later defined an expanded Bonanza King Formation that is presently used by most southern Great Basin workers; and Barnes and Palmer (1961) defined the lower, Papoose Lake and upper, Banded Mountain Members. The Papoose Lake Member does not crop out in northern Potosi Valley because of cut-out by the Keystone thrust.

The exposed thickness of the Banded Mountain Member on the west side of northern Potosi Valey is about 1.485 to 1,650 ft (450 to 500 m). The most distinctive lithology is a dark gray, mottled dolomite that occurs in the lower and middle parts of the section. Blue-gray chert nodules and silicified domal stromatolites also occur in the lower half of the section. Other lithologies include cyclic intervals of light and dark gray-weathered dolomicrites and cherty, pinkish-gray, fine- to medium-crystalline dolomite. A distinct medium gray peloidal dolomite provides a good marker horizon some 130 ft (40 m) below the top of the Bonanza King. The uppermost part of the formation is predominantly light gray, medium-bedded, medium-crystalline, sucrosic dolomite, about 130 ft (40 m) thick, which forms a prominent, generally light-colored ridge below the contrasting orange-brown-weathered, recessed Dunderberg Shale interval (Figs. 3, 4) above and the darker-colored dolomites below. Although pervasively dolomitized, the Banded Mountain Member reveals a wealth of primary depositional features including bioclastic, peloidal, and bioturbated fabrics, domal, digitate, and stratiform

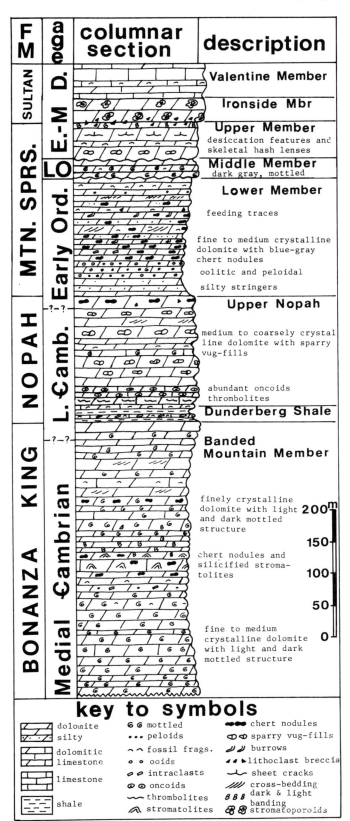

Figure 4. Columnar section, east face of ridge, west side of northern Potosi Valley.

cryptalgal structures, and some current-bedded features. Little detailed facies work has been done on the Bonanza King Formation of the Goodsprings district; however, Kepper (1972) has conducted regional analysis of Middle Cambrian paleoenvironmental patterns.

Nopah Formation

Conformably overlying the Banded Mountain Member of the Bonanza King Formation is the Nopah Formation, named by Hazzard (1937) for the type locality in the northern Nopah Range (Fig. 1). Here the basal unit of the Nopah consists of 99 ft (30 m) of brown-weathered, interbedded shale and bioclastic-intraclastic-oolitic carbonates comprising the Dunderberg Shale member (Cooper and others, 1982, p. 114-115). Barnes and Christiansen (1967) divided the Nopah Formation into the lower, Dunderberg Shale member; the middle, Halfpint Member; and the upper, Smoky Member. Regionally, its characteristic saddle topographic expression and orange-brown-weathering color permit recognition of the Dunderberg Shale throughout the Spring Mountains and a number of nearby ranges. It is one of the most widespread, convenient, and reliable markers in the Paleozoic section of the southern Great Basin. In the central Great Basin of Nevada and Utah, the Dunderberg Shale is generally regarded as a separate formation (e.g., Palmer, 1960); however, in the southern Great Basin, the member terminology of Barnes and Christiansen (1967) is most commonly used. In the Goodsprings district, the Dunderberg Shale is clearly represented, but the distinction between the Halfpint and Smoky members is not so clear. Cooper and others (1981, 1982) have documented the regional lithofacies and biostratigraphic zonation of the Dunderberg-Halfpint interval. The overlying Smoky Member, comprising the thickest part of the Nopah, has not been studied in detail.

Along the west side of northern Potosi Valley (Fig. 3), immediately above the uppermost light-colored, ledge-forming dolomite of the Bonanza King Formation, an orange-brown-weathered, recessed, 50- to 66-ft- (15- to 20-m-) thick interval forms a distinct band across the landscape. This is the Dunderberg Shale, which consists of interbedded 0.33- to 1.65-ft- brown to green shale that weathers into small thin flakes, and ledge-forming 0.33-1.65 ft (0.1-0.5 m) thick, orange-brown, crinkly weathered dolomitic limestone beds. The shale intervals are generally covered, being exposed only in small scars. The bench-forming carbonate beds contain large (up to 4.8 in; 12 cm) laminated intraclasts in a trilobite-eocrinoid grainstone host. Trilobites of the Upper Cambrian (Dresbachian Stage; Pterocephaliid biomere of Palmer, 1965; Steptoean Stage of Ludvigsen and Westrop, 1985) *Dunderbergia* biozone have been recovered from some of these beds (Miller and others, 1981; Cooper and others, 1982). This section lies within the peritidal eastern facies belt of Cooper and others (1982).

Overlying the Dunderberg Shale is approximately 548 ft (175 m) of medium- to thick-bedded generally light-tan- to light-gray-weathered dolomite and dolomitic limestone. The lower 66 to 99 ft (20 to 30 m) consists of beds of cross-stratified eocrinoid grainstone overlain by pale yellowish-gray dolomitic limestone with a characteristic clotted texture believed to represent cryptalgal thrombolites, succeeded by somewhat darker gray dolomite beds containing abundant oncoids. The main part of the upper Nopah section consists of crystalline dolomitic limestone and dolomite with sparse recognizable depositional textures. The predominant lithology is a light gray, sugary, medium-crystalline, thick-bedded to massive dolomite that forms jointed, somewhat bouldery outcrops developed as several prominent cliff-forming ledges (Fig. 4). This lithology contains abundant coalescing patches of coarsely crystalline sparry calcite and dolomite possibly representing either solution vug fills or stromatactis-like networks of bird's eye enlargements. A few medium to dark gray intervals are present in the section and consist of fine- to medium-crystalline dolomite with occasional fossil fragments and chert nodules and stringers. Locally, the upper 33 ft (10 m) of the Nopah Formation contains abundant brown-gray chert, some in the form of silicified lithoclasts and intraclasts. The upper Nopah maintains a rather constant thickness and light color band along strike in the ridge face. Based on stratigraphic position and regional relationships, the upper Nopah is considered to be Late Cambrian, although the Cambro-Ordovician boundary probably lies below the top.

Mountain Springs Formation

In subdividing the Goodsprings Dolomite, Gans (1974) named the Mountain Springs Formation, whose type section is in the east face of hill 6646 in the northeastern corner of the U.S.G.S. Shenandoah Peak, Nevada-California 15-minute Quadrangle (Fig. 2). The Mountain Springs is the stratigraphic interval between the Nopah Formation and the Ironside Member of the Sultan Formation (Fig. 4). Based on sparse megafossil control and little physical evidence, together with regional stratigraphic comparisons, Gans (1974) postulated the existence of post-Early Ordovician/pre-Late Ordovician and post-Late Ordovician/pre-Middle Devonian disconformities in the Mountain Springs Formation.

Miller and Zilinsky (1981) and Miller and Cooper (1982) corroborated the existence of these two disconformities and, based on tight conodont biostratigraphic control and detailed field descriptions (Cheek and others, 1983), these disconformities have been documented more precisely (Fig. 4). Hintzmann (1983) subdivided the Mountain Springs Formation in the Spring Mountains into three informal members separated by the two disconformities.

Following the subdivision scheme of Hintzmann (1983), the Mountain Springs Formation on the west side of northern Potosi Valley is developed as three distinct units separated by disconformities (Fig. 4). The thickest, lower unit is conformable with the underlying Nopah Formation and is Early Ordovician to early Medial Ordovician in age (Miller and Cooper, 1982). It consists of approximately 660 ft (200 m) of light gray, fine- to

medium-crystalline dolomite that forms several bold ledges. The basal 49.5 to 66 ft (15 to 20 m) is developed as a topographic saddle within which thinly bedded orange-tan-weathered bioclastic silty dolomite contrasts with the sugary, light gray, ledge-forming uppermost Nopah. Above this basal recessed interval is medium to dark gray peloidal and oolitic dolowackestone and dolopackstone succeeded by a thick interval of light gray, thin- to thick-bedded dolomite containing blue-gray chert nodules, stringers, and thin beds, and occasional thin beds of dolomite lithoclast breccia. This cherty interval gives way upsection to light gray, thin-bedded finely crystalline dolomite with locally abundant horizontal feeding traces (Fig. 4). The upper part of the lower unit is light to medium gray, fine- to medium-crystalline dolomite with some gray to white chert nodules and coarse bioclastic beds.

Despite the pervasive dolomitization, the lower unit of the Mountain Springs contains abundant recognizable allochems including skeletal grains, ooids, peloids, and intraclasts; in many places primary depositional textures and fabrics can be observed. Sedimentary structures include cross-stratification in dolarenite, occasional cryptalgal structures, bioturbation, and small-scale scours. Complete calcareous body fossils are rare; recognizable fossil fragments include crinoid, gastropod, articulate brachiopod, and ostracode. Edwards (1983) interpreted the lower unit of the Mountain Springs Formation as the product of peritidal environments including upper, middle, and lower intertidal flats, and both agitated and restricted shallow subtidal. Secondary dolomitization is judged to be early diagenetic; shallow burial (Edwards and Cooper, 1984b) based on stable oxygen isotope values, dolomite crystal size and morphology, and retention of primary depositional fabrics.

Approximately 660 ft (200 m) above the base of the Mountain Springs Formation is a distinctive 40-50 ft (12-15 m) thick, ledge-forming massive bed of highly mottled, dark olive gray, fetid, finely crystalline dolomite. Conodonts from this bed indicate a late Medial to Late Ordovician age (Miller and Cooper, 1982). Although the contact with underlying strata is generally poorly exposed, physical evidence for a disconformity can be observed in a few places along strike. Where exposed, the contact is irregular, highly calichified, and shows small-scale solution features (microkarst). Conodonts suggest a hiatus embracing much of medial Ordovician. Overlying this dark gray mottled bed is a thin interval of light gray dolomicrite with locally abundant small intraclasts and small chert nodules.

The upper member of the Mountain Springs Formation is approximately 165 ft (50 m) thick and consists of alternating medium- to thick-bedded intervals of fine- to coarsely crystalline, light to dark gray dolomite. In this part of the section, the topographic profile steepens and the slope becomes more vegetated with cedar and pinon pine. The contact with the underlying middle member is generally well exposed and is developed as thick-bedded, boudery weathered, coarsely crystalline, medium gray dolomite on light gray, finely crystalline, thin- to medium-bedded dolomite (Fig. 4). The disconformable contact is sharp and undulating and locally contains a thin band of paleocaliche

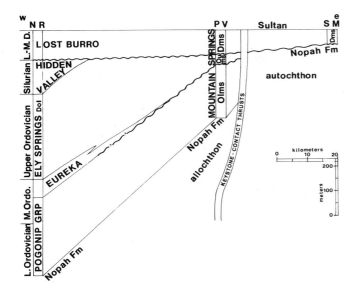

Figure 5. West-to-east stratigraphic panel showing thinning and unit pinchouts related to regional disconformities in post-Nopah/pre-Sultan section. NR, Nopah Range; PV, Potosi Valley; SM, Sheep Mountain; Olms, lower member of Mountain Springs Formation; Oums, middle member of Mountain Springs Formation; Dms, upper member of Mountain Springs Formation. See Figure 1 for line of section.

and microkarst solution features. Conodonts suggest an age of late Early to Medial Devonian for this upper unit.

Several dark gray, fetid, thin intervals containing sparse gastropod, brachiopod, and crinoid fragments are present. Also, several coarsely crystalline stringers and lenses containing large pentamerid brachiopod valve fragments occur in the middle part of the unit. Another prominent lithology includes thinly banded intervals of dark and light gray, fine- to medium-crystalline dolomite, particularly in the uppermost part of the unit.

Desiccation features are abundant in the upper member and are represented by sheet cracks and bird's eye (fenestrae) enlargements filled with coarsely crystalline calcite and dolomite, some of which coalesce to form a stromatactis-like structure. Occasional sediment-floored solution cavities are also present. Cheek and others (1983) interpret this upper, Devonian part of the Mountain Springs to represent a complex of supratidal, intertidal, and shallow subtidal environments. Harrington (1982) has described the transition from the upper Mountain Springs into the lower part of the overlying Sultan Formation. East of Potosi Valley, in the parautochthonous and autochthonous blocks at Stateline and Sheep Mountain, only the Devonian part of the Mountain Springs Formation is present (Fig. 5).

Sultan Formation

Ironside Member. Hewett (1931) named the Sultan Formation and its lower member, the Ironside Dolomite, in the Goodsprings district. The Ironside Member is a dark brown to dark gray-brown, fine- to medium-crystalline dolomite that averages about 132 to 148.5 ft (40 to 45 m) thick and forms a distinct

color band and cliff expression throughout its outcrop extent. In most places, it is a clear marker and no doubt inspired Hewett (1931) to use the clearly mappable base of the Ironside as the upper boundary of the undifferentiated Goodsprings interval. Along the west side of northern Potosi Valley, dark brown, ledge-forming, thick-bedded to massive dolomite of the basal Ironside overlies the generally light gray dolomite of the uppermost Mountain Springs. One of the characteristic features of the Ironside Member is the occurrence of several levels containing abundant globular stromatoporoids (Fig. 4). The contact between the upper member of the Mountain Springs and the Ironside is, in places, an interval of spectacular breccia containing angular blocks from both units, and probably representing post-Ironside intrastratal solution. Physical relationships and conodonts suggest an essentially conformable contact. A distinctly light and dark gray banded succession near the top of the Mountain Springs may represent a transition between the two units.

Regional Stratigraphic Relationships

The Goodsprings interval along the west side of northern Potosi Valley represents the transition between the much thinner cratonic section at Sheep Mountain and the thicker miogeoclinal section at the Nopah Range (Fig. 5). Figure 5, whose line of section is oriented generally perpendicular to paleodepositional strike, shows the eastward thinning and pinching out of units in relationship to two disconformities. Of particular note is the loss of Silurian section between the northern Nopah Range and southern Spring Mountains, as well as the loss of the entire Ordovician section between the eastern Spring Mountains and Sheep Mountain, where the Devonian upper part of the Mountain Springs Formation rests on Upper Cambrian strata of the Nopah Formation. This Goodsprings section at Sheep Mountain is equivalent to the Muav Limestone of the classic Grand Canyon cratonic section.

Although Gans (1974) recommended abandoning the name Goodsprings as a formation, its elevation to group status would seem to have some justification.

SUMMARY

Along the west side of northern Posoti Valley, between Potosi Pass and Potosi Wash, one can climb through more than 3,300 ft (1000 m) of craton margin to miogeoclinal transition strata belonging to subdivisions of the Goodsprings Dolomite exposed in the upper plate of the Keystone thrust. Less than one hour's drive from Las Vegas, this easily accessible area can serve as a half- to full-day field excursion to examine shallow-marine depositional facies and diagenetic fabrics of this transition section. It could also provide an instructive stop on a more regionally oriented field trip designed to focus on the east-to-west stratigraphic and facies changes from cratonic, autochthonous sections such as those at Sheep and Frenchmen Mountains, to miogeoclinal sections such as in the northern Nopah Range and Death Valley region.

REFERENCES CITED

Barnes, H., and Palmer, A. R., 1961, Revision of stratigraphic nomenclature of Cambrian rocks, Nevada Test Site and vicinity, Nevada: U.S. Geological Survey Professional Paper 424-C, p. C11–C103.

Barnes, H., and Christiansen, R. L., 1967, Cambrian and Precambrian rocks of the Groom District, Nevada, Southern Great Basin: U.S. Geological Survey Bulletin, 1244–G, p. G1–G34.

Cheek, K. N., Campbell, A. G., Conner, F. J., Flynn, P. E., Goodman, R. W., McKee, M. E., Tucker, C. A., and Cooper, J. D., 1983, Petrology and paleoenvironments of the Mountain Springs Formation (Lower Ordovician-Middle Devonian), Potosi Valley, Spring Mountains, Southern Nevada: Geological Society of America Abstracts with Programs, v. 15, no. 6, p. 305.

Cooper, J. D., Miller, R. H., and Sundberg, F. A., 1981, Upper Cambrian depositional environments, Southeastern California and Southern Nevada: U.S. Geological Survey Open File Report 81-743, Short Papers for the Second International Symposium on the Cambrian System, Golden, Colorado, August, 1981, M. E. Taylor, ed., p. 57–62.

——1982, Environmental stratigraphy of the lower part of the Nopah Formation (Upper Cambrian), southwestern Great Basin: Geological Society of America, Cordilleran Section, Guidebook for Field trip no. 9, p. 97–117.

Edwards, J. C., 1983, Depositional environments and diagenesis of Member A (Lower to Middle Ordovician), Mountain Springs Formation, Southern Great Basin [M.S. thesis]: San Diego, California, San Diego State University, 188 p.

Edwards, J. C., and Cooper, J. D., 1984a, Dolomitization of Lower Ordovician carbonate rocks, Southern Great Basin: Society of Economic Paleontologists and Mineralogists, Pacific Section, Abstracts with Programs, Annual Meeting, San Diego, California, p. 60.

——1984b, Diagenetic history of Lower Ordovician Mountain Springs Formation, Southern Great Basin: Geological Society of America Abstracts with Programs, v. 16, no. 5, p. 280.

Gans, W. T., 1974, Correlation and redefinition of the Goodsprings Dolomite, southern Nevada and eastern California: Geological Society of America Bulletin, v. 85, no. 2, p. 189–200.

Harrington, R. J., 1982, Depositional environments and ecological gradients of the Upper Devonian Sultan Formation (Ironside Dolomite and Valentine Limestone members), and subjacent beds of the uppermost Mountain Springs Formation, near Mountain Springs, Clark County, Nevada [M.S. thesis]: Riverside, California, University of California, Riverside, 147 p.

Hazzard, J. C., 1937, Paleozoic section in the Nopah and Resting Springs Mountains, Inyo County, California: California Journal of Mines and Geology, v. 33, p. 273–339.

Hazzard, J. C., and Mason, J. F., 1936, Middle Cambrian formations of the Providence and Marble Mountains, California: Geological Society of America Bulletin, v. 47, p. 229–240.

Hewett, D. F., 1931, Geology and ore deposits of the Goodsprings quadrangle, Nevada: U.S. Geological Survey Professional Paper, 72 p.

Hintzmann, K. J., 1983, Stratigraphic correlation of the Mountain Springs Formation (Ordovician-Devonian), southern Great Basin [M.S. thesis]: San Diego, California, San Diego State University, 125 p.

Kepper, J. C., 1972, Paleoenvironmental patterns in Middle to lower Upper Cambrian interval in eastern Great Basin: American Association of Petroleum Geologists Bulletin, v. 56, p. 503–527.

Longwell, C. R., Pampeyan, E. H., Bowyer, B., and Roberts, R. J., 1965, Geology and Mineral deposits of Clark County, Nevada: Nevada Bureau of Mines Bulletin, v. 62, 218 p.

Ludvigsen, R., and Westrop, S. R., 1985, Three new Upper Cambrian stages for North America: Geology, v. 13, no. 2, p. 139–144.

Miller, R. H., and Zilinsky, G. H., 1981, Lower Ordovician through Lower Devonian cratonic margin rocks of the southern Great Basin: Geological Society of America Bulletin, v. 92, p. 255–261.

Miller, R. H., Cooper, J. D., and Sundberg, F. A., 1981, Upper Cambrian faunal distribution in southeastern California and southern Nevada: U.S. Geological Survey Open File Report 81-743, Short Papers for the Second International Symposium on the Cambrian System, Golden, Colorado, August 1981, M. E. Taylor, ed., p. 138–143.

Miller, R. H., and Cooper, J. D., 1982, Ordovician-Devonian conodonts from southeastern California and southern Nevada: Geological Society of America Abstracts with Programs, v. 14, no. 4, p. 217.

Palmer, A. R., 1960, Trilobites of the Upper Cambrian Dunderberg Shale, Eureka District, Nevada: U.S. Geological Survey Professional Paper 334-C, p. C53–C109.

——1965, Trilobites of the Late Cambrian Pterocephaliid biomere in the Great Basin, United States: U.S. Geological Survey Professional Paper 493, 105 p.

Palmer, A. R., and Hazzard, J. C., 1956, Age and correlation of Cornfield Springs and Bonanza King Formations in southeastern California and southern Nevada: American Association of Petroleum Geologists Bulletin, v. 40, p. 2494–2499.

Steamboat Springs, Nevada

Donald E. White, *U.S. Geological Survey, 345 Middlefield Road, Menlo Park, California 94025*
David B. Slemmons, *Center for Neotectonic Studies, Mackay School of Mines, University of Nevada, Reno, Nevada 89557*

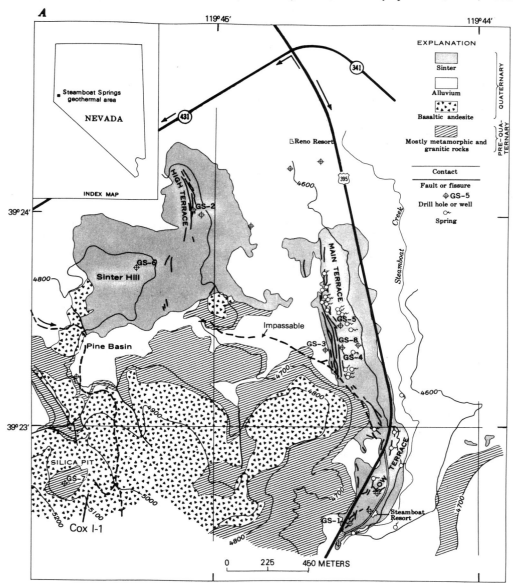

Figure 1. Map of locality

LOCATION

Steamboat Springs geothermal area, Nevada, is at the boundary between the Sierra Nevada and the Great Basin, and 10 mi (16 km) south of downtown Reno (Fig. 1). Access to the thermal area is convenient by bus or passenger car, from Nevada 431 at the northern edge of the area and U.S. 395 at the eastern edge.

SIGNIFICANCE OF SITE

The spring system has the longest known period of activity

and is regarded as the surface expression of an epithermal gold-silver–depositing hydrothermal system. A second feature of this system is that it is among the first geothermal areas in Nevada that will have commercial development of electrical energy.

SITE INFORMATION

The spring system consists of the youngest still-discharging area at the Main and Low Terraces next to Steamboat Creek; the older, pre–Lake Lahontan High Terrace with subsurface discharge (up to 30,000 years old); and the old silica sinter and clay deposits (up to 3 m.y. old), about 1.25 mi (2 km) westward in the Pine Basin–Silica Pit area of the main part of Steamboat Hills (Thompson and White, 1964; White and others, 1964; White, 1968).

This geothermal area has attracted attention for many years for deposition of ore minerals and development of spas. More recent investigations of this area as a potential geothermal resource include deep exploratory drilling to the southwest. Here, Phillips Petroleum Company reached the highest measured temperature, 228°C about 1.5 mi (2.5 km) southwest of the Main Terrace, which is close to the 180° to 230°C predicted from geochemical thermometers.

The importance of Steamboat Springs to economic geologists is indicated by a quote from White (1984, pp. 256, 258): "The Steamboat area has been of long-standing interest to economic geologists for its bearing on hydrothermal ore deposits, and is now viewed as the present-day equivalent of geothermal systems of Tertiary age that formed epithermal gold-silver deposits throughout the Great Basin of the western United States and elsewhere. . . .At Steamboat, hot spring sinter deposits, chemical sediments in spring vents, and veins intersected in drill holes all contain significant concentrations of gold, silver, mercury, antimony, arsenic, thallium, and boron."

The springs are an excellent area for opaline and chalcedonic sinter deposits, with various replacement and depositional textures that formed at the surface. The older, generally more chalcedonic silica sinter is commonly pigmented with cinnabar and black replacements of cinnibar. Native sulfur is commonly observed as an incrustation on surface springs and steaming areas. The minerals observed in the springs or drill core include stibnite, metastibnite, and pyrargyrite.

The longevity of the Steamboat area is indicated by the earliest hot-spring sinter (about 3 Ma) that was deposited below the basaltic andesite (2.52 Ma). Intermittent activity is indicated by the 3-Ma sinter, hydrothermal adularia of 1.1 Ma, and evidence for continuous activity during the past about 0.1 m.y. Although the water chemistry favors a partly magmatic origin of the fluids, continuous activity would require about 720 mi^3

(3,000 km^3) of magma, a volume that appears to be unlikely, judging from mass-and heat-flow constraints. The youthful and local magmatic activity is shown by a 3-Ma rhyolitic dome, the basaltic andesite of 2.52 Ma, and five rhyolite domes clustering near 1.2 Ma. This suggests that the magmatic and geothermal activity are intermittent, representing only about 10 percent of the total time elapsed (Silberman and others, 1979).

The Main and Low Terraces are accessible from near the Steamboat Resort (Fig. 1). This area is actively changing in aspect by silica sinter deposition of porous vuggy opal and many minor minerals. Dissolution and disintegration of old sinter also occurs along the partly open and steaming fissures of this zone (White, 1984).

The High Terrace is about 30,000 years old and is still thermally active, but the water level is 40 ft (12 m) below the surface and is discharging into the subsurface. All three terrace areas have been drilled and are currently being developed for low- or moderate-temperature conversion into electrical energy and for nonelectrical uses.

The older deposits tend to be of chalcedonic sinter, often stained red to pink from trace amounts of cinnabar. The black surfaces develop from exposure of fine-grained cinnabar to sunlight.

REFERENCES CITED

Silberman, M. L., White, D. E., and Keith, T.E.C., and Dockter, R. D., 1979, Duration of hydrothermal activity at Steamboat Springs, Nevada, from ages of spatially associated volcanic rocks: U.S. Geological Survey Professional Paper 458-D, 64 p.

Thompson, G. A., and White, D. E., 1964, Regional geology of the Steamboat Springs area, Washoe County, Nevada: U.S. Geological Survey Professional Paper 458-A, 51 p.

White, D. E., 1968, Hydrology, activity, and heat flow of the Steamboat Springs thermal system, Washoe County, Nevada: U.S. Geological Survey Professional Paper 458-C, 109 p.

—— , 1984, Summary of Steamboat Springs geothermal area, Nevada, with attached roadlog commentary, *in* Lintz, J., Jr., ed., Western geological excursions, v. 3: Reno, University of Nevada, Department of Geological Sciences, p. 256–265.

White, D. E., Thompson, G. A., and Sandberg, C. A., 1964, Rocks, structure, and geologic history of Steamboat Springs thermal area, Washoe County, Nevada: U.S. Geological Survey Professional Paper 458-B, 63 p.

Complex fold and thrust relationships, Mac Canyon, northern Pilot Mountains, west-central Nevada

John S. Oldow, Department of Geology and Geophysics, Rice University, Houston, Texas 77251

LOCATION

Mac Canyon is located on the northwestern flank of the Pilot Mountains, approximately 3.6 mi (6 km) east of Mina, in west-central Nevada. Mina is situated on U.S. Highway 95 approximately 33 mi (50 km) southeast of Hawthorne and about 76 mi (120 km) northwest of Tonopah. Mac Canyon is accessible by a good gravel road that leaves the southeastern quadrant of the Mina townsite (Fig. 1). The road ends at an old mine dump (suitable for parking) approximately 0.15 mi (0.25 km) south of the mouth of the canyon. A fair trail leads north along the range-front fault of the Pilot Mountains and is an easy walk to the canyon mouth. Some problem may be encountered in locating the appropriate road from Mina to the northwestern flank of the Pilot Mountains, because it is not marked. If local directions are sought, ask for the road to Water or Spearmint Canyon (site of the town water supply). Location of the town airstrip and use of the accompanying copy of parts of the Mina and Sodaville quadrangles should be sufficient direction. It is important to note that some of the roads indicated on the map have moved since publication of the quadrangle maps. Roads are good and passenger cars can be used. It is doubtful that buses could negotiate the road to the parking site, but they probably could approach to within about 2.4 mi (4 km) of the trail head.

SIGNIFICANCE OF SITE

The excellent exposures in Mac Canyon allow for detailed observation of complex fold and fault relationships in the upper plate of part of the Luning thrust system. The Luning thrust represents the southern extension of the late Mesozoic Luning-Fencemaker fold and thrust belt of the northwestern Great Basin (Oldow, 1983, 1984). Lithologic units exposed consist of the volcanic and volcanogenic rocks of the Permian Black Dyke Formation (Speed, 1977) and the carbonate and clastic rocks of the Upper Triassic Luning and Upper Triassic and Lower Jurassic Sunrise-Gabbs Formations (Muller and Ferguson, 1939; Oldow, 1981a). The Black Dyke Formation represents part of the controversial terrane, Sonomia (Speed, 1979), which may be

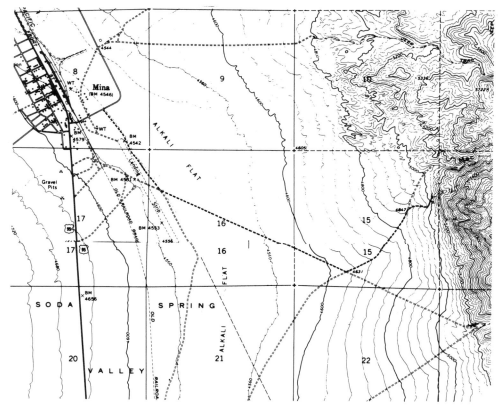

Figure 1. Location map based on the Mina and Sodaville 7.5′ quadrangles. Access road to Mac Canyon marked by dark line.

part of an accreted microplate which was attached to the western margin of western North America in the Permo-Triassic. The Mesozoic carbonate and clastic rocks were deposited in shallow marine environments supplied with continentally derived debris and minor amounts of volcanic detritus. They are representative of part of the Mesozoic marine province of the northwestern Great Basin (Speed, 1978; Oldow, 1983, 1984), which consists of rocks that are largely allochthonous but undoubtedly of North American affinity (Oldow, 1984).

Three phases of folds are recognized, designated D_1, D_2, and D_3 in order of decreasing age, and are kinematically related to multiple phases of thrust faulting. The relationships between folds and thrusts are deduced by the existence of folded faults and faulted folds and yield a remarkably complete relative timing scheme. Careful examination of the field relations offers an excellent field-laboratory example of complex structures.

SITE INFORMATION

The complex structural relationships in Mac Canyon were originally recognized by Ferguson and Muller (1949), whose classic structural interpretation is used as an example of progressive deformation in the laboratory text by Ragan (1968). The main points presented by Ferguson and Muller (1949) involve a complex history of folding and refolding accompanied by synorogenic sedimentation. Their interpretation is substantially modified by detailed mapping and structural analysis by Oldow (1981a, 1981b), the results of which are depicted in the accompanying map and cross sections and discussed in the text below.

The thrust sheets in Mac Canyon (Fig. 2) and adjacent areas are composed of three lithologic units, consisting of the Permian Black Dyke, the Upper Triassic Luning, and the Upper Triassic and Lower Jurassic Sunrise-Gabbs Formations. The units are lithologically distinct and easily recognized in the field. No interformational contacts are preserved within the Luning allochthon in the Pilot Mountains; all formational contacts are thrust faults.

The Black Dyke Formation is composed of interbedded volcanic and volcanogenic sedimentary rocks. Volcanic rocks consist of flows and breccias of basic to intermediate composition and are interbedded with feldspathic wackes and arenites and locally with thin distal turbidites. This unit is easily recognized in Mac Canyon as dark-green massive outcrops exposed on the north wall of the canyon.

The Luning Formation is divided into three members: the lower member is composed of interbedded calcareous argillite and thin bedded limestone; the middle member consists of interleaved argillite, quartz and chert wacke and arenite, and chert-pebble conglomerate; and, the upper member is composed of medium to thick bedded micrite, calcarenite, and crystalline limestone. Locally, the upper member contains a carbonate sandstone and breccia unit which Ferguson and Muller (1949) originally mapped as a younger unit unconformably overlying the Luning Formation. The unit is exposed high on the south wall of the canyon, where it presents an irregular light-brown to buff outcrop pattern. The unit is important because it was correlated with the

Jurassic Dunlap Formation, exposed south of the Luning allochthon in the Pilot Mountains. On the basis of this correlation and their interpretation of the significance of the unit, Ferguson and Muller (1949) argued that synorogenic deposition of the Jurassic Dunlap could be proved and thus used to date the age of emplacement of the Luning thrust belt. They interpreted the carbonate sandstone and breccia unit as being a subaerial accumulation of debris shed from the leading edge of the advancing Luning thrust system. This argument stemmed largely from the interpretation that the unit was unconformably overlying deformed rocks of the Luning Formation. Careful, detailed mapping and structural analysis of the outcrops on the south wall of Mac Canyon clearly show that the carbonate clastic unit is an interbed within the Luning Formation and does not overlie it. The earlier interpretation probably arose from the complex outcrop pattern of the carbonate clastic unit. The apparently chaotic distribution of the carbonate clastic unit is the product of a complex superposition of three phases of folds. Moreover, the unit does not have a laterally continuous thickness, which further complicates the distribution of the exposures. The origin of the clastic unit is argued to be the product of collapse and subterranean sedimentation in a karst environment formed prior to late Mesozoic deformation. This interpretation is based on several lines of evidence, and the interested reader is referred to Oldow (1981a).

The Sunrise-Gabbs Formation consists of interbedded silty calcarenite, mudstone, and feldspathic arenite and wacke. The unit weathers brown to buff and occupies the lower parts of the north and south wall of Mac Canyon.

Three phases of folds, developed throughout the Mesozoic rocks of the Pilot Mountains, are exposed in Mac Canyon. First folds, D_1, consist of northeast to east-west trending tight to isoclinal major and minor folds with a well-developed penetrative axial-planar cleavage. The axial surfaces of these folds are generally south vergent to recumbent. Much variability in axial surface orientation is observed and is the product of later deformation. Only a few first-generation folds are exposed in Mac Canyon and the best examples of these structures are major folds, one of which folds a thrust fault. Second-phase folds, D_2, are northwest striking close to tight major and minor folds which locally exhibit a spaced cleavage. The axial surfaces of second-generation folds range from upright to southerly vergent. Near the base of major thrust sheets, second-phase folds are recumbent and locally coaxially folded by later stage second-phase folds (Oldow, 1981a). D_2 folds are easily recognized in that they fold the pervasive D_1 cleavage. Third-generation structures, D_3, are sporadically developed north striking gentle to close folds that result in minor reorientation of the earlier fold sets.

Several generations of thrust faults are developed, and Mac Canyon exposes the boundary between two major nappes, both of which consist of several thrust sheets that have experienced a complex history of imbrication. Near the mouth of the canyon, Tertiary high-angle faults complicate the distribution of the Mesozoic thrust faults, but within the canyon the spatial relationships among the thrusts are well preserved.

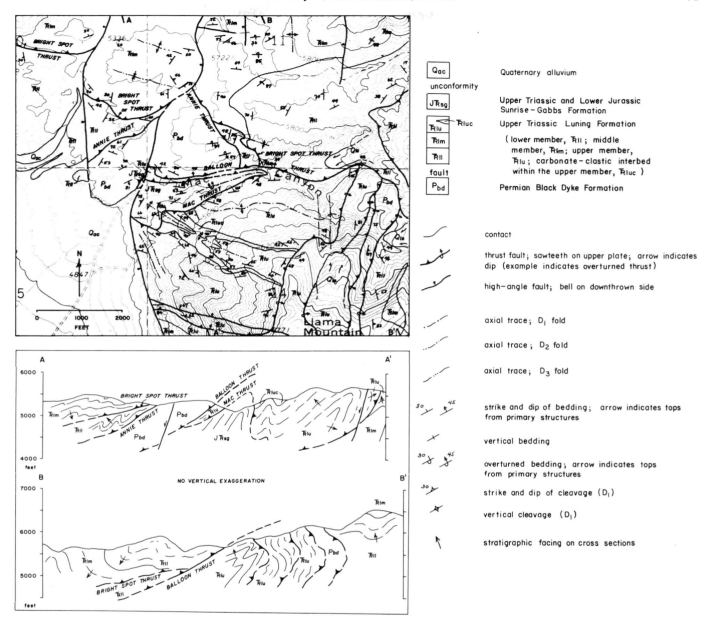

Figure 2. Generalized geologic map and cross sections of the Mac Canyon area of the northwestern Pilot Mountains, west-central Nevada.

The sequential development of thrust faults in Mac Canyon is established by cross cutting relationships. The thrust with the most recent displacement in the canyon is the Balloon thrust which is subplanar and has an east-west strike with a 35° to 50° north dip. The Balloon thrust represents the boundary between the two nappes exposed in Mac Canyon and has had a complex history of motion. The Balloon thrust is exposed on the north wall of the canyon and underlies rocks of the Black Dyke Formation in the west and the Luning Formation in the east. In the upper plate of the thrust, the contact between the Black Dyke and Luning Formations is also a fault, the Annie thrust. Unlike the Balloon thrust, the Annie thrust is not planar and is involved in major D₂ folds. In Mac Canyon, the northerly striking Annie thrust is exposed on the north wall and is nearly vertical. Farther to the north, the trace of the Annie thrust swings westerly and it has a moderate north dip. The intersection of the Annie and Balloon thrusts is obscured by talus, but sufficient outcrop exists to demonstrate that the folded Annie thrust is truncated by the subplanar Balloon thrust. East of the Annie thrust, the complexly deformed Bright Spot thrust separates two sheets of Luning Formation and is also truncated by the Balloon thrust. The contact between the Luning and Sunrise-Gabbs Formations, exposed on both walls of the canyon, is also a thrust, the Mac thrust (Ferguson and Muller, 1949), which lies in the lower of the two nappes.

The Mac thrust has a complicated outcrop pattern which is the product of a complex structural history. On the north wall of the canyon, the Mac thrust is subparallel to the overlying Balloon thrust, but toward the east the two faults diverge. The Mac thrust is folded in a major D_1 isoclinal fold and dips to the north on the northern flank of the canyon and steeply south on the southern wall of the canyon.

The Mac thrust is a key to understanding the relationship between first-generation folds and thrust development. A major D_1 fold in the Luning Formation on the southern flank of the canyon is truncated by the Mac thrust, which itself is folded in a major D_1 fold. This relationship demonstrates that thrusting was activated during D_1 folding. Near the mouth of Mac Canyon, the trace of the Mac thrust is additionally folded in major northwest trending D_2 folds and minor D_3 flexures. Elsewhere in the lower nappe, a major D_3 fold deforms several thrust sheets but does not deform the overlying Balloon thrust. The internal thrust sheets within the lower nappe are all deformed in major D_2 folds and were imbricated prior to the development of D_2 folds and the emplacement of the Balloon thrust.

In the upper nappe, the Bright Spot thrust juxtaposes a thrust sheet composed of stratigraphically upright rocks of the lower member of the Luning Formation beneath a sheet composed of an inverted section of the Luning. The fault is deformed in major D_2 folds and is truncated at its lower contact by the subplanar Balloon thrust. This relationship illustrates the relative timing of the internal imbrication of the thrusts in the upper plate of the Balloon thrust and the development of second-phase folds. The thrusts within the upper nappe were developed either during or after D_1 folding, as indicated by the truncation of first-generation cleavage by the fault contacts, and were obviously emplaced prior to D_2 folding.

The displacement history of the Balloon thrust is complex and is deduced from data derived from Mac Canyon and elsewhere in the Pilot Mountains. The fault was a major structural feature during D_1 deformation and accommodated several tens of kilometers of displacement. Subsequent displacement was smaller and involved only a few kilometers of movement and occurred during and after D_2 deformation. The inferred displacement history of the Balloon thrust, particularly during the first phase of deformation, is not obvious from relationships that can be directly observed in Mac Canyon but is required by stratigraphic and structural relations established elsewhere in the Pilot Mountains (Oldow, 1981a, 1981b).

In summary, the deformation of upper Paleozoic and lower Mesozoic rocks in Mac Canyon involves an intricate history of folding and associated thrusting. Two nappes, separated by a laterally extensive subplanar thrust fault (Balloon thrust), are each composed of a highly deformed stack of imbricate thrust sheets. Internal structural relationships shed light on the kinematics of thrust imbrication and fold development. The existence of first-generation folds truncated by a thrust which is itself folded in a first-phase isoclinal fold illustrates the synchronous development of thrusts and folds during D_1 deformation. Based on observations other than those that can be made in Mac Canyon (Oldow, 1981a, 1981b), it is possible to demonstrate that most of the thrust sheets within the two nappes are the limbs of dismembered first-phase major folds. The imbrication of the thrust sheets preceeded development of second-generation folds as indicated by the folds of the bounding fault surfaces in D_2 structures and the map-scale involvement of the imbricate stacks in major second folds (Oldow, 1981a, 1981b). In the Pilot Mountains, the Balloon thrust has played an important role in the evolution of the Luning allochthon and exhibits a complex displacement history associated both with D_1 folding and with D_2 deformation. The deduced transport history of the Balloon thrust records a change in transport direction. During D_1 deformation, tectonic transport was from the northwest to the southeast (Oldow, 1981a, 1981b; Seidensticker and others, 1982) and was associated with displacements of on the order of tens of kilometers. Subsequent deformation, D_2, resulted in relatively small displacements (a few kilometers or less) and thrust transport was from the northeast to the southwest. The last episode of contractional deformation is recorded by D_3 folds which are found to be truncated by the Balloon thrust in Mac Canyon, but elsewhere they demonstrably fold the thrust. Based on total shortening estimates for D_3 folds, it is unlikely that this deformational event was accompanied by substantial thrust displacement.

REFERENCES CITED

Ferguson, H. G., and Muller, S. W., 1949, Structural geology of the Hawthorne and Tonopah quadrangles, Nevada: U.S. Geological Survey Professional Paper 216, 55 p.

Muller, S. W., and Ferguson, H. G., 1939, Mesozoic stratigraphy of the Hawthorne and Tonopah quadrangles, Nevada: Geological Society of America Bulletin, v. 50, p. 1573–1624.

Oldow, J. S., 1981a, Structure and stratigraphy of the Luning allochthon and the kinematics of allochthon emplacement, Pilot Mountains, west-central Nevada: Geological Society of America Bulletin, v. 92: Part I, p. 889–911, Part II, p. 1647–1669.

——1981b, Kinematics of late Mesozoic thrusting, Pilot Mountains, west-central Nevada, USA: Journal of Structural Geology, v. 3, p. 39–49.

——1983, Tectonic implications of a late Mesozoic fold and thrust belt in northwestern Nevada: Geology, v. 11, p. 542–546.

——1984, Evolution of a late Mesozoic back-arc fold and thrust belt, northwestern Great Basin, USA: Tectonophysics, v. 102, p. 245–274.

Ragan, D. M., 1968, Structural Geology: An Introduction to Geometrical Techniques: New York, John Wiley and Sons, 393 p.

Seidensticker, C. M., Oldow, J. S., and Ave Lallement, H. G., 1982, Development of bedding-normal boudins: Tectonophysics, v. 90, p. 335–349.

Speed, R. C., 1977, Excelsior Formation, west-central Nevada: stratigraphic appraisal, new divisions, and paleogeographic interpretation, *in* Stewart, J. H., Stevens, C. H., and Fritsche, A. E., Paleozoic Paleogeography of the Western United States: Society of Economic Paleontologists and Mineralogists, Pacific Section, Pacific Coast Paleogeography Symposium, v. 1, p. 325–336.

——1978, Paleogeographic and plate tectonic evolution of the early Mesozoic marine province of the western Great Basin, *in* Howell, D. G., and McDougall, K. A., eds., Mesozoic Paleogeography of the Western United States: Society of Economic Paleontologists and Mineralogists, Pacific Section, Pacific Coast Paleogeography Symposium, v. 2, p. 253–270.

——1979, Collided Paleozoic microplate in the western United States: Journal of Geology, v. 87, p. 279–292.

1954 Fairview Peak earthquake area, Nevada

David B. Slemmons, Center for Neotectonic Studies, Mackay School of Mines, University of Nevada, Reno, Nevada 89557
John W. Bell, Nevada Bureau of Mines and Geology, University of Nevada, Reno, Nevada 89557

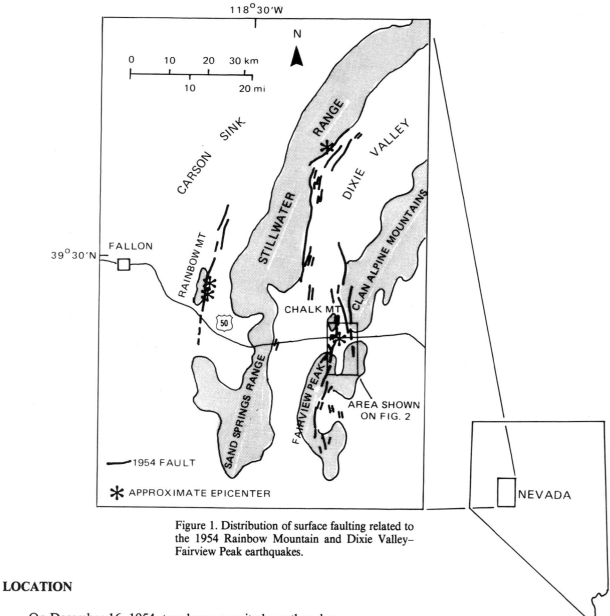

Figure 1. Distribution of surface faulting related to the 1954 Rainbow Mountain and Dixie Valley–Fairview Peak earthquakes.

LOCATION

On December 16, 1954, two large-magnitude earthquakes occurred nearly simultaneously in the Fairview Peak–Dixie Valley area of western Nevada (Fig. 1) (Slemmons, 1984). The first earthquake (Ms = 7.25) was centered near the middle of the Fairview Peak fault zone, where it crosses U.S. 50. It was followed four minutes later by a second shock (Ms = 7–) centered on the east flank of the Stillwater Range in the northwest part of Dixie Valley, about 37 mi (60 km) to the north. Both earthquakes were accompanied by extensive and spectacular surface faulting (Slemmons, 1957). The combined rupture zone is about 62 mi

(100 km) long and 18 mi (30 km) wide. The most spectacular scarps of this zone are at the main site (Figs. 2 and 3) on the southeast side of Fairview Peak, an area that is about 5.5 mi (9 km) south of U.S. 50 and is accessible by passenger car on a new graded road.

SIGNIFICANCE OF SITE

The Fairview Peak fault zone is the most readily accessible area of 14 examples of historical surface faulting in the Basin and

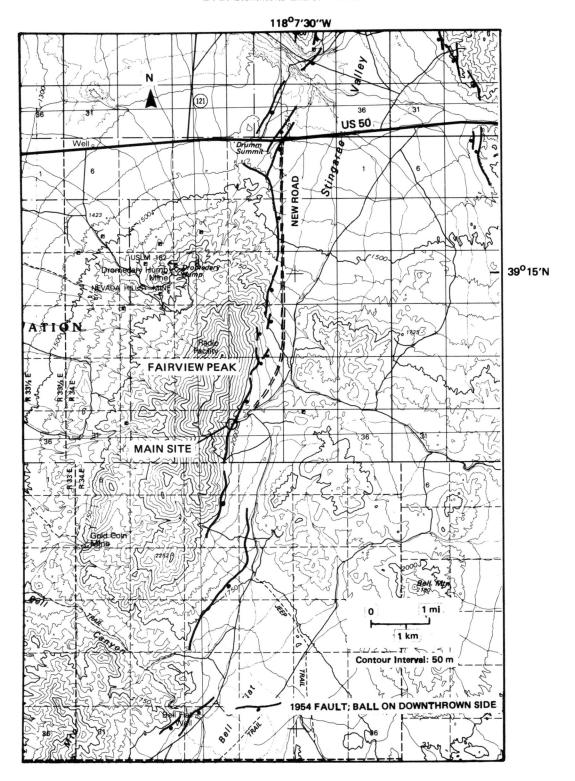

Figure 2. Location map showing distribution of 1954 faulting along
Fairview Peak and Stingaree Valley. Base is the Fallon 30 × 60 minute
Topographic Quadrangle, scale 1:100,000.

Figure 3. Ground view looking south along 1954 scarp at main site (photograph by D. B. Slemmons, 1966).

Figure 4. Ground view looking south along 1954 scarp at main site. The scarp is more than 23 ft (7 m) in height, although the tectonic displacement is less. The right-oblique slip shown by offset tree roots on the ridgecrest is expressed by the 7 ft (2.1 m) vertical separation, the 11.3 ft (3.4 m) of right-lateral separation, and the net slip is 13.3 ft (4.1 m) (photograph by D. B. Slemmons, 1966).

Range Province. It also provides a classic example of the kinds of seismotectonic events that, with many repetitions in the last 10 or 20 million years, have developed the geologic features of this province.

SITE INFORMATION

There were two other large earthquakes in this region in 1954 with associated surface faulting: the July 6 earthquake of $Ms = 6.3$ and the August 24 earthquake of $Ms = 6.95$ in the Rainbow Mountain region about 25 mi (40 km) northwest of Fairview Peak (Fig. 1).

The Fairview Peak earthquake was felt over an area of about 220,000 mi^2 (570,000 km^2). The maximum MMI was at least VII and would have been closer to X if it were not located in such a remote area. The focus of the main shock was about 9 mi (15 km) deep on a fault striking N11°W and dipping 62°E, with an epicenter between Chalk Mountain and Fairview Peak (Fig. 1).

Although the engineering damage was relatively light, due to the remoteness of the region, the geological effects were dramatic and are still visible. Bell and others (1984) commented as follows about this area,

The ruptures from the December 16, 1954, earthquake were frantically noticed by motorists along this section of U.S. Highway 50. Three main en echelon scarps broke the asphalt with a cumulative displacement of 1.7 m of vertical and 1.8 m of right-lateral displacement. The vertical effects appear to be asymmetric, with the graben of Stingaree Valley to the east tilted westward with activation from surface faults on both sides of the valley. The eastern side of the graben was ruptured with about 30 cm of vertical and 1 m of right-lateral offset. The horizontal rupturing may show elastic rebound effects with 5 minutes of counterclockwise bending to the west of the fault (Slemmons, 1957).

Large, fresh-appearing scarps and fissures can be seen in the Fairview Peak earthquake area (Fig. 4). They are up to 23 ft

(7.0 m) in height, particularly along the narrow grabens that are aligned along some of the fault traces. Right-slip offsets of at least 12 ft (3.7 m) were reported (Slemmons, 1957) and can still be seen where the fault offsets the alluvial-fan ridgeline just south of the main site.

Fairview Peak adjoins the Walker Lane, a major northwest-trending feature within the Basin and Range Province. It divides the main northern and eastern part of the province of elongate north-south ranges from the less elongate structures of the western part of the province. Faults within the Walker Lane, which is of similar trend to the San Andreas fault, have mainly right-lateral offsets and thus appear to influence the offsets along the Fairview Peak zone. The maximum measured fault offset is where the fault zone narrows to a single break on the ridge crest shown in Figure 4. Here, tree roots were offset with a vertical separation of 7 ft (2.1 m), a horizontal separation of 11.3 ft (3.4 m), and a net slip of about 13.3 ft (4.1 m). Most of the previously surveyed geodetic stations were not close to the fault, but the data of Whitten (1957) indicates that the right-lateral geodetic separation was more than 12 ft (3.7 m), and the vertical separation along U.S. 50 was about 5 ft (1.5 m). The ratio of right-slip to vertical separation decreases on the north-south to N15°W scarps in the Dixie Valley earthquake area; some faults show little or no right-lateral offsets. On the west side of the Dixie Valley graben, the scarps are also conspicuous and continuous along the base of the range front, but the scarps are somewhat lower—about 15 ft (4.6 m) in maximum height.

During these two earthquakes the main tectonic effect was to tilt both the Dixie Valley graben and the Stingaree Valley graben to the west through activation of faults on either side of the Dixie Valley graben, on the eastern side of the Chalk Mountain horst, and on either side of the Stingaree Valley graben.

Wallace and Whitney (1984) describe a zone of discontinuous surface historical faulting that extends southward from the 1915 Pleasant Valley earthquake zone, through the 1954 Fairview Peak–Dixie Valley–Rainbow Mountain area (which also includes the 1903 Wonder earthquake), and the 1932 Cedar Mountain, 1934 Excelsior Mountain, and 1872 Owens Valley earthquake and surface faulting zones. Three important gaps have been recognized in this discontinuous zone (Fig. 5). The Stillwater seismic gap is at the north end of Dixie Valley (Wallace, 1984; Wallace and Whitney, 1984).

REFERENCES CITED

Bell, J. W., Slemmons, D. B., and Wallace, R. E., 1984, Roadlog Reno to Dixie Valley–Fairview Peak earthquake areas, *in* Lintz, J., Jr., ed., Western geological excursions: Guidebook, 1984 Geological Society of America Annual Meeting, v. 4, p. 425–472.

Slemmons, D. B., 1957, Geological effects of the Dixie Valley–Fairview Peak, Nevada, earthquakes of December 16, 1954: Seismological Society of America Bulletin, v. 47, n. 4, p. 353–375.

———, 1984, Dixie Valley–Fairview Peak earthquake areas, *in* Lintz, J., Jr., ed., Western geological excursions: Guidebook, 1984 Geological Society of America Annual Meeting, v. 4, p. 418–420.

Wallace, R. E., 1984, Notes on surface faulting in Dixie Valley, Nevada, *in* Lintz, J., Jr., ed., Western geological excursions: Guidebook, 1984 Geological Society of America Annual Meeting, v. 4, p. 402–407.

Wallace, R. E., and Whitney, R. A., 1984, Late Quaternary history of the Stillwater seismic gap: Seismological Society of America Bulletin, v. 74, n. 1, p. 301–314.

Whitten, C. A., 1957, Geodetic measurements of the Dixie Valley–Fairview Peak, Nevada, earthquakes of December 16, 1954: Seismological Society of America Bulletin, v. 47, n. 4, p. 321–326.

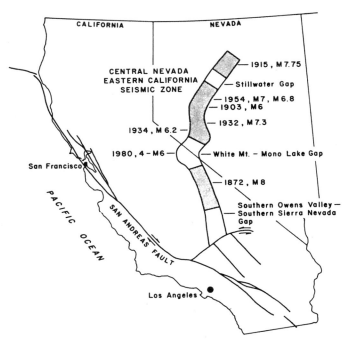

Figure 5. Seismic zones showing the dates of earthquakes accompanied by surface faulting. Three gaps between the faulted segments are shown (from Wallace, 1984).

The Snake Range decollement, eastern Nevada

Elizabeth L. Miller, Phillip B. Gans, and Jeffrey Lee, Department of Geology, Stanford University, Stanford, California 94305

LOCATION AND ACCESSIBILITY

The northern Snake Range is located in east-central Nevada, along the Utah-Nevada border. It can be reached by driving approximately 50 mi (80 km) east from Ely, Nevada, on U.S. 50 (Fig. 1). The closest town of Baker, Nevada has food, accommodations (The Silver State motel), and nearby camping. The field trip stops described below are some of the most easily accessible; for a more complete field trip guide and road log, the reader is referred to Gans and Miller (1983).

SITE SIGNIFICANCE

The northern Snake Range provides excellent exposures of both brittle and ductile structures formed as a result of large magnitude crustal extension. The most prominent structural feature of the range is the northern Snake Range decollement (NSRD), a subhorizontal surface that separates Paleozoic carbonate units extended by normal faulting above, from ductilely attenuated Cambrian and Precambrian metasedimentary and igneous rocks below. Similar detachment faults are exposed throughout the western U.S. within "metamorphic core complexes" (Crittenden and others, 1980), and their origin and amount of displacement are controversial topics. We have argued that the NSRD represents an uplifted Oligocene–Miocene extensional ductile-brittle transition or zone of decoupling between imbricate normal faulting above and ductile flow below (Miller and others, 1983; Gans and others, 1985). As such, it is analogous to the present-day midcrustal ductile-brittle transition zone inferred beneath the modern normal fault systems that define the Basin and Range Province.

GEOLOGIC BACKGROUND

The evolution of ideas about the NSRD charts our growing understanding of the geologic history of the Basin and Range Province. Until the early 1970s, most low-angle faults in the western U.S. were mapped as thrust faults. The NSRD was first described by Hazzard and others (1953) and then discussed by Misch (1960), who speculated that it was the basal decollement or shearing-off surface for the Mesozoic Sevier thrust belt exposed to the east in Utah. Later mapping in the Snake Range by Whitebread (1969), Hose and Blake (1976), and Hose (1981) documented drastic thinning of the upper plate by normal faulting and led Hose and Danes (1973) and Hintze (1978) to propose that east-central Nevada and Utah represented an uplifted, extended hinterland behind a gravity-driven Mesozoic thrust belt to the east. In his classic synthesis of east-central Nevada geology, Armstrong (1972) suggested that many of the low-angle faults, including the NSRD, were Tertiary, rather than Mesozoic in age. This

interpretation was accepted by Coney (1974) who argued that mesoscopic data from immediately below the NSRD indicated radial gravity sliding of the upper plate from the top of the range. Our recent and ongoing studies (Gans and Miller, 1983; Miller and others, 1983; Gans and others, 1985; and Lee and others, 1986) support a Tertiary age for most of the deformation in this region but have also emphasized the importance of Mesozoic metamorphism and plutonism at deep structural levels.

Despite the extremely complex map pattern (Fig. 1), a systematic structural style is evident in the upper plate of the NSRD. Structural sections of Middle Cambrian to Permian strata (and locally, overlying Tertiary) that "young" to the west are repeated eastward on east-dipping normal faults that merge with, but do not offset the NSRD. Older, west-dipping faults within these structural sections omit units as well, and are interpreted as originating as steep, down-to-the-east, normal faults that once also shallowed into the NSRD. These older normal faults rotated to low angles as they moved until cut by the younger faults, which rotated them past horizontal and into their present westward dips. Ductile deformation, warping (normal drag), folding, and low-angle fault splays helped alleviate space problems at the toes of normal faults as they merged with the NSRD.

The geometry of upper plate normal faulting indicates a WNW-ESE direction of extension. Our palinspastic reconstructions indicate that normal faulting caused several hundred percent extension, thinning the upper plate from 4.3 mi (7 km) to less than 1.2 mi (2 km) thick (Figs. 1 and 2). Stratigraphic relations in the Sacramento Pass area indicate that faulting began after extrusion of 35 Ma volcanic rocks and was ongoing at 32 Ma (Grier, 1984). Unfortunately, no younger stratigraphic age brackets exist for faulting in this area.

In marked contrast to the upper plate, rocks beneath the NSRD are flat-lying, are not cut by faults, and have been dramatically thinned. The lower plate units are parallel to the NSRD and define a gentle N-S trending antiform (Fig. 1). A conformable sequence of metasedimentary rocks in the lower plate ranges in age from late Precambrian (McCoy Creek Group) at the deepest levels of exposure, to Middle Cambrian (Pole Canyon Limestone) immediately beneath the NSRD (Fig. 1). The deformational, metamorphic, and intrusive history of these rocks is still incompletely understood, but clearly includes both Mesozoic and Cenozoic events.

Lower plate metasedimentary rocks are typically upper greenschist to amphibolite facies. Jurassic plutons intrude these rocks along the south flank of the range (Miller and Wright, unpublished U-Pb zircon data), and, like their counterparts in adjacent ranges, have narrow amphibolite-facies contact aureoles with low pressure assemblages that include andalusite and cor-

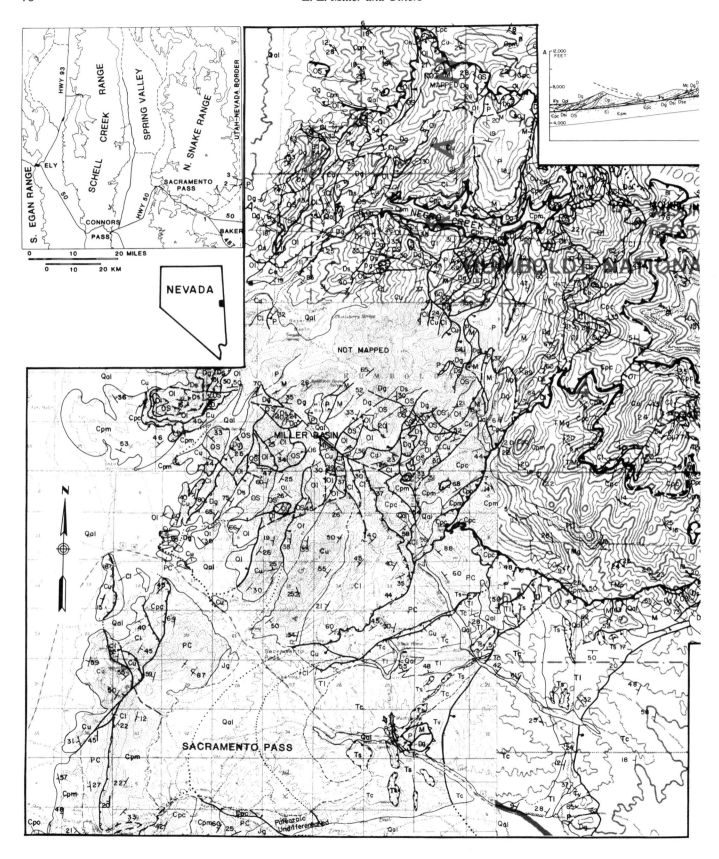

GEOLOGIC MAP OF PART OF THE NORTHERN SNAKE RANGE

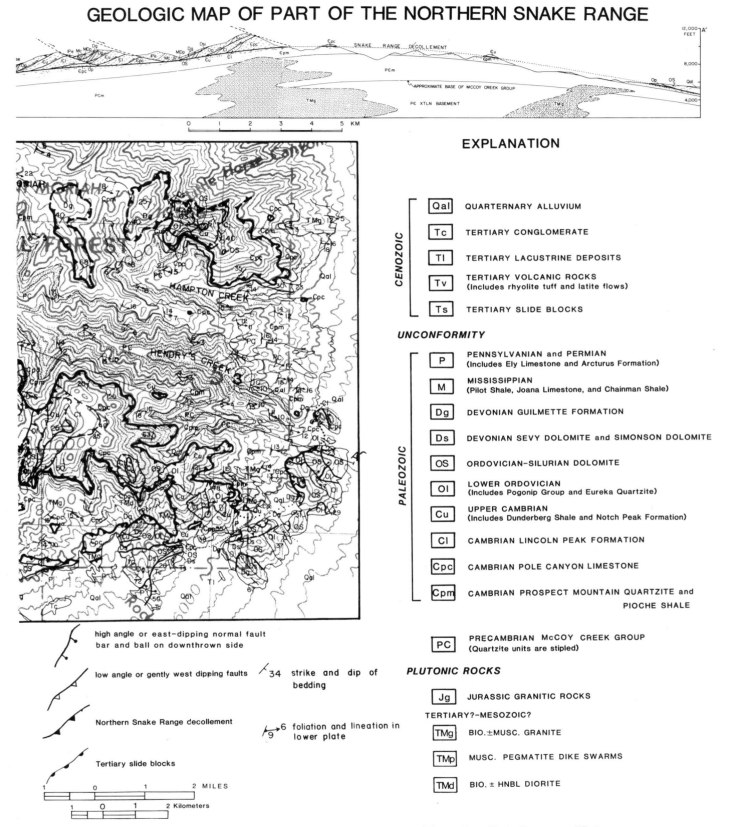

EXPLANATION

CENOZOIC

Qal	QUARTERNARY ALLUVIUM
Tc	TERTIARY CONGLOMERATE
Tl	TERTIARY LACUSTRINE DEPOSITS
Tv	TERTIARY VOLCANIC ROCKS (Includes rhyolite tuff and latite flows)
Ts	TERTIARY SLIDE BLOCKS

UNCONFORMITY

PALEOZOIC

P	PENNSYLVANIAN and PERMIAN (Includes Ely Limestone and Arcturus Formation)
M	MISSISSIPPIAN (Pilot Shale, Joana Limestone, and Chainman Shale)
Dg	DEVONIAN GUILMETTE FORMATION
Ds	DEVONIAN SEVY DOLOMITE and SIMONSON DOLOMITE
OS	ORDOVICIAN-SILURIAN DOLOMITE
Ol	LOWER ORDOVICIAN (Includes Pogonip Group and Eureka Quartzite)
Cu	UPPER CAMBRIAN (Includes Dunderberg Shale and Notch Peak Formation)
Cl	CAMBRIAN LINCOLN PEAK FORMATION
Cpc	CAMBRIAN POLE CANYON LIMESTONE
Cpm	CAMBRIAN PROSPECT MOUNTAIN QUARTZITE and PIOCHE SHALE

| PC | PRECAMBRIAN McCOY CREEK GROUP (Quartzite units are stipled) |

PLUTONIC ROCKS

| Jg | JURASSIC GRANITIC ROCKS |

TERTIARY?-MESOZOIC?

TMg	BIO.±MUSC. GRANITE
TMp	MUSC. PEGMATITE DIKE SWARMS
TMd	BIO. ± HNBL DIORITE

high angle or east-dipping normal fault bar and ball on downthrown side

low angle or gently west dipping faults

Northern Snake Range decollement

Tertiary slide blocks

34 strike and dip of bedding

6 foliation and lineation in lower plate

Figure 1. Geologic map and cross section of the southern part of the northern Snake Range, modified from Miller and others (1983) showing the location and accessibility of localities described in text.

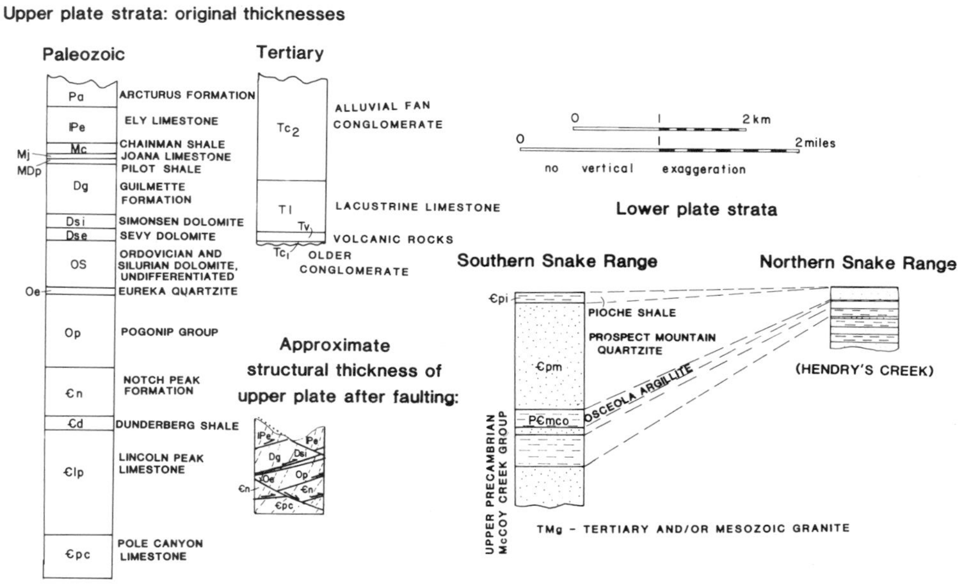

Figure 2. Simplified stratigraphic columns for upper and lower plate strata in the northern Snake Range, showing structural thicknesses before and after Tertiary extensional deformation. Modified from Miller and others (1983).

dierite. Metamorphic grade increases northward and with depth along the east flank of the range. Garnet, staurolite, and kyanite-in isograds have been mapped in the Precambrian schist units in the vicinity of Hendry's and Hampton creeks (Fig. 1). On the basis of new ^{40}Ar-^{39}Ar radiometric studies, Lee and others (1986) have argued, as did Lee and others (1981), for a Late Cretaceous age for peak amphibolite facies metamorphism in the Hendry's and Hampton creek region.

Superimposed on all lower plate rocks is a penetrative, subhorizontal foliation and a consistent WNW mineral elongation lineation. Strain related to the development of this foliation is highest on the east flank of the range, and dies out along the very westernmost flank of the range. Everywhere, this foliation (the XY plane of the finite strain ellipsoid) is parallel to the NSRD and to lower plate stratigraphic contacts. Mineral stretching lineations and long axes of pebbles consistently trend N55°W to N70°W and are subparallel to the extension direction inferred from the geometry of upper plate faults. The penetrative deformation in the lower plate has resulted in drastic thinning of units compared to correlative sections in the adjacent southern Snake Range and Schell Creek Range (Fig. 2). We estimate that units in the lower plate have been thinned to as little as one tenth of their original thickness on the east flank of the range (Miller and others, 1983; Lee and others, 1986).

The present controversy surrounding the origin and movement history of detachment faults such as the NSRD is centered on several related issues: the timing of lower plate deformation (Mesozoic versus Tertiary), the lower plate strain path (pure shear versus simple shear), and the amount of displacement represented by these surfaces.

Petrographic studies of metamorphic rocks from the lower

plate of the Snake Range indicate that a distinctly younger metamorphic event accompanied the high strain deformation and retrograded older amphibolite facies assemblages. Temperatures during this deformation increase with depth as suggested by growth of syntectonic chlorite at shallow levels and by biotite at the deepest levels of exposure. Pervasive Tertiary re-setting of K-Ar mica ages has been documented in lower plate rocks by Lee and others (1980) and new ^{40}Ar-^{39}Ar studies support a Tertiary age for deformation (Lee and others, 1986). These new data argue that at least the latest stages of lower plate ductile deformation occurred at greater than 320° (the argon closure temperature of muscovite) and was ongoing, ending shortly after 23–24 Ma, in the Miocene.

The geometry and the history of lower plate strain do not provide an unambiguous answer to the question of whether lower plate strain occurred by pure or simple shear. The observed drastic thinning of units, the parallelism of foliation, and the NSRD throughout the range, coupled with the fact that deformation does not appear to die out with depth, are compatible with a large component of "pure shear" deformation characterized by vertical shortening and horizontal extension. However, assymmetric quartz C-axis fabrics and consistent "S-C plane" relationships (Lister and Snoke, 1984) indicate at least a late stage component of top to the east simple shear (Lee and others, 1986).

Although some amount of displacement could have occurred along the NSRD, the comparable magnitudes and orientation of upper and lower plate strain axes and our inability to trace the NSRD into adjacent ranges suggest very strongly that this translation has been minor (less than 6 mi; 10 km; Miller and others, 1983; Gans and others, 1985). Thus the NSRD probably originated as a local dislocation between a brittle extending upper

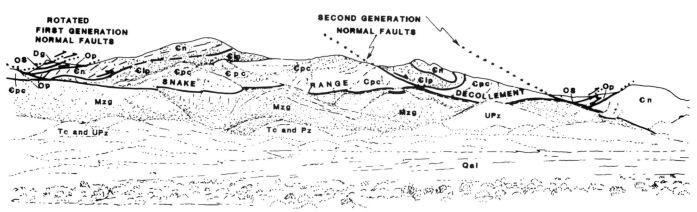

Figure 3. View looking north towards the southern flank of the northern Snake Range. The NSRD lies above a thin but conspicuous ledge of white tectonized marble. It is folded about a WNW-ESE trending axis and dips south beneath the hills in foreground, which are underlain by faulted upper Paleozoic (UPz) and Tertiary strata. Two generations of normal faults can be seen in the upper plate (see text for discussion). Lower plate rocks on this flank of the range consist almost entirely of strongly deformed Mesozoic plutons (Mzg). Unit labels are the same as those shown in Figures 1 and 2.

plate and a plastically stretching lower plate. Furthermore, it appears that ductile strain was concentrated in the thermally weakened part of the lower plate, presumably coincident with elevated isotherms due to the intrusion of plutons at depth.

Locality 1. View Stop of the Northern Snake Range. A good overview of the northern Snake Range and the NSRD can be seen from the parking lot of Moriah's Great Basin Inn at the junction of U.S. 50 and Nevada 487, the turnoff south to Baker (Fig. 3).

Locality 2. The Snake Range Decollement. From Moriah's Great Basin Inn go east on U.S. 50 about 100 yd (90 m) and turn left on graded Snake Valley Road. Slow down through ranch. After 7.1 mi (11.4 km), turn left on dirt road (turnoff is 0.9 mi (1.4 km) north of Utah-Nevada State Line and before turnoff to Hatch Rock Quarry (Loc. 3, below) and may have a sign pointed to Wright Quarry). Drive west toward the Snake Range

about 2.6 mi (4.2 km) passing through small hills of highly faulted and brecciated upper plate rocks. Smooth, light colored hills straight ahead are underlain by ductilely deformed Cambrian Prospect Mountain Quartzite. Park near trailer. If inhabited, please request permission to pass. Walk about 0.5 mi (0.8 km) due west up hill towards the white cliffs of marble.

This short hill will allow one to see the top of a highly deformed Jurassic(?) pluton intruded into the Cambrian Prospect Mountain Quartzite. In stratigraphic continuity above the Prospect Mountain Quartzite are the slope-forming Cambrian Pioche Shale and the prominent ledge-forming, intricately folded Middle Cambrian Pole Canyon Limestone. The NSRD is located at the top of the marble mylonite ledge and is overlain by a thin fault-bound sliver of variably foliated and brecciated upper plate Cambrian Pole Canyon Limestone, which is in turn overlain by relatively unmetamorphosed cherty limestone of the Late Cambrian Notch Peak Formation. Note that what we have defined as the NSRD here is simply the last surface of movement within what was probably a broader, highly complex zone of interaction between upper plate normal faults and ductile flow below.

Locality 3. Lower Plate Fabrics and Metamorphism in Hendry's Creek. Continue north along Snake Valley Road about 0.3 mi (0.5 km) past the turnoff to locality 2. Turn left on the road up Hendry's Creek, which is well marked by a stone monument and maintained by Hatch flagstone quarry. The NSRD projects under the dark hills of faulted upper plate carbonates in the foreground. The light-colored slopes beyond are underlain by dramatically thinned Prospect Mountain Quartzite. The NSRD dips about 15° to the east, to that the slope of the eastern flank of the range approximates its exhumed surface. After 3.0 mi (4.8 km), take left fork; another 0.1 mi (0.2 km) take right fork; another 0.3 mi (0.5 km) take left fork. Road is washed out just a little farther on so you will have to park and hike. An excellent view of the Snake Range decollement and a detailed tour of lower plate units, their fabrics, and their metamorphism can be had by hiking straight up the north canyon wall of Hendry's Creek. We suggest walking up the canyon road, across the major washout, the bog, and up to the first conspicuous stand of pine trees before hiking up the canyon wall. Although it is a steep hike, it is easy to avoid cliffs by walking along their base until you find a safe passage up. The view from the ridge above is one of the best, as are the joint-bound "swords" of flaggy Prospect Mountain Quartzite.

Exposures of metasedimentary rocks in the deeply incised canyons along the eastern flank of the Snake Range, such as Hendry's Creek, best exhibit the ductile thinning of lower plate strata. Here, the Precambrian McCoy Creek Group, consisting of interbedded quartzite and schist, and the overlying Cambrian Prospect Mountain Quartzite and Pioche Shale have been thinned to about 1,600 ft (500 m) from an original thickness of about 10,000 ft (3 km). In particular, the thinning of the Prospect Mountain Quartzite from an original thickness of 4,000 ft (1,200 m) to only several hundred ft (m) is most impressive. This unit forms the highest quartzite cliff in the canyon walls of Hendry's Creek.

Ductile stretching of the quartzites formed classic "Type II S-C mylonites" as described by Lister and Snoke (1984), while interlayered schists behaved in a more brittle fashion and exhibit spaced "extensional crenulation cleavage" (Platt and Vissers, 1980). Micro and mesoscopic conjugate normal faults and subvertical joints perpendicular to the direction of extension overprint these earlier ductile fabrics. This progressive extensional deformation affected amphibolite facies rocks but occurred at lower greenschist grade conditions as exhibited by the synkinematic growth of retrograde chlorite and white mica as pressure shadows on older metamorphic minerals. Older metamorphic minerals such as garnet, staurolite, muscovite, and biotite have been mechanically kinked, bent, broken, or rotated into parallelism with the younger subhorizontal foliation and lineation, while quartz, behaving in a ductile fashion, has flowed about them.

REFERENCES CITED

Armstrong, R. L., 1972, Low-angle (denudational) faults, hinterland of the Sevier Orogenic Belt, eastern Nevada and western Utah: Geological Society of America Bulletin, v. 83, p. 1729–1754.

Bartley, J. M., and Wernicke, B. P., 1984, The Snake Range decollement interpreted as a major extensional shear zone: Tectonics, v. 3, p. 647–657.

Coney, P. J., 1974, Structural analysis of the Snake Range decollement east-central Nevada: Geological Society of America Bulletin, v. 85, p. 973–978.

Crittenden, M. D., Jr., Coney, P. J., and Davis, G. H., 1980, Cordilleran Metamorphic Core Complexes: Geological Society of America Memoir 153, 490 p.

Eaton, G. P., 1982, The Basin and Range Province; Origin and tectonic significance: Annual Review Earth and Planetary Sciences, v. 10, p. 409–440.

Gans, P. B., and Miller, E. L., 1983, Style of mid-Tertiary extension in east-central Nevada: Utah Geological and Mineral Survey Special Studies 59, p. 107–160.

Gans, P. B., Miller, E. L., McCarthy, J., and Ouldcott, M. L., 1985, Tertiary extensional faulting and evolving ductile-brittle transition zones in the northern Snake Range and vicinity; New insights from seismic data: Geology, v. 13, p. 189–193.

Grier, S. P., 1984, Alluvial fan and lacustrine carbonate deposits in the Snake Range; A study of Tertiary sedimentation and associated tectonism [M.S. thesis]: Stanford, California, Stanford University, 61 p.

Hazzard, J. C., Misch, P., Wieses, J. H., and Bishop, W. C., 1953, Large-scale thrusting in the northern Snake Range, White Pine County, northeastern Nevada [abs.]: Geological Society of America Bulletin, v. 64, p. 1507–1508.

Hintze, L. F., 1978, Sevier orogenic attenuation faulting in the Fish Springs and House ranges, western Utah: Brigham Young University Geologic Studies, v. 25, part 1, p. 11–24.

Hose, R. K., 1981, Geologic map of the Mount Moriah further planning (Rare II) area, eastern Nevada: U.S. Geological Survey Map MF-1244A.

Hose, R. K., and Blake, M. C., Jr., 1976, Geology and mineral resources of White Pine County, Nevada; Part I, Geology: Nevada Bureau of Mines and Geology Bulletin 85, 105 p.

Hose, R. K., and Danes, Z. E., 1973, Development of late Mesozoic to early Cenozoic structures of the eastern Great Basin, *in* DeJong, K. A., and Scholten, R., eds., Gravity and tectonics: New York, John Wiley and Sons, p. 429–441.

Lee, D. E., Marvin, R. F., and Mehnert, H. H., 1980, A radiometric age study of Mesozoic-Cenozoic metamorphism in eastern White Pine County, Nevada and nearby Utah: U.S. Geological Survey Professional Paper 1158C, p. C17–C28.

Lee, D., Stern, T. W., and Marvin, R. F., 1981, Uranium-thorium-lead isotopic ages of metamorphic monazite from the northern Snake Range, Nevada: Isochron/West, no. 31, p. 23.

Lee, J., Miller, E. L., Marks, A.M.B., Lister, G. S., and Sutter, J. F., 1986, Ductile strain and metamorphism in an extensional tectonic setting; A case study from the northern Snake Range, Nevada, U.S.A.: Journal of the Geological Society of London, Syposium Volume for Special Meeting on Continental Extensional Tectonics (in press).

Lister, G. S., and Snoke, A. W., 1984, S-C mylonites: Journal of Structural Geology, v. 6, p. 617–638.

Miller, E. L., Gans, P. B., and Garing, J., 1983, The Snake Range decollement; An exhumed mid-Tertiary ductile-brittle transition: Tectonics, v. 2, p. 239–263.

Misch, P., 1960, Regional structural reconnaissance in central-northeast Nevada and some adjacent areas: Observations and interpretations: Intermountain Association of Petroleum Geology 11th Annual Field Conference Guidebook, p. 17–42.

Platt, J. P., and Vissers, R.L.M., 1980, Extensional structures in anisotropic rocks: Journal of Structural Geology, v. 2, p. 397–410.

Whitebread, D. H., 1969, Geologic map of the Wheeler Park and Garrison Quadrangles, Nevada and Utah: U.S. Geological Survey, Map I-578.

Fault scarps formed during the earthquakes of October 2, 1915, Pleasant Valley, Nevada

Robert E. Wallace, U.S. Geological Survey, 345 Middlefield Road, Menlo Park, California 94025

LOCATION

Four fault scarps that formed during the earthquakes of October 2, 1915, are located in north-central Nevada, 150 mi (240 km) northeast of Reno, and 30 to 55 mi (48 to 88 km) south of Winnemucca (Fig. 1). A site along the largest of the four scarps, the Pearce scarp, at Golconda Canyon on the west flank of the Tobin Range (Sec. 5,T28N,R.39E, Mt. Tobin 15-minute Quadrangle) is easily reached by driving south about 45 mi (72 km) from Winnemucca along the Grass Valley–Pleasant Valley road, which is paved for about 7 mi (11.2 km) after which it is graded gravel. At the Golconda Canyon road, go east 1.5 mi (2.4 km) to the fault scarp at the range front. The Pearce scarp is also clearly visible from the Pleasant Valley road for about 15 mi (24 km) and excellent photo sites abound. Afternoon lighting emphasizes the scarp especially well. The Sou Hills scarp (3–6 ft) (1–2 m) high, can be seen in the distance to the south from near the junction of the Pleasant Valley–Dixie Valley roads. Additional map coverage is provided by the Cain Mountain and Fencemaker 15-minute and the China Mountain 7½-minute Topographic Maps.

SIGNIFICANCE

The fault scarps that formed during the earthquakes of October 2, 1915, are spectacular examples of surface faulting along a seismogenic fault. Block faulting characteristic of the Basin and Range province is demonstrated. Displacement is predominantly normal and dip slip. The upthrown Tobin Range block represents a horst, and the downthrown Pleasant Valley block is a graben. The Tobin Range, China Mountain, and Sou Hills blocks are all tilted to the east. At Golconda Canyon and elsewhere along the scarps, the interaction of active tectonism and geomorphic processes is well displayed.

SITE INFORMATION

The main earthquake of Richter magnitude 7¾ struck at 2252 PST, October 2, 1915. Two strong foreshocks occurred at 1541 and 1750 PST, and aftershocks continued for months. Four separate scarps formed during the earthquakes; from north to south they are the China Mountain, Tobin, Pearce, and Sou Hills scarps (Fig. 1).

The Golconda Canyon site lies midway on the Pearce scarp, and in the section a few miles (kilometers) long to the north are found the largest displacements (up to 19 ft; 5.8 m). Scarp heights are commonly greater than this because of slope and fault geometry and postearthquake erosion.

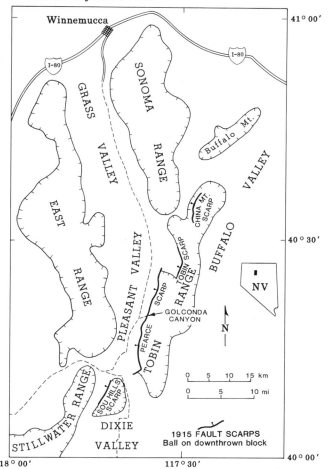

Figure 1. Index map showing location of fault scarps formed during earthquakes of October 2, 1915, Pleasant Valley, Nevada.

The 1915 fault scarps are generally characterized by a sharp crest, below which is a free face, a steep face from which sand and gravel spall and fall to a debris slope below. The sand and gravel of the debris slope stand at the angle of repose of about 30° to 35°. Commonly, coarser clasts of cobbles and boulders accumulate at the base of the debris slope. Where the original colluvium or alluvium is poorly indurated, the free face may have been covered by debris slope. Note the steep, almost unchanged free face at the end of the fence a few tens of meters south of the Golconda Canyon road (Fig. 2). Apparently, the gravels underlying the swale are slightly more cemented than the colluvium to the north and south. Assuming that all of the original scarp face shown in Figure 2 stood at a position equivalent to that of the steep face at the fence line, one can estimate the amount (several feet; meters) of free face retreat of the rest of the scarp. I estimate

Figure 2. Fault scarp at Golconda Canyon site. Note characteristic parts of this young scarp; a sharp "crest" above a steep "free face," and a "debris slope" sloping about 30° to 35° in the lower half.

that a free face will survive somewhere along the 36-mi (60-km) length of the 1915 scarps for at least several hundred years and possibly for as long as 2,000 years.

About 0.3 mi (0.5 km) north of Golconda Canyon is an old fence that was displaced right laterally about 6.6 ft (2 m). Given an axis of regional extension of about N.65°W., those segments that strike northwest, such as here, have a significant right-lateral component of displacement. Those segments that strike N.25°E., perpendicular to regional extension, have simple dip-slip displacement. On the hillside above the site can be seen arcuate scarps that represent slope-failure or landslide-like features.

From many points along the main Pleasant Valley road 3 to 4 mi (5 to 7 km) south of Golconda Canyon Road are excellent views of faceted spurs and wine glass–shaped valleys characteristic of fault-generated range fronts. During each displacement event, a steep scarp a few feet (meters) high is added to the base of the faceted spur. The planar nature of the entire facet persists for several million years, but note that erosion has caused the upper part of each triangular facet to decline to a lower slope angle than the basal part of the facet.

In many places the 1915 scarps formed along the line of older scarps, and remnants of the older scarps are commonly left as bevels at the crest of the 1915 scarp. Analysis of the profiles of these older scarps, uplift rates of the range blocks, and dating by radiocarbon and other methods indicate that larger earthquakes (M 7) accompanied by surface faulting occur at intervals of several thousand years. An average displacement rate of a few tenths of a millimeter per year is indicated along the range front faults in the region.

SELECTED REFERENCES

Jones, J. C., 1915, The Pleasant Valley, Nevada, earthquake of October 2, 1915: Seismological Society of America Bulletin, v. 5, no. 4, p. 190–205.

Page, B. M., 1934, Basin-range faulting of 1915 in Pleasant Valley, Nevada: Journal of Geology, v. 43, no. 7, p. 690–707.

Wallace, R. E., 1977, Profiles and ages of young fault scarps, north-central Nevada: Geological Society of America Bulletin, v. 88, no. 9, p. 1267–1281.

——— , 1978, Geometry and rates of change of fault-generated range fronts, north-central Nevada: U.S. Geological Survey Journal of Research, v. 6, no. 5, p. 637–650.

——— , 1984, Fault scarps formed during the earthquakes of October 2, 1915, in Pleasant Valley, Nevada, and some tectonic implications: U.S. Geological Survey Professional Paper 1274-A, 33 p.

Mesozoic structural features in Sonoma Canyon, Sonoma Range, Humboldt and Pershing counties, Nevada

Stephen D. Stahl, *Department of Geology, Central Michigan University, Mt. Pleasant, Michigan 48859*

LOCATION

Sonoma Canyon is on the west flank of the Sonoma Range, directly south of Winnemucca, Nevada (Fig. 1A, B). Coming from either the east or west on I-80, take the second Winnemucca exit ("to U.S. 95 north"), turn west on Business I-80 at the base of the ramp, and continue on Business I-80 through the U.S. 95 junction. If entering Winnemucca via U.S. 95, turn right (west) at the junction with Business I-80 (highlighted by "to I-80" signs). Travelers from all directions are now on the same road and heading in the same direction and should get into the left lane of Business I-80 west. Approximately 0.5 mi (0.8 km) west of the Business I-80/U.S. 95 junction, turn left (southeast) on Hanson Street (at sign for "Hanson Street Overpass") and follow this road over the railroad tracks. At the base of the viaduct, turn right (southwest) on Grass Valley Road. Follow Grass Valley Road southwest and then south (at a "Y" in the road after passing the Grass Valley school and Star City deli/laundromat) for approximately 11.5 mi (18.5 km). The pavement ends at the Pershing County line. Turn left (east) onto the Sonoma Canyon access road (no sign; Fig. 1C). This turnoff is approximately 0.75 mi (1.3 km) north of Sonoma Ranch, which is well marked. The total distance from the center of Winnemucca is about 13 mi (21 km).

Sonoma Canyon is serviced by a rough canyon bottom road requiring a four-wheel-drive vehicle. The approach road from Grass Valley Road to the canyon mouth is always passable by two-wheel drive, but visitors with such vehicles will have to walk in the canyon. Visitors in late July and August should pack in an ample water supply. The entire canyon is Bureau of Land Management (BLM) property, so no permission from the local ranchers is required to gain access. Enjoyment of this site will be greatly enhanced if one carries binoculars in addition to the usual field implements.

SIGNIFICANCE

The Sonoma Range offers the best evidence of middle Mesozoic compressional reshuffling of the Triassic-Paleozoic tectonostratigraphy of north-central Nevada. Lower Paleozoic rocks of the Roberts Mountains allochthon, upper Paleozoic rocks of the Roberts overlap sequence, and upper Paleozoic rocks of the Golconda allochthon were thrust from east to west over Upper Triassic shelf rocks (Ferguson and others, 1951; Gilluly, 1967; Speed and others, 1982) during the Jurassic Winnemucca orogeny (Stahl and Speed, 1983). Recent interpretation of structural fabric elements in the range suggests that the sole thrust of the Winnemucca thrust system is exposed in the Sonoma Range

(Stahl and Speed, 1983), and that the Winnemucca thrust system developed in overstep fashion (Stahl and Speed, 1984).

SITE INFORMATION

Introduction. The geology of Sonoma Canyon will prove challenging to both the interested undergraduate and the professional geologist. To the undergraduate, the site offers textbook quality examples of structural windows with fold-and-fault vergence and stratigraphic relationships leading to the interpretation of east-to-west overthrusting during the Mesozoic Winnemucca orogeny. To the professional geologist, the interpretation of structural fabric elements yields significant constraints on Jurassic regional tectonics.

Regional geology. The Sonoma Range lies within the Winnemucca fold-and-thrust belt (Speed and others, 1982) and displays relations characteristic of this belt: west-verging folds of Triassic shelf strata structurally beneath lower and upper Paleozoic rocks (Fig. 1C). To the east of the range lies an enclave of little Mesozoic deformation. The Fencemaker fold-and-thrust belt lies 12 mi (20 km) to the west of the Winnemucca fold-and-thrust belt. Relationships characteristic of the Fencemaker fold-and-thrust belt are exposed in the East Range, where Triassic basinal rocks are thrust eastward over Triassic shelf strata (Elison and Speed, 1985).

Pertinent literature. This paper is an offshoot of the third generation of geologic mapping in the Sonoma Range. The earlier works (Ferguson and others, 1951; Gilluly, 1967; Silberling, 1975) were primarily stratigraphic efforts, whereas this latest approach is principally structural in nature (Speed and others, 1982; Stahl and Speed, 1983, 1984). The significant interpretive differences between the map by Ferguson and others (1951) and the Gilluly (1967) and Silberling (1975) maps are in the number and extent of Mesozoic thrusts. Ferguson and others (1951) proposed a single Mesozoic thrust at the top of the Triassic rocks, with the Triassic and structurally lower rocks being Mesozoic autochthonous; Gilluly (1967) and Silberling (1975) proposed three Mesozoic thrusts and no Mesozoic autochthonous rocks. This author's interpretations are most succinctly explained as a combination of those of Ferguson and others (1951) and Silberling (1975), plus the recognition of low-angle normal faulting at the base of the Triassic package; these interpretations are summarized on Figures 1C, 2 and 3.

The site consists of excellent outcrops of Triassic shelf rocks and poorly exposed lower Paleozoic Harmony "formation" rocks on the upper plate between the windows. The Triassic lithologies

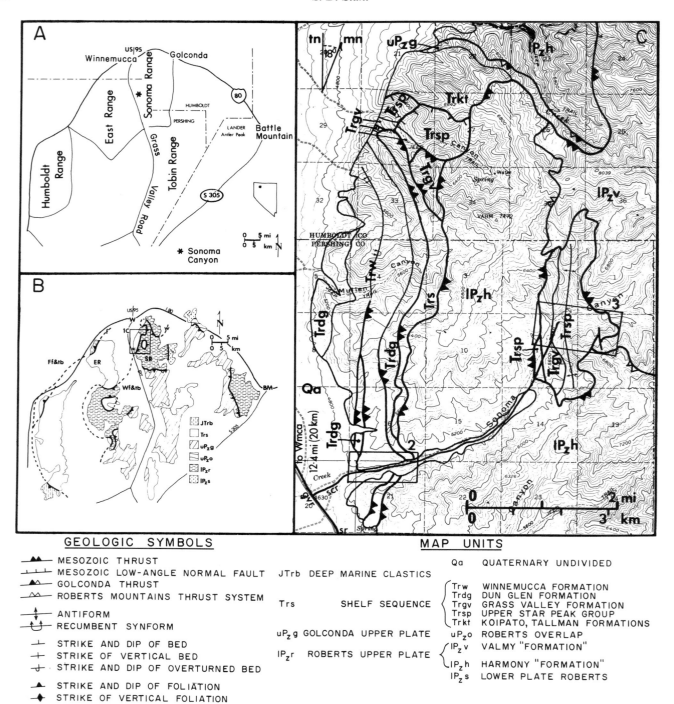

GEOLOGIC SYMBOLS

- ▲▲ — MESOZOIC THRUST
- ⊥⊥⊥ — MESOZOIC LOW-ANGLE NORMAL FAULT
- ▲▲ — GOLCONDA THRUST
- △△ — ROBERTS MOUNTAINS THRUST SYSTEM
- ⊥ — ANTIFORM
- ⊥ — RECUMBENT SYNFORM
- ⊥ — STRIKE AND DIP OF BED
- + — STRIKE OF VERTICAL BED
- ⊥ — STRIKE AND DIP OF OVERTURNED BED
- ▲ — STRIKE AND DIP OF FOLIATION
- ◆ — STRIKE OF VERTICAL FOLIATION

JTrb DEEP MARINE CLASTICS

Trs   SHELF SEQUENCE

uPzg GOLCONDA UPPER PLATE

IPzr ROBERTS UPPER PLATE

MAP UNITS

Qa   QUATERNARY UNDIVIDED

Trw WINNEMUCCA FORMATION
Trdg DUN GLEN FORMATION
Trgv GRASS VALLEY FORMATION
Trsp UPPER STAR PEAK GROUP
Trkt KOIPATO, TALLMAN FORMATIONS

uPzo ROBERTS OVERLAP

IPzv VALMY "FORMATION"

IPzh HARMONY "FORMATION"

IPzs LOWER PLATE ROBERTS

Figure 1. A. Reference map of north-central Nevada, inset map shows location within Nevada. S 305, Nevada State Route 305; county lines indicated by dashed lines with names in all caps; selected ranges as indicated. B. Generalized geologic map of north-central Nevada at same scale as Figure 1A. Box shows approximate area of Figure 1C. Geography as before (gvr, Grass Valley Road; I 80, Interstate 80; US 95, U.S. Highway 95), geologic symbols and map units (Tr, Triassic) at bottom of figure: w, Winnemucca thrust; f, Fencemaker thrust; Wf&b, Winnemucca fold and thrust belt; Ff&tb, Fencemaker fold and thrust belt; ER, East Range; SR, Sonoma Range; BM, Battle Mountain. C. Generalized geologic map of the Sonoma Range (after Ferguson and others, 1951; Gilluly, 1967; Silberling, 1975), Boxes show approximate area of Figures 2 and 3. Geologic symbols and map units (Tr, Triassic) at bottom of figure; gvr, Grass Valley road; scr, Sonoma Canyon Road; sr, Sonoma Ranch.

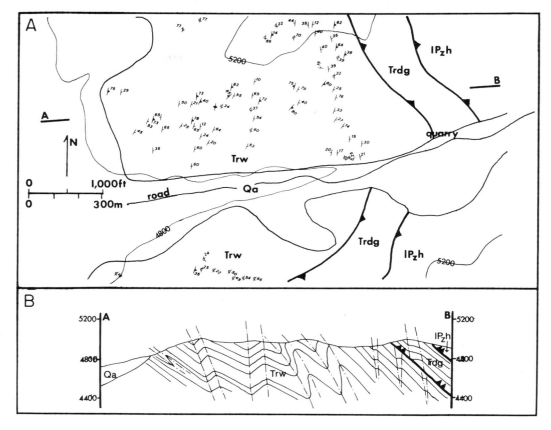

Figure 2. A. Geologic map of the range-front window in Sonoma Canyon. For explanation of symbols, see Figure 1. Road and quarry indicated. Contour interval is 400 ft (122 m). B. Cross section along line A–B.

are best summarized by Burke and Silberling (1973) and Nichols and Silberling (1977); the Harmony rocks are amply described by Ferguson and others (1951), Hotz and Willden (1964), Roberts (1964), and Stewart and Suczek (1977).

The Triassic shelf sequence exposed in Sonoma Canyon is composed of, from youngest to oldest, the Winnemucca, Dun Glen, and Grass Valley formations of the Auld Lang Syne Group (Burke and Silberling, 1973) and the uppermost unit of the Star Peak Group (Nichols and Silberling, 1977). The Winnemucca Formation is predominantly thin to medium beds of arkose to subarkose and thick sequences of multicolored shale with minor thin to thick beds of impure carbonate and pure quartz arenite. The sandstones commonly show cross-stratification and ripple marks. The Dun Glen Formation is a monotonous succession of thick to very thick massive beds of dark gray crystalline carbonate with local sand stringers and chert nodules. The Grass Valley Formation consists of very thin to thin interbeds of arkose (cross-bedding common) and olive gray to black siltstone and mudstone. The Star Peak Group rocks are medium to thick beds of massive nonfossiliferous light gray carbonate. The lower Paleozoic Harmony rocks exposed in Sonoma Canyon are distinctive quartz- and potassium feldspar–rich pebble conglomerate,

yellow-brown arkose, and yellowish muscovite-rich silty mudstone.

Five generations of Winnemucca orogenic folding are recognized in the Triassic rocks of the Sonoma Range. First-phase folds (F1) are minor recumbent isoclines in the Grass Valley and Winnemucca formations and west-verging inclined open folds in the Star Peak Group rocks and Dun Glen Formation. Foliation is seen only in the Grass Valley Formation, where it is bedding parallel. Second-phase folds (F2) are major folding of the entire Triassic sequence into a tight recumbent synform and tight inclined antiform; this generation is the result of blind thrusting within the Triassic rocks (Stahl and Speed, 1984). Third-phase folds (F3) are seen solely in the Star Peak rocks proximal to the overthrusted lower Paleozoic rocks in the central window (Fig. 3). Fourth-phase folds (F4) are minor inclined to upright west-verging tight folds of beds and earlier structural surfaces with near-vertical axial-plane foliation. Unlike earlier generations (F1 to F3), which are only present in the Triassic rocks, a correlative to this phase is also present in the lower Paleozoic rocks of the Mesozoic allochthon. Fifth-phase folds (F5) are open upright major folds with vertical axial-planar foliation. Correlatives to this phase are seen in both the Mesozoic autochthonous rocks

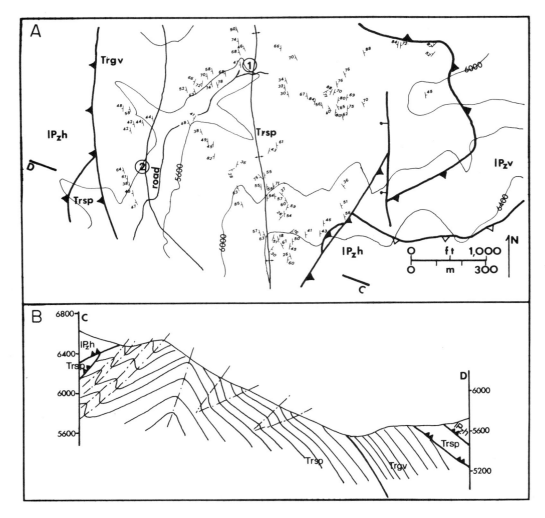

Figure 3. A. Geologic map of the southern portion of the central window. For explanation of symbols, see Figure 1. Contour interval is 400 ft (122 m). Locations 1 and 2 indicated by circled numerals, road as indicated. B. Cross section along line C–D as viewed from location 1 (the east and west ends of the section are opposite those of Fig. 2B).

(unlike Triassic rocks F1 to F4) and lower Paleozoic rocks of the Mesozoic allochthon.

SITE DESCRIPTION

The primary features of this site are two structural windows of upper Triassic shelf rocks beneath lower Paleozoic rocks (Fig. 1C). While both the Triassic and Paleozoic rocks are nonfossiliferous in Sonoma Canyon, the thrust nature of the contact is evident by the superposition of deep-water clastics on shallow-water carbonates and deltaic deposits. The range-front window (Fig. 2) exposes the uppermost rocks of the Triassic sequence on the overturned limb of a recurrent synform (F2), and in the central window (Fig. 3) the oldest units of the shelf strata define an inclined to upright antiform (also F2). Evidence for the east-to-west overthrusting of the lower Paleozoic rocks is apparent when one considers the implications of the stratigraphy of the

windows; the Winnemucca sole thrust cuts upsection (ramps) from within the Star Peak Group (the base of the Triassic shelf sequence) at the eastern margin of the central window to the top of the Winnemucca Formation (the youngest of the Triassic rocks) at the range-front window.

Range-front window. Figure 2A shows the geology of the range-front window. The traverse described below parallels cross section line A–B and shows Winnemucca deformations F1, F2, F4, and F5. Utilizing a Brunton compass and keeping an eye on sedimentary evidence for tops would be helpful in identifying the different phases of folding. To reach the start of the traverse, park at the turnout for a small quarry in the north wall of the canyon, climb the slope to the west (down canyon) of the quarry to a height of 20 to 40 ft (6 to 12 m) above the quarry, and walk due east (up canyon) to the last exposure of the massive, thickly to very thickly bedded carbonates of the upper Triassic Dun Glen

Formation. The talus of lower Paleozoic Harmony rocks east and uphill of the Dun Glen marks the sole thrust of the Winnemucca thrust system. The described traverse is due west from this point, ending at the barbed-wire fence.

Beds of the Dun Glen show no internal organization, so evidence for their overturned nature is solely the stratigraphic observation that they are overlying the younger, overturned beds of the Winnemucca Formation. In Sonoma Canyon these units are bounded by a thrust fault, but farther to the north it is a gradational contact. Asymmetric ("S" when axes plunge north) open folds of beds (F1) are evident in the Dun Glen at the beginning of the traverse.

The thrust contact with the Winnemucca Formation, easily seen by the truncation of Winnemucca fine clastic beds against the contact, is not folded by these F1 folds. Minor interbeds of sandstones within these fine clastics show cross lamination, which indicates the overturned nature of the unit. As one continued west, carbonate interbeds become more common. The medium to thick carbonate beds show open asymmetric folding (F1) not unlike those folds seen in the Dun Glen.

Careful attention to bedding attitude and facing over the next 600 ft (200 m) of the traverse will reveal a series (2 antiforms and 2 synforms) of tight west-verging asymmetric folds (F2). Axes plunge less than 10° north, hence these are "S" folds. The axial traces of these folds are folded about the major F2 recumbent synform suggesting that these folds are an early F2 phase. No foliation was developed during this deformation.

Within approximately 300 ft (100 m) of the last synformal hinge region of the fold train referred to above, look for near-vertical foliation in exposures of mudstone. This foliation is found in the hinge region of F4 close asymmetric antiformal and synformal folds. These folds are S-asymmetric as their axes plunge north.

Minor F4 folds and some open F1 folds are evident over the next 1,500 ft (500 m) of the traverse. Near the barbed-wire fence at the end of the traverse, minor isoclinal folding (F1) is evidenced by juxtaposition of parallel upright and overturned beds. No foliation is present and it will take a careful eye to spot these folds.

Central window. Figure 3A shows the geology of the central window, which spectacularly exposes the Winnemucca sole thrust as the boundary between the unexposed lower Paleozoic rocks and the white outcrops of the uppermost unit of the Star Peak Group. The traverse described below consists of an orientation stop at Location 1 and a brief transect of the Grass Valley rocks exposed at Location 2. F1, F2, F3, and F4 folds are observable in the window. The visitor should park his/her vehicle at the base of Location 2.

Looking due south from Location 1 at the high ridge two valleys away, one views the geology drawn on cross section C–D (Fig. 3B). Two interesting features are shown by the thick to very thick massive beds of the uppermost unit of the Star Peak Group. These are a horse (small thrust slice) of Star Peak rocks and a deformation gradient in the Star Peak rocks adjacent to the Win-

nemucca sole thrust. The horse is easily recognizable by the discordance of bedding across its faulted lower boundary; beds within the horse are parallel to the thrust and those outside are truncated by the thrust.

The deformation gradient is composed of tight asymmetric folds ("Z" as one looks south) in which apical angles increase and short limb lengths decrease with increasing distance from the thrust. The folds are west verging, and axial surface dip increases systematically from parallel to the thrust adjacent to the fault (recumbent) to more steeply east dipping (inclined) away from the thrust). All folds show significant amounts of hinge thickening. This deformation gradient is modeled as syn-emplacement of the Winnemucca allochthon (F3).

Another interesting structural feature clearly seen from Location 1 is the attitude of the Winnemucca sole thrust. Where the sole thrust crosses the small ridge, immediately to the south of Location 2 at the east edge of the window, the fault is dipping steeply to the east. It shallows in dip as it cuts above the rocks showing the deformation gradient. On the west margin of the window it is dipping shallowly to the west. In part, this antiformal structure is primary; that is, due to the change in the declivity of the sole thrust during emplacement, from a ramp in the east to flat in the central portion of the window. The west-dipping attitude at the western margin indicates that to some extent this structure must also represent post-emplacement deformation (F4 or F5).

Finally, one can trace the axial surface of the major F2 antiform exposed in the central window. It is roughly parallel to the canyon floor and trends south as one looks north, but it trends to the east of the canyon floor as one looks south. Outcrop patterns of the Star Peak Group show the inclined, close, west-verging nature of the fold. Clearly the F2 fold of Triassic rocks is not related (harmonic) to the F4 fold defined by the Winnemucca sole thrust.

Location 2 is a series of exposures of interbedded sandstones and mudstones of the Grass Valley Formation. F1 folds are recognized by the juxtaposition of upright and overturned beds; facing is indicated by load casts, flame structures, and cross-laminations in the sandstones. Hinges of these isoclines are generally faulted, but show much hinge thickening where observable. In marked contrast to the F1 isoclines of the fine clastics of the Winnemucca Formation seen in the range-front window, the folded Grass Valley rocks of the central window show a strongly developed bedding subparallel foliation.

Above and to the west of the Grass Valley rocks is a small horse of Star Peak Group rocks. This block was translated from the east to its present position on the western margin of the central window during the emplacement of the Winnemucca allochthon. Above and to the west of the Star Peak thrust slice is the Winnemucca allochthon; again the rocks are poorly exposed, but the fault is readily visible from the change in vegetation from grasses over the Grass Valley Formation to sage brush above Harmony rocks. The Winnemucca sole thrust dips gently to the west.

DISCUSSION

There are two general end-member models for the development of thrust systems, the piggyback and overstep (Fig. 4; Dahlstrom, 1970; Elliot and Johnson, 1980). In the piggyback model, thrusting jumps in the direction of fault displacement with older thrusts incorporated into the allochthon. In the overstep model, thrusting jumps opposite the sense of fault displacement; hence the sole thrust is the oldest fault of the thrust system. The observations of the trends of the sequential structural fabric elements of the Triassic rocks enumerated in the previous section, coupled with the interpretation that the sole thrust of the Winnemucca thrust system is exposed in the Sonoma Range, yields the interpretation that the Winnemucca thrust system developed in overstep fashion. The arguments for this interpretation are made elsewhere (Stahl and Speed, 1984); a brief treatment follows.

Two assumptions are necessary to develop the argument. First, the style of folding reflects the amount of tectonic overburden during deformation. Folding occurring in rocks under little overburden should show little penetrative deformation. The marker for the degree of penetrative deformation in the Triassic rocks exposed in the Sonoma Range is the degree to which foliation is developed; early generations (F1 to F3) show virtually no foliation where subsequent phases (F4 and F5) have well-developed foliations. The interpretation is that post-Winnemucca allochthon emplacement folding (F4 and F5) was under significant overburden, and preemplacement (F1 and F2) folding was not. Second, axial-plane attitude reflects the relative proximity of thrusting during deformation. Folding near thrusting would have axial planes parallel to the thrust (recumbent) and folding in rocks more distal to thrusting would have axial planes increas-

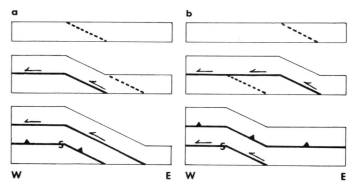

Figure 4. Schematic east-west cross sections for models of thrust system development: a, overstep; b, piggyback. The location of the Sonoma Range indicated by "S," faults by thick lines, active thrusts by arrows, inactive thrusts by sawteeth, and future thrusts by dashed lined.

ingly discordant (inclined to upright) to the attitude of the thrust. The observations in the Triassic rocks of the Sonoma Range are that postemplacement folds (F4 and F5) are steeply inclined (F4) and upright (F5), whereas pre- and synemplacement folds (F1 to F3) are recumbent to shallowly inclined. These observations document the migration of thrusting away from the rocks exposed in the Sonoma Range with time.

The observations may be restated as follows: the sole thrust of the Winnemucca thrust system is exposed in the Sonoma Range and any other thrusting must, by definition, be structurally higher than the sole thrust. Fabric evidence indicates that this thrusting is younger than the sole thrust. In essense, this is the definition of an overstep thrust system.

REFERENCES CITED

Burke, D. B., and Silberling, N. J., 1973, The Auld Lang Syne Group of Late Triassic and Jurassic(?) age, north-central Nevada: U.S. Geological Survey Bulletin 12394-E, 14 p.

Dahlstrom, C.D.A., 1970, Structural geology in the eastern margin of the Canadian Rocky Mountains: Bulletin of Canadian Petroleum Geology, v. 18, p. 332–406.

Elison, M., and Speed, R. C., 1985, Kinematics of the Fencemaker allochthon, East Range, north-central Nevada: Geological Society of America Abstracts with Programs, v. 17, p. 353–354.

Elliot, D., and Johnson, M.R.W., 1980, Structural evolution in the northern part of the Moine Thrust belt, northwest Scotland: Transactions of the Royal Society of Edinburgh, Earth Sciences, v. 71, p. 69–96.

Ferguson, H., Muller, S., and Roberts, R., 1951, Geologic map of the Winnemucca Quadrangle, Nevada: U.S. Geological Survey Map GQ-11, scale 1:125,000.

Gilluly, J., 1967, Geologic map of the Winnemucca Quadrangle, Pershing and Humboldt counties, Nevada: U.S. Geological Survey Map GQ-656, scale 1:62,500.

Hotz, P. E., and Willden, R., 1964, Geology and mineral deposits of the Osgood Mountain Quadrangle, Humboldt County, Nevada: U.S. Geological Survey Professional Paper 431, 128 p.

Nichols, K. M., and Silberling, N. J., 1977, Stratigraphy and depositional history of the Star Peak Group (Triassic), northwestern Nevada: Geological Society

of America Special Paper 178, 73 p.

Roberts, R., 1964, Stratigraphy and structure of the Antler Peak Quadrangle, Humboldt and Lander counties, Nevada: U.S. Geological Survey Professional Paper 459A, 93 p.

Silberling, N. J., 1975, Age relations on the Golconda thrust fault, Sonoma Range, northwestern Nevada: Geological Society of America Special Paper 72, 58 p.

Speed, R., Elison, M., Engeln, J., Stahl, S., Vaverka, E., and Wiens, D., 1982, Mesozoic Winnemucca fold and thrust belt: Geological Society of America Abstracts with Programs, v. 14, p. 236.

Stahl, S. D., and Speed, R. C., 1983, Tectonic implications of pre-Cenozoic structures, Sonoma Range, northern Nevada: Geological Society of America Abstracts with Programs, v. 15, p. 282.

—— , 1984, Overstep thrust system development in the Winnemucca fold and thrust belt, north-central Nevada: Geological Society of America Abstracts with Programs, v. 16, p. 334.

Stewart, J. H., and Suczek, C. A., 1977, Cambrian and latest Precambrian paleogeography and tectonics in the western United States, in Stewart, J. H., Stevens, C. H., and Fritsche, A. E., eds., Paleozoic Paleogeography of the western United States; Pacific Ocean paleogeography symposium 1: Los Angeles, Pacific section, Society of Economic Paleontologists and Mineralogists, p. 337–347.

21

Post–Early Triassic, pre–middle Eocene folds and thrust faults, northern Adobe Range, Nevada

Keith B. Ketner, U.S. Geological Survey, Box 25046, MS 939, Denver Federal Center, Denver, Colorado 80225

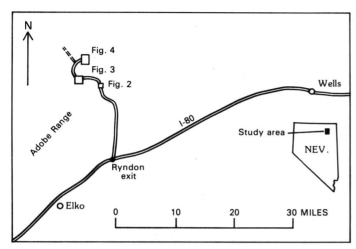

Figure 1. Map showing route to sites in the northern Adobe Range, northeastern Nevada.

LOCATION

The Adobe Range is in northeastern Nevada near the city of Elko (Fig. 1). The northern end of the range can be reached in less than an hour from Elko via I-80 and a gravel and dirt road that extends northward from the Ryndon Exit, 10 mi (16 km) east of Elko. (From the Ryndon Exit, proceed 7 mi (11.2 km) north to intersection; 2.7 mi (4.3 km) northwest to house; then right 3 mi (4.8 km) up Coal Mine Canyon to area shown on Fig. 2.) The areas indicated on Figures 2 through 4 are shown on the Coal Mine Basin 7½-minute Quadrangle, and the geology of the northern Adobe Range is shown in Ketner and Ross (1983). In dry weather, these areas can be reached by two-wheel drive pickup truck or by vans, but four-wheel drive vehicles are recommended. These locations are inaccessible by bus.

SIGNIFICANCE

The northern Adobe Range is significant because major folds and thrust faults there are certainly bracketed between Early Triassic and middle Eocene and are probably of Jurassic or Cretaceous age. Major tectonism of this age in northern Nevada has been described (Coats and Riva, 1983; Ketner, 1984), but most of the emphasis in the older literature is on Paleozoic to Early Triassic tectonism, with recent interest centering on middle-to-late Tertiary extensional tectonics.

In the northern Adobe Range, a thrust plate composed of Silurian and Permian rocks (Plate II) and one composed largely of Triassic rocks (Plate III) were thrust over Mississippian autochthonous or parautochthonous rocks (Plate I); all three were

then folded to form the conspicuous Adobe Range syncline and the less well exposed Garamendi anticline. These structures were overthrust by Plates IV and V, composed mainly of lower Paleozoic rocks of a type such as the Ordovician Vinini Formation, which would customarily be assigned to the Roberts allochthon. The youngest rocks involved in the folding and thrusting are of Early Triassic age. The oldest dated rocks that depositionally cut across, and therefore postdate, the structures are of middle Eocene age.

Coal Mine Canyon. This area (Fig. 2) displays a profound unconformity between the Paleozoic and the Paleogene. The Paleozoic rocks of Plate I are Lower Mississippian sandstones that dip steeply southwest, west, and northwest. The Tertiary sequence here includes a basal conglomerate, the overlying Elko Formation, and a rhyolitic welded tuff, all dipping south-southeast. The Elko Formation, a lake deposit rich in oil shale, has been dated radiometrically near Elko and is Eocene and possibly Oligocene in age. Conglomerate underlying the Elko Formation near Elko is middle Eocene (Solomon and Moore, 1982). Tertiary units exposed at the Coal Mine Canyon site can be traced 6 mi (9.6 km) northeast, where they depositionally cut across the axis of the Adobe Range syncline and several thrusts. The principal structures of the northern Adobe Range must therefore be older than middle Eocene and, allowing time for the deep erosion represented by the unconformity, may be as old as Cretaceous.

Coal Mine Basin. This area (Fig. 3) lies along the Coal Mine Canyon road about 5 mi (8 km) west of the area shown in

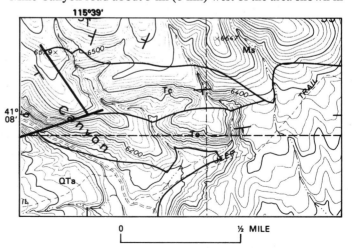

Figure 2. Coal Mine Canyon. Angular unconformity between Lower Mississippian sandstone (Plate I) and Lower Tertiary rocks. QTa, Quaternary and Tertiary alluvium, colluvium; Te, Eocene and Oligocene(?) Elko Formation; Tc, middle Eocene conglomerate; Ms, Lower Mississippian sandstone.

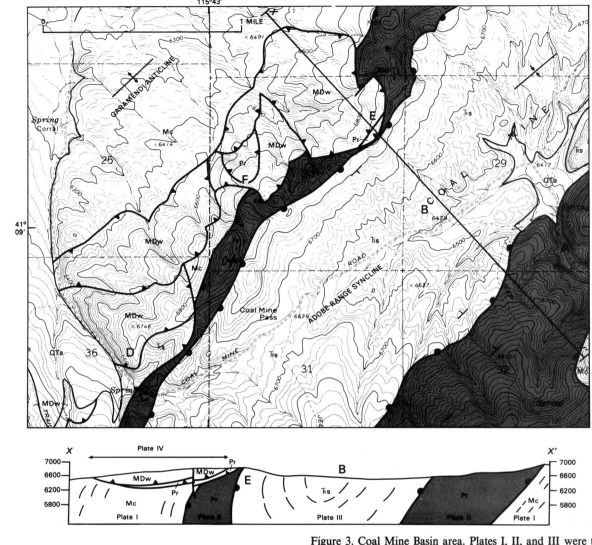

MAP SYMBOLS

———— Contact

———— High angle fault

●——●—— Pre-fold thrust fault,
teeth on upper plate

▲——▲—— Post-fold thrust fault,
teeth on upper plate

↕ Axis of anticline

↕ Axis of syncline

⊥ Generalized attitude of beds

Figure 3. Coal Mine Basin area. Plates I, II, and III were tectonically stacked, folded, and overthrust by Plate IV. QTa, Quaternary and Tertiary alluvium; Ŧs, Triassic shale and limestone; Pr, Permian rocks; Mc, Mississippian Chainman Shale; MDw, Mississippian and Devonian western (siliceous) facies rocks. Plate II is shaded.

Figure 2. It displays the relation of Plate IV to Plates I, II, III, and the folds. The road crosses the northwest-dipping limb of the Adobe Range syncline in Plate II at A, crosses the axis of the syncline in Plate III at B, and crosses vertical beds in the northwest limb of the syncline at C. Here, at C, a tectonically attenuated sequence of Permian beds in Plate II forms a conspicuous outcrop next to the road. Just to the northwest of the outcrop, at D, Plate IV lies on a low-angle thrust across steeply dipping beds of Plates I and II. In the area of Figure 3, this plate is composed mainly of uppermost Devonian and lowermost Mississippian black chert and argillite. Permian and Triassic rocks form the sole of the allochthon at D and E, and a slice of Permian rock is enclosed within it at F.

Garamendi Mine. This abandoned lead mine (Fig. 4) is 2.5 mi (4 km) north of the Coal Mine Basin locality on a branch of the Coal Mine Canyon road. Plate II, composed of Silurian rocks at A and Permian rocks at B, was thrust over the Mississippian Chainman Shale of Plate I. Plates I, II, and III were then folded to form the Adobe Range syncline and the Garamendi anticline, whose axis can be seen at C. Basal Silurian beds are composed of thick-bedded chert that may be black or white, or brilliant shades of red and green. Some beds within the basal Silurian chert sequence near the mine and elsewhere are porous and brecciated owing to solution of syngenetic, stratiform iron, lead, and zinc sulfides (Ketner, 1983).

Badger Spring. Badger Spring (Fig. 4) is located about 3

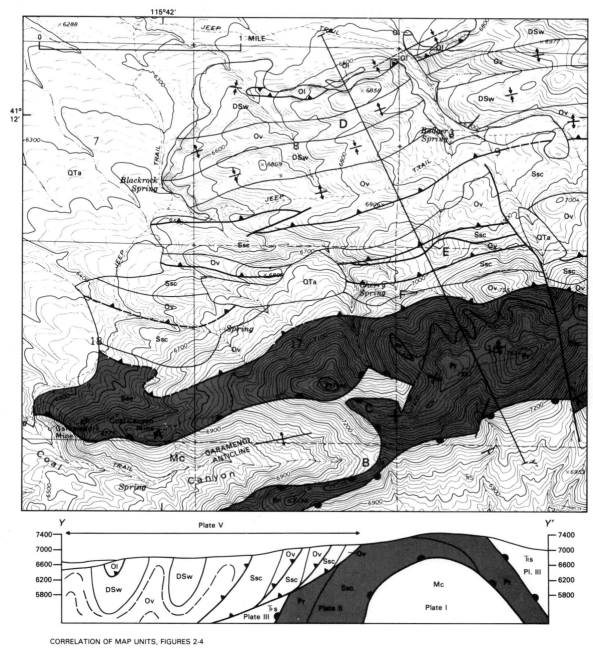

Figure 4. Garamendi Mine and Badger Spring areas. Plates I, II, and III were tectonically stacked, folded and overridden by Plate V. QTa, Quaternary and Tertiary alluvium; Ᵽs, Triassic shale and limestone; Pr, Permian rocks; Mc, Mississippian Chainman Shale; DSw, Devonian and Silurian western (siliceous) facies; Ssc, Silurian siltstone and chert; Ov, Ordovician Vinini Formation; Ol, Ordovician limestone buildup. Plate II is shaded.

mi (4.8 km) northeast of the Garamendi Mine along a branch of the Coal Mine Canyon road. The spring is within Plate V, which here overlies the Garamendi anticline and part of the Adobe Range syncline. The presence of a large number of fossil collections has made it possible to work out the essential features of the stratigraphy and structure. Ordovician, Silurian, and Devonian western (siliceous) facies rocks are here tightly folded along axial planes steeply dipping north-northwest near D and are sliced at intervals by thrust faults steeply dipping north-northwest near E. Upper Middle to lower Upper Ordovician beds in this plate consist of black shale rich in graptolites. The uppermost Ordovician consists of thick-bedded black chert and thin, shaly interbeds with sparse graptolites. Generally, the basal Silurian is thick-bedded, varicolored chert; the Middle Silurian consists mostly of light-colored micaceous siltstone. Silurian beds in most of Plate V, as exposed in a trench and outcrops at F, represent a some-what different facies than those in Plate II, as exposed near the Garamendi Mine. In Plate V the basal chert is relatively thin, discontinuously exposed, and only weakly mineralized in comparison with equivalent beds of Plate II near the mine. Unconformable on the Silurian is an Upper Devonian unit (and sporadically Lower to Middle Devonian erosional remnants). The Devonian is extremely heterogeneous, consisting of chert, shale, sandstone, limestone, and conglomerate.

About 0.4 mi (0.64 km) north of Badger Spring is the axial remnant of a tight syncline composed of lower Middle Ordovician shallow-water shelf limestone overlain unconformably by Lower Silurian graptolite-bearing black chert, and Middle Silurian limy siltstone of slope and basin facies. This unique sequence is separated from the underlying Devonian rocks by a folded thrust and represents an unusual depositional history entirely different from that of other parts of Plate V.

REFERENCES CITED

Coats, R. R., and Riva, J. F., 1983, Overlapping overthrust belts of late Paleozoic and Mesozoic ages, northern Elko County, Nevada, *in* Miller, D. M., Todd, V. R., and Howard, K. A., eds., Tectonic and stratigraphic studies in the eastern Great Basin: Geological Society of America Memoir 157, p. 305–327.

Ketner, K. B., 1983, Strata-bound, silver-bearing iron, lead, and zinc sulfide deposits in Silurian and Ordovician rocks of allochthonous terranes, Nevada and northern Mexico: U.S. Geological Survey Open-File Report 83-792, 6 p.

——, 1984, Recent studies indicate that major structures in northeastern Nevada and the Golconda thrust in north-central Nevada are of Jurassic or Cretaceous age: Geology, v. 12, p. 483–486.

Ketner, K. B., and Ross, R. J., Jr., 1983, Preliminary geologic map of the northern Adobe Range, Elko County, Nevada: U.S. Geological Survey Open-File Map 83-290, scale 1:24,000.

Solomon, B. J., and Moore, S. W., 1982, Geologic map and oil shale deposits of the Elko East quadrangle, Elko County, Nevada: U.S. Geological Survey Miscellaneous Field Studies Map, MF-1421.

Lamoille Canyon nappe in the Ruby Mountains metamorphic core complex, Nevada

Keith A. Howard, U.S. Geological Survey, 345 Middlefield Road, Menlo Park, California 94025

LOCATION AND ACCESSIBILITY

The Ruby Mountains form a high and rugged range in northeastern Nevada. Lamoille Canyon in the Ruby Mountains is an easy half-hour drive by passenger car from the town of Elko (Fig. 1). From Elko, located on I-80, take Nevada 227 southeast for 19 mi (30 km) to a sign marking a right turn onto the Lamoille Canyon road. This paved road leads south into the Ruby Mountains and ascends Lamoille Canyon for 10 mi (16 km) to an elevation of 8,730 ft (2,661 m). The site is best visited in summer, for the canyon is generally snowbound from November to May. The Lamoille Canyon road offers spectacular mountain scenery and a U.S. Forest Service campground. Geologic features are well exposed in roadside outcrops and in glaciated canyon walls 2,000 to 4,000 ft (600 to 1,200 m) high. To enhance your visit, bring a camera, a geologic road log (Snoke and Howard, 1984), a colored 1:125,000-scale geologic map (Howard and others, 1979), and the Lamoille, Nevada, 15-minute topographic quadrangle published by the U.S. Geological Survey.

SIGNIFICANCE

The Ruby Mountains and the adjoining East Humboldt Range form one of more than a score of North American Cordilleran metamorphic core complexes (Crittenden and others, 1980). Like many of the core complexes, the Ruby–East Humboldt Range exposes deep-seated metamorphic rocks surmounted by mylonitic zones and low-angle normal faults. Cordilleran core complexes are commonly considered to expose the middle crust, tectonically denuded by Cenozoic extensional faulting (Hamilton, 1982). Lamoille Canyon cuts through the heart of one of these complexes and offers cross-sectional mountainside views of sillimanite-grade Paleozoic miogeoclinal strata, folded in a large nappe and thoroughly migmatized by two-mica granite. A mylonitic carapace in which stratigraphic units are thinned to 5 percent of their original thickness lies above the migmatitic core.

The formation of metamorphic core complexes is widely held to include metamorphism and plutonism at deep levels followed by younger ductile and brittle tectonic extension that denuded them (Armstrong, 1982). Lamoille Canyon features a record of both processes. The chronology of events in the Ruby Mountains is debated, but a history consistent with available data is the following: Mesozoic westward thrusting was followed later in the Mesozoic by eastward-directed recumbent folding that accompanied metamorphism and migmatization. These Mesozoic events likely relate to convergent shortening and crustal thickening across the Cordillera, but specific tectonic models remain

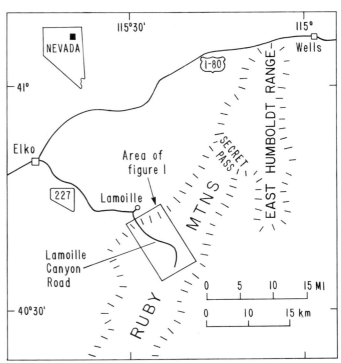

Figure 1. Location map of the Ruby Mountains and Lamoille Canyon.

highly speculative. Cenozoic crustal extension resulted in west-directed mid-Tertiary mylonitic shearing and then Miocene to Quaternary brittle low-angle extensional faulting as the terrane was tectonically unloaded.

Lamoille Canyon presents one of the clearest drive-in views of the interior of a Cordilleran metamorphic core complex, where the style of deformation, metamorphism, and plutonism can be examined from roadside or on foot. This view is a window into the Cordilleran orogen at mid-crustal levels.

GEOLOGY

Lamoille Canyon exposes a migmatite terrane, with half or two-thirds the total volume of bedrock consisting of granite and gneissic granite. The granitic rocks occur in sills, dikes, phacoliths, and irregular bodies intimately injected into a framework of marble, quartzite, and schist. With rare exception, remnants of metasedimentary rocks surrounded by granite are concordant with foliation in the granitic rocks, and the remnants outline relict stratigraphy and structure even in areas where granitic rocks greatly predominate. Relicts of the strata, commonly swimming

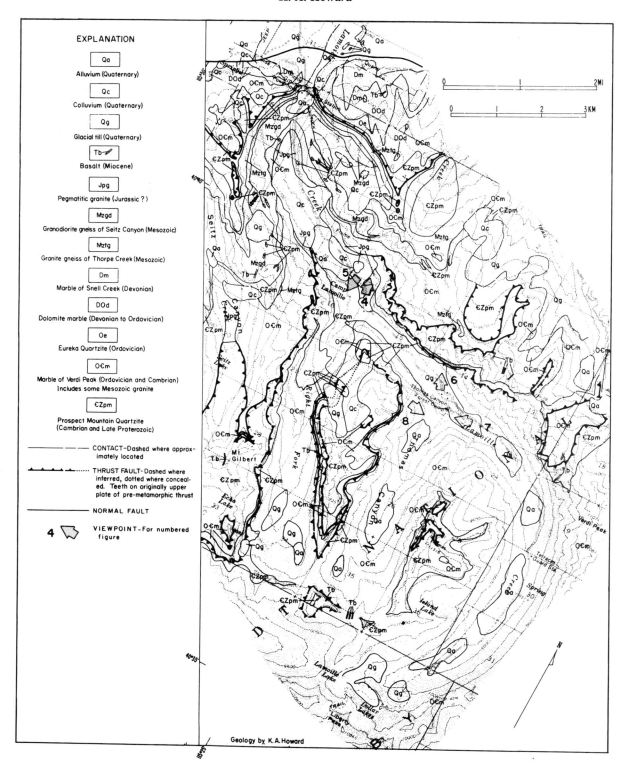

EXPLANATION

Qa
Alluvium (Quaternary)

Qc
Colluvium (Quaternary)

Qg
Glacial till (Quaternary)

Tb
Basalt (Miocene)

Jpg
Pegmatitic granite (Jurassic ?)

Mzgd
Granodiorite gneiss of Seitz Canyon (Mesozoic)

Mztg
Granite gneiss of Thorpe Creek (Mesozoic)

Dm
Marble of Snell Creek (Devonian)

DOd
Dolomite marble (Devonian to Ordovician)

Oe
Eureka Quartzite (Ordovician)

OЄm
Marble of Verdi Peak (Ordovician and Cambrian)
Includes some Mesozoic granite

ЄZpm
Prospect Mountain Quartzite
(Cambrian and Late Proterozoic)

———— ——— CONTACT-Dashed where approximately located

╌╌╌╌ THRUST FAULT-Dashed where
inferred, dotted where concealed. Teeth on originally upper
plate of pre-metamorphic thrust

———————— NORMAL FAULT

4 ⬦ VIEWPOINT-For numbered
figure

Geology by K. A. Howard

Figure 2. Geologic map of Lamoille Canyon, showing numbered viewpoints for the scenes shown in Figures 4–8. The stratigraphic units as mapped include much (commonly 50 to 80 percent) intrusive pegmatitic granite. The Lamoille Canyon Road was paved since the base map was prepared (Lamoille 15-minute Quadrangle, contour interval 80 ft; 24.4 m). Geologic maps of the area are available at 1:24,000 scale (Howard, 1966) and in color at 1:125,000 scale (Howard and others, 1979).

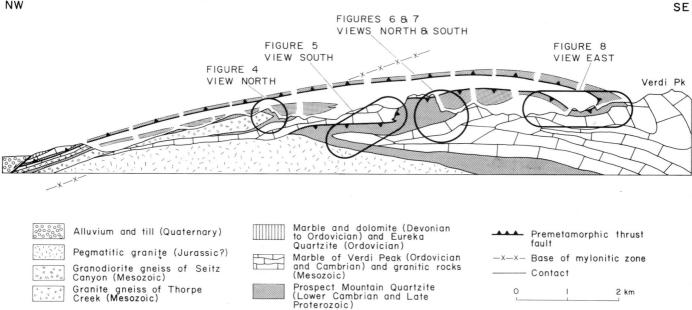

Figure 3. Geologic cross section along the north side of Lamoille Canyon. Granite that pervades the section is shown only where metasedimentary relicts are absent. A large recumbent anticline, the Lamoille Canyon nappe, closes to the right and folds the older inferred Ogilvie thrust fault.

in a sea of gneissic granitic sills and dikes, outline such structures as large recumbent folds with a wondrous variety of vergence directions (Howard, 1980) and prefolding thrust faults (Figs. 2 to 8). A mylonitic overprint on the west flank of the range displays thinned strata and structures.

The stratigraphic age and the chronology of structural and metamorphic events in the Ruby Mountains have long been debated. Geologists of the Fortieth Parallel Survey in the nineteenth century had presumed the terrane was Archean in age. When mid-Tertiary potassium-argon ages were discovered, it became clear that the terrane had experienced geologically young heating and uplift (Armstrong and Hansen, 1966). Detailed mapping demonstrated that the lithologic sequence of metamorphosed strata matches the sequence of unmetamorphosed Paleozoic miogeoclinal formations found elsewhere in eastern Nevada (Howard, 1966, 1971). This match argues strongly for a post-Paleozoic age of metamorphism and migmatization, although Willden and Kistler (1969, 1979) and Kistler and others (1981) alternatively interpret most of the metamorphosed rocks as Precambrian, based mainly on Rb-Sr isotopic data that they interpret as indicating a Cambrian age for some of the orthogneisses.

The metamorphic complex is considered by Snoke and Howard (1984) and by Snoke and Lush (1984) to consist of Paleozoic and latest Proterozoic strata and local older basement, all comingled with pervasive Mesozoic and Tertiary granitoids. A series of geochronologic studies summarized by those authors and by Dallmeyer and others (1986) indicates that most of the prograde metamorphism, recumbent folding, and granites are of Mesozoic age, but that some granite is Tertiary. Early Mesozoic

Figure 4. Photo of the north wall of Lamoille Canyon above Camp Lamoille, showing the nose of a body of orthogneiss, the granodiorite gneiss of Seitz Canyon (Mzgd), that occupies the core of the Lamoille Canyon nappe. Vertical relief in the photo is 2,500 ft (about 750 m). Symbols are the same as in Figure 2.

westward thrusting on the Ogilvie thrust fault before metamorphism and recumbent folding explains a folded duplication of the stratigraphic section.

The migmatitic terrane grades upward into a mylonitic carapace in which structures and strata are extremely attenuated and sheared (Snelson, 1957; Howard, 1966, 1968, 1980; Snoke, 1980). The mylonites dip about 10–30° off the west flank of the Ruby Mountains and form dip slopes and flatirons that appear

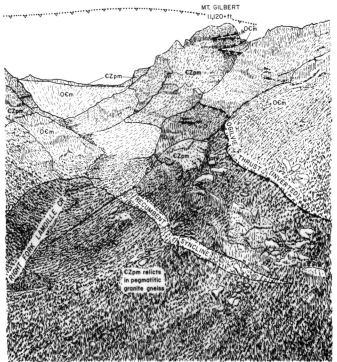

Figure 5. View looking south over Camp Lamoille. Granitic rocks pervade the section. Strata above the recumbent syncline are stratigraphically inverted on the lower limb of the Lamoille Canyon nappe. Symbols are the same as in Figure 2.

prominent as the range is approached. An elongation lineation trends approximately N80°W and characterizes the mylonitic zone for a distance of 50 mi (80 km) north-south. Shear was top to the west in the mylonitic zone as indicated by S-C structures (Snoke and Lush, 1984; Snoke and Howard, 1984) and by noncylindrical disharmonic sheathlike folds analyzed by the Hansen technique (Howard, 1966, 1968, 1980). Although structural studies had suggested that the mylonitization was coeval with Mesozoic nappe formation, the mylonitization is now known to affect Tertiary granite and is considered to represent a deep stage of mid-Tertiary extensional shearing, on which Miocene low-angle faulting was superposed (Snoke and Lush, 1984; Snoke and Howard, 1984).

The low-angle extensional faults, which juxtapose allochthonous upper Paleozoic and Tertiary strata against the metamorphic and mylonitic rocks, are present at structural levels higher than those exposed in Lamoille Canyon. The faults and allochthons can be visited easily 20 mi (32 km) north of Lamoille Canyon at road cuts along Nevada 229 in Secret Pass (Fig. 1), described by Snoke and Howard (1984), Snelson (1957), Snoke (1980), and Snoke and Lush (1984). Late Pleistocene or Holo-

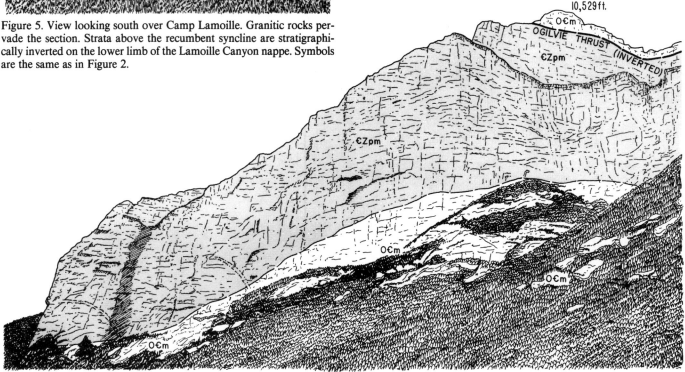

Figure 6. View of the north wall of Lamoille Canyon above the Thomas Canyon campground. The high brown cliff is composed of the Prospect Mountain Quartzite, above intricately folded rusty-weathering graphitic calc-silicate rock at the inverted stratigraphic base of the marble of Verdi Peak. Light-colored pegmatitic granite forms sills in the calc-silicate rock and irregular dikes in the quartzite. For scale, two curved rainbow-shaped dikes of biotite monzogranite in the left part of the cliff near the waterfall stain are each 20 to 25 ft (7 to 8 m) thick. Symbols are the same as in Figure 2. The folded calc-silicate rock at the base of the rainbow dikes is extraordinarily photogenic and is well worth a hike up the mountainside to see.

cene fault scarps along the west side of the range suggest that low-angle normal faulting down to the northwest has continued into modern times. The sinuous scarps parallel the mylonitic map patterns (Howard and others, 1979) and project on seismic reflection profiles into gently dipping listric faults (Effimoff and Pinezich, 1981; Smith, 1984). The Lamoille Canyon road crosses one of these scarps where it offsets late Pleistocene moraines at the range front (Fig. 2; Sharp, 1938, 1939).

Structures in the mylonitic zone are easily visited at the mouth of Lamoille Canyon on the hillsides above the Powerhouse Picnic Grounds where the road enters the Ruby Mountains (Fig. 2). The stratigraphic sequence is partly duplicated by an inferred west-directed prefolding thrust fault (Ogilvie thrust) and intruded by granite, including a sill of the garnet-two-mica granite gneiss of Thorpe Creek, the rock exposed in the first road cut. Stratigraphic units here are thinned to 5 percent of original thickness. Tan micaceous quartzite assigned to the Prospect Mountain Quartzite, of Late Proterozoic and Cambrian age, is a flaggy quartzitic schist in the mylonitic zone. The overlying marble of Verdi Peak, correlated with formations of Cambrian and Ordovician age, consists of siliceous marble, calc-silicate rock, and local pelitic schist. The overlying white Eureka Quartzite of Ordovician age is present locally. Above the Eureka is sugary white dolomite marble correlated with formations of Ordovician to Devonian age; it is succeeded upward by the marble of Snell

Creek, a gray metalimestone correlated with the Devonian Guilmette Formation.

The mylonitic fabrics fade and disappear as one proceeds structurally downward (Fig. 3, line of *x*'s), going up Lamoille Canyon, and are virtually absent along the road beyond about 1 mi (1.6 km) up canyon from the mouth. Instead the rocks beyond have coarsely crystalline, generally annealed textures.

A visitor driving up Lamoille Canyon will see mostly gneissic two-mica pegmatitic leucogranite, commonly containing garnet or sillimanite. Kistler and others (1981) reported that the granites have strontium and oxygen isotopic characteristics similar to their wall rocks. The compositional and isotopic character of the granites suggest they were melted from crustal sources, perhaps including clastic rocks such as the Late Proterozoic McCoy Creek Group, which in nearby ranges underlies the Prospect Mountain Quartzite. In addition to two-mica granite, granodiorite gneiss (Fig. 4) and biotite granite are also common, and locally metagabbros are present. The abundance of granite in this migmatite terrane is particularly high in the deep structural levels encountered along the canyon bottom.

In spite of all the granitoids, a traverse almost anywhere will encounter the metasedimentary relicts that outline the ghost stratigraphy and structure mapped in Figure 2 and shown in section in Figure 3. High on the canyon walls, a visitor can make out brownish cliffs of quartzite assigned to the Prospect Mountain

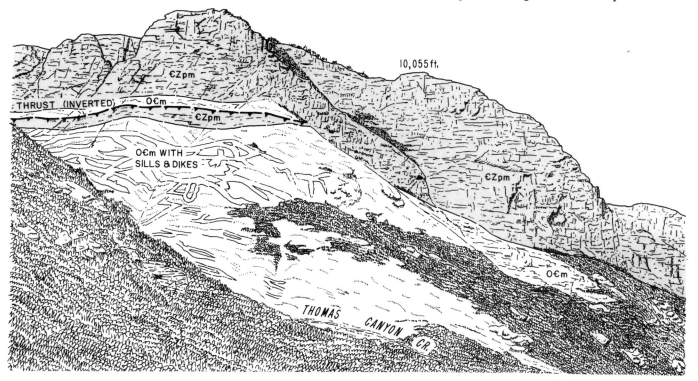

Figure 7. View of the southwest wall of Lamoille Canyon above the Thomas Canyon campground. Prominent white sills and dikes of pegmatitic granite in the lower part of the canyon wall intrude marble and calc-silicate rocks. The section is inverted. Below the high brown cliffs of Prospect Mountain Quartzite is a sliver of the quartzite that is interpreted as an inverted thrust slice on a splay of the Ogilvie thrust fault. Symbols are the same as in Figure 2.

Quartzite (Figs. 4–8). In the marble of Verdi Peak, pods and layers of greenish and rusty calc-silicate rock, light-gray marble, and dark-gray schist comingle with the ubiquitous white granites (Figs. 7 and 8).

The Lamoille Canyon fold nappe is prominently visible from the road. The nappe and parasitic folds and mineral lineations trend north-northeast. Some parasitic folds are coaxially refolded, such as at photogenic exposures to which the visitor is directed by the caption for Figure 6. The nappe is shown in cross section in Figure 3 and in scenes pictured in Figures 4 to 8. The positions of these views are indicated in Figure 2 (arrows) and in Figure 3. All the views except Figure 8 are from the road.

Granodiorite orthogneiss cores the nappe in the scene shown in Figure 4. A view south up Right Fork (Fig. 5) includes the recumbent syncline that underlies the nappe. In the central part of Lamoille Canyon, stratigraphic sections are largely inverted on the lower limb of the nappe (Figs. 6 and 7). Quartzite outlines the nose of the nappe on the northeast wall of Lamoille Canyon as viewed in Figure 8. The nappe where seen in Lamoille Canyon is overturned to the east; nappes elsewhere in the range show all other vergence directions.

Figure 8. Photo looking east at Lamoille Canyon from the mountain shown in Figure 7. Symbols are the same as in Figure 2. Sills of pegmatitic granite form bold white outcrops intermixed with the marble of Verdi Peak. The hinge of the Lamoille Canyon fold nappe is outlined by brown cliffs composed of the Prospect Mountain Quartzite, which noses out eastward where metacarbonate rocks wrap around it. The quartzite nose envelopes a repetition of the stratigraphically higher marble of Verdi Peak in the inferred prefolding lower plate of the Ogilvie thrust. The nose can be viewed from roadside, in the upper part of Lamoille Canyon between the Terraces Picnic Area and the road head, by looking north.

REFERENCES

Armstrong, R. L., 1982, Cordilleran metamorphic core complexes; From Arizona to southern Canada: Annual Review of Earth and Planetary Science, v. 10, p. 129–154.

Armstrong, R. L., and Hansen, E., 1966, Cordilleran infrastructure in the eastern Great Basin: American Journal of Science, v. 264, p. 112–154.

Crittenden, M. D., Jr., Coney, P. J., and Davis, G. H., eds., 1980, Cordilleran metamorphic core complexes: Geological Society of America Memoir 153, 490 p.

Dallmeyer, R. D., Snoke, A. W., and McKee, E. H., 1986, The Mesozoic-Cenozoic tectonothermal evolution of the Ruby Mountains, East Humboldt Range, Nevada; A Cordilleran metamorphic core complex: Tectonics, v. 5, p. 931–954.

Effimoff, I., and Pinezich, A. R., 1981, Tertiary structural development of selected valleys based on seismic data; Basin and Range province, northeastern Nevada: Royal Society of London Philosophical Transactions, Ser. A 300, p. 435–442.

Hamilton, W., 1982, Structural evolution of the Big Maria Mountains, northeastern Riverside County, southeastern California, in Frost, E. G., and Martin, D. L., eds., Mesozoic-Cenozoic tectonic evolution of the Colorado River region, California, Arizona, and Nevada: San Diego, Cordilleran Publishers, p. 1–27.

Howard, K. A., 1966, Structure of the metamorphic rocks of the northern Ruby Mountains, Nevada [Ph.D. thesis]: New Haven, Connecticut, Yale University, 170 p.

——, 1968, Flow direction in triclinic folded rocks: American Journal of Science, v. 266, p. 758–765.

——, 1971, Paleozoic metasediments in the northern Ruby Mountains, Nevada: Geological Society of America Bulletin, v. 82, p. 259–264.

——, 1980, Metamorphic infrastructure in the northern Ruby Mountains, Nevada, in Crittenden, M. D., Jr., Coney, P. J., and Davis, G. H., eds., Cordilleran metamorphic core complexes: Geological Society of America Memoir 153, p. 335–347.

Howard, K. A., Kistler, K. W., Snoke, A. W., and Willden, R., 1979, Geologic map of the Ruby Mountains, Nevada: U.S. Geological Survey Misc. Investigations Series Map I-1136, scale 1:125,000.

Kistler, R. W., Ghent, E. D., and O'Neil, J. R., 1981, Petrogenesis of garnet two-mica granites in the Ruby Mountains, Nevada: Journal of Geophysical Research, v. 86, p. 10591–10606.

Sharp, R. P., 1938, Pleistocene glaciation in the Ruby–East Humboldt Range, northeastern Nevada: Journal of Geomorphology, v. 1, p. 296–323.

——, 1939, Basin-Range structure of the Ruby–East Humboldt Range, northeastern Nevada: Geological Society of America Bulletin, v. 50, p. 881–919.

Smith, J. F., Jr., and Howard, K. A., 1977, Geologic map of the Lee 15-minute quadrangle, Elko County, Nevada: U.S. Geological Survey Map GQ-1393, scale 1:62,500.

Smith, K. A., 1984, Normal faulting in an extensional domain; Constraints from seismic reflection interpretation and modeling [M.S. thesis]: Salt Lake City, University of Utah, 165 p.

Snelson, S., 1957, The geology of the northern Ruby Mountains and the East Humboldt Range, Elko County, Nevada [Ph.D. thesis]: Seattle, University of Washington, 268 p.

Snoke, A. W., 1980, The transition from infrastructure to suprastructure in the northern Ruby Mountains, Nevada, in Crittenden, M. D., Jr., Coney, P. J., and Davis, G. H., eds., Cordilleran metamorphic core complexes: Geological Society of America Memoir 153, p. 287–333.

Snoke, A. W., and Howard, K. A., 1984, Geology of the Ruby Mountains–East Humboldt Range, Nevada; A Cordilleran metamorphic core complex (includes road log), in Lintz, J., Jr., Western geological excursions: Reno, Nevada, Department of Geological Sciences of the Mackay School of Mines, v. 4, p. 260–303.

Snoke, A. W., and Lush, A. P., 1984, Polyphase Mesozoic-Cenozoic deformational history of the northern Ruby–East Humboldt Range, Nevada, in Lintz, J., Jr., Western geological excursions: Reno, Nevada, Department of Geological Sciences of the Mackay School of Mines, v. 4, p. 232–260.

Willden, R., and Kistler, R. W., 1969, Geologic map of the Jiggs quadrangle, Elko County, Nevada: U.S. Geological Survey Map GQ-859, scale 1:62,500.

——, 1979, Precambrian and Paleozoic stratigraphy in central Ruby Mountains, Elko County, Nevada, in Newman, G. W., and Goode, H. D., eds., Basin and Range symposium and Great Basin field conference: Rocky Mountain Association of Geologists–Utah Geological Association, p. 221–243.

The Algodones dunes, Imperial Valley, southern California

Lisa A. Rossbacher, Department of Geological Sciences, California State Polytechnic University, Pomona, California 91768

LOCATION AND ACCESS

The Algodones dunes are located along the eastern edge of the Imperial Valley, south and east of the Salton Sea (Fig. 1). This area is covered by U.S. Geological Survey 1:250,000 topographic maps of El Centro and Salton Sea, California.

The northwestern end of the dune field can be reached easily by passenger car via California 78 east from Brawley or southwest from Blythe. Just west of the town of Glamis, California 78 crosses the dunes. South of the highway, Hugh Osborn County Park offers ample parking, although on weekends it may be filled by trailers carrying off-road vehicles or film crews shooting Saharan scenes. An unpaved desert road (Ted Kipf Road) parallels the eastern edge of the entire dune field; this is generally passable with a passenger car, but changing conditions, including drifting sand, may make four-wheel-drive vehicles advisable or necessary. I-8 crosses the southeastern end of the dune field. The dunes are accessible by foot or dune buggy.

SIGNIFICANCE

The Algodones dunes are one of the largest and most accessible dune fields in the United States, and they offer an excellent analog for the large aeolian landforms that have been observed on the surface of Mars. The origin of the sand has been debated for more than 20 years, but the sediment was probably derived from the Colorado River sand to the southeast. The dune field is also noted for the variety of landform scales, including longitudinal dunes and "megabarchans" that are kilometers in length, stabilized dunes and intradune flats that are measured in meters, and centimeter-scale ripples that occur on the dune surfaces. Unresolved questions about the dune field include interpretation of paleowind directions and an explanation of the linearity and uniform width of the dune chain.

SITE INFORMATION

Early interpretations of the origin of the Algodones dunes were based on the nearby availability of beach sand from Quaternary Lake Cahuilla (Fig. 1), on the southeastern edge of the Salton Sea (Norris and Norris, 1961). Subsequent studies of the mineralogy of the dune sand have indicated a Colorado River source, based on analyses of feldspars, rock fragments, detrital foraminifera, and carbonate material (Merriam, 1969; Van de Kamp, 1973). Although the Colorado River now seems to be the most likely sand source, both present and paleowind directions are poorly understood (Smith, 1978).

The large-scale dune forms in the Algodones chain can be divided into three morphologic sections along its length (Sharp, 1979). The northern third exhibits a series of complex coalesced dome-shaped or barchanoid dunes (Smith, 1978). Sharp (1979) has described these as "disheveled dunes." The southwestern edge

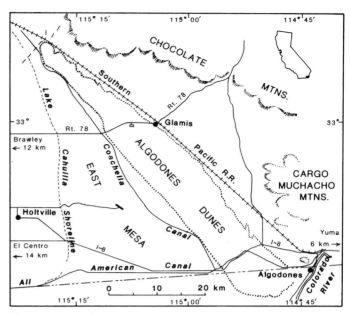

Figure 1. Location map for Algodones Dunes, Imperial Valley, California.

also includes sinuous longitudinal dunes that extend up to 12 mi (20 km) to the southeast. The central part of the dune chain has wide ridges of sand perpendicular to the axis of the chain (Smith, 1978). The southern third contains large, isolated "megabarchan" dunes (Norris and Norris, 1961). All of these large-scale features indicate that the predominant direction of sand drift is toward the southeast (Sharp, 1979).

Intermediate-scale features include barchans and transverse dunes that average 2 to 20 ft (0.5 to 6 m) in height. The southern end of the dune field is characterized by gravel-floored intradune flats between the larger dunes (Norris, 1966; Sharp, 1979). Average rates of migration for these dunes have been measured from 14 to 16 in/yr (35 to 40 cm/yr; Sharp, 1979) to 60 ft/yr (20 m/yr; Smith, 1972). Small-scale aeolian features include small sand ripples (Fig. 2), large granule ripples on the dune surface, and sand streamers extending outward from clumps of vegetation.

Laboratory studies have shown that the dune sands have a 3:1 ratio of potassium feldspar to plagioclase, and they also contain abundant fine-grained rock fragments. Grain sizes are very uniform, with at least 70 percent of the particles in the 0.125–0.25 mm range (Merriam, 1969). Abraded, calcite-filled detrital foraminifera occur in the sands; these fossils are identical to those of the upper Cretaceous Mancos Shale of the Colorado Plateau, which supports a Colorado River origin for the dune sand (Merriam and Bandy, 1965; Merriam, 1969).

Figure 2. Algodones dunes field, showing small-scale ripples in foreground and peak-and-hollow topography. The four small dark objects are dune buggies, and the Cargo Muchacho Mountains are in the background. (View to east-southeast from Hugh Osborn County Park; October 1981.)

The similarity between the Algodones dunes and the large dune fields on the surface of Mars was noted soon after the first close-up images of that planet became available (Cutts and Smith, 1973; Breed, 1977). Sharp (1979) has also pointed out that both the dunes and the intradune flats of the Algodones chain are large enough that they would be visible on satellite images with the same resolution as the Viking images of the Martian surface. Thus, the Algodones dune field is one of the few terrestrial landforms that approximate the Martian features in size. Most terrestrial landforms are an order of magnitude smaller than the analogous features on Mars.

Several important questions about the Algodones dunes remain to be answered. The linearity of the dune field may be controlled by a fault, perhaps a trace of the San Andreas fault, but geophysical studies have only hinted at this (Sharp, 1979). The field also has a remarkably uniform width, ranging from 4 to 7 mi (6 to 12 km) over its 43-mi (70-km) length. This width may be controlled by fault traces, particularly where predune vegetation along a fault could have trapped sand (Sharp, 1979). As already noted, modern and ancient wind regimes are poorly understood, but the field width could also be affected by interaction of winds from varying directions or the "bow-wave" effect of the Cargo Muchacho and Chocolate mountains, 3 to 6 mi (5 to 10 km) to the east, on the prevailing westerly winds (Sharp, 1979).

REFERENCES CITED

Breed, C. S., 1977, Terrestrial analogs of the Hellespontes dunes, Mars: Icarus, v. 30, p. 326–340.

Cutts, J. A., and Smith, R.S.U., 1973, Aeolian deposits and dunes on Mars: Journal of Geophysical Research, v. 78, p. 4139–4154.

Merriam, R., 1969, Source of sand dunes of southeastern California and northwestern Sonora, Mexico: Geological Society of America Bulletin, v. 80, p. 531–534.

Merriam, R., and Bandy, O. L., 1965, Source of upper Cenozoic sediments in Colorado delta region: Journal of Sedimentary Petrology, v. 35, p. 911–916.

Norris, R. M., 1966, Barchan dunes of Imperial County, California: Journal of Geology, v. 74, p. 292–306.

Norris, R. M., and Norris, K. S., 1961, Algodones dunes of southeastern California: Geological society of America Bulletin, v. 72, p. 605–620.

Sharp, R. P., 1979, Intradune flats of the Algodones chain, Imperial Valley, California: Geological Society of America Bulletin, v. 90, p. 908–916.

Smith, R.S.U., 1972, Barchan dunes in a seasonally reversing wind regime, southeastern Imperial County, California: Geological Society of America Abstracts with Programs, v. 4, p. 240–241.

——, 1978, Guide to selected features of the aeolian geomorphology of the Algodones dune chain, Imperial County, California, in Greeley, R., Womer, M. B., Papson, R. P., and Spudis, P. D., eds., Aeolian features of southern California: Washington, D.C., National Aeronautics and Space Administration, p. 74–98.

Van de Kamp, P. C., 1973, Holocene continental sedimentation in the Salton basin, California; A reconnaissance: Geological Society of America Bulletin, v. 84, p. 827–848.

Structure section in Painted Canyon, Mecca Hills, southern California

Arthur G. Sylvester, *Department of Geological Sciences, University of California, Santa Barbara, California 93106*
Robert R. Smith, *Shell Oil Company, Houston, Texas 77001*

LOCATION

The Mecca Hills lie on the northeast margin of the Salton Sea astride the San Andreas fault zone near its southern terminus (Fig. 1). Painted Canyon bisects the Mecca Hills and is reached by driving 5 mi (8 km) eastward from Mecca on State Highway 195, and 0.2 mi (500 m) across and beyond the Coachella Canal, to a graded dirt road which follows a powerline northwestward 2 mi (3 km) to the mouth of the canyon (Fig. 1).

A few logistical remarks are worth emphasizing. Visitors should be aware of the hazards associated with flash floods, washed out roads, soft sand, off-road vehicles, and shooters. Temperatures soar to more than 104° (40° C) in late spring, summer, and early fall. Two plants, the smoke tree and desert holly, are protected by law and should not be disturbed. Rattlesnakes and scorpions are among the endemic fauna. Nearly all aspects of the general geology can be seen adequately from the canyon floor or can be reached by short treks up side canyons. Due caution should be exercised when climbing the friable rocks of the canyon walls. In the courses of many little canyons are abrupt, vertical dry falls, around most of which are no detours.

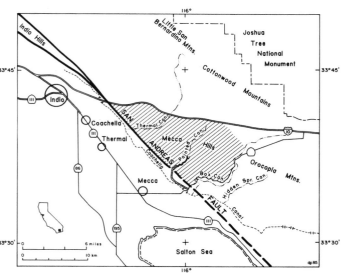

Figure 1. Index map showing the location of the Mecca Hills (ruled pattern) and Painted Canyon relative to local highways and towns.

SIGNIFICANCE

The Mecca Hills are the surficial expression of a "palm tree structure" (Sylvester and Smith, 1976), or "flower structure" in

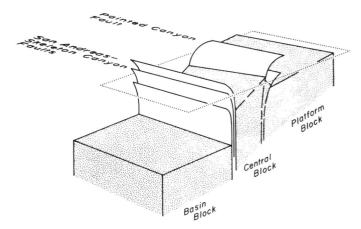

Figure 2. Idealized block diagram of the basement and principal faults in the Painted Canyon part of the Mecca Hills. Dotted parallelogram represents the surface. By permission of American Association of Petroleum Geologists.

the terminology of Wilcox and others (1973). This refers to an arrangement of faults, folds and, in the case of the Mecca Hills, basement blocks, which forms as a result of convergent strike-slip faulting (Fig. 2). It is well exposed and quite accessible because basement rocks are exposed in their structural relations with the overlying sedimentary rocks, and because there is little vegetation and alluvial cover.

STRUCTURAL AND LITHOLOGIC OVERVIEW

Within the part of the Mecca Hills crossed by Painted Canyon, three structural domains or blocks are distinguished by the style and degree of deformation as well as by the type and thickness of late Cenozoic sedimentary rocks; the three are informally designated the platform block, the central block, and the basin block (Fig. 2; Table 1). They are separated from one another by the Painted Canyon and San Andreas-Skeleton Canyon faults which flatten upward, carrying rocks of the central block outward upon the adjacent blocks (Fig. 3). It is this geometrical arrangement of faults relative to the three blocks which led us to designate this a "palm tree structure" (Sylvester and Smith, 1976). The structural and lithologic contrasts among the three domains are summarized in Table 1.

Basement Rocks

The basement comprises two main rock units: (1) the Chuckawalla Complex (Miller, 1944) which is chiefly Precam-

TABLE 1.   LITHOLOGIC AND STRUCTURAL CONTRASTS AMONG THE THREE STRUCTURAL BLOCKS OF THE MECCA HILLS

Basin Block	Central Block	Platform Block
Pre-Cenozoic Basement Rocks		
Not exposed	Highly sheared gneiss and granite of the Chuckawalla Complex.	Moderately sheared to unsheared gneissic and and plutonic rocks of Chuckawalla Complex; Orocopia Schist.
	Basement-sediment surface steeply tilted to the southwest.	Basement-sediment surface gently inclined to to southwest.
Cenozoic Sedimentary Rocks		
Alluvium	Arkose and conglomeratic arkose.	Conglomeratic arkose and conglomerate.
Thickness: 12,000-15,000 ft (3,000- 5,000 m)	Thicker stratigraphic sequence than in eastern block--approximately 5,000 ft (1750 m).	Relatively thin stratigraphic sequence (<2,000 ft or <750 m).
Structure of sediments beneath allivial cover is not known.	Broad, open folds, locally appressed and overturned, with axes oblique to traces of major faults.	Virtually unfolded except for minor drag folds with axes slightly oblique to fault trends.
	Steep, west-trending, normal cross-faults.	Steep-to-gently inclined, northwest-trending normal faults.

brian gneiss, migmatite, and anorthosite and related rocks intruded by Mesozoic plutonic granitic rocks, and (2) the Orocopia Schist which is thought to have been regionally metamorphosed during late Mesozoic time (Ehlig, 1981). The Chuckawalla Complex is thrust upon the Orocopia Schist in the Orocopia Mountains (Crowell, 1975), but in the Mecca Hills the two rock units are separated by the high-angle Platform fault (Figs. 2, 3).

Cenozoic Stratigraphy

Late Tertiary and Quaternary nonmarine sedimentary rocks (Table 2), including intercalated alluvial fan, braided stream, and lacustrine deposits, rest unconformably upon the Precambrian basement. Stratigraphic thicknesses, age relations and correlation of various rock units across faults are not well-known in the area because of numerous depositional discontinuities, abrupt lateral and vertical facies changes, and lack of fossils and distinctive marker beds. The overall nature of the stratified sequence, however, records a period of nonmarine deposition near a tectonically active basin margin. Clast lithology and sedimentary structures show that the sedimentary detritus was derived from the Cottonwood, Little San Bernardino, and Orocopia Mountains to the northeast and east, just as it is today.

The Mecca Formation (Table 2) is the oldest unit of the Tertiary sedimentary sequence. Composed chiefly of angular, dark red-weathering detritus derived locally from the Chuckawalla Complex and Orocopia Schist, it forms a nonconformable blanket 6–15 ft (2–5 m) thick upon the basement pediment northeast of the Painted Canyon fault. It is much thicker and coarser southwest of that fault where the contact with the basement in Painted Canyon is a buttress unconformity.

The Palm Spring Formation (Table 2) marks an abrupt change in provenance in that it was derived almost entirely from a granitic terrane. Its deposition in the Mecca Hills area marks the spreading of alluvial fans from the Cottonwood and Little San Bernardino Mountains across the pediment of the platform block. Like the Mecca Formation, the Palm Spring Formation thickens abruptly across the Painted Canyon fault and is progressively finer-grained basinward. Numerous diastems within the formation southwest of Painted Canyon indicate depositional interruptions reflecting Plio-Pleistocene episodes of folding and faulting at the margin of Salton Trough. Today, in Painted Canyon, the thicker sedimentary sequence of the central block is uplifted relative to that on the platform block. Thus, apparent tectonic inversion implied by this reversal may be resolved by vertical *separation* associated with strike-slip.

STRUCTURAL PROFILE

Basin Block

Geophysical studies by Biehler, Kovach, and Allen (1964) indicate that the depth to basement ranges from 6,000 ft (2,000 m) to as much as 15,000 ft (5,000 m) beneath Coachella Valley. A steep gravity gradient across the San Andreas fault along the southwest edge of the Mecca Hills probably indicates a near-vertical step of the basement of at least 12,000 ft (4,000 m). Thus, the San Andreas fault is the principal structural boundary between the Salton Trough and the high-standing terrane to the northeast in the Mecca Hills.

A thin strip of sedimentary rocks is upturned along the northeast edge of the basin block. Comprising distal fanglomerate

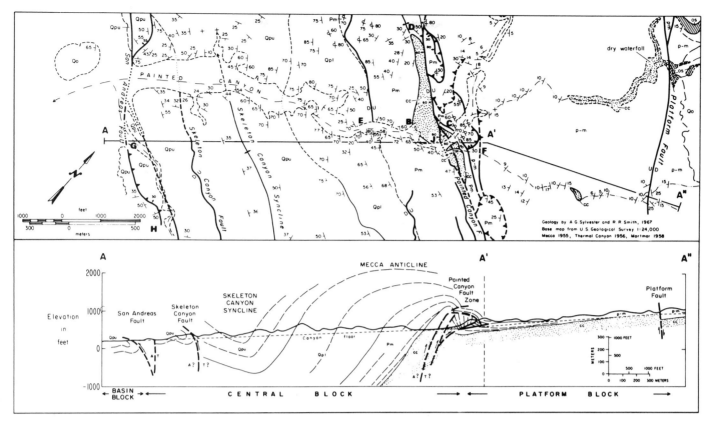

Figure 3. Geologic strip map and structural cross section parallel to Painted Canyon in the Mecca Hills. Letters B–J are localities referred to in the text. Stippled pattern is Chuckawalla Complex basement; ruled pattern is Orocopia Schist.

and lacustrine equivalents of the Palm Spring and younger Ocotillo Formations, the strata are strongly folded, especially southeast of the mouth of Painted Canyon. Here, fold axes trend west-northwest, oblique to the strike of the San Andreas fault, and locally plunge up to 70° in the same direction (locality H in Fig. 3).

Nearly monolithologic beds of Orocopia Schist clasts in the base of the Ocotillo Conglomerate (Pleistocene) show that these strata have been displaced right-laterally at least 15 mi (22 km) from their source in the Orocopia Mountains (Ware, 1958).

San Andreas–Skeleton Canyon Fault Zone

The southwest side of the central block is bounded by a complex zone of faults and folded sedimentary rocks. At the mouth of Painted Canyon, the relatively low structural and topographic relief precludes good exposures of these structures, but they may be studied in Skeleton Canyon, a major tributary marked by low hills of brick-red, phacoid-bearing fault gouge of the San Andreas fault on the southeast side of Painted Canyon (locality G in Fig. 3). The faults are steep in the bottoms of

TABLE 2.   THICKNESSES, AGES AND LITHOLOGY OF CENOZOIC FORMATIONS IN THE MECCA HILLS
(after Dibblee, 1954)

Formation	Lithology
Canebrake-Ocotillo Conglomerate (Pleistocene) 0-5,000 ft (0-750 m).	Gray conglomerate of granitic debris in central Mecca Hills, reddish conglomerate of schist in eastern Mecca Hills.
Palm Spring Formation (Pliocene[?] and Pleistocene) 0-4,800 ft (0-1,200 m).	Upper member: thin-bedded buff arkosic sandstone, grading basinward into light-greenish sandy siltstone. Lower member: thick-bedded buff arkosic conglomerate and arkose with thin interbeds of grey-green siltstone.
Mecca Formation (Pliocene) 0-800 ft (0-225 m).	Reddish arkose, conglomerate, claystone; chiefly metamorphic debris in basal strata.

canyons, and they flatten upward on the sides of the canyon walls. Locally, tight and nearly vertical folds are beneath low-angle segments of the gouge zones, such as at locality H.

The most recently active trace of the San Andreas fault is marked northwest of Painted Canyon by aligned gulches and ridge notches, deflected stream courses, fault gouge, nearly vertical shear surfaces with horizontal slickensides, and en echelon fractures and fault scarps in alluvium (Clark, 1984). Interpretations of several of these features are complementary and consistent, indicating right-slip movement with local vertical uplift.

Central Block

The central block is a 1 mi (1.5 km)-wide, northwest-trending zone of broad, open folds and relatively minor, high-angle faults, bounded by the Painted Canyon and San Andreas faults (Figs. 2, 3). Northwest of Painted Canyon, the axial traces of most folds trend about N70°W and define a step-right en echelon pattern. However, the folds are appressed—overturned in some instances—and trend parallel to, or are truncated by the Painted Canyon and San Andreas faults. The largest and most prominent of these folds is the Mecca anticline which forms the topographically highest part of the hills northwest of Painted Canyon. The slivers of basement exposed along the Painted Canyon fault represent the core, and are structurally the deepest exposures of the anticline.

The lower part of Painted Canyon, in the central block, is a structural depression exposing a thick, nearly flat-lying sequence of sandy siltstone and silty sandstone (upper member of Palm Spring Formation) in the trough of the shallow Skeleton Canyon syncline (Fig. 3). As one proceeds up the canyon, the stratigraphic progression is downsection into increasingly steeper tan sandstone and pebbly sandstone strata with thin, gray beds of micaceous siltstone (lower member of Palm Spring Formation) on the south flank of Mecca anticline.

A small anticline and syncline are prominently exposed in the northwest wall of Painted Canyon at locality E (Fig. 3). They are relatively minor structures and are not shown on the map because they are so small and die out laterally and vertically in very short distances: they do not project across the canyon to the southeast wall and are only gentle flexures in the next canyon to the northwest. These folds and others similar in style and position along the flank of the anticline formed in response to layer-parallel shortening in the fold limb shared by Mecca anticline and Skeleton Canyon syncline.

Painted Canyon probably received its name from the varicolored exposures of basement rocks and overlying Mecca Formation in the central part of the canyon around localities B, J, and C in Figure 3. There, dark migmatitic gneiss, intricately intruded by small, irregularly-shaped bodies of white granitic rocks of Mesozoic age, and light orange and yellow felsite dikes (K-Ar age of about 24 m.y.), is overlain by a very coarse, bouldery facies of dark red-brown-weathering Mecca Formation. The contact is a low-angle buttress unconformity that is best observed on the west wall of the canyon at locality B, where it dips 60° to the southwest. The contact and overlying beds form the core of the northwest-plunging Mecca anticline at locality D. There the northeast limb of the anticline is truncated by the Painted Canyon fault; elsewhere, however, structurally higher parts of the northeast limb are overturned and thrust short distances upon the platform block (locality C, Fig. 3).

In contrast to the relatively unsheared basement in the platform block, the basement in the anticline at locality D and adjacent to the Painted Canyon fault, such as at locality J, is pervasively fractured and sheared into a granulated mass of rock fragments ranging typically from 0.5 cm to 5 cm in diameter. The degree of fracturing is highest next to the fault. The overlying sedimentary rocks, however, are strongly fractured only within a meter or so of the fault surface; the basement-sedimentary rock contact is not a surface of slip. These field observations show that in response to contractile strain, the basement adjusted cataclastically by slip on old fractures and shear surfaces that we assume formed during a long history of pre-Mecca Formation deformation in the San Andreas fault zone; the sedimentary cover responded to deformation at the basement level by folding passively, partly by intergranular slip and partly by flexural slip concentrated in thin mudstone and siltstone beds. This mechanism is analogous to passive warping of a pliable material over a deformed mass of buckshot.

Painted Canyon Fault

The Painted Canyon fault is a major structural discontinuity at least 15 mi (24 km) long and is defined by a zone of crushed rock and fault gouge from a few centimeters to several meters wide. The fault surface dips as steeply as 70° in canyon bottoms, and it flattens upward to nearly horizontal attitudes beneath slabs of the central block that have been carried up to 330 ft (100 m) out, over the southwest edge of the platform block (locality C, Fig. 3). Beneath the low-angle segments, footwall strata of the platform block are dragged abruptly to vertical and overturned attitudes. Right horizontal separation on the Painted Canyon fault is about 2 mi (3 km), whereas the vertical separation of the basement-Mecca Formation contact locally exceeds 490 ft (150 m).

The geometry of Painted Canyon fault and of its associated structures is displayed best in the walls of central Painted Canyon as shown diagrammatically in Figure 4. In general, the structure is a faulted anticline in the hanging wall and an overturned syncline in the footwall, but it is a structure in which the two fault blocks are juxtaposed by a fault having about 2 mi (3 km) of right separation. The contact between the Mecca Formation and the crushed migmatite basement in the footwall is also a buttress unconformity, but it is tilted northeastward almost 90° from its initial gentle southwest dip (Fig. 4, see esp. section A-A'). A sequence of beds in the overturned syncline is buckled between older and younger strata in the way that the pages of a flat-lying book might be shoved and folded between their covers. The buckled beds are bounded by a triangular arrangement of high- and low-angle faults that are best observed in Little Painted Canyon at locality F (Fig. 3). The thrust faults and associated folds are additional manifestations of contraction and uplift of parts of the central block with respect to the platform and basin blocks in transpressional deformation.

Platform Block

The upper part of Painted Canyon is incised into the northeastern structural domain: the platform block. Nearest the Painted Canyon fault, the basement rocks are overlain nonconformably by strata of the Mecca and Palm Spring Formations that are much thinner, and typically composed of coarser and more angular detritus, in this block than in the central block. The contact is a nearly planar, pre-Mecca Formation erosion surface into which channels up to 16 ft (5 m) deep were incised and filled with locally-derived, very coarse and angular Mecca Formation detritus. The erosion surface and overlying strata dip gently southwestward when mapped from canyon to canyon. Except for faulting and minor drag folds adjacent to the faults, however, strata on the platform block are undeformed.

The dry waterfall prevents further access up the canyon by motor vehicle. Near it (Fig. 3) is the best place to observe the relatively undeformed character of the basement and overlying sedimentary rocks, the details of the nonconformable contact, and the geometry of subsidiary faults and associated minor drag folds. The dry waterfall is cut into migmatite of the Chuckawalla Complex that is massive and relatively unfractured in contrast to that in the central block. Smooth and contorted flow folds in the migmatite are products of high temperature and pressure-ductile deformation in Precambrian time. About 660 ft (200 m) up the canyon from the dry waterfall are exposures of anorthosite and related rocks that Crowell and Walker (1962) described and correlated with similar rocks on the west side of the San Andreas fault in the Transverse Ranges. Farther up the canyon, these and other rocks of the Chuckawalla Complex are juxtaposed against the Orocopia Schist by the high-angle Platform fault (Fig. 3). Nearly horizontal slickensides show that the latest movement on that fault was horizontal, but drag folds with nearly horizontal axes indicate that a significant component of vertical displacement has occurred as well.

REFERENCES CITED

Biehler, S., Kovach, R. L., and Allen, C. R., 1964, Geophysical framework of the northern end of the Gulf of California structural province: American Association of Petroleum Geologists Memoir 3, p. 126–143.

Clark, M. M., 1984, Map showing recently active breaks along the San Andreas fault and associated faults between Salton Sea and Whitewater River-Mission Creek, California: U.S. Geological Survey Map I-1483, scale 1:24,000.

Crowell, J. C., 1975, Geologic sketch of the Orocopia Mountains, southeastern California, *in* Crowell, J. C., ed., San Andreas Fault in California—A Guide to San Andreas Fault from Mexico to Carrizo Plain: California Division of

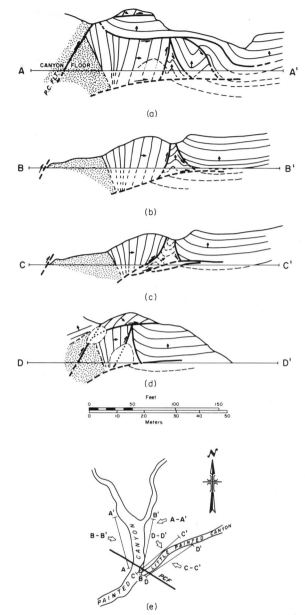

Figure 4. Generalized cross-sections of buckled strata and low- to high-angle faults in the footwall of the Painted Canyon fault. (a) Northwest wall, Painted Canyon; (b) southeast wall, Painted Canyon; (c) northwest wall, Little Painted Canyon; (d) southeast wall, Little Painted Canyon; (e) index map showing locations of cross sections. In (a), (b), (c) and (d) arrows indicate tops of beds. In (e) open arrows indicate locations of view points and view directions for each cross section.

Mines and Geology Special Report 118, p. 99–110.

Crowell, J. C., and Walker, J.W.R., 1962, Anorthosite and related rocks along the San Andreas fault, southern California: University of California Publications in the Geological Sciences, v. 40, no. 4, p. 219–288.

Dibblee, T. W., Jr., 1954, Geology of the Imperial Valley, California: California Division of Mines Bulletin 170, Chapter 2, p. 21–28.

Ehlig, P. L., 1981, Origin and tectonic history of the basement terrane of the San Gabriel Mountains, central Transverse Ranges, *in* Ernst, W. G., ed., The

Geotectonic Development of California, Rubey Volume I: Englewood Cliffs, New Jersey, Prentice-Hall, p. 253–283.

Miller, W. J., 1944, Geology of the Palm Springs-Blythe strip, Riverside County, California: California Journal of Mines and Geology, v. 40, no. 1, p. 11–72.

Sylvester, A. G., and Smith, R. R., 1976, Tectonic transpression and basement-controlled deformation in the San Andreas fault zone, Salton Trough, California: American Association of Petroleum Geologists Bulletin, v. 60, no. 12, p. 2081–2102.

Ware, G. C., 1958, The geology of a portion of the Mecca Hills, Riverside County, California [M.A. thesis]: Los Angeles, University of California, 60 p.

Wilcox, R. E., Harding, T. P., and Seely, D. R., 1973, Basic wrench tectonics: American Association of Petroleum Geologists Bulletin 57, p. 74–96.

The following 1:250,000-scale geologic maps in southern California show the Mecca Hills and their relation to the regional geology:

Jennings, C. W., 1967, Salton Sea sheet; Geologic map of California (Olaf P. Jenkins edition): California Division of Mines and Geology.

Rogers, T. H., 1965, Santa Ana sheet; Geologic map of California (Olaf P. Jenkins edition): California Division of Mines and Geology.

ACKNOWLEDGMENTS

Previous versions of the manuscript were reviewed by R. V. Sharp, J. C. Crowell and G. L. Meyer, and we published other versions of it for the Geological Society of America, the National Association of Geology Teachers, and the American Association of Petroleum Geologists. We thank the latter organization for permission to reproduce Figure 2 and the tables. The Shell Development Company (Houston, Texas) supported the original study.

Blackhawk landslide, southwestern San Bernardino County, California

Ronald L. Shreve, *Department of Earth and Space Sciences and Institute of Geophysics and Planetary Physics, University of California, Los Angeles, California 90024*

LOCATION AND ACCESSIBILITY

The Blackhawk landslide is located in southeastern Lucerne Valley at the southern edge of the Mojave Desert 85 mi (135 km) east of the Los Angeles, California, civic center. It lies across the eastern half of the line separating the Cougar Buttes and Big Bear City, California, U.S. Geological Survey 7½-minute Topographic Quadrangles. Its distal (lower) end is readily apparent on both the map and the ground near 34°25′N, 116°47′W.

The easiest way to reach the landslide is to take I-15 to Victorville, then California 18 east through Apple Valley, approximately 25 mi (40 km) to the center of the town of Lucerne Valley, and finally County Highway 247 (CH247, Fig. 1) east 8.7 mi (13.9 km) to its intersection with Santa Fe Road (SFR, Fig. 1). The steep, gray, 50-ft (15-m) scarp approximately 2,000 ft (600 m) to the southeast is the distal edge of the landslide. It can most conveniently be visited on foot either from the prospectors' road (moderate clearance or four-wheel drive advisable) that enters the highway at Santa Fe Road or from the highway itself about 0.6 mi (1.0 km) farther east. Features along the west side of the landslide can be reached along Blackhawk Canyon Road (BCR, Fig. 1; moderate clearance or four-wheel drive advisable), which crosses the highway 1.05 mi (1.7 km) west of the Santa Fe Road intersection. Features within the landslide lobe can be reached by taking Blackhawk Canyon Road south from the highway 1.05 mi (1.7 km) to the remains of a jeep trail that goes eastward, following it (four-wheel drive strongly advisable) to the dry wash nearest the landslide, and driving up the wash. Alternatively, these features can be visited on foot by hiking from the roads to the west and south. Features of the proximal (upper) part of the landslide can best be reached by taking County Highway 247 east from the Santa Fe Road intersection 3.25 mi (5.2 km) to a well-graded mine road marked by a large white marble boulder, then taking the mine road (low hill on right 4.4 mi (7.1 km) from highway is eastern lateral ridge of Silver Reef landslide) southwest 5.3 mi (8.5 km) to a prospectors' road (intersection about 1,400 ft (425 m) northeast of Round Mountain), next taking the prospectors' road (moderate clearance or four-wheel drive advisable) west approximately 0.25 mi (0.4 km) to where it crosses another similar road, and finally taking the other road northwest 1.6 mi (2.6 km) to the mouth of Blackhawk Canyon (BCM, Fig. 2). Features in the vicinity can be visited either on foot or, in most cases if road conditions permit, by vehicle (moderate to high clearance and low gear or four-wheel drive advisable). Good views of the landslide lobe on the alluvial slope and of its source area on Blackhawk Mountain can be obtained from the mine road (high clearance and low gear or four-wheel drive strongly advisable) that climbs the ridge just east of Blackhawk Canyon (VP, Fig. 2).

An ample supply of fuel and water should always be carried in the area, regardless of season. Only County Highway 247 and Santa Fe Road are regularly maintained. All other roads and trails are unpatrolled and are only sporadically maintained by the local ranchers, miners, and prospectors, so they can, and often do, change drastically in condition or even location.

SIGNIFICANCE OF SITE

The Blackhawk is among the largest landslides known on Earth, although it is small compared to some on Mars. It is the type example for a class of relatively rare large landslides that Shreve (1966, p. 1642; 1968a, p. 37–38) proposed slid on a layer of trapped, compressed air, in order to explain not only their high speed, low friction, and long runout but also their special peculiarities of form and structure, which are nearly all outstandingly exemplified at this site. The mechanism is still controversial, and a variety of alternatives to air-layer lubrication have been proposed: simple unlubricated sliding (McSaveney, 1978, p. 232; Cruden, 1980, p. 299); sliding lubricated not by air but by water-saturated mud (Buss and Heim, 1881, p. 145; Voight and Pariseau, 1978, p. 31; Johnson, 1978, p. 502–503), by frictionally melted ice, snow, or rock (Lucchitta, 1978, p. 1607; Erismann, 1979, p. 34), by frictionally vaporized ice, snow, or ground water (Habib, 1975, p. 194; Goguel, 1977, p. 697–698), or by carbon dioxide from disassociated carbonate rock (Erismann, 1979, p. 34); and "thixotropic" fluidization of the debris by interstitial water, air, or dust (Heim, 1932; Kent, 1966, p. 82; Hsü, 1975, p. 135; Lucchitta, 1979, p. 8111), by intergranular impacts (mechanical fluidization, or inertial grain flow; Heim, 1882, p. 83; Hsü, 1978, 1975, p. 134; Davies, 1982, p. 14), or even by intense sound waves (acoustic fluidization; Melosh, 1983, 1979, p. 7513). The Blackhawk-type landslides and many of these mechanisms have also been invoked, in most cases equally controversially, in connection with large volcanic debris avalanches (Siebert, 1984, p. 180), pyroclastic flows (Sparks, 1976, p. 175), underwater debris flows (Foley and others, 1978, p. 115), Tertiary megabreccias (Brady, 1984, p. 145, 152; Kerr, 1984, p. 239), the cryptic Heart Mountain structure in northeastern Wyoming (Hsü, 1969, p. 945; Kehle, 1970, p. 1649), the Tsiolkovsky Crater ejecta lobe on the Moon (Guest, 1971, p. 99; Guest and Murray, 1969, p. 133), and the huge landslides (as much as 60 mi (100 km) of runout) and rampart craters of Mars (Lucchitta, 1978, 1979; Schultz and Gault, 1979, p. 7681).

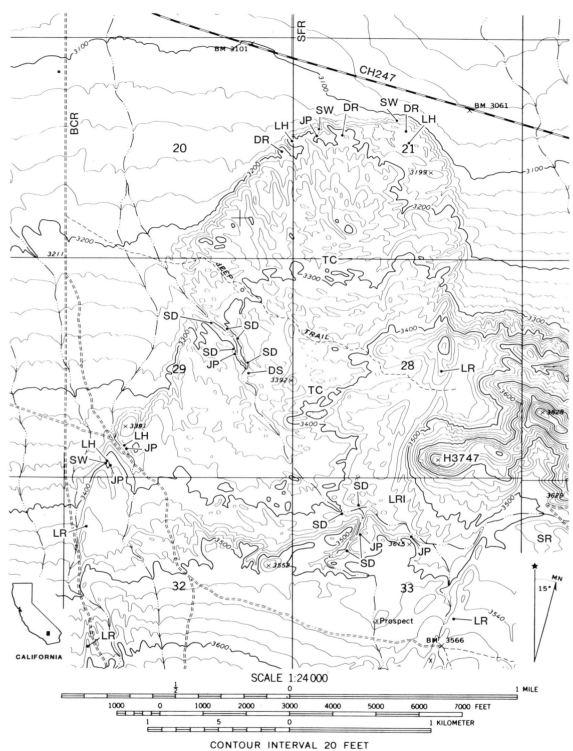

SCALE 1:24 000

CONTOUR INTERVAL 20 FEET

Figure 1. Distal part of Blackhawk landslide. DR, distal rim; DS, dated shells; JP, three-dimensional jigsaw-puzzle structure; LH, local homogeneity of debris; LR, lateral ridge; LRI, interior lateral ridge; SD, crushed-sandstone dike; SW, crushed-sandstone wedge; TC, transverse corrugations. JP, LH, LR, and TC are also present at many unlabelled localities. BCR, Blackhawk Canyon Road; CH247, County Highway 247; H3747, Hill 3747; SFR, Santa Fe Road; SR, Silver Reef landslide. Base map from U.S. Geological Survey Cougar Buttes, California, 7½-minute Topographic Quadrangle, 1971. Contour interval 20 ft (6.1 m).

SITE INFORMATION

The Blackhawk landslide was first recognized by Woodford and Harriss (1928), who described many of its peculiarities of form and structure and suggested that it was a debris outrush like the landslide of 1881 at Elm, Switzerland (Buss and Heim, 1881; Heim, 1882). It was studied in detail by Shreve (1968a), who likened it not only to the Elm landslide but also to the landslide of 1903 from Turtle Mountain at Frank, Alberta, Canada (McConnell and Brock, 1904; Daly and others, 1912), to the landslide of 1964 on the Sherman Glacier near Cordova, Alaska (Shreve, 1966; McSaveney, 1978), and to an adjacent older landslide, the Silver Reef (SR, Figs. 1 and 2; Shreve, 1968a, p. 16, 30, and Plate 2, facing p. 28). It was briefly restudied by Johnson (1978), who confirmed the previous observations and discovered sandstone dikes intruded upward into the distal parts of the landslide lobe, and by Stout (1977, p. 102–103), who obtained a radiocarbon age of 17,400 ±550 yr from fresh-water gastropod and pelecypod shells found in calcareous-mudstone pond deposits on the landslide surface (DS, Fig. 1).

The landslide originated as a huge rockfall from the summit of Blackhawk Mountain, which consists of resistant marble thrust northward over uncemented sandstone and weathered gneiss. It fell 2,000 ft (600 m) vertically and 7,000 ft (2,000 m) northward to the mouth of Blackhawk Canyon (BCM, Fig. 2), where it debouched onto the alluvial slope to form a narrow, symmetrical lobe of nearly monolithologic marble breccia 30 to 100 ft (10 to 30 m) thick (estimated), 1 to 2 mi (2 to 3 km) wide, and 5 mi (8 km) long. Its volume is 1×10^{10} ft^3 (3×10^8 m^3).

The edges of the proximal 3 mi (5 km) of the lobe (that is, the part nearest the source) are bounded by straight, narrow lateral ridges (LR, Figs. 1 and 2), like levees, that rise 50 to 100 ft (15 to 30 m) above the surrounding terrain. In places, the major ridge is accompanied, usually on its interior side, by parallel subsidiary lateral ridges (LRS, Fig. 2). The edge of the distal 2 mi (3 km) of the lobe is bounded by a somewhat sinuous scarp about 50 ft (15 m) high, whose crest generally rises a few feet (2–3 m) above the nearby landslide surface to form a definite distal rim (DR, Fig. 1). The surface of the lobe, where it is not buried by later alluvial fan gravels, is covered with low rounded hills and small closed basins with about 10 to 30 ft (3 to 10 m) of local relief. In the distal 2 mi (3 km) of the lobe these hills and valleys are elongate and form a strong pattern of transverse corrugations (TC, Fig. 1). The higher hills on the landslide lobe probably reflect underlying gneiss knobs that projected above the former alluvial surface. The most prominent of these is Hill 3747 (H3747, Fig. 1) near the eastern edge of the lobe about 2 mi (3 km) from its distal end, which blocked the progress of the landslide across a zone 1,500 ft (500 m) wide. The subparallel ridges on the western and southwestern slopes of the hill doubtless are interior lateral ridges (LRI, Fig. 1) that were successively formed and abandoned as the unimpeded middle region of the landslide adjusted to the arrested motion of the debris ascending the hill.

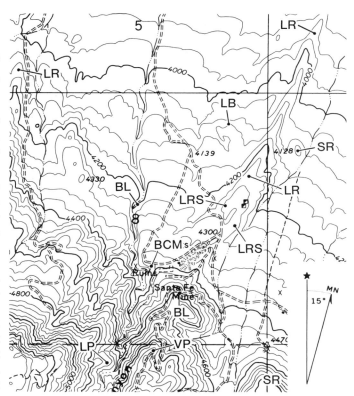

Figure 2. Proximal part of Blackhawk landslide. BL, base of landslide debris; LB, largest block on landslide; LP, launch point; LR, lateral ridge; LRS, subsidiary lateral ridge. BCM, mouth of Blackhawk Canyon; SR, Silver Reef landslide; VP, view point. Base map from U.S. Geological Survey Big Bear City, California, 7½-minute Topographic Quadrangle, 1971, photorevised 1979. Map scale same as in Figure 1; contour interval 40 ft (12.2 m).

The landslide lobe consists almost exclusively of crushed marble. The individual fragments are roughly equant, and range in size from powder to about 10 in (250 mm), the most common (that is, the modal) diameter being approximately 1 in (25 mm). A few exceptionally large blocks of a well-cemented older breccia range up to 35 ft (11 m) in maximum dimension (LB, Fig. 2). Neither the size distribution nor the lithologic characteristics of the fragments varies systematically with position on the landslide lobe. Locally, however, the debris tends to consist predominantly of a certain size of fragment or variety of rock; that is, it displays local homogeneity (LH, Fig. 1). In many places the fragments in roughly lenticular zones up to 20 ft (6 m) across are all pieces of a single source block that are loosely fitted together, giving a distinctive three-dimensional jigsaw-puzzle structure (JP, Fig. 1), in which color bands, for example, continue from fragment to fragment without significant offset.

Along the distal end and western edge of the landslide, the marble debris is nearly everywhere underlain by a wedge of crushed sandstone (SW, Fig. 1) and minor gneiss that was transported more than 4 mi (7 km) down the alluvial slope. The contact between the two generally dips toward the interior of the landslide, normally is marked by up to 6 in (150 mm) of clayey green gouge, and is quite sharp, although in places scattered angular fragments of marble are mixed with the crushed sandstone within about 1 to 2 ft (0.5 m) of the contact. At the distal end of the lobe, several mildly contorted layers of marble debris as much as several feet (a few meters) thick, alternating with thinner layers of crushed sandstone and gneiss in sheared contact, overlie the sandstone wedge and extend at least 150 ft (50 m) toward the interior of the landslide lobe. Crushed-sandstone dikes (SD, Fig. 1) intrude the marble debris in arroyo walls at several localities 1 to 2 mi (2 to 3 km) from the edges of the lobe (Johnson, 1978, p. 492–493), which suggests that the whole landslide may be underlain by a layer of crushed sandstone. Unfortunately, the base of the landslide debris is exposed in very few places (BL, Fig. 2).

No remnants of crushed sandstone are present anywhere on the upper surface of the landslide, however. This implies that, barring a remarkable coincidence, the landslide cannot have crossed the alluvial slope simply as an unusually large debris flow. In debris flows, material at the forward edge necessarily arrives there by way of the upper surface, because the surface material in any relatively wide flow moves faster than the forward edge, eventually overtakes it, and is rolled under. Thus, the presence of crushed sandstone at the distal edge of the Blackhawk landslide unaccompanied by similar material anywhere on its upper surface means that it could not have been a flow in the normal sense. Instead, it must have had essentially a sliding mode of movement.

The landslide overtopped Hill 3747 (H3747, Fig. 1), which would require a minimum speed of 75 mph (35 m/s or 120 km/hr) if accomplished solely by conversion of kinetic energy to potential energy. The actual minimum would of course differ to the extent that the debris ascending the hill was pushed from behind by that descending the alluvial slope (Johnson, 1978, p. 497), which is a process that cannot be treated quantitatively in general terms. Support for the approximate correctness of the conservation-of-energy estimate, however, comes from the Elm and Frank landslides, which are the two historical cases that most closely resemble the Blackhawk. Conservation of energy gives minimum speeds of 65 mph (30 m/s or 105 km/hr) and 110 mph (50 m/s or 175 km/hr) for the two, in reasonable agreement with the average speeds of 110 mph (50 m/s or 175 km/hr) and 90 mph (40 m/s or 145 km/hr) estimated from eyewitness accounts (Heim, 1932, p. 93; McConnell and Brock, 1904, p. 8). The same witnesses also reported that both landslides decelerated from high speeds to a complete halt with remarkable abruptness. Thus, the Blackhawk probably also traveled at a high speed and stopped abruptly.

The assemblage of characteristics just described defines a remarkable genre of landslides that Shreve (1966, p. 1639) termed Blackhawk type. These landslides are exemplified not only by the Blackhawk but also by the Silver Reef, Elm, Frank, and Sherman landslides, and doubtless others. Indeed, had the Elm not been entirely obliterated (Shreve, 1968a, p. 33) by subsequent agricultural rehabilitation, it would have been an even better type example, because it not only had all the same characteristics but also was seen in action; and, as Voight and Pariseau (1978, p. 28) noted, it has precedence.

Not all large, high-speed, long-runout landslides are Blackhawk type; a notable exception is the Huascarán, Peru, landslide of 1970 (Plafker and Ericksen, 1978). Nor does Blackhawk type imply air-layer lubrication, as Voight and Pariseau (1978, p. 32) mistakenly assumed. Rather, it implies an assemblage of characteristics that the air-layer lubrication hypothesis was put forward to explain.

Other landslides of the Blackhawk type have additional characteristics either originally lacking or subsequently lost in the Blackhawk. In the Silver Reef, the Elm, and possibly the Sherman, but not the Blackhawk or Frank, for example, the preserved sequence of lithologic and other characteristics demonstrates that the lower part of the source block became the distal part of the landslide lobe (Heim, 1882, p. 102–103; Shreve, 1966, p. 1641, 1968a, p. 30, 36) and gives further support to the inference that the mode of movement was more akin to sliding than to flowing. A different mode of failure of the source block could explain the lack of this characteristic in the Blackhawk and Frank. The surfaces of the Elm, Frank, and Sherman were also dotted by scattered debris cones, which would long since have disappeared if originally present on the Blackhawk and Silver Reef. These cones consist of finer debris usually piled at the angle of repose atop single large blocks (Heim, 1882, p. 101, 104; McConnell and Brock, 1904, p. 9; Shreve, 1966, p. 1642). About a third of the cones on the Sherman are xenolithologic debris cones, in which both the cone and its underlying block have identical peculiarities not present in the surrounding surface debris, such as an uncommon rock type or distinctive quartz veining (Shreve, 1966, p. 1642).

A striking feature of the Sherman landslide, which sets it apart from the Blackhawk, Silver Reef, Elm, and Frank landslides, is the pattern of hundreds of parallel shallow V-shaped longitudinal grooves that covers almost its entire surface (Shreve, 1966, p. 1641). Although some of the more prominent grooves were formed by shear between substreams of the debris, the vast majority were not, and are probably the result of lateral spreading. Similar longitudinal grooves are present on many contemporaneous landslides in the vicinity, as well as on the Martian landslides (Lucchitta, 1978, p. 1602–1603; 1979, p. 8098), on the ejecta blankets of certain Martian rampart craters (Mouginis-Mark, 1979, p. 8013), and on the ejecta lobe associated with Tsiolkovsky Crater on the Moon (Guest, 1971, p. 99). All these grooved features seem to have formed in the presence of strong ground shaking. The cause was a magnitude-8.5 earthquake in the case of the Sherman and its contemporaries; possibly earthquake or impact in the Martian landslides, as indicated by widespread synchroneity (Lucchitta, 1979, p. 8106); and almost certainly impact in the case of the crater ejecta.

According to the hypothesis of air-layer lubrication proposed by Shreve (1966, p. 1642, 1968a, p. 37–38), landslides of the Blackhawk type start as huge rockfalls which acquire so much momentum in their fall that at a projecting rib of rock or abrupt steepening of slope they leave the ground, overriding and trapping a cushion of compressed air, upon which they slide with little friction. The Elm and Frank definitely left the ground and overrode substantial volumes of air (Buss and Heim, 1881, p. 145; McConnell and Brock, 1904, p. 8), as hypothesized; and the Blackhawk and Sherman (Shreve, 1966, p. 1642) almost certainly did so, the launch point for the Blackhawk being a projecting spur ridge about 2,000 ft (600 m) upstream of the mouth of Blackhawk Canyon (LP, Fig. 2; see Johnson, 1978, p. 501, for photograph taken from the source area).

The high speed, the long runout, the local homogeneity of the debris, the distal wedge of crushed sandstone bulldozed from sources near the proximal end of the landslide, and, where present, the preserved sequence of lithologic and other characteristics result because the air layer is so easily sheared. The dikes (also present in the Sherman landslide; Shreve, 1966, p. 1641) originate from a relatively thin basal layer of lower permeability crushed sandstone (or, in the other cases, mud or snow) that necessarily had to be present in order for the pressure drop in the leaking air to be sufficient to counterbalance the weight of the loose fragments at the underside of the debris sheet (Shreve, 1968b, p. 655–656). The lateral ridges form where leakage allows the sides of the sliding sheet to fall and stop, forming levees that stand higher than the thinner debris that arrives later. The subsidiary lateral ridges form where air escapes along the interior side of an existing lateral ridge or ridges; and the interior lateral ridges form where it escapes at shear zones between sublobes of debris. The original ridge in some cases may not be the main one, because it can be smaller than a later one, and hence be subsidiary (Fig. 2). The debris cones form where large blocks, carrying on their tops finer debris from within the landslide, emerge from the moving debris as it spreads and thins, so that its surface settles around them (Heim, 1882, p. 101). They are xenolithologic where the debris at depth is lithologically homogeneous and differs from that at the surface.

The abrupt stop occurs and the distal rim and scarp form when the air layer becomes spread so thin that the leading edge of the sheet of debris hits the ground, slows rapidly, and causes the debris behind to pile up. The pattern of transverse corrugations reflects an imbricate internal structure that forms as the zone of impact propagates rearward up the landslide lobe, in the process destroying any lontigudinal grooves present. However, when the initial impact occurs toward the proximal end of the sheet of debris, the zone of impact propagates forward down the landslide lobe, in which case the stop, though still quick, is somewhat less abrupt, no distal rim or transverse corrugations form (unless the zone of impact meets a second one propagating rearward), and longitudinal grooves are preserved. The three-dimensional jigsaw-puzzle structure forms where the impact shatters large blocks into fragments that remain loosely fitted together because of the near lack of further movement.

The hypotheses advanced as alternatives to air-layer lubrication are motivated by doubts, predominantly intuitive, as to whether the initial speed of the Blackhawk was sufficient for the necessary launch (Johnson, 1978, p. 502; Voight and Pariseau, 1978, p. 31), whether sufficient air can be trapped (Johnson, 1978, p. 502; Voight and Pariseau, 1978, p. 30–31; McSaveney, 1978, p. 235; Erismann, 1979, p. 30, 31), whether it can be retained sufficiently long (Erismann, 1979, p. 31), and whether debris supported by a layer of compressed air is sufficiently stable (Bishop, 1973, p. 360–361), among other concerns. They are also motivated by purported long-runout landslides on the Moon (Hsü, 1975, p. 129, 132, 1978, p. 88, 91–92; Erismann, 1979, p. 32; Davies, 1982, p. 10, 11; Melosh, 1983, p. 158), which has almost certainly always lacked an atmosphere (and water), and by the Sherman-like landslides on Mars (Lucchitta, 1979, p. 8110; Davies, 1982, p. 11; Melosh, 1983, p. 158) which, except locally during impacts, probably always lacked sufficient atmosphere (Lucchitta, 1979, p. 8110), although not definitely (Schultz and Gault, 1979, p. 7681; Mouginis-Mark, 1979, p. 8020).

Actually, only two lunar candidates are known. One is the ejecta lobe at Tsiolkovsky Crater (Guest, 1971, p. 98–100), which formed at the same time as the crater (Guest and Murray, 1969, p. 133; Guest, 1971, p. 102), so that high energy (Lucchitta, 1978, p. 1605) and possibly released gas (Guest, 1971, p. 99; Sparks, 1976, p. 184) from the impact could have been involved. The other is the light-mantle deposit at the Apollo-17 site (Howard, 1973), which definitely is not Blackhawk type in form and probably is not even a landslide (Lucchitta, 1977). No landslide forms have been seen on Mercury (Lucchitta, 1979, p. 8097). Thus, the Martian examples are the only known apparently Blackhawk-type landslides from airless or nearly airless environments; and even they are open to debate.

Not surprisingly, inasmuch as they are intended to explain the same things, the alternative hypotheses have much in common with air-layer lubrication. They can be classified into three groups: unlubricated sliding, lubricated sliding, and "thixotropic" fluidization. Proponents of the first group believe that the apparent friction is not in fact unusually low or that it does not need to be. In order to account for the abrupt stop, they have to assume, with Heim (1882, p. 106), that friction increases considerably with decreased speed. Proponents of the second group, which includes air-layer lubrication, agree as to the necessity for a lubricant but differ as to its nature or source. Those of the third group insist, on the other hand, that the motion is flow, not sliding, but state (or imply) that the fluidized debris has special (as yet unconfirmed) rheologic properties such that a relatively thin basal layer shears much more rapidly than the rest, producing a result closely akin to sliding.

Virtually all the proposed processes are likely to occur in large landslides somewhere and sometime; and in many cases, if not most, some of them undoubtedly occur simultaneously. The question, therefore, is not whether they occur but what is their relative importance in specific classes of landslides, such as those of the Blackhawk type. Although eyewitness reports, as at Elm and Frank, have produced much essential information, direct observation—especially by trained observers—is severely limited by the rarity and unpredictability of all large landslides. Observation of experimentally induced landslides is possible, albeit not very practical for the Blackhawk type, because large size seems to be a prerequisite for long runout. More promising are experimental studies aimed at elucidating and quantifying the proposed processes, which would provide fundamental information on which to base theoretical calculations or computer simulations of the resultant landslides. Equally important are further field investigation and interpretation of the details of form and structure of existing Blackhawk-type landslides, including the Blackhawk itself as the premier example of the genre.

REFERENCES CITED

Bishop, A. W., 1973, The stability of tips and spoil heaps: Quarterly Journal of Engineering Geology, v. 6, p. 335–376.

Brady, R. H., 1984, Neogene stratigraphy of the Avawatz Mountains between the Garlock and Death Valley fault zones, southern Death Valley, California; Implications as to late Cenozoic tectonism: Sedimentary Geology, v. 38, p. 127–157.

Buss, E., and Heim, A., 1881, Der Bersturz von Elm den 11. September 1881: Zurich, Wurster, 163 p.

Cruden, D. M., 1980, The anatomy of landslides: Canadian Geotechnical Journal, v. 17, p. 295–300.

Daly, R. A., Miller, W. G., and Rice, G. S., 1912, Report of the commission appointed to investigate Turtle Mountain, Frank, Alberta: Geological Survey of Canada Memoir 27, 34 p.

Davies, T.R.H., 1982, Spreading of rock avalanche debris by mechanical fluidization: Rock Mechanics, v. 15, p. 9–24.

Erismann, T. H., 1979, Mechanisms of large landslides: Rock Mechanics, v. 12, p. 15–46.

Foley, M. G., Vessell, R. K., Davies, D. K., and Bonis, S. B., 1978, Bed-load transport mechanisms during flash floods, *in* Conference on flash floods; Hydrometeorological aspects: Boston, American Meteorological Society, p. 109–116.

Goguel, J., 1977, Scale-dependent rockslide mechanisms, with emphasis on role of pore-fluid vaporization, *in* Voight, B., ed., Rockslides and avalanches, 1; Natural phenomena: Amsterdam, Elsevier Scientific Publishing Co., p. 693–705.

Guest, J. E., 1971, Geology of the farside crater Tsiolkovsky, *in* Fielder, G., ed., Geology and physics of the Moon: Amsterdam, Elsevier Scientific Publishing Company, p. 93–103.

Guest, J. E., and Murray, J. B., 1969, Nature and origin of Tsiolkovsky Crater, lunar farside: Planetary and Space Science, v. 17, p. 121–141.

Habib, P., 1975, Production of gaseous pore pressure during rock slides: Rock Mechanics, v. 7, p. 193–197.

Heim, A., 1882, Der Bergsturz von Elm: Zeitschrift der Deutschen Geologischen Gesellschaft, v. 34, p. 74–115.

—— , 1932, Bergsturz und Menschenleben: Zurich, Fretz und Wasmuth, 218 p.

Howard, K. A., 1973, Avalanche mode of motion; Implications from lunar examples: Science, v. 180, p. 1052–1055.

Hsü, K. J., 1969, Role of cohesive strength in mechanics of overthrust faulting and of landsliding: Geological Society of America Bulletin, v. 80, p. 927–952 (reprinted 1977 *in* Voight, B., ed., Mechanics of thrust faults and décollement [Benchmark Papers in Geology, v. 32]: Stroudsburg, Pennsylvania, Dowden, Hutchinson, and Ross, p. 314–339.

—— , 1975, Catastrophic debris streams (sturzstroms) generated by rockfalls: Geological Society of America Bulletin, v. 86, p. 129–140.

—— , 1978, Albert Heim; Observations on landslides and relevance to modern interpretations, *in* Voight, B., ed., Rockslides and avalanches, 1; Natural phenomena: Amsterdam, Elsevier Scientific Publishing Company, p. 71–93.

Johnson, B., 1978, Blackhawk landslide, California, U.S.A., *in* Voight, B., ed., Rockslides and avalanches, 1: Natural phenomena: Amsterdam, Elsevier Scientific Publishing Company, p. 481–504.

Kehle, R. O., 1970, Analysis of gravity sliding and orogenic translation: Geological Society of America Bulletin, v. 81, p. 1641–1663 (reprinted 1977 *in*

Voight, B., ed., 1977, Mechanics of thrust faults and décollement [Benchmark Papers in Geology, v. 32]: Stroudsburg, Pennsylvania, Dowden, Hutchinson, and Ross, p. 412–434.

Kent, P. E., 1966, The transport mechanism in catastrophic rock falls: Journal of Geology, v. 74, p. 79–83.

Kerr, D. R., 1984, Early Neogene continental sedimentation in the Vallecito and Fish Creek Mountains, western Salton Trough, California: Sedimentary Geology, v. 38, p. 217–246.

Lucchitta, B. K., 1977, Crater clusters and light mantle of the Apollo 17 site; A result of secondary impact from Tycho: Icarus, v. 30, p. 80–96.

—— , 1978, Large landslide on Mars: Geological Society of America Bulletin, v. 89, p. 1601–1609.

—— , 1979, Landslides in Valles Marineris, Mars: Journal of Geophysical Research, v. 84, p. 8097–8113.

McConnell, R. G., and Brock, R. W., 1904, The great landslide at Frank, Alberta: Canada Department of the Interior Annual Report, 1902–1903, pt. 8, Report of the Superintendent of Mines, Appendix, 17 p. (also Canada Parliament Sessional Papers, v. 38, no. 10, Sessional Paper no. 25, pt. 8, Report of the Superintendent of Mines, Appendix, 17 p.)

McSaveney, M. J., 1978, Sherman Glacier rock avalanche, Alaska, U.S.A., *in* Voight, B., ed., Rockslides and avalanches, 1: Natural phenomena: Amsterdam, Elsevier Scientific Publishing Company, p. 197–258.

Melosh, H. J., 1979, Acoustic fluidization; A new geologic process: Journal of Geophysical Research, v. 84, p. 7513–7520.

—— , 1983, Acoustic fluidization: American Scientist, v. 71, p. 158–165.

Mouginis-Mark, P., 1979, Martian fluidized crater morphology; Variations with crater size, latitude, and target material: Journal of Geophysical Research, v. 84, p. 8001–8022.

Plafker, G., and Ericksen, G. E., 1978, Nevados Huascarán avalanches, Peru, *in* Voight, B., ed., Rockslides and avalanches, 1: Natural phenomena: Amsterdam, Elsevier Scientific Publishing Company, p. 277–314.

Schultz, P. H., and Gault, D. E., 1979, Atmospheric effects on Martian ejecta emplacement: Journal of Geophysical Research, v. 84, p. 7669–7687.

Shreve, R. L., 1966, Sherman landslide, Alaska: Science, v. 154, p. 1639–1643.

—— , 1968a, The Blackhawk landslide: Geological Society of America Special Paper 108, 47 p.

—— , 1968b, Leakage and fluidization in air-layer lubricated avalanches: Geological Society of America Bulletin, v. 79, p. 653–658.

Siebert, L., 1984, Large volcanic debris avalanches; Characteristics of source areas, deposits, and associated eruptions: Journal of Volcanology and Geothermal Research, v. 22, p. 163–197.

Sparks, R.S.J., 1976, Grain-size variations in ignimbrites and implications for transport of pyroclastic flows: Sedimentology, v. 23, p. 147–188.

Stout, M. L., 1977, Radiocarbon dating of landslides in southern California: California Geology, v. 30, p. 99–105.

Voight, B., and Pariseau, W. G., 1978, Rockslides and avalanches; An introduction, *in* Voight, B., ed., Rockslides and avalanches, 1: Natural phenomena: Amsterdam, Elsevier Scientific Publishing Company, p. 1–67.

Woodford, A. O., and Harriss, T. F., 1928, Geology of Blackhawk Canyon, San Bernardino Mountains, California: University of California Publications in the Geological Sciences, v. 17, p. 265–304.

Relationship of the Jurassic volcanic arc to backarc stratigraphy, Cowhole Mountains, San Bernardino County, California

John E. Marzolf and Robert D. Cole, Department of Geology, Southern Illinois University, Carbondale, Illinois 62901

LOCATION

Geologic relationships observable at this site are displayed in Paleozoic sedimentary and Mesozoic sedimentary and volcanic rocks preserved beneath a cap of Jurassic rhyolite on a ridge herein referred to as Rhyolite Ridge. Rhyolite Ridge lies at the northeastern corner of the Cowhole Mountains approximately 13 mi (21 km) south of Baker, California, comprising portions of sections 12 and 13 of T12N,R9E and 7 and 18 of T12N,R10E, San Bernardino Principle Point. Rhyolite Ridge (not labeled) can be found in the southwest corner of the Seventeenmile Point Quadrangle, 7½-minute Topographic Series of the U.S. Geological Survey. The ridge may be reached by taking the paved Kelbaker Road south from the four-way stop in central Baker or by turning south from either of the central Baker off-ramps from I-15. Figure 1 indicates mileage between points mentioned in the following text.

Proceed as follows: From the four-way stop in central Baker (arrow), proceed southeast on the paved Kelbaker Road to the County Refuse-Disposal Road. Turn right (south). Proceed until road turns toward refuse disposal. Take the road that bears to the right. After following the road around toward the southeast and passing through a gate, a road will fork back to the left. Continue to the right (south). Continue past crossroad, along the edge of Soda Dry Lake, to intersection at corral (tank). Turn left (southeast). Continue past fork in road. Bear to right. Continue through gate in fence. Bear left at next road to a triangular flat area bounded by three roads and designated by a signpost as Fox Hills Camp.

Although the site may be reached by passenger car at most times of the year, a high ground-clearance vehicle is desirable. After the initial draft of this paper was written (May 1986), the roads were severely damaged by a summer storm, making high ground-clearance vehicles a necessity and four-wheel drive advisable. Check on road conditions at the San Bernardino County Sheriff's substation in Baker. If the site is visited in the summertime, the daytime temperature will probably be over 100°F (~38°C), so take ample water along.

SIGNIFICANCE

Rhyolite Ridge displays, in almost 100 percent exposure, the development of the Jurassic Cordilleran volcanic arc on a deformed Paleozoic carbonate terrain. A similar carbonate and volcanic terrain, extending from the Big Maria Mountains (Hamilton, 1982) to the southern Inyo Mountains (Oborne and others, 1983), is the source of coarse clastics in Moenave/Kayenta-equivalent strata in southern Nevada, thus indicating the Early

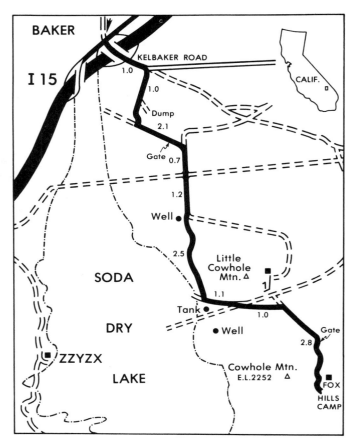

Figure 1. Location map: Baker to Rhyolite Ridge (near Cowhole Mountain). Distances in miles. Mileage begins at four-way stop in central Baker (arrow).

Jurassic initiation of the arc. The unconformity beneath the coarse clastics extends well into the Colorado Plateau, separating overlying Sinnemurian to Pliensbachian strata from underlying strata of Norian age. In the Cowhole Mountains, the unconformity is angular and is overlain by Mesozoic volcanics filling topographic lows eroded in the carbonate terrain (Marzolf, 1983a). The Aztec Sandstone overlapped the Moenave/Kayenta strata to rest on volcanics and bury the carbonate topographic highs. The volcanics, interbedded with and overlying the Aztec Sandstone, document the relationship of the Nugget-Navajo eolian sand sea to the volcanic arc (Marzolf, 1982, 1983b). Subsequently, the Aztec Sandstone and volcanics were homoclinally folded and eroded. Renewed silicic volcanism produced the rhyolite flow and silicic plugs that unconformably overlie and com-

Post-Aztec volcanic facies

Extrusive and
intrusive rhyolites

Rhyolite flow and
associated plug

Aztec Sandstone and
syndepositional volcanics

Aztec Sandstone

Andesite to
rhyodacite

Paleozoic
carbonates

Fault

Thrust-Fault
- Upper Plate

Attitude of
Sedimentary or
Volcanic Rocks

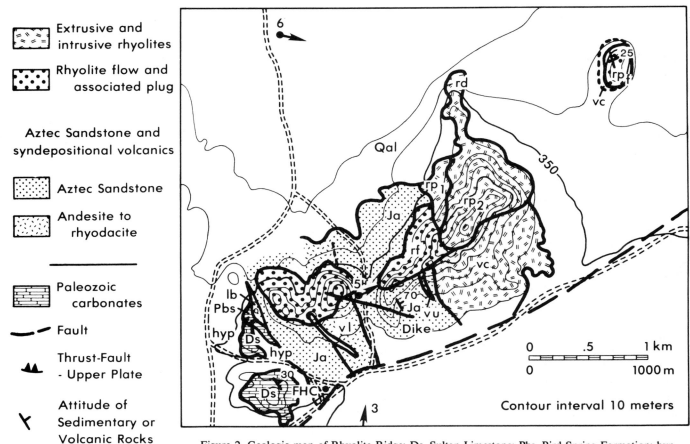

Figure 2. Geologic map of Rhyolite Ridge: Ds, Sultan Limestone; Pbs, Bird Spring Formation; hyp, hypabyssal mafic rocks; 1b, limestone breccia; vl, lower volcanics; vu, upper volcanics; vc, capping volcanics; rf, rhyolite flow; rp_1, flow facies of brecciated vent; rp_2, vent breccia; rd, radial dike; rp_f, columnar-jointed rhyolite plug, main source of rhyolite flow; Qal, quaternary alluvium and wind-blown sand; FHC, Fox Hills Camp. Fault has left-lateral separation. Arrows with numbers indicate figure numbers and direction of view.

pose the eastern end of Rhyolite Ridge, respectively. The unconformity probably correlates with the J1-J2 unconformities of Peterson and Pipiringos (1979) which separate the Middle Jurassic Temple Cap and Carmel formations from the underlying Navajo Sandstone on the Colorado Plateau. Bentonites in both the Temple Cap and Carmel formations (Schultz and Wright, 1962; Marvin and others, 1965) and silicic volcanic clasts in the Carmel Formation were derived from this or a similar terrain.

SITE INFORMATION

When you arrive at Fox Hills Camp, FHC on the geologic map of Rhyolite Ridge (Fig. 2), you will be standing on the lower red bed section of the Aztec Sandstone. As you stand looking north at Rhyolite Ridge, your view is similar to that in Figure 3, a photograph taken from higher elevation approximately 2,300 ft (700 m) behind you. At the west end of the ridge, the Early

Jurassic(?) Cowhole Mountain thrust places the Pennsylvanian to Permian Bird Spring Formation on the Devonian Sultan Limestone (Novitsky-Evans, 1978). The Aztec Sandstone, homoclinally dipping 70° to the east, overlies the Paleozoic carbonates that dip approximately 30° to the east. At this location, the contact relationships between the Aztec Sandstone and the Sultan Limestone and Cowhole Mountain thrust are somewhat obscured by mafic hypabyssal rocks, which have invaded the contact and thrust as well as the carbonate rocks. At the southern end of the mountains, the Aztec Sandstone depositionally crosscuts the Cowhole Mountain thrust, which places the Bird Spring Formation on the Goodsprings Dolomite (Novitsky-Evans, 1978).

The basal Aztec Sandstone is marked by a lenticular limestone breccia containing angular cobbles and boulders of the Bird Spring Formation (Fig. 2, 1b). The limestone blocks contain fusulinids and productid brachiopods. The breccia is lenticular and

a

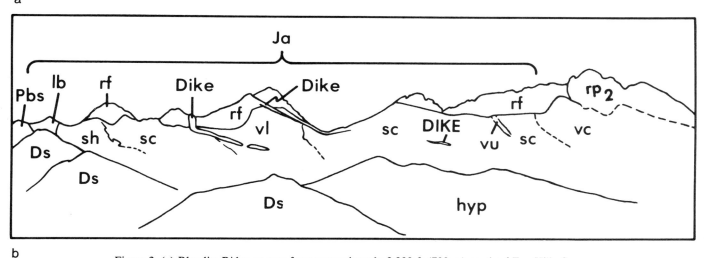

b

Figure 3. (a) Rhyolite Ridge as seen from approximately 2,300 ft (700 m) south of Fox Hills Camp. Base of the Aztec Sandstone is to the far left; volcanics overlying the Aztec Sandstone are to the right. (b) Geologic sketch of (a): Ds, Sultan Limestone; Pbs, Bird Spring Formation; lb, limestone breccia; sh, horizontally stratified sandstone; sc, crossbedded sandstone; vl, lower volcanics; vu, upper volcanics; vc, capping volcanics; Ja, Aztec Sandstone; rf, rhyolite flow; rp_2, rhyolitic vent breccia, and hyp, hypabyssal mafic rocks.

fills a broad channel cut in the deformed carbonate terrain. The breccia is overlain by 200 ft (60 m) of dark reddish-brown volcanic lithlutites and very fine grained volcanic litharenites, which are horizontally stratified or cross-stratified on a scale of a few inches (Fig. 3, sh). Lenses of very pale orange to grayish orange, very fine-grained to fine-grained volcanic sublitharenite to quartz arenite, cross-stratified on a large scale (3 to 10 ft; 1 to 3 m), are enclosed in the dark reddish brown siltstones and sandstones. The siltstone and sandstone composite set is overlain by a coset of cross-stratified, yellowish gray quartz arenite to subarkose (Fig. 3, sc). The trough and wedge planar sets are cross-stratified on a scale of 3 to 10 ft (1 to 3 m). The cross-stratified set is overlain by a horizontally stratified coset of similar lithology containing dispersed coarse sand-sized to granule-sized volcanic clasts.

The purple-gray slopes overlying the lower Aztec sandstones are developed on the lower of two volcanic units interbedded with the Aztec Sandstone at Rhyolite Ridge (Fig. 3, vl). The lower contact is cut diagonally by an aplite dike of later Mesozoic or Cenozoic age. Because of low grade metamorphism and weathering, the rock types composing the lower volcanic unit, which is 226 ft (69 m) thick, are difficult to decipher but include flows of porphyritic pyroxene-plagioclase andesite and dacite, welded tuffs, and possibly flow breccias and debris flows of similar composition. Cross-stratified Aztec sandstone immediately overlying the volcanics contains pebbles and, in places, cobbles of the volcanic rocks, demonstrating the extrusive, rather than intrusive, origin of these igneous rocks.

The section overlying the lower volcanic unit comprises 1,300 ft (400 m) of fine-grained yellow-brown, quartz arenite containing large-scale trough and wedge-planar cross-stratification consisting of sets 6 to 16 ft (2 to 5 m) in amplitude (Fig. 3, sc). Volcaniclastic, horizontally stratified sandstone increases in

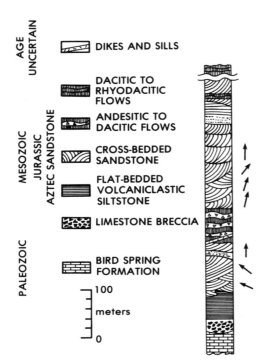

a

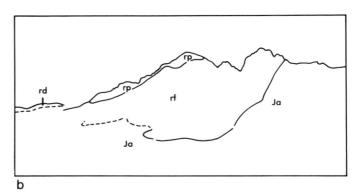

b

Figure 4. Measured stratigraphic section of the Aztec Sandstone at Rhyolite Ridge showing relationship to underlying Bird Spring Formation and overlying volcanics. Arrows indicate vector-resultant of crossbed-dip direction. For orientation of each resultant, north, rather than east, is to top of section.

Figure 5. (a) View looking east at contact between Aztec Sandstone and rhyolite flow. Cross-sectional shape of contact geometry is shown in (b). Note: Aztec Sandstone (Ja) underlies the rhyolite (rf) across the entire exposure. The rhyolite flowed toward the viewer; the Aztec Sandstone dips away from the viewer at approximately 70°. The rhyolitic vent breccia and associated flow facies (rp) and a radial dike (rd) can be seen in the background.

proportion toward the top of the sandstone section. About 140 ft (43 m) below the top, a single porphyritic rhyodacitic flow, 40 ft (12 m) thick, pinches out to the south (Fig. 3, vu).

Cross-bed orientation in the Aztec Sandstone in the Cowhole Mountains, as elsewhere from southwestern Utah to southeastern California, rotates clockwise from bottom to top of the section. Unlike most other localities where the rotation is from southeast to southwest, rotation in the Cowhole Mountains is from northwest to northeast, suggesting that the Cowhole Mountain block has been rotated tectonically approximately 180° about a vertical axis.

The contact of the sandstone with the overlying volcanic rocks (Fig. 3, vc) is not well exposed on Rhyolite Ridge but appears to be conformable. According to Novitsky-Evans (1978), a short distance to the south, the volcanic rocks overlying the Aztec Sandstone comprise at least 1,900 ft (575 m) of porphyritic dacitic, latitic, and rhyolitic flows and ash-flow tuffs; however, a presently unknown thickness of the volcanics assigned by her to the volcanic sequence conformably overlying the Aztec Sandstone includes at least one radial dike and flow facies of a brecciated plug that intrudes the volcanic sequence. Like the Aztec Sandstone, conformably overlying volcanics dip homoclinally 70° to the east. A stratigraphic section of the Aztec Sandstone measured at Rhyolite Ridge is illustrated in Figure 4.

Rhyolite Ridge is capped by an aphanitic, intensively flow-banded rhyolite (Fig. 3, rf). The contact is highly discordant. Viewed from the south, the line of contact is irregular but, on average, plunges to the west at approximately 5°, whereas the underlying sandstone and volcanic beds dip 70° to the east. Toward the east end of the ridge, near the top of the Aztec Sandstone, the contact dips steeply to the north and flattens down dip. Toward the west end of the ridge, the contact dips gently north. At three locations, the west end of Rhyolite Ridge and the east and west sides of the pass through the ridge, the Aztec Sandstone can be seen to underlie the rhyolite flow (Fig. 5). These relationships may be interpreted in two ways. The contact of the rhyolite flow with the Aztec Sandstone may represent the walls and floor of a canyon, deeper and steeper walled on the east and shallower and broader on the west, down which the rhyolite flowed. Alternatively, the contact may be in part intrusive, representing upward-expanding vent plugs and coalesced rhyolite domes aligned along an east-west fissure. According to the latter interpretation, the differing geometry of the contact is the result of the intersection of the contact by the present topographic surface that has not intersected any of the small vent plugs. On the south side of Rhyolite Ridge where the rhyolite flow (rf) is in contact with

a

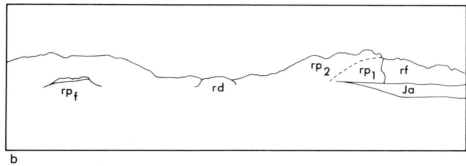

b

Figure 6. (a) View from north side of Rhyolite Ridge looking east. (b) Geologic sketch of (a): Ja, Aztec Sandstone; rp_2, rhyolitic vent breccia; rp_1, flow facies associated with rp_2; rd, radial dike; rp_f, columnar-jointed rhyolite plug, the main source of the rhyolite flow that caps Rhyolite Ridge.

the lower volcanic unit (vl) of the Aztec Sandstone, concentrically flow-banded cylinders, 6 to 10 ft (2 to 3 m) in diameter and inclined at a high angle to the base of the flow, possibly indicate the presence of such small vents. Although not mutually exclusive, which of these two interpretations is more nearly correct has yet to be resolved. In either case, the rhyolite flowed from east to west (present coordinates). The concentrically flow-banded cylinders, which now plunge east, opposite the direction of flow, plunged at a high angle in the direction of flow at time of emplacement.

The highest topographic point on Rhyolite Ridge is composed of a silicic vent breccia of probable rhyolitic composition (rp_2, Figs. 2 and 3). The rhyolitic breccia not only intrudes the Aztec Sandstone and overlying concordant volcanics but also clearly intrudes the rhyolite (rf) that caps Rhyolite Ridge to the west. The climb to the top of the ridge to observe this contact is worth the effort. The location of this vent breccia (rp_2), centered along the long axis of the rhyolite body (rf) it intrudes, supports the interpretation of a fissure-eruption origin of the rhyolite (rf) capping Rhyolite Ridge. Radial dikes extend both northward and southward from the vent breccia. A single distinct radial dike (rd) extending from the north side of the breccia body is illustrated in Figure 6. The vent breccia and radial dikes are surrounded by porphyritic rhyolitic flows (rp_1) that represent flow facies of the intrusive phases. The breccia and associated flows truncate the rhyolite flow capping Rhyolite Ridge, separating it from a columnar-jointed rhyolite plug (rp_f) lying along the projected axis

of the rhyolite (rf) to the east-northeast (Figs. 2 and 6). Field relationships and preliminary petrographic and geochemical data strongly suggest the columnar-jointed plug was the main source of the rhyolite (rf) now capping Rhyolite Ridge (Marzolf and Cole, in preparation). The columnar joints in the plug now plunge west at 25°. Restoration of the columns to the vertical implies the rhyolite (rf) now capping Rhyolite Ridge flowed down a surface dipping at approximately 30°. At this time, the Aztec Sandstone and interbedded and conformably overlying volcanics dipped at approximately 45°. Tectonic rotation of 180°, suggested by cross-bed orientation in the Aztec Sandstone, implies the rhyolite flowed eastward over eroded Aztec Sandstone dipping 45° to the west. The timing of this postulated rotation and the tilting that has brought the entire sequence to its present attitude, are unknown but probably are related to Tertiary tectonic events.

SUMMARY OF IMPORTANT RELATIONSHIPS

1. Paleozoic carbonates were deformed and thrust faulted prior to deposition of the Aztec Sandstone.

2. The Aztec Sandstone was deposited on a surface of significant relief developed on Paleozoic carbonates, cobbles and boulders of which were incorporated into a basal breccia of the sandstone.

3. The Aztec Sandstone is interbedded with and conformably overlain by volcanics of varied genesis and composition.

4. The Aztec Sandstone and associated volcanics were deformed and eroded prior to intrusion and extrusion of rhyolitic vent and flow facies.

5. Regional relationships and the presence of clasts of rhyolitic flows and tuffs and beds of altered tuff in the late Bathonian to Callovian Carmel Formation of southwestern Utah suggest that the rhyolitic volcanics in the Cowhole Mountains are approximately of the same age, and that the unconformity between these volcanics and the Aztec Sandstone correlates with the J1 and J2 unconformities of the Colorado Plateau.

REFERENCES

Hamilton, W., 1982, Structural evolution of the Big Maria Mountains, northeastern Riverside County, California, *in* Frost, E. G., and Martin, D. L., eds., Mesozoic-Cenozoic Tectonic Evolution of the Colorado River Region, California, Arizona, and Nevada; San Diego, California, Cordilleran Publishers, p. 1–28.

Marvin, R. F., Wright, J. C., and Walthall, F. G., 1965, K-Ar and Rb-Sr ages of biotite from the Middle Jurassic part of the Carmel Formation, Utah: U.S. Geological Survey Professional Paper 525-B, p. B104–B107.

Marzolf, J. E., 1982, Paleogeographic implications of the Early Jurassic (?) Navajo and Aztec Sandstones, *in* Frost, E. G., and Martin, D. L., eds., Mesozoic-Cenozoic Tectonic Evolution of the Colorado River Region, California, Arizona, and Nevada; San Diego, California, Cordilleran Publishers, p. 493–501.

—— , 1983a, Early Mesozoic eolian transition from cratonal margin to orogenic-volcanic arc, *in* Marzolf, J. E. and Dunne, G. C., eds., Evolution of early Mesozoic tectonostratigraphic environments—southwestern Colorado Plateau to southern Inyo Mountains: Utah Geological and Mineral Survey Special Studies 60, Guidebook Part 2, Annual Meeting of the Rocky Mountain and Cordilleran Sections of the Geological Society of America, Salt Lake City, Utah, p. 39–46.

—— , 1983b, Changing wind and hydrologic regimes during deposition of the Early Jurassic (?) Navajo and Aztec Sandstones, southwestern United States, *in* Brookfield, M. E. and Ahlbrandt, T. S., eds., Eolian Sediments and Processes: Developments in Sedimentology, v. 38, Amsterdam, Elsevier Science Publishing Company, p. 583–612.

Novitsky-Evans, J. M., 1978, Geology of the Cowhole Mountains, southeastern California; structural, stratigraphic, and geochemical studies [Ph.D. thesis]: Houston, Texas, Rice University, 95 p.

Oborne, Mark, Fritsche, A. E., and Dunne, G. C., 1983, Stratigraphic analysis of Middle (?) Triassic marine-to-continental rocks, southern Inyo Mountains, east-central California, *in* Marzolf, J. E. and Dunne, G. C., eds., Evolution of early Mesozoic tectonostratigraphic environments—southwestern Colorado Plateau to southern Inyo Mountains: Utah Geological and Mineral Survey Special Studies 60, Guidebook Part 2, Annual Meeting of the Rocky Mountain and Cordilleran Sections of the Geological Society of America, Salt Lake City, Utah, p. 54–57.

Peterson, F., and Pipiringos, G. N., 1979, Stratigraphic relations of the Navajo Sandstone to Middle Jurassic Formations, southern Utah and northern Arizona: U.S. Geological Survey Professional Paper 1035F, 43 p.

Schultz, L. G., and Wright, J. C., 1962, Bentonite beds of unusual composition in the Carmel Formation, southwest Utah: U.S. Geological Survey Professional Paper 450-E, p. E67–E72.

ACKNOWLEDGMENTS

We would like to thank E. I. Smith for insightful comments on the interpretation of the volcanic rocks, Mason Hill for guidance in preparation of the manuscript and figures, colleagues J. Zimmerman and R. H. Fifarek for critical reading of same, and the Southern Illinois University Department of Geology and Office of Research Development and Administration for financial assistance in support of field work.

Tertiary extensional features, Death Valley region, eastern California

Bennie W. Troxel, *University of California, Davis, California 95616*
Lauren A. Wright, *Pennsylvania State University, University Park, Pennsylvania 16803*

INTRODUCTION

The southeastern part of the Death Valley region (Fig. 1) displays two remarkable structural features: turtlebacks (Curry, 1938), and the Amargosa chaos (Noble, 1941). The changing ideas during the past half-century about the origin of these features reflect the growth of understanding of the major aspects of Basin and Range tectonics.

Although these features were initially believed to be related to thrust faulting, a consensus now exists that they are different aspects of widespread Tertiary extension associated with the development of the Basin and Range province. The evidence upon which this historical debate is based is discussed in the site descriptions presented herein.

SITE 27. AMARGOSA CHAOS
B. W. Troxel
L. A. Wright

LOCATION AND ACCESS

From Shoshone, California, a small community at the intersection of California 127 and 178, follow California 127/178 north one mile; then turn west on California 178 about 15 mi (18 km) to the edge of the area of concern, hereafter called the Virgin Spring area. All stops are on California 178 (Fig. 2). Shoshone is about 60 mi (96 km) north of Baker, California, a community situated on I-15 between Las Vegas and Los Angeles.

This text is abstracted from Wright and Troxel (1984). It, as well as sections of two guidebooks (Troxel, 1974 and 1982, various pages), are useful supplements to this guide.

SIGNIFICANCE

Noble (1941) observed a style of faulting in the subject area so intricate and complex that he referred to the faulted rock units as "chaos." He referred to these, as well as other similarly faulted terranes in the Death Valley region, as the "Amargosa chaos." He selected the Virgin Spring area in the west-central part of the Black Mountains (Fig. 1) as the type locality for the Amargosa chaos.

Noble (1941) interpreted the terrane of the Virgin Spring area as broadly divisible into three lithologic-structural units (Fig. 2): (1) an autochthonous, relatively intact basement complex composed mostly of Precambrian quartzo-feldspathic metamorphic rocks and containing subordinate intrusive bodies variously of Precambrian, Mesozoic (?), and Tertiary age; (1) a

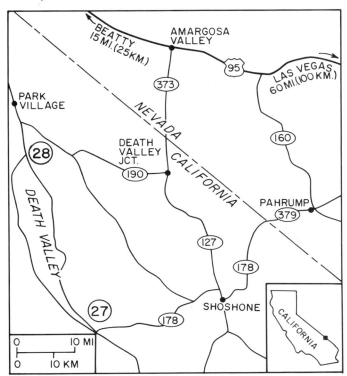

Figure 1. Map showing access to Amargosa chaos (Site 27) and turtlebacks (Site 28), Death Valley, southeastern California.

composite allochthonous plate, comprising the various elements of the Amargosa chaos, much more deformed than the underlying complex, and composed largely of nested fault blocks of later Precambrian sedimentary rocks and diabase, Cambrian sedimentary rocks, and Tertiary volcanic, plutonic, and sedimentary rocks; and (3) an autochthonous cover of Late Cenozoic fanglomerate, basalt, and alluvium.

Noble (1941) interpreted a faulted contact between the basement rocks and the overlying Amargosa chaos as a regional thrust fault and the dominant structural feature of the Virgin Spring area. He named it the "Amargosa thrust" (Fig. 2). As Tertiary volcanic and sedimentary rocks are involved in the chaos, he held that most or all of the movement on the proposed thrust occurred in Tertiary time. Noble later questioned the regional thrust concept.

SITE INFORMATION

Noble (1941) recognized three phases of the Amargosa chaos and named them the Virgin Spring, Calico, and Jubilee

121

phases (Fig. 2). The Virgin Spring phase is composed almost entirely of units of the Pahrump Group (Fig. 3) and of the overlying latest Precambrian and Cambrian units. The Calico phase consists mostly of Tertiary volcanic units. The Jubilee phase comprises Tertiary conglomerate, finer-grained strata, and bodies of monolithologic breccia. He visualized the Virgin Spring phase as emplaced first and the Calico and Jubilee phases as moving over the Virgin Spring phase and semi-independently of it. Noble (1941) noted that much of the Calico phase "is intricately broken up by faulting, but not entirely chaotic" (p. 970) and that the Jubilee phase "presents a more confused picture than the other two phases" (p. 972).

The contact between the basement complex and the overlying Virgin Spring phase of the Amargosa chaos dips southwestward in some places and northeastward in others, delineating southwest-plunging antiforms and synforms. The largest antiform is termed the "Desert Hound anticline." The Malpais Hill syncline, Graham anticline, and Rhodes anticline appear in the eastern third of Noble's mapped area (Fig. 2). Noble cited these foldlike features as evidence that the Amargosa thrust was folded after most or all of the thrusting had ceased.

Since then, various persons have expressed views on the origin of the Amargosa chaos. Some have supported Noble's initial (regional thrust) interpretation; others have held that the constituent rock units of the chaos have remained close to their original sites of deposition.

Curry (1954) considered the turtleback surfaces of the Black Mountain front as marking northern extensions of the Amargosa thrust. Hunt and Mabey (1966) concurred with Noble (1941) that the dominant structural features of the Panamint Range, west of Death Valley, may be an anticline in a thrust plate like the Amargosa chaos. Hunt and Mabey suggested that the Amargosa chaos is a gravity-propelled detachment feature that began to move westward in Mesozoic time, was later folded, and then broken up by late Cenozoic normal faults.

Sears (1953) proposed that bodies of Tertiary granite and the various anticlines and synclines formed simultaneously, being effects of vertical forces related to rising magma, and that the chaos formed by gravity sliding off the flanks of the anticlines.

Bucher (1956) suspected that the Virgin Spring phase was caused by gravity sliding, but he related the sliding to a violent disruption. Drewes (1963), like Noble and Wright (1954), was inclined to limit the chaos to the vicinity of the Black Mountain block east of Death Valley and to attribute it to "repeated adjustments to large movements on the steep faults that bound the block." As alternate possibilities he suggested "near-surface bifurcation of a thrust fault" and "gravity sliding off a rising structural block."

In our mapping of the chaos (Wright and Troxel, 1984), the following features of the Virgin Spring and Calico phases became obvious:

(1) Nearly all of the faults that feature the internal structure of the chaos are either normal or strike-slip; rarely do older rocks rest upon younger; (2) Where traceable downdip, the normal faults flatten with depth; some of them join along detachment surfaces within the chaos; others join along fault contacts between the Precambrian basement complex and the overlying later Precambrian units. Still others offset the contact and penetrate the complex; (3) The Virgin Spring phase is most chaotic within several tens of meters of the contact with the underlying complex.

We thus interpreted the chaos as an extensional feature, which formed on the underside of rotated fault blocks (Wright and Troxel, 1969) and also in the vicinity of low-angle detachment surfaces where normal faults flatten and join at shallow depths (Wright and Troxel, 1973). We also suggested that, in some areas, the crustal extension was accommodated by normal faulting in the basement, and by the emplacement of dikes and plutons (Wright and Troxel, 1973).

The basement complex is involved in the chaos and chaos-related faulting to a greater degree than Noble (1941) implied. Basement involvement is particularly obvious on the southwestern flank of the Desert Hound anticline (Fig. 2). The fault surface extends beyond the most westerly exposures of the Pahrump Group (Fig. 3) and splays into the complex. Southwest of the fault, the basal strata of the Crystal Spring Formation rest depositionally upon the complex. Many low-angle normal faults cut the crystalline complex (Wright and Troxel, 1984). The existence of these are important to a consideration of the origin of the chaos, as such faults permit extension of the basement concurrently with the formation of the chaos, and unaccompanied by the intrusion of bodies of igneous rock.

High-angle faults of relatively small displacement, apparently lateral, and commonly closely spaced, were mapped at several localities in the Virgin Spring chaos. Some lie entirely within the chaos; others offset the contact between the chaos and the underlying complex. All contribute to the disordered appearance of the chaos, but we interpret them as being superimposed upon the characteristic fault patterns of the chaos.

Some folds are Mesozoic or Early Tertiary in age, formed concurrently with the Desert Hound anticline and strongly modified by movement along the principal fault and by innumerable smaller faults.

ORIGIN OF THE AMARGOSA CHAOS

The geologic features of the Virgin Spring area record four major deformational events. The first, occurring as early as 1,700 Ma, accompanied and followed the metamorphism of the crystalline complex. It contributed to the angular discordance between planar and linear features in the complex and bedding planes in the overlying later Precambrian sedimentary units.

The second began with the deposition of the arkosic-conglomeratic strata low in the Crystal Spring Formation and continued through Noonday time (Fig. 3), spanning a poorly bracketed interval of time that probably lasted about 400 m.y. This event was accompanied by vertical crustal shifts (Wright and Troxel, 1984) causing facies changes in the Pahrump Group and Noonday Dolomite, and the angular unconformity beneath

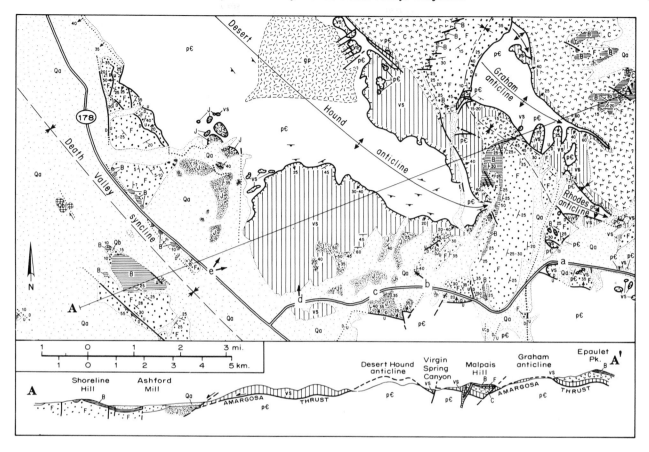

Explanation		Structural units	Age of component material	Character of material	Symbols
Qa	Alluvial deposits		Quaternary	Sand, gravel, silt, and clay; rock salt in Death Valley.	Strike and dip of beds _____
Qb	Basaltic ash		Quaternary	Dissected cinder cone and stratified ash.	Strike and dip of schistosity in Precambrian rocks _____
	—UNCONFORMITY—				
F / B	Fanglomerate Interbedded basalt flows	Funeral fanglomerate	Pliocene ?	Interlayered basalt, breccia, and fanglomerate.	Amargosa thrust; hachures on over-thrust side (dotted where concealed) _____
	—UNCONFORMITY—				
J		Jubilee phase	Precambrian to Tertiary	Sedimentary and volcanic rocks and breccias of granitic, sedimentary, and metamorphic rocks.	Klippe _____
C	Amargosa chaos	Calico phase	Almost wholly Tertiary	Rhyolitic lava and tuff.	Fenster _____
					POST-THRUST STRUCTURES
vs		Virgin Spring phase	Almost wholly Cambrian and later precambrian	Dolomite, limestone, sandstone, quartzite, shale, and slate.	Normal fault; U, upthrown; D, downthrown; arrow indicates relative direction of horizontal component (dotted where concealed) _____
	—Amargosa overthrust—				
pЄ / gp	Metamorphosed rocks Granite and granite porphyry intrusive	Autochthonous block	Precambrian metamorphic rocks, intruded by granite and granite porphyry of Tertiary ? age.	Granitic gneiss and greenstone sills.	Axis of anticline and direction of plunge _____ Axis of syncline _____

Figure 2. Noble's (1941) original map and cross section of the Virgin Spring area, redrafted and slightly modified for reduction and black and white reproduction. Small letters identify vantage points along paved road discussed in text.

the Noonday. These features, as expressed in the Virgin Spring area, indicate the presence of a major Precambrian discontinuity.

We interpret foldlike features preserved in the later Precambrian and Cambrian sedimentary rocks as actual folds forming before intricate faulting that produced the chaotic appearance of the Pahrump and younger units. We suggest that this folding occurred in Mesozoic or Early Tertiary time.

We continue to attribute the formation of the Virgin Spring and Calico phases of the chaos, the fourth deformational event, to faulting related to crustal extension in Cenozoic time. When the Death Valley region was deeply eroded, within the late Mesozoic–early Cenozoic interval, and then severely extended in later Cenozoic time, the resulting pattern of faulting led to the illusion of a single Cenozoic dislocation surface, originally planar and later folded.

To our earlier interpretations that related the telescoping of the later Precambrian and Cambrian strata in the chaos largely or wholly to movement on normal faults, and that involve the underlying crystalline complex in the chaos-related faulting (Wright and Troxel, 1973), we add the following interpretations. (1) The complex and younger cover rocks have responded differently to severe crustal extension, thus creating the appearance of a single Tertiary thrust fault bringing the younger units over the complex without involving the complex. The complex has been broken and extended by normal faults. (2) The chaos-forming event has consisted of a continuum featured by normal faulting accompanied by intervals of erosion, basinal sedimentation, and volcanism. Thus, the Virgin Spring phase of the chaos is more intricately faulted than the Calico phase and the Calico phase more so than the Funeral Formation. (3) The high-angle faults of apparent lateral slip we interpret as genetically and temporally related to the normal faults.

SUGGESTED FIELD EXCURSION

Depart from Shoshone, travel one mile (1.2 km) north, then turn west on California 178. Five stops are shown on Figure 2 and discussed below. General features of the geology from Shoshone into Death Valley are described by Troxel (1974, p. 2–16 and 1982, p. 37–42, 71–74).

The best single panorama of chaos exposures available from the highway is provided at a point in Bradbury Wash 3.5 mi (about 5.5 km) west of Salsberry Pass and 0.5 mi (about 0.8 km) east of the east boundary of Death Valley National Monument (Fig. 4).

The first ridge toward the viewer from the Panamint Range exposes the major features of Noble's (1941) Desert Hound anticline. The central part of the anticline is marked by exposures of the gray crystalline complex beneath Desert Hound Peak. Its limbs are identifiable by exposures of the varicolored, younger Precambrian and Cambrian units that compose the Virgin Spring phase of the chaos.

The near low ridge is underlain by east-tilted conglomerate and basalt of the late Cenozoic Funeral Formation. They are

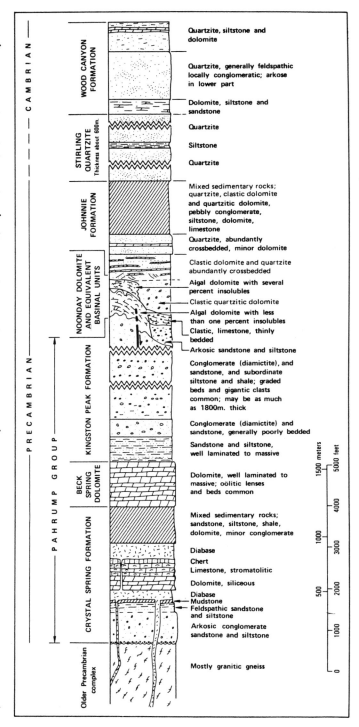

Figure 3. Generalized columnar section of Precambrian to Lower Cambrian strata, Death Valley region. Equivalent basinal units of Noonday Dolomite are now known as the Ibex Formation. From Wright and others (1974).

much less deformed than the rock units of the chaos and thus postdate the formation of the chaos.

Rhodes Hill, in the near foreground north of the highway, is underlain by gray gneiss of the Precambrian complex. The overridden part of the complex is exposed on the crest of the low ridge that limits Bradbury Wash on the south. Jubilee Peak also is underlain by the Precambrian complex.

Epaulet Peak, identifiable by a capping and fringelike talus slopes of dark brown- to black-weathering basalt, dominates the skyline north of Bradbury Wash. The basalt and a thin, discontinuous, underlying layer of conglomerate apparently are correlative with the Funeral Formation.

Exposed over most of the southwest slope of Epaulet Peak are rhyolitic volcanic rocks, varicolored, but mostly in shades of yellow. These are the Shoshone Volcanics of Pliocene age. They are faulted considerably more than the overlying Funeral Formation, and form the principal exposures in the Virgin Spring area of Noble's Calico phase of the chaos.

Exposed in a belt still lower on the southwest slope of Epaulet Peak are highly faulted latest Precambrian and Cambrian units in an occurrence of the Virgin Spring phase of the chaos. Within the belt are fault-bounded segments of the Noonday Dolomite, Johnnie Formation, Stirling Quartzite, Wood Canyon Formation, and Zabriskie Quartzite. Viewed collectively, they are darker—mostly in shades of red—than the overlying volcanic units. They are more colorful and resistant and much more faulted than the gray underlying crystalline complex, which is barely in view from here. This contact is a segment of Noble's Amargosa thrust and marks the northeast limb of the Graham anticline (Fig. 2). He also interpreted the contact between the Virgin Spring and Calico phases of the chaos as a surface of movement, but less than the movement on the lower contact.

Exposures of the Virgin Spring phase of the Amargosa chaos in lower Bradbury Wash. Upon entering Death Valley Monument, and for the next 4 mi (about 6.5 km) westward, the road is close to exposures of the Virgin Spring phase of the chaos and its contact with the underlying complex. Here, as elsewhere, the contact is marked by an abrupt change from the gray of the complex to the brighter and more varied colors of the chaos. In this area, only isolated erosional remnants of the chaos remain, but they show features much like those that characterize larger bodies of the Virgin Spring phase. Of the chaos-forming units at this locality, the Noonday Dolomite is the easiest to identify. It is the yellowish gray, resistant unit that supports most of the knobs within 0.5 mi (0.8 km) of the highway. At numerous places, one can observe details of the faulted lower surfaces and the intensely fractured nature of the various overlying rock units.

The best exposure of the Virgin Spring chaos along California 178 lies adjacent to and south of the highway and west of the Monument boundary (point a, Fig. 2). There the chaos underlies the steep north face of a hill about 300 ft (90 m) high and displays most of the features that are commonly ascribed to the lower part of the chaos in general.

The lower part of this face is underlain by the gray-weathering, locally red-stained crystalline complex. Within it are sheared masses of dark green diabase dikes and nearly white granitic pegmatite dikes. All are thoroughly sheared and become progressively more so upward to the nearly horizontal contact with the overlying chaos. The strong evidence of dislocation along this contact, together with the deformation recorded in the chaos, impressed Noble to the extent that he identified it as an occurrence of his Amargosa thrust.

The pale gray to dark lavender, thin fault-bounded lenses at the base of the overlying chaos consist of arkosic sandstone and siltstone of the dominantly clastic lower part of the Crystal Spring. The dark green lenses higher on the face are slices of the diabase sill that, regionwide, separates the lower clastic members from the carbonate member. The latter, in turn, is represented by the still higher, dark reddish brown lenses. This hill, like other hills in the vicinity, is upheld by yellowish gray dolomite of the Noonday Dolomite. Strata of the Johnnie Formation are exposed on the south side of the hill crest. Both the Noonday and Johnnie, like the Crystal Spring, occur as fault-bounded lenses and thus also qualify as chaotic.

The full thickness of the Crystal Spring ordinarily ranges between 2,500 and 4,000 ft (750 and 1,200 m; Fig. 3). The fault-bounded slices of Crystal Spring exposed on the nearby vertical north face of the hill in the lower Bradbury Wash are limited to about a 200-ft-segment (60 m) of the face. The Beck Spring Dolomite and Kingston Peak Formation may have been eroded away from this location in Precambrian time before the Noonday Dolomite was deposited, but most of the Crystal Spring has been faulted out in the formation of the chaos. Each slice retains its proper stratigraphic position, younger over older.

View of the southwest limb of the Desert Hound anticline from the west side of Jubilee Pass. A point about 0.5 mi (0.8 km) west of Jubilee Pass (point b, Fig. 2) affords an excellent distant view of the crest and southwest limb of the Desert Hound anticline and of the southwestern body of the Virgin Spring chaos (Fig. 5). From Desert Hound Peak eastward is exposed the gray-weathering, earlier Precambrian crystalline complex. The dark green patches within it are exposures of parts of an anastomosing system of Precambrian diabase dikes; the lighter patches are exposures of prediabase pegmatite bodies and Tertiary acidic dikes.

The contact between the complex and the Virgin Spring phase of the chaos is about halfway down the slope southward and is identifiable by the characteristic change in color, from the gray of the complex to the warmer colors of the later Precambrian and Cambrian units. The dark green unit near the skyline is the sill of diabase in the Crystal Spring Formation. The lightest-colored rock, which tends to form topographic highs, is the yellowish gray dolomite of the Noonday Dolomite. The post-Noonday formations are more difficult to distinguish from one another. Of these, the most distinctive are the pale orange to pale lavender, well-layered units of the Johnnie Formation.

The contact between this body of Virgin Spring chaos and the underlying complex is everywhere strongly faulted, but is unbroken by later faults. It dips moderately to steeply southwest-

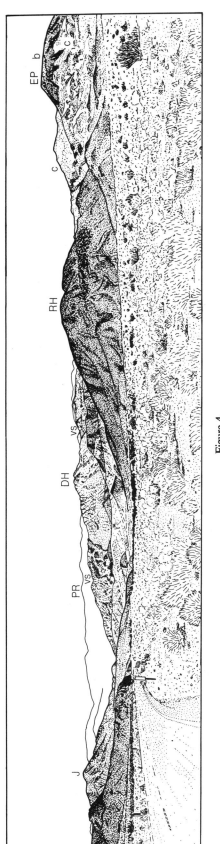

Figure 4.

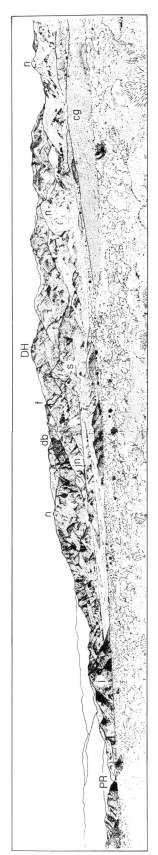

Figure 5.

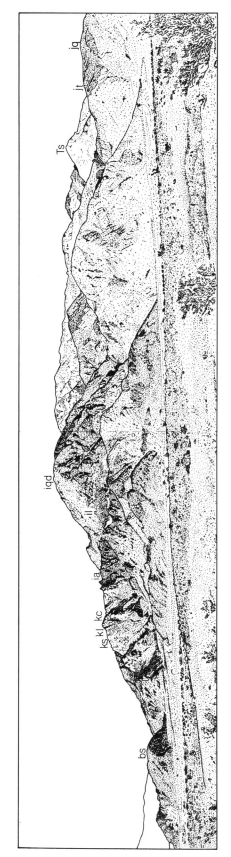

Figure 6.

ward and resembles in detail the contact and associated overlying and underlying rock units observed at point a along the highway in lower Bradbury Wash. This contact is the most continuously exposed segment of Noble's Amargosa thrust (Fig. 2). The overlying, younger units compose the thickest and most extensively exposed body of the Virgin Spring phase of the chaos in the map area. Most of the chaos in this body is much less intricately faulted than the chaos observed near the highway.

Jubilee phase of the Amargosa chaos exposed near Point of Rocks. The most accessible and some of the best examples of Noble's (1941) Jubilee phase of the Amargosa chaos underlie three hills just north of the highway and opposite Point of Rocks, about 2 mi (3.2 km) west of Jubilee Pass (point c, Fig. 2). Noble (1941) distinguished this phase from the other two phases because, unlike them, it contains abundant conglomerate and siltstone of Tertiary age, as well as bodies of Tertiary volcanic and granitic rock and various rock units of the Pahrump Group and latest Precambrian and Cambrian formations. In addition, many of the bodies of Tertiary volcanic and granitic rock and all of the bodies of the older units are truly breccia layers, which, although monolithologic, are interlayered with Tertiary conglomerate, siltstone, and tuff. Noble and Wright (1954) reinterpreted most or all of these bodies of breccia as sedimentary units

---

Figure 4. Sketch of westward view of the terrane of the Amargosa chaos from a point on California 178, 3.5 mi (5.6 km) west of Salsberry Pass and 0.5 mi (0.8 km) east of the eastern boundary of Death Valley National Monument. Topographic features are indicated by capital letters, geologic features by small letters. DH, Desert Hound Peak; EP, Epaulet Peak; J, Jubilee Peak; PR, Panamint Range; RH, Rhodes Hill; b, basalt of Funeral Formation; c, Calico phase of chaos; vs, Virgin Spring phase of chaos as distributed along both sides of Desert Hound Peak.

Figure 5. Sketch of the terrane of the Amargosa chaos as viewed northward and westward from the vicinity of Jubilee Pass (point b, Fig. 2). Exposed in succession from the northern skyline toward the viewer are (1) the Precambrian crystalline complex, underlying the highest part of the landscape; (2) the Virgin Spring phase of the chaos forming a continuous belt along the intermediate slopes; and (3) the Jubilee phase of the chaos discontinuously exposed in low hills and ridges surrounded by alluvium. DH, Desert Hound Peak; PR, Point of Rocks; cg, conglomerate of Funeral Formation; db, diabase of Crystal Spring Formation; f, fault contact between the Precambrian crystalline complex and the overlying Virgin Spring phase of the chaos; j, Jubilee phase of the chaos; jn, Johnnie Formation; n, Noonday Dolomite; s, Stirling Quartzite.

Figure 6. Sketch of a part of the Black Mountains, looking northward from lower Jubilee Wash (Point c, Fig. 2). The rock units underlying the prominent slopes are mostly of Precambrian age and were included by Noble (1941) in his Virgin Spring phase of the Amargosa chaos. They form, in general, an east-tilted fault block, broken by many normal faults of relatively small displacement and which cause repetitions of the sedimentary units; bs, Beck Spring Dolomite; ia, il, and iqd, arkose, limestone, and quartz-dolomite sandstone members of the Ibex Formation; jt and jg, transitional and quartzite members of the Johnnie Formation; ks, kl, and kc, siltstone, limestone, and conglomerate members of Kingston Peak Formation; Ts, Tertiary sedimentary rock.

in the Tertiary section, and our mapping has reinforced this view.

Monolithologic breccia of quartz monzonite underlies most of the east and middle hills and displays a cavernous type of weathering. The west hill is underlain mostly by breccia derived from the Crystal Spring Formation, including the diabase (green), carbonate member (red), and arkose of the lower units (gray). The evenly bedded conglomerate and sandstone exposed at Point of Rocks are representative of the other sedimentary rocks associated with the bodies of breccia.

Pahrump Group and latest Precambrian formations exposed on north wall of lower Jubilee Wash. As the traverse continues still farther westward and down Jubilee Wash, the Virgin Spring chaos north of the highway assumes a progressively less chaotic appearance. Viewed from a point about 1.5 mi (2.4 km) west of Point of Rocks (point d, Fig. 2), the north wall of the wash provides a cross section through the upper part of the Beck Spring Dolomite, the Kingston Peak Formation, and the basin facies of the Noonday Dolomite (Ibex Formation). Although faulted, these formations retain most of their original thickness and dip moderately eastward (Fig. 6).

Perhaps the simplest to identify is the unit of dark lavender strata in the middle part of the face. This is the arkose member of the Ibex Formation. It consists of arkosic sandstone and siltstone and composes the lowest part of the formation (Williams and others, 1976). It is underlain by small, discontinuous lenses of yellowish gray dolomite. These lenses are remnants of the southward-thinning lower dolomite member of the Noonday Dolomite (platform facies). Successively above the arkose member are well-bedded yellow limestone and limestone conglomerate, then massive dolomite-quartz sandstone.

The low hill at the western end of the face is underlain by the gray-appearing Beck Spring Dolomite. The generally orange to reddish orange strata between the Beck Spring and the lenses of Noonday Dolomite are units of the Kingston Peak Formation. A thin layer of thinly-bedded, black limestone separates the fine-grained lower siltstone member of the Kingston Peak from the conglomeratic middle member (diamictite). The upper member consists of a relatively evenly bedded unit of mixed conglomerate, sandstone, siltstone, and sedimentary breccia. At the Jubilee Wash locality, the part of the Kingston Peak that overlies the limestone member consists mostly of diamictite and includes only a thin occurrence of the upper member. All of the bodies of conglomerate contain abundant debris from the Beck Spring Dolomite and Crystal Spring Formation. We cite this as evidence that the Beck Spring and Crystal Spring once extended well to the north of their most northerly exposures in the Confidence Hills Quadrangle.

View of the Black Mountains escarpment from Ashford Mill site; Pahrump Group and Noonday Dolomite. The Pahrump Group and the overlying Noonday Dolomite, where exposed on the Black Mountains escarpment, are much less faulted and more completely exposed than they are in the chaos of lower Bradbury Wash. When viewed from Ashford Mill site (point e, Fig. 2) in the afternoon sun or on a cloudy day, the escarpment clearly shows the differences in color that permit identification of the various Precambrian units. The yellowish gray unit, supporting the highest point, is the Noonday Dolomite. The change from the platform to the basin facies (Ibex Formation) occurs abruptly near lower Jubilee Wash. The gray unit, beneath the Noonday and traceable diagonally up the escarpment, south to north, is the Beck Spring Dolomite. The Kingston Peak Formation is missing along all but the southernmost part of the escarpment, as it wedges out a short distance north of Jubilee Wash. Detectable even from this distance, however, is an interlayering of gray dolomite typical of the Beck Spring, and orange strata like those of the siltstone member of the Kingston Peak.

Successively exposed beneath the Beck Spring along the rest of the escarpment are the various members of the Crystal Spring Formation. Especially obvious are the dark green diabase sills at various positions within the formation. The upper sedimentary units are varicolored; the dolomite, which here forms the carbonate member, is orange, and the lower arkosic units are various shades of gray and lavender.

SITE 28. TURTLEBACK SURFACES
B. W. Troxel

LOCATION AND ACCESS

Turtleback surfaces are exposed along the west front of the Black Mountains between about 15 and 35 mi (24 and 56 km) south from Furnace Creek Ranch, Death Valley, California. They lie within a few miles of the paved road that extends southward from Furnace Creek Inn to Shoshone, California. Access to the northernmost turtleback, the Badwater turtleback, is obtained by driving to the parking area at the east end of a gravel road identified by a sign that denotes "Natural Bridge Canyon." The northwestern tip of the turtleback is cut by Natural Bridge Canyon. The southwest wall of the Badwater turtleback is well exposed and easily accessible by hiking from the parking area.

The next turtleback to the south is the Copper Canyon turtleback. Access to it is gained by parking near the mountain front at the south edge of the Copper Canyon fan and hiking north along the mountain front to the point where the crystalline rocks beneath the turtleback surface plunge northwestward beneath the faulted Tertiary sedimentary rocks. A moderately steep, but short, climb affords excellent detailed exposures of the turtleback fault.

The Mormon Point turtleback, a few miles farther southwest from the Copper Canyon turtleback, plunges northwestward beneath Quaternary gravel. Details of the bedrock beneath the turtleback surface can be observed at many points along the west flank of the mountain front south from Mormon Point.

SIGNIFICANCE

The turtleback surfaces were recognized and named by Curry (1938). Since then they have been the subject of consider-

able debate as to their origin and significance. Five significantly different origins have been proposed for the surfaces. Parts of the surfaces are moderately easily accessible; these features invite intense field discussions. The features are important in that they have been involved at least in Tertiary Basin and Range extension and perhaps in Mesozoic compression.

SITE INFORMATION

Background information. Curry's pioneer work (1938) led him to attribute the origin of the three turtleback surfaces to compressional folding of a regional thrust fault (Curry, 1954). Noble (1941) and Hunt and Mabey (1966) likewise related them to thrust faulting. Drewes (1959) proposed that differential erosion produced an "undulating topographic surface upon which the Cenozoic rocks were deposited and from which they later slid, propelled by gravity" (Wright and others, 1974). Sears (1953) related the arching to the intrusion of shallow plutons. Hill and Troxel (1966) stated that the turtleback surfaces were formed during regional compression and that the Tertiary cover rocks essentially moved as the basement rocks folded. Wright and others (1974) and Otton (1974) stated that the turtleback surfaces "were colossal fault mullion resulting from severe crustal extensions which were localized along undulating and northwest-plunging zones of weakness that were in existence prior to this deformation." Stewart (1983) considered the turtleback surfaces to be gigantic mullions related to the detachment and transport of the overlying rocks 50 mi (80 km) northwestward.

Noble (1941) related his "Amargosa thrust" to the "turtleback fault" of Curry (1954) but later doubted the existence of the Amargosa thrust (Noble and Wright, 1954). The turtleback folds and metamorphism of mantling carbonate rocks are now considered to be analogous to core complexes in that they are domal, consist of a core of gneiss that dips away from the domes, have a mantle of metamorphosed rocks, and are covered by deformed but unmetamorphosed rocks separated from the mantled core by mylonitized rocks beneath the detachment surface. The three turtlebacks are overlain by Cenozoic sedimentary rocks cut by abundant listric normal faults that flatten and merge with the detachment faults atop the turtlebacks. Similar fault patterns are characteristic of the Virgin Spring area farther south (see Wright and Troxel, this guidebook).

Physical features of the Death Valley turtlebacks. The three turtlebacks as identified by Curry (1938) are, from north to south, the Badwater, Copper Canyon, and Mormon Point turtlebacks. The antiformal and topographic axes of the turtleback surfaces trend northwest, and the crests plunge northwest.

Slickensides on the southwest flanks of the turtleback surfaces and on many of the frontal faults on the west side of the Black Mountains trend northwest and plunge 10° to 15° to the northwest (unpublished data). Each of the turtleback surfaces is underlain by a mantle composed of discontinuous carbonate rocks, which are internally highly deformed. The rocks beneath the detachment surface are usually mylonitized and commonly

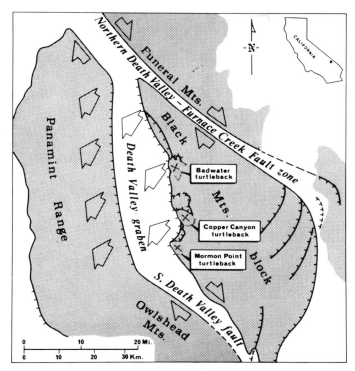

Figure 7. Generalized structural map of Death Valley region, showing position of three turtleback surfaces of Black Mountains. Hachured lines mark positions of major normal faults; full arrows show inferred direction of crustal extension; half arrows show relative displacement on strike-slip fault zones. Figure from Wright and others (1974).

enriched in iron. The mantle of carbonate rocks is underlain by foliated gneissic rock of Precambrian age (Drewes, 1963) and intruded by Mesozoic (?) dioritic rocks (Otton, 1974). The southeastern extension of the Mormon Point turtleback (the Desert Hound anticline of Noble, 1941) is intruded by Miocene (?) quartz monzonite (Drewes, 1963; Wright and Troxel, 1984). Tertiary intrusive rocks crop out also southeast of the domal crests of the other turtleback surfaces.

The common northwest trend of the antiformal axes of the three turtleback surfaces and the continuation of the Mormon Point turtleback surface into the Desert Hound anticline of Noble (1941) are significant. Moreover, the trend of these features is remarkably coincident with the trend of the Northern and Southern Death Valley fault zones, the other anticlines in the Virgin Spring area (Noble, 1941; Wright and Troxel, 1984), and the trend of the domed surface formed beneath the Boundary Canyon and Keene Wonder faults in the Funeral Mountains (Troxel and Wright, unpublished data) situated farther north. Most of these trends are apparent on the Geologic Map of California (Jennings, 1977).

The geology of the Death Valley turtlebacks is shown on various geologic maps. These include Curry (1954), Drewes (1959, 1963), Otton (1974), the Death Valley 1:250,000-scale

map (Streitz and Stinson, 1974), and the Geologic Map of California (Jennings, 1977). Noble (1941) published a map of the Virgin Spring area of chaos, and Noble and Wright (1954) published a general structural map of Death Valley.

Some important differences exist between the turtlebacks of Curry (1938, 1954), the anticlines in the Virgin Spring area (Noble, 1941; Noble and Wright, 1954; Wright and Troxel, 1984), and the domal detachment surface exposed in the Funeral Mountains (Troxel and Wright, unpublished data). The common trend of these features and of the major strike-slip faults is obvious. Its meaning is less so. The subtle to obvious differences in the rocks and their fabric is also important and incompletely understood at this time. The following field traverse is suggested to stir interest and acquaint you with features of Curry's (1938, 1954) original observations.

FIELD EXCURSION

A traverse from south to north in the floor of Death Valley is suggested. A review of the discussion of the Amargosa chaos by Wright and Troxel (1984, and this volume) is recommended before progressing northward from the Virgin Spring area into central Death Valley, where the Death Valley turtlebacks are exposed.

Proceed on California 178 and 127 for 1 mi (1.6 km) north from Shoshone, California. Shoshone is about 60 mi (97 km) north of Baker, California, which is situated on I-15 that connects Las Vegas, Nevada, and Los Angeles, California. Proceed west on California 178, into the floor of Death Valley (about 30 mi; 48 km), then north along the paved road that follows the east side of Death Valley to the intersection of California 190 at Furnace Creek Inn.

When you obtain the position of Mormon Point (the northwestern promontory of the Mormon Point turtleback (Fig. 7), you are at the point where lateral motion on the southern Death Valley fault zone gives way to transtension in a pull-apart region that lies between the Southern Death Valley fault zone and the Northern Death Valley–Furnace Creek fault zone. The Death Valley turtlebacks lie within this transtension zone (Fig. 7). This part of Death Valley has been identified as a pull-apart basin (Burchfiel and Stewart, 1966; Wright and others, 1974). The direction of motion is implied to trend parallel with the orientation of the crests of the Death Valley turtlebacks (Curry, 1938, 1954) as shown on Figure 7. The topographic low of central Death Valley occupies a half-graben that lies between the Panamint Mountains to the west and the Black Mountains to the east. For the most part, the Black Mountains are devoid of Precambrian and Paleozoic rocks that are exposed on nearly all sides of the Black Mountains (e.g., see Jennings, 1977). The lack of the Precambrian and late Paleozoic strata in most of the Black Mountains block (Fig. 7), and other phenomena, led Stewart (1983) to postulate a 50-mi transport (80 km) of the Panamint Mountains northwestward from a position above the Black Mountains block along a fault plane (or planes) related to the turtleback surfaces.

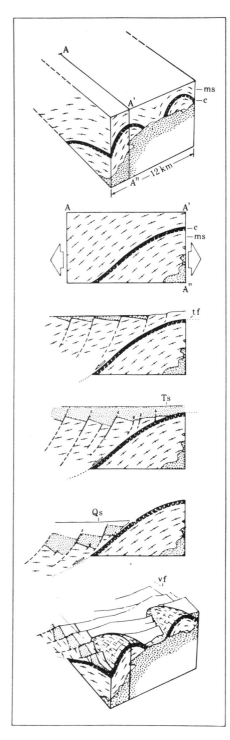

Figure 9. Copper Canyon turtleback from west. Left arrow denotes place to observe turtleback surface, transported overriding rocks, and fractured bedrock. Central arrow denotes suggested place to park. Third arrow marks crest of turtleback. Copper Canyon fan in lower left foreground. Photo by L. A. Wright.

Figure 8. Idealized block diagrams and cross sections, illustrating pull-apart concept of turtleback formation; based on observations of Copper Canyon and Mormon Point turtlebacks, Death Valley. c, Carbonate layers; ms, mixed metasedimentary rock; Qs, Quaternary sediments; tf, turtleback fault; Ts, Tertiary sedimentary rock; vf, valley floor. Figure from Wright and others (1974).

Figure 8, idealized block diagrams and cross sections, demonstrates the pull-apart concept of Wright and others (1974). Each of the three turtlebacks is discussed below.

Mormon Point turtleback. The Mormon Point turtleback, mapped most recently and in most detail by Otton (1974), is easily accessible from the paved road in Death Valley that follows closely the west flank of the turtleback. Many small west-flowing stream channels afford access into the flank of the ridge. In some of the channels, one can observe the turtleback fault preserved beneath Quaternary gravel that has been deposited upon the fault surface and subsequently moved essentially down the dip of the fault surface. Normal faults that dip more steeply to the west cut the Quaternary gravel and merge with the renovated turtleback fault (see Troxel, 1986). Beneath the turtleback fault, the bedrock, most commonly Precambrian carbonate rocks (Otton, 1974), is intensely brecciated. The degree of brecciation diminishes downward away from the fault surface. The Quaternary gravel has been rotated downward to the east during slip on the main fault plane and subsidiary fault planes that merge downward into it. This pattern is typical of listric faults that merge downward into major extensional fault planes in many parts of the Death Valley region. A particularly good exposure of the

Figure 10. Badwater turtleback. White line on fan surface is Natural Bridge Canyon road to parking area (lower arrow). Arrows denote approximate location of natural bridge (left), exhumed turtleback surface (center), and a triangular-shaped erosional remnant of Tertiary rocks transported over turtleback fault (right). North is to the left. Photo scale 1:12,000. Low sun-angle photograph; courtesy of D. B. Slemmons.

Quaternary faults that join the rejuvenated turtleback fault is in a small canyon situated 1 mi (1.6 km) south from Mormon Point. It can also be found by parking 0.3 mi (0.5 km) north of highway mileage marker 36. The mouth of the canyon is crossed by west-facing fault scarps in very young stream gravel. The north wall of the mouth of the canyon contains gently east-tilted fine-grained sediments overlain and underlain by coarse gravel.

From many points on the paved road can be seen pale-colored marble intruded by dark green dioritic rock. The marble forms a coating or shell beneath the turtleback fault surface (Otton, 1974). It must be assumed that the southwest downdip continuation of the turtleback surface has been downdropped beneath the Death Valley floor along Quaternary (and older?) normal (oblique slip?) faults that abound along the range front.

Copper Canyon turtleback. Mormon Point affords an excellent view of the profile of the Copper Canyon turtleback crest and the overlying Tertiary sedimentary and volcanic rocks that have been transported northwestward over the turtleback surfaces. From Mormon Point the view to the northeast is almost at a right angle to the northwest plunge and trend of the crest of the turtleback surface.

Not visible from this viewpoint is the relatively thin skin of carbonate rock directly beneath the turtleback surface. This is mainly because the southwest flank of the turtleback surface has been cut by younger faults along the mountain front. Precambrian foliated gneissic rock and flow-banded or foliated Mesozoic dioritic rock form most of the rangefront in view beneath the turtleback surface (Fig. 9). A few small patches of carbonate rock are exposed at the crest of the turtleback, which is nearly coincident with the ridge crest visible from Mormon Point.

A few dikes of Tertiary intrusive rocks cross the turtleback mass at nearly right angles to the trend of the ridge crest. The dikes are oriented in a proper direction to be fractures that were filled as the basement mass extended during northwest transport of the rocks above the turtleback fault surface. Proceed east, then north, along the road from Mormon Point to the point where the road begins to veer northwest from the mountain front. Park here if you wish to hike to the point where the turtleback fault is exposed beneath the transported Tertiary rocks.

In a distance of less than 1 mi (1.6 km) one can hike northward, parallel to the mountain front, to a small canyon that cuts across the northwestward-plunging nose of the Copper Canyon turtleback. At this point, it is advisable to climb the steep (and moderately difficult) surface on the north side of the narrow stream channel. After a few tens of yards, the topography becomes less steep and one is rewarded with excellent exposures of the multiple faults that separate the Tertiary rocks from the underlying Precambrian rocks. One would probably want to spend one to two hours at this locality noting the details of mylonitization, fault imbrication, iron enrichment, and brecciation associated with the Copper Canyon turtleback fault.

Badwater turtleback. The Badwater turtleback, some 12 to 15 mi (19 to 24 km) north of the Copper Canyon turtleback, is more easily accessible than the Copper Canyon turtleback, and, in addition, contains remnants of the Tertiary cover rocks preserved along the southwest flank of the turtleback surface (Fig. 10). After leaving the parking lot at Badwater and traveling north along the highway, the patches of Tertiary rocks can easily be distinguished at a distance from the underlying drab Precambrian bedrock by the distinct bright and pale colors of the Tertiary rocks. Proceed to the turnoff denoting Natural Bridge Canyon, then east to the end of the gravel road.

From the parking area at the end of the road it is recommended that you proceed east across the fan and deep channel that cuts it to observe the patches of Tertiary rocks preserved above the Badwater turtleback fault (Fig. 10). The footpath up Natural Bridge canyon permits you to see Quaternary gravel in the nearer canyon walls cut by many faults. Up-canyon, beyond the natural bridge, are exposures of Tertiary rocks in fault contact with the underlying Precambrian rocks, however, access is more difficult than to the remnants of Tertiary rocks preserved along the mountain front, where the exhumed surface of the turtleback is exposed and details of bedding in the Tertiary strata are preserved.

REFERENCES CITED

Bucher, W., 1956, Role of gravity in orogenesis: Geological Society of America Bulletin, v. 67, p. 1295–1318.

Burchfiel, B. C., and Stewart, J. H., 1966, "Pull-apart" origin of the central segment of Death Valley, California: Geological Society of America Bulletin, v. 77, p. 439–442.

Curry, H. D., 1938, "Turtleback" fault surfaces in Death Valley, California [abs.]: Geological Society of America Bulletin, v. 49, p. 1875.

——, 1954, Turtlebacks in the central Black Mountains, Death Valley, California, *in* Jahns, R. H., ed., Geology of southern California: California Division of Mines Bulletin 170, p. 53–59.

Drewes, H., 1959, Turtleback faults of Death Valley, California—a reinterpretation: Geological Society of America Bulletin, v. 70, p. 1497–1508.

Drewes, H., 1963, Geology of the Funeral Peak Quadrangle, California, on the east flank of Death Valley: U.S. Geological Survey Professional Paper 413, 78 p.

Hill, M. H., and Troxel, B. W., 1966, Tectonics of Death Valley region, California: Geological Society of America Bulletin, v. 77, p. 441–444.

Hunt, C. B. and Mabey, D. R., 1966, General geology of Death Valley, California—Stratigraphy and structure: U.S. Geological Survey Professional Paper 494-A, 165 p.

Jennings, C. W., 1977, Geologic map of California: California Division of Mines and Geology, Geologic Data Map Series, scale 1:750,000.

Noble, L. F., 1941, Structural features of the Virgin Spring area, Death Valley, California: Geological Society of America Bulletin, v. 52, p. 942–1000.

Noble, L. F., and Wright, L. A., 1954, Geology of the central and southern Death Valley region, California, *in* Jahns, R. H., ed., Geology of southern California: California Division of Mines Bulletin 170, p. 143–160.

Otton, J. K., 1974, Geologic features of the central Black Mountains, Death Valley, California, *in* Guidebook, Death Valley region, California and Nevada: Shoshone, California, Death Valley Publishing Company, p. 65–72.

Sears, D. H., 1953, Origin of the Amargosa chaos, Virgin Spring area, Death Valley, California: Journal of Geology, v. 61, p. 182–186.

Stewart, J. H., 1983, Extensional tectonics in the Death Valley area, California; Transport of the Panamint Range structural block 80 km northwestward: Geology, v. 11, p. 153–157.

Streitz, R., and Stinson, M. C., 1974, Geologic map of California, Death Valley sheet: California Division of Mines and Geology, scale 1:250,000.

Troxel, B. W., 1974, Geologic guide to the Death Valley region, California and Nevada, *in* Guidebook, Death Valley region, California and Nevada: Shoshone, California, Death Valley Publishing Company, p. 2–16.

——, 1982, Geologic road guide; Day 2, Baker–southern Death Valley–Shoshone, and Day 3, Segment A, *in* Cooper, J. D., ed., Geology of selected areas in the San Bernardino Mountains, western Mojave Desert, and southern Great Basin, California: Shoshone, California, Death Valley Publishing Company, p.37–42 and 71–76.

——, 1986, Significance of Quaternary fault pattern, west side of the Mormon Point turtleback, southern Death Valley, California; A model of listric normal faults, *in* Quaternary tectonics of southern Death Valley, California, field trip guide: Shoshone, California, B. W. Troxel, publisher, p. 37–40.

Williams, E. G., Wright, L. A., and Troxel, B. W., 1976, The Noonday Dolomite and equivalent stratigraphic units, southern Death Valley region, California, *in* Troxel, B. W., and Wright, L. A., eds., Geologic features, Death Valley, California: California Division of Mines and Geology Special Report 106, p. 45–50.

Wright, L. A., and Troxel, B. W., 1969, Chaos structure and Basin and Range normal faults; Evidence for genetic relationship: Geological Society of America Abstracts with Programs, v. 1, no. 7, p. 242.

——, 1973, Shallow-fault interpretation of Basin and Range structure, southwestern Great Basin, *in* DeJong, R., and Scholten, R., eds., Gravity and tectonics: Amsterdam, Elsevier Scientific Publishing Company, p. 397–407.

——, 1984, Geology of the northern half of the Confidence Hills 15-minute Quadrangle, Death Valley region, eastern California; The area of the Amargosa chaos: California Division of Mines and Geology Map Sheet 34, 21 p., scale 1:24,000.

Wright, L. A., Otton, J. K., and Troxel, B. W., 1974, Turtleback surfaces of Death Valley viewed as phenomena of extensional tectonics: Geology, v. 2, p. 53–54.

Wright, L. A., Troxel, B. W., Williams, E. G., Roberts, M. T., and Diehl, P. E., 1974, Precambrian sedimentary environments of the Death Valley region, eastern California, *in* Guidebook, Death Valley region, California and Nevada: Shoshone, California, Death Valley Publishing Company, p. 27–36.

Quaternary fault-line features of the central Garlock fault, Kern County, California

Bruce A. Carter, Pasadena City College, Pasadena, California 91106

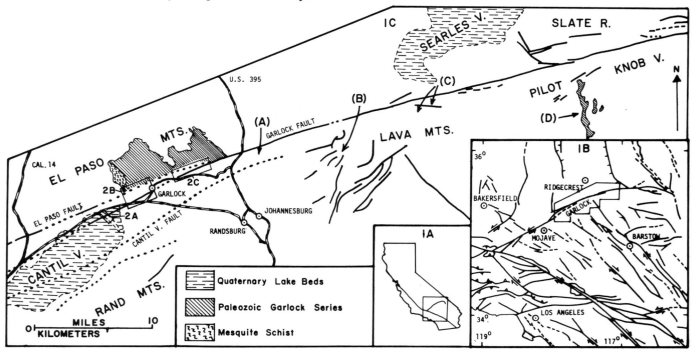

Figure 1. Index maps to the central Garlock fault, California. Figure 1C. (A) Eastern limit of Pleistocene gravels containing Mesquite Schist clasts south of the fault and 12 mi (19 km) east of their probable site of deposition. (B) Mesquite schist clasts in sediments interlayered with the Pliocene Almond Mountain volcanics about 20 mi (32 km) east of their probable site of deposition. (C) Large blocks of Garlock series and Mesquite Schist within the latest Miocene Bedrock Springs Formation about 29 mi (47 km) east of their probable site of deposition. (D) Paleozoic Garlock series rocks displaced a total of about 40 mi (64 km) from similar rocks in El Paso Mountains.

LOCATION

The part of the Garlock fault near the town of Garlock can be reached by turning east on the Randsburg road from California 14 about 20 mi (32 km) north of Mojave or by turning west from U.S. 395 about 6 mi (10 km) north of Johannesburg (Fig. 1C). From California 14, it is about 10 mi (16 km) east to the offset gravel bar, 1 mi (1.6 km) farther to the Mesquite Canyon Road, and then 4.5 mi (7.2 km) to the Goler area. The Goler area is about 4.5 mi (7.2 km) west of U.S. 395.

SIGNIFICANCE

The Garlock fault is one of the major faults of southern California, running about 160 mi (257 km) from the San Andreas fault near the town of Gorman to the south end of Death Valley. This left-lateral fault separates the Basin Ranges and Sierra Nevada provinces on the north from the Mojave Desert on the south, and has been described as an intracontinental trans-

form fault (Davis and Birchfiel, 1973). Total displacement on the Garlock fault is about 40 mi (64 km); Smith, 1962; Davis and Birchfiel, 1973), and the well-defined fault-line features in the central and eastern parts of the fault suggest very recent activity. Along the central part of the fault, near the town of Garlock, a variety of well-preserved fault line features includes an offset Pleistocene gravel bar, scarps, shutter ridges, tension graben, offset streams, and others (Clark, 1973). More important than the prominance of these features is the fact that some of them can be used to deduce the recent age and rate of movement on the Garlock fault. El Paso Mountains, north of the fault in this area, contain exposures of several distinctive bedrock lithologies, of which the Mesquite Schist and the Garlock Series are the most important (Fig. 1C). Gravels derived from these lithologies were transported southward across the fault and deposited in Cantil Valley. In several instances, highly distinctive deposits south of the Garlock fault now lie east of their probable areas of deposition, and therefore demonstrate left-lateral displacement on the fault. Striking examples of these offset deposits can be seen near

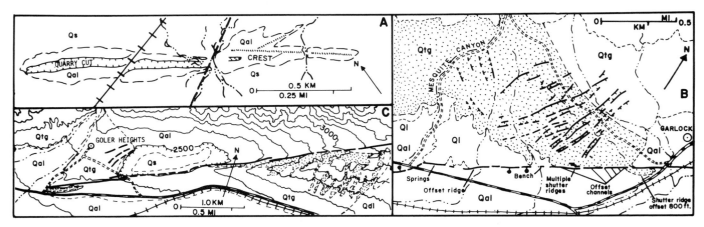

Figure 2. Detailed geologic relationships along the central part of the Garlock fault, California. Figure 2A. Well-preserved shoreline bar offset about 260 ft (79 m) by the Garlock fault. Figure 2B. Fault-line features along the Garlock fault near the mouth of Mesquite Canyon. Normal fault scarps up to 25 ft (7.6 m) high cut the older uplifted fan surface east of Mesquite Canyon. Ball on relatively downthrown side of scarps along break on which movement has been predominantly horizontal; hachures on relatively downthrown side of scarps along break on which movement has been predominantly vertical. Figure 2C. Goler graben. Qs, fine sand and silt; Qal, coarse sand and gravel; Qtg, older uplifted gravel; Ql, tilted Pleistocene lake beds. Stippled pattern, older uplifted gravels containing abundant clasts of Mesquite Schist (in Fig. 1C clasts are abundant only in gravel along the base of the scarp).

the town of Garlock, although much greater offsets can be demonstrated using deposits exposed in the Lava Mountains farther east. Matching offset deposits of different ages with their bedrock source areas in El Paso Mountains shows that the Garlock fault has averaged about 7 mm/year displacement since latest Miocene, and that strike slip on the Garlock fault probably began about 9 ma.

LOCALITY

Lake-shore Bar. On the south side of the Randsburg Road about 10 mi (16 km) east of California 14, a well-preserved gravel shoreline bar is present on the northeast side of Koehn Lake, and about 30 ft (9 m) above the lake. Although its northwestern part has been quarried for gravel, its geomorphic form is well-preserved about 0.5 mi (0.8 km) south of the railroad and suggests that the bar is relatively young. About 1 mi (1.6 km) south of the road (you can drive as far as the railroad) the Garlock fault crosses the bar and offsets it about 260 ft (80 m) left-laterally, and about 21 ft (7 m), north side down (Fig. 2A). Based on an age of about 11,000 years obtained on tufa deposits on gravel (Clark and Lajoie, 1974), this indicates an average displacement rate on the Garlock fault of about 7 mm/year. A trench across the fault trace, about 300 ft (90 m) west of the bar, exposed beds deposited during the most recent high water stand of the lake, which are about 14,000 years old (Burke, 1979).

Mesquite Fan. About 2 mi (3.2 km) east of the bar the Mesquite Canyon Road is on the left side of the highway. From here to the town of Garlock, about 1.5 mi (2.4 km) to the east, the fault cuts the old Mesquite Canyon alluvial fan, exposed on the north side of the fault, and a variety of Quaternary fault-line features are well-preserved (Fig. 2B). North of the fault, the fan has been broken by a group of tension graben that have been moderately modified by subsequent erosion, but are probably Holocene in age. The uplifted Mesquite fan (as well as present stream courses) is covered by coarse gravels derived from the Mesquite Schist, a distinctive quartz-sericite-albite schist nearly always containing porphyroblasts of chlorite, albite, or chloritoid, which is readily distinguished from any other schist in this part of the Mojave Desert. This distinctive schist crops out only in a small area near Mesquite Canyon, and gravels containing this schist could therefore only have been deposited south of the Garlock fault in the vicinity of Mesquite Canyon. Gravels containing clasts of this lithology are exposed south of the fault and east of this area, and therefore must have reached their present locations due to left-lateral displacement on the Garlock fault (Carter, 1980). Just above this intersection, the fault scarp exposes a group of tilted lacustrine strata that contain vertebrate fossils which are Pleistocene in age (Dibblee, 1952). The terrace gravels containing distinctive Mesquite Schist clasts unconformably overlie these Pleistocene lake beds, and therefore offset of the terrace gravels must have occurred entirely in the Pleistocene. Mesquite Schist clasts have also been identified in Pliocene and Miocene formations south of the fault in the Lava Mountains and therefore indicate amounts of Pliocene and Miocene displacement as well.

Goler Graben. About 3.5 mi (5.6 km) east of the town of Garlock and 0.5 mi (0.8 km) east of Goler Wash, a dirt road on the north side of the highway runs to Goler Heights. East of this intersection a depression is formed where the bank of lower Goler Wash has been offset about 0.5 mi (0.8 km) and sealed by alluvium deposited from the west (Fig. 2C). About 0.5 mi

(0.8 km) farther east, a large graben is formed by the north-facing scarp of the Garlock fault facing El Paso Mountains on the north side of the concealed El Paso fault. A few poorly defined tension graben cut the terrace gravels between Goler Wash and the west end of the graben.

The surface of the terrace gravels on the south side of the fault is covered by gravels that include large boulders deposited at the mouth of Golder Wash. The mountain front north of the fault is underlain by distinctive rocks of the Paleozoic Garlock series, consisting of slightly metamorphosed chert, shale, quartzite conglomerate, sandstone, basalt, andesite porphyry, and tuff. Gravels south of the fault contain boulders derived from the Black Mountain Basalt, and from the Goler Formation, which could only have been deposited at the mouth of Goler Wash, and indicate at least about 4 mi (6.4 km) of left-lateral offset of the gravels. Terrace gravels containing abundant clasts of Mesquite Schist are exposed on the face of the fault scarp east of the graben where they underlie the surface gravels deposited by Goler Wash (Fig. 2C). These deposits now lie about 7 mi (11 km) east of their depositional area near the mouth of Mesquite Canyon. A small fault bench exposes gravels that underlie the schist-bearing gravels and contain abundant clasts of hornblende quartz diorite, which is exposed west of Mesquite Canyon.

Quaternary terrace gravels containing abundant clasts of Mesquite Schist are found as much as 12 mi (19 km) east of Garlock. Clasts of Mesquite Schist are also found in Pliocene sediments in the Lava Mountains about 20 mi (32 km) east of their probable site of deposition. Large blocks of Garlock series and Mesquite Schist are found within the latest Miocene (mid to late Hemphillian) Bedrock Springs Formation in the eastern Lava Mountains about 29 mi (47 km) east of their probable site of deposition (Fig. 1C). These offsets indicate an average long-term slip rate of 7 mm/year on the Garlock fault, and suggest that the 40 mi (64 km) total left-lateral slip on this fault could have accumulated entirely within the last 8 to 9 m.y.

REFERENCES CITED

Burke, D. B., 1979, Log of a trench in the Garlock fault zone, Fremont Valley, California: U.S. Geological Survey Map MF-1028, scale 1:20.

Carter, B. A., 1980, Quaternary displacement on the Garlock fault, California, *in* Fife, D. L., and Brown, A. R., eds., Geology and mineral wealth of the California desert: South Coast Geological Society, p. 457–466.

Clark, M. M., 1973, Map showing recently active breaks along the Garlock and associated faults, California: U.S. Geological Survey Map I-741, scale 1: 24,000.

Clark, M. M., and Lajoie, K. R., 1974, Holocene behavior of the Garlock fault: Geological Society of America Abstracts with Programs, v. 6, p. 156–157.

Davis, G. A., and Burchfiel, B. C., 1973, Garlock fault; An intracontinental transform structure, southern California: Geological Society of America Bulletin, v. 84, p. 1407–1422.

Dibblee, T. W., 1952, Geology of the Saltdale Quadrangle, California: California Division of Mines Bulletin 160, p. 1–43.

Smith, G. I., 1962, Large lateral displacement on the Garlock fault, California, as measured from offset dike swarm: American Association of Petroleum Geologists Bulletin, v. 46, p. 85–104.

Searles Valley, California: Outcrop evidence of a Pleistocene lake and its fluctuations, limnology, and climatic significance

George I. Smith, U.S. Geological Survey, MS 902, 345 Middlefield Road, Menlo Park, California 94025

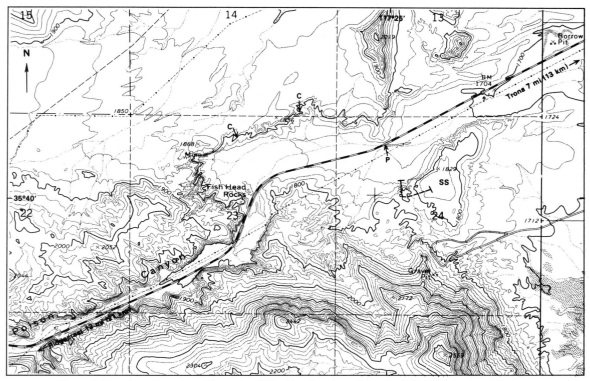

Figure 1. Location of stratigraphic section (SS). Also, location of suggested (dry-weather) parking area (P) and channels (C) that produce unconformities in lake beds north of highway. California 178 bisects map. From Westend 7½-minute Quadrangle, section boundaries are 1-mile apart.

LOCATION

Searles Valley, the site of Pleistocene Searles Lake, is 125 mi (200 km) north-northeast of Los Angeles and 40 mi (65 km) east of the south end of the Sierra Nevada. The main locality described here is 15 mi (24 km) east of Ridgecrest on California 178, at the mouth of Poison Canyon, and 0.7 mi (1.1 km) east of Fish Head Rocks (Fig. 1). Park well off the highway on the south side where the wash crosses the road—unless there is a possibility of rain! A short walk south, over a sandy fan surface, brings one to the outcrops of Pleistocene lake sediments (Fig. 2). Ascend and descend along the ridge marked on Figure 1 or the more gentle slopes 650 ft (200 m) southwest of it. It is advisable to wear boots or shoes with good traction.

SIGNIFICANCE

At this exposure of Pleistocene Searles Lake sediments, 11 undeformed beds composed of lacustrine gravel, sand, silt, marl, and tufa are exposed between the base and top of the 80-ft (25-m) high ridge. They were deposited on the floor of Searles Lake during a period that started about 30,000 B.P. and ended 10,000 B.P. Sediments representing the period between about 24,000 and 13,000 B.P. are absent from this section but are preserved on the north side of this amphitheater and elsewhere in Searles Valley. At this locality, these sediments were removed by sublacustrine—lake floor—erosion prior to deposition of the overlying beds. Dates on small snails and clams from sediment layers stratigraphically equivalent to those exposed near the base of this cliff, using ^{14}C methods, range from about 28,000 to 35,000 B.P.; dates from the next-to-top unit are as young as 11,000 B.P.

BACKGROUND

During much of the past 3.2 m.y., Searles Valley and its ancestral basins were sites of large lakes that fluctuated in size in response to regional climate (Smith and others, 1983). Most of the water came from the east side of the Sierra Nevada via the Owens River. Owens Lake and China Lake, upstream from

137

Figure 2. Pleistocene lake sediments described in this field guide. Lower three-fourths of stratigraphic section (Fig. 5) aligned with arrow.

Searles Valley, first had to fill and overflow before Searles received any water from that source (Fig. 3). When Searles Lake received large inflow volumes, it too filled and overflowed into Panamint Valley. When that valley filled, water overflowed into Death Valley. During the last 130,000 years, Searles Lake expanded and overflowed several times, but the lake in Panamint Valley expanded to only a fraction of its maximum area (Smith and Street-Perrott, 1983, p. 198).

The series of lakes in Searles Valley left evidence of their existence and character in the middle of the valley and around its edges. Each time the valley was inundated by deep lakes, sediments were deposited on its floor. Marl (calcium carbonate plus subordinate sand, silt, and clay) slowly accumulated on the floor of the deeper parts of the basin where currents were minimal, calcareous sand and silt accumulated on the shallower flanks of the basin where currents were stronger, and gravel, sand, and tufa (calcium carbonate masses deposited by algae) accumulated in the shallow turbulent water around the lake margin. Maximum water depth in Searles Valley was about 660 ft (200 m), and maximum lake area was about 400 mi^2 (1,000 km^2), twice the present area of Lake Tahoe.

During periods of increasing aridity, inflow diminished or ceased, and the lake became smaller or desiccated. When the lake became smaller but not dry, monomineralic salt layers sometimes accumulated in the center of the basin where they are now preserved as subsurface deposits, and coarse-grained lake deposits and tufa were deposited low on the valley sides. When the lake desiccated, beds composed of many salt species crystallized in the center of the basin. Upon exposure to subaerial conditions, the lacustrine sediments deposited previously on the flanks of the basin either underwent erosion or developed soils.

GEOLOGIC AND LIMNOLOGIC INTERPRETATIONS

The site described here is on the flanks of Searles Valley at an elevation of about 1,800 ft (550 m), a third of the distance between the present lake floor and the highest shoreline, and near the inlet that received water overflowing from China Lake (Fig. 3). After one climbs the ridge composed of lake sediments, the high shorelines can be seen on the lower half of the bedrock ridge 0.6 mi (1 km) to the south, and the present dry lake can be seen 3

to 6 mi (5–10 km) to the northeast where it appears as a large white area.

The history of lake fluctuations (Fig. 4), derived from geologic mapping of exposures around the entire valley plus subsurface data, shows that a site at this elevation was one of lacustrine deposition much of the time between 40,000 and 10,000 B.P. During the brief intervals within this span when the lake receded, those lacustrine deposits were exposed to subaerial erosion, but its effect at this site was minor because the catchment area was small. However, a major disconformity caused by sublacustrine erosion in a zone near the edge of the readvancing lake is evident at this site. This process is a result of wind-driven currents that first erode the poorly-indurated older sediments in the near-shore zone, smoothing any ridges and filling any channels, and then

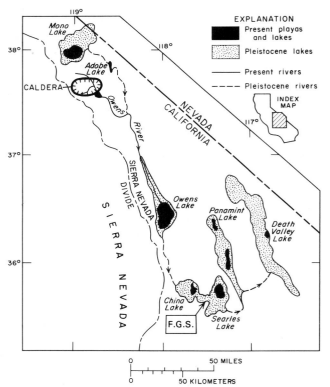

Figure 3. Pleistocene chain of lakes that included Searles Lake. Location of field-guide site (FGS) also indicated.

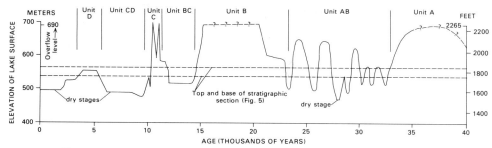

Figure 4. Reconstructed history of lake levels in Searles Valley, 0–40,000 yr ago.

deposit new sediments on this conspicuously planar surface—the most distinguishing characteristic of this form of erosion. The planar horizons that represent missing sediments or gaps in deposition are mostly in coarse-grained units, and they are *within* those units rather than at boundaries between differing lithologies. Many of these coarser-grained units display a change in bedding, color, grain size, or some other characteristic, near the middle of the bed, and this probably represents the actual hiatus.

Along the north side of this amphitheater, channels cut between periods of lacustrine deposition document episodes of subaerial erosion (Fig. 1), but evidence of this type is rare. More commonly, when the lake receded, the new channels reoccupied the approximate paths of earlier channels, eroding away evidence of previous subaerial erosion. Why? Most readvancing lakes rose rapidly to their maximum levels, and the deep-water lake deposits tended to drape themselves over the previous topography, including the channels; after the next cycle of lake sedimentation ceased, the sites of most old channels thus emerged as topographic troughs, and the new stream channels developed in them.

During geologic mapping of Searles Valley, deposits were grouped into units designated by letters. The sedimentary sequence exposed at this locality, and the informal designations of its units, are shown in Figure 5. There is also a unit A that is older than the sediments exposed here, units B and BC are missing because of sublacustrine erosion, and unit C4 was apparently not deposited at this elevation. Unit D, a small Holocene lake, is represented at this site by its beach, the smooth firm apron of sand that extends out from the base of the exposure of older sediments shown diagrammatically on Figure 5. This explains why this and other exposures at about this elevation in this amphitheater are good—they are all wave-cut cliffs.

A notable characteristic of unit AB, the lower two-thirds of this outcrop section, is the orange hue of the sand-and-gravel layers and the upward increase in its intensity. Unit AB7 represents the culmination of this characteristic; it thins to little more than a lag gravel where it is exposed on the south edge of this ridge, and the gravel fragments are stained a distinctive dark orange. What caused this? Some element of climate apparently promoted rapid deposition of iron-bearing desert varnish at the time this unit's surface was exposed to the atmosphere, and partial hydration of that varnish after burial then produced limonite, the mineral whose distinctive orange color now characterizes the zone. A buried desert varnish thus has the same geologic significance as a buried soil.

The contacts between the tops of outcropping sand layers and the bases of fine-sediment or marl layers are typically very sharp (Fig. 5). These contacts mark the time when the lake's "chemocline," which was migrating up the slopes of the lake floor during lake expansion, reaches this level. A chemocline is the horizontal boundary between discrete lake-water masses that have contrasting densities caused by differing salinities and thus remain stratified. Such lakes become two nearly-independent water bodies, each with its individual current, chemical, and biological regime. Where the lake floor was covered by the more dense water mass below the chemocline, wind-driven currents were minimal because wind energy was mostly absorbed by the water mass above it; those parts of the floor were also free from fresh-water swimming and burrowing organisms. The chemocline was also responsible for the deposition of annual(?) layers of aragonite, now observable as thin white laminae in both outcrop and subsurface deposits (Fig. 6). The laminae seem to have been products of the annual floods of fresh—but calcium-bearing—water caused by the early-summer influx of meltwaters from the Sierra Nevada snowpack. Each year, those waters initially flowed over the entire surface of the stratified lake. Later, however, because of late-summer evaporation or wind action, they mixed with the alkaline waters below the chemocline, causing rapid supersaturation with calcium carbonate and a "snowstorm" of microscopic white aragonite crystals on the lake floor.

Unit C1 is well exposed on the surface at the bend in stratigraphic section SS (Fig. 1). A thin layer of white tufa 3 ft (1 m) above the base of this unit is characteristic of the unit over most of this amphitheater area. Note, too, the gypsum crystals in the layer that are indicative of an increasing sulfate concentration in the stratified lake's lower brine body. Similar crystals are observed in this unit elsewhere in the valley, and this illustrates a general principle: Any mineralogic or lithologic character that reveals a quirk in lake-water chemistry is a more reliable basis for basin-wide stratigraphic correlation than any lithologic property related to lake energy levels or clastic composition; water chemistry tends to be uniform throughout a basin because waters within a lake or a stratified water mass tend to be well mixed, whereas distinctive energy regimes and sediment sources tend to be localized characteristics. The laminated character of units C1 and C3 is also found

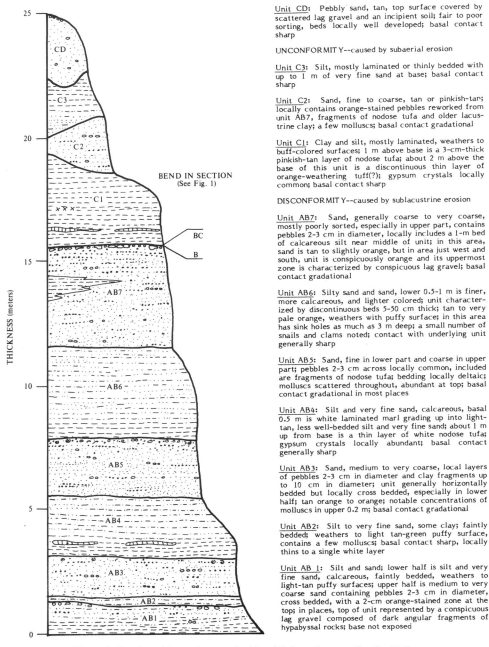

Figure 5. Stratigraphic section, south side of Poison Canyon, Searles Valley.

Unit CD: Pebbly sand, tan, top surface covered by scattered lag gravel and an incipient soil; fair to poor sorting, beds locally well developed; basal contact sharp

UNCONFORMITY--caused by subaerial erosion

Unit C3: Silt, mostly laminated or thinly bedded with up to 1 m of very fine sand at base; basal contact sharp

Unit C2: Sand, fine to coarse, tan or pinkish-tan; locally contains orange-stained pebbles reworked from unit AB7, fragments of nodose tufa and older lacustrine clay; a few molluscs; basal contact gradational

Unit C1: Clay and silt, mostly laminated, weathers to buff-colored surfaces; 1 m above base is a 3-cm-thick pinkish-tan layer of nodose tufa; about 2 m above the base of this unit is a discontinuous thin layer of orange-weathering tuff(?); gypsum crystals locally common; basal contact sharp

DISCONFORMITY--caused by sublacustrine erosion

Unit AB7: Sand, generally coarse to very coarse, mostly poorly sorted, especially in upper part, contains pebbles 2-3 cm in diameter, locally includes a 1-m bed of calcareous silt near middle of unit; in this area, sand is tan to slightly orange, but in area just west and south, unit is conspicuously orange and its uppermost zone is characterized by conspicuous lag gravel; basal contact gradational

Unit AB6: Silty sand and sand, lower 0.5-1 m is finer, more calcareous, and lighter colored; unit characterized by discontinuous beds 5-50 cm thick; tan to very pale orange, weathers with puffy surface; in this area has sink holes as much as 3 m deep; a small number of snails and clams noted; contact with underlying unit generally sharp

Unit AB5: Sand, fine in lower part and coarse in upper part; pebbles 2-3 cm across locally common, included are fragments of nodose tufa; bedding locally deltaic; molluscs scattered throughout, abundant at top; basal contact gradational in most places

Unit AB4: Silt and very fine sand, calcareous, basal 0.5 m is white laminated marl grading up into light-tan, less well-bedded silt and very fine sand; about 1 m up from base is a thin layer of white nodose tufa; gypsum crystals locally abundant; basal contact generally sharp

Unit AB3: Sand, medium to very coarse, local layers of pebbles 2-3 cm in diameter and clay fragments up to 10 cm in diameter; unit generally horizontally bedded but locally cross bedded, especially in lower half; tan orange to orange; notable concentrations of molluscs in upper 0.2 m; basal contact gradational

Unit AB2: Silt to very fine sand, some clay; faintly bedded; weathers to light tan-green puffy surface, contains a few molluscs; basal contact sharp, locally thins to a single white layer

Unit AB1: Silt and sand; lower half is silt and very fine sand, calcareous, faintly bedded, weathers to light-tan puffy surfaces; upper half is medium to very coarse sand containing pebbles 2-3 cm in diameter, cross bedded, with a 2-cm orange-stained zone at the top; in places, top of unit represented by a conspicuous lag gravel composed of dark angular fragments of hypabyssal rocks; base not exposed

throughout the valley because these structures, as interpreted, are products of basin-wide lake stratification—another expression of lake-water chemistry. Unit CD rests on a subaerially eroded surface and appears to be a fine-grained alluvium that was deposited immediately after the lake retreated below this elevation for the last time. Its lithology also suggests reworking by wind and colluvial processes.

PALEOCLIMATIC INTERPRETATIONS

Several types of paleoclimatic data are derived from the record of fluctuations in the size and depth of Searles Lake. The record chiefly reflects variations in the volume of runoff from its tributaries, although it does not differentiate quantitatively between changes in regional precipitation and temperature. Limits to these uncertainties can be estimated, however, and it appears that for Searles Valley to receive any contribution of water from the Owens River, its flow had to increase by at least 30% (and possibly by 160%); for Searles Lake to fill and overflow, Owens River flow had to increase by a minimum of 200% (and possibly by 500%) (Smith and Street-Perrott, 1983, table 10-2).

However, even if reconstructed precisely, the streamflow volumes that controlled the depth of water and sedimentation character in Searles Lake cannot be translated linearly into a regional record of climate-induced streamflow variations. This is because the sensitivity of lake sedimentation to climate change varied as a function of the regional hydrologic regime prevailing at the time. As discussed below, hydrologic regimes that produced intermediate levels of regional runoff (relative to the driest and wettest recorded) allowed lake levels and sedimentation in Searles Lake to respond most dramatically to moderate changes in hydrology, and the paleoclimatic records of those periods are therefore the most detailed. The driest and wettest regimes produced less sensitive records. During very arid periods like the present, which led to the desiccation of Searles and China lakes, sedimentation in both lakes changes little in response to climates that produced moderate variations in regional runoff; those variations only altered the frequency, size, and duration of their intermittent lakes, while fluctuations of Owens Lake represented the major hydrologic response. Conversely, during very wet periods, when Owens, China, and Searles lakes were full and overflowing, variations in regional runoff affected their overflow volumes but produced no sedimentation changes, though it was reflected downstream in the levels of the last lake in the chain.

The evidence from this outcrop indicating small-magnitude, short-period fluctuations in lake depth seems to indicate the existence at the time of an intermediate-level hydrologic regime. Evidence observed at this site also provides a more detailed and informative history of short (e.g., less than 2,000 yr) climate variations than many geologic records. Two examples: (1) The seven fluctuations indicated by units AB1 to AB7 cover a period about 10,000 yr long, allowing an average of 1,500 yr for each climatic scenario; (2) the five fluctuations recorded by units C1, C2, C3, C4, and CD required only about 1,500 years, allowing an average of only 300 years for the deposition of each unit. The fairly precise correlation of outcrops with subsurface deposits also allows the combining of data that suggest possible links between short-lived climatic phenomena and geologic processes. An example is the evidence from subsurface salt-mineral assemblages that the mean annual temperatures 24,000 to 26,000 B.P. were about 4°C below modern averages (Smith, 1979, p. 89–90). The possibility arises that the lower temperatures could be part of the explanation for the abnormally rapid development of desert varnish inferred to be the cause of the intense orange coloration of the top of unit AB7, the time-correlative of these saline assemblages.

OTHER AREAS OF INTEREST

About 5 highway-mi (8 km) northeast of this field-guide site is the first of three chemical plants in Searles Valley. Their existence is a result of the geologic history of Searles Lake. During several long periods in its geologic past, Searles was a large—but not overflowing—lake. When that situation existed, large ton-

Figure 6. Exposure showing horizontal, thin, white aragonite layers, north side of Poison Canyon.

nages of relatively rare dissolved elements accumulated in the lake because thermal springs related to a large caldera were active and were introducing those elements into the headwaters of the Owens River (Fig. 3). When the lake desiccated, those components crystallized as beds of brine-filled salts. Those brines now are the source of soda ash (Na_2CO_3), salt cake (Na_2SO_4), borate ($Na_2B_4O_7$), and potash (KCl), chemicals produced by the Kerr-McGee Chemical Corporation at its Trona, Argus, and Westend plants (Moulton, 1981). To date, the sales values of those products exceed $2 billion.

Because of the economic exploitation of the present-day Searles (dry) Lake itself, subsurface data have been obtained by the company for operational purposes, and many of the data have been released to researchers. A study of more than 100 cores of the top 165 ft (50 m) of lake deposits allowed a detailed reconstruction of the subsurface stratigraphy of the deposit; almost 100 ^{14}C dates on those cores provide time control for that history (Smith, 1979). One 3,050-ft (930-m) core, which extends to bedrock in the center of the basin, is especially informative (Smith and others, 1983). It provides documentation of lacustrine sedimentation in Searles Valley for the last 3.2 m.y. and is the basis for speculation that the most extreme shifts in the hydrologic regimes of this area have been responses to the globe's 400,000 yr orbital-eccentricity cycle, which also appears responsible for long cyclic changes in tropical sea-surface conditions (Smith, 1984). Unfortunately, few paleoclimatic records of this length are available to test this hypothesis. Figure 7 is a diagrammatic history of that long record and also illustrates why the wettest and driest hydrologic regimes produced a relatively uninformative record of paleoclimatic fluctuations in Searles Valley.

A nearby area of geologic and scenic interest is The Pinnacles, tufa towers, as much as 100 ft (30 m) high, constructed of algal-precipitated calcium carbonate beneath the surface of the Pleistocene lake (Fig. 8). To drive to them from the locality shown in Figure 1, follow the highway 1.6 mi (2.6 km) northeast, turn right on the gravel road, turn right again just prior to crossing the railroad tracks (the road crosses them eventually), and follow the road that parallels the tracks south for about 4 mi (6.5 km). Passenger cars have no difficulty on this road unless it is wet, but after heavy rains it is impassable to all vehicles. The light-colored lacustrine sediments along this route are mostly assigned to units B or C. About 1.3 mi (2 km) prior to reaching The Pinnacles, the road ascends a flat, dark, gravel-covered slope that dips 1°–3° northeast and extends 0.6 mi (1 km) to the southeast. This is a gravel bar built into Searles Lake when its surface remained near that level for a long period, ending probably about 35,000 B.P. The bar was later covered by finer lacustrine sediments, but they were mostly eroded during Holocene time, exhuming the original bar surface. Both pre- and post-bar lacustrine sediments are well exposed where the railroad tracks cut through this bar and in the stream-cut at its southeast end.

The Pinnacles probably document the sites of sublacustrine springs that injected calcium-bearing water into the slightly alkaline lake when its surface stood at or slightly above the present level of The Pinnacles, and remnants of aquatic algae (Scholl, 1960, Figs. 4–6) in the tufa show that they also participated in the

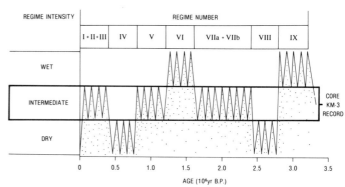

Figure 7. Diagrammatic portrayal of fluctuations in a hypothetical lake during last 3.2 million years. Record in Searles Valley resembles history in "Intermediate" hydrologic-regime box; fluctuations during "Dry" and "Wet" regimes recorded in upstream and downstream valleys, respectively.

carbonate-precipitation process. The main mass of the tufa towers is a light-tan, porous material that weathers to a lattice-like surface. Stratigraphic relations show that this tufa was deposited between about 130,000 and 35,000 B.P., prior to deposition of the lacustrine sediments visited at the main study site. Nearly 3 ft (1 m) of darker brown, nodular tufa coats the older tufa, and most of it was deposited during a short period about 16,000 B.P., near the end of deposition of unit B. A short time later, much of that tufa was eroded, and fragments of it are present throughout the basin at the base of unit C.

Figure 8. The Pinnacles and dissected lacustrine sediments, southwest of Searles Lake (light area between The Pinnacles and distant mountains).

REFERENCES CITED

Moulton, G. F., Jr., 1981, Compendium of Searles Lake operations: Transactions, American Institute of Mining, Metallurgical, and Petroleum Engineers, Society of Mining Engineers AIME, v. 270, p. 1918–1922.

Scholl, D. W., 1960, Pleistocene algal pinnacles at Searles Lake, California: Journal of Sedimentary Petrology, v. 30, no. 3, p. 414–431.

Smith, G. I., 1979, Subsurface stratigraphy and geochemistry of late Quaternary evaporites, Searles Lake, California: U.S. Geological Survey Professional Paper 1043, 130 p.

——— , 1984, Paleohydrologic regimes in the southwestern Great Basin, 0–3.2 m.y. ago, compared with other long records of "global" climate: Quaternary Research, v. 22, p. 1–27.

Smith, G. I., and Street-Perrott, F. A., 1983, Pluvial lakes of the western United States, *in* Porter, S. C., ed., Late Quaternary of the United States: University of Minnesota Press, p. 190–212.

Smith, G. I., Barczak, V. J., Moulton, G. F., Liddicoat, J. C., 1983, Core KM-3, a surface-to-bedrock record of late Cenozoic sedimentation in Searles Valley, California: U.S. Geological Survey Professional Paper 1256, 24 p.

Red Cinder Mountain and Fossil Falls, California

Pierre Saint-Amand, Code 013, Naval Weapons Center, China Lake, California 93555

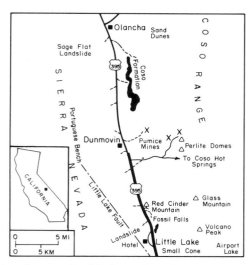

Figure 1. Locality map. Fossil Falls and Red Cinder Mountain are found on the east side of U.S. 395 about 3 mi (4.8 km) north of the town of Little Lake.

Figure 2. Red Cinder Mountain viewed from the northwest. The partially breached, phreatic cone is mantled with reddish brown vesicular cinders. Small hydrolaccoliths surround the cone. The background hills are granodiorite with dikes of aplite and alaskite. Flows of basalt cascade from the fault trough.

LOCATION AND SIGNIFICANCE

Turn right (east) on Cinder Road, 3 mi (4.8 km) north of Little Lake Hotel on U.S. 395 in Rose Valley, Inyo County, California (Fig. 1). Red Cinder Mountain lies just north of Cinder Road. After 0.5 mi (0.8 km), the road bifurcates. Turn south (right) and continue for 0.6 mi (1 km) over the asperate surface of a lava flow, passing hydrolaccoliths 50 ft (15 m) high and several hundred feet (>100 m) across, their upper surfaces domed and broken like crusts of over-risen bread. The road ends in a parking area. A trail to the southeast leads to the dry bed of the Owens River and Fossil Falls. These sites tell of interplay between climate change and volcanic activity.

SITE DESCRIPTIONS

Red Cinder Mountain (Fig. 2) is a partially breached phreatic cone, developed by vaporization of ground water beneath the flow upon which it stands. The cinders contain small, lath-shaped crystals of labradorite and augite, and dark brown inclusions of basaltic glass. The lava has more sodium, potassium, and titanium than is common in basalts (Chesterman 1956, p. 66–67). The light-weight vesicular scoria is used for making roofing granules and concrete bricks.

The cone appears to be but a few tens of thousand years old. Duffield and Smith (1978) suggest that it is 22,000 years old. If it is phreatic it should be contemporaneous with the flow upon which it stands. The flow has, however, a potassium-argon date of 140,000 ± 90,000 years BP, from a sample taken a mile to the south (Duffield and others, 1980). The flow was present during the Tioga glacial epoch.

Figure 3. Fossil Falls and the course of the Owens River, looking east. The river drained from the playas in the background, through Fossil Falls and into the flat playa below the falls. Relics of Indian occupation abound in this area.

The cone lies on the projection of a mile-wide zone of basaltic cones and rhyolite plugs that extends along a lineal depression to the vicinity of Airport Lake. Volcano Peak, a symmetrical cone of dark basalt, flanked by adventitious cones and accompanied by plugs of perlitic rhyolite, is surrounded with

Figure 4. Large potholes at the head of the canyon (one of them filled by a geochemist).

Figure 5. Water running from a crack at the base of Fossil Falls following a heavy local rain in 1985. Water rarely, if ever, runs over the top of the falls these days, but following rains it often descends through fissures in the body of the basalt. [Photo by David C. H. Saint-Amand]

lapilli from the eruption of the plugs. A basalt flow descends toward us from Volcano Peak. Lava from Volcano Peak is dated at 38,000 years (Duffield and others, 1980, p. 2396).

Fossil Falls formed when Owens River was dammed by the lava flow about 0.5 mi (0.8 km) east of the falls. Water accumulated in a shallow lake, now a playa (Fig. 3). An abandoned channel, 100 ft (30 m) wide, is cut 6 to 10 ft (2 to 3 m) into the lava flow. Water-carved boulders are strewn about the river bed. Numerous chips of obsidian from Glass Mountain, 3 mi (5 km) to the east, lie in the river bed. Many have been worked by Indians. House rings are found along the sides of the stream.

At Fossil Falls, a narrow canyon abruptly incises the flat bed of the river and descends steeply to the west. It soon reaches a depth of 40 ft (12 m) in the deepest part of the canyon. After 300 ft (100 m), the canyon widens into a smooth, flat-bottomed channel. Below this flat area, the canyon floor drops abruptly for about 30 ft (10 m), without the elaborate carving noted at the top. Blocks of rock lie below the lip. Water once flowed over the lip to a depth of at least 3 ft (1 m), as may be seen by scour marks on the sides of the canyon. In the flat river bed below, puddles of sun-warmed water have dissolved and undercut the sides of the boulders. All through the gorge, the rocks are carved into fantastic shapes. Large potholes abound (Fig. 4); one vertical hole, open at the bottom, is 2 to 3 ft (0.6 to 1 m) in diameter and 12 ft (4 m) deep. The carving action of the water is ascribed to semipermanent eddies formed during the flow, with erosive effects being due to suspended silt, sand and rocks; the basalt itself is slightly soluble in warm water. The lava is traversed by fractures through which water descends freely. Figure 5 shows water issuing from fractures during a recent rain. Water descending through such holes can produce rotational flow, like water draining from a sink. Heavily covered with desert varnish, the dense basalt gleams blackly in the sunlight; deep shadows within the canyon are but poorly illuminated by reflected light. It is a striking site for

photography. Footing is good, but polished rocks at the edges of the gorge are slippery when damp.

An enormous mass of water coursed through Fossil Falls until between 2,000 and 4,000 years ago (Gale 1914, p. 264; Smith and Street-Perrot, 1983). At the time of maximum filling, China and Searles lakes were joined and covered 384 mi^2 (1000 km^2). Owens River must have maintained a flow of at least 3,500 cubic feet a second over Fossil Falls, to supply the evaporational demands of the lakes.

REFERENCES CITED

Chesterman, C. W., 1956, Pumice, pumicite, and volcanic cinders in California: California State Division of Mines and Geology Bulletin 174, 93 p.

Duffield, W. A., and Bacon, C. R., 1981, Geologic map of the Coso Volcanic Field and adjacent areas, Inyo County, California: U.S. Geological Survey Miscellaneous Investigations Series Map I-1200, scale 1:50,000 with notes.

Duffield, W. A., and Smith, G. I., 1978, Pleistocene history of volcanism and the Owens River near Little Lake, California: U.S. Geological Survey Journal of Research, v. 6, p. 395–408.

Duffield, W. A., Bacon, C. R., and Dalyrymple, B., 1980, Late Cenozoic volcanism, geochronology, and structure of the Coso Range, Inyo County, California: Journal of Geophysical Research, v. 85,

Gale, H. S., 1914, in Salines in the Owens, Searles, and Panamint Basins, Southeastern California: U.S. Geological Survey Contributions to Economic Geology, 1913, Part I-L, Bulletin 580-L, 323 p.

Smith, G. I., and Street-Perrott, A., 1983, in Wright, H. E., ed., Pluvial lakes in the Western United States; Late Quaternary environments of the United States: Minneapolis, University of Minnesota Press, v. 1, p. 190–212.

32

Owens Lake, an ionic soap opera staged on a natric playa

P. Saint-Amand and C. Gaines, Code 013, Naval Weapons Center, China Lake, California 93555
D. Saint-Amand, Saint-Amand Scientific Consultants, P.O. Box 532, Ridgecrest, California 93555

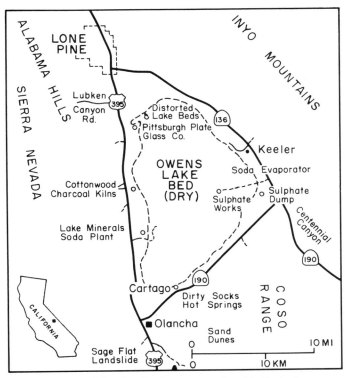

Figure 1. Map showing features mentioned in the text.

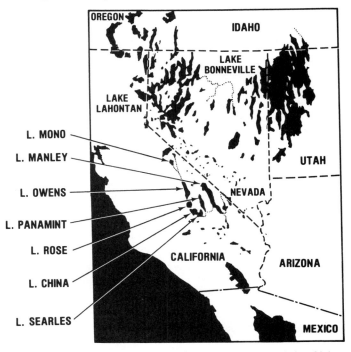

Figure 2. Pleistocene Lakes. Owens Lake was a part of a chain of lakes that stretched from Mono Lake to Death Valley. All the lakes except Mono Lake are now dry, and the streams no longer run.

LOCATION AND SIGNIFICANCE

Owens Lake, 17 mi (27 km) south of Lone Pine in Inyo County, California, is reached via U.S. 395, (Fig. 1). This 110-mi^2 (285-km^2) lake bed, once part of a chain of Pleistocene lakes, was full of saline water until the 1920s. Now dry, it shows the processes at work in a wet, natric playa and tells of climatic change and of the effect of man on the desert.

HISTORICAL SYNOPSIS

The chain of Pleistocene lakes that extended from Mono Lake to Lake Manly in Death Valley (Fig. 2) was full during most of the Pleistocene; Gale (1914, p. 264) concluded that the flow to the south of Owens Lake ceased 3,500 to 4,000 years ago. A well 920 ft (280 m) deep in the central part of Owens Lake revealed a continuous series of clays and silts but no buried salines (Smith and Pratt, 1957). This led Smith and Street-Perrot (1983, p. 198) to conjecture that the lake had not desiccated for several hundred thousand years. However, some episodic drying must have taken place, because extensive layers of coarser sediments beneath the clays give rise to artesian wells. Studies of salt bodies in Searles Lake reveal fillings and dryings not easily seen

in the other lakes in the series (Smith, 1979). The latest desiccation of Owens Lake was initiated by climatic change, accelerated by irrigation, and finished by export of water (Chalfant, 1933; Nadeau, 1974; Krahl, 1982).

In 1862, farmers began to move to Owens Valley because water was abundant. By 1917, 62,000 acres were in cultivation and 160,000 fruit trees had been planted (Newcomb, 1917); 450,000 acre ft (555 hm^3) of water were being used annually. Although the level of the lake dropped 16 ft (4.9 m) between 1894 and 1905, it later rose, due to above normal rainfall, and by 1911 had regained 18 ft (5.5 m), despite irrigation and export of water. The Department of Water and Power of Los Angeles (DWP) began to convey water from Owens Valley to San Fernando Valley in November 1913, and now exports 350,000 acre feet (430 hm^3) annually. The lake level began to fall in 1917. By 1926, the lake was dry, and most of the artesian wells and flowing springs within Owens Valley had dried up.

SITE DESCRIPTION

Start at Olancha, California (Fig. 1), and take California

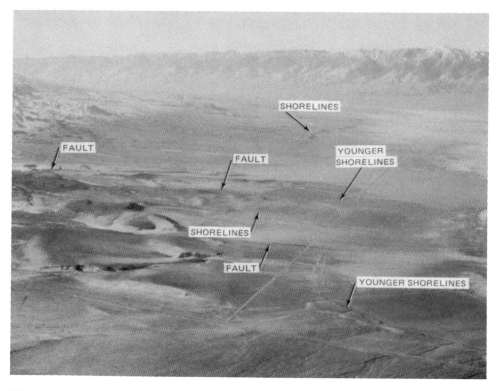

Figure 3. Shorelines around the south side of Owens Lake, looking southwest. A series of south-southeasterly trending faults offsets the Tioga Shoreline 1 mi (1.6 km) left laterally and 80 ft (24 m) vertically, up on the west side. A few younger shorelines can be seen between the Tioga lines and the 1913 shorelines at the extreme right of the photograph.

190 east to the junction with California 136. Shorelines of Tioga age ring the basin 300 ft (100 m) above the lake bottom (Carver, 1967), and are incised by small quebradas that have built subaerial fans onto the sloping surface beneath, but are otherwise uneroded. Thirteen mi (21 km) east of Olancha, faults multiply offset the Tioga shorelines 1 mi (1.6 km) left laterally (Fig. 3). Turn north on California 136 and continue for 4 mi (6 km). Turn west 1.4 mi (2 km) south of Keeler to the center of Owens Lake playa. In winter, the playa surface is covered with mirabilite; in summer, mirabilite melts, leaving thenardite.

A large pile of sodium sulphate lies to the south of the road, which leads to the ruins of a loading ramp. An artesian well, surrounded by grass, waters the surface. Ponds near artesian wells are saturated with sodium sulphate. The soda has disintegrated ties from an abandoned railway; the soft parts of the wood have been destroyed and only the hard, lignitic, winter rings remain. The poles of a tripod standing in a pool of water have an expanded section for 2 ft (0.6 m) above the maximum reach of the brine (Fig. 4).

The clay playa. The playa surface is hard in summer but soft in winter, when vehicles sink several inches into an efflorescent crust, and it cannot be traversed after rain or snow. When the surface temperature of the playa first reaches 95°F in the spring, polyhydrates lose water of hydration and wet the playa surface; this process cycles diurnally.

The playa is covered with a thin layer of windblown sand mixed with clay, and an alkali crust. The clay beneath is free of sand. When the surface is dry, the first few centimeters are a loose, fluffy layer of aggregated clay particles. The crust when dry arches above the clays, but when wet it collapses and the fluffy material reverts to a reconstituted plastic clay. The clays are illite and montmorillinite, with some chlorite. The clay beneath contains 40 to 50 percent water in about a 3 percent brine.

In summer, the upper clays dry. Desiccation polygons, tens of feet (10+ m) across, form on the surface. The edges of the polygons are covered with salts. Open cracks appear in the edges (Fig. 5). The clay, to a depth of 2 or 3 ft (~1 m), breaks into blocks a foot or less (<30 cm) in size, with lesser cracks an inch or so (2–3 cm) wide throughout the polygons. The cracks at the edges of the polygons fill with sand and form clastic dikes. The clay near the top is khaki to greenish gray in color. At depth, it is a bluish black, solonetz containing carbonized leaves and twigs. Near the clastic dikes the color is bleached to that of the surface (Fig. 6).

Efflorescent crust. Alkali crusts form on playas where the water table is less than 10 ft (3 m) below the surface. Crusts on

Figure 4. The effects of crystallizing salts are seen where the legs of the tripod are greatly expanded, just above the water. The town of Keeler is in the background; the dark areas are springs, where grasses grow in the fresher water. The surface is covered with a winter crust of halite, mirabilite, and natron.

Figure 5. Cracks forming on the surface of the playa in November 1985. A melting snow has dissolved most of the early crust, leaving holes in the surface developed by piping. The soil beneath is a loose aggregation of clay particles.

playas containing only sodium chloride, or nonhydrated minerals are usually damp and hard. The crust on Owens Lake contains sodium chloride, carbonate, sulphate, and minor amounts of borates, nitrates, potassium, and lithium.

The chemistry of the crust varies with the seasons as shown in Figure 7. The left side of the diagram explains the winter, the right side the summer situation. Begin at the "start" block. The first question asks if the temperature is greater than 18°C. If so, follow the yes arrow to the right—trona, thenardite, and halite form. If enough rain falls, the precipitates dissolve and move below the surface by the Soret effect, the forces of osmosis opposing those of capillarity. If the temperature rises above 65°C, thenardite and halite remain, but trona converts to thermonatrite. The crust is hard and not easily dislodged by wind.

In winter, when the temperature is below 18°C, trona and halite form, and thenardite changes to mirabilite in the presence of water. A transient dihydrate phase may precede the formation of mirabilite (Bernasovskii, 1953); it is likely that a septahydrate also forms. The mirabilite occupies 4.1 times the volume of the thenardite. This breaks the crust and separates the clay grains.

With enough rain, it all dissolves and starts over; with less rain, more mirabilite forms. Upon exposure to dry air, trona and halite are stable, but the uppermost mirabilite dehydrates to amorphous thenardite with a volume decrease. If the wind blows, a dust rich in sulphate-bicarbonate with particles of clay is ablated.

If the temperature drops below 10°C., mirabilite, amorphous sodium sulphate, and halite are stable but trona converts the natron with a 4.8 times volume increase. The hydrates grow on the bottom, encouraged by osmotic pressure operating in the same direction as the capillary forces, to bring water to the surface through the clay. They dehydrate as they grow, encouraged in part by an osmotic gradient of the water, to halite. Natron dehydrates to amorphous sodium carbonate, with a concomitant volume decrease. Anhydrous sodium carbonate and sodium sulphate are thus present in a thin farinaceous surface layer. Samples of the crust show no structure in X-ray crystallography, except for halite and silica, but chemical tests reveal the presence of carbonate and sulphate ions. A 15 knot wind will ablate a carbonate-sulphate–rich dust and particles of clay.

These conditions lead to dust storms when the wind blows

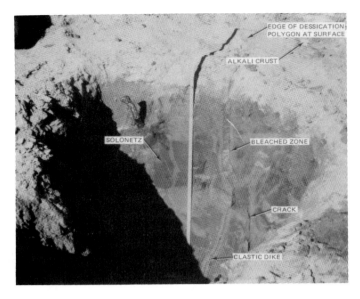

Figure 6. The crack with the piping in Figure 5 has been excavated. The thin, alkali crust on the surface is almost gone. The crack extends downward into the clays and fills with windblown sand, and with silts and clays washed in during the piping. These clastic dikes give the clays a vertical permeability. At about 1 ft (30 cm) depth, the clay is a deep blue-black solonetz. Numerous cracks develop from just above the vadose layer and extend upward into the clay.

in late fall, winter, and early spring. The storms interfere with air and surface transport and cause health problems (Saint-Amand and others, 1986). Dust from the Owens Lake playa has been tracked for 250 mi (400 km) south. Several tons of material are removed per second during large storms.

SODA PRODUCTION AT KEELER

This town of 100 souls had 5,000 people, schools, churches, two hotels, theatres, and other adult amusements in the early days. A soda extraction plant used the seasonal changes just described until 1904. Lake water was concentrated in shallow basins by solar evaporation. In hot weather, trona (summer soda) precipitated, was collected, dried, heated to drive off excess CO_2, and sold as sodium carbonate monohydrate. In winter, when the temperature fell below the stability field of trona, icelike crystals of natron (winter soda) formed in the pools, were collected, dehydrated, and sold as anhydrous sodium carbonate. Large amounts were used in China for making china. This operation was abandoned when the lake water became so concentrated that trona precipitated in the lake. In 1911, carbon dioxide from limestones and dolomites of the Inyo Mountains was used to precipitate trona. The operation continued until 1937 (Dub, 1947).

TECTONICS OF THE LAKE BASIN

In the northwestern "corner" of the lake, an area of distorted lake sediments has been uplifted, truncated, and dextrally offset by north-south faulting (Fig. 8). The site is accessible by foot starting about 1.2 mi (1.9 km) south of Lubken Canyon Road. This fault is dextral; the one on the other side of the lake is sinistral. They diverge slightly, hinting that the Coso Mountains are being displaced southward, leaving an area of low tectonic relief into which the main part of the lake is dropping, perhaps explaining the formation of this deep basin.

As the lake continued to dry, the brine became concentrated on the west side of the lake. The Pittsburg Plate Glass Company plant at Bartlett Point, one of the world's largest soda producers in its time, operated from 1929 to the 1950s. The brine was concentrated in evaporators; trona was precipitated by carbonation, calcined, and CO_2 recycled; and borax was recovered. Production ceased because of changes in the water level of the lake. A 6,920 ft (2,110 m) hole, drilled near the plant, passed through lake bed sediments and gravels but did not reach bedrock!

Lake Mineral Company controls the roads onto the playa and along the west bank of the lake. Responsible persons requesting access are welcomed. Trona is collected for use at Boron. Complex sodium minerals, including burkeite, and mirabilite are found in the area. Collectors should seal their samples in airtight containers and transport them in ice chests.

A short drive (<2 mi; 3 km) along the shore to the north of the Lake Minerals Company staging area leads to outcrops of an older formation of unknown provenance. This well-sorted, cemented, and folded pebble conglomerate is found along the west side of the lake from here northward. The nearest similar outcrop is in Darwin Wash, 5 mi (8 km) to the south-southeast on top of the Coso Mountains. The formations may not be related, but both are anomalous in location and similar in appearance.

BRINE POOL

One can walk to the water body from the Lake Minerals Company operational area, from the Pittsburgh Glass Plant, from the charcoal kilns, or from Cottonwood Spring. In dry years, it is 1 or 2 ft (30 to 60 cm) deep, and extends for about 1 mi (1.6 km) north-south and a few hundred feet (meters) east-west. Salts form on the bottom, and float on the top, as the many possible compounds precipitate. When water has been dumped in the lake, or following rains, the pond is larger. The properties change as salines are redissolved (Smith, 1979; Friedman and others, 1976). Owens Lake contains 1.6×10^8 tons of anhydrous salts (Gale, 1914) of which 6.7×10^7 tons are sodium chloride, 6.3×10^7 tons are sodium carbonate, 2.3×10^7 tons are sodium sulphate, and 7.4×10^6 tons are potassium, boron, iron, aluminum, lithium, and other elements.

On the shore of the brine pool, the surface temperature at noon often exceeds 165°F. Just below the surface, a gelatinous material, probably sodium silicate, thickens the brine. At a depth of a few centimeters, the temperature is usually 75°F. Fresh spring water enters the playa, mixes with the brine and forms

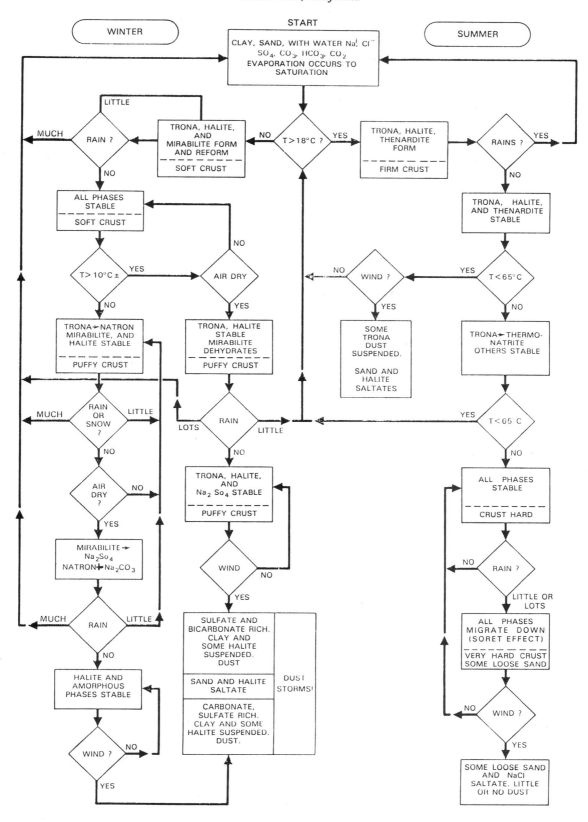

Figure 7. A logic diagram of the processes by which the alkali crust is formed on Owens Lake, and how the surface is conditioned to yield dust storms.

quicksand. It is unwise to wade in this area. Scratches produced by falling through the crust into the brine heal slowly because the carbonates saponify subcutaneous fats.

Hydrobiology. Algae such as *Dunaliela salina, D. viridis,* and halophilic bacteria such as *Halobacterium, Chromatium, Ectothiorhodospira,* and *Halococcus* (Larsen, 1980, p. 25 and 30), which have red carotenoids in their cells, color the water red. The microbiota metabolize sulphates and modify the carbonate dioxide content of the brine (Tew, 1980).

CLIMATES CHANGE

During the aluvial episodes, rainfall of at least 20 in (50 cm) per year prevailed over most of the desert. The rain probably came from the Pacific Ocean to the south, rather than from frontal storms of northern provenance that are now prevalent.

Under the influence of the present-day semipermanent high pressure area, a strong diurnal heating develops. Heated air rises at the sides of the valleys and descends in the central parts, being reheated by compression during descent. The heating produces a low-level, dry, thermal low that advects surface air from the desert to the south and east, further warming the region. If the Great Basin were damp enough to preclude the formation of a permanent high, the climate would become much more benign.

The drying trend exacerbates the desertification, which in turn reinforces the permanent high. Deserts are selfpropagating. The propagation is often accelerated by human activity, as has happened in the Sahara and the Sahel of Africa, the Caspian Basin, and India. The world has become much drier since biblical times. We are seeing this elsewhere in the arid west, but that is another story.

REFERENCES CITED

Bernasovskii, V. Y., 1953, Natural dehydration of mirabilite: Vestnik Akademii Nauk Kazakhskoi SSR, v. 10, no. 12, (whole no. 105), p. 87–89.

Carver, G. A., 1967, Shoreline deformation at Owens Lake: California Geology, v. 28, p. 111.

Chalfant, W. A., 1933, The story of Inyo: Lone Pine, California, Chalfant Press, 430 p.

Dub, G. D., 1947, Owens Lake; Source of sodium minerals: American Institute of Mining and Metallurgical Engineers Technical Publication 2235, p. 1–13.

Friedman, I., Smith, G. I., and Hardcastle, K. G., 1976, Studies of Quaternary saline lakes; Isotopic and compositional changes during desiccation of the brines in Owens Lake, California 1969–1971: Geochimica et Cosmochimica Acta, v. 40, p. 501–511.

Gale, H. S., 1914, Salines in the Owens, Searles, and Panamint basins, southeastern California: U.S. Geological Survey Contributions to Economic Geology, 1913, pt. I-L, Bulletin 580-L, 323 p.

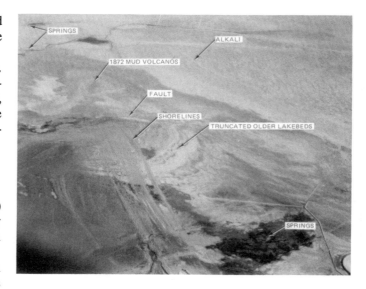

Figure 8. Looking northeasterly from the northwest corner of Owens Lake. The fault called out is a branch of the fault that produced the 1872 earthquake. Abandoned shorelines overlie a series of folded and faulted lake bed sediments truncated so that the edges are exposed. These lakebeds are repeatedly offset to the right by faulting. Thus, the Coso Mountains may move relatively southward between the two slightly divergent fractures.

Kahrl, W. L., 1982, Water and power: Berkeley and Los Angeles, University of California Press, 574 p.

Larsen, H., 1980, Ecology of hypersaline environments, *in* Nissenbaum, A., ed., Hypersaline brines and evaporitic environments, Proceedings of the Bat Sheva Seminar on Saline Lakes and Natural Brines: Amsterdam, Elsevier Scientific Publishing Company, p. 23–39.

Nadeau, R. A., 1974, The water seekers: Santa Barbara, California, Peregrine Smith, Incorporated, 278 p.

Saint-Amand, P., Mathews, L. A., Gaines, C., and Reinking, R., 1986, Dust storms from Owens and Mono valleys, California: China Lake, California, Naval Weapons Center Technical Publication 6731, 79 p.

Smith, G. I., 1979, Subsurface stratigraphy and geochemistry of late Quaternary evaporites, Searles Lake, California: U.S. Geological Survey Professional Paper 1043, 1130 p.

Smith, G. I., and Pratt, W. P., 1957, Core logs from Owens, China, Searles, and Panamint basins, California: U.S. Geological Survey Bulletin 1045-A, p. 1–62.

Smith, G. I., and Street-Perrott, A., 1983, Pluvial lakes in the western United States, *in* Wright, H. E., ed., Late Quaternary environments of the United States: Minneapolis, University of Minnesota Press, p. 190–212.

Smith, G. I., Barczak, V. J., Moulton, G. F., and Liddicott, J. C., 1983, Core KM-3, a surface-to-bedrock record of late Cenozoic sedimentation in Searles Valley, California: U.S. Geological Survey Professional Paper 1256, 24 p.

Tew, R. W., 1980, Halotolerant *Ectothiorhodospira* survival in mirabilite; Experiments with a model of chemical stratification by hydrate deposition in saline lakes: Geomicrobiology Journal, v. 2, no. 1, p. 13–20.

Late Quaternary fault scarp at Lone Pine, California; Location of oblique slip during the great 1872 earthquake and earlier earthquakes

Lester K. C. Lubetkin, U.S. Forest Service, Placerville, California 95667
Malcolm M. Clark, U.S. Geological Survey, 345 Middlefield Road, Menlo Park, California 94025

LOCATION

To reach this site, drive west on Whitney Portal Road about 0.7 mi (1.2 km) from U.S. Highway 395 in the center of Lone Pine (Fig. 1). About 0.15 mi (0.2 km) west of the Los Angeles Aqueduct, park in the paved area north of the road. Walk north and then northeast about 0.3 mi (0.5 km) along the dirt roads that lead to the area just east of the main scarp.

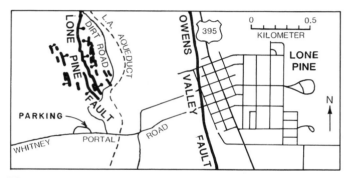

Figure 1. Index map showing main trace of Owens Valley fault and field site at Lone Pine fault. Ball on downthrown side of fault scarps. Location shown on Figure 2. Base from U.S. Geological Survey Lone Pine quadrangle, 1:24,000 scale, 1982.

INTRODUCTION

The great earthquake of March 26, 1872, in Owens Valley (M~8) was one of the three largest historic shocks in California. The earthquake was associated with extensive surface ruptures along the Owens Valley fault zone and caused strong ground shaking throughout a vast region (Whitney, 1872; Oakeshott and others, 1972). At this site, a prominent fault scarp preserves some of the clearest evidence in the Owens Valley fault zone of slip during this great earthquake and earlier events. Here, the fault scarp crosses an abandoned alluvial and outwash fan of Lone Pine Creek. The fan preserves a record of both horizontal and vertical fault displacement after abandonment. This record has been interpreted from scarp morphology, weathering characteristics of the fan and scarp, and history of the fan. Study of this scarp is important to the interpretation of tectonic history and assessment of earthquake hazard in this part of California.

Each of us must try to make a minimum impact upon this exceptional site. Because relatively fragile surface morphology is a critical source of information here, we request that you "tread lightly" to avoid disturbing the fault scarp and fan surface. Please

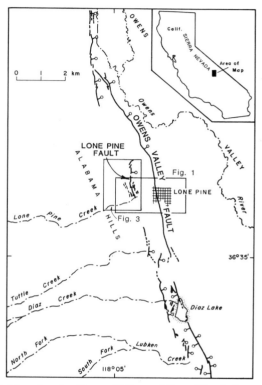

Figure 2. Map showing Owens Valley and Lone Pine faults near Lone Pine. Ball on downthrown side of fault scarps; faults dashed where approximately located.

do not climb on the scarp at any place where you will dislodge material. We hope that future investigators will not be thwarted by our carelessness. Please do not accelerate the natural erosional processes at work on the scarp and fan surface.

THE LONE PINE FAULT

The Owens Valley fault zone near Lone Pine consists of a main trace (Owens Valley fault), several prominent secondary traces, and many smaller traces (Fig. 2). The main fault trace extends across the western part of Lone Pine, and its scarp forms the east side of Diaz Lake to the south. Early investigators reported as much as 16 ft (4.8 m) of right-lateral displacement along the main trace at Lone Pine in 1872 (Hobbs, 1910). Vertical displacement also occurred along the main trace to the north and south. Prominent secondary traces lie as much as 1 mi (1.4 km) west of the main trace at Lone Pine and extend southward

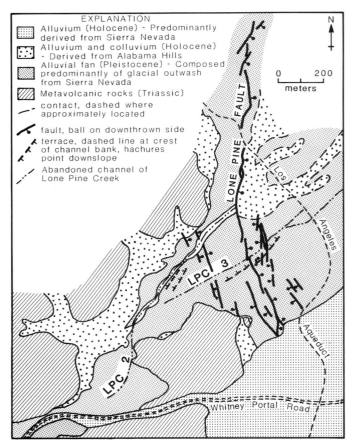

Figure 3. Geologic map showing scarp of Lone Pine fault across abandoned fan of Lone Pine Creek. LPC 2 and LPC 3 are former channels of Lone Pine Creek; LPC 3 is older. Location shown on Figure 2.

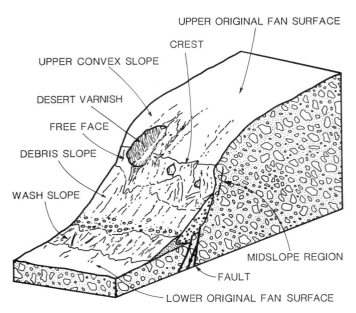

Figure 4. Idealized profile of fault scarp, modified from Wallace (1977).

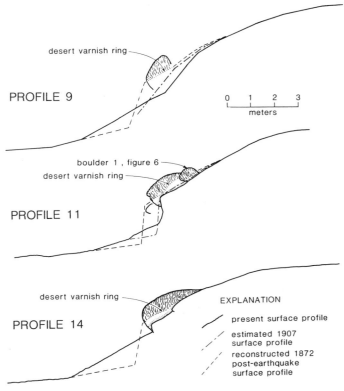

Figure 5. Present and reconstructed 1872 post-earthquake and 1907 profiles across Lone Pine fault scarp. Locations of profiles shown on Figure 7.

discontinuously to the west side of Diaz Lake. Lubetkin (1980) informally named the most westerly of these prominent secondary faults the Lone Pine fault. This field guide is adapted in part from Lubetkin and Clark (1985).

At this locality, scarps of the Lone Pine fault offset the surface of the abandoned fan of Lone Pine Creek. The fan is composed of alluvium and glacial outwash from the Sierra Nevada to the west. The fan surface is inactive now except for encroachment of local colluvium and infrequent debris flows from the flanking Alabama Hills along one of the relict channels on the fan surface. This inactivity has helped to preserve the record of fault displacement. If you go westward to the head of the fan, you will see the cross-cutting relationships of the modern and abandoned channels of Lone Pine Creek (Fig. 3).

The fault scarp has an upper convex slope, a steep midslope, and a lower concave portion (Fig. 4). The scarp is compound, the result of more than one slip event. The steep midslope is the erosionally modified scarp of the most recent slip event (1872), whereas the upper convex slope is the modified scarp of one or more older slip events. The lower concave portion is a sedimentary apron that conceals the original fault scarp and consists of the debris slope and wash slope, derived from the upper part of the scarp. Several north-facing ramps separate left-stepping en echelon scarps. Such ramps are common features of faults with both right-lateral and vertical displacement (Bateman, 1961).

Figure 6. Lone Pine scarp in 1907 (left) and 1978 (right), showing changes in free face and debris and wash slopes during 70 years. Little or no change has occurred on the upper convex slope (above A-A'). Individual cobbles and boulders (numbered) are recognizable in both photographs, although camera positions were not identical. View west at profile 11, 115 ft (35 m) north of LPC 3 (Figs. 5, 7). 1907 photo is W. D. Johnson, no. 685.

1872 DIP SLIP

To understand how this fault scarp developed, the amount of dip slip during the 1872 earthquake must first be determined. Although an early report suggested that the entire scarp developed in 1872 (Hobbs, 1910), later investigators recognized that the scarp is the product of multiple slip events (Oakeshott and others, 1972). We have estimated 1872 dip slip from scarp weathering and profiles at many places along the scarp, as well as exposures in a backhoe trench (Lubetkin and Clark, 1985).

We estimate a dip slip component of 3–7 ft (1–2 m) along the Lone Pine fault in 1872 by measuring the slip on reconstructed 1872 post-earthquake profiles (Fig. 5). To reconstruct the 1872 post-earthquake scarp profiles, we used features of the present fault scarp that record, or are remnants of, the scarp that was here before the 1872 earthquake. These features include the upper convex slope, the wash slope, exposed desert varnish rings, caliche coatings on clasts, and weathering of cobbles and boulders. The exposed desert varnish rings show the actual position of an earlier ground surface (Smith, 1979), whereas caliche coatings and disintegrated clasts, which develop close to the ground surface, commonly indicate only a lower limit for the

position of the earlier surface. The reconstructed post-earthquake profiles closely approximate the scarp just after the earthquake, before erosional or depositional modification. Notice that many of the larger clasts and boulders in the midportion of the scarp have caliche coatings on the bottom sides or have desert varnish rings.

Detailed comparisons between 1907[1] and 1978 conditions at various places along the fault scarp show little degradation or aggradation of the upper convex slope during this period, except for incision by some narrow rills (Fig. 6). In contrast, material from the midslope has buried the lower portion of the scarp. This nearly stable upper convex slope, and its downward projection to a surface represented by the varnish rings, closely mark the true surface profile west of the Lone Pine fault before and immediately after the 1872 earthquake.

The lower portion of the reconstructed profile includes the original lower fan surface, plus the wash-controlled slope and its projection into the fault (Fig. 5). A large debris slope now covers

---

[1] The earliest detailed photos of the scarp were taken by W. D. Johnson of the USGS in 1907 and are in the USGS library, Denver. The estimated 1907 surface profiles of Figure 5 were reconstructed from these photographs. Johnson's investigation of the 1872 earthquake was reported by Hobbs (1910).

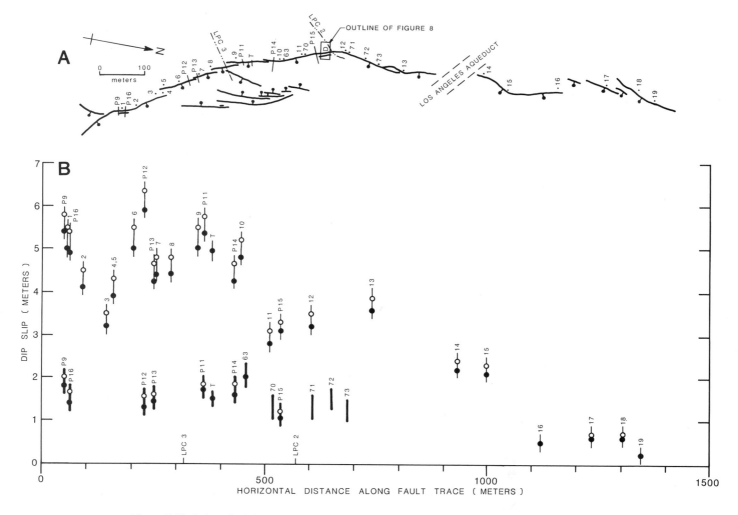

Figure 7. Variation of 1872 and total dip slip along Lone Pine fault. From Lubetkin and Clark (1985). A) Plan view of Lone Pine fault showing locations of numbered profiles (P), displacement measurement sites (numerals), trench (T), and offset debris flow (D), along abandoned fan of Lone Pine Creek. LPC 2 and LPC 3 are abandoned channels of Lone Pine Creek. See Figure 3 for location. B) Estimated dip slip in 1872 (heavy lines) and cumulative dip slip since abandonment of fan (light lines) at locations shown in A), for assumed fault dips of 70° (open circles) and 90° (solid circles). Length of lines approximates uncertainty in measurements.

the base of the scarp and the older wash slope. Between slip events, material eroded from this debris slope moves down to form the wash slope.

The position of the 1872 rupture surface in the fault scarp can be approximated from the position of boulders in the midslope of the scarp, together with the observed position of the fault-plane in the backhoe trench and other exposures. Figure 7 shows estimated 1872 and total dip slip for these profiles and measurements.

LATE QUATERNARY SLIP, RECURRENCE INTERVAL, AND SLIP RATE

As you follow the scarp northward, notice the change in

lithology of the material in the scarp where a relict channel (LPC 2, Fig. 3) crosses the fault. A debris flow complex fills much of this channel and has been offset by the Lone Pine fault. This complex consists of many individual, tongue-like masses of debris derived from the Alabama Hills. At the south edge of the complex, the fault displaces one of the most recent debris flows right-laterally 35–40 ft (10–12 m) and vertically 6–8 ft (1.9–2.4 m), east side down (Fig. 8). Horizontal offset of this debris flow and the older relict channel 700 ft (220 m) to the south (LPC 3, Fig. 3) is under investigation. Margins of the older channel indicate horizontal offset of about 50 ft (16 m). This horizontal offset, combined with total vertical offset of about 20 ft (6 m) near the older channel yields a minimum oblique offset of the fan surface of about 55 ft (17 m) at the older channel.

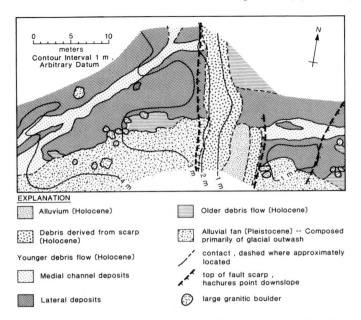

EXPLANATION

Alluvium (Holocene)	Older debris flow (Holocene)
Debris derived from scarp (Holocene)	Alluvial fan (Pleistocene) -- Composed primarily of glacial outwash
Younger debris flow (Holocene)	contact , dashed where approximately located
Medial channel deposits	top of fault scarp , hachures point downslope
Lateral deposits	large granitic boulder

Figure 8. Geologic map of offset debris flow along Lone Pine fault. Location shown on Figure 7.

Figure 9. Boulder in scarp of Lone Pine fault that apparently records 3 slip events. Line A separates areas of different weathering, and coincides with the projection of the original upper fan surface. The part of this boulder above line A apparently lay above the ground surface before scarp development. Two rings of desert varnish, B and C, may record later stable positions of the ground surface, each following prehistoric faulting. Rod is 5 ft (1.5 m) long. View north between profiles 12 and 13 (Fig. 7).

Several lines of evidence indicate that displacement from three earthquakes created the scarp. The maximum height of the southern part of the scarp is about 16–20 ft (5–6.5 m) in the same area where the maximum component of dip slip in 1872 was about 5–7ft (1.5–2 m) (Fig. 7). If the 1872 earthquake was a characteristic event (Schwartz and Coppersmith, 1984) for this fault, then the scarp was created by displacement during three such earthquakes. Desert varnish on the large boulder shown in profile 11 (Fig. 5) indicates that dip slip during the earthquake before 1872 was at least 3 ft (1 m) (Lubetkin and Clark, 1985). That at least three earthquakes created the scarp is indicated by coatings on a 10 × 15 ft (3 × 5 m) boulder exposed in the scarp between profiles 12 and 13 (Fig. 7). These coatings record at least three slip events (Fig. 9). Stratigraphic evidence in a trench excavated north of profile 11 also suggests three slip events (Lubetkin and Clark, 1985).

The postulated 1872-type earthquakes recorded by the scarp permit calculations of average late Quaternary earthquake recurrence intervals and fault slip rates. The maximum age of the fault scarp is limited by that of the fan surface, which in turn is limited by a radiocarbon date of about 21,000 years on tufa from a nearby shoreline of ancient Lake Owens. Short stretches of ancient shorelines survive along the eastern flank of the Alabama Hills at a maximum elevation of 3750 ft (1,144 m), about the same elevation as the Los Angeles Aqueduct. On pre-aqueduct photos, the old shoreline does not appear to cut the surface of the abandoned fan; moreover, geomorphic and stratigraphic evidence suggests that the fan was not built into a lake at this high-water stand (Lubetkin and Clark, 1985).

The minimum age of the fan surface appears to be about 10,000 years. The degree of weathering of granitic boulders, soil oxidation color, and surface morphology of the fan, relative to that of glacial deposits along the eastern slopes of the Sierra Nevada, suggest the abandoned fan is contemporary with the Tioga glaciation (Lubetkin and Clark, 1985), which ended about 10,000 years ago (Burke and Birkeland, 1979).

Three slip events along the Lone Pine fault during the past 10,000–21,000 years give an average recurrence interval that ranges from 3,300 to 10,500 years. The 3,300-year interval assumes that the fan surface was deposited 10,000 years ago, just after an earthquake. The 10,500-year interval assumes the fan was deposited 21,000 years ago just before an earthquake. The estimate of average recurrence interval for 1872-type earthquakes in turn leads to an estimate of late Quaternary slip rate for the Lone Pine fault. Total oblique slip of about 55 ft (17 m) at the older offset channel (LPC 3) yields an average oblique slip of about 19 ft (5.7 m)/event. Combined with an average earthquake recurrence of 3,300–10,500 years, this average oblique slip gives slip rates that range from .02 to .07 in (0.5 to 1.7 mm)/yr.

These recurrence intervals and slip rates are tentative, awaiting further confirmation of the dates of the fan surface and postulated horizontal slip. Some investigators think the fan may be older; others think the young debris flow at LPC 2 may have been offset by only one event. Furthermore, the Lone Pine fault is but one of several strands within the Owens Valley fault zone, and the average recurrence interval for major late Quaternary earthquakes, and the nature of slip for the Lone Pine fault, may not define activity over the entire fault zone. Thus, investigations of the Lone Pine fault, and its relation to other faults of the Owens Valley fault zone, continue.

NOTE ADDED IN PROOF

Our further research on this fault[1] and recent work on the Owens Valley fault zone by Sarah Beanland (written communication, 1986) modify some of these conclusions. Most important, fan morphology near LPC 2 indicates that the first earthquake occurred before abandonment of the fan. This change in our interpretation would increase minimum recurrence interval[1] somewhat and comparably decrease slip rate.

---

[1]Lubetkin, L.K.C., and Clark, M. M., Late Quaternary activity along The Lone Pine fault, eastern California: Geological Society of America Bulletin (in press).

REFERENCES CITED

Bateman, P. C., 1961, Willard D. Johnson and the strike-slip component of fault movement in the Owens Valley, California earthquake of 1872: Bulletin of the Seismological Society of America, v. 51, p. 483–493.

Burke, R. M., and Birkeland, P. W., 1979, Reevaluation of multiparameter relative dating techniques and their application to the glacial sequence along the eastern escarpment of the Sierra Nevada, California: Quaternary Research, v. 11, p. 21–51.

Hobbs, W. H., 1910, The earthquake of 1872 in the Owens Valley, California: Beitrage zur Geophysik, v. 10, p. 352–385.

Lubetkin, L.K.C., 1980, Late Quaternary activity along the Lone Pine fault, Owens Valley fault zone, California [unpublished M.S. thesis]: Stanford University, California, 85 p.

Lubetkin, L.K.C., and Clark, M. M., 1985, Late Quaternary activity along the Lone Pine fault, eastern California, *in* Stein, R. S., and Bucknam, R. C., eds., Proceedings of workshop XXVII on the Borah Peak, Idaho, earthquake: U.S. Geological Survey Open-File Report 85-290A, p. 118–140.

Oakeshott, G. B., Greensfelder, R. W., and Kahle, J. E., 1972, 1872–1972—one hundred years later: California Geology, v. 25, p. 55–61.

Schwartz, D. P., and Coppersmith, K. J., 1984, Fault behavior and characteristic earthquakes: Examples from the Wasatch and San Andreas faults: Journal of Geophysical Research, v. 89, p. 5681–5698.

Smith, R.S.U., 1979, Holocene offset and seismicity along the Panamint Valley fault zone, western Basin and Range Province, California: Tectonophysics, v. 52, p. 411–415.

Wallace, R. E., 1977, Profiles and ages of young fault scarps, north-central Nevada: Geological Society of America Bulletin, v. 88, p. 1267–1281.

Whitney, J. D., 1872, The Owens Valley Earthquake: Overland Monthly, v. 9, p. 130–140, 266–278; *Reprinted in 1888 in* California Division of Mines, 8th Annual Report of the State Mineralogist, p. 224–309.

Papoose Flat pluton, Inyo Mountains, California

C. A. Nelson, Department of Earth and Space Sciences, University of California, Los Angeles, California 90024

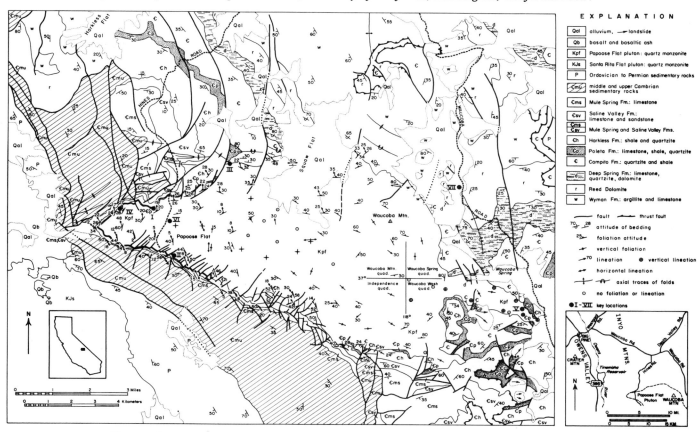

Figure 1. Geologic map of Papoose Flat pluton, Inyo Mountains, California. From Sylvester and others, 1978. Key locations are indicated by Roman numerals.

LOCATION AND ACCESS

Papoose Flat pluton, a late Cretaceous (75–81 Ma) granitic body in the Inyo Mountains, California, is one of several Mesozoic intrusives in the White–Inyo Range that are "satellites" of the large Sierra batholith to the west. The western portion of the pluton can be reached by way of the Hines Road (Waucoba Mountain 15-minute Quadrangle) (Fig. 1, insert); the eastern exposures by way of the Waucoba Road leading to Saline Valley (Waucoba Spring and Waucoba Wash 15-minute quadrangles). The published geologic maps of these quadrangles (Nelson, 1966, 1971; Ross, 1967) and the geologic map of the pluton (Nelson and others, 1978) are important adjuncts to this report. Only the Hines Road requires a four-wheel drive vehicle.

SIGNIFICANCE OF SITE

The pluton is composed chiefly of quartz monzonite, with large megacrysts of K-feldspar very abundant in the foliated border zone of the pluton. It crops out as an elongated E-W

trending dome, 9.9 mi (16 km) long by 4.9 mi (8 km) wide, emplaced along the south-west flank of the Inyo anticline, a major SE plunging structure (Fig. 2), into a succession of relatively unmetamorphosed late Precambrian to Cambrian strata.

Several structural features associated with the pluton are particularly notable and unusual. The contact with the country rocks is everywhere sharp; along the western two-thirds of the body, the contact is also concordant; the contact along the eastern (deeper) portion of the pluton is gently to strongly discordant. Along the concordant western portion of the body the stratigraphic units have been tectonically thinned to as little as 10% of their regional thicknesses, without loss of stratigraphic identity or continuity; there is little or no attenuation of the section near the discordant contact in the east. The rocks in the narrow metamorphic aureole in both the pluton and adjacent units are strongly foliated and lineated near the concordant (western) contact and have the aspect of regionally metamorphosed rocks, whereas those near the discordant (eastern) part of the pluton have textures and structures more characteristic of hornfelses. Also, metamorphic foliation and lineation are less common in the eastern

157

than in the western portion of the body. The pluton contains zoned potash feldspar megacrysts, especially in the border zone of the concordant west portion. Identical megacrysts have been found in the adjacent country rocks at the discordant east end, suggesting the possibility of a metasomatic origin for the megacrysts in the pluton. The presence of a thin garnet-epidote skarn at many of the contacts and the cross-cutting nature of the eastern contacts indicate that the principal mass is of magmatic origin.

STRUCTURAL AND STRATIGRAPHIC FRAMEWORK

The geologic maps (Figs. 1 and 2) illustrate the position of the pluton athwart the southwest side of the Inyo anticline. Initial emplacement into the Campito Formation, massive fine-grained sandstone and siltstone, was along pre-granite faults; continued upwelling of magmatic material caused the formation of a large westerly directed "blister" that is expressed as the major disruption of the regional strike of the initial homocline. This westward deflection affects all stratigraphic units from the middle of the Campito Formation upward. Although palinspastic restoration of post-emplacemen structures reduces the apparent westward bulge, the disruption of the general structure is a striking feature of the pluton.

Along the north contact of the pluton, the lowermost Campito Formation and the strata beneath the Campito—the Deep Spring, Reed, and Wyman formations—have also been affected by the emplacement of the pluton, but to the northeast. North of the pluton these units have the characteristic regional southeasterly strike; as they approach the pluton contact, they steepen and eventually turn to a more easterly strike and are overturned to dips as low as 30°. This represents a significant "blistering" in a north-northeasterly direction as well as to the west, and supports the suggestion that it was the faulted Campito Formation that provided the avenue for upwelling of magmatic material, and which was "split" and shouldered aside by the pluton.

A major significant feature of the Papoose Flat pluton is the striking concordance around its western contact. From a point on the northern border, to a point on the southern border directly across the pluton (Fig. 1), the pluton-metasedimentary rock contact lies within the Poleta Formation and within approximately 50 ft (15 m) of section, over an outcrop distance of 9 mi (15 km). Beyond these limits, the contact is only slightly discordant and for long distances remains parallel to stratigraphic contacts within other formational units. Foliation in the border zone of the pluton is parallel in strike and dip to foliation and stratigraphic bedding in the metasedimentary rocks.

Over the western two-thirds of the contact, granitic apophyses are nearly lacking. With the exception of thin granitic sills in the Campito Formation within a few feet (a meter) of the contact (key location III), the contact between the pluton and the country rocks is not violated by granitic material over this large distance, even though the pluton contains numerous aplite dikes, which largely parallel the foliation in the pluton. This absence of

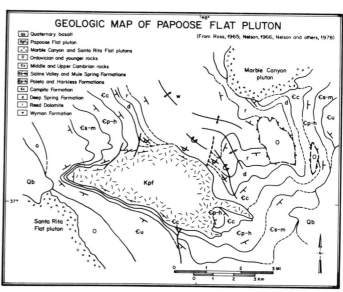

Figure 2. Generalized geologic map of Papoose Flat pluton, illustrating position of pluton on southwest flank of the Inyo Mountain anticline.

apophyses may be attributed in part to the character of the strata at the contact. Over long distances the pluton is either adjacent to or very close to limestone units of the Poleta Formation which, upon being heated by the granitic magma, were decarbonated. This resulted in the liberation of CO_2, which entered the vapor phase of the magma, promoting rapid isothermal crystallization and the development of a "solid" border zone, preventing penetration into the country rock (Sylvester and others, 1978). This is shown well along the north border of the pluton at Squaw Flat and to the west. There, three faults (one beneath Squaw Flat) cut the contact; the stratigraphic throw on each of these faults is more than twice the throw of the fault as measured by the offset of the pluton contact, indicating renewal of movement on pre-plutonic faults. Thus, it is difficult to explain the absence of granitic material as apophyses in the vicinity of the faults, unless the magma were essentially a viscous solid when emplaced and the CO_2 pressure along the faults, produced from the carbonate masses undergoing metamorphism nearby, was sufficient to prevent migration of magmatic material into the fault zone.

The possible role of CO_2 pressure from the decarbonation of nearby carbonate bodies in the style of intrusion is well illustrated adjacent to the second of the faults mentioned above. There, although no igneous material found its way along the fault, intrusion along the foliation in the Campito Formation— a unit containing no carbonate—was relatively easy and numerous sills occur (key location III).

Along the southwestern border of the pluton, over a distance of 6 mi (10 km), the granite-metasedimentary contact is off-set by numerous small faults; 58 of the 59 mapped (Nelson and others, 1978) show horizontal left-lateral separation. The fault traces are parallel in trend and inclination to a set of closely spaced joints in the pluton and probably bear a genetic relationship to the joint

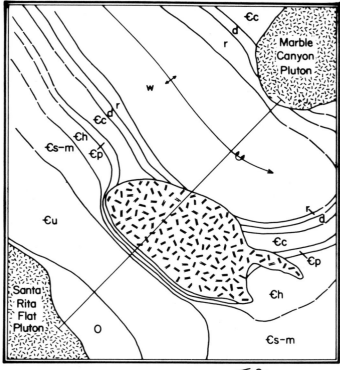

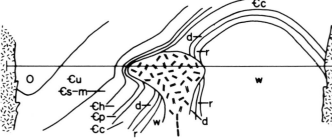

Figure 3. Generalized geologic map and cross section of Papoose Flat pluton, showing conceptualized geometry of attenuated strata around and beneath pluton after emplacement. w, Wyman Formation; r, Reed dolomite; d, Deep Spring Formation; Єc, Campito Formation; Єp, Poleta Formation; Єh, Harkless Formation; Єs-m, Saline Valley, Mule Spring, and Monola formations; Єu, undifferentiated Middle and Upper Cambrian Formations; O, undifferentiated Ordovician formations. (After Sylvester and others, 1978.)

set. Where the faults pass from within the pluton across the contact, the separations are clearly displayed and the massive granitic rocks and the immediately adjacent marble and meta-siltstone have fractured in a brittle fashion. Higher into the stratigraphic succession, however, the faults rapidly diminish in separation, and the beds overlying the brittle limestone bend and "wrap around" the terminations of the faults. In most cases, the faults show no separation beyond 330 ft (100 m) stratigraphically from the contact (key location II).

STRATIGRAPHIC ATTENUATION

In addition to the remarkable concordance of the pluton

contact with the bedding in the country rocks, a striking feature of the metasedimentary rocks is the tremendous attenuation produced during the emplacement of the pluton. Even though stratigraphic identity and continuity, except for faulting, have been maintained around the western two-thirds of the body, individual units have been thinned to as little as one-tenth of their regional stratigraphic thickness.

The Poleta Formation, the unit that in most cases is adjacent to the pluton in the western portion, is represented by gray to black schist with interbedded calc-silicate layers overlain by a persistent unit of buff-yellow to blue crystalline limestone. Detailed stratigraphic recognition within the schist unit is difficult, but the identity of the overlying crystalline limestone is clear. It is the uppermost unit of the Poleta Formation, which in numerous exposures in the White–Inyo region, averages 50 to 65 ft (15 to 20 m) in thickness. It is distinguished by an alternation from base to top, of buff-yellow, blue, and buff-yellow limestone. In proximity to the pluton, generally no more than 30 ft (10 m) above the contact, this unit has been reduced to 7 ft (2 m) of recrystallized limestone. Even at this degree of attenuation, it is possible to recognize the same stratigraphic color variations and similar relative thicknesses as in the unaltered limestone (key location II).

The overlying Harkless Formation, 1,870 ft (570 m) thick 4.3 mi (7 km) north of the pluton, has been similarly affected by the emplacement of the pluton. Regionally, it is characterized by a basal gray shale, a middle orthoquartzite, and an upper brown shale. Around the western half of the pluton, it is from 200 ft (60 m) to less than 150 ft (45 m) thick and is composed of a basal fine-grained black schist, a middle meta-quartzite, and an upper coarse-grained brown biotite-andalusite schist.

The Saline Valley Formation has been reduced from a thickness of 785 ft (240 m) several mi (km) from Papoose Flat to approximately 165 ft (50 m) near the contact. The overlying Mule Spring Limestone, generally from 850 to 1,0000 ft (260 to 300 m) in thickness over the White–Inyo Range, has been reduced to approximately 100 ft (30 m) near the contact, and has been transformed to a medium- to coarse-grained well-foliated marble. The Monola Formation has been reduced to about one-half its normal thickness, but the overlying massive Bonanza King Dolomite has been affected little by the emplacement of the pluton.

STYLE OF EMPLACEMENT

The pluton was intruded diapirically as a rising dike, probably along pre-granite faults, truncating the Wyman, Reed, Deep Spring, and Campito formations. The large eastern apophysis is probably a vestige of the initial stage of intrusion, as shown by vertical lineations in the granite and the higher grade of metamorphism of the wall rocks. Upon reaching the upper portion of the Campito and the overlying Poleta carbonates, the pluton ceased rising as a discordant body and expanded laterally and concordantly to the west and to the northeast. It uplifted and plastically stretched and flattened the overlying strata, producing the well-developed foliation in the border zone of the pluton and the adjacent metasedimentary shell (Fig. 3).

KEY LOCATIONS (I–VII OF FIG. 1)

Location I. Immediately SE of hill 8878 (Waucoba Mountain 15-minute Quadrangle). Excellent exposure of border zone of pluton, illustrating foliation and lineation, aplite dikes, and K-feldspar megacrysts. Reached by side road 0.8 mi (1.3 km) SW of summit of Hines Road (at north side of Papoose Flat).

Location II. At contact of pluton and Poleta Formation. Exposure of border zone of pluton, calc-silicated meta-sediments, attenuated Upper Poleta and Harkless formations, NE trending faults, and joint complex. 0.9 mi (1.4 km) SE of west edge of Papoose Flat along road to Badger Flat. Park in alluvial area and take trail 500 ft (150 m) to the north.

Location III. Thin fine-grained quartz monzonite sills in upper Campito Formation. 1.5 mi (2.4 km) SE of Hines Road, along north contact. Reached by jeep and foot trail.

Location IV. Large (10 × 3 ft; 3 × 1 m) calc-silicate boudins at pluton-Poleta contact. 0.8 mi (1.3 km) west of hill 8878, reached by road and trail down west slope of range.

Location V. Large granitic apophysis, showing vertical lineation and abundant zoned K-feldspar megacrysts. West of Waucoba–Saline Valley Road, 1.4 mi (2.2 km) south of Waucoba Spring, at north end of Saline Valley (Waucoba Wash Quadrangle).

Location VI. View from west end of Papoose Flat 6 mi (10 km) NE to saddle north of Waucoba Mountain, showing contact of pluton (gray) and overturned Wyman marble and siltstone.

Location VII. View from Whippoorwill Flat (Waucoba Spring Quadrangle) west 2.5 mi (4 km) to saddle, showing contact of pluton (gray) and overturned Wyman marble.

REFERENCES CITED

Nelson, C. A., 1966, Geologic map of the Waucoba Mountain Quadrangle, Inyo County, California: U.S. Geological Survey Geological Quadrangle Map GQ-528, scale 1:62,500.

—— , 1971, Geologic map of the Waucoba Spring Quadrangle, Inyo County, California: U.S. Geological Survey Geological Quadrangle Map GQ-921, scale 1:62,500.

Nelson, C. A., Oertel, G., Christie, J. M., and Sylvester, A. G., 1978, Geologic map, structure sections and palinspastic map of the Papoose Flat pluton, Inyo Mountains, California: Geological Society of America Map and Chart Series MC-20.

Ross, D. C., 1965, Geology of the Independence Quadrangle, Inyo County, California: U.S. Geological Survey Bulletin 1181-0, 64 p.

—— , 1967, Geologic map of the Waucoba Wash Quadrangle, Inyo County, California: U.S. Geological Survey Geological Quadrangle Map GQ-612, scale 1:62,500.

Sylvester, A. G., Oertel, G., Nelson, C. A., and Christie, J. M., 1978, Papoose Flat pluton; A granitic blister in the Inyo Mountains, California: Geological Society of America Bulletin, v. 89, p. 1205–1219.

Big Pumice cut, California: A well-dated, 750,000-year-old glacial till

Robert P. Sharp, *Division of Geological and Planetary Sciences, California Institute of Technology, Pasadena, California 91125*

LOCATION

Big Pumice is a roadcut alongside U.S. Highway 395 at the east base of the Sierra Nevada in Mono County, east-central California. The site lies within the NW¼ of sec. 34, T.4S., R.30E. of Casa Diablo, 15′ quadrangle, at 118°39′W Long., 37°33′E Lat. It is on the north side of an east-west highway reach, 1.2 mi (2 km) east of Tom's Place, 0.1 mi (0.2 km) east of the crossing of Rock Creek, and 22.8 mi (38 km) by highway northwest of the city of Bishop (Fig. 1).

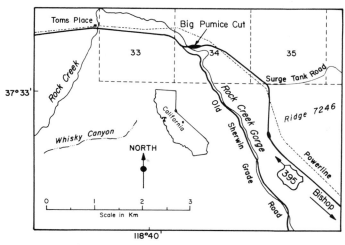

Figure 1. Locality map for Big Pumice cut.

ACCESSIBILITY

The site is easily accessible, essentially year-round, by bus or normal motor car via Highway 395. A large, graded flat immediately south of the cut provides ample parking for large fleets of cars or buses as well as a safe viewing point. A heavily traveled, 4-lane highway has to be crossed on foot for hands-on inspection of deposits and relationships. Parties making this somewhat hazardous transit should treat the cut with gentle respect and be prepared with a convincing science-educational statement of objectives in case they are interviewed by officers of the California Highway Patrol, who maintain a paternalistic interest in the welfare of the cut as well as pedestrians crossing the highway. The locality is within the highway right-of-way across Forest Service domain.

SIGNIFICANCE

Big Pumice cut exposes what must be one of the best-dated older Pleistocene glacial deposits of the world. A weathered and eroded bouldery glacial till stratigraphically underlies layers of a feldspar-crystal- (sanidine-) bearing pumice that composes the base of extensive Bishop Tuff deposits (Gilbert, 1938). Both tuff and pumice have now been accurately dated by repeated K/Ar measurements on feldspar (sanidine) crystals extracted from pumice fragments, hence uncontaminated by foreign inclusions in the tuff. The preferred age is 725,000 years (Dalrymple, 1980, p. 3), although Sarna-Wojcicki and others (1984, p. 22) use 730,000 years. Polarity of tuff magnetization is normal, and it has served as a major reference point in dating the Brunhes-Matuyama polarity boundary (Dalrymple and others, 1965; Dalrymple, 1972). There is little reason to think that the pumice age is significantly different from the preferred age of the entire tuff as they were emplaced by short-lived catastrophic events. Comparison with the weathering and erosion of younger surficial glacial deposits in this region suggests that the extent of pre-pumice weathering and erosion of the buried till in Big Pumice cut involved a conservatively estimated 25,000 to 40,000 years, making its age at least 750,000 years.

Detailed geological mapping (Sharp, 1968) shows that this buried till is Blackwelder's (1931, p. 895–900) classical Sherwin till of Ridge 7246 (Fig. 1) and further that it is correlative with glacial deposits known in other nearby sites to underlie the Bishop Tuff (Gilbert, 1938, p. 1860; Rinehart and Ross, 1957; Putnam, 1960, p. 233; Wahrhaftig and Birman, 1965, p. 310). Increasing recognition of occurrences of the contemporary Bishop ash in terrestrial and even marine deposits of the western United States (Izett and others, 1970; Merriam and Bischoff, 1975; Sarna-Wojcicki and others, 1984, p. 19-26) is making the relationships exposed in Big Pumice cut of concern and interest to a variety of Quaternary geologists. Clastic dikes cutting all units in the cut are an added feature of interest.

DESCRIPTION

Big Pumice is a large highway cut about 100 ft high by 460 ft long (30 m by 140 m). Relationships exposed are described in greater detail elsewhere (Sharp, 1968, p. 352-355) and are best treated by reference to Figure 2. Extending from the middle of the cut to its west end, and rising westward from the highway level, is a buried hillock of bouldery glacial till. The till is only modestly oxidized to a depth of 6 to 10 ft (2 to 3 m), but granitic boulders are deeply disintegrated to a depth of about 25 ft (8 m) below the contact with overlying pumice. The east flank of the buried hillock bears a mantle of brownish, soil-like, disintegrated grus, thickening to 1.5 ft (0.5 m) down the hill's slope, beneath the pumice mantle. These relationships indicate considerable weathering of the till before burial beneath the pumice. Configuration of the till-pumice contact over a wider area also suggests that the glacial depositional topography had undergone considerable erosion prior to pumice emplacement.

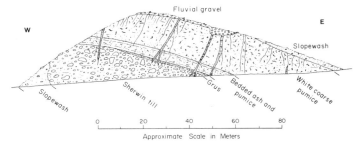

Figure 2. Details of relationships exposed in Big Pumice cut, looking north. Scale bar is horizontal; vertical and horizontal scales the same.

Boulder lithologies, predominantly granodiorite and quartz monzonite, indicate the Rock Creek drainage to the west-southwest as the principal source of ice and debris, coming into the area via Whisky Canyon rather than by way of the younger present course of lowermost Rock Creek (Fig. 1). The till matrix is tight, with considerable silt, and some boulders are striated. The silty rather than clayish matrix, the many square mi (km) covered, a thickness in places approaching or even exceeding 660 ft (200 m), and faint remnants of possible glacial topographic forms suggest that the deposit is a glacial till, not just a diamicton of some other origin. Exposures in roadcuts are excellent for more than 2 mi (3 km) southeast along Highway 395.

The 76 ft (23 m) of overlying, predominantly white, rhyolite pumice consist of two units. The lower unit is a well-layered, faintly brownish sequence of fine ash and small pumice fragments, 15 ft (4.5 m) thick, dipping gently east in conformity with the 10° slope of the underlying till surface. This is regarded as airborne tephra laid down as a mantle over a rolling landscape. The upper pumice unit consists of 60 ft (18 m) of coarser, more poorly sorted, looser, poorly bedded, white pumice layers in near-horizontal attitude. These layers were probably emplaced as a sequence of pumice flows. The pumice and the cut are capped

unconformably by 10 ft (3 m) of much younger fluvial or glacio-fluvial oxidized gravels consisting of smooth roundstones up to small boulder size. The entire Big Pumice cut sequence is truncated on the west by the current hill slope graded to the downcutting bed of Rock Creek.

A series of clastic dikes dipping steeply west cuts through the entire sequence into the underlying till where they terminate. Some are composed solely of pumice, but many contain roundstones from the capping gravels, indicating a significant contribution to the dike filling from above and a date of origin later than gravel deposition. Small clastic dikes of matrix cut through boulders in the till in other exposures along Highway 395 to the southeast. Ground cracking associated with seismic events in this tectonically active area would seem a possible explanation for the dikes.

Under proper antecedent conditions of weathering, erosion, eolian winnowing, and road scraping, concentrations of glassy sanidine feldspar crystals can be found along the ditch at the cut's base in its eastern part.

DISCUSSION

Eliot Blackwelder (1931, p. 895-900), in his superb classical paper on east-side Sierra Nevada Pleistocene glaciation, recognized and named the Sherwin till. He regarded it as younger than the Bishop Tuff, although no fragments of that formation have ever been found in the till. Other workers, listed above, recogized that deposits of presumed glacial origin also underlie the Bishop Tuff. Detailed geological mapping was required to resolve the question of whether they were related to the Sherwin glaciation or to an earlier glacial episode. Such mapping succeeded in demonstrating that Blackwelder's type Sherwin, composing Ridge 7246 (Fig. 1), was indeed the till buried beneath Bishop pumice at the Big Pumice cut, making this site a benchmark on the time scale of Quaternary events.

REFERENCES CITED

Blackwelder, Eliot, 1931, Pleistocene glaciation in the Sierra Nevada and Basin Ranges: Geological Society of America Bulletin, v. 42, p. 865–922.

Dalrymple, G. B., 1972, Potassium-argon dating of geomagnetic reversals and North American glaciations, in Bishop, W. S., and Miller, J. A., eds., Calibration of hominoid evolution: Wenner-Gren Foundation for Anthropological Research, New York, Scottish Academic Press, p. 107–134.

——1980, K-Ar ages of the Friant pumice member of the Turlock Lake Formation, the Bishop Tuff, and the tuff of Reds Meadow, central California: New Mexico Bureau of Mines and Mineral Resources, Isochron/West, no. 28, p. 3–5.

Dalrymple, G. B., Cox, Allan, and Doell, R. R., 1965, Potassium-argon age and paleomagnetism of the Bishop Tuff, California: Geological Society of America Bulletin, v. 76, p. 665–674.

Gilbert, C. M., 1938, Welded tuff in eastern California: Geological Society of America Bulletin, v. 49, p. 1829–1862.

Izett, G. A., and others, 1970, The Bishop ash bed, a Pleistocene marker bed in the western United States: Quaternary Research, v. 1, p. 121–132.

Merriam, Richard, and Bischoff, J. L., 1975, Bishop ash: A widespread volcanic ash extended into southern California: Journal of Sedimentary Petrology, v. 45, p. 207–211.

Putnam, W. C., 1960, Origin of Rock Creek and Owens River gorges, Mono County, California: University of California Publications in the Geological Sciences, v. 34, p. 221–280.

Rinehart, C. D., and Ross, D. C., 1957, Geology of the Casa Diablo Mountain quadrangle, California: U.S. Geological Survey Geological Quadrangle Map, GQ–99, scale 1:62,500.

Sarna-Wojcicki, A. M. and others, 1984, Chemical analyses, correlations, and ages of Upper Pliocene and Pleistocene ash layers of east-central and southern California: U.S. Geological Survey Professional Paper 1293, 40 p.

Sharp, R. P., 1968, Sherwin till-Bishop Tuff geological relationships, Sierra Nevada, California: Geological Society of America Bulletin, v. 79, p. 351–364.

Wahrhaftig, Clyde, and Birman, J., 1965, The Quaternary of the Pacific mountain system in California, in Wright, H. E., and Frey, D. G., eds., The Quaternary of the United States: Princeton, Princeton University Press, p. 299–340.

Long Valley caldera, eastern California

Roy A. Bailey, U.S. Geological Survey, 345 Middlefield Road, Menlo Park, California 94025

LOCATION

Long Valley caldera (Fig. 1) is in eastern California, at the eastern base of the Sierra Nevada between Mono Lake and Owens Valley. Traversed by U.S. 395, the caldera is about 310 mi (500 km) north of Los Angeles and 165 mi (265 km) south of Reno. It is midway between the towns of Bishop and Lee Vining and contains the resort town of Mammoth Lakes.

SIGNIFICANCE

Long Valley caldera is a 10 by 20 mi (17 by 32 km) elliptical volcanic depression that formed at 0.73 Ma as a consequence of catastrophic pyroclastic eruptions from a large, shallow, rhyolite magma chamber. The deposits of these eruptions, the Bishop Tuff, occur as thick, welded ash flows surrounding the caldera and as thin plinian ash falls over much of western United States. The caldera is one of the largest centers of Quaternary rhyolite volcanism in North America. It is underlain at depth by a residual, partially molten magma chamber, and is a potential site for future volcanic activity. The caldera contains an active hydrothermal system that is being explored as a geothermal energy resource.

STOP 1. LOOKOUT MOUNTAIN OVERVIEW

[Refer to *Old Mammoth, California 7½-minute Topographic Quadrangle.*] One of the better vantage points for viewing the caldera is from the summit of Lookout Mountain (Fig. 2), a high point in the northwestern part of the caldera. The summit is accessible by a gravel road that leads east from U.S. 395, 5.2 mi (8.4 km) north of the Mammoth Lakes (Route 203) exit and 9.5 mi (15.3 km) south of June Lake junction. The road (not plowed in winter) is passable to ordinary passenger vehicles from May through October. It terminates 3.1 mi (5.0 km) from U.S. 395 in a summit loop at an elevation of 8,343 ft (2,543 m). The panoramas of Figures 3 and 4 serve as a guide to the views, and the following geologic summary provides background information.

Geologic Summary. Long Valley caldera, located at the western edge of the Basin and Range Province, straddles the Eastern Sierran frontal fault escarpment, forming a reentrant or offset commonly referred to as the "Mammoth embayment." The floor of the caldera ranges in elevation from 6,700 ft (2,040 m) in its eastern half to 8,500 ft (2,600 m) in its western half. The caldera walls rise steeply to elevations of 10,000 to 12,000 ft (3,000 to 3,500 m) on all sides except the east and southeast. The prevolcanic basement in the area is mainly Mesozoic granitic rock of the Sierra Nevada batholith and Paleozoic metasedimentary and Mesozoic metavolcanic rocks of the Mount Morrison and Ritter Range roof pendants.

Volcanism associated with Long Valley caldera (Bailey and others, 1976; Bailey and Koeppen, 1977) began about 3.2 Ma with widespread eruption of trachybasaltic-trachyandesitic lavas (Fig. 1), erosional remnants of which are scattered over a 1,500

mi^2 (4,000 km^2) area around the caldera—a distribution that suggests an extensive mantle source region for these initial eruptions.

Associated quartz-latite domes and flows erupted between 3.0 and 2.6 Ma near the north and northwest rims of the present caldera and represent the onset of deep-crustal magmatic accumulation and differentiation that eventually culminated in formation of the large shallow Long Valley magma chamber, from which subsequent, more silicic eruptions originated. The first eruptions from this silicic chamber were on the northeast rim of the present caldera at Glass Mountain, a 3,200-ft-thick (1,000 m) complex of high-silica rhyolite domes, flows, and tuffs that formed between 2.1 and 0.8 Ma (Metz and Mahood, 1985).

At 0.73 Ma, catastrophic rupturing of the roof of the magma chamber caused the expulsion of 140 mi^3 (600 km^3) of rhyolite magma as plinian ash falls and hot incandescent ash flows. This partial emptying of the chamber caused collapse of its roof to form the 1- to 2-mi-deep (2 to 3 km) oval depression of Long Valley caldera. The resulting ash-flow deposits, the Bishop Tuff (Gilbert, 1938; Hildreth, 1979), inundated 580 mi^2 (1,500 km^2) of the surrounding countryside and accumulated locally to thicknesses approaching 660 ft (200 m) in upper Owens Valley, Adobe Valley, and Mono Basin. A large volume of Bishop Tuff also ponded within the caldera, where drill holes have confirmed that as much as 4,500 ft (1,400 m) of Bishop Tuff is buried beneath younger caldera fill. Associated plinian ash clouds drifted thousands of miles downwind and deposited an ash layer recognized as far east as Kansas and Nebraska (Izett, 1982).

After collapse of the roof of the magma chamber, volcanism continued on the caldera floor with early postcaldera tephra eruptions followed by extrusion of thin, hot, fluid obsidian flows. Simultaneous renewal of magma pressure uplifted, arched, and faulted the early rhyolite flows and tephra, forming a resurgent dome with a northwest-trending medial graben (Figs. 1 and 3). This resurgent dome rose within 100,000 years after caldera collapse, between 0.73 and 0.65 Ma, as indicated by K-Ar ages of the contemporaneous obsidian flows. The early rhyolite that erupted during this episode is aphyric to sparsely porphyritic, contains less than 5 percent crystals of plagioclase, hypersthene, biotite, and Fe-Ti oxides, and contrasts strikingly with the preceding crystal-rich Bishop Tuff.

After an interlude of quiescence, crystal-rich rhyolite began erupting in the caldera moat, probably from ring fractures peripheral to the resurgent dome. This moat rhyolite, typically containing up to 20 percent phenocrysts of plagioclase, quartz, sanidine, biotite, hornblende, and Fe-Ti oxides, formed thick, steep-sided domes and flows, suggesting higher viscosity and lower temperature than the early rhyolite and signalling probable onset of cooling and crystallization of the magma chamber. Moat

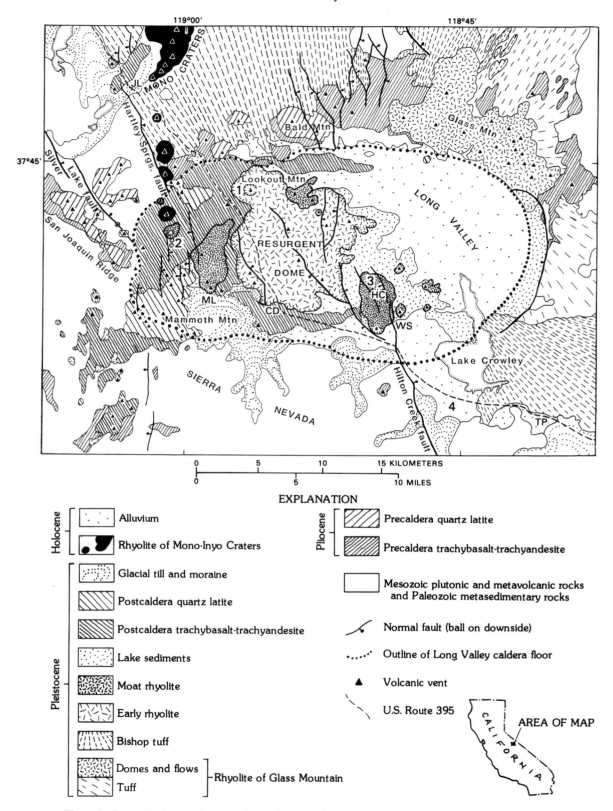

Figure 1. Generalized geologic map of Long Valley caldera. JL, June Lake junction; ML, Mammoth Lakes; CD, Casa Diablo Hot Springs; HC, Hot Creek; WS, Whitmore Springs; TP, Toms Place. Numbers 1–4 refer to stops in the roadlog.

rhyolite erupted at about 200,000-year intervals, at 0.5, 0.3, and 0.1 Ma, in clockwise succession around the resurgent dome, in the northern, southeastern, and western sectors of the moat, respectively. The 0.1-Ma west-moat rhyolites appear to be the youngest eruptives so far derived from the Long Valley chamber. However, seismological and geodetic studies suggest that a large body of partially molten magma still underlies the resurgent dome, providing a source for possible future eruptions (see below).

Overlapping both spatially and temporally this 3.2 to 0.1 Ma mafic-to-silicic sequence of volcanism centered on Long Valley caldera is a younger mafic-to-silicic sequence localized along a 30-mi (50-km), north-south–trending fissure system extending from Mammoth Mountain through the western caldera moat to Mono Lake. This younger sequence began about 0.3 to 0.2 Ma with the eruption of trachybasalt-trachyandesite lavas in and near the west moat, where they accumulated to depths of at least 820 ft (250 m) and poured around the resurgent dome, sending lava tongues into both the north and south moat (Fig. 1). Successively younger mafic vents erupted farther north near June Lake and at Black Point on Mono Lake, suggesting localization along a northward-opening fissure system. Intermittently during these mafic eruptions, quartz-latite domes and flows erupted locally in the western part of the caldera, as well as farther north in and near Mono Lake; the greatest accumulations, however, are on the northwest and southwest caldera margins, where the north-south fissure system intersects caldera ring fractures. On the southwest rim, 150,000 to 50,000 years ago, extrusion of quartz-latite domes and flows built Mammoth Mountain, an imposing cumulovolcano straddling the southwest caldera wall.

Rhyolites began erupting along the north-south fissure system 38,000 to 40,000 years ago—first at the Mono Craters, northwest of the caldera, and more recently at the Inyo Craters, which span the northwest caldera rim and extend into the west moat. The Mono Craters form an 11-mi-long (17 km), arcuate chain of 30 or more coalesced rhyolite domes, flows, and craters, which consist predominantly of aphric to sparsely porphyritic obsidian and pumice and are remarkably homogeneous chemically. The most recent Mono Craters eruptions occurred about 550 yrs B.P. (A.D. 1325 to 1365) at the north end of the chain (Sieh and Bursik, 1986).

The Inyo Craters form a discontinuous 7.5-mi-long (12 km) chain of rhyolite dome-flows and craters that range in age from about 6,000 to 500 years. The youngest Inyo Craters eruptions (A.D. 1369–1472) blanketed the surrounding terrane with thick pumice deposits and concluded with extrusion of Obsidian, Glass Creek, and Deadman Creek domes (Fig. 4), all apparently fed by a shallow, 5-mi-long (8 km), rhyolite dike (Miller, 1985).

Recent Seismicity and Ground Deformation. Based on the moat-rhyolite periodicity of 200,000 years and the most recent eruption of 0.1 Ma, future eruptions from the residual Long Valley magma chamber would seem only a remote possibility, not to be expected for another 100,000 years. However, an unusual sequence of magnitude-6 earthquakes in May 1980 and

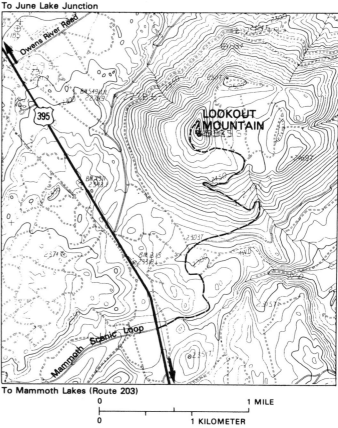

Figure 2. Location map showing access route to Lookout Mountain. Elevations in meters; contour interval 10 m.

another of magnitude-5 earthquakes in January 1983, accompanied by 20 in (50 cm) uplift of the resurgent dome, suggested that new magma was being injected into the chamber and possibly into the south moat ring fracture zone. While the intensity of seismicity and the rate of uplift declined through 1985 and 1986, the potential for future activity remains (Hill and others, 1985). However, statistically, the more likely site for future eruptions is the Mono–Inyo Craters volcanic chain, where, on the basis of ^{14}C and obsidian hydration-rind dating (Wood, 1977), eruptions have occurred at about 500-year intervals for the past 3,000 years, and where the last eruption was 550 years B.P. Miller and others (1982) have outlined the nature of the potential hazards associated with possible future eruptions in the area.

STOP 2. INYO CRATERS

[Refer to ***Old Mammoth and Mammoth Mountain, California 7½-minute Topographic Quadrangles.***] Access to the Inyo Craters is via the Mammoth Scenic Loop, which leads west from U.S. 395 directly opposite the Lookout Mountain Road (Fig. 2). Follow the paved Mammoth Loop to the Inyo Craters sign (3.3 mi; 5.3 km); turn right onto the winding unpaved road (not plowed in winter) and continue following Inyo Craters signs to the visitors' parking area (4.5 mi; 7.2 km). Park and follow the well-worn foot path west 0.3 mi (500 m) to the craters.

The Inyo Craters are three north-south–alined phreatic ex-

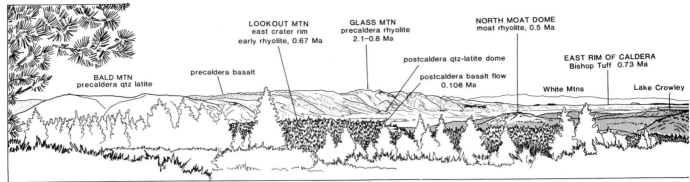

Figure 3 (this and facing page). Panorama looking generally east from the summit of Lookout Mountain, from Bald Mountain on the left (north) to Mammoth Mountain on the right (southwest). [In both Figures 3 and 4 a few foreground trees, which can be avoided by slightly shifting one's viewpoint, have been deleted in the sketches.] The present viewpoint on Lookout Mountain, a cratered stratovolcano of early rhyolite, is on the western crater rim; the obsidian underfoot is typical of the early rhyolite. The steep south-facing escarpment between Bald and Glass mountains is the north wall of Long Valley caldera. Wheeler Crest, McGee Mountain, Mount Morgan, Mount Morrison, Laurel Mountain, Bloody

plosion craters on the summit and south flank of Deer Mountain, a 113,000-year-old rhyolite dome in the west moat. The two southernmost craters average about 660 ft (200 m) in diameter and contain small lakes; the summit crater is smaller, breeched on its south side, and dry. The lake in the southern crater is yellowish green, suggesting suspended sulfur, but the water is cold (11°C), and other evidence of thermal activity is lacking. In spite of differences in appearance, all three craters formed at nearly the same time, probably within hours or days. However, the summit crater was the first to erupt, as its lighter-colored deposits, derived from

the rhyolite of Deer Mountain, can be seen in the northeast wall of the middle crater underneath the darker deposits of the middle and southern craters, derived mainly from trachyandesitic flows well exposed in the southern crater. In the north wall of the southern crater, exposed in succession above the dark trachyandesite flows, are trachyandesitic cinder (6.7 ft; 2 m), glacial till (26 ft; 8 m), pumice tephra from the Inyo domes eruptions (3 ft; 1 m), and coarse, crudely bedded phreatic deposits (26 ft; 8 m). The latter, which extends as far as 0.3 mi (0.5 km) from the crater, consist mainly of trachybasalt-trachyandesite blocks in a greyish

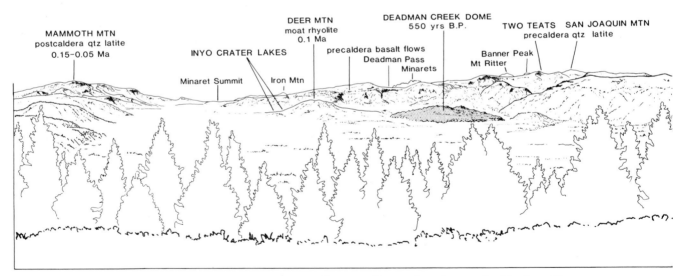

Figure 4 (this and facing page). Panorama looking generally west from the summit of Lookout Mountain, from Mammoth Mountain on the left (southwest) to the Mono Craters on the right (north). From left to right on the distant skyline (interrupted) are: Iron Mountain, The Minarets, Mount Ritter, Banner Peak, Carson Peak, Mount Wood, Mount Lewis, Mount Gibbs, Mount Dana, and Mount Warren, peaks along the crest of the Sierra Nevada underlain by Mesozoic granitic plutons and metavolcanic and metasedimentary roof-pendant rocks. The skyline from Minaret Summit through Deadman Pass, Two Teats, and San Joaquin Mountain to June Mountain constitute the San Joaquin Ridge. The steep scarp from Minaret Summit northward in front of San Joaquin Ridge and White Wing Mountain continuing

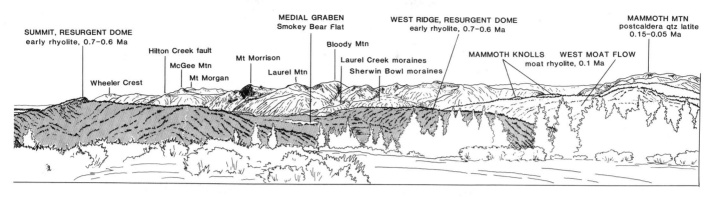

Mountain, and the skyline west to Mammoth Mountain are peaks along the crest of the Sierra Nevada underlain by Mesozoic granitic plutons and metamorphic roof-pendant rocks; the steep north-facing scarp immediately in front of these peaks is the south wall of the caldera. The forested ridges in the central part of the panorama (shaded) are the resurgent dome. The northwest-trending medial graben transects the resurgent dome, its east-bounding fault forming the steep west-facing cliff on the summit peak of the dome and its west-bounding fault the east-facing forested scarp just west of Smokey Bear Flat.

brown compact to semi-indurated fine-grained matrix; blocks and boulders of granite and metamorphic rocks also occur. Battered logs incorporated in the deposits give ^{14}C ages of 500 years B.P. The terrane around the craters is broken by many north-trending faults and fissures, with 0 to 100 ft (0 to 30 m) displacement, that define a graben 2,000 ft (600 m) wide and 1.2 mi (2 km) long. The graben and craters most likely formed when the rhyolite dike feeding the Inyo domes, upon continued southward and upward injection, encountered groundwater at shallow depth, causing it to flash explosively to steam, ejecting hot mud and angular blocks. Surprisingly, no juvenile rhyolite magma reached the surface during the eruptions.

STOP 3. HOT CREEK GORGE

[Refer to *Whitmore Hot Spring, California 7½-minute*

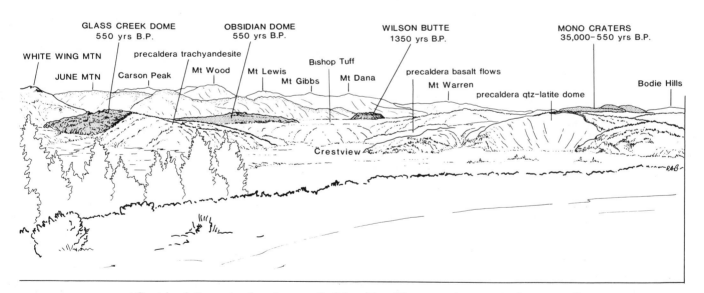

past Crestview is the west and northwest wall of the caldera. The tree- and sage-covered area extending west from the base of Lookout Mountain to the caldera wall is the west and northwest moat, underlain by postcaldera trachybasalt-trachyandesite flows 150,000 to 90,000 years old. Extending north from Deer Mountain are the 550-yr-old Inyo domes (shaded), which include Deadman Creek dome, within the caldera, Glass Creek dome, which erupted on the caldera rim and flowed down the northwest wall, Obsidian Dome, which erupted outside the caldera, and Wilson Butte, which erupted on the same north-south fissure system but is older (1,350 yrs B.P.) and chemically identical to the Mono Craters, visible on the skyline to the north.

Topographic Quadrangle.] Access to Hot Creek is via the Mammoth Lakes Airport exit from U.S. 395, 3.0 mi (4.9 km) east of the Mammoth Lakes (Route 203) exit. After turning left (north) onto the Airport exit, pass the airport turnoff (first right) and take the second right (0.5 mi; 0.8 km); continue to Hot Creek Recreational Area parking lot (on the left, 3.4 mi; 5.4 km), which overlooks Hot Creek Gorge.

Hot Creek, a segment of Mammoth Creek, has several vigorous hot springs that form a popular bathing area. Hot Creek Gorge has been incised into the Hot Creek rhyolite flow, a sanidine-augite-phyric moat rhyolite that erupted 280,000 years ago from a nearby vent on the shore of Pleistocene Long Valley Lake; it flowed northward into the lake, becoming a subaqueous flow in the vicinity of the present gorge. The initially hot glassy rhyolite was pervasively hydrated by interaction with the lake water and later partly hydrothermally altered by local hot-spring activity. Outcrops of partly altered, perlitized flow breccia are exposed along the paved trail leading to the bottom of the gorge. The several boiling (93°C) pools on the sides and floor of the gorge commonly change in vigor and configuration in response to local earthquakes and are also significantly affected by small changes in the water level of Hot Creek. The white deposits lining the pools are calcium carbonate (travertine). Older travertine deposits surround extinct hot springs elsewhere on the gorge floor.

Hot Springs, fumaroles, and areas of active hydrothermal alteration are prevalent in many parts of Long Valley caldera, particularly in the south and southeast moat and on adjacent flanks of the resurgent dome. Most of these features are localized along faults (here, the intracaldera extension of the Hilton Creek fault). The heat source for these features is presumably the magma chamber beneath the resurgent dome, but drill holes show that hot waters are confined to relatively shallow aquifers (<2,300 ft; 700 m deep) and that except in the west moat, temperatures decline or do not increase substantially with depth. Maximum temperatures measured in drill holes are about 220°C, but chemical geothermometry indicates deep reservoir temperatures near 280°C. In recent years substantial efforts have been made to develop and utilize this geothermal resource. At Casa Diablo Hot Springs several shallow drill holes supply hot water to a 7-megawatt binary electrical power plant; nearby, a California State fish hatchery utilizes local warm-spring waters to breed trout; and at Whitmore Hot Springs, thermal waters are used in a public swimming pool.

STOP 4. LAKE CROWLEY VIEWPOINT

[Refer to *Convict Lake and Toms Place, California 7½-minute Topographic Quadrangles.*] An excellent vantage point for viewing the eastern part of the caldera is from the Lake Crowley Volunteer Fire Station on *old* U.S. 395. Exit U.S. 395 at the Lake Crowley exit, 4.0 mi (6.5 km) southeast of the Mammoth Lakes Airport exit, turning right onto *old* U.S. 395 and continuing southeast to fire station (on the left, 3.6 mi; 5.8 km). Park beyond and well clear of the station and walk to the rear, where a sweeping view to the north and east of Long Valley and Lake Crowley can be seen.

During early postsubsidence (600,000 years ago), the caldera was filled by a large lake; terraces and strand lines are well preserved along the eastern caldera wall and locally on the flanks of the resurgent dome. Rise of the resurgent dome gradually raised the lake level above the low southeastern rim (east of Lake Crowley), where it overflowed and cut the Owens River Gorge. Gradual downcutting, accompanied by tectonic lowering of the southeast rim along the Sierran front, eventually drained the lake, probably between 100,000 and 50,000 years ago. (Lake Crowley is a modern, dammed reservoir.) The downwarped terraces of Pleistocene Long Valley Lake can be traced from 6,900 ft (2,100 m) elevation east of Lake Crowley northward to 7,600 ft (2,300 m) at the foot of Glass Mountain. (They are best seen in the low-angle light of early morning or late afternoon.) During the early life of the caldera lake, glaciers flowing into it from the High Sierra generated debris-laden icebergs that drifted across it, depositing large Sierran granite erratics on the flanks of the resurgent dome, which stood as an island, and on lake terraces along the eastern caldera wall.

The view west from the fire station toward the steep Sierran front is of the Hilton Creek fault escarpment near the mouth of McGee Creek canyon. At the base of the escarpment south (left) of the canyon can be seen the light-colored jagged trace of the slickensided fault surface cutting Paleozoic Hilton Creek marble. The Holocene fault displacement on the late Pleistocene moraines at the canyon mouth is 50 ft (15 m), but the total displacement on the escarpment exceeds 3,900 ft (1,200 m).

REFERENCES CITED

Bailey, R. A., and Koeppen, R. P., 1977, Preliminary geologic map of Long Valley caldera, Mono County, California: U.S. Geological Survey Open-File Report 77-468, 2 sheets, scale 1:62,500.

Bailey, R. A., Dalrymple, G. B., and Lanphere, M. A., 1976, Volcanism, structure, and geochronology of Long Valley caldera, Mono County, California: Journal of Geophysical Research, v. 81, p. 725–744.

Gilbert, C. M., 1938, Welded tuff in eastern California: Geological Society of America Bulletin, v. 49, p. 1829–1862.

Hildreth, W., 1979, The Bishop tuff; Evidence for the origin of compositional zoning in silicic magma chambers: Geological Society of America Special Paper 180, p. 43–75.

Hill, D. P., Bailey, R. A., and Ryall, A. S., 1985, Active tectonic and magmatic processes in Long Valley caldera, eastern California; An overview: Journal of Geophysical Research, v. 90, p. 11111–11120.

Izett, G. A., 1982, The Bishop Ash Bed and some older compositionally similar ash beds in California, Nevada, and Utah: U.S. Geological Survey Open-File Report 82-582, 30 p.

Metz, J. M., and Mahood, G. A., 1985, Precursors to the Bishop Tuff eruption, Glass Mountain, Long Valley, California: Journal of Geophysical Research, v. 90, p. 11121–11126.

Miller, C. D., 1985, Holocene eruption at the Inyo volcanic chain, California: implications for possible eruptions in Long Valley: Geology, v. 13, p. 14–17.

Miller, C. D., Mullineaux, D. R., Crandell, D. R., and Bailey, R. A., 1982, Potential hazards from future volcanic eruptions in the Long Valley–Mono Lake area, east-central California and southwestern Nevada; A preliminary assessment: U.S. Geological Survey Circular 77, 10 p.

Sieh, K., and Bursik, M., 1986, Most recent eruptions of the Mono Craters, eastern central California: Journal of Geophysical Research (in press)

Wood, S. H., 1977, Chronology of late Pleistocene and Holocene volcanics, Long Valley and Mono Basin geothermal areas, eastern California: U.S. Geological Survey Open-File Report 83-747, 76 p. [1983].

37

The Silver Strand; A unique tombolo, San Diego, California

William J. Elliott, P.O. Box 541, Solana Beach, California 92075

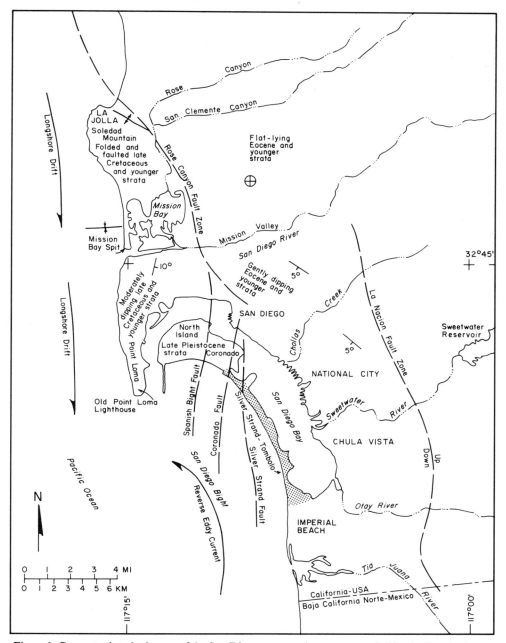

Figure 1. Conceptual geologic map of the San Diego area, southwesternmost California. Standard map symbols have been used.

LOCATION AND ACCESS

The Silver Strand tombolo is located in southwesternmost California, in the city of Coronado, approximately 115 mi (185 km) south of Los Angeles. It can be reached from the city of San Diego by crossing over San Diego Bay on the Coronado Toll Bridge (California 75), or from the city of Imperial Beach, to the south, via the same highway. A standard road map of San Diego and vicinity will provide a ready guide to the geology and physiography of the area as shown conceptually in Figure 1. From the tip of Point Loma, at The Old Point Loma Lighthouse, one can enjoy a world-class view of the Silver Strand, Coronado, North Island, San Diego Bay, and the city of San Diego.

SIGNIFICANCE OF SITE

Net annual longshore beach sand transport along most of the San Diego County coastline is from north to south. This normal drift current pattern is interrupted, however, by the seaward projections of Soledad Mountain (La Jolla) and Point Loma. South of Point Loma, a reverse (counterclockwise) eddy current carries sand in a northerly direction, from the mouth of the Tia Juana River toward Coronado and North Island. Net northward transport of sand across the bight provides a natural depositional connection between the mainland and Coronado/North Island—The Silver Strand tombolo.

SITE DESCRIPTION

A tombolo is a depositional feature built of beach sand and/or gravel that connects two islands, or connects an island with the mainland. By comparison to the east coast of the United States, where tombolos are a relatively common occurrence, there are only a few good examples on the west coast. One such tombolo, in central California, projects roughly perpendicularly outward to connect Morro Rock with the mainland. Another, the Silver Strand, projects northerly, roughly parallel to the coastline to connect Coronado and North Island with the mainland (Fig. 1). This approximately 6-mi-long (9.6 km) stretch of narrow (average 0.3-mi; 0.5-km wide) sandy beach, because of geology and physiography unique to the San Diego area, closes off a portion of the San Diego Bight to form the western boundary of San Diego Bay (Ross and Dowlen, 1973; Abbott and O'Dunn, 1981).

Coastal San Diego has been at the edge of the North American continent, that is, at the edge of the Pacific Ocean, since Cretaceous time. Late Cretaceous, Eocene, Oligocene, Miocene, Pliocene, and Pleistocene sedimentary rocks, exposed in coastal San Diego outcrop, silently attest to this ancient, nearly static coastline (Kennedy, 1975; Kennedy and Tan, 1977). Most of these clastic, nonmarine, marginal-marine, and near-shore-marine strata are flat-lying, or nearly so; the principal exceptions being in outcrops west of the Rose Canyon fault zone (Fig. 1).

West of the right-oblique Rose Canyon fault, Late Cretaceous, Eocene, and younger sedimentary strata have been uplifted, tilted, and exposed in the asymmetrical, faulted Soledad Mountain anticline, and in the long, linear, eastward-tilted Point Loma fault block. These same strata are below sea level in the intervening Mission Bay syncline. Protected from Pacific Ocean wave attack, late Pleistocene-age sedimentary strata are exposed on the Coronado/North Island, topographic/structural island situated east of the Point Loma topographic promontory.

East of the Rose Canyon fault zone, and roughly north of Mission Valley, Eocene and younger strata are nearly flat-lying (Fig. 1). However, south of Mission Valley, Eocene and younger strata are gently tilted and down-faulted into a structural/topographic depression—now occupied by San Diego Bay, and the offshore San Diego Bight.

In 1542, fifty years after Columbus discovered America, the Portuguese navigator and explorer, Juan Rodriguez Cabrillo, sailed between North Island and Point Loma into a previously uncharted, natural deep-water harbor—San Diego Bay (Brown, 1981). He was doubtless unaware of the dynamic chain of geologic events that had created this unexpected gift—a safe refuge where he could rest, mend his sails, and take on fresh water and food. Almost 450 years later, San Diego Bay is a key link in a thriving coastal economy. And, for the geologist, the bay and its surrounding geology provide an unparalleled opportunity to understand and appreciate the complex set of events making this beautiful and scenic geologic/physiographic setting possible.

REFERENCES CITED

Abbott, P. L., and O'Dunn, S., eds., 1981, Geologic investigations of the coastal plain, San Diego County, California: San Diego Association of Geologists annual field trip, 166 p.

Brown, J. E., 1981, Cabrillo National Monument: Cabrillo Historical Association, 44 p.

Kennedy, M. P., 1975, Geology of the San Diego metropolitan area, California: California Division of Mines and Geology Bulletin 200, 39 p.

Kennedy, M. P., and Tan, S. S., 1977, Geology of National City, Imperial Beach, and Otay Mesa quadrangles, southern San Diego metropolitan area, California: California Division of Mines and Geology, Map Sheet 29, scale 1:24,000.

Ross, A., and Dowlen, R. J., eds., 1973, Studies on the geology and geologic hazards of the greater San Diego area, California: San Diego Association of Geologists annual field trip, 152 p.

The Cristianitos fault and Quaternary geology, San Onofre State Beach, California

Roy J. Shlemon, P.O. Box 3066, Newport Beach, California 92663

LOCATION AND ACCESSIBILITY

The site is on the San Onofre Bluff Quadrangle 7½-minute Series, California, in San Diego County (SE¼,sec.30,T.9S., R.6W.,SB). The site, located within the San Onofre State Beach, is bounded on the east and the southeast by the Camp Pendleton Marine Corps Base and on the northwest by the City of San Clemente. The Cristianitos fault and adjacent wave-cut platforms are best exposed in the sea cliffs (San Onofre Bluff area, approximately 0.8 mi [1.3 km] downcoast [southeast] from the San Onofre Nuclear Generating Station [SONGS; Fig. 1]).

The site is readily reached by passenger vehicle from either Los Angeles or San Diego. Exit I-5 (San Diego Freeway) at the Basilone Road Interchange (approximately 1 mi [1.6 km] south of the San Diego County–Orange County line) and follow the frontage road southeastward (parallel to the coast), past SONGS, 3. 2 mi (5.1 km) to the entrance of the San Onofre State Beach.

From the parking lot, follow a well-marked trail from the top of the San Onofre Bluffs to the beach (approximately 15-minute walk). Partially down the bluffs, the trail merges into a graded access road and, at the Echo Arch Trail camping area, is gravel covered (private vehicles not permitted).

At the beach, follow the sea cliffs about 330 ft (100 m) to the north. The Cristianitos fault is clearly visible in the bluffs where it juxtaposes light-colored, massive silts and sands of the Pliocene San Mateo Formation on the northwest and dark-colored, laminated siltstones and claystones of the late Miocene Monterey Formation on the southeast (Fig. 2).

SIGNIFICANCE OF SITE

The San Onofre State Beach and bluffs provide near-textbook illustrations of the Cristianitos fault, of marine terraces, and of related Quaternary geological features, all fortuitously preserved in this increasingly urbanized area. The site has been well studied because of its proximity to the nearby nuclear generating station (McNey, 1979). The site is particularly significant because: (1) the Cristianitos normal-slip fault is overlain by unbroken, regressive marine terrace sediments dated by U/Th, amino acid, faunal association, and soil-stratigraphic techniques as pertaining to marine oxygen isotope stage 5e (ca. 125,000 years b.p.), thus providing a minimum age for last displacement; (2) overlying, prograded continental deposits contain at least seven intercalated buried paleosols, some of which are significant stratigraphic markers that, where dated by soil-stratigraphic and radiocarbon techniques, provide information about late Quaternary climatic and vegetation change; and (3) multiple marine terraces occur both onshore and offshore and, with nearly continuous sea-cliff exposures of the 125,000-year-old wave-cut plat-

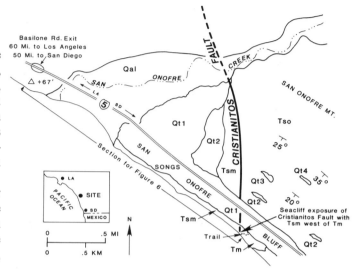

Figure 1. Generalized location and geologic map of the San Onofre State Beach and adjacent area. Geographic symbols: LA = Los Angeles; SD = San Diego; SONGS = San Onofre Nuclear Generating Stations. Geologic symbols: Qal = alluvium; Qt1 (younger), Qt2, etc. = marine terraces and fluvial terraces of San Onofre Creek; Tm = Tertiary Monterey Formation; Tsm = Tertiary San Mateo Formation; Tso = Tertiary San Onofre Breccia; + 67′ = Medio triangulation station (modified from Moyle, 1973; Hunt and Hawkins, 1975).

form, yield excellent data for assessing vertical rates of neotectonic movement in this part of coastal southern California.

SITE INFORMATION

From its exposure in the San Onofre Bluffs, the Cristianitos fault is traced about 21 mi (35 km) inland where it displaces the Tertiary Capistrano, San Mateo, and Monterey formations and the San Onofre Breccia (Morton and Miller, 1981; Moyle, 1973; Vedder and others, 1957). The approximate age of last displacement is constrained by the presence of overlying, unbroken Quaternary marine sediments at the coast and fluvial terrace deposits bordering the nearby San Onofre Creek (Figs. 1 and 2).

At the sea cliff the Cristianitos fault strikes approximately north 32° west and dips about 58° northwest, separating the San Mateo and Monterey formations (Fig. 2; Ehlig, 1977). The San Mateo Formation, west of the fault, ranges in age from late Miocene to early Pliocene. It is entirely of marine origin and is thought to have been laid down as backfill in submarine channels, possibly localized in a structural trough along the downthrown side of the Cristianitos fault (Ehlig, 1977). The greyish-green, thin-bedded siltstones, east of the fault, have previously been called the Capistrano Formation (Fig. 1). Recently, however, microfossil assemblages have been dated as early to

Figure 2. The Cristianitos fault at the San Onofre Bluffs juxtaposing Tertiary San Mateo (left) and Monterey (right) formations and overlain by basal marine gravels and sands and continental sediments with intercalated buried paleosols.

Figure 3. Small faults displacing beds within the San Mateo Formation, adjacent to Cristianitos fault, San Onofre sea cliffs.

middle upper Miocene in age, and thus the beds are now assigned to the Monterey Formation (Anderson, Warren Assoc., 1977; Ehlig, 1977).

At the sea cliff, the San Mateo Formation is cut by many conjugate faults with only several inches (cm) displacement. They increase in frequency near the Cristianitos fault, indicating that the fault has involved dip-slip and therefore is an extensional or normal-slip fault (Fig. 3).

Where exposed in the sea cliff, the Cristianitos fault is overlain by about 3.3 ft (1 m) of marine gravels and sands and 40 ft (12 m) of continental deposits (Fig. 2). These sediments were laid down on a widely traceable, wave-cut platform visible here and for several miles on either side of the Cristianitos fault (Fig. 4).

Mollusks collected from basal terrace deposits about 2.5 mi (4 km) southeast of the Cristianitos fault (near the Camp Pendleton–San Onofre State Beach boundary) pertain to a warm-water fauna, indicative of a late Sangamon (marine isotope substage 5e) sea that transgressed to about 20 to 33 ft (6 to 10 m) above the persent (Bloom and others, 1974; Shackleton and Opdyke, 1973, 1976). The platform age is also indicated by amino-acid and uranium-series dates from this and other localities on the southern California coast (Kern, 1977; Ku and Kern, 1974; Lajoie and others, 1979; Shlemon, 1978a; Wehmiller and others, 1977). Accordingly, because these terrace deposits are clearly not offset by the Cristianitos fault, last displacement took place at least 125,000 years ago and most likely well before that time.

The San Onofre Bluffs also expose several buried paleosols (Fig. 2). These soils (pedogenic profiles) are slightly to moderately developed. They are best exposed in steep-walled canyons incised within continental deposits overlying the 125,000-year-old wave-cut platform but are also visible in trailside exposures between the top of the bluffs and the beach.

The buried paleosols have been best documented in road

Figure 4. The 125,000-year-old marine platform cut across the Tertiary San Mateo Formation, San Onofre Bluffs area, looking northwest from the Cristianitos fault to the San Onofre Nuclear Generating Station.

cuts on Camp Pendleton, several miles southeast of the Cristianitos fault (Shlemon, 1978a). Here, seven moderately developed profiles were identified, each capping a fining-upward sequence of fan gravels, sands, and silts. Radiocarbon dates from sediments bracketing the soils, obtained at Camp Pendleton and in the San Diego area, indicate that moderate profile development (with argillic B horizon) in this area can take place in about 3,000 to 6,000 years (Carter, 1957; Shlemon, 1978a). This rate of coastal soil development, which is much faster than that in inland parts of California, is apparently due to the presence of locally derived, fine-grained parent material increasing soil moisture content, to the influx of eolian-derived salt causing rapid dispersal and migration of illuvial clays, and to the influence of summer fog decreasing evapotranspiration (Shlemon and Hamilton, 1978).

Although unbroken by the Cristianitos fault, the 125,000-year-old wave-cut platform, as exposed between about Oceanside on the south and Laguna Beach on the north, is tilted up to the north. The platform has been uplifted at a rate of about 3.6 in/ 1,000 yr (9 cm) at San Onofre and 10 in/1,000 yr (26 cm) at Dana Point. In comparison, uplift rates elsewhere on the southern California coast are about 6.4 to 8 in/1,000 yr (16 to 20 cm) at San Diego, 12 to 18 in/1,000 yr (30 to 45 cm) at Malibu, and more than 80 in/1,000 yr (200 cm) at Ventura (Shlemon, 1978b). These data thus suggest that neotectonic uplift in the San Onofre area is one of the lowest in coastal southern California.

Visible from the top of the San Onofre bluffs (from the parking lot area) are remnants of marine terraces, particularly well displayed a few miles (km) to the south on Camp Pendleton. These terraces are perhaps the last classic mainland sequences remaining on the southern California coast. Others, for example those on the Palos Verdes Peninsula (Woodring and others, 1946), have been generally covered as urbanization takes place.

Onshore terraces older than the 125,000-year-old wave-cut platform are expressed mainly by veneers of gravel bars (Ehlig, 1977). Their ages are estimated by association with the marine isotope stage chronology (Shlemon, 1978a; Fig. 5). The terraces are judged to reflect glacio-eustatic sea levels superimposed on a tectonically rising block. Accordingly, the highest terrace remnants, at about 1,250 ft (370 m), are conservatively estimated to be almost 800,000 years old (Fig. 5). It is possible, however, that this and other high terraces are even older, perhaps well in excess of a million years (Ehlig, 1979).

Offshore geophysical surveys show the persence of three well-defined platforms cut across the Monterey Formation. These submarine terraces are capped by sediments informally designated "younger" cover (Y.C.) and "older" cover (O.C.), respectively (Fig. 5). Five samples of wood dredged from the base of the younger cover yielded radiocarbon ages ranging from about 13,000 years offshore to 8,500 years nearshore, documenting particularly well the late Holocene (Flandrian) rise of sea level (Barneich and Brogan, 1980). The absolute ages of the underlying submarine platforms are not known but are likewise estimated by

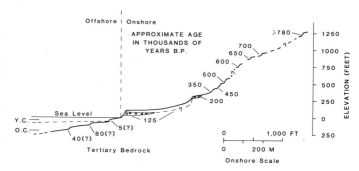

Figure 5. Schematic representation, onshore and offshore terraces, elevations and estimated ages (marine isotope stage chronology), San Onofre area.

association with the marine isotope stage chronology and judged to be about 5,000, 40,000, and 80,000 years, respectively (Fig. 5).

The complex interfingering and age relationships of fluvial and marine terrace deposits are visible in the sea cliffs between the Cristianitos fault and the mouth of San Onofre Creek (Fig. 1). At the Cristianitos fault and at SONGS, the San Mateo Formation is covered by younger, transgressive marine deposits (depicted as Qt5e for isotope substage 5e; Fig. 6). Exposures at SONGS, however, show that well-rounded and "clean" marine gravels and sands laterally merge into angular gravels and clayey sediments laid down in an ancient channel of San Onofre Creek (deposits Qc4; Fig. 6). These gravels underlie a San Onofre Creek fluvial terrace and, in turn, are overlain by overbank and locally derived distal fan deposits.

The continental deposits also bear several intercalated buried paleosols. As shown schematically (Fig. 6, unit Qc3), one of the paleosols is very strongly developed (with argillic horizon) and is typified by multiple calcic horizons. Soil carbonates here yield radiocarbon dates of about 30,000 to 35,000 years (R. Berger, 1983, personal communication). Accordingly, based on

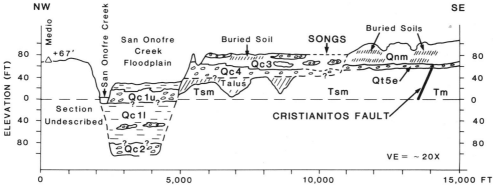

Figure 6. Stratigraphic relationship of late Quaternary channel, fluvial, and marine terrace deposits between the Cristianitos fault and San Onofre Creek as inferred from sea-cliff exposures and from well-log and bridge-boring data. Section location shown on Fig. 1. Symbols: Tm = Monterey Formation; Tsm = San Mateo Formation. Quaternary units identified by inferred age association with the marine isotope-stage chronology: e.g., Qt5e = terrace gravels of stage 5e age, ca. 125,000 years old; Qc4 and Qc2 = stage 4 and 2 channel gravels laid down by San Onofre Creek; Qc3 = stage 3 overbank and fan deposits; Qc1l and Qc1u = lower and upper stage 1 deposits (latest Pleistocene and Holocene) of San Onofre Creek (after Shlemon, 1979).

soil dates and on stratigraphic relations to the 125,000-year-old platform, the San Onofre Creek terrace gravels and their covering deposits are accorded ages of about 30,000 to 50,000 and 60,000 to 70,000 years, equated to marine isotope stage 3 and 4, respectively (Fig. 6).

Shown also in Figure 6 are fluvial gravels of San Onofre Creek encountered at the coastline about 100 ft (30 m) below sea level (Shlemon, 1979). As indicated by water well and bridge borings, these deposits rise to the surface a few miles inland but deepen considerably offshore, likely grading to the last major, low

stand of sea level, about 15,000 to 20,000 years ago (isotope stage 2; unit Qc2, Fig. 6).

In sum, the bluffs at San Onofre State Beach expose not only the Cristianitos fault but also marine and fluvial terrace deposits and numerous buried paleosols. In essence, many Quaternary geomorphic and soil-stratigraphic markers useful to assess paleo- and neotectonic rates in coastal southern California are well preserved here. A geologic textbook is thus open to all wishing to read it by taking a pleasant walk along this beautiful beach.

REFERENCES CITED

Anderson, Warren and Assoc. Inc., 1977, Results of microfossil identification and geologic age correlation, vicinity of San Onofre Generating Station, California, *in* Geotechnical studies, northern San Diego County, California; San Onofre Nuclear Generating Station, Units 1, 2, and 3, for Southern California Edison Company (Rosemead) and San Diego Gas & Electric Company (San Diego), Appendix A, Enclosure 3, Consultants' Technical Reports (Aug. 22, 31, and Sept. 20, 1977), variously paginated.

Barneich, J. A., and Brogan, G. E., 1980, Summary report on basic data from two onshore and six offshore geologic borings, SONGS Units 2 and 3, San Onofre, California; for Southern California Edison Company (Rosemead): Woodward-Clyde Consultants Technical Report, Project No. 41299I, 8 p., tables, appendices.

Bloom, A. L., Broecker, W. S., Chappell, J. M., Matthews, R. K., and Mesollela, K. J., 1974, Quaternary sea level fluctuations on a tectonic coast; New Th-230/U-234 dates from the Huon Peninsula, New Guinea: Quaternary Research, v. 4, p. 185–205.

Carter, G. F., 1957, Pleistocene Man at San Diego: Baltimore, Johns Hopkins Press, 400 p.

Ehlig, P. L., 1977, Geologic report on the area adjacent to the San Onofre Nuclear Generating Station, northwestern San Diego County, California, *in* Geotechnical studies, northern San Diego County, California; San Onofre Nuclear Generating Station, Units 1, 2, and 3; *for* Southern California Edison Company (Rosemead) and San Diego Gas & Electric Company (San Diego), Appendix A, Enclosure 3, Consultant's Technical Report, 40 p., tables, plates.

—— , 1979, Late Cenozoic evolution of the Capistrano Embayment, *in* Keaton, J. R., ed., Guidebook to selected geologic features, coastal areas of southern Orange and northern San Diego counties, California: South Coast Geological Society, October 20, 1979, Field Trip, p. 26–34.

Hunt, G. S., and Hawkins, H. G., 1975, Geology of the San Onofre area and portions of the Cristianitos fault, *in* Ross, A., and Dowlen, R. J., eds., Studies on the geology of Camp Pendleton and western San Diego County, California: San Diego Association of Geologists, p. 7–14.

Kern, J. P., 1977, Origin and history of upper Pleistocene marine terraces, San Diego, California: Geological Society of America Bulletin, v. 88, p. 1553–1556.

Ku, T. L., and Kern, J. P., 1974, Uranium-series age of the upper Pleistocene Nestor terrace, San Diego, California: Geological Society of America Bulletin, v. 85, p. 1713–1716.

Lajoie, K. R., Kern, J. P., Wehmiller, J. F., Kennedy, G. L., Mathieson, S. A., Sarna Wojcicki, A. M., Yerkes, R. F., and McCrory, P. F., 1979, Quaternary terraces and crustal deformation in southern California, *in* Abbott, P. L., ed., Geological excursions in the southern California area, Geological Society of America Annual Meeting, November 1979: San Diego, California, Department of Geological Sciences, San Diego State University, p. 3–15.

McNey, J. L., 1979, General geology, San Onofre area, *in* Keaton, K. R., ed., Guidebook to selected geologic features, coastal areas of southern Orange and northern San Diego counties, California: South Coast Geological Society, October 20, 1979, Field Trip, p. 90–104.

Morton, P. K., and Miller, R. V., compilers, 1981, Geologic map of Orange County, California showing mines and mineral deposits: California Division of Mines and Geology, Department of Conservation, scale 1:48,000.

Moyle, W. R., 1973, Geologic map of western part of Camp Pendleton, southern California: U.S. Geological Survey Open-File Map, 2 plates, scale 1:62,500.

Shackleton, N. J., and Opdyke, N. D., 1973, Oxygen isotope and palaeomagnetic stratigraphic of equatorial Pacific core V28-238; Oxygen isotope temperatures and ice volumes on a 10^5 and 10^6 year scale: Quaternary Research, v. 3, no. 1, p. 39–55.

—— , 1976, Oxygen isotope and paleomagnetic stratigraphy of Pacific core V28:239, late Pliocene to latest Pleistocene: Geological Society of America Memoir 145, p. 449–464.

Shlemon, R. J., 1978a, Late Quaternary evolution of the Camp Pendleton–San Onofre State Beach coastal area, northwestern San Diego County, California, for Southern California Edison Company (Rosemead) and San Diego Gas & Electric Company (San Diego), Consultant's Technical Report, 128 p.

—— , 1978b, Late Quaternary rates of deformation, Laguna Beach–San Onofre State Beach, Orange and San Diego counties, California, for Southern California Edison Company (Rosemead) and San Diego Gas & Electric Company (San Diego), Consultant's Technical Report, 40 p.

—— , 1979, Late Cenozoic stratigraphy, Capistrano Embayment coastal area, Orange County, California, *for* Southern California Edison Company (Rosemead) and San Diego Gas & Electric Company (San Diego), Consultant's Technical Report, 27 p.

Shlemon, R. J., and Hamilton, P., 1978, Late Quaternary rates of sedimentation and soil formation, Camp Pendleton–San Onofre State Beach coastal area, southern California, U.S.A.: Jerusalem, Israel, Tenth International Congress on Sedimentology, Abstract volume II, p. 603–604.

Vedder, J. G., Yerkes, R. F., and Schoellhamer, J. E., 1957, Geologic map of San Joaquin–San Juan Capistrano area, Orange County, California: U.S. Geological Survey Oil and Gas Investigations Map, OM 193.

Wehmiller, J., Lajoie, K. R., Kvenvolden, K. A., Peterson, E., Belknap, B. F., Kennedy, G. L., Addicott, W. O., Vedder, J. G., and Wright, R. W., 1977, Correlation and chronology of Pacific marine terrace deposits of continental United States by fossil amino-acid stereochemistry-technique evaluation, relative ages, kinetic model ages, and geologic implications: U.S. Geological Survey Open-File Report 77-680, 106 p.

Woodring, W. P., Bramlette, N. N., and Kew, W. S., 1946, Geology and paleontology of Palos Verdes Hills, California: U.S. Geological Survey Professional Paper 207, 145 p.

ACKNOWLEDGMENTS

Most data presented here stem from investigations commissioned by the Southern California Edison Company (SCE) for siting the San Onofre Nuclear Generation Station, Units 2 and 3. Helpful discussions and logistical support were generously made available by SCE geologists, by private consultants, and by technical reviewers.

Late Pleistocene angular unconformity at San Pedro, California

J. Douglas Yule and Donald H. Zenger, Department of Geology, Pomona College, Claremont, California 91711

LOCATION

This outcrop is located on Second Street between Pacific Avenue and Mesa Street in central San Pedro (Fig. 1), approximately 25 mi (40 km) south of downtown Los Angeles, California. To reach the "Second Street locality" from downtown, take the Harbor Freeway (I-110) south to its termination at Gaffey Street in San Pedro. Proceed south on Gaffey Street approximately 0.3 mi (0.5 km) to Second Street. Turn left (east) on Second Street and proceed 0.2 mi (0.3 km) to Pacific Avenue. Cross Pacific Avenue and park along Second Street. The outcrop extends along the north side of Second Street for 360 ft (110 m) east of the alleyway adjacent to the Tropicana Bakery. (The authors wholeheartedly recommend a visit to this authentic *panaderia.*)

SIGNIFICANCE OF LOCALITY

Exposed at Second Street are the tilted (17° to 23° NE), fossiliferous (marine) lower(?) Pleistocene Lomita Marl and Timms Point Silt, unconformably overlain by the nearly horizontal Palos Verdes Sand, an upper Pleistocene marine terrace deposit that grades upward into alluvial and colluvial(?) sand and silt. The Lomita is one of the most famous fossiliferous Pleistocene "marls" in the world (Bandy and Emery, 1954). Unfortunately, the classic Lomita quarry and Hilltop quarry localities have been destroyed, leaving the Second Street locality and a small outcrop of uppermost Lomita marl on the south side of Third Street between Pacific Avenue and Mesa Street as the only remaining surface exposures of the unit in San Pedro (Fig. 1). The significant late Pleistocene angular unconformity strikingly exhibited at Second Street is present throughout the Los Angeles Basin, Ventura Basin, and California Coast Ranges.

GENERAL BACKGROUND

Lower(?) Pleistocene. The classical type section for marine lower(?) Pleistocene sediments in southern California is located in the southwest part of the Los Angeles Basin along the north and east borders of the Palos Verdes Hills. Sediments here range in thickness from 100 to 600 ft (30 to 180 m). These sediments consist of three formations: Lomita Marl, Timms Point Silt, and San Pedro Sand. Central San Pedro is the only area where all three units occur together at the surface, in an upward sequence as listed above. Contacts between these units have been described as gradational in places (Woodring and others, 1946), and discordant at others (Arnold, 1903; Valentine, 1961).

Depositional environments of these units have been described to varying degrees by Bandy (1967), Bandy and others (1971), Clark (1931), Kennedy (1975), Valentine (1961), Valentine (1976), Woodring and others (1946), and Woodring (1957). The Lomita is thought to have been deposited in water depths of

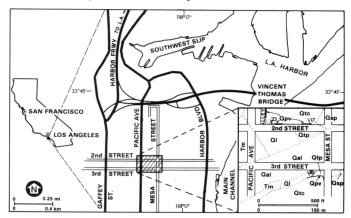

Figure 1. Location map showing central San Pedro, California, and the Second Street locality. Geology modified from Woodring and others (1946, Plate 14). The Torrance 7½-minute Quadrangle is north of 33° 45′ latitude; the San Pedro 7½-minute Quadrangle lies south of this line. Qsp = San Pedro Formation; other Quaternary symbols explained on Figure 2.

151 to 600 ft (46 to 183 m) (see summary in Kennedy, 1975, p. 8). An abundance of pelagic foraminifera, glauconite, and phosphate nodules in the lowermost beds indicates slow deposition. Nonglauconitic layers higher upsection, such as those at Second Street, contain an outer sublittoral fauna indicative of 151- to 302-ft (46- to 92-m) depths. The Timms Point Silt is the most uniform facies of the three units, representing a 151- to 600-ft (46- to 183-m) deep, cool-water depositional site. The San Pedro Sand represents a more shallow, 59- to 302-ft (18- to 92-m) depositional environment.

The base of the Lomita Marl has, in the past, been considered to represent the basal Pleistocene in southern California. Obradovich (1965, 1968), using the K/Ar dating method, reported an age of 3.04 ~0.09 Ma for the Lomita Marl. On the other hand, an age of 155 ~30 ka was determined by Fanale and Schaeffer (1965) using the He/U dating method. Bandy and others (1971) suggested an age of 700 to 800 ka for the Lomita based upon planktonic foraminiferal datum planes, represented in the Lomita by the appearance of the cold water, sinistrally coiled *Globerigerina pachyderma,* and their correlation with radiometric dates of Pleistocene volcanic rocks. Isotope ratios of strontium-87 to strontium-86 in a bivalve shell from the Santa Barbara Formation, which has been correlated with the Lomita Marl (Valentine, 1961; Valentine, 1976), and in three samples from the Lomita are consistent with a 700 ka age and certainly indicate an age not more than 1 Ma (DePaolo and Ingram, 1985; D. J. DePaolo, oral communication, 1985). The base of the Calabrian in Italy, considered to be the base of the Pleistocene, has been dated at 1.79 Ma (Bandy and Wilcoxon, 1970; Berggren and Van Couvering, 1974, p. 52–56). Amino acid racemization data suggest a middle Pleistocene age (i.e., <700 ka) for the Lomita at the

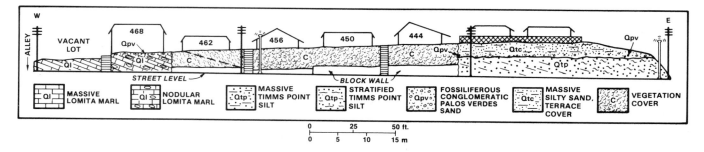

Figure 2. Sketch of the outcrop along the north side of Second Street. Houses, staircases, telephone poles, and light posts are shown schematically. House numbers refer to their respective street addresses. The dashed line below the house at 462 Second Street represents the approximate location of the Lomita/Timms Point contact (see Woodring and others, 1946, Plate 15). The actual slope of Second Street is about 2° to the east. Refer to text for further details.

old Hilltop quarry (J. F. Wehmiller, unpublished data, 1985). Naeser and others (1981, p. 37–41) present a discussion of the various proposed ages of the Lomita. In view of the uncertainty of the age of the Lomita at this time, we have chosen to consider it as lower Pleistocene with qualification; it does appear that the age of the Lomita is much less than the 3 Ma figure of Obradovich (1965, 1968) and very likely less than 1 Ma. Subsequent work may substantiate a middle Pleistocene age.

Upper Pleistocene. Middle and upper Pleistocene units exposed in the Palos Verdes Hills are represented primarily by 13 marine terraces that range in elevation from 100 to 1,300 ft (30 to 400 m). Fossils have been reported from marine deposits on nine of these terraces (Marincovich, 1976; Valentine, 1962; Woodring, 1957; Woodring and others, 1946). The name "Palos Verdes Sand" was introduced by Woodring and others (1946) and restricted to the marine deposits on the lowermost terrace. The name has subsequently been used for upper Pleistocene terrace deposits around the Los Angeles basin, including the upper terraces on the Palos Verdes Hills (Marincovich, 1976), a convention that should be dropped. Bandy and Emery (1954, p. 3) reported a minimal radiocarbon date of 30 ka for the Palos Verdes on the lowest emergent terrace. Szabo and Rosholt (1969), using uranium-series dating methods, determined an average "open system age" of 86 ka for mollusks in the Palos Verdes Sand on the first terrace. Numerous uranium-series determinations (e.g., Fanale and Schaeffer, 1965) and amino-acid racemization analyses (e.g., Wehmiller and others, 1977) support an age of 120 to 130 ka for the Palos Verdes Sand, correlative with substage 5e of the marine oxygen isotope record.

The Palos Verdes Sand unconformity overlies Miocene, Pliocene, and lower(?) Pleistocene units at various localities. Along the north flank of the Palos Verdes Hills, the Palos Verdes and underlying Pleistocene sediments have undergone deformation such that they are dipping 25° to 30°N (Woodring and others, 1946, p. 109). Elsewhere, vertical uplift has affected the Palos Verdes Sand (such is the case of the raised marine terrace at Second Street). Thus, post-Palos Verdes uplift is of latest Pleistocene or Holocene age, attesting to ongoing tectonism in the Los Angeles Basin.

SITE INFORMATION

History of the Second Street Locality. Only a portion of the original exposure at Second Street survives today. The uppermost 5 to 10 ft (1.5 to 3.0 m) of Timms Point Silt and lowermost San Pedro Sand were removed east of the present outcrop and the Lomita-Timms Point contact has become obscured by vegetation subsequent to work of Woodring and others (1946). The exposure of Lomita and Timms Point along the south side of Second Street was destroyed by urban renewal construction in the latest 1970s and earliest 1980s. What remains of this significant outcrop on the north side of Second Street is shown in Figure 2.

Lomita Marl. Approximately 38 ft (12 m) of the middle part of the Lomita Marl are exposed along Second Street and the alley adjacent to the bakery at the west end of the outcrop (Fig. 2). The basal Lomita is covered beneath the bakery, and the upper 8–9 ft (2.4–2.7 m) are covered by vegetation. The lower 18 ft (5.5 m) of exposed Lomita, designated "massive Lomita Marl" in Figure 2, consist of zones of light-gray to slightly olive-gray coquina in a less fossiliferous limestone that varies laterally with respect to shell content. The upper 20 ft (6 m), designated "nodular Lomita Marl" in Figure 2, are characterized by an upward increase in coquina and nodular layers. Thin layers of coarse calcareous sand without nodules form re-entrants and alternate with more resistant nodular coquinoid layers. The exposure is probably equivalent to units 3 and 4 of Woodring and others (1946, p. 46), described from the south side of Second Street. Petrographically, the Lomita is predominantly an algal-foraminiferal-mollusk-bryozoan grainstone to rudstone. More highly indurated nodular layers contain a micritic matrix that locally supports grains. The Lomita at Second Street contains very little argillaceous material and is not a marl; it is best described as a calcareous sand. X-ray diffraction (XRD) reveals, in order of decreasing abundance, calcite, Mg-calcite, Ca-rich dolomite, minor aragonite, and very minor quartz silt.

Larger invertebrate fossils are mainly mollusks and generally are smaller individuals than at other, now destroyed, outcrops. Very common bivalves here are *Cyclocardia* sp., aff. *C. occidentalis, Epilucina californica, Lucinoma annulata, Psephidia*

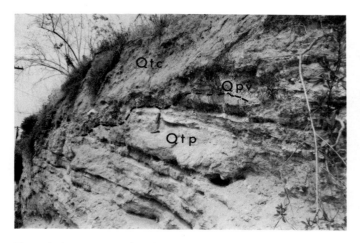

Figure 3. Angular unconformity at the east end of the exposure (view to NW). The stratified Timms Point Silt (Qtp) is truncated by the coarse, pebbly Palos Verdes Sand (Qpv) that in turn is overlain by a terrestrial terrace cover (Qtc). Note the thin white layers in the Timms Point Silt (at hammer handle) and the irregularity of the unconformity (dashed line).

Figure 4. Palos Verdes Sand (Qpv, top of photograph) unconformably overlying Timms Point Silt (Qtp, bottom of photograph). The irregularity of the unconformity is evidenced by the vertical burrows filled with Palos Verdes material (dashed line); note the molds of large bivalves at the tops of two burrows.

sp., and *Semele incongrua.* Commonest gastropods include *Amphissa versicolor, Bittium* spp., *Homalopoma carpenteri, Nassarius* spp., and *Turritella cooperi.* Foraminifera are common, particularly the genera *Bolivina, Cassidulina, Cibicides, Dentalina, Discorbis, Elphidium, Globigerina, Quinqueloculina, Triloculina,* and *Uvigerina* (Emiliani and Epstein, 1953). Also present are fragments of *Strongylocentrotus* sp., bryozoans, and chiton plates.

Timms Point Silt. Approximately 40 ft (12 m) of the middle part of the Timms Point Silt are exposed at the eastern end of the exposure (Fig. 2). The lowermost 33 ft (10 m) of the unit are covered by vegetation (Fig. 2), and the uppermost 5–10 ft (1.5–3.0 m) have been removed. The lower 30 ft (9 m) of exposed Timms Point, designated "massive Timms Point Silt" in Figure 2, consist of friable grayish-orange to dark yellowish-orange, massive sandy silt. Conical concretions occur along the bases of some bedding planes and there are a few fossiliferous horizons. The uppermost 10–12 ft (3.0–3.5 m), designated "stratified Timms Point Silt" in Figure 2, consist of thin bedded (3–12 in; 7.5–30 cm) orange-brown silty sand and less resistant gray sandy silt. Three conspicuous thin (up to 2 in; 5 cm) white limy beds occur in a 12-in (30 cm) interval near the top of the section (Fig. 3). The exposure is probably equivalent to units 3 through 6 of Woodring and others (1946, p. 46). Petrographically, the Timms Point Silt is mainly a sandy, nonlaminated quartz-feldspar-mica siltstone to a silty, very fine-grained, micaceous, feldspathic wacke with very angular grains. A "blue-green" amphibole is a surprisingly common constituent. Also present are a few radiolarians, foraminifera, and diatoms. XRD reveals, in order of decreasing abundance, quartz and plagioclase, microcline, mica, amphibole, and minor Ca-rich dolomite, calcite, and chlorite. One sample was 7.7% soluble in hydrochloric acid. The Timms Point is immature and probably was derived from material shed off the Coast Ranges (Valentine, 1976).

The Timms Point is much less fossiliferous at this locality than the Lomita Marl. According to Woodring and others (1946), the most common mollusks are the bivalves *Mya truncata, Panomya beringiana, Patinopecten caurinus,* and the gastropod *Mitrella carinata.* According to Valentine (1961, p. 405), the fauna of the Timms Point Silt on Second Street is a *Cyclocardia barbarensis–Antiplanes perversa* association representing the deep, outer sublittoral zone, 300 to 600+ ft (90 to 180+ m).

Lomita/Timms Point contact. As mentioned above, the contact between the Lomita Marl and Timms Point Silt at Second Street is concealed (but see Woodring and others, 1946, Pl. 15, Fig. A). Arnold (1903), Clark (1931), and Valentine (1961) interpreted the Timms Point to lie unconformably on the Lomita. The contact at Second Street, however, has been described as conformable by Woodring and others (1946). We observed an upward decrease in carbonate content through about a 1 ft (0.3 m) zone along the former exposure on the south side of Second Street, suggestive of a gradational conformable contact. The contact at the Third Street exposure (Fig. 1) is also gradational.

Palos Verdes Sand. The Palos Verdes Sand lies unconformably above the lower(?) Pleistocene units, and is exposed and accessible above the Timms Point Silt at the eastern end of the outcrop (Figs. 2, 3, and 4). The Palos Verdes Sand is largely absent or inaccessible above the Lomita, except in an exposure along the alley to the west (Fig. 5). The Palos Verdes Sand is a 1–2 ft (30–61 cm) unit of pale yellowish-brown to dark yellowish-brown fossiliferous silty sand with a basal conglomerate. Pebble and cobble-sized clasts are composed of metaquartzite, Tertiary sedimentary rocks (Monterey and Repetto), and Lomita that are commonly bored by pholadid bivalves. Fossils are primarily mollusks; vertebrate remains are rare. The large (4–5 in; 10–12 cm) infaunal bivalves *Saxidomus nuttalli* and *Tresus nuttallii* are most conspicuous. The Palos Verdes grades vertically

Figure 5. Palos Verdes Sand (Qpv, darker) just above the unconformity in the alleyway at west end of exposure. Note the light, irregular erosional remnant of Lomita Marl (Ql) in pebbly, fossiliferous Palos Verdes Sand. Pen knife (2.7 in long; 7 cm) for scale.

into friable light-brown to moderately yellowish-brown, sandy and silty continental terrace deposits that reach a thickness of approximately 10 ft (3 m) at the eastern end of the outcrop, and a thickness of 4 ft (1.2 m) along the alley.

    Nature of the Unconformity. The angular unconformity is an irregular surface. Vertical burrows filled with Palos Verdes material extend 1 to 2 ft (0.3–0.6 m) downward into the Timms Point Silt, and were made by *Tresus nuttallii,* commonly preserved whole in burrowing position (Fig. 4). In the alleyway, clasts and erosional remnants of Lomita Marl are incorporated in Palos Verdes siliciclastics (Fig. 5). The sequence of geologic events was as follows: (1) deposition of the Lomita Marl, Timms Point Silt, and San Pedro Sand (now removed at Second Street); (2) subsequent uplift, deformation, and erosion; (3) marine transgression and terrace cutting in the late Pleistocene time, and deposition of the marine terrace deposits (Palos Verdes Sand); and (4) continued uplift and deposition of the continental terrace cover.

REFERENCES CITED

Arnold, R., 1903, The paleontology and stratigraphy of the marine Pliocene and Pleistocene of San Pedro, California: California Academy of Science Memoirs, v. 3, 420 p.

Bandy, O. L., 1967, Foraminiferal definition of the boundaries of the Pleistocene in southern California, U.S.A., *in* Sears, M., ed., Progress in Oceanography, Volume 4: London and New York, Pergamon Press, p. 27–49.

Bandy, O. L., and Emery, K. O., 1954, Geologic Guide No. 4, Southwestern part of the Los Angeles Basin: California Division of Mines Bulletin 170, 14 p.

Bandy, O. L., and Wilcoxon, J. A., 1970, The Pliocene-Pleistocene boundary, Italy and California: Geological Society of America Bulletin, v. 81, p. 2939–2948.

Bandy, O. L., Casey, R. E., and Wright, R. C., 1971, Late Neogene planktonic zonation, magnetic reversals, and radiometric dates, Antarctic to the tropics: Antarctic Research Series (Oceanology I), v. 15, p. 1–26.

Berggren, W. A., and Van Couvering, J. A., 1974, The late Neogene; Biostratigraphy, geochronology and paleoclimatology of the last 15 million years in marine and continental sequences: Palaeogeography, Palaeoclimatology, Palaeoecology, v. 16, p. 1–216.

Clark, A., 1931, The cool-water Timms Point Pleistocene horizon at San Pedro, California: Transactions of the San Diego Society of Natural History, v. 7, p. 25–42.

DePaolo, D. J., and Ingram, B. L., 1985, High-resolution stratigraphy with strontium isotopes: Science, v. 227, p. 938–941.

Emiliani, C., and Epstein, S., 1953, Temperature variations in the lower Pleistocene of southern California: Journal of Geology, v. 61, no. 2, p. 171–181.

Fanale, F. P., and Schaeffer, O. A., 1965, Helium-uranium ratios for Pleistocene and Tertiary fossil aragonites: Science, v. 149, p. 312–317.

Kennedy, G. L., 1975, Paleontologic record of areas adjacent to the Los Angeles and Long Beach harbors, Los Angeles County, California, *in* Soule, D. F., and Oguri, M., eds., Marine studies of San Pedro Bay, California, Part 9, paleontology: Los Angeles, Alan Hancock Foundation, 119 p.

Marincovich, L. N., Jr., 1976, Late Pleistocene molluscan faunas from upper terraces of the Palos Verdes Hills, California: Natural History Museum of Los Angeles County, Contributions in Science 281, 28 p.

Naeser, C. W., Brigg, N. D., Obradovich, J. D., and Izett, G. A., 1981, Geochronology of Quaternary tephra deposits, *in* Self, S., and Sparks, R.S.J., eds., Tephra Studies: Boston, D. Reidel Publishing Company, p. 13–47.

Obradovich, J. D., 1965, Age of the marine Pleistocene of California [abs.]: American Association of Petroleum Geologists Bulletin, v. 49, p. 1087.

Obradovich, J. D., 1968, The potential use of glauconite for late-Cenozoic geochronology, *in* Morrison, R. B., and Wright, H. E., Jr., eds., Means of correlation of Quaternary successions: Proceedings VII Congress, International Association for Quaternary Research, Volume 8, Salt Lake City, University of Utah Press, p. 267–279.

Szabo, B. J., and Rosholt, J. N., 1969, Uranium-series dating of Pleistocene molluscan shells from southern California—An open system model: Journal of Geophysical Research, v. 74, p. 3253–3260.

Valentine, J. W., 1961, Paleoecologic molluscan geography of the Californian Pleistocene: University of California Publications in Geological Sciences, v. 34, no. 7, p. 309–442.

——, 1962, Pleistocene molluscan notes. No. 4. Older terrace faunas from Palos Verdes Hills, California: Journal of Geology, v. 70, p. 92–101.

Valentine, P. C., 1976, Zoogeography of Holocene Ostracoda off western North America and paleoclimatic implications: U.S. Geological Survey Professional Paper 916, 47 p.

Wehmiller, J. F., Lajoie, K. R., Kvenvolden, K. A., Peterson, E., Belknap, D. F., Kennedy, G. L., Addicott, W. O., Vedder, J. G., and Wright, R. W., 1977, Correlation and chronology of Pacific Coast marine terrace deposits of continental United States by fossil amino acid stereochemistry—Technique evaluation, relative ages, kinetic model ages, and geologic implications: U.S. Geological Survey Open File Report 77-680, 196 p.

Woodring, W. P., 1957, Marine Pleistocene of California: Geological Society of America Memoir 67, p. 589–597.

Woodring, W. P., Bramlette, M. N., and Kew, W.S.W., 1946, Geology and paleontology of Palos Verdes Hills, California: U.S. Geological Survey Professional Paper 207, 145 p.

ACKNOWLEDGMENTS

    We are indebted to D. B. McIntyre for formally proposing the Second Street site for "commemoration." J. D. Obradovich discussed the age of the Lomita. We are especially grateful to G. L. Kennedy who thoroughly reviewed the manuscript and checked the names of the molluscan species. D. J. DePaolo and R. C. Capo kindly provided data on the age of a Lomita sample based on $^{87}Sr/^{86}Sr$ ratios. The Graphics Department, Union Oil Research, aided in the drafting of Figures 1 and 2. J. MacKay typed the various drafts of the manuscript and G. Ott did the darkroom phase of the photography.

Portuguese Bend landslide complex, southern California

Perry L. Ehlig, *Department of Geology, California State University, Los Angeles, California 90032*

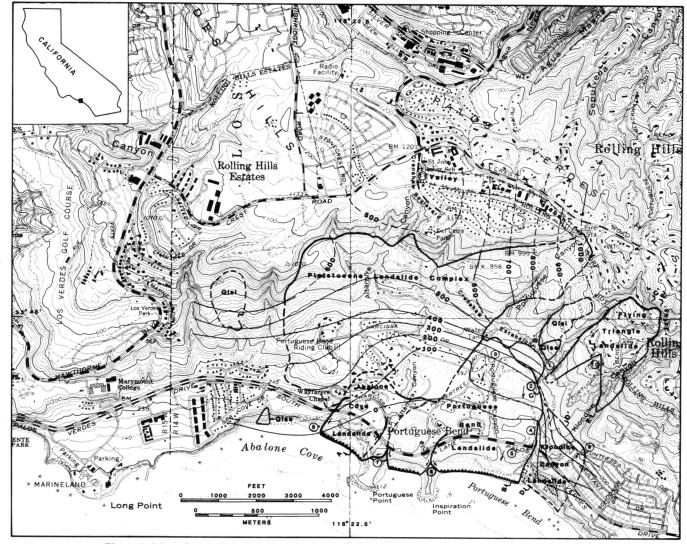

Figure 1. Map of landslides in Portuguese Bend landslide complex. Qlsa, minor landslides active in historic time; Qlsi, minor landslides inactive during historic time. Structure contours show elevation in feet above sea level at base of Portuguese Tuff. Numbers in circles are locations referred to in text.

LOCATION AND ACCESSIBILITY

The Portuguese Bend landslide complex occupies a 2.5 mi^2 (6 km^2) amphitheater-shaped topographic embayment on the south flank of the Palos Verdes Peninsula, 22 mi (35 km) south of the Los Angeles civic center in southern California. One-third of the complex consists of the active Portuguese Bend and Flying Triangle landslides and the recently active Abalone Cove and Klondike Canyon landslides (Fig. 1). The complex is in the northwest corner of the San Pedro, California, 7½-minute Topo-

graphic Quadrangle and adjoining parts of the Torrance and Redondo Beach 7½-minute quadrangles.

As part of the greater Los Angeles metropolitan area, the Portuguese Bend area is easily reached by automobile or by Rapid Transit District (RTD) buses. The most direct route from the Los Angeles civic center is south on the Harbor Freeway (I-110) to its terminus at Gaffey Street in San Pedro. Continue south on Gaffey Street for 1.7 mi (2.8 km), turn right on 25th Street and go 2.8 mi (4.5 km) west to where 25th Street merges with Palos Verdes Drive South (PVDS). The east edge of the

Portuguese Bend landslide is 1.5 mi (2.4 km) farther west on PVDS. The area can also be approached from the north by way of Crenshaw and Hawthorne Boulevards (Fig. 1).

An excellent overview of the area is obtained from the ridge crest on the south side of Del Cerro Park, located near the southern terminous of Crenshaw Boulevard (Fig. 1). Hiking and riding trails and unpaved roads are present within undeveloped parts of the complex, but most of this area is private land and much of the area is within gated communities where entry requires special permission. The only public land is the beach area seaward of PVDS and west of Inspiration Point (Fig. 1). Public parking along PVDS is restricted to a lot on the south side of PVDS, 0.1 mi (0.2 km) west of Wayfarers Chapel (Fig. 1) and costs $3.00 per day.

Geologists wishing current information on the landslides or assistance in gaining access to the area should contact Tom Bandy, Community Services Director, City of Rancho Palos Verdes, 30940 Hawthorne Boulevard, Rancho Palos Verdes, CA 90274; telephone (213) 377-0360. Mr. Bandy is also the Executive Director of the Abalone Cove Landslide Abatement District and the Klondike Canyon Geologic Hazard Abatement District.

SIGNIFICANCE OF SITE

The area contains the 260 acre (105 ha) Portuguese Bend landslide, which has moved continuously since 1956; the 80 acre (32 ha) Abalone Cove landslide, which began moving in 1978 but has been stabilized by dewatering; the 50 acre (20 ha) Klondike Canyon landslide, which began moving in 1979 but is not moving at present; and the 88 acre (36 ha) Flying Triangle landslide, which began moving in a 5 acre (2 ha) area in 1980 but expanded greatly in 1983 and may continue to enlarge. Slippage occurs on gently dipping bentonite beds at rates ranging from about an inch to 35 ft (0.02 to 11 m) per year. Block glide is a dominant mode of sliding, but lateral spreading on listric normal faults is also common. Slide toes consist of fault-propagated folds and thrust ramps. Groundwater plays an important role in all of the slides. The low angle of internal friction (as low as 6°) along the bases of these slides, as well as their style of deformation, make them excellent models for some types of large detachment faults and gravity glide overthrusts. In addition to their scientific value, the slides have played an important role in the growth of engineering geology. Although the area had been mapped as a landslide, extensive residential development was permitted prior to the activation of the Portuguese Bend landslide, because Los Angeles County lacked geologic requirements in their grading ordinance. After movement began, both the County and the City of Los Angeles added geologic requirements, ushering in the mandatory role of engineering geologists in land development in southern California. In 1961 a legal precedent was set when the County of Los Angeles was held liable for causing the slide and subsequent destruction of about 130 homes (because the County loaded the head of the slide with fill during construction for Crenshaw Boulevard). This decision enhanced the role of engi-

neering geology in public works. Another first was the establishment of the Abalone Cove Landslide Abatement District in 1980, following passage of enabling legislation. It is the first such abatement district in California and is a highly successful model for similar districts. The district is responsible for mitigating the landslide using funds obtained by taxation of the benefiting properties. The Klondike Canyon Geologic Hazard Abatement District was established in 1982 as the second such district in California. In 1984, the Rancho Palos Verdes Redevelopment Agency was established to stabilize the Portuguese Bend landslide, the first such use of a redevelopment agency. Stabilization efforts are under way with the aid of a $2,000,000 grant from the state.

SITE INFORMATION

The Palos Verdes Peninsula is a dome-shaped ridge whose configuration reflects its origin as a northwest-trending, doubly plunging anticline. Its uplift began in the Pliocene while the area was still below sea level and it spent most of the Pleistocene as an island before it became part of the mainland near the end of the Pleistocene. While it was an island, a series of marine terraces was eroded into its flanks. Marine erosion was greatest on its ocean-facing south and west sides where wave erosion continues to occur today. Removal of down-slope support by marine erosion has played a major role in most large landslides within the peninsula.

Mesozoic Catalina Schist forms the basement beneath the peninsula and is locally exposed in the northeastern part. The schist is unconformably overlain by the marine Monterey Formation, which is subdivided into the middle Miocene Altamira Shale, the upper Miocene Valmonte Diatomite, and the upper Miocene to lower Pliocene Malaga Mudstone. The Altamira Shale is the thickest and most widely exposed part of the Monterey Formation and the host for most landslides. All but the upper part of the Altamira Shale was deposited simultaneously with local volcanism, and it contains tuff and basalt intrusions and extrusions along with siliceous shale and dolostone of largely biogenic origin, and siltstone and sandstone derived from Catalina Schist. This part of the Altamira Shale is referred to as the tuffaceous member and has an exposed thickness of about 1,000 ft (300 m). The tuff has been partially to entirely altered to bentonite, including a distinctive unit referred to as the Portuguese Tuff. The upper part of the Altamira Shale includes a cherty member containing distinctive chert concretions and consisting of about 52 ft (16 m) of siliceous shale, porcellanite, and chert, and an overlying phosphatic member of variable thickness and lithology, but consisting of about 80 ft (24 m) of phosphatic siltstone, diatomaceous shale, and laminated sandstone in the area upslope from Portuguese Bend.

Those wishing to learn more about the geology of the peninsula should refer to Woodring and others (1946). Rowell (1982) updates the chronostratigraphy of the Monterey Formation, and Conrad and Ehlig (1983) update the lithostratigraphy of the

Monterey Formation. Bryant (1982) provides an update on the Pleistocene marine terraces. Jahns and Vonder Linden (1973) discuss space-time relationships of landsliding on the south flank of the Peninsula. The Portuguese Bend landslide is described by Merriam (1960) and Vonder Linden and Lindval (1982); the mechanics of the Portuguese Bend landslide and plans to stabilize it are discussed by Ehlig (1986). The mechanics of the Abalone Cove landslide are described by Ehlig (1982), and groundwater conditions and dewatering of the Abalone Cove landslide by Ehlig and Bean (1982). The Klondike Canyon landslide is described by Kerwin (1982), and the Flying Triangle landslide is described by Keene (1986).

The Portuguese Bend landslide complex occupies a structurally simple part of the south flank of the Palos Verdes anticline. Like most landslides in the peninsula, those within the complex have their basal slip surfaces along bentonite beds in the tuffaceous member of the Altamira Shale. The stratigraphic positions of slip surfaces are usually referenced to the Portuguese Tuff because it is an easily identified lithologic unit. The Portuguese Tuff has an average thickness of about 60 ft (18 m) and has its base about 530 ft (162 m) below the top of the tuffaceous member. Structure contours on the base of the Portuguese Tuff are shown on Figure 1.

The stratigraphically deepest slip surfaces are along the base of the Flying Triangle landslide and may be as much as 460 ft (140 m) below the base of the Portuguese Tuff. Studies in progress indicate the active Flying Triangle landslide is part of a much larger Pleistocene landslide that extends down slope beneath the Klondike Canyon landslide and the eastern margin of the active Portuguese Bend landslide. The down-slope movement of this hypothesized slide was accommodated by anticlinal uplift near its toe. The earliest movement of the Flying Triangle landslide and its hypothesized down-slope extension predate a marine terrace whose shoreline angle is at an elevation of 410 ft (125 m) along the upslope edge of the Klondike Canyon landslide. Correlation of marine terraces by Bryant (1982) indicate this terrace is about 400,000 years old.

Within the main part of the Portuguese Bend landslide complex, the deepest slip surfaces generally range from slightly above the top of the Portuguese Tuff to slightly below the base of the tuff. The lowest known position is 70 ft (21 m) below the base of the tuff in the Klondike Canyon landslide. Studies by Jahns and Vonder Linden (1973) indicate some sliding occurred in the up slope part of the complex prior to erosion of a marine terrace at an elevation of about 660 ft (200 m), or prior to about 600,000 years ago, using correlations by Bryant (1982). However, most of the complex is much younger as suggested by the fact that the deepest slip surface is continuous between the up-slope and down-slope parts of the complex and by the absence of marine terraces within the slide affected area. Most of the complex probably formed about 120,000 years ago when the sea occupied the prominent terrace that caps Inspiration and Portuguese points and has a shoreline angle at an elevation of about 130 ft (40 m). The seaward edge of this Pleistocene landslide is exposed in the cliff on the east side of Inspiration Point (locality 6, Fig. 1). Here, the Portuguese Tuff was bent upward to a nearly vertical position by up thrusting along the toe of the slide and was subsequently unconformably overlain by marine terrace deposits. The toe of the Pleistocene landslide probably developed as an anticlinal uplift above a slip surface at the base of the Portuguese tuff, about 200 ft (60 m) below the base of the marine terrace. As movement progressed, the slip surface propagated upward as a thrust ramp. In the area west of Portuguese Canyon, the base of the presently active slide overrides the deep toe of the Pleistocene landslide in a manner similar to duplex structure at the toe of an overthrust.

There appears to have been no major movement within the main part of the Portuguese Bend landslide complex during the latter part of the Pleistocene, probably because the shoreline receded during glacial lowering of sea level and was not eroding the toe of the landslide. One or more major slide events occurred in the Holocene prior to historic time, probably as a result of wave erosion undermining the toe of the Pleistocene slide. The movement occurred less than 4,800 years ago, a carbon-14 date on pond deposits tilted during the Holocene event (Emery, 1967). During this event the western edge of the prehistoric Abalone Cove landslide truncated the Holocene beach and shoreline angle (locality 8, Fig. 1). The locations of scarps and closed depressions that predate historic movement indicate this Holocene slide had nearly the same boundaries as the present Abalone Cove and Portuguese Bend landslides and show that the two slides were interconnected. The two slides became separated after a large mass of slide material was thrust onto the landward edge of Portuguese Point (locality 7, Fig. 1). This formed a buttress that resisted movement, although there was minor movement in the zone between the two slides from 1979 to 1984.

It is difficult to establish an itinerary for those who wish to visit the active slides because conditions change through time, particularly with efforts under way to stabilize the Portuguese Bend landslide. Figure 1 shows locations where important features are likely to be preserved for many years to come. Features to be observed at each location are briefly described below.

Location 1. On the upslope side of Crenshaw Extension to the west of Paintbrush Canyon, southwest-dipping beds of siliceous and tuffaceous shale and pillow basalt began sliding after the beds were undercut during road construction by the Los Angeles County Road Department. The slide appears as a light area in Figure 2. The County Road Department was in the process of disposing of the slide material by enlarging the road fill directly downslope from the slide when the road fill began to move in August, 1956. This was the beginning of the active Portuguese Bend landslide.

Location 2. This is the area where movement of the Portuguese Bend landslide was first observed. Since 1956, part of this road fill has been down dropped as much as 120 ft (37 m), leaving a scarp exposing road fill above the head of the slide. In this area the base of the Portuguese Bend landslide is at the base of the Portuguese Tuff, and the up-slope edge of the landslide occurs where the base of the tuff rises to the surface on the south

Figure 2. View of Portuguese Bend area looking north on September 26, 1956, one month after the Portuguese Bend landslide reactivated. Note Crenshaw Boulevard under construction on right side of photo. Slide movement was first observed in road fill at right center of photo; white scar on hill northeast of fill is landslide caused by road excavation.

flank of an anticlinal flexure (cross section BB', Fig. 3). East of location 2, in-place bedrock is exposed along the channel of Paintbrush Canyon. Pressure at the toe of the active Flying Triangle landslide is causing this bedrock to uplift very slowly (cross section DD', Fig. 3).

Location 3. The striated base of the slide is exposed along the up-slope edge of the slide mass. A few inches of sheared bentonite generally separates the slip surface from underlying bedrock, which consists of siliceous and silty shale and dips about 20° southwest in this area.

Location 4. A steep west-facing dip-slope forms the edge of the Portuguese Bend landslide. Subhorizontal striations occur in areas where slide material has recently moved across the face of this slope. The beds that form this slope may have been bent upward by movement of the Pleistocene Flying Triangle landslide.

Location 5. A segment of the 1956 roadway of PVDS is still preserved but has been displaced 600 ft (180 m) from its original position.

Location 6. On the east side of Inspiration Point, the Portuguese Tuff was pushed upward to a nearly vertical position at the toe of the Pleistocene Portuguese Bend landslide and subsequently overlain by marine terrace deposits. Also, observe how the active slide bifurcates at the landward edge of the point. Inspiration and Portuguese points are both buttressed at their seaward ends by a basalt sill. Mélange, slip surfaces, and folds can be observed in slide material exposed along the beach and cliffs at the toe of the slide on either side of Inspiration Point. Exposures are usually best during the winter.

Location 7. The toe of the prehistoric Holocene landslide was thrust onto the marine terrace at the landward edge of Portuguese Point, creating the hill between the terrace and PVDS. A surveying monument at the crest of the hill was uplifted 1.2 ft (0.4 m) by slide movement from 1979 to 1984.

Location 8. The west edge of the prehistoric Holocene Abalone Cove landslide truncated the beach and shoreline angle, which are on the landward side of the beach house. This is shown by the topography on Figure 1, which predates recent movement

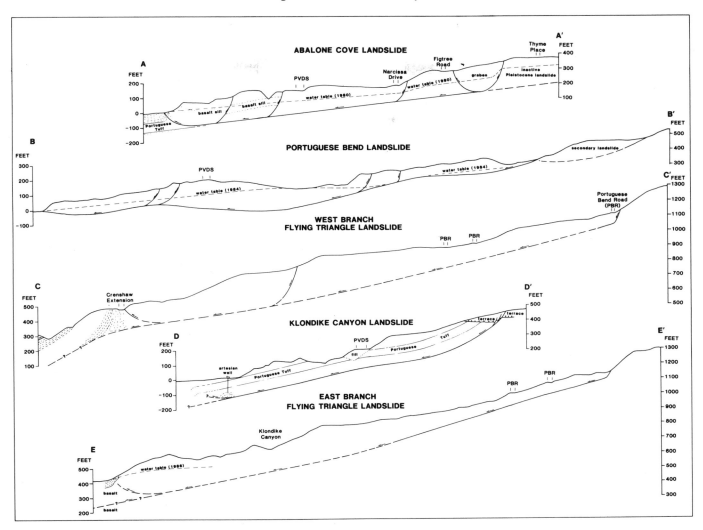

Figure 3. Vertical cross sections in the direction of movement through the major active and recently active landslides. Same horizontal and vertical scale. See Figure 1 for locations of cross sections.

of the Abalone Cove landslide. A short distance to the east, the pavement of the abandoned tennis courts shows interesting deformational features, including strike-slip and thrust faults. An assemblage of slide-related features is also visible at the toe of the slide along the beach. Remnants of the pre-1978 beach have been uplifted about 10 ft (3 m) near the mouth of Altamira Canyon. In this same area, the Portuguese Tuff is exposed in the surf zone along the toe of the slide and can be observed during low tide. The tuff contains steeply inclined gouge zones and slip surfaces and is overlain by displaced bedrock, which dips steeply landward. In-place bedrock exposed in the surf zone on either side of the slide has gentle seaward dips. The structure of the Abalone Cove landslide is shown in cross section AA′ on Figure 3.

Location 9. Extensive cracks are visible in the road and curbs where the up-slope edge of the Klondike Canyon landslide crosses the intersection of Dauntless and Exultant drives. Overall, this landslide has a subtle expression because its total historic displacement is less than 3 ft (1 m) and its prehistoric displacement is only a few tens of feet. Recent movement was facilitated by artesian water pressure at its toe. An artesian well was drilled near the beach in 1981 and has subsequently been flowing at a rate of 15 to 20 million gallons per year. The structure of the slide is shown in cross section DD′ on Figure 3.

Location 10. At "Flat Top" hill, the Flying Triangle landslide abruptly bifurcates at a 50° angle (Fig. 1). One branch moves southwest and converges at a low angle with Paintbrush Canyon. Its toe rises to the surface along the north side of Crenshaw Extension, although some movement is accommodated by uplift of beds south of the road (cross section CC', Fig. 3). The other branch moves southward and is laterally displacing the deeply incised channel of Klondike Canyon instead of toeing up into the canyon (cross section EE', Fig. 3). Anticlinal uplift is occurring at the down-slope termination of this branch.

Location 11. A Late Pleistocene graben occurs along the crest of the peninsula at this location. Valley View Road extends eastward from Crenshaw Boulevard along the axis of the graben. The graben probably formed by slippage along the top of the Portuguese Tuff during the initial stages of the development of the Portuguese Bend landslide as described by Ehlert (1986).

REFERENCES CITED

Bryant, M. E., 1982, Geomorphology, neotectonics, and ages of marine terraces, Palos Verdes Peninsula, *in* Cooper, J. D., compiler, Landslides and landslide mitigation in southern California: Association of Engineering Geologists, Southern California Section, p. 15–25.

Conrad, C. L., and Ehlig, P. L., 1983, The Monterey Formation of the Palos Verdes Peninsula, California; An example of sedimentation in a tectonically active basin within the California Continental Borderland, *in* Larue, D. K., and Steel, R. J., eds., Cenozoic marine sedimentation, Pacific margin, U.S.A.: Pacific Section, Society of Economic Paleontologists and Mineralogists, p. 103–116.

Ehlert, K. W., 1986, Origin of a mile long valley located northerly of the ancient Portuguese Bend landslide, Palos Verdes Peninsula, southern California, *in* Ehlig, P. L., compiler, Landslides and landslide mitigation in southern California: Guidebook and volume prepared for 82nd meeting of Cordilleran Section, Geological Society of America, p. 167–176.

Ehlig, K. A., and Bean, R. T., 1982, Dewatering of the Abalone Cove landslide, City of Rancho Palos Verdes, Los Angeles County, California, *in* Cooper, J. D., compiler, Landslides and landslide mitigation in southern California: Association of Engineering Geologists, Southern California Section, p. 67–79.

Ehlig, P. L., 1982, Mechanics of the Abalone Cove landslide including the role of groundwater in landslide stability and a model for development of large landslides in the Palos Verdes Hills, *in* Cooper, J. D., compiler, Landslides and landslide mitigation in southern California: Association of Engineering Geologists, Southern California Section, p. 57–66.

—— , 1986, The Portuguese Bend landslide; Its mechanics and a plan for its stabilization, *in* Ehlig, P. L., compiler, Landslides and landslide mitigation in southern California: Guidebook and volume prepared for 82nd meeting of Cordilleran Section, Geological Society of America, p. 167–172.

Emery, K. O., 1967, The activity of coastal landslides related to sea level: Revuew de Géographie Physique et de Géologie Dynamique, v. 9, p. 177–180.

Jahns, R. H., and Vonder Linden, K., 1973, Space-time relationships of landsliding on the southerly side of Palos Verdes Hills, *in* Moran, D. E., Slosson, J. E., Stone, R. O., and Yelverton, C. A., eds., Geology, seismicity, and environmental impact: Association of Engineering Geologists, Special Publication, p. 123–138.

Keene, A. G., 1986, Flying Triangle landslide, City of Rolling Hills, California, *in* Ehlig, P. L., compiler, Landslides and landslide mitigation in southern California: Guidebook and volume prepared for 82nd meeting of Cordilleran Section, Geological Society of America, p. 173–176.

Kerwin, C. T., 1982, Land stability in the Klondike Canyon area, *in* Cooper, J. D., compiler, Landslides and landslide mitigation in southern California: Association of Engineering Geologists, Southern California Section, p. 39–48.

Merriam, R., 1960, Portuguese Bend landslide, Palos Verdes Hills, California: Journal of Geology, v. 68, p. 140–153.

Rowell, H. C., 1982, Chronostratigraphy of the Monterey Formation of the Palos Verdes Hills, *in* Cooper, J. D., compiler, Landslides and landslide mitigation in southern California: Association of Engineering Geologists, Southern California Section, p. 7–13.

Vonder Linden, K., and Lindvall, C. E., 1982, The Portuguese Bend landslide, *in* Cooper, J. D., compiler, Landslides and landslide mitigation in southern California: Association of Engineering Geologists, Southern California Section, p. 49–56.

Woodring, W. P., Bramlette, M. N., and Kew, W.S.W., 1946, Geology and paleontology of the Palos Verdes Hills, California: U.S. Geological Survey Professional Paper 207, 145 p.

A section through the Peninsular Ranges batholith, Elsinore Mountains, southern California

Margaret S. Woyski, California State University, Fullerton, California 92634
Astrid H. Howard, California Institute of Technology, Pasadena, California 91125

LOCATION

The Elsinore Mountains in the Cleveland National Forest of southern California are situated in the northwestern Peninsular Ranges. They lie southwest of and overlook Lake Elsinore at elevations of 2,600 to 3,600 ft (790 to 1,100 m). The northeastern front of the mountains is a precipitous escarpment that rises 1,350 to 2,350 ft (410 to 720 m) above the lake at its base. The site is reached by taking the Ortega Highway (California 74) from either I-15 at Lake Elsinore or I-5 at San Juan Capistrano to the Main Divide Truck Trail (Fig. 1) where it intersects the Ortega Highway 0.2 mi (0.3 km) west of the summit. Road cuts along the Main Divide Truck Trail (a graded dirt road) are accessible by passenger car and, with caution, by bus. Two campgrounds are located at the intersection of the Ortega Highway with the Main Divide Truck Trail and group camping sites (limited to 70 people) are located just off the West Extension of the Main Divide Truck Trail. Make reservations with the Cleveland National Forest Trabuco District Office, Corona Ranger Station, 1147 East Sixth Street, Corona, California 91719, (714) 736-1811. There is a daily charge for all campgrounds.

GEOLOGIC SIGNIFICANCE OF SITE

The major units of the northwestern Peninsular Ranges batholith that Larsen (1948) described are exposed along the Main Divide Truck Trail in the Elsinore Mountains. Pre-batholith country rock of folded Bedford Canyon metasediments is intruded by three units of the western batholith: San Marcos gabbro, Bonsall tonalite, and Woodson Mountain granodiorite. The San Marcos gabbro here shows its two chief facies (calcic hornblende norite and quartz-biotite hornblende norite) and its two major structures (large poikilitic hornblende crystals and rhythmic banding). It is intruded by both the Bonsall tonalite and Woodson Mountain granodiorite. The Bonsall tonalite contains distinctive ellipsoidal inclusions. The Woodson Mountain granodiorite exhibits a wide range of border effects.

Pegmatite and aplite dikes cut all the plutons and porphyry dikes cut the Woodson Mountain granodiorite. Spheroidal weathering is well displayed. Paleocene sediments, which rest on weathered granodiorite, are covered by Tertiary basalts. Remnants of the Perris erosion surface persist on the divide where they have been uplifted by motion on the Elsinore fault. The truck trail parallels the scarp of the active Elsinore fault zone and overlooks Lake Elsinore, a pull-apart basin filled by disrupted drainage.

The Peninsular Ranges batholith is interpreted as the root system of a Late Jurassic–Early Cretaceous volcanic arc off the

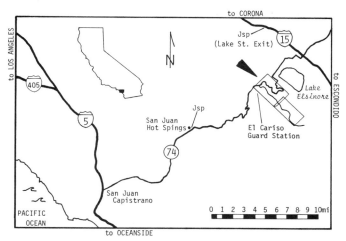

Figure 1. Location of Main Divide Truck Trail. Jsp = Santiago Peak Volcanics.

southwest edge of the North American craton. The extrusive rocks of this arc (Santiago Peak volcanics) were erupted upon the earlier Mesozoic flysch deposits of the Bedford Canyon Formation. From 130 to 105 Ma the island arc system remained static while plutons, represented by those in the Elsinore Mountains, were intruded across the 31 mi (50 km) width of the arc. From 105 to 80 Ma the locus of plutonism migrated steadily eastward to the craton.

PENINSULAR RANGES BATHOLITH

The Peninsular Ranges batholith extends from the Transverse Ranges through southern California and the length of Baja California in a SSE trending belt 995 mi (1,600 km) long and spreads 68 mi (110 km) wide from the foothills along the Pacific coast to the Salton trough along the San Andreas fault. The batholith consists of hundreds of plutons that were intruded side by side, leaving a few screens of country rock between them.

Like many batholiths and volcanic arc systems, the Peninsular Ranges batholith is zoned longitudinally by differences in lithology, chronology, and chemistry. The plutons to the west are more varied and consist of gabbros and quartz gabbros, tonalites, and leucogranodiorites whereas to the east the plutons are mostly tonalites and closely related low K_2O granodiorites; nevertheless, all of the rocks appear to have the chemistry and mineralogy of I-type granites, indicating an igneous mantle source. An igneous mantle source is also indicated by the calcic alkali-lime index of 63% for the batholith. The zircon U-Pb ages of the western plutons range from 122–105 Ma and have no preferred distribution within the belt; on the other hand, the U-Pb ages of the eastern plutons range from 105–89 Ma and become progressively younger to the east. Hornblende and biotite K-Ar ages follow the same general pattern (Silver and others, 1979).

Several geochemical parameters are zoned across the batholith without regard for rock type indicating a common but changing magma source. The initial Sr isotopic ratio becomes increasingly radiogenic eastward; $\delta^{18}O$ values increase eastward from +6 to +12 with a sharp step that closely matches the 105 Ma isochron; Light Ion Lithophile (LIL) elements (e.g., Rb, Sr, Ba, Pb, U, Th) increase in the tonalites eastward (Silver and others, 1979); and Nd isotope ratios decrease eastward +8.2 to −6.4 (DePaolo, 1981). All these trends suggest that the magmas for the western part of the batholith are primarily from a primitive source region, probably the upper mantle, whereas the magmas in the eastwardly migrating belt involved small, but increasing eastward, assimilation of subducted sediments or highly altered upper portions of the oceanic crust. On the eastern edge, in the San Jacinto block, the still lower $\delta^{18}O$ and still higher $^{87}Sr/^{86}Sr$ values are attributed to the involvement of the southwestern edge of the North American craton in the melting process that produced the magmas (Taylor and Silver, 1978). Of the major elements, the alkalis increase eastward and Mg, Fe, and Ca decrease eastward but K varies independently (Baird and others, 1979).

The country rock on the west consists of low-grade metamorphic Triassic-Jurassic flysch-type sediments of the Bedford Canyon Formation and andesitic volcanics of the Santiago Peak Formation, whereas to the east it is medium-grade metamorphic Mesozoic (Paleozoic?) clastic sediments and minor limestones.

L. T. Silver and co-workers consider the Peninsular Ranges batholith to be the result of two stylistically different but related, and essentially continuous, magmatic events that formed a Late Jurassic–Early Cretaceous arc. In Late Jurassic times a plate to the west of the North American plate (possibly the Kula or Farallon plate) began to subduct beneath the North American plate creating a volcanic arc that formed to the southwest of the Precambrian craton (Silver, 1979). Rising magmas were intruded into and erupted onto the sediments of the Bedford Canyon Formation. The spatially random distribution of emplacement ages of plutons of all types on the western side of the batholith implies a static arc system from 130 to 105 Ma. At 105 Ma there was a change in the tectonic regime, possibly an increase in the rate of convergence. The arc then moved progressively eastward from 105 to 89 Ma as demonstrated by the regularly decreasing emplacement ages of the plutons in the eastern half of the batholith (Silver and others, 1979).

By Late Cretaceous time isostatic uplift and erosion, much greater to the east, had exposed the root system of these two related arcs in the Peninsular Ranges. This is deduced from: (1) the increasing grade of metamorphism in country rocks presumedly due to greater depth from west to east (Gastil, 1975); (2) the increasing divergence of the hornblende and biotite K-Ar ages from the zircon emplacement ages to the east that probably re-

sulted from increased cooling time at greater depth; (3) the increase in size and homogeneity and therefore depth of the plutons to the east; and (4) the $\delta^{18}O$ evidence for a roof zone on the western edge (Taylor and Silver, 1978). The Peninsular Ranges batholith was separated from its initial position contiguous to mainland Mexico when the Gulf of California developed as an extension of the East Pacific spreading center, perhaps beginning as early as Miocene times (Silver and others, 1979). The tectonics of the Lake Elsinore region are now dominated by right oblique movement along the en echelon faults of the Elsinore fault zone (Weber, 1983).

ROCK UNITS OF THE ELSINORE MOUNTAINS

The folded and faulted **Bedford Canyon Formation** is the predominant country rock in the western zone of the batholith. It forms large masses, screens between plutons, numerous roof pendants, and abundant inclusions within plutons. The total thickness of the Bedford Canyon Formation cannot be determined because it is overlain unconformably by the Santiago Peak volcanics and the base is not known. It weathers to boulder-free tan soil that supports denser vegetation than the other rock units.

The formation is dark gray argillite or slate interbedded with fine arkosic or lithic quartzite, resulting in its distinctive alternation of dark and lighter beds from 0.4 to 59 in (1 to 150 cm) thick. Rare lenses of dark limestone contain poorly preserved fossils that have been dated as Jurassic (Callovian). No good horizon markers have been found in this thick sequence. Criscione and others (1978) have found two different ages by Rb/Sr isotope analysis, 176 Ma (Middle Jurassic) and 228 Ma (Permo-Triassic). Even close to contacts the Bedford Canyon metasediments are only mildly metamorphosed although inclusions within the plutons have been converted to hornfels. The foliation everywhere parallels the bedding, implying tight isoclinal folds. The beds strike NW, with local deviations to N–S, and dip steeply. Where crossbedding or graded bedding is detectable, it appears that the majority of beds are overturned.

Santiago Peak volcanics and hypabyssal rocks of Late Jurassic age occur along the west flank of the Peninsular Ranges and around the northwestern tip. Outcrops are most accessible in road cuts on I-15 or on California 74. (See Jsp on Fig. 1.) The volcanics resist erosion, producing steep slopes and high peaks, and weather to a rocky clay soil, supporting dense vegetation. The Santiago Peak Formation consists of andesites interbedded with quartz latites and minor rhyolites, as flows, dikes, tuffs, and volcanic breccias. The rocks are aphanitic, altered dark gray or green, with feldspar phenocrysts. Where intensely metamorphosed near contacts, the rock is converted to a dense hornfels with mere traces of phenocrysts (Larsen, 1948).

The **San Marcos gabbro** occurs as small plutons and as screens, roof pendants, and swarms of inclusions in the other plutonic rocks. The gabbro, especially the quartz-bearing facies, is prone to spheroidal weathering, reducing it to a red soil with a few round boulders. On divides the quartz-free gabbro weathers to sharp peaks with distinctive conical profiles. The outstanding characteristic of the San Marcos gabbro is its variability even within single outcrops. The grain size ranges from 0.5 to 20 mm although the average is about 1 mm. The mafites range from 20% to 55%. The composition ranges from typical gabbronorite (augite and hypersthene with labradorite, an_{55-70}) to calcic olivine norite (hypersthene and olivine with very calcic plagioclase, an_{89-95}) or else to quartz biotite norite (quartz, biotite, and hypersthene with intermediate plagioclase, an_{46-59}). Hornblende is conspicuously erratic. Some hornblende is an early alteration product, some is a late hydrothermal mineral, but most is a primary mineral that occupies interstices with poikilitic texture. It may constitute as little as several percent of the rock to as much as 50%. Miller (1937) suggests that irregular distribution of water in the magma may explain its erratic character. Miller (1937) grouped all the various gabbros into one unit, "San Marcos," because their ages and occurrences are similar and the variations in mineralogy, although great, show continuous gradation from one extreme to the other.

One common internal structure is rhythmic banding in which layers of hornblende and pyroxene alternate with layers of plagioclase. Another structure is a penetration of finer gabbro by coarser, more hornblende-rich gabbro, as an angular network or uncommonly, as matrix around nodules. Orbicular structures of radiating olivine, hypersthene, and plagioclase occur locally (Miller, 1938).

The **Bonsall tonalite** and other tonalites intrude the gabbro and are intruded in turn by the granodiorites. The gold veins of the Pinacate district are almost exclusively associated with the Bonsall tonalite. This tonalite is noted for its abundant and well-oriented inclusions but few schlieren. Most field geologists have used the nature of schlieren, inclusions, and xenocrysts to distinguish the tonalite plutons. That the size, shape, and abundance of inclusions is distinctive for each pluton, regardless of country rock, suggests that these inclusions are related to the source of magma.

The tonalite is medium gray, coarse-grained from 0.5 to 10 mm, composed of 50%–60% andesine, 20%–25% quartz, 5%–20% biotite, 5%–15% hornblende, and less than 10% orthoclase. The inclusions contain the same minerals as the tonalite but with a greater proportion of mafites.

The **Woodson Mountain granodiorite** forms large plutons that intrude the tonalite and gabbro in the western Peninsular Ranges. Engel (1959) reports that, where vertical sections are distinguishable, contacts dip gently as if the plutons flared outward upward. The Woodson Mountain granodiorite is light gray to tan in fresh specimens with medium-coarse texture. It is composed of 30%–35% oligoclase, 10%–30% microcline perthite, 30%–40% quartz, and 1%–10% mafites (chiefly biotite). The granodiorite appears to have intruded late in the igneous sequence when the surrounding rock was cool enough to induce chilled, streaky, more mafic borders and brittle enough to shatter at the contacts.

The **Paleocene** was an epoch of profound **lateritic weath-ering** during which alumina-rich clays and iron-rich red soils were developed in situ. These were covered by fresh and brackish water clays and sandstones of the lower **Silverado Formation.** Deep red soil that was developed during the unique Paleocene climatic conditions serves as a time-marker in the Peninsular Ranges. It is well displayed in road cuts at the summit of the Ortega Highway. Paleocene clays in the Elsinore Trough, which are used for ceramics, constitute one of the major mineral resources of southern California.

Four or five horizontal basalt flows of the **Santa Rosa volcanics** form extensive northeast-trending mesas close to the Elsinore fault zone. The basal flow is alkali basalt; overlying flows are tholeiitic. Elsinore Peak and adjacent hills are composed of basalt that weathers to brown soil covered by grassy vegetation. Aphanitic porphyritic dikes with chilled margins that cut the youngest of the plutonic rocks may be associated with the volcanic activity that produced the Santa Rosa basalts.

The **Perris Surface** is one of the oldest and most extensive of the six erosion surfaces on the Perris block, which is bounded by the Elsinore and San Jacinto faults. It lies beneath the Santa Rosa volcanics that were erupted at 8.3 Ma, before the other surfaces were developed (Woodford and others, 1971). On this basis, the erosion surface beneath the volcanics of Elsinore Peak can be correlated with the Perris surface.

The **Elsinore fault zone** extends along the southwest edge of the Elsinore Trough running from Corona southeastward about 50 mi (80 km). The trough is bordered by low hills to the northeast and by the steep Santa Ana and Elsinore Mountains to the southwest and parallels the more seismically active San Jacinto Fault to the northeast and Newport-Inglewood Fault to the southwest. All along the Elsinore fault zone there is geomorphic evidence of oblique-slip movement through the late Quaternary. Weber (1983) has proposed 5.6 to 6.8 mi (9 to 11 km) of right lateral displacement based on offsets of thin bodies of gabbro, aberrant foliation in the Bedford Canyon Formation, and the contact between the Santiago Peak volcanics and the Bedford Canyon Formation. Uplift of the Elsinore Mountains along the Elsinore fault zone has cut off the San Jacinto River, which is now impounded in a pull-apart basin forming Lake Elsinore.

ROAD LOG

Individual outcrops along the Main Divide Truck Trail (Fig. 2) are described in road-log format with mileages given to the nearest 0.05 mi (0.1 km).

0.0 mi (0.0 km). Ortega Highway California 74. At El Cariso campgrounds, take Forest Service road 6S07 southward; this is the Main Divide Truck Trail.

0.85 mi (1.4 km) Tonalite/Gabbro/Granodiorite Contact. Overlook. On the NE face of the north corner of the truck trail, Bonsall tonalite is in sharp contact with the San Marcos gabbro, with no change in grain size; within a few centimeters of the contact, however, the tonalite shows a very faint foliation

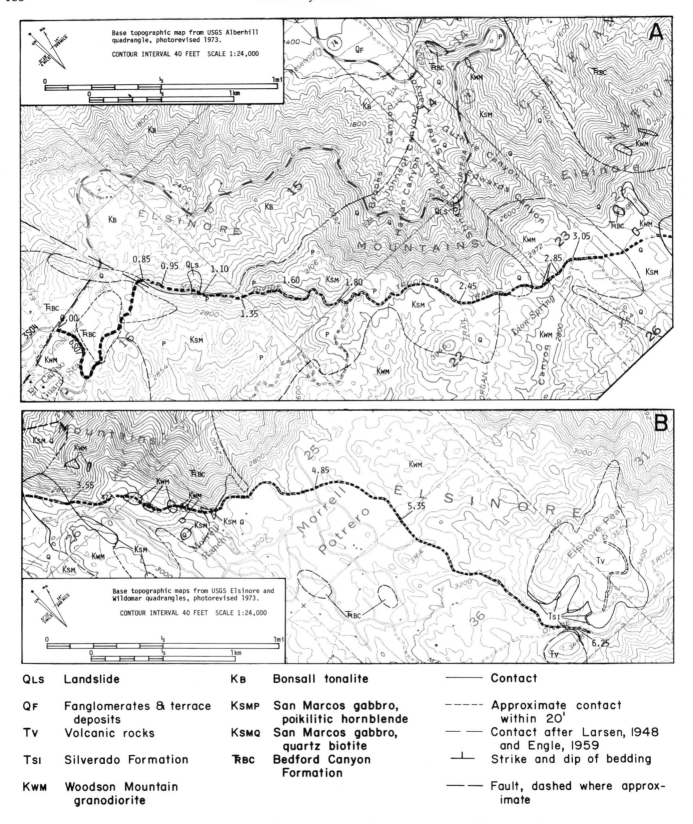

QLS	Landslide	KB	Bonsall tonalite	——	Contact
QF	Fanglomerates & terrace deposits	KSMP	San Marcos gabbro, poikilitic hornblende	- - - -	Approximate contact within 20'
Tv	Volcanic rocks	KSMQ	San Marcos gabbro, quartz biotite	— —	Contact after Larsen, 1948 and Engle, 1959
TSI	Silverado Formation	ŦBC	Bedford Canyon Formation	⊥	Strike and dip of bedding
KWM	Woodson Mountain granodiorite			— ——	Fault, dashed where approximate

Figure 2. Geologic Map. Main Divide Truck Trail. A. Miles 0.0 to 3.53 (0.0 to 5.7 km); B. Miles 3.53 to 6.25 (5.7 to 10.1 km).

parallel to the contact. An apophysis of tonalite cuts the gabbro. Absence of chilling suggests that the gabbro had solidified but not yet cooled when the tonalite was intruded. (A similar contact can be seen at 1.1 mi [1.8 km]). On the NW face Woodson Mountain granodiorite is in contact with the gabbro. It is coarse and mafic with large inclusions of Bedford Canyon metasediments and gabbro. The San Marcos gabbro here is the calcic poikilitic hornblende facies of this highly variable formation. At the top of the cut, the gabbro shows its typical red clay weathering product. Cutting the gabbro along an aplite dike is a shear zone about 1.6 ft (0.5 m) wide that has been thoroughly altered. This shear zone lies along the trend of a presumed fault valley north of the Ortega Highway.

From this vantage point, one can look northeast over the 1,600-ft (490-m) scarp of the Willard Fault across Lake Elsinore to the scarp of the Glen Ivy North Fault. These two faults are strands of the Elsinore fault zone that defines the Elsinore Trough. In the middle distance are clay pits in Paleocene deposits at Alberhill. Beyond, on the low plateau of the Gavilan, the Steele Valley pluton of granodiorite is outlined by its blanket of large boulders. The Pinacate gold district borders the pluton to the right and behind it. In the distance, the Lakeview Mountains rise from the Perris plain, showing their arcuate structure. On a clear day, Mount San Jacinto, Mount San Gorgonio, and Mount San Antonio can be seen on the horizon.

0.95 mi (1.5 km). Bonsall Tonalite. Here the typical inclusions in the Bonsall tonalite are conspicuous. They are small (3–6 in; 8–15 cm), elliptical and compose several percent of the rock. About half the inclusions pitch to the SE about 30°. A few pegmatite stringers parallel this flow direction also. Toward the gabbro contact, larger angular inclusions become more abundant. Inclusions in the Bonsall tonalite appear to be of two generations: an early ubiquitous set, with granoblastic texture, of uniform size, elliptical shape, and oriented in parallel that apparently came from considerable depth; and a late localized set, of heterogeneous size, irregular shape, without orientation, and composed of identifiable local rock that apparently came from the country-rock roof. The early set of inclusions distinguishes the Bonsall tonalite from other tonalites.

1.35 mi (2.2 km). Poikilitic Hornblende Gabbro. At this wide parking area all three major pluton types of the western Peninsular Ranges batholith can be seen in close proximity. To the north, scattered boulders of gray Bonsall tonalite in poor sandy soil outline an elliptical pluton about 1.9 mi (3 km) long that lies between this overlook and the base of the scarp. Even from a distance, the ubiquitous dark inclusions make the boulders look spotted. To the south, far hillsides are littered with the huge pinkish-white exfoliation boulders in sandy grus by which the Woodson Mountain granodiorite can be recognized. This pluton extends for 6 mi (10 km) to the southwest and 12 mi (20 km) to the southeast. In the road cuts at the parking area is San Marcos gabbro.

At this location, the gabbro contains poikilitic hornblende crystals up to 5 in (12 cm) across that enclose small grains of calcic plagioclase (anorthite) and olivine now altered to brown spots of iddingsite and iron oxides. Most of the original hypersthene has been replaced. Between the hornblende crystals is labradorite and red biotite. The small grains of calcic plagioclase and olivine were probably precipitated in response to assimilation of small amounts of aluminous sediments as predicted by Bowen's reaction principle.

1.6 mi (2.6 km). Hornblende Streaks in Gabbro. This road cut contains narrow pegmatite dikes that cut black dike-like streaks of poikilitic hornblende in the hornblende gabbro. The association of the hornblende streaks with pegmatite dikes and the dike-like character of the streaks suggests that some poikilitic hornblende may have developed in response to introduction of fluids from invading magma.

1.8 mi (2.9 km). Rhythmically Banded Poikilitic Gabbro. Here the gabbro contains a zone of rhythmic banding 23 ft (7 m) wide where dark hornblende-rich layers 0.5 in (1 cm) wide alternate with light plagioclase-rich layers. The composition change is probably due to sequential depletion of constituents as crystallization progressed.

2.45 mi (3.9 km). Quartz-Biotite Gabbro. On the hillside near the top of the rise are light gray spheroidal boulders of quartz-biotite gabbro that contain biotite flakes and quartz as well as hornblende, labradorite, and minor hypersthene. This gabbro, unlike the poikilitic gabbro, contains inclusions of fine-grained hornfels derived from the Bedford Canyon metasediments. Quartz-biotite gabbro occurs close to metasediment contacts both in this area and northwest of the Ortega Highway. The presence of biotite and quartz, in total amounts from 5% to 25%, could well be the result of assimilation of Bedford Canyon silty shales. The finer grain size (suggesting rapid heat loss), the association with the Bedford Canyon formation, the presence of metasediment inclusions, and the increase in minerals (quartz and biotite) that occur in metasediment all point to this origin.

2.85 mi (4.6 km). Granodiorite/Gabbro/Metasediment/ Porphyry Contacts. At the head of Morrell Canyon are three road cuts in the space of 0.1 mi (0.2 km). The western cut contains a border facies of the Woodson Mountain granodiorite that is more mafic and finer-grained than the typical granodiorite. (A larger exposure can be seen at mile 2.6 [4.2 km].) It contains small inclusions of gabbro so numerous that it forms an intrusion breccia. Large inclusions of metasediment occur in both the granodiorite and the quartz biotite gabbro on the east end of the cut. Both gabbro and metasediment inclusions have been metamorphosed to fine hornfels. Cutting the border facies granodiorite is a porphyritic dike about 3 ft (1 m) wide. The feldspar phenocrysts range from 1 mm at the chilled margin to 4 mm in the center of the dike. The groundmass is dark gray, fine aphanitic, drawn into platy foliation parallel to and adjacent to the contacts. All these relationships are shown as well in the middle road cut. The east cut contains a very coarse facies of the Woodson Mountain granodiorite with grain size of 5–10 mm. The outcrop is cut by two porphyry dikes 12 and 50 in (30 and 130 cm) wide, faulted at top and bottom. Inclusions of gabbro in the granodiorite are few, rounded, and average 4 in (10 cm) in diameter. Wide-set joints are deeply weathered.

These outcrops show that gabbro intruded metasediments, incorporating large blocks into the magma. Subsequently, granodiorite intruded along the gabbro-metasediment contact. Near the contact the granodiorite was chilled to a more mafic border facies and incorporated inclusions of both gabbro and metasediment. An alternate interpretation would attribute the mafic border to assimilation of gabbro; however, the gabbro inclusions consistently show sharp contacts with no disintegration zones or schlieren. Although the east outcrop is near the contact, the granodiorite has an uncommonly coarse grain size, perhaps due to the addition of water to the magma from the Bedford Canyon Formation, or more likely, due to offset on a fault. Much later, probably during the late Tertiary volcanic activity of the region, porphyry dikes were introduced into the cold rock.

3.55 mi (5.7 km). Granodiorite/Metasediment Contact. At this road cut (Fig. 2B), just past a small gully running northwest, granodiorite is in intrusive contact with the Bedford Canyon metasediments. The Bedford Canyon Formation shows its distinctive bedding of alternating layers 3 to 9 in (8 to 24 cm) thick. One set of layers is fine sandstone, metamorphosed to non-foliated quartzite that resists weathering. The other set is clayey siltstone, metamorphosed to fine-grained mica schist that weathers readily along the foliation. The foliation is parallel to the bedding except where it is deformed around a bend (fault?) in an aplite dike.

5.35 mi (8.6 km). Morrell Potrero. Here and at mile 3.05 (4.9 km) are former river beds on an uplifted erosion surface.

6.25 mi (10.1 km). Basalt. Paleocene Surface. Silverado Formation. At this large parking area, Elsinore Peak, with its microwave relay station, rises to the northeast. Elsinore Peak and the hill to the southwest are composed of basalt with glassy-matrix; some is vesicular at the top of the SW hill. On weathered surfaces, the 1 mm phenocrysts of olivine make brown spots on the otherwise smooth surface. The base of the flows, which were extruded onto the Perris erosion surface, lies at about the level of the parking area. At the foot of the hillside to the southwest is a layer of white medium-grained, moderately-sorted micaceous feldspathic sandstone about 4 ft (1.2 m) thick. It is lithologically similar to the white sands of the Silverado Formation that crop out in the Elsinore Trough. Across the meadow to the north can be seen a ledge of the same flat-lying sandstone. About 0.05 mi (0.1 km) down the road to the southeast is mottled red and white residual clay that grades downward into weathered granodiorite. This clay marks the Paleocene erosion surface and basal unconformity. About 0.2 mi (0.3 km) up the service road to the microwave station is a pull-out from which the basalt is easily accessible.

REFERENCES CITED

Baird, A. K., Baird, K. W., and Welday, E. E., 1979, Batholithic rocks of the northern Peninsular and Transverse Ranges, southern California: Chemical composition and variation: *in* Abbott, P. L., and Todd, V. R., eds., Mesozoic crystalline rocks, Manuscripts and road logs prepared for the Geological Society of America Annual Meeting, San Diego, 1979: Department of Geological Sciences, San Diego State University, p. 111-132.

Criscione, J. J., Davis, T. E., and Ehlig, P., 1978, The age of sedimentation/diagenesis for the Bedford Canyon Formation and the Santa Monica Formation in southern California: A Rb/Sr evaluation, *in* Howell, D. G., and McDougall, K. A., eds., Mesozoic paleogeography of the western United States: Pacific Section, Society of Economic Paleontologists and Mineralogists, Symposium 2, p. 385–396.

DePaolo, D. J., 1981, A neodymium and strontium isotopic study of the Mesozoic calc-alkaline granitic batholiths of the Sierra Nevada and Peninsular Ranges, California: Journal of Geophysical Research, v. 86, p. 10470–10488.

Engel, R., 1959, Geology and mineral deposits of the Lake Elsinore Quadrangle, California: California Division of Mines Bulletin, v. 149, p. 9–58.

Gastil, R. G., 1975, Plutonic zones in the Peninsular Ranges of southern California and northern Baja California: Geology, v. 3, p. 361–363.

Larsen, E. S., Jr., 1948, Batholith and associated rocks of Corona, Elsinore, and San Luis Rey Quadrangles, southern California: Geological Society of America Memoir 29, 182 p.

Miller, F. S., 1937, Petrology of the San Marcos gabbro, southern California: Geological Society of America Bulletin, v. 48, p. 1397–1426.

——, 1938, Hornblendes and primary structures of the San Marcos gabbro: Geological Society of America Bulletin, v. 49, p. 1213–1232.

Silver, L. T., 1979, Peninsular Ranges batholith; A case study in continental margin magmatic arc setting, characteristics, and evolution: Geological Society of America Abstracts with Programs, v. 11, p. 517.

Silver, L. T., Taylor, H. P., Jr., and Chappell, B. W., 1979, Some petrological, geochemical and geochronological observations of the Peninsular Ranges batholith near the international border of the U.S.A. and Mexico, *in* Abbott, P. L., and Todd, V. R., eds., Mesozoic crystalline rocks, Manuscripts and road logs prepared for the Geological Society of America Annual Meeting, San Diego: Department of Geological Sciences, San Diego State University, p. 83–110.

Taylor, H. P., Jr., and Silver, L. T., 1978, Oxygen isotope relationships in plutonic igneous rocks of the Peninsular Ranges batholith, southern and Baja California: *in* Zartman, R. E., ed., Short Papers of 4th International Conf. Geochronology, Cosmochronology, Isotope Geology, U.S. Geological Survey Open-File Report 78-701, p. 423–426.

Weber, F. H., Jr., 1983, Geology and seismicity of the Elsinore and Chino fault zones in northwestern Riverside County and very small parts of adjoining Orange and San Bernardino counties, California, *in* Steller, D. L., Bryant, M., and Gath, E., eds., Geology of the northern Elsinore trough: South Coast Geological Society Annual Fall Field Trip, p. 39–43.

Woodford, A. O., Shelton, J. S., Doehring, D. O., and Morton, D. K., 1971, Pliocene-Pleistocene history of the Perris Block, southern California: Geological Society of America Bulletin, v. 82, p. 3421–3448.

Banning fault, Cottonwood Canyon, San Gorgonio Pass, southern California

D. M. Morton, J. C. Matti, and J. C. Tinsley, U.S. Geological Survey, 345 Middlefield Road, Menlo Park, California 94025

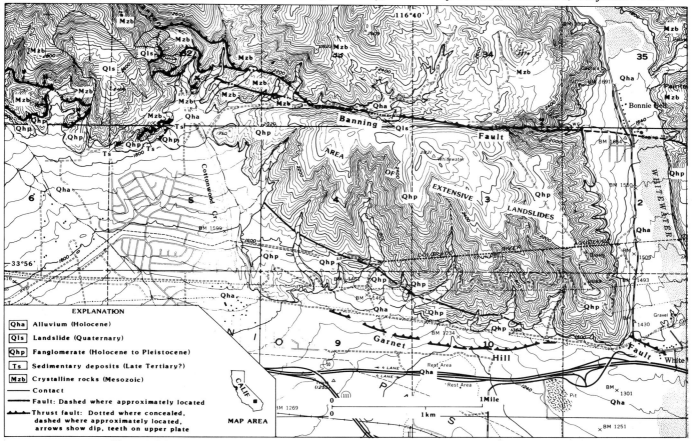

Figure 1. Map showing geologic relations along the Banning Fault in the Cottonwood Canyon area.

LOCATION AND ACCESSIBILITY

The Banning fault site is located at the mouth of Cottonwood Canyon, at the eastern end of San Gorgonio Pass. San Gorgonio Pass is the low divide separating the high steep terrain of the Peninsular Ranges Province to the south, from the higher but less steep Transverse Ranges Province to the north. The San Gorgonio Pass area is reached by taking I-10 east from the Los Angeles–San Bernardino–Riverside region toward Indio.

Most places with spectacular exposures of the Banning fault are, unfortunately, on property with restricted access, but one location is readily accessible to the public in Cottonwood Canyon near the eastern end of the Banning fault. Cottonwood Canyon is reached by taking the Verbenia exit from I-10 about 45 mi (72 km) east of Riverside and San Bernardino. Cross north over I-10 and turn left on the first road, Tamarack. Proceed west 0.3 mi (0.5 km) on Tamarack to Cottonwood Canyon Road. Go north 1.8 mi (2.9 km) on Cottonwood Canyon Road to a sharp right turn. Park in a parking area to the north of the turn, which is at the mouth of Cottonwood Canyon. Part of Cottonwood Can-

yon Road is paved; the remainder is an improved dirt road easily passable by passenger cars and buses.

SIGNIFICANCE OF SITE

Banning fault zone in the San Gorgonio Pass area is an old (about 5 Ma) complex and composite fault zone currently characterized by compressional oblique-slip movement. The seismicity includes some of the deepest seismic activity of the entire San Andreas fault system (Nicholson and others, 1986). It is also the zone along the San Andreas fault where evidence of compressional tectonism is by far the most pronounced.

SITE INFORMATION

The Banning fault zone constitutes a major lithologic boundary in southern California. It separates Peninsular Ranges batholith and prebatholithic rocks to the south from a complex of Transverse Ranges basement rocks to the north. The Banning fault extends about 40 mi (64 km) westward from Cottonwood Canyon to its inferred termination against the San Jacinto fault. It extends 30 mi (48 km) southeast of Cottonwood where it merges with the Mission Creek branch of the San Andreas (Rogers, 1965; Matti and others, 1985). Vaughan (1922) first mapped faults belonging to the Banning fault zone but applied no names to the faults. Hill (1928) reinterpreted fault relationships in San Gorgonio Pass area and introduced the name Banning Fault. Allen (1957) produced the first detailed geologic map of the San Gorgonio Pass area and clarified many of the complex geologic and nomenclatural problems of the area. A more recent regional geologic synthesis of the pass area was produced by Matti and others (1985).

The Banning fault zone had an initial period of left-lateral displacement prior to late Miocene time, followed by right-lateral displacement during latest Miocene and early Pliocene time (Matti and others, 1985). The left-lateral displacement juxtaposed Transverse Ranges basement against Peninsular Ranges basement. Assuming that the fault is the eastward extension of the Cucamonga-Malibu fault zone, there has been at least 55 mi (88 km) of left slip on the Banning fault (Yerkes and Campbell, 1971; Jahns, 1973; Campbell and Yerkes, 1976). Clasts within Tertiary conglomerates in the San Gorgonio Pass area have been right-laterally displaced 10 to 15 mi (6 to 24 km) from their source areas (Matti and others, 1985). Stratigraphic successions in the area suggest the right-lateral displacement occurred between 4 and 7.5 Ma, when the Banning fault was the active strand of the San Andreas transform fault system (Matti and others, 1985).

During Quaternary time, the eastern two-thirds of the Banning fault (Fig. 1) was reactivated to once again form an active strand of the San Andreas system. The western third of the Banning fault appears to have been dormant since the Pliocene. The renewed displacement on the eastern third of the Banning fault appears to be mostly right lateral, with a total of 1 to 2 mi (1.6 to 3.2 km) displacement. The central part of the Banning fault in the San Gorgonio Pass area underwent reverse displacement during the Quaternary producing a complex of low-angle thrust faults (Allen, 1957; Matti and others, 1985). This complex is well exposed in Cottonwood Canyon. The lateral component of this renewed displacement is not known. For a different interpretation of the geologic relationship of the Cottonwood Canyon area see Rasmussen and Reeder (1986).

Cottonwood Canyon is located at the eastern end of the complex zone of north-dipping thrust faults that extends at least 18 mi (29 km) west from the canyon. West of the mouth of Cottonwood Canyon, the lowest major thrust places basement (mainly a biotite-hornblende gneiss cut by ilmenite-bearing granitic pegmatite dikes) over deformed late Tertiary and Quaternary gravels. The thrust and gravels are overlain by younger Quaternary gravels (Fig. 1). The dip of the fault ranges from shallow north and locally south dips to horizontal (Allen, 1957). It is likely that creep has lowered the initial dip of the faults to account for the horizontal and south-dipping attitudes. Farther west, where the thrust faults cut alluvial fan surfaces, the faults dip about 25° to the north.

Readily visible on the steep mountain face above the basement-gravel contact are major north-dipping thrusts in in-

tensely deformed, green-colored basement rock. A large landside deposit, which forms the stepped topography west of the well-exposed basement thrusts, overrides both the basement and gravels farther to the west. East of the mouth of Cottonwood Canyon, several splays of the Banning fault merge to form an apparent single, north-dipping fault. Splays just east of Cottonwood Canyon dip north 35°. East of the merged splays the fault dips about 40° to 45° to the north.

Two mi (3.2 km) east of Cottonwood Canyon, the Banning fault is well exposed on the west side of Whitewater Canyon. Here, the fault dips 40° to the north and juxtapose a thick section of conglomeratic Pleistocene fluvial deposits on the south and gneissic basement on the north. Whitewater Canyon is reached by taking the Whitewater exit off of I-10 about 2.5 mi (4 km) east of the Verbenia exit. The paved Whitewater Road leads to the Banning fault, which crosses Whitewater Canyon 1.5 mi (2.4 km) north of I-10. Here it forms a prominent groundwater barrier at the community of Bonnie Bell. A rough, unimproved dirt roads leads from Bonnie Bell to the west side of Whitewater Canyon, where the Banning fault is easily seen. Entry is restricted during summer months owing to fire closure. It is an easy 0.25 mi (0.4 km) walk from the Whitewater Road to the fault.

REFERENCES CITED

Allen, C. R., 1957, San Andreas fault zone in San Gorgonio Pass, southern California: Geological Society of America Bulletin, v. 68, p. 319–350.

Campbell, R. H., and Yerkes, R. F., 1976, Cenozoic evolution of the Los Angeles basin area; Relation to plate tectonics, *in* Howell, D. G., ed., Aspects of the geologic history of the California Continental Borderland: Pacific section, American Association of Petroleum Geologists Miscellaneous Publication 24, p. 541–558.

Hill, R. T., 1928, Southern California geology and Los Angeles earthquakes: Los Angeles, Southern California Academy of Sciences, 232 p.

Jahns, R. H., 1973, Tectonic evolution of the Transverse Ranges Province as related to the San Andreas fault system, *in* Kovach, R. L., and Nur, A., eds., Proceedings of the Conference on tectonic problems of the San Andreas fault system: Stanford University Publications in Geological Sciences, v. XIII, p. 149–170.

Matti, J. C., Morton, D. M., and Cox, B. F., 1985, Distribution and geologic relations of faults systems in the vicinity of the Central Transverse Ranges, southern California: U.S. Geological Survey Open-File Report 85-365, 27 p.

Nicholson, C., Seeber, L., Williams, P., and Sykes, L. R., 1986, Seismicity and fault kinematics through the Eastern Transverse Ranges, California; Block rotation, strike-slip faulting, and low angle thrusts: Journal of Geophysical Research, v. 91, no. B5, p. 4891–4908.

Rasmussen, G. S., and Reeder, W. A., 1986, What happens to the real San Andreas fault at Cottonwood Canyon, San Gorgonio Pass, California? *in* Kooser, M. A., and Reynolds, R. E., eds., Geology around the margins of the eastern San Bernardino Mountains: Publications of the Inland Geological Society, v. 1, p. 57–62.

Rogers, T. H., compiler, 1965, Geologic map of California, Olaf P. Jenkins edition—Santa Ana sheet: California Division of Mines and Geology, scale, 1:250,000.

Vaughan, F. E., 1922, Geology of the San Bernardino Mountains north of San Gorgonio Pass: University of California Publications of the Department of Geological Sciences Bulletin, v. 13, p. 319–411.

Yerkes, R. F., and Campbell, R. H., 1971, Cenozoic evolution of the Santa Monica Mountains–Los Angeles Basin area; I. Constraints on tectonic models: Geological Society of America Abstracts with Programs, v. 3, no. 2, p. 222–223.

San Andreas fault, Cajon Pass, southern California

Ray J. Weldon, *U.S. Geological Survey, 345 Middlefield Road, Menlo Park, California 94025*

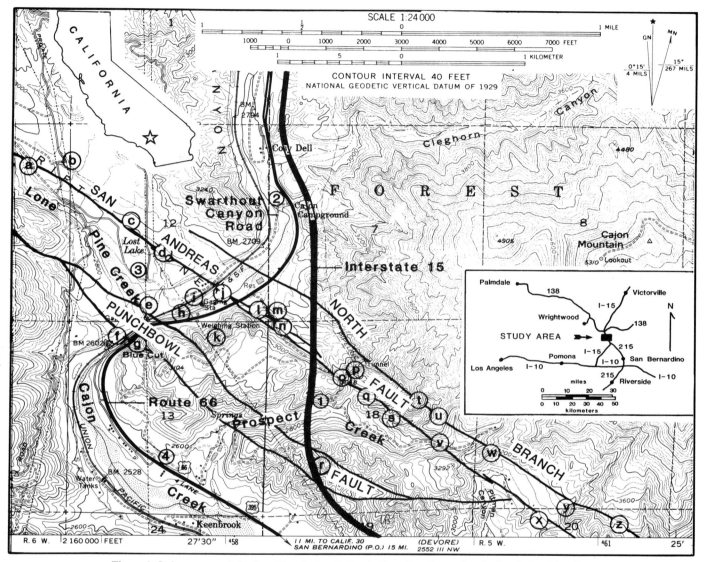

Figure 1. Index map and site localities. Lettered sites indicate locations of geologic relationships discussed in text. Numbers are access localities: (1) a freeway pullout that provides parking and access to the localities east of I-15; (2) Swarthout Canyon Road joins Route 66, providing access to the area west of Cajon Creek; (3) powerline road joins Swarthout Canyon Road, providing access to the Lost Lake area; (4) dirt road to the top of Blue Cut from Route 66. Route 66 can be reached from I-15 at Cleghorn Road (2.5 mi; 4 km north of the San Andreas fault) or Kenwood Road (2.5 mi; 4 km south of the San Andreas fault).

LOCATION AND ACCESSIBILITY

The field area discussed here straddles I-15 and old Route 66 (here part of U.S. 395) where they cross the Transverse Ranges, about 50 mi (80 km) northeast of Los Angeles, California (Fig. 1). Most of the sites can be reached by car on paved or good dirt roads. Sites east of I-15 are accessible only with four-wheel drive or a short walk from a paved road. The land not obviously occupied by structures or surrounded by fences belongs to the Forest Service, the Highway Department, or the railroads, so access does not require special permission.

SIGNIFICANCE AND HISTORICAL PERSPECTIVE

A 3 mi (5 km) stretch of the San Andreas fault, centered at Cajon Creek (Figs. 1 and 2), contains excellent examples of the style of faulting and the geomorphology associated with the San

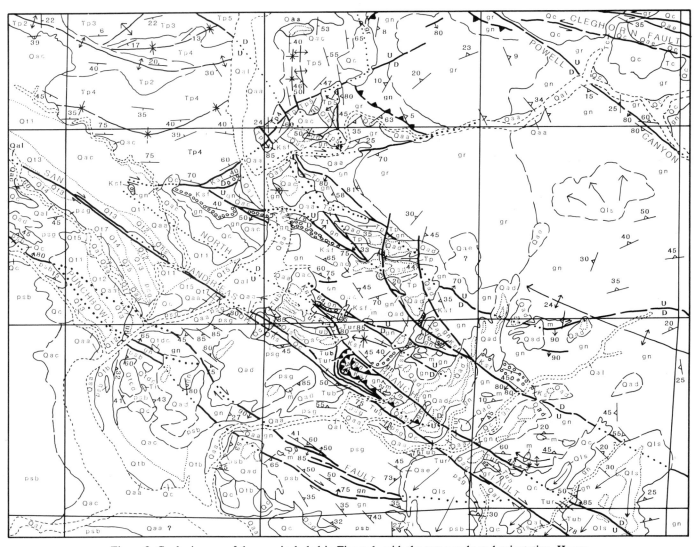

Figure 2. Geologic map of the area included in Figure 1, with the same scale and orientation. Heavy lines are faults; medium lines, fold axes; light lines, depositional or intrusive contacts. Short dashes mark the limit of the active wash, and lines of small dots define geomorphic surfaces (where different from depositional contacts). The units are: gn, gneiss; gr, granitic rocks; m, metasedimentary rocks in pendants; psg and psb, Pelona Schist; Ksf, San Francisquito sandstone; Ti, Tertiary dikes; Tp 1–5, Cajon (Punchbowl) Formation; Tc, Crowder Formation; Tub and Tur, unnamed Tertiary(?) buff and red sandstones and conglomerates; Qaa through Qae, Quaternary alluvium ("a" is youngest and "e" is oldest); Qtl through Qt7, Quaternary terraces ("1" is oldest and "7" is youngest, Qt6 is the aggradational surface of Qaa, Qtl is the original surface of Qac, and Qt3=Qtb outside the Lost Lake area); Qhs, Holocene sag pond deposits; Qc, colluvium (where mapped); Qls, landslides. The units are described in the text and Figure 3 shows a section through the late Quaternary deposits.

Andreas fault. Abandoned traces of the San Andreas fault, evidence for lateral and vertical deformation across the fault zone, and late Quaternary offset river terraces, landslides, and sag ponds can all be found in this small area. The relationships seen here have been used to characterize the uplift of the Transverse Ranges and the Quaternary slip rate and recurrence interval of earthquakes on the San Andreas fault.

Padre Garces was the first European to cross Cajon Pass

(called El Cajon de los Mejicanos), on his way to the San Gabriel Mission in 1776, and Jedediah Smith was the first American to follow him in 1826 (Hill, 1980). The site was settled by Mormons between 1846 and 1851 (Blake, 1856). The first railroad, the Santa Fe, was completed through the area in 1885 (Hill, 1980). In the nineteenth century, Marcou correctly inferred from the geology of Cajon Pass that the Transverse Ranges were uplifted in the Quaternary or even the "Modern Epoch" (Marcou,

1876), and Schuyler (1896) described a great "earthquake crack" caused by the 1857 earthquake, running more than 180 mi (300 km) through southern California, ending near Cajon Pass. The first description of the San Andreas fault at this site was made by Campbell (1902). He noted both the straight course of Lone Pine Canyon and the different rocks across the fault; the fault was subsequently named the San Andreas following the 1906 San Francisco earthquake (Lawson and others, 1908). The first suggestion of many miles of right-lateral slip on the San Andreas fault was made by Levi Noble, based on his mapping in the Cajon Pass area (Noble, 1926). Thus, the central location of Cajon Pass on the southern San Andreas fault and in the Transverse Ranges has made it an important site both historically and geologically.

SITE DESCRIPTION

Pre-Holocene geology. Near Cajon Creek, three major faults of the San Andreas system are well exposed in deep, cross-cutting canyons; they are the San Andreas fault, the North Branch (of the San Andreas fault), and the Punchbowl fault (Fig. 1). These faults juxtapose different rock types, but only the San Andreas fault has experienced significant lateral displacements during the late Quaternary. The North Branch is nearly everywhere a high angle fault separating unnamed Tertiary(?) sandstones and conglomerates (Tub + Tur, Fig. 2) from generally dioritic gneiss (gn) of the San Bernardino Mountains and, locally, shales, sandstones, and conglomerates of the Cretaceous and Paleocene(?) (Kooser, 1985) San Francisquito Formation (Ksf). The North Branch is especially well exposed at sites w and y (Fig. 1). The gneiss and San Francisquito sediments were thrust over the initially strike-slip North Branch, and the North Branch has subsequently been reactivated as a dip-slip fault, uplifting and exposing the overlying thrust faults (well exposed at site p). Small scarps formed by the current dip-slip motion of the North Branch can be seen running across the surface of Qad (sites t, u, and w) and the landslide at the southeast edge of the map (site z).

The Punchbowl fault juxtaposes two facies of Pelona Schist (Psb and Psg), locally separated by gneiss, intrusives, and marble of unknown affinity (Fig. 2). Psb is dominantly a well-layered muscovite-quartz-plagioclase metasedimentary schist, intruded by Tertiary dikes, and is part of the Pelona Schist exposed beneath the Vincent thrust in the San Gabriel Mountains (Ehlig, 1981). Psg consists mainly of albite-epidote-chlorite schist (derived from volcanic or volcano-clastic sediments) and is believed to be offset from the larger mass of Pelona Schist at Sierra Pelona (Blake, 1856; Dibblee, 1975). Based on evidence to the northwest, the Punchbowl fault was most active during the Pliocene (Noble, 1954; Dibblee, 1975; Barrows and others, 1985); here it is overlain by Quaternary sediments. Correlations of rock types and structures outside of the area indicate that the Punchbowl fault has about 28 mi (45 km) of right-lateral offset (Dibblee, 1975; Barrows, 1979). This fault is well exposed where it is crossed by old Route 66 (sites f and g) and I-15 (site r).

Penetrative deformation related to San Andreas system extends only tens of yards (meters) beyond the Punchbowl and North Branch faults (sites g, p, and w). In contrast, the deformation in the slivers in the fault system is extreme. The foliation in the Pelona Schist between the San Andreas and the Punchbowl faults is generally vertical and discontinuous (site k) and the Tertiary sediments between the North Branch and the San Andreas faults have shattered clasts and discontinuous vertical bedding (sites m, o, and w). The currently active San Andreas fault that separates these deformed units, however, is commonly less than 3 ft (1 m) wide and locally is only tens of centimeters wide (site o). The intense deformation associated with the fault zone appears to be related to the generation of new strands in the system, not distributed shear once a fault strand assumes the total deformation across the zone.

Late Quaternary deposits in the area are right-laterally offset up to 0.9 mi (1.4 km) by the San Andreas fault and up to 30 ft (10 m) vertically by the North Branch and several other unnamed dip-slip faults northeast of the San Andreas fault (Fig. 2). Restoring the 0.9 mi (1.4 km) offset of Qad straightens the courses of Cajon and Prospect creeks across the San Andreas fault (Fig. 1); this indicates that the post-Qad incision set the major streams into their courses for the rest of the Pleistocene and Holocene. Incision dominated until the latest Pleistocene, and Cajon Creek was near its current level between 25,000 and 15,000 years ago (Weldon and Sieh, 1985). Aggradation by Cajon Creek where it crosses San Andreas fault occurred between about 14,400 and 12,400 years ago, depositing Qac. The last 12,000 years were again dominated by incision, returning the creek to its current level.

Holocene geology. The Holocene record at the site is preserved in a flight of terraces cut into Qac or bedrock (Fig. 3); minor fluvial deposits, sag pond sediments, small landslides, and colluvium complete the Holocene sedimentary record. Most of the geomorphic features associated with the San Andreas fault that are visible on the ground postdate the Qac fill and are Holocene in age. The young fault features are discussed from southeast to northwest along the fault. East of I-15 all three tributaries of Prospect Creek are offset about 950 ft (290 m; sites q, s, v). These streams were set into their courses by the rapid incision that postdated Qac, and have been progressively offset by the fault since then (Weldon and Sieh, 1985). These offsets are best seen from a trail that runs up the ridge northeast of site p. In this area, small gullies and steep slopes are offset as little as 13 ft (4 m), which is believed to be the offset associated with the last earthquake rupture on the San Andreas fault in Cajon Pass at the beginning of the eighteenth century (Weldon and Sieh, 1985). This offset can be seen at site o. The fault can be located there by the contrast between the blue-green Pelona Schist and the buff-pink sedimentary rocks; southeast of the saddle, the wall of the canyon is conspicuously offset. Slumping has modified the expression of this offset, but the amount of rupture associated with the latest earthquake can be determined by comparing many such offsets (e.g., Sieh, 1978). A similar 13-ft (4-m) offset channel wall is at the northwest end of the area (site a), and several small gullies between sites o and q are offset the same amount.

R. J. Weldon

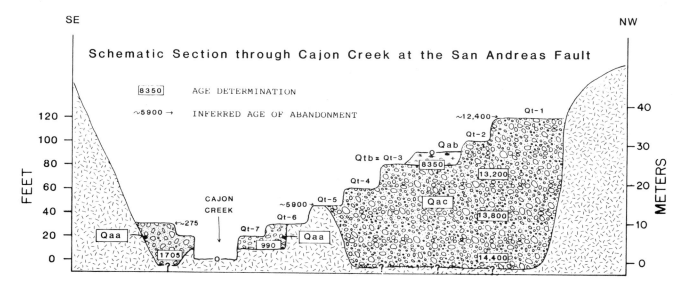

Figure 3. Schematic section through Cajon Creek, showing late Pleistocene and Holocene deposits and terraces. The ages in boxes are radiocarbon dates in stratigraphic position and the unboxed dates are abandonment ages of terraces extrapolated from the ages of deposits overlying and underlying particular surfaces (Weldon and Sieh, 1985). See Figure 2 for definition of units.

Between I-15 and Route 66 there are two traces of the San Andreas fault. A small sag pond (site n) sits between the two traces and a small landslide at its western end (Fig. 2). About 820 ft (250 m) northwest of the sag, the southern trace of the fault is exposed in the northwest facing cliff (site 1; due to the poison oak–infested chaparral, this exposure is best seen from the west late in the day). The slope above the cliff and the contact between colluvium (reddish brown) and Pelona Schist–landslide (green) are dropped to the north by the fault.

Northwest of Route 66, Cajon Creek and Lone Pine Creek have formed a broad flight of terraces across the San Andreas fault. Given the time (and an unlocked gate where the dirt road leaves Route 66, locality 4) these terraces can be seen from the top of the hill between Blue Cut and the San Andreas fault (Fig. 1). The road up to this overview crosses several strath terraces formed during downcutting that occurred between the deposition of Qad and Qac.

Several of the younger terraces and a fault-line scarp associated with the San Andreas fault can be seen from Route 66, looking across Cajon Creek (between sites h and j; a stop here would be best for observers with time for only one stop). Different strath terraces, cut into the vertically foliated Pelona Schist and overlain by Holocene sand and gravel (Qc and Qaa, Fig. 2), create spectacular angular unconformities. The Pelona Schist ends abruptly at the San Andreas fault (site j). Lateral slip along the San Andreas fault has offset the pre-Qac channel, north of the fault, against Pelona Schist, south of the fault (Weldon and Sieh, 1985, Fig. 9). Differential erosion into the softer Qac fill north of

the fault has created the fault-line scarp. The southeastern edge of Cajon Creek when Qac was deposited can be seen in the railroad cut (site i), the wall of Lone Pine Canyon (site e), and the edge of Cajon Creek today (site f).

The terraces with the best preserved geomorphic surfaces were cut by Lone Pine Creek. These terraces can be reached from Swarthout Road, which leaves Route 66 at locality 2 (Fig. 1; the campground on the map here no longer exists). There are seven terraces recognized near the mouth of Lone Pine Creek (Fig. 3). The oldest five terraces are offset between 475 and 950 ft (145 and 290 m) by the San Andreas fault (between sites c and d). The geomorphic development of this area, called the Lost Lake depression, involves a complex interaction of faulting, lateral offset of terraces of different heights, and erosion by minor streams, and is portrayed on Figure 4. Offsets of the terraces are measured by the separation of their terrace risers (the abandoned river bank when the creek flowed at that level) across the fault. Because Lone Pine Creek flowed across the fault at an oblique angle, the surface expression of the fault drops downstream except where it rises to a higher and older terrace (Weldon and Sieh, 1985, Fig. 8). This creates an illusion of ridges and troughs, not terraces and risers, encountering the fault. However, approaching the fault by walking along a terrace riser from a distance (e.g., from locality 3 or site b) yields the proper perspective at the fault. The Lost Lake depression is best viewed from a small knoll just north of the powerline tower at site d. If time permits, the offsets of the terraces (shown in Fig. 4f) can be seen by walking along the many roads and trails in this area.

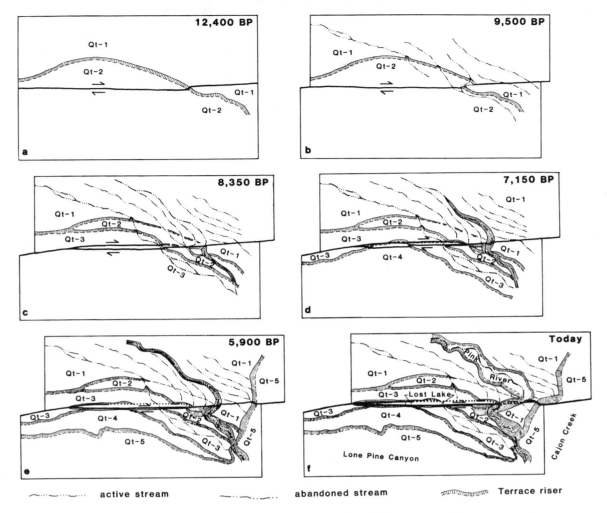

Figure 4. Development of the Lost Lake depression and associated offset terraces and streams across the San Andreas fault. This figure covers the area between sites b and j (Fig. 1). A flight of Quaternary terraces (Qt-1 through Qt-5) are progressively offset by the San Andreas fault. a, c, d, and e represent times when younger terraces were cut across the fault and had not yet been offset.

SLIP RATE AND EARTHQUAKE RECURRENCE

A Holocene history of sedimentation and terrace formation has been combined with the numerous offsets of geomorphic features to calculate an average slip rate of 24.5 ± 3.5 mm/yr on the San Andreas fault (Weldon and Sieh, 1985). Four separate determinations of the slip rate, for different periods of the Holocene, indicate that the fault slips at a uniform rate when measured over thousands of years. However, trenching at site d revealed that the fault has experienced repeated large earthquakes during the last 1,000 years (Weldon and Sieh, 1985). The data from this site, coupled with the distribution of shaking intensities associated with the 1857 earthquake (Agnew and Sieh, 1978), suggest that rupture during that earthquake terminated northwest of Cajon

Pass. The observations that the fault is locked here and probably has been for more than the 215 years of historical record, and that rupture occurs in discrete displacements of about 13 ft (4 m) every few hundred years require a large earthquake in the near future to relieve the more than 16 ft (5 m) of strain that has accumulated since the last earthquake.

REFERENCES CITED

Agnew, D., and Sieh, K., 1978, A documentary study of the felt effects of the great California earthquake of 1857: Seismological Society of America Bulletin, v. 68, p. 1717–1729.

Barrows, A. G., 1979, Geology and fault activity of the Valyermo segment of the San Andreas fault zone, Los Angeles County, California: California Division of Mines and Geology Open-File Report 79-1 LA, 49 p.

Barrows, A. G., Kahle, J. E., and Beeby, D. J., 1985, Earthquake hazards and tectonic history of the San Andreas fault zone, Los Angeles County, California: Final Technical Report to the USGS Earthquake Hazard Reduction Program, 126 p.

Blake, W. P., 1856, Geologic reconnaissance in California: Pacific Railroad Reports, 33rd Congress Senate Document no. 78, v. 5, 370 p.

Campbell, M. R., 1902, Reconnaissance of the borox deposits of Death Valley and the Mojave Desert: U.S. Geological Survey Bulletin 200, 23 p.

Dibblee, T. W., Jr., 1975, Tectonics of the western Mojave Desert near the San Andreas fault, in Crowell, J. C., ed., San Andreas fault in southern California: California Division of Mines and Geology, Special Report 118, p. 155–161.

Ehlig, P. L., 1981, Origin and tectonic history of the basement terrane of the San Gabriel Mountains, central Transverse Ranges, in Ernst, W. G., ed., The geotectonic development of California, Rubey Volume no. 1: New Jersey, Prentice-Hall, p. 253–283.

Hill, M. L., 1980, History of geologic explorations in the deserts of southern California, in Fife, D. L., and Brown, A. R., eds., Geology and mineral wealth of the California desert, Dibblee volume: Santa Ana, California, South Coast Geological Society, p. 2–12.

Kooser, M. A., 1985, Paleocene pleisiosaur(?), in Reynolds, R. E., ed., Geologic investigations along Interstate 15, Cajon Pass to Manix Lake, California: San Bernardino County Museum Publication, p. 43–48.

Lawson, A. D., and others, 1908, The California earthquake of April 18, 1906; Report to the State Earthquake Investigation Commission: Carnegie Institution of Washington Publication 87, v. 1, 451 p.

Marcou, J., 1876, Report on the geology of southern California: U.S. Geological Surveys West 100th Meridian, Annual Report for 1876, p. 378–392.

Noble, L. F., 1926, The San Andreas rift and some other active faults in the desert region of southeastern California: Carnegie Institute of Washington Year Book, no. 25, p. 415–422.

—— , 1954, Geology of the Valyermo Quadrangle and vicinity, California: U.S. Geological Survey Quadrangle Map GQ-50, scale 1:24,000.

Schuyler, J. D., 1896, Reservoirs for irrigation: U.S. Geological Survey, 18th Annual Report, Part 4, p. 617–740.

Sieh, K. E., 1978, Slip along the San Andreas fault associated with the great 1857 earthquake: Seismological Society of America Bulletin, v. 68, p. 1421–1448.

Weldon, R. J., and Sieh, K. E., 1985, Holocene rate of slip and tentative recurrence interval for large earthquakes on the San Andreas fault, Cajon Pass, southern California: Geological Society of America Bulletin, v. 96, p. 793–812.

Cucamonga fault zone scarps, Day Canyon alluvial fan, eastern San Gabriel Mountains, southern California

D. M. Morton, J. C. Matti, and J. C. Tinsley, U.S. Geological Survey, 345 Middlefield Road, Menlo Park, California 94025

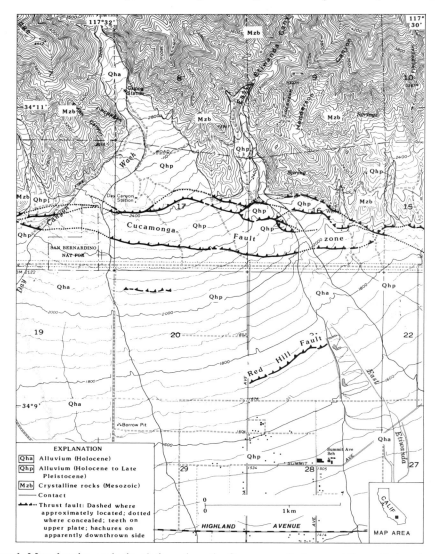

Figure 1. Map showing geologic relations along the Cucamonga fault zone at the south margin of the eastern San Gabriel Mountains.

LOCATION AND ACCESSIBILITY

The Cucamonga fault zone divides the high eastern San Gabriel Mountains from a broad alluvial apron to the south. The site is located on the Day Canyon alluvial fan where strands of the Cucamonga fault zone have produced impressive scarps crossing the fan surface (Fig. 1). This site is readily accessible by taking the Baseline Road exit from I-10 approximately 10 mi (16 km) east of the city of Ontario. Drive 0.25 mi (0.4 km) west on Baseline Road to Etiwanda Avenue; turn north and go 3 mi (5 km) to the end of Etiwanda Avenue. At the end of Etiwanda

Avenue, an east-west oriented dirt power-line service road is parallel to and just south of the southern fault scarp of the Cucamonga fault zone. Proceeding 0.5 mi (0.8 km) west from Etiwanda Avenue on the power-line service road, a dirt road leads north up the Day Canyon alluvial fan crossing the southern scarp of the Cucamonga fault zone. This road ends at a locked gate about 0.5 mi (0.8 km) from the power-line road. This point is situated on top of one of the northern scarps of the Cucamonga fault zone. The U.S. Forest Service buildings, referred to as the Day Canyon Station, are located south of the locked gate at the top of a 130 ft (40 m) high scarp. The dirt roads can be rough, but are generally easily passable for passenger cars.

SIGNIFICANCE OF THE SITE

The Cucamonga fault zone is one element of a composite frontal fault system, commonly referred to as the Malibu–Sierra Madre–Santa Monica–Cucamonga fault system, or zone, which forms the southern boundary of the Transverse Ranges Province west of the San Jacinto fault. The fault zone is considered to be the result of convergence of the Peninsula Ranges to the south, with the Transverse Ranges to the north (Morton and Matti, 1987). The minimum convergence rate at Day Canyon is approximately 5 mm/yr for the last 13,000 years (Matti and others, 1985). Earthquakes producing the scarps are estimated to have had surface wave magnitudes of M 6.5 to M 7.2, and produced vertical surface displacements of about 6 ft (2 m) with an average recurrence interval of 625 years. The Cucamonga fault zone has a width of 0.5 mi (0.8 km) and has the greatest number of youthful fault scarps cutting the alluvial fans flanking the southern margin of the high San Gabriel Mountains. The Day Canyon alluvial fan contains the greatest assortment of young scarps and the most detailed Quaternary tectonic record along the Cucamonga fault (Eckis, 1928).

SITE INFORMATION

The Cucamonga fault zone here consists of a series of anastamosing, north-dipping reverse and thrust faults, with an abundance of discontinuous south-facing scarps. Scarps extend discontinuously over a distance of 14 mi (22 km) between San Antonio Canyon, 8 mi (13 km) west of Day Canyon, and Lytle Creek, 6 mi (10 km) to the east (Morton, 1976).

Of the ten larger alluvuial fans along the Cucamonga fault zone between San Antonio Canyon and Lytle Creek, all except one, the Deer Canyon fan 2 mi (3 km) west of Day Canyon, have entrenched fan heads. Fault scarps cut all ages of alluvium except on Deer Canyon fan and the active alluvium covering the entrenched channels. In the area of Day Canyon fan, the older faults are primarily located in the northern part of the zone where they are within mylonitic basement rock or are thrust contacts between mylonitic basement and Pleistocene alluvium. The upper plates of the thrusts commonly show evidence of backward rotation. The younger faults are located in the southern part of the zone where they cut 13,000 yr old and younger alluvium.

The oldest and highest scarp on the Day Canyon fan is near the Day Canyon Station where the maximum height is 131 ft (40 m). The scarp is preserved here because it is an isolated bedrock knoll. The knoll is capped and flanked by alluvium; the steep (36°) scarp face is basement covered by a thin veneer of colluvium. The age of the alluvium capping the knoll is estimated to be no older than 13,000 years on the basis of soil profile development. On the west side of the buildings are a lower composite scarp 40 ft; (12 m) and a paleo-stream channel with a

rather gentle gradient (14°), which may reflect incomplete degradation of an initially much steeper scarp.

The southernmost scarp paralleling the power-line road on Day Canyon is the most instructive in deducing a sequence of surface-rupturing thrust-faulting events. The age of the alluvial surface cut by this composite scarp is estimated to be 5,000 yrs. The length of this nearly linear scarp is 2.5 mi (4 km). Topographic profiles constructed across the scarp show heights ranging from 6.5 to 52 ft (2 to 16 m). The highest part of the scarp is readily seen just west of where the road leading to Day Canyon Station crosses the fault scarp. The lowest scarps are best seen on both sides of East Etiwanda Canyon east of Day Canyon fan. They can be reached by taking the power-line road east from Etiwanda Avenue 0.3 to 0.5 mi (0.5 to 0.8 km). Of particular interest is the presence of three different height scarps just to the west of the entrenched fanhead of East Etiwanda Canyon.

Scarp heights are interpreted to have been produced in multiples of 6.6 ft (2 m). A trench cut across the southernmost scarp of the Cucamonga fault on Day Canyon fan exposed a 35° north dipping, 6.6 ft (2 m) thick fault zone at essentially the midpoint of the scarp face. The scarps formed from an initially unfaulted alluvial fan surface, preserved now as erosional remnants at the top of the 52 ft (16 m) high scarp. This surface was subjected to a series of thrust-faulting events along a realtively thin shear zone. Each event produced vertical offsets of about 6.6 ft (2 m) at the surface. Erosion between faulting episodes removed segments of the scarp, producing a stepped transverse profile. Each step represents one or possibly more thrust-faulting events. Thus, the southernmost composite Cucamonga fault scarp appears to record evidence of eight earthquakes, each producing a 6.6-ft (2-m) component of vertical displacement with a recurrence interval of approximately 625 years. If movement on this fault is all dip-slip, the displacement for each event would be approximately 11 ft (3.5 m). If 6.6 ft (2 m) vertical offsets formed the 131-ft-high (40 m) scarp at Day Canyon Station, the recurrence interval for ground-rupturing events is about the same as determined for the southernmost scarp, 650 years.

A 0.5-mi (0.8-km) hike up East Etiwanda Canyon above the power-line road leads to an exposure of mylonitic basement thrust over Pleistocene gravels. The fault at this exposure, on the east side of the wash, dips 35° to the north, which is typical for the dip of the Cucamonga fault.

REFERENCES CITED

Eckis, R., 1928, Alluvial fans in the Cucamonga district, southern California: Journal of Geology, v. 36, p. 224–247.
Matti, J. C., Morton, D. M., and Cox, B. F., 1985, Distribution and geologic relations of fault systems in the vicinity of the Central Transverse Ranges, southern California: U.S. Geological Survey Open-File Report 85-365, 27 p.
Morton, D. M., 1976, Geologic map of the Cucamonga fault zone between San Antonio Canyon and Lytle Creek, southern California: U.S. Geological Survey Open-File Report 76-726, scale 1:24,000.
Morton, D. M., and Matti, J. C., 1987, The Cucamonga fault zone; Geologic setting and Quaternary history: U.S. Geological Survey Professional Paper 1339 (in press).

Hog Back: A grossly stable prehistoric translatory rock block slide, San Antonio Canyon, southern California

Lawrence J. Herber, Geological Sciences, California State Polytechnic University, Pomona, California 91768

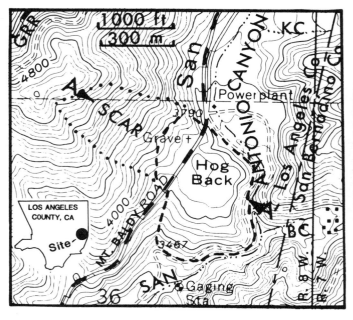

Figure 1. Index Map of Hog Back slide. BC, Barrett Canyon; KC, Kerkhoff Canyon; GRR, Glendora Ridge Road; Cross section A–A'.

Figure 2. South face of Hog Back looking north. Mt. Baldy Road cut at upper left corner.

LOCATION AND ACCESSIBILITY

Hog Back, as named on the 7½-minute Mt. Baldy, California, Quadrangle, lies in San Antonio Canyon (NE¼Sec.36, T.2N.,R.8W.) in the Angeles National Forest 1.1 mi (1.8 km) south of Mt. Baldy Village (Fig. 1). Hog Back appears as a large obstruction nearly choking off the canyon when looking upstream (Fig. 2). Looking south from Glendora Ridge Road, it appears as a low, rounded lump in the canyon floor (Fig. 3; Shelton, 1966).

Mt. Baldy Road crosses the head of Hog Back, providing easy access, and a roadside parking area suitable for large buses is near the head of the slide mass. A switchback footpath descends from the parking area to the old Mt. Baldy Road (abandoned) excavated across the toe of the slide. A 1-mi (1.6-km) hike along the footpath, abandoned road, and return to parking area via Mt. Baldy Road yields good cross-sectional views of the toe and head of the slide mass.

SIGNIFICANCE

Hog Back is a well-preserved translatory rock block slide in crystalline rock with good exposures of a thick shear zone across its toe and intact rock at its head, and is a natural for studying geomorphic effects of large slides, including major changes of the valley profile, a displaced stream, narrow bedrock gorge, and a 30-ft (10-m) waterfall with plunge pool.

SITE INFORMATION

Hog Back is one of hundreds of major slides (Morton and Streitz, 1969) in the rugged, early mature San Gabriel Mountains. The bedrock slope along the south side of Hog Back rises 1,600 ft (490 m) at a slope angle of 31° from the canyon floor to the ridgeline above the scar. Three other large bedrock slides with a combined scar/slide surface area from 55 to 320 acres occur within 2 mi (3.5 km) of Hog Back.

Estimated slide parameters are: surface area, 40 acres; length, 1,870 ft (570 m); width, 1,600 ft (490 m); thickness, 300 ft (90 m) maximum; volume, 5×10^6 m³; forward movement, 1,150 ft (350 m); elevation drop, 1,000 ft (300 m) maximum.

The fresh, randomly and pervasively fractured source area rocks for Hog Back consist largely of coarsely layered hornblende-plagioclase gneiss and lesser biotite-plagioclase gneiss. Both are locally migmatized and part of the pre-Tertiary San Antonio Canyon migmatite unit (Baird, 1956). Poorly layered gneissic diorite and a few dikes of latite and basalt also occur. Green epidote and randomly oriented slicks are common on fracture surfaces. Layering in the gneiss strikes N30° to 55°W and dips 30° to 45°S, defining a homoclinal structure and a corresponding neutral slope. Bedrock exposures are limited largely to the head and flanks of the scar and show no basal shear surface nor discontinuities dipping out of slope that could correspond to a slide-slip surface. The rock comprising the head of the slide mass along the Mt. Baldy Road cut correlates with the source area rock but has not been significantly disrupted or rotated. Thus, Hog Back is a coherent slide, but with slight internal disruption suggested by its jumbled surface layer.

Figure 4 shows the inferred four-layer anatomy of Hog

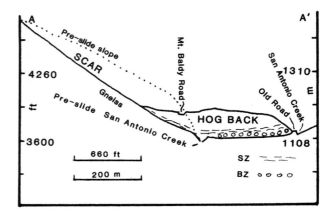

Figure 4. Cross section A–A'. BZ, Bulldozed zone; SZ, Shear zone.

Figure 3. Hog Back looking south from Glendora Ridge Road is framed by Mt. Baldy Road at lower middle right and S-shaped road at center. Broadened valley floor at lower left, and Barrett Canyon Road at left.

Back. Layers 1 and 2 are visible along the toe, and layer 3 at the head of the slide mass along Mt. Baldy Road.

1. Bulldozed zone. Loose, gray, unlayered, fresh stream gravels 15 ft (5 m) thick with rounded cobbles and boulders dragged along during forward motion of the slide. It overlies undeformed old brown silty terrace gravels.

2. Shear zone. Gray to greenish gray crushed and sheared gneiss up to 65 ft (20 m) thick. The basal section, 3 to 7 ft (1 to 2 m) thick, is the main shear zone and has been crushed to a fine gravel size with a rock flour matrix. Angular blocks show upward coarsening to 10 ft (3 m) near the top of the zone.

3. Body of slide. Central coherent mass of slide.

4. Jumbled surface layer. This layer forms most of the slide surface, except for the head area, and is a jumble of angular gneiss blocks with little soil development. The surface at the head, including the higher elevations of the slide mass, is relatively smooth, dips gently toward the slide scar, and shows a few small, gentle depressions. It represents the shortest transport away from the source rock area, and minimal surface and internal deformation. The jumbled layer is characterized by disoriented fresh angular rock blocks up to 15 ft (5 m), which locally enclose spaces to form small caves, or are loosely linked forming small steep depressions. Decreased lateral confining pressures, high basal shearing resistance, and severe vibrations accompanied catastrophic slide movement away from the source rock area across the irregular alluviated valley floor. The resulting deformation (rotation and small translations) caused slight internal disruptions leading to the jumbled surface with its small depressions.

Earthquakes from nearby active faults or saturation of the steep, highly fractured bedrock slopes may have triggered sliding. The active San Andreas, San Jacinto, and Cucamonga faults are only 9 to 4 mi (15 to 7 km) from Hog Back.

San Antonio Creek was forced to cut into its west slope by Kerkhoff Canyon Creek 1,800 ft (550 m) upstream and by the protruding north wall of Barrett Canyon (Fig. 1). A steep, unstable 60° slope perhaps 410 ft (125 m) high was cut into the toe of a 30° slope 1,800 ft (550 m) high (Fig. 4).

The slide moved southeasterly across the canyon floor ramping part way up a bedrock ridge projecting southwesterly into the canyon (the north wall of Barrett Canyon, Fig. 1). Hog Back dammed San Antonio Creek, and up to 200 ft (60 m) of coarse alluvial gravels were deposited against its upstream face. The gravels raised and broadened the valley floor (Fig. 3), and lake waters flowed across the toe of Hog Back at its lowest point when surface flows exceeded underflow through and below the slide. This nearly doubled the gradient to perhaps 1,000 ft/mi (190 m/km), dramatically increasing the erosive powers of the newly displaced San Antonio Creek. Rejuvenated, it rapidly sawed headward through the buried bedrock ridge without detour or deflection. The deep narrow bedrock gorge and remaining 30-ft (10-m) waterfall attest to vigorous downward and headward erosion, which in turn exposed the toe anatomy of Hog Back. Grading for the old Mt. Baldy Road significantly enlarged the exposure without obvious major damage to the gorge or falls, and provides a comfortable observation platform.

Despite its location across a major canyon, Hog Back is remarkably well preserved except for erosion of its toe and burial of its upstream edge by stream gravel. The silicate rock composition and porous surface layer of Hog Back preclude rapid weathering and erosion. In addition, San Antonio Creek confined itself to a bedrock gorge that protects the critical, easily erodible upstream toe area—a fortunate circumstance as this is the only spot where Hog Back's anatomy is on display.

REFERENCES CITED

Baird, A. K., 1956, Geology of a portion of San Antonio Canyon, San Gabriel Mountains [M.S. thesis]: Claremont, California, Pomona College, 91 p.

Morton, D. M., and Streitz, R., 1969, Preliminary map of major landslides, San Gabriel Mountains, California: California Division of Mines and Geology Map Sheet 15, scale 1:63,000.

Shelton, J. S., 1966, Geology illustrated: San Francisco, W. H. Freeman, 434 p.

The San Gabriel anorthosite-syenite-gabbro body, San Gabriel Mountains, California

Bruce A. Carter, Department of Physical Sciences (Geology), Pasadena City College, Pasadena, California 91106

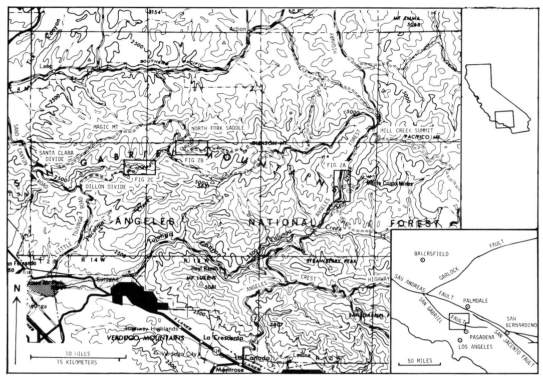

Figure 1. Index maps to rocks of the anorthosite-syenite-gabbro body of the western San Gabriel Mountains, Los Angeles County, California.

LOCATION

The San Gabriel anorthosite-syenite-gabbro body is exposed in the western San Gabriel Mountains about 30 mi (48 km) north of Los Angeles (Fig. 1). The most abundant lithologies of this body, which constitute the anorthosite-leucogabbro unit, are easily accessible via Angeles Forest Highway (N 3) along Mill Creek about 18 mi (28.8 km) south of Palmdale (Figs. 1, 2A). However, many of the most interesting lithologies of this body can be seen only by driving several mi (km) on U.S. Forest Service roads into the central part of the body west of Angeles Forest Highway. These localities can be reached from Angeles Forest Highway by driving west from Mill Creek Summit, past Mt. Gleason, to North Fork Saddle, and on down Pacoima Canyon. Alternately, they can be reached by driving east either from Santa Clara Divide or from Dillon Divide on the Sand Canyon–Little Tujunga Road (Fig. 1).

A second important group of lithologies constitutes the syenite unit and can be seen near "Sold" on the North Fork Saddle–Magic Mountain–Mount Gleason road (Figs. 1, 2B). This road is usually in good condition and is easily passable by con-

ventional automobile except between December and February, when it is occasionally covered by snow at higher elevations.

A third important group of lithologies constitutes the Jotunite unit and can be seen along Pacoima Canyon (Fig. 2C). The road through Pacoima Canyon, reached from either Dillon Divide or North Fork Saddle (Fig. 1), is rough and may require use of a four-wheel-drive vehicle during winter and spring.

SIGNIFICANCE

The San Gabriel anorthosite-syenite-gabbro body, which underlies about 100 mi^2 (250 km^2) of the western San Gabriel Mountains constitutes a well-defined Precambrian massif-type anorthosite pluton (Anderson, 1969; DeWaard, 1969; Emslie, 1973) that is one of the most distinctive units in the crystalline terrains of the Transverse Ranges of southern California (Carter and Silver, 1972; Silver and others, 1963; Crowell, 1962; Oakeshott, 1958). Although only a part of an original larger body, the rocks presently exposed in the San Gabriel Mountains exhibit nearly all of the characteristic lithologies, structures, and textures of other anorthosite massifs, and they have not been subjected to

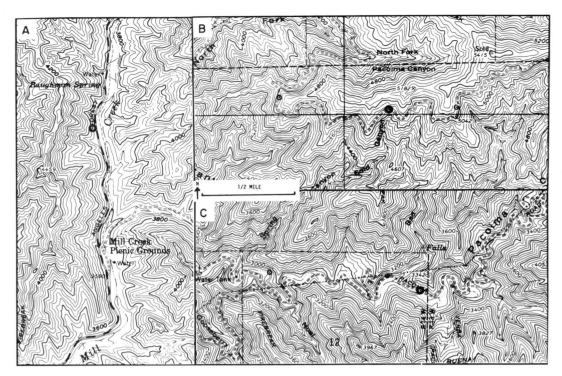

Figure 2. Index maps to localities in the San Gabriel anorthosite-syenite-gabbro body. A. Mill Creek area, Chilao Flat 7½-minute Quadrangle. Rocks of the anorthosite-leucogabbro unit are exposed along Mill Creek; best exposures are at the point indicated (34 21′9″N, 118 6′36″W). B. Sold Ridge area, Agua Dulce, Sunland, Acton and Condor Peak 7½-minute Quadrangles. Rocks of the syenite unit are exposed along the road. Best exposures of syenite are at the point indicated by the larger circle (34 22′32″N, 118 14′51″W). Other excellent exposures occur along the road about 1 mi (1.6 km) in either direction as indicated by smaller circles. C. Pacioma Canyon area, Sunland 7½-minute Quadrangle. Rocks of the jotunite unit are exposed along Pacioma Canyon. Best exposures are at the larger circle (34 21″46″N, 118 17′26″W). Other exposures of different lithologies occur near the mouths of Bad and Spring Canyons to the west (small circles).

post-emplacement, high grade regional metamorphism. Thus this body provides an excellent opportunity to study rocks of the anorthosite association.

Primary cumulate structures and textures in these rocks have greatly aided the structural reconstruction of this body and provide strong evidence of its origin by bottom crystal accumulation. The portion now exposed was at least 6 mi (10 km) in thickness and about 9 mi (15 km) in diameter. It probably had the form of an inverted cone, with a subhorizontal, concordant upper contact. Primary quartz is rare in rocks of this suite; the following lithologies have been distinguished on the basis of the percentage and composition of their constituent feldspars: anorthosite, leucogabbro, gabbro, ferrogabbro (all with calcic andesine), ultramafite (olivine, augite, ilmenite and apatite), jotunite (predominantly antiperthitic sodic andesine), mangerite (antiperthite and mesoperthite), and syenite (predominantly mesoperthite). The body has been subdivided into three main stratigraphic units (from oldest to youngest): anorthosite-leucogabbro unit, syenite unit, and jotunite unit (Fig. 3). The overall lithologic variation, as well as compositional variations of plagioclase, suggests that the anorthosite-leucogabbro unit and the jotunite unit were each

formed in open systems in which major influxes of new magma were responsible for producing individual "differentiation units" 300 to 3,000 ft (100 to 1,000 m) in thickness.

LOCALITY
Anorthosite-Leucogabbro Unit

The anorthosite-leucogabbro unit consists of alternating subunits of massive anorthosite and cyclically layered leucogabbro; it is at least 4 mi (7 km) in thickness and becomes more gabbroic near its top. Exposures of rocks of this unit are best seen along Mill Creek; Angeles Forest Highway crosses several thick (average about 2,600 ft; 800 m) anorthosite and leucogabbro subunits in the vicinity of the Monte Cristo campground (Mill Creek picnic grounds, Fig. 2A). The best outcrop is about 1 mi (1.6 km) north of the campground, where exposures consist of gabbroic rocks of one of the leucogabbro subunits that display distinct cyclic compositional layering with a vertical orientation. This layering probably was originally subhorizontal and oriented so that up was to the north (right).

Anorthosite is present near the northern end of the roadcut; it is much more abundant farther south along the highway and is

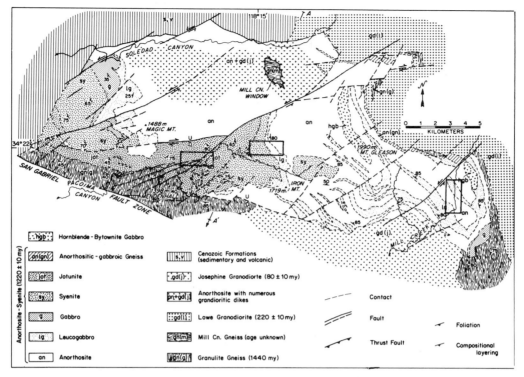

Figure 3. Simplified geologic map of the San Gabriel anorthosite-syenite-gabbro body. Boxes indicate locations of maps in Figure 2A-C.

estimated to constitute about 70 percent of the anorthosite-leucogabbro unit. Anorthosite consists of large equant anhedral crystals (1 cm to 3 cm) of gray to white calcic andesine with 0 to 10 percent ferromagnesian minerals. Most of the outcrop consists of leucogabbro with 10 to 35 percent ferromagnesian minerals. Most leucogabbro consists of a coarse anorthosite matrix surrounding 1- to 40-cm pyroxene crystals that ophitically enclose numerous tabular plagioclase crystals (0.5 to 2 cm) of the same composition as the andesine of the anorthosite matrix (Fig. 4). This texture is best seen near the south end of the exposure to the left of a large, south-dipping mafic dike (which is younger than and unrelated to rocks of the anorthosite-syenite-gabbro body). This texture is produced by postcumulous recrystallization, which has produced the extremely coarse-grained textures of anorthosite and leucogabbro by recrystallization of original "cumulate" andesine tablets whenever they were not mantled by the pyroxene that crystallized from intercumulate magma (forming anorthosite and the anorthositic part of leucogabbro; Wager and others, 1960; Carter, 1982).

One "crescumulate" layer near the center of the roadcut consists of perpendicular plagioclase crystals on the left that grew upward into magma (on the right) and were then buried by subsequent crystallization of pyroxene. Several wedge-shaped bodies of very coarse anorthosite cross-cutting layered leucogabbro point to the right and have their bases in thin layers of coarse plagioclase and pyroxene (the largest of these wedges was at the bottom of the roadcut and has been covered by fill). These wedges are interpreted to have formed by flow of lower density plagioclase crystal mush (0.5 to 2-cm tablets) upward into overlying higher density leucogabbro crystal mush. Subsequent postcumulous recrystallization produced the extremely coarse textures now observed.

Pervasive deuteric alteration has affected all primary ferromagnesian minerals in these rocks; the original pyroxenes are now alteration aggregates (uralite). Because of the alteration, the predominance of orthopyroxene in these rocks (as in many anorthosite suites elsewhere in the world) cannot be confirmed, so that the rocks have been called gabbros rather than norites. This uralitization of nearly all rocks of the entire body (with the exception of a few in the youngest jotunite unit) suggests that the magmas from which they formed probably had a relatively high water content.

Syenite Unit

The syenite unit is adjacent to (overlies) the anorthosite-leucogabbro unit and locally attains thicknesses of at least 2 to 3 mi (3 to 5 km), but is commonly much thinner or absent. In some areas, the basal 300 to 3,000 ft (100 to 1,000 m) of this unit is extremely mafic, with pronounced layering (the ultramafic syenite subunit), but the remainder of the unit is fairly homogeneous and massive (the normal syenite subunit). Rocks of this unit are most accessible along the road near "Sold" (Fig. 2B).

Along about 2 mi (3.2 km) of the road, there are good exposures of rocks of the ultramafic syenite subunit (gabbro, ferrogabbro, ferrojotunitic gabbro, ferrosyenite, ultramafite, jotunite, and mangerite). Compositional layering 1 to 20 in (3 to 50 cm) is very common and is primarily defined by differences in color index. Rocks of this subunit commonly contain large 3.3 to 100 ft; (1-30 m) angular blocks of anorthosite. Individual layers are commonly asymmetric, with contrasting feldspar-enriched and feldspar-depleted sides. Layers sometimes drape across anorthosite blocks. These rocks formed by bottom accumulation of mostly ferromagnesian minerals on the floor and side of the syenite magma chamber, and reach about 0.9 mi (1.5 km) in thickness in the upper Pacoima Canyon southeast of "Sold."

The main locality indicated on Figure 2B is an exposure of massive syenite of the normal syenite subunit. The rock consists of equant alkali feldspar (mesoperthite) crystals (2 to 5 mm) and 15 to 35 percent alteration aggregates (micas, amphiboles, epidote, and quartz) after primary ferromagnesian minerals. Although very little massive syenite is present on this road, about 1.5 mi (2.5 km) to the south, along the ridge south of Pacoima Canyon, this lithology comprises the normal syenite subunit, which is about 1.8 mi (3 km) in thickness.

Jotunite Unit

The uppermost jotunite unit is a highly compositionally variable unit that intruded the overlying granulite gneiss; it has been subdivided into five subunits based on contrasting lithologies or structures. Rocks of this unit are best exposed in Pacoima Canyon (Fig. 2C). Along the road southwest of North Fork Saddle, rocks of the lower two "anorthosite block" and "lower jotunite" subunits are exposed. At the main point indicated, east of the mouth of Bad Canyon, rocks of the fourth "ultramafic jutonite" subunit are exposed (the rock is a ferrogabbro consisting of about equal amounts of olivine, augite, plagioclase [calcic andesine], ilmenite, and apatite, with minor biotite or hornblende).

Just west of the mouth of Bad Canyon, rocks of the third "layered jotunite" subunit display conspicuous 2- to 12-in (5 to 30-cm) often asymmetric compositional layering with individual layers which may persist over tens of meters. About one mile to the west, near the mouth of Spring Canyon (Fig. 2C), gabbroic rocks of the fifth "upper jotunite" subunit contain numerous gabbro and gneiss xenoliths and display abundant evidence of slumping in a crystal mush, and of bottom accumulation of ferromagnesian minerals.

These localities include many classical "cumulate" structures, including crescumulate layers, a variety of slump structures, density (and sometimes size)-graded layering, cross-layering, and other current structures. Although many of these features have traditionally been explained as the result of crystals sinking (and less commonly, floating) in a magma chamber (Wager and Brown, 1967), many of them could also have formed by direct crystallization in a static boundary layer on the floor of a magma chamber (McBirney and Noyes, 1979). It is clear that settling of ferromagnesian minerals did occur during crystallization of some rocks of the ultramafic syenite subunit and the jotunite unit, and it is concluded that nearly all rocks of this body formed near the base of the magma chamber. However, it is possible, even likely, that crystal settling was not an important process during much of

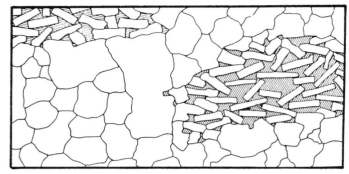

Figure 4. Ophiolitic leucogabbro. This texture resulted when cumulate plagioclase tablets were overgrown by both plagioclase and pyroxene crystallized from the intercumulate magma (heteradcumulous crystallization) followed by post-cumulous recrystallization of the plagioclase matrix. All plagioclase is unzoned calcic andesine.

its crystallization, and that many of its rocks (anorthosite, syenite, and many gabbroic rocks) formed by direct crystallization on the floor of the magma chamber. The classical "cumulate" textures and structures in these rocks must be carefully interpreted in light of their probable origin by direct bottom crystallization, which occurred without significant concurrent crystal settling in many cases.

REFERENCES CITED

Anderson, A. T., 1969, Massif-type anorthosite; A widespread Precambrian igneous rock, in Isachsen, Y. W., ed., Origin of Anorthosites and Related Rocks: New York State Museum and Science Service Memoir 18, p. 47–55.

Carter, B., 1982, Geology and structural setting of the San Gabriel anorthosite-syenite body and adjacent rocks of the western San Gabriel Mountains, Los Angeles County, California, in Cooper, J. D., ed., Geologic Excursions in the Transverse Ranges: Volume and Guidebook, Geological Society of America Cordilleran Section, 78th Annual Meeting, p. 1–53.

Carter, B., and Silver, L. T., 1972, Structure and petrology of the San Gabriel anorthosite-syenite body, California: 24th International Geological Congress, Montreal Report, Section 2, p. 303–311.

Crowell, J. C., 1962, Displacement along the San Andreas Fault, California: Geological Society of America Special Paper 71, 61 p.

DeWaard, D., 1969, The anorthosite problem; The problem of the anorthosite-charnockite suite of rocks, in Isachsen, Y. W., ed., Origin of Anorthosite and Related Rocks: New York State Museum and Science Service Memoir 18, p. 71–91.

Emslie, R. F., 1973, Some chemical characteristics of anorthositic suites and their significance: Canadian Journal of the Earth Sciences, v. 10, p. 54–71.

McBirney, A. R., and Noyes, R. M., 1979, Crystallization and layering of the Skaergaard intrusion: Journal of Petrology, v. 20, p. 487–554.

Oakeshott, G. B., 1958, Geology and Mineral Deposits of the San Fernando Quadrangle, Los Angeles County, California: California Division of Mines Bulletin, v. 172, 147 p.

Silver, L. T., McKinney, C. R., Deutsch, S., and Bolinger, J., 1963, Precambrian age determinations in the western San Gabriel Mountains, California: Journal of Geology, v. 71, p. 196–214.

Wager, L. R., and Brown, G. M., 1967, Layered igneous rocks: Edinburgh, Scotland, Oliver & Boyd, 587 p.

Wager, L. R., Brown, G. M., and Wadsworth, W. J., 1960, Types of igneous cumulates: Journal of Petrology, v. 1, p. 73–85.

Geology of the Devil's Punchbowl, Los Angeles County, California

T. W. Dibblee, Jr., 316 E. Mission St., Santa Barbara, California 93105

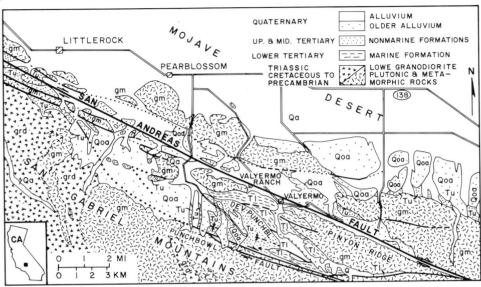

Figure 1. Small geologic map showing road from Pearblossom, California, to the Devil's Punchbowl overlook.

LOCATION AND ACCESS

The Devil's Punchbowl, now a county park, is a spectacular exposure of Tertiary sandstone strata on the southwest margin of the Mojave Desert adjacent to the San Gabriel Mountains.

This unusual feature is located south of Valyermo Ranger Station, about 7 mi (11 km) southeast of Pearblossom in the Mojave Desert in northern Los Angeles County (Fig. 1). It is between the high San Gabriel Mountains to the south, and lower hills to the north, which hide it from the adjacent Mojave Desert. The Devil's Punchbowl is on and within the north border of the Angeles National Forest. This site is accessible via a paved road (L.A. County Road N6) direct from Pearblossom on California 138.

The Devil's Punchbowl is shown on the Valyermo 7½- and 15-minute Quadrangles. The geology described here is summarized from Dibblee (1967).

SIGNIFICANCE OF SITE

The Devil's Punchbowl is an unusual exposure of a folded, thick upper Miocene terrestrial formation; it is unconformable on an equally thick, more steeply tilted lower Tertiary marine formation on crystalline basement that is within a 1.8-mi (3-km) wide slice within the San Andreas fault zone. These thick formations are exotic to the adjacent blocks, because the San Gabriel Mountain block southwest of this slice is composed only of pre-Tertiary crystalline basement, and the Mojave Desert block northeast of this slice is composed of crystalline basement with a thin strip of a late Tertiary formation along the San Andreas fault zone unlike that of the Devil's Punchbowl area. The slice that includes the Devil's Punchbowl is here referred to as the Punchbowl slice. It was probably juxtaposed between these blocks from a distant area by large right-slip movements on the San Andreas and Punchbowl faults that bound it on the northeast and southwest, respectively.

Geomorphology

The Punchbowl slice is a strip of moderate relief, as compared to the adjacent Mojave block of low relief and the San Gabriel Mountain block of very high relief. The Devil's Punchbowl divides the Punchbowl slice into two parts. The part to the east is semi-mountainous, of which Pinyon Ridge of basement, with a subdued summit, is the dominant feature. The part to the west is an area of dissected piedmont alluvial fans sloping northward from the San Gabriel Mountains toward the Mojave Desert, with several low hills of granitic basement in its northern part.

The Punchbowl slice was eroded to an area of low relief in Pleistocene time and in part covered by piedmont alluvial fans derived from the San Gabriel Mountains. Renewed uplift since that episode caused rejuvenated erosion, during which streams deepened their channels in the hill areas and caused severe dissection of the alluvial fans west of the Devil's Punchbowl.

Rock Units

The exotic Tertiary sedimentary formations of the Punchbowl fault slice were deposited in the Soledad-Ridge basin in the Transverse Ranges to the west, from which this slice was evidently derived and emplaced along the San Andreas fault zone.

Alluvial Fan Deposits. In the flat areas of the Punchbowl slice west of the Devil's Punchbowl, the beveled erosion surface of the late Miocene Punchbowl Formation is covered by extensive late Pleistocene piedmont alluvial fan deposits as thick as 140 ft (50 m). These deposits, which slope northward from the base of the San Gabriel Mountains, are composed entirely of basement detritus derived from that range. These fan deposits are terminated downslope by the San Andreas fault. Remnants of alluvial fans north of that fault are composed of detritus derived from low mountains farther west, but these alluvial fan remnants have since been shifted southeastward from that source area by right slip on the fault.

Late Tertiary Strata North of San Andreas Fault. North of the San Andreas fault is a strip about 0.5 mi (0.8 km) wide of late Tertiary terrestrial friable sandstone and gray to reddish claystone. This unit is largely covered by alluvial deposits and is exposed in only a few places. It is bounded on the northeast by granitic basement, which is in large part covered by alluvial deposits.

Punchbowl Formation. The Miocene sandstone that forms the Devil's Punchbowl was named and mapped as the Punchbowl Formation by Noble (1954a, b). It consists of about 5,000 ft (1,500 m) of semi-friable conglomeratic arkosic sandstone; which is light gray to nearly white and medium to coarse grained; it is composed of thick, massive strata separated by thin layers of greenish to reddish silty mudstone. The lower part includes conglomerate of subrounded granitic and gneissic cobbles in a light gray sandstone matrix. Sandstone of the upper part contains pebbles of granitic rocks and of andesitic porphyries. Cross-bedding in some strata dips generally to the southwest.

The Punchbowl Formation lies with angular discordance on more steeply tilted strata of the marine San Francisquito Formation. In Sandrock Canyon south of Valyermo, the Punchbowl Formation contains a basal lens as thick as 200 ft (60 m) of reddish-brown breccia of large, angular fragments of sandstone eroded from the San Francisquito Formation.

The Punchbowl Formation was deposited in late Miocene time by streams in a valley or plain that was then part of the Soledad basin to the west. The sediments were derived from granitic and gneissic rocks exposed probably northeast of the San Andreas fault, such as those now exposed in the vicinity of the Cajon Pass area. Some may have been derived from the Pinyon Ridge area. It is doubtful that they were derived from the San Gabriel Mountains, as inferred by Noble (1954a, b), because detritus of distinctive basement rocks of that area is absent in the Punchbowl Formation. The clasts of porphyritic andesites in the upper part were either derived from the metavolcanic porphyries now exposed in central parts of the Mojave Desert, or they may have been redeposited from conglomerates of the San Francisquito Formation.

The Punchbowl Formation yielded mammalian fossil remains

diagnostic of the Clarendonian and Hemphillian Stages of the vertebrate time scale (Woodburne, 1975) or late Miocene–early Pliocene. Similar fossils were found in the upper part of the Mint Canyon Foundation of the Soledad basin near Saugus to the west. The Punchbowl Formation was thought to be correlative with the "Punchbowl" Formation, of similar lithology and thickness, that is superbly exposed in Cajon Canyon about 25 mi (32 km) to the east on the northeast side of the San Andreas fault (Noble, 1954a). However, in that area, mammalian fossils found are diagnostic of the Barstovian and late Hemingfordian Stages, middle Miocene (Woodburne and Golz, 1972), and thereafter must be older.

San Francisquito Formation. On the east and north sides of the Devil's Punchbowl, the Punchbowl Formation is unconformably underlain by a marine formation of inferred Eocene-Paleocene age. It was assigned to the Martinez Formation on the basis of paleontologic evidence by Dickerson (1914) and Noble (1954a, b), but Dibblee (1967) later referred it to the San Francisquito Formation of the San Francisquito Canyon area about 20 mi (32 km) west, the type section. In both areas this formation consists of hard, light buff arkosic sandstone and thin interbeds of dark gray micaceous shale. In many places the sandstones contain lenses of coarse cobble conglomerate with rounded cobbles and pebbles of mostly hard granitic rocks and andesitic porphyries.

In the Devil's Punchbowl area the San Francisquito Formation is estimated to be about 4,700 ft (1,500 m) in exposed thickness, and unconformably overlies granitic and gneissic basement. In exposures east of the Devil's Punchbowl, the upper part is predominantly sandstone and the lower part mostly shale, but in western exposures north of the Devil's Punchbowl, the lower part is composed of nearly equal amounts of sandstone and shale.

The San Francisquito Formation of the Devil's Punchbowl area yielded molluscan fossils diagnostic of Paleocene age (Dickerson, 1914). Most of these were found in a thin basal sandstone and granitic conglomerate east of the Devil's Punchbowl.

The San Francisquito Formation was deposited on the continental shelf submerged under the eastern Pacific margin. The sandstones and conglomerates were deposited as proximal submarine fans, and were derived from granitic terrane that included hypabyssal and metavolcanic porphyritic andesites such as those now exposed in central parts of the Mojave Desert far to the northeast.

Crystalline Basement Rocks. The basement rocks of the Punchbowl slice exposed on Pinyon Ridge and west of Valyermo consist of mostly late Mesozoic biotite-granodiorite and older banded gneiss. These rocks are much shattered and not very coherent.

In the adjacent part of the Mojave Desert block north of the San Andreas fault gray-white quartz monzonite of late Mesozoic age comprises most of the basement. North of Pinyon Ridge the rock is intrusive into pre-existing north-dipping gneiss that includes lenses of white marble.

The basement complex exposed in the San Gabriel Moun-

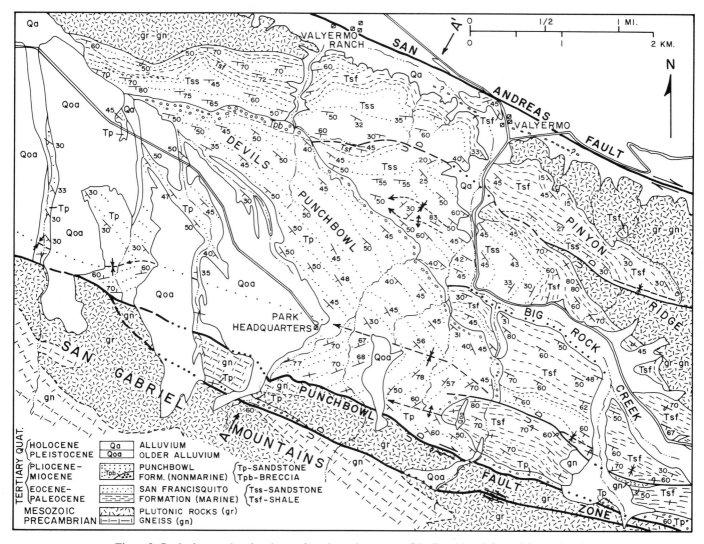

Figure 2. Geologic map showing the stratigraphy and structure of the Punchbowl site and the road to the Los Angeles County Park headquarters.

tains south of the Punchbowl slice is composed of a variety of crystalline rocks, as shown on Figure 2. The oldest is biotite gneiss of Precambrian(?) age. This is intruded complexly by variable gray biotite diorite–quartz diorite of Mesozoic age, and injected by younger white aplitic dikes. Associated with the gneiss in one area is a strip of distinctive, nearly white, hornblende granodiorite (Lowe Granodiorite) of Permo-Triassic age. This whole complex is intruded by nearly white quartz monzonite of late Mesozoic age.

Geologic Structure

Punchbowl Slice. The geologic structure of the rocks within the Punchbowl slice (Figs. 2, 3) is generally unlike that of the adjacent blocks. The gneissic rocks of Pinyon Ridge and ridge west of Valyermo strike nearly parallel to the San Andreas fault and dip steeply south to vertical. This basement strip, together

with the overlying San Francisquito Formation, was apparently tilted steeply southwestward, away from the San Andreas fault, prior to late Miocene time.

The San Francisquito Formation dips generally southwestward, in some places vertically or even overturned. This steeply tilted formation is repeated by three near-vertical southwest-side-up faults (Fig. 2). There are no major thrust faults as erroneously interpreted by Noble (1954b).

The Punchbowl Formation that unconformably overlies the San Francisquito Formation is also tilted southwestward but much less than that formation, generally about 45°. Two of the faults that disrupt that formation do not break the Punchbowl Formation, indicating that they were active before deposition of this Miocene formation. However, the most southeasterly fault that breaks the San Francisquito Formation breaks only the basal part of the Punchbowl Formation, where it is deformed into a small west-plunging anticline (Fig. 2).

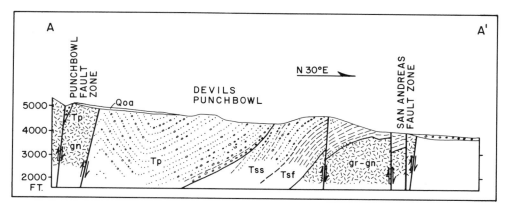

Figure 3. Geologic section across the Punchbowl between the San Andreas and Punchbowl faults.

Near the Punchbowl fault, the Punchbowl Formation is deformed into a west-plunging syncline with a very steep south flank (Fig. 2). This structure must be the effect of right lateral and south-side-up movement on the Punchbowl fault.

San Andreas Fault. In the Valyermo area the great active San Andreas fault is a single straight vertical fault that displays many physiographic features characteristic of this fault, such as scarplets, pressure ridges, sag ponds, and laterally offset or deflected stream channels. All these features are the effects of Holocene right lateral movements on the fault. The last major slip and earthquake on this segment was in 1857. An estimated cumulative Cenozoic right slip on this segment is about 130 mi (208 km) (Crowell, 1962, p. 36).

Punchbowl Fault. The Punchbowl fault is the high-angle fault along which the major part of the San Gabriel Mountains was elevated against the Punchbowl slice. This fault was formerly assumed to be the northwest extension of the San Jacinto fault (Noble, 1954a, b), but was later found to join southeastward into the San Andreas fault in Cajon Canyon (Dibblee, 1967, 1968, 1975), so it is a strand of the San Andreas fault zone. Apparent movement on this segment of the Punchbowl fault was up on the southwest. However, from regional geology it is evident that there has been about 25 mi (40 km) of right slip on the fault during late Cenozoic time (Dibblee, 1967, 1968).

In the Devil's Punchbowl area, the Punchbowl fault is composed of two closely spaced strands that eventually converge in both directions. The north strand is very conspicuously exposed where it separates the Punchbowl Formation from dark gray dioritic basement injected by white aplitic dikes. This strand dips steeply southwest. The south strand is within the basement rocks and forms an alignment of notches. It contains a thin slice of basal Punchbowl conglomerate.

The Punchbowl fault does not break any of the late Pleistocene alluvial fans that cover it except at two places about 2 mi (3 km) west of park headquarters (Fig. 2; Noble, 1954b), where it forms low scarplets. Therefore this fault has been largely inactive in late Quaternary time.

CONCLUSION

The Punchbowl slice which includes the Devil's Punchbowl is within the San Andreas fault zone as defined by Crowell (1962) and Dibblee (1968). It has been juxtaposed by large right-slip movements in late Cenozoic time on the San Andreas and Punchbowl faults between the Mojave Desert block and the San Gabriel Mountains block. The Tertiary sedimentary formations of this exotic slice were deposited in the Soledad-Ridge basin area far to the west and are therefore exotic to the adjacent blocks of this area.

REFERENCES CITED

Crowell, J. C., 1962, Displacements along the San Andreas fault, California: Geological Society of America Special Paper 71, 61 p.

Dibblee, T. W., Jr., 1967, Areal geology of the western Mojave Desert, California: U.S. Geological Survey Professional Paper 552, 153 p., map scale 1:125,000.

——, 1968, Displacements on the San Andreas fault system in the San Gabriel, San Bernardino, and San Jacinto Mountains, Southern California, in Dickinson, W. R., and Grantz, A., eds., Proceedings of conference on geologic problems, San Andreas fault system: Stanford University Publications in Geological Sciences, v. XI, p. 260–268.

——, 1975, Tectonics of the western Mojave Desert near the San Andreas fault, in Crowell, J. C., ed., San Andreas fault in Southern California: California Division of Mines and Geology Special Report 118, p. 155–161.

Dickerson, R. E., 1914, The Martinez Eocene and associated formations of Rock Creek on the western border of the Mojave Desert area: California University Department of Geology Bulletin, v. 8, no. 14, p. 289–298.

Noble, L. F., 1954a, The San Andreas fault zone from Soledad Pass to Cajon Pass, California, in Jahns, R. H., ed., Geology of southern California: California Division of Mines Bulletin 170, ch. IV, contr. IV, p. 37–48, pl. 5, scale 1:124,000.

——, 1954b, Geology of the Valyermo (7½') quadrangle and vicinity, California: U.S. Geological Survey Map GQ 50, scale 1:24,000.

Woodburne, M. O., 1975, Cenozoic stratigraphy of the Transverse Ranges and adjacent areas, southern California: Geological Society of America Special Paper 162, 91 p.

Woodburne, M. O., and Golz, D. J., 1972, Stratigraphy of the Punchbowl Formation, Cajon Valley, southern California: University of California Publications in Geological Sciences, v. 92, 71 p.

Roadcut exposure of the San Andreas fault zone along the Antelope Valley Freeway near Palmdale, California

Allan G. Barrows, California Division of Mines and Geology, 107 S. Broadway, Room 1065, Los Angeles, California 90012

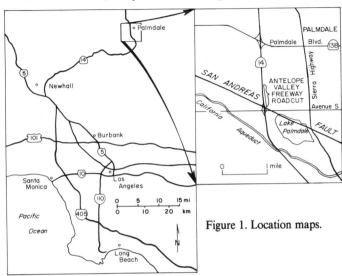

Figure 1. Location maps.

LOCATION AND ACCESS

The spectacular, trench-like roadcut where the Antelope Valley Freeway (California 14) crosses the San Andreas fault zone not only excellently exposes the geology within the zone but is also the most readily accessible point on the San Andreas fault for residents of the Los Angeles region. The locality is 60 mi (100 km) from the Los Angeles Civic Center via I-5 and California 14, and is about 1 mi (1.6 km) southwest of the city of Palmdale between the offramps for Palmdale Boulevard and Avenue S (Fig. 1).

The Antelope Valley Freeway is a limited-access, divided highway; stopping along the shoulder to view the geology is not permitted. Slowing down while traveling through the cut is also very dangerous because of high-speed freeway traffic. The best way to view the roadcut is to exit the freeway at Avenue S and climb the hill outside the fence along the freeway.

SIGNIFICANCE OF THE SITE

The great 1857 Fort Tejon earthquake that occurred along the south-central stretch of the San Andreas fault resulted in surface faulting across northern Los Angeles County where the fault passes within 35 mi (56 km) of downtown Los Angeles. Intensive geologic studies of the fault in recent years have led to the conclusion that a similar great earthquake, potentially much more catastrophic due to vast growth in the region, is likely to recur before the end of the twentieth century (Davis and others, 1982). Widespread publicity of this anticipated earthquake has, naturally, stimulated concern about its effects and curiosity about the nature of the San Andreas fault. The roadcut exposure of contorted rocks within the fault zone provides visual proof of the existence of the fault and dramatically displays the effects of powerful forces that have acted along it for a long time.

SITE INFORMATION

The San Andreas fault, in the vicinity of Palmdale, consists of a modern "main trace" along which the most recent surface rupture has taken place. The fault is expressed as an alignment of youthful geomorphic features such as scarps, troughs, and linear ridges that comprise a belt typically 50 to 100 ft (15 to 30 m) wide. The modern fault lies at the center of the San Andreas fault zone, which includes all the subparallel fault strands that are spatially and tectonically associated with the main trace proper. The San Andreas fault zone is 1 mi (1.6 km) wide where the Antelope Valley Freeway crosses it west of Lake Palmdale (a reservoir developed within a closed depression along the fault). Approximately 2 mi (3.2 km) to the east, the fault zone is 2 mi (3.2 km) wide.

Excavation of the roadcut across the uplifted area north of the San Andreas fault exposed complexly folded and faulted strata of the Anaverde Formation for 2,400 ft (730 m) (Fig. 2) between the main trace and another major strand called the Little Rock fault (Smith, 1976). With minor exceptions, the entire exposure consists of interbedded buff, arkosic sandstone, and dark brown, gypsiferous, clay shale of the lower-middle Pliocene, nonmarine, Anaverde Formation. A map of the distribution and generalized structure of the Anaverde Formation near Palmdale was first published by Wallace (1949), who also presented a geologic section that, fortuitously, coincides with the freeway alignment. Dibblee's map of the Lancaster Quadrangle (1960) also covers the area surrounding the roadcut.

As many as eight members of the Anaverde Formation have been mapped in the 30-mi (48 km) stretch between Elizabeth Lake and Juniper Hills along the north side of the San Andreas fault (Barrows and others, 1976, 1985), but only the clay shale (Tac) and buff arkose (Tab) members are present in the roadcut. At its northern boundary the Anaverde Formation is juxtaposed against granitic rocks along the Little Rock fault, which is not exposed in the roadcut but is visible along the base of the northern slope of the hill. The Little Rock fault, a possible ancestral trace of the San Andreas fault, parallels the San Andreas for more than 37 mi (60 km) and has been the site of substantial lateral slip, in excess of 13 mi (20 km), since the deposition of the Anaverde Formation.

A distinct bump in the freeway pavement, presumably a result of swelling of the clayey gouge beneath the roadbed, reveals to motorists the location of the San Andreas fault 400 ft (120 m) north of Avenue S. At the southern end of the roadcut, the San Andreas fault manifests itself as a 50-ft (15-m) wide zone of dark

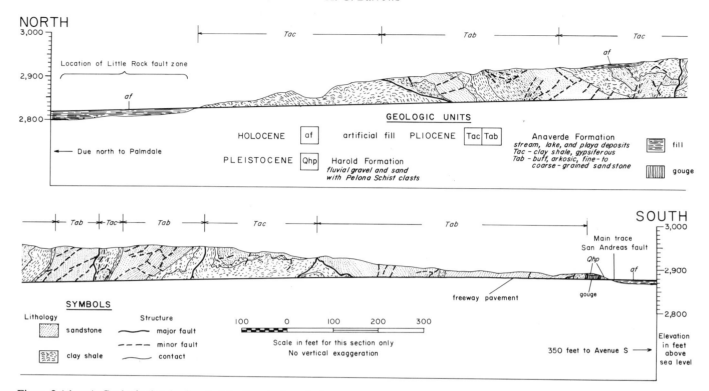

NORTH

Location of Little Rock fault zone

Due north to Palmdale

GEOLOGIC UNITS

HOLOCENE [af] artificial fill PLIOCENE [Tac][Tab]

PLEISTOCENE [Qhp] Harold Formation
fluvial gravel and sand
with Pelona Schist clasts

Anaverde Formation
stream, lake, and playa deposits
Tac - clay shale, gypsiferous
Tab - buff, arkosic, fine- to
coarse-grained sandstone

[image] fill

[image] gouge

SOUTH

Main trace
San Andreas fault

freeway pavement   gouge

SYMBOLS

Lithology   Structure
[image] sandstone   ——— major fault
                    - - - - minor fault
[image] clay shale ～～～ contact

Scale in feet for this section only
No vertical exaggeration

350 feet to Avenue S →

Elevation
in feet
above
sea level

Figure 2 (above). Geologic sketch of part of the San Andreas fault zone exposed in the east wall of a roadcut along the Antelope Valley Freeway (California 14) north of Avenue S near Palmdale, California. (Adapted from Smith, 1976.)

Figure 3. (right) Deformed Anaverde Formation sandstone and gypsiferous clay shale (dark) layers exposed near the center of the east wall of the Antelope Freeway roadcut near Palmdale. Note fault on right side of photo that extends to surface. Height of cut is 90 ft (27 m). (Photo by Drew Smith.)

gray gouge. Buff, arkosic sandstone is folded into a possible, although minor, anticline. Internal faulting disrupts the symmetry of the fold, which extends along the cut for about 600 ft (180 m) north of the San Andreas fault. Proceeding northward, a prominent fault separates the sandstone from an assemblage of intricately folded shale and sandstone that extends for 250 ft (75 m) to another prominent fault. The faults bounding this section have several inches of gouge and may be sites of important lateral movement. The remainder of the cut, beyond the fault that is 850 ft (260 m) north of the San Andreas, consists of alternating, internally faulted shale and sandstone sequences that have been deformed into an asymmetrical synclinal fold whose axis lies within the tightly folded and faulted area near the deepest part of the cut (Fig. 3). Plastic deformation is graphically demonstrated within this area by the complex folding of dark clay shale and interbedded resistant, thin, light-colored layers of nearly pure gypsum. Numerous pits on the surface of the hillside nearby attest to the mining of the abundant gypsum during the early part of this century.

REFERENCES CITED

Barrows, A. G., Kahle, J. E., and Beeby, D. J., 1976, Geology and fault activity of the Palmdale segment of the San Andreas fault zone, Los Angeles County, California: California Division of Mines and Geology Open-File Report 76-6 LA, 30 p., scale 1:12,000.

——, 1985, Earthquake hazards and tectonic history of the San Andreas fault zone, Los Angeles County, California: California Division of Mines and Geology Open-File Report 85-10 LA, 236 p., scale 1:12,000.

Davis, J. F., Bennett, J. H., Borchardt, G. A., Kahle, J. E., Rice, S. A., and Silva, M. A., 1982, Earthquake planning scenario for a magnitude 8.3 earthquake on the San Andreas fault in southern California: California Division of Mines and Geology Special Publication 60, 128 p.

Dibblee, T. W., Jr., 1960, Geologic map of the Lancaster Quadrangle, Los Angeles County, California: U.S. Geological Survey Miscellaneous Field Studies Map MF-76, scale 1:62,500.

Smith, D. P., 1976, Roadcut geology in the San Andreas fault zone: California Geology, v. 29, no. 5, p. 99–104.

Wallace, R. E., 1949, Structure of a portion of the San Andreas rift in southern California: Geological Society of America Bulletin, v. 60, p. 781–806.

Potrero Canyon fault, landslides and oil drilling site, Pacific Palisades, Los Angeles, California

J. T. McGill, (retired) *U.S. Geological Survey, 11705 West 24th Place, Lakewood, Colorado 80215*
D. L. Lamar, Lamar-Merifield Geologists, Inc., 1318 Second Street #25, Santa Monica, California 90403
R. L. Hill, Geological Consultant, 3155 S. Barrington Avenue #B, Los Angeles, California 90066
E. D. Michael, Consulting Geologist, 6225 Bonsall Drive, Malibu, California 90265

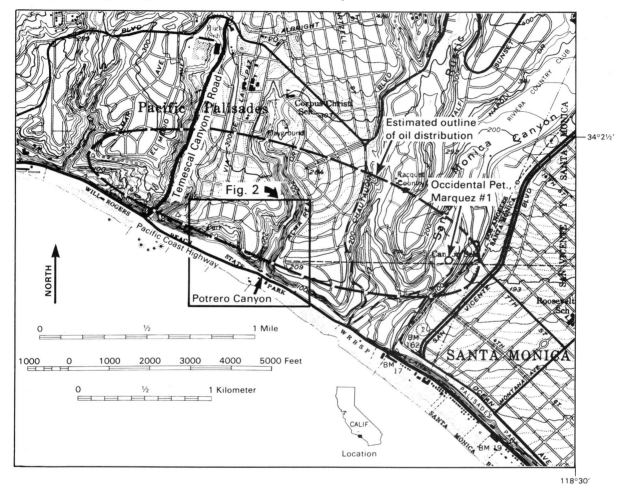

Figure 1. Index map of area showing access roads, limits of oil prospect (Anonymous, 1982), and outline of Figure 2; from U.S.G.S. Topanga Quadrangle.

LOCATION

This stop is located in the Pacific Palisades area of the City of Los Angeles, directly west of the Sunspot Restaurant and Motel, 15145 Pacific Coast Highway, between traffic signals at Chautauqua Boulevard, 0.5 mi, (0.8 km) to the southeast, and Temescal Canyon Road, 0.6 mi (1.0 km) to the northwest (Fig. 1). The mouth of Potrero Canyon is situated behind the motel. The site can be reached from central Los Angeles to the east via the Santa Monica Freeway (I-10), which continues west of the McClure Tunnel in the City of Santa Monica as the Pacific

Coast Highway (California 1), or from the Ventura-Oxnard area to the west via the Pacific Coast Highway through Malibu. Parking is available directly west of the restaurant-motel, or, for a fee, directly across the highway; during periods of beach use on summer weekends, parking may be difficult and heavy traffic may be a problem. Extreme care should be taken if it is necessary to cross Pacific Coast Highway.

SIGNIFICANCE

The mouth of Potrero Canyon is one of the most frequently visited sites in southern California for engineering geology field trips because the following may be seen: (1) Potrero Canyon fault exposed in a cliff north of the stop and in Potrero Canyon to the east; (2) warm-air ventilating system installed to stabilize the Huntington Palisades landslide on the southeastern side of the canyon at its mouth; (3) Via de las Olas landslide about 1,500 ft (460 m) to the northwest; and (4) proposed Occidental Petroleum Corporation Highway Drill Site directly southeast of the Via de las Olas landslide (Fig. 2).

The Potrero Canyon fault is an excellent example of a fault with evidence of late Pleistocene and possibly Holocene displacement. This fault is considered capable of generating an earthquake, which could trigger landslides on the nearby slopes. Rapid erosion and soil falls in the Huntington Palisades have been a hazard to the Pacific Coast Highway and residences overlooking the cliff. The Via de la Olas landslide is one of the largest landslides in the area. Because of concern over geologic hazards and public resistance to oil production in this exclusive suburban area, the drilling site has been a matter of political controversy. Thus, the area is of interest not only to geologists, but to all concerned with a safe environment in this seaside residential and recreational area of metropolitan Los Angeles.

SITE INFORMATION

The Potrero Canyon fault and adjacent area have been mapped by Hoots (1931), Johnson (1932; several of the remarkable plates in this classic engineering geology study are reproduced by Hill, 1979), and McGill (1973, 1982a, 1982b). Portions of the following are modified from Lamar (1978).

Potrero Canyon fault. The Potrero Canyon fault is a strand of the Santa Monica fault system, which forms the boundary of the Santa Monica Mountains of the Transverse Ranges Province on the north and the Los Angeles basin of the Peninsular Ranges Province on the south (Yerkes and others, 1965). The fault projects offshore at this location and is the probable eastward continuation of the Malibu Coast fault, which appears onshore 7.3 mi (11.7 km) to the west (Campbell and others, 1966, 1970; Yerkes and Campbell, 1980; McGill, 1980). The Dume (Anacapa) fault, a major strand of the Santa Monica fault system, is located offshore south of and parallel to the Malibu Coast and Potrero Canyon strands of the fault system (Green and others, 1975; Junger and Wagner, 1977).

The name Potrero Canyon and the fault at the mouth of the canyon are shown on Hoots' (1931) geologic map, but Hoots did not refer to the fault by name. Johnson (1932) called it the Potrero fault. The term Potrero Canyon fault was first used in a study of landslides along the Pacific Coast Highway (Anonymous, 1959). This name is preferred because a fault in the Newport-Inglewood zone is named the Potrero fault (Jennings, 1962). The following descriptions are from Johnson (1932), McGill (1980, 1982a), and Hill (1979).

The fault is well exposed on the cliff directly northwest of the restaurant-motel as a contact between Pliocene marine siltstone on the northwest and upper Pleistocene deposits on the southeast; there the fault strikes E–W to N80°E and dips 43° to

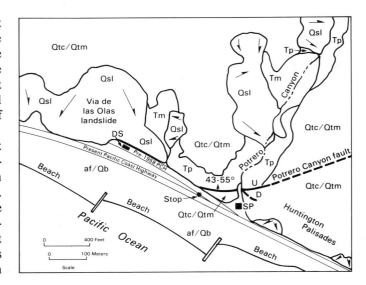

Figure 2. Geologic map of area adjacent to stop; generalized from McGill (1982a). See Figure 1 for location. Symbols: af/Qb, undifferentiated artificial fill and Quaternary beach deposits; Qsl, Quaternary landslide debris; Qtc/Qtm, undifferentiated upper Pleistocene marine and nonmarine terrace deposits; Tp, Pliocene siltstone; Tm, upper Miocene Modelo Formation; SP, Sunspot Motel, 15145 Pacific Coast Highway; DS, proposed Occidental Petroleum drill site.

55° N. Cobbles and pebbles along the fault plane have polished and striated flat surfaces; the striations are generally oriented parallel to the dip of the fault plane. The striations and relative ages of the rocks indicate that most recent displacement was reverse, northern side up. On the eastern wall of Potrero Canyon a near-vertical upper strand splits from the fault and dips 81° N.

The upper Pleistocene deposits consist of marine terrace sand and gravel overlain by nonmarine alluvial fan deposits. North of the fault the marine deposits rest on the Pacific Palisades wave-cut platform (Davis, 1933). Vertical separation of the marine deposits is about 16 ft (5 m) on the north-dipping main fault and about 95 ft (29 m) on the vertical branch fault. The total vertical shift, including fault drag adjacent to the main fault is about 154 ft (47 m). Strands of the vertical segment of the fault have been traced to within 13 ft (4 m) of the ground surface by McGill (1980) and, according to Johnson (1932), could be traced to the ground surface in 1932.

The marine terrace deposits are estimated to be about 125,000 years old and probably correlate with the highest eustatic late Quaternary sea level (Bloom and others, 1974). A fossil assemblage from the basal conglomerate of the marine deposits near the head of Potrero Canyon (Woodring in Hoots, 1931, p. 121–122; Valentine, 1956) is considered by George L. Kennedy (written communication, 1977) of the Los Angeles County Museum to represent a warm-water fauna correlative with the Palos Verdes Sand (Woodring and others, 1946). The overlying alluvial deposits probably were deposited during the following eustatic sea level fall, which may have lasted until about 116,000 yr B.P. (Bloom and others, 1974).

A well-defined but gentle scarp east of Potrero Canyon is aligned for 0.7 mi (1.2 km) with the inferred trace of the N75°E–trending vertical strand of the Potrero Canyon fault. The scarp height decreases eastward from about 26 ft (8 m) to 10 ft (3 m) or less; the width decreases from about 330 ft (100 m) to 98 ft (30 m). The scarp slopes about 5° south, whereas the terrace surface in the vicinity slopes about 2° south. From the physiographic evidence, it is estimated that as much as 26 ft (8 m) of the 111 ft (34 m) total vertical separation occurred after the alluvial cover was deposited. McGill (1980) believes that the remaining 85 ft (26 m) of vertical separation occurred between 125,000 and 116,000 yr B.P., which yields an average displacement rate of about 0.1 in/yr (3 mm/yr) during that period.

The scarp suggests the possibility of some Holocene displacement, and the Dume fault was the source of the 1973 M5.3 Point Mugu earthquake (Ziony and Yerkes, 1985). According to Greensfelder (1974), M7.5 is the maximum expected earthquake along the Santa Monica fault system. Thus, strands of the Santa Monica fault system, including the Potrero Canyon fault, are capable of generating significant earthquakes.

Huntington Palisades. Occasionally, especially during some rainy seasons, the cliffs of the Huntington Palisades experience rapid erosion. The cliffs are 110 to 180 ft (34 to 55 m) high and extend along Pacific Coast Highway from Chautauqua Boulevard in Santa Monica Canyon to Potrero Canyon. The materials forming the cliffs are poorly stratified upper Pleistocene alluvial fan deposits (McGill, 1982a), which are well exposed in the cliff adjacent to the highway at the mouth of the canyon on its southeastern side. The materials have a high coefficient of friction, low cohesion, and relatively high hydraulic conductivity (Anonymous, 1959). The deposits vary in strength resulting in ledges and overhangs. Failure of the cliffs is attributed to near-vertical joints and excess groundwater. The highway has been relocated several times due to soil falls and debris avalanches, and traffic disruption is common during and after intense rainfall.

The cliff east of Potrero Canyon failed in the fall of 1932, and 100,000 yd³ (76,500 m³) of debris blocked Pacific Coast Highway; an expensive residential estate on the edge of the cliff was damaged (Hill, 1934). The landslide developed along a southerly shallow-dipping clay bed. This is one of the few landslides known to be photographed during failure. Motion pictures were taken of the landslide from the roof of a restaurant, no longer in existence, on the opposite side of the highway.

To prevent further failure, a network of interconnected tunnels and drill holes were driven into a clay stratum at the base of the cliff, along which failure occurred; air heated by a gas furnace was blown through the network to dry out the clay and increase the shear strength (Hill, 1934). The mouth of the main tunnel is situated behind the metal building at the eastern side of the parking lot adjacent to the highway. Behind this metal building is a small concrete structure that houses the original heating equipment. Concrete portals of several of the tunnels still exist. As late

as 1956, the tunnels were open; however, now they probably are collapsed. The heater was turned off in 1941 and never reactivated. Extensive damp zones have been present at the base of the cliff at the site of the slide for the past 10 years; dampness in the clay may again be approaching levels where failure may occur.

Via de las Olas Landslide. The historical record of landsliding in the area south of Via de las Olas began about 1889 or earlier; at that time, available maps suggest a slide with vertical displacement of 100 ft (30 m). Development of the mesa above the slide area began in 1922, and the Pacific Coast Highway was regraded and realigned initially at that time. Movement at the toe of the slide began in 1930. Between 1931 and 1937, cut slopes along the highway failed and were regraded. In 1938, cracks developed along the roadway above the main slide mass. Cracks formed in 1942 as the result of the firing of artillery from an emplacement in the upper surface of the slump mass. Intermittent movement of the main slide continued until 1952, when movement accelerated; major landslides of 150,000 yd³ (115,000 m³) occurred in 1956 and 65,000 yd³ (50,000 m³) on March 27, 1958. The most spectacular movement occurred March 31, 1958, when 780,000 yd³ (600,000 m³) of debris buried the Pacific Coast Highway to a depth of 100 ft (30 m); the toe of the landslide reached the ocean. A highway maintenance supervisor was killed and several trucks and other pieces of heavy equipment were destroyed or damaged while removing debris from the previous landslide of March 27 (U.S. Army Corps of Engineers, 1976). The hapless highway employee and his truck are still entombed in the debris. This somewhat macabre note, although of little geologic relevance, should give pause to geologists working in the area, especially, one may imagine, to those engaged in exploratory drilling on the slide mass. The Pacific Coast Highway was rerouted to the south in 1958 and 1959.

Occidental Petroleum Drilling Site. In August 1966, Occidental Petroleum Corporation drilled an exploratory hole, the Marquez No. 1, to a depth of 9,271 ft (2826 m) near Elder Street and Entrada Drive in Santa Monica Canyon (Fig. 1; Anonymous, 1982). Natural gas and 26° gravity low-sulfur (0.7%) crude oil were discovered. In May 1970 the "Highway Drill Site" near the mouth of Potrero Canyon (Fig. 2) was acquired in a land exchange with the State of California and the City of Los Angeles. In July 1970 an application for a conditional use permit authorizing the drilling of an exploratory core hole on the Highway Drilling Site was granted by the Zoning Administrator, City of Los Angeles. However, upon appeal to the Board of Zoning Appeals, the right to drill the exploratory hole was denied.

In June 1972, Occidental Petroleum Corporation applied for the three drilling districts for production of oil and gas. In July 1972 the Planning Commission, after conferring with the Planning staff, City Attorney, and Assistant Petroleum Administrator, recommended approval of the three districts with special conditions providing for the drilling of two informational bore holes without commercial production. In October 1972 the City Council adopted the three drilling district ordinances with special conditions set by the Planning Commission.

After appeals and challenges in the courts by four nonprofit corporations, including No Oil, Incorporated, representing persons opposed to oil drilling in the Pacific Palisades, the California Supreme Court, in December 1974, said that the Council erroneously failed to render a written determination respecting the environmental effects of the drilling project before it approved the project, and that the City Council followed an erroneous test in deciding the drilling project did not require an environmental impact report. The State Supreme Court further stated: "The Superior Court shall set aside the ordinances establishing the oil drilling districts on the ground that the City in enacting these ordinances, failed to comply with the provisions of the California Environmental Quality Act."

To satisfy the California Environmental Quality Act, Occidental Petroleum Corporation prepared an Environmental Impact Report (EIR; Anonymous, 1982), which indicates the following. A 1.80mi-long (2.9 km) symmetrical anticline closure of about 600 acres lies beneath an elliptical-shaped area north of the Pacific Coast Highway and northwest of Santa Monica Canyon (Fig. 1); the sandstone reservoir is calculated to contain 60 million barrels of 26° gravity low-sulfur crude oil and natural gas in a 500-ft-thick (152 m) section at a depth of 9,500 ft (2896 m). Also offered in support of the EIR (Baker, 1982) are reservoir values for voidage, voidage-pore volume, and subsidence prior to repressuring. Michael (1982) has strenuously challenged the validity of extrapolating data from the EIR and from Baker (1982) to the entire postulated reservoir, because they are based on information from a single exploration core hole, Marquez No. 1 (Fig. 1) and because they are presented as demonstrated fact. At the time of this writing, the matter is yet to be presented to the California Coastal Commission.

REFERENCES CITED

Anonymous, 1959, Final report, Pacific Palisades landslide study: Moran, Proctor and Mueser and Rutledge, Consulting Engineers, New York, Report to California Department of Public Works, v. 1, text, 203 p.; v. 2, drawings, 72 sheets; v. 3, technical data, 194 p.

Anonymous, 1982, Draft Environmental Impact Report; Riviera oil drilling districts and proposed alternate drill sites: Ultrasystems report for Occidental Petroleum Corporation, Appendices A–X, 216 p.

Baker, R. H., 1982, Comments on the various subsidence reports and associated criticisms for the Riviera drilling districts and alternate drill sites EIR: California Division of Oil and Gas unpublished report, 17 p.

Bloom, A. L., Broecker, W. S., Chappell, J.M.A., Mathews, R. K., and Mesolella, K. J., 1974, Quaternary sea level fluctuations on a tectonic coast; New ^{230}Th/^{234}U dates from the Huon Peninsula, New Guinea: Quaternary Research, v. 4, p. 185–205.

Campbell, R. H., Yerkes, R. F., and Wentworth, C. M., 1966, Detachment faults in the central Santa Monica Mountains, California: U.S. Geological Survey Professional Paper 550-C, p. C1–C11.

Campbell, R. H., Blackerby, B. A., Yerkes, R. F., Schoellhamer, J. E., Birkeland, P. W., and Wentworth, C. M., 1970, Preliminary geologic map of the Point Dume Quadrangle, Los Angeles County, California: U.S. Geological Survey Open-File Map, scale 1:12,000.

Davis, W. M., 1933, Glacial epochs of the Santa Monica Mountains, California: Geological Society of America Bulletin, v. 44, p. 1041–1133.

Green, H. G., Clarke, S. H., Jr., Field, M. E., Linker, F. I., and Wagner, H. C., 1975, Preliminary report on the environmental geology of selected areas of the southern California Continental Borderland: U.S. Geological Survey Open-File Report 75-596, p. 50–66.

Greensfelder, R. W., 1974, Maximum credible rock acceleration from earthquakes in California: California Division of Mines and Geology, Map Sheet 23, 12 p.

Hill, R. A., 1934, Clay statum dried out to prevent landslips; Heated air forced through tunnels and drill holes to control earth movement: Civil Engineering, v. 4, no. 8, p. 403–407.

Hill, R. L., 1979, Potrero Canyon fault and University High School escarpment, *in* J. R. Keaton, chairman, Field guide to selected engineering geologic features Santa Monica Mountains: Association of Engineering Geologists, Southern California Section Annual Field Trip, May 19, 1979, p. 83–103.

Hoots, H. W., 1931, Geology of the eastern part of the Santa Monica Mountains, Los Angeles County, California: U.S. Geological Survey Professional Paper 165c, p. 83–134.

Jennings, C. W., 1962, Long Beach sheet, Geologic map of California: California Division of Mines and Geology, scale 1:250,000.

Johnson, H. R., 1932, Folio of plates to accompany geologic report, Quelinda Estate: Harry R. Johnson, Consulting Geologist, unpublished report, 25 pls.

Junger, A., and Wagner, H. C., 1977, Geology of the Santa Monica and San Pedro basins, California Continental Borderland: U.S. Geological Survey Miscellaneous Field Studies Map MF-820, scale 1:250,000.

Lamar, D. L., chairman, 1978, Geologic guide and engineering geology case histories, Los Angeles Metropolitan Area: Association of Engineering Geologists First Annual California Section Conference, May 12–14, 1978, p. 33–37.

McGill, J. T., 1973, Map showing landslides in the Pacific Palisades area, City of Los Angeles, California: U.S. Geological Survey Miscellaneous Field Studies Map MF-471, scale 1:4800.

——, 1980, Recent movement on the Potrero Canyon fault, Pacific Palisades area, Los Angeles, *in* Geological Survey research 1980; U.S. Geological Survey Professional Paper 1175, p. 258–259.

——, 1982a, Preliminary geologic map of the Pacific Palisades area, City of Los Angeles, California: U.S. Geological Survey Open-File Report 82-194, 15 p., map scale 1:4,800.

——, 1982b, Map showing relationship of historic to prehistoric landslides, Pacific Palisades area, City of Los Angeles, California: U.S. Geological Survey Miscellaneous Field Studies Map MF-1455, scale 1:4,800.

Michael, E. D., 1982, Review of geologic data, Final EIR No. 731-80-SUD (O) SUP, Occidental drilling project, City of Los Angeles: unpublished report for No Oil, Incorporated, September 13, 41 p.

U.S. Army Corps of Engineers, Los Angeles District, California, 1976, Report on landslide study, Pacific Palisades area, Los Angeles County, California: U.S. Army Corps of Engineers, Los Angeles District, California, main report, 30 p.; appendix 1 (prepared by J. T. McGill, U.S. Geological Survey), table, 89 p., map, scale 1:4,800.

Valentine, J. W., 1956, Upper Pleistocene mollusca from Potrero Canyon, Pacific Palisades, California: San Diego Society of Natural History Transactions, v. 12, no. 10, p. 181–205.

Woodring, W. P., Bramlette, M. N., and Kew, W.S.W., 1946, Geology and paleontology of Palos Verdes Hills, California: U.S. Geological Survey Professional Paper 207, 145 p.

Yerkes, R. F., and Campbell, R. H., 1980, Geologic map of east-central Santa Monica Mountains, Los Angeles County, California: U.S. Geological Survey Miscellaneous Investigations Series Map I-1146, scale 1:24,000.

Yerkes, R. F., McCulloh, T. H., Schoellhamer, J. E., and Vedder, J. G., 1965, Geology of the Los Angeles basin, California; An introduction: U.S. Geological Survey Professional Paper 420-A, p. A1–A57.

Ziony, J. I., and Yerkes, R. F., 1985, Evaluation of earthquake and surface-faulting potential, *in* Ziony, J. I., ed., Evaluating earthquake hazards in the Los Angeles region, An earth-science perspective: U.S. Geological Survey Professional Paper 1360, p. 56.

Sierra Madre thrust fault, Arcadia, California

Richard J. Proctor, *Richard J. Proctor, Inc., Engineering Geology, 327 Fairview Avenue, Arcadia, California 91006*
Richard Crook, Jr., *Consulting Geologist, 93 N. Sunnyside Avenue, Sierra Madre, California 91024*

LOCATION

One of the best exposures of the Sierra Madre thrust fault can be seen in the west wall of Santa Anita Canyon in Wilderness Park, city of Arcadia, 15 mi (25 km) northeast of downtown Los Angeles. Figure 1 shows how one can drive to the fault exposure by turning left into the first available parking area in Wilderness Park. The park gates are open 9:00 am to 5:00 pm daily.

The 54-mi-long (90 km) Sierra Madre fault zone forms the southern base of the San Gabriel Mountains along which they have been thrust over the valleys to the south. The 1971 San Fernando M 6.4 earthquake, caused by this fault, created 15 mi (25 km) of surface ruptures beginning 24 mi (40 km) west of this exposure.

At the west side of Arcadia's Wilderness Park (Fig. 2), banded gneiss is thrust over old alluvium containing large boulders. The fault consists of several feet of gouge and crushed rock generated from the gneiss; it dips 35 degrees north. The fault cannot be traced into the upper part of the old alluvium and probably has been inactive since it was deposited, roughly 2,000 to 5,000 years ago (Crook and others, 1978).

Two of the earliest geologists to describe the San Gabriel Mountains (Davis, 1927; Miller, 1928) believed the range was bounded on the south by normal faults. Mason Hill (1980) was first to show that the western part of the range was not created by the typical basin and range-type faulting.

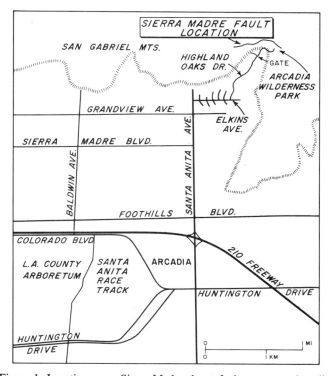

Figure 1. Location map, Sierra Madre thrust fault exposure, Arcadia, California.

Figure 2. Sierra Madre thrust fault exposure in Wilderness Park, Arcadia, California. Fault plane (35 dip) below man's arms. Gneiss at right overrides bouldery old alluvium.

In the late 1930s, John P. Buwalda (1940) mapped portions of the fault zone between La Canada and Monrovia. He was the first to describe the Sierra Madre fault as "a rather wide zone complexly braided as to pattern of fracture lines" rather than a single line trace.

It was not until the 1960s that portions of the fault zone were mapped in detail. This mapping was done by a group of Metropolitan Water District (MWD) of Southern California geologists of which the authors were a part. The 1:12,000 scale mapping was carried out between the Arroyo Seco on the west and Sycamore Canyon on the east and established 11 localities where crystalline basement rocks could be seen thrust over Quaternary alluvium. The Wilderness Park site was, to our best recollection, discovered during this project by Daniel C. Kalin and is judged by us to be the best easily accessible example.

The first published map showing the fault zone and the Wilderness Park site was produced by Douglas M. Morton (1973). His map shows this fault trace to branch into the interior of the mountain range rather than follow the southern base as does the main fault zone.

From 1975-78 the authors, along with C. R. Allen, B. Kamb, and C. M. Payne mapped the fault zone in greater detail, and trenched the fault, as part of a USGS grant to Caltech (Crook and others, 1978). The area mapped is a strip 1.2 to 2.4 mi (2 to 4 km) wide extending 24 mi (40 km) from Big Tujanga to San Gabriel Canyons.

The ^{14}C dates obtained from the Caltech work reveal that the central part of the Sierra Madre fault has not moved in 5,000 years. This contrasts with the 1971 earthquake to the west and the recent evidence that the Cucaonga branch to the east has a recurrence interval of about 800 years (Morton and others, 1982). Is the central part of the fault zone truly orders of magnitude less active, or has it accumulated enough strain energy to be currently worrisome?

REFERENCES CITED

Association of Engineering Geologists, 1983, Field Trip Guidebook Sierra Madre and associated faults: Association of Engineering Geologists, Southern California Section, Los Angeles (various pagination).

Buwalda, J. P., 1940, Geology of the Raymond Basin: Report to the Pasadena Water Department, Pasadena, California, 131 p.

Crook, Richard, Jr., Relative dating of Quaternary Deposits based on P wave velocities in weathered granitic clasts: Quaternary Research (in press).

Crook, R., Jr., Allen, C. R., Kamb, B., Payne, C. M., and Proctor, R. J., 1978, Quaternary geology and seismic hazard of the Sierra Madre and associated faults, western San Gabriel Mountains, California: Contribution No. 3191, Division of Geology and Planetary Science, California Institute of Technology, 117 p. (In press, USGS Professional Paper, tentative title: Thrust faults along the south front of the Transverse Ranges Province, Southern California.)

Davis, W. M., 1927, The rifts of southern California: American Journal of Science, v. 13, p. 57–72.

Ehlig, P., 1975, Geologic framework of the San Gabriel Mountains, in Oakeshott, G. B., ed., San Fernando, California, Earthquake of 9 February 1971: California Division of Mines and Geology Bulletin 196, p. 7–18.

Hill, M. L., 1930, Structure of the San Gabriel Mountains north of Los Angeles, California: University of California Publications in the Geological Sciences, v. 19, p. 137–170.

Jahns, R. H., 1973, Tectonic evolution of the Transverse Ranges Province as related to the San Andreas Fault System: Proceedings of the Conference on Tectonic Problems of the San Andreas Fault System, Stanford University Publication, v. XIII, p. 149–170.

Jahns, R. H., and Proctor, R. J., 1975, The San Gabriel and Santa Susana-Sierra Madre fault zones in the western and central San Gabriel Mountains, Southern California: Geological Society of America Abstracts with Programs, v. 7, no. 3, p. 329.

Lamar, D. L., Merifield, P. M., and Proctor, R. J., 1973, Earthquake recurrence intervals on major faults in southern California, in Geology, Seismicity, and Environmental Impact: Los Angeles, Association of Engineering Geologists, p. 265–276.

Miller, W. J., 1928, Geomorphology of the San Gabriel Mountains of California: University of California Publications in Mathematics and Physical Sciences, v. 1, no. 1, p. 1–114.

Morton, D. M., 1973, Geology of parts of the Azusa and Mt. Wilson quadrangles, San Gabriel Mountains, Los Angeles County, California: California Division of Mines and Geology Special Report 105, 21 p.

Morton, D. M., Matti, J. C., and Tinsley, J. C., 1982, Quaternary history of the Cucamonga fault zone, California: Geological Society of America Abstracts with Programs, v. 14, p. 184.

Proctor, R. J., Payne, C. M., and Kalin, D. C., 1970, Crossing the Sierra Madre fault in the Glendora Tunnel, San Gabriel Mountains, California: Engineering Geology, Amsterdam, Elsevier Publishing Co., v. 4, p. 5–63.

Proctor, R. J., Crook, R., Jr., McKeown, M. H., and Moresco, R. L., 1972, Relation of known faults to surface ruptures, 1971 San Fernando earthquake, southern California: Geological Society of America Bulletin, v. 83, p. 1601–1618.

Ventura Avenue anticline: Amphitheater locality, California

Robert S. Yeats and F. Bryan Grigsby, Department of Geology, Oregon State University, Corvallis, Oregon 97331-5506

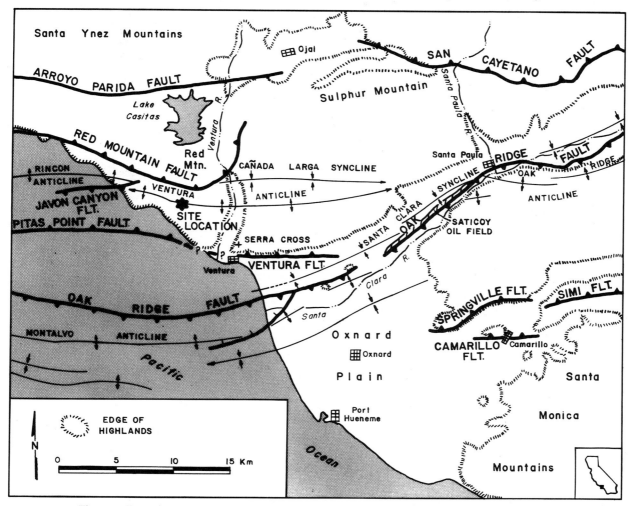

Figure 1. Tectonic map of the lower Santa Clara Valley in Ventura basin showing Ventura Avenue anticline and site location.

INTRODUCTION

The Ventura basin in the Transverse Ranges of southern California (Fig. 1) contains evidence for major folding and faulting in Quaternary time, deformation that is still in progress. Much of the evidence for age and deformation rates comes from the Ventura Avenue anticline that began to form about 200 Ka. The locality exposes the axis of the western continuation of this anticline and a folded reverse fault similar to larger ones known from subsurface information.

LOCATION AND ACCESS

The locality is in the San Miguelito oil field on a lease operated by Conoco Incorporated. Permission to take this trip

must be obtained from the Operations Director, Onshore for Conoco Incorporated, 290 Maple Court, Suite 284, Ventura, California 93003, phone 805-658-5800. The geologic map of the locality (Fig. 2) uses the Ventura 7½-minute Quadrangle as a topographic base.

From the junction of California 33 and U.S. 101 at Ventura, turn west on U.S. 101 toward Santa Barbara (Mileage 0). After crossing the Ventura River, exposures of stream terrace gravels of the lowest large terrace lie unconformably over south-dipping San Pedro (or Saugus) Formation on the right side of the road. Here, the San Pedro Formation contains massive conglomerate beds that resemble the terrace gravels. Continue on U.S. 101 1.2 mi (1.9 km) to the first exit, which is on the right, and continue beneath the freeway to the old Coast Highway, following it for another 2.9 mi (4.7 km). This route travels obliquely across the

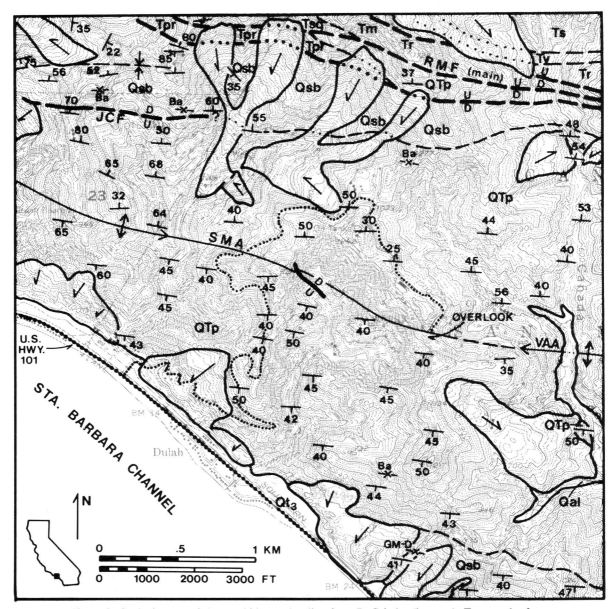

Figure 2. Geologic map of the amphitheater locality, from B. Grigsby (in prep.). Topography from Ventura 7½-minute Quadrangle. Ts, Sespe Formation (Oligocene); Tv, Vaqueros Formation (Oligocene); Tr, Rincon Formation (Miocene); Tm, Monterey Formation (Miocene); Tsq, Sisquoc Formation (Miocene-Pliocene); Tpr, Repetto Formation (Pliocene); QTp, Pico Formation (Pliocene-Pleistocene); Qsb, Santa Barbara Formation (Pleistocene); Qt₃, Seacliff terrace (Holocene); arrows show landslides. -x- marks volcanic ash exposures: Ba, Bailey ash; GM-D, Mono-Glass Mountain D ash, both from A. Sarna-Wojcicki. JCF, Javon Canyon fault; RMF, Red Mountain fault; SMA, San Miguelito anticline; VAA, Ventura Avenue anticline. Dots show field trip route to site.

south limb of the Ventura Avenue anticline; exposures to the right are San Pedro Formation. The route passes into the Santa Barbara Formation, predominantly mudstone, and eventually the Pico Formation, predominantly interbedded turbidite sandstone and mudstone (stratigraphic thickness and age in Fig. 3). Continue beyond Solimar to the road leading to the Conoco production office at San Miguelito oil field. Stop at this office to indicate

that you have permission to visit the locality. A Conoco employee may be assigned to accompany you.

Mileage 4.5 (7.2 km): Turn right onto the Conoco road, across the railroad tracks, through the freeway underpass, and follow oil company roads to the anticline overlook, about 3.6 mi (5.8 km) from the guardhouse. Watch for oncoming traffic; roads are narrow, steep and tortuous.

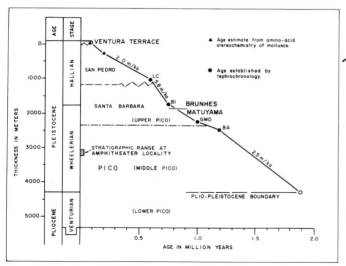

Figure 3. Stratigraphic thickness on south limb, Ventura anticline versus numerical ages, modified slightly from Lajoie and others (1982). BA, Bailey ash (1.2 m.y.); GMD, Glass Mountain-D ash (0.8–0.9 m.y.); BI, Bishop ash (0.7 m.y.); LC, Lava Creek ash (0.6 m.y.). Brunhes-Matuyama chron boundary from Liddicoat and Opdyke (1981). Stages are those of Natland (1952) with ages from Van Eysinga (1975). Stratigraphic nomenclature from Weber and others (1973) except for subsurface units (in parentheses).

SIGNIFICANCE OF SITE

The anticline contains the largest oil field in the Ventura basin, with recoverable reserves close to one billion barrels. The anticline trends east, oblique to the trend of the main Pliocene-Pleistocene trough of the Ventura basin, which is east-northeast from the coastline to the town of Santa Paula (Fig. 1). The site is important because the age of folding is closely constrained and is shown to be very young.

Figure 3, modified slightly from Lajoie and others (1982), shows the stratigraphic thickness of the south limb of the Ventura Avenue anticline, which is very close to the thickness measured at the San Miguelito site. These strata have been dated radiometrically, paleomagnetically, and by tephrochronology (Lajoie and others, 1979, 1982; Yeats, 1983). The Bailey ash (BA of Fig. 3) was dated as 1.2 ± 0.2 Ma using fission tracks in zircon (Izett and others, 1974). Blackie and Yeats (1976) determined that the Bailey ash datum at the Saticoy oil field near Santa Paula occurs in a reversely magnetized sequence between the Olduvai and Jaramillo events; the Jaramillo has not yet been recognized at Ventura. An ash bed near the top of the Santa Barbara Formation was correlated to the Bishop Tuff, 0.7 Ma, and an ash bed above the base of the San Pedro Formation was correlated to the Lava Creek ash (formerly the Pearlette Type O ash) dated as 0.6 Ma (Lajoie and others, 1982). The boundary between the normally magnetized Brunhes Chron and the reversely magnetized Matuyama Chron, 0.73 Ma in age, was identified by Liddicoat and Opdyke (1981) just below the Bishop Tuff. Amino-acid racemization age estimates from fossil shells in the San Pedro Formation

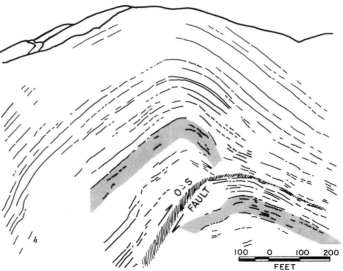

Figure 4. Line drawing of amphitheater site by B. Grisby (in prep.). Arrows show separation of reverse fault that curves over anticlinal crest. Shaded area marks a sequence of turbidites correlated across fault to determine offset. View is west.

show that the age of the top of the San Pedro is about 0.2 Ma (Lajoie and others, 1979, 1982).

Sometime after 0.2 Ma, folding of the Ventura Avenue anticline took place. The Ventura Terrace of Lajoie and others (1982) overlies the San Pedro with angular unconformity and is itself tilted southward 9°–15° (Lajoie and others, 1982; Sarna-Wojcicki and others, 1976). This relationship can be seen on the road to Junipero Serra Cross north of the San Buenaventura Mission in downtown Ventura (cf. Sarna-Wojcicki and others, 1976, and field trip guide by Yeats and others, 1982). Amino acid age estimates of this terrace are 85–105 Ka. A terrace 45 Ka in age, dated by the uranium-series method and by amino-acid racemization age estimates on molluscs (Lajoie and others, 1982) can be seen from U.S. 101 5.6 mi (9 km) northwest of the Conoco turnoff; it rests with angular unconformity on steeply dipping Plio-Pleistocene strata (Yeats and others, 1982).

Based on this age calibration, sedimentation rates calculated by Lajoie and others are 2 to 5.8 mm/y, after which time the anticline began to fold. If folding began immediately after the end of deposition, the south flank of the anticline would have rotated to its present south dip at an average rate of 3.4 μrad/y, and horizontal shortening during this time would have taken place at a rate of 20 mm/y (Yeats, 1983). Uplift rates of the axis of the anticline were calculated by Rockwell (1983) based on a soils chronosequence, in part calibrated by ^{14}C dating, developed on warped terraces of the antecedent Ventura River as it cut through the rising anticline. Uplift rates were 15 to 16 mm/y from 0.2 Ma to 80 Ka, 10.5 to 11.5 mm/y from 80 to 29.6 Ka, and 4.3 to 5.2 mm/y since 29.6 Ka. These rates are reasonable for a rootless fold that buckles and slides southward on a ductile decollement of fine-grained Miocene strata (Yeats, 1983); the

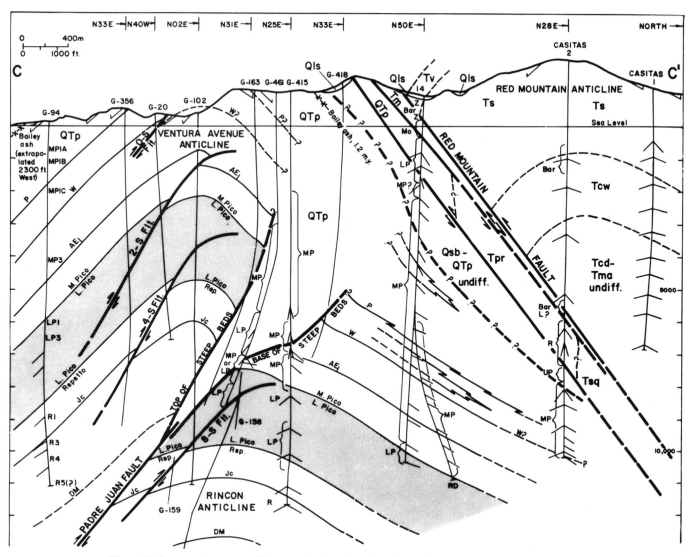

Figure 5. Cross section at amphitheater site showing subsurface control, stratigraphic position of site with respect to Bailey ash, and structural relations to deeper faults and to Red Mountain fault (from B. Grigsby, in prep.). Formation abbreviations in order of decreasing age: Tma, Matilija Formation (Eocene); Tcd, Cozy Dell Shale (Eocene); Tcw, Coldwater Sandstone (Eocene); Ts, Sespe Formation (Oligocene); Tv, Vaqueros Formation (Oligocene); Tm, Monterey Formation (Miocene); Tsq, Sisquoc Formation (Miocene-Pliocene); Tpr or Rep., Repetto Formation (Pliocene); QTp, Pico Formation (Pliocene-Pleistocene); Qsb, Santa Barbara Formation (Pleistocene); Qls, landslide deposits. Microfaunal zones: Z, Zemorrian; L, Luisian; Mo, Mohnian; R, Repettian; LP, MP, UP, Lower, Middle, Upper Pico. Other abbreviations are electric-log markers.

anticlinal crest would rise rapidly in the early stages of buckling but more slowly in the later stages (see also Currie and others, 1962; Adams, 1984).

The high rate of folding may account for two anomalous characteristics of the Ventura Avenue anticline. (1) It is overpressured, and the ratio of fluid pressure to lithostatic pressure approaches 1 with increasing depth (Duggan, 1964; McCulloh, 1969; Yeats, 1983). (2) Oil-water contacts are nonplanar and nonhorizontal (Schneider, 1972), suggesting that tilting occurs

too rapidly to allow the oil-water interface to reach equilibrium (Yeats, 1983).

SITE INFORMATION

The view is west toward the San Miguelito anticline, western continuation of the Ventura Avenue anticline. Folded strata are alternating sandstone and mudstone of the middle Pico of early Pleistocene age 1970 to 2460 ft (600–750 m) below the Bailey ash bed horizon (Fig. 3). Sandstones were deposited by turbidity currents in an elongate west-trending trough that shoaled to the north (Hsü, 1977) against a seaknoll on the site of the younger Red Mountain anticline (Yeats and others, 1986). Folding is by flexural slip; there is no change in thickness of strata between axis and limbs. Figure 4 is a sketch of the fold. A reverse fault dips south about 55° and has been found in the subsurface by B. Grigsby (in prep.; Fig. 5). The fault is curved over the crest of the anticline and becomes a bedding thrust on the north flank with a separation of 377 ft (115 m), measured along the fault plane. This fault is typical of many south-dipping reverse faults that curve over the crest of the Ventura Avenue anticline (Barnard fault set of Yeats, 1983). These faults formed during early stages of folding and were themselves folded across the crest during later stages of folding. The largest of these is the Padre Juan fault, which formed at the same time as the San Miguelito anticline such that the structure should be described as a fault-propagation fold (Fig. 5; B. Grigsby, in prep.; Suppe, 1985).

To the north, the Red Mountain reverse fault juxtaposes Oligocene redbeds of the Sespe Formation against fine-grained Miocene and early Pliocene strata of the Rincon, Monterey, and Sisquoc formations (Fig. 5). The fault is largely obscured by landslides. Farther south, there is a narrow exposure of north-dipping Santa Barbara Formation, and this formation is also exposed south of the anticlinal crest. The 1.2 Ma Bailey ash is preserved on both sides of the anticline, north and south of the site (A. Sarna-Wojcicki, unpublished map).

REFERENCES CITED

Adams, J., 1984, Active deformation of the Pacific Northwest continental margin: Tectonics, v. 3, p. 449–472.

Blackie, G. W., and Yeats, R. S., 1976, Magnetic reversal stratigraphy of Pliocene-Pleistocene producing section of Saticoy oil field, Ventura basin, California: American Association of Petroleum Geologists Bulletin, v. 60, p. 1985–1992.

Currie, J. B., Patnode, H. W., and Trump, R. P., 1962, Development of folds in sedimentary rocks: Geological Society of America Bulletin, v. 73, p. 655–674.

Duggan, D. E., 1964, Porosity variations in two deep Pliocene zones of Ventura Avenue anticline as a function of structural and stratigraphic position [M.A. thesis]: Riverside, University of California, 57 p.

Grigsby, F. B., in prep., Late Pleistocene deformation associated with lateral compression, western Ventura basin, California: To be submitted to American Association of Petroleum Geologists Bulletin.

Hsü, K. J., 1977, Studies of Ventura field, California, I, Facies geometry and genesis of lower Pliocene turbidites: American Association of Petroleum Geologists Bulletin, v. 61, p. 137–168.

Izett, G. A., Naeser, C. W., and Obradovich, J. D., 1974, Fission track age of zircon from an ash bed in the Pico Formation (Pliocene-Pleistocene) near Ventura, California: Geological Society of America Abstracts with Programs, v. 6, p. 197.

Lajoie, K. R., Kern, J. P., Wehmiller, J. F., Kennedy, G. L., Mathieson, S. A., Sarna-Wojcicki, A. M., Yerkes, R. F., and McCrory, P. F., 1979, Quaternary marine shorelines and crustal deformation, San Diego to Santa Barbara, California, *in* Abbott, P. L., ed., Geological excursions in the southern California area: San Diego, Department of Geological Sciences, San Diego State University, p. 3–15.

Lajoie, K. R., Sarna-Wojcicki, A. M., and Yerkes, R. F., 1982, Quaternary chronology and rates of crustal deformation in the Ventura area, California, *in* Cooper, S. D., compiler, Neotectonics in southern California: Anaheim, California, Cordilleran Section Field Trip Guidebook, 51, Geological Society of America, volume and guidebook, p. 43–51.

Liddicoat, J. C., and Opdyke, N. D., 1981, Magnetostratigraphy of sediments in the Atlantic Coastal Plain and Pacific Coast of the United States as an aid for dating tectonic deformation: Menlo Park, California, Technical Report Summary, Contract 14-08-0001-18377, U.S. Geological Survey.

McCulloh, T. H., 1969, Geologic characteristics of the Dos Cuadras offshore oil field: U.S. Geological Survey Professional Paper 679-C, p. 29–46.

Natland, M. L., 1952, Pleistocene and Pliocene stratigraphy of southern California [Ph.D. thesis]: Los Angeles, University of California, 165 p.

Rockwell, T. K., 1983, Soil chronology, geology and neotectonics of the north central Ventura basin [Ph.D. thesis]: Santa Barbara, University of California, 425 p.

Sarna-Wojcicki, A. M., Williams, K. M., and Yerkes, R. F., 1976, Geology of the Ventura fault, Ventura County, California: U.S. Geological Survey Miscellaneous Field Studies Map MF-781.

Schneider, J. J., 1972, Geological factors on the design and surveillance of water-floods in the thick structurally complex reservoirs in the Ventura field, California: Paper presented at the American Institute of Mining, Metallurgical, and Petroleum Engineers 47th Annual Fall Meeting, San Antonio, Texas.

Suppe, J., 1985, Principles of Structural Geology: Englewood Cliffs, New Jersey, Prentice-Hall, Incorporated, 537 p.

Van Eysinga, F.W.B., 1975, Geologic time table, 3rd edition: Amsterdam, Elsevier Scientific Publishing Company.

Weber, F. H., Jr., Cleveland, G. B., Kahle, J. E., Kiessling, E. F., Miller, R. V., Mills, M. F., Morton, D. M., and Cilweck, B. A., 1973, Geology and mineral resources study of southern Ventura County, California: California Division of Mines and Geology, Preliminary Report 14, 102 p.

Yeats, R. S., 1983, Large-scale Quaternary detachments in Ventura basin, southern California: Journal of Geophysical Research, v. 88, p. 569–583.

Yeats, R. S., Keller, E. A., Rockwell, T. K., Lajoie, K. R., Sarna-Wojcicki, A. M., and Yerkes, R. F., 1982, Field trip number 3; Neotectonics of the Ventura basin—road log, *in* Cooper, J. D., compiler, Neotectonics in southern California: Geological Society of America Cordilleran Section, 78th Annual Meeting April 19–21, 1982, Anaheim, California, volume and guidebook, p. 61–76.

Yeats, R. S., Lee, W. H. K., and Yerkes, R. F., 1986, Geology and seismicity of the eastern Red Mountain fault, Ventura County, California, *in* Morton, D. M., and Yerkes, R. F., eds., Recent reverse faulting in the Transverse Ranges, California: U.S. Geological Survey Professional Paper 1339 (in press).

Turbidite features in the Pico Formation (Pliocene) at Santa Paula Creek, California

E. J. Baldwin, *Department of Earth Sciences, El Camino College, Torrance, California 90506*

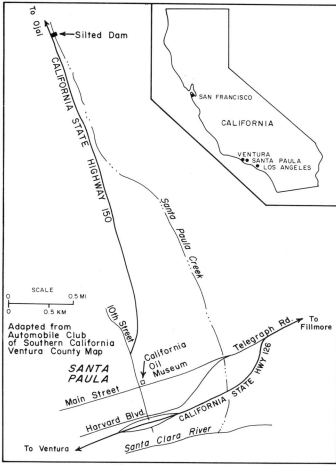

Figure 1. Location map.

Figure 2. Pliocene strata outcrops in Santa Paula Creek.

GEOLOGY

Much of the 13,000-ft (3,900 m) thick Pliocene section of the Ventura Basin consists of alternating mudstones, sandstones, and conglomerates. In their classic paper of 1951, Natland and Kuenen described the fauna and sedimentary features of these strata and presented evidence in favor of transportation and deposition of the sand from shallow water to bathyal depths by turbidity currents.

Foraminifera in the mudstones indicate a water depth at the time of deposition of 900-2,000 ft (270-600 m) (Natland and Kuenen, 1951; Natland, 1957). Some of the interbedded sandstones contain fossil shells of shallow-water molluscan fossils, yet do not exhibit other features typical of shallow-water deposits (Natland and Kuenen, 1951). The sedimentary structures exhibited in these strata that suggest emplacement by turbidity currents are: graded bedding (Fig. 3), load casts (Fig. 4), flame structures

LOCATION AND ACCESSIBILITY

Exposures of turbidite features in the Pico Formation are located in Santa Paula Creek along the east side of California State Highway 150, 3.1 mi (5 km) north of Main Street, Santa Paula (Fig. 1). Santa Paula is 14 mi (22.4 km) east of the city of Ventura, Ventura County, southern California.

At the roadside entrance to the site, there is a chain and a no trespassing sign. Permission to enter may be obtained from the Santa Paula Water Department (805-525-5591). The best exposures of the strata start at a silted dam and gauging station and continue south for about 600 ft (180 m). Santa Paula Creek crosses the steeply dipping strata at a right angle, thus providing an excellent cross section for study (Fig. 2).

In order to obtain quality close-up photographs of the features, take a brush, a knife or trowel, and a bucket (water is usually available from the stream) to clean up the exposures.

Figure 3. Graded bedding, Santa Paula Creek.

Figure 4. Load casts, Santa Paula Creek.

Figure 5. Flame structures, Santa Paula Creek.

Figure 6. Convolute bedding, truncated beds, and abrupt upward transition of mudstone to coarse sandstone, Santa Paula Creek.

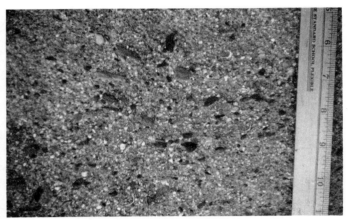

Figure 7. Shale pebbles in sandstone, Santa Paula Creek.

(Fig. 5), convolute bedding, truncated beds and abrupt upward transition of mudstone to coarse sandstone (Fig. 6), shale pebbles in the sandstones (Fig. 7), pull aparts and warped clay slabs (Fig. 8), fossil shells with preferred orientation (statistical study by Crowell and others, 1966), and partial Bouma sequences. These features are documented by sketches and photographs in Natland and Kuenen, 1951; Baldwin, 1959; Crowell and others, 1966.

For any geologist interested in turbidites, the Santa Paula Creek section is a worthwhile visit because of its accessibility, excellent exposure, and variety of features.

REFERENCES CITED

Baldwin, E. J., 1959, Pliocene turbidity current deposits in the Ventura Basin, California [M.A. thesis]: Los Angeles, California, University of Southern California, 66 p.

Crowell, J. C., Hope, R. A., Kahle, J. E., Ovenshine, A. T., and Sams, R. H., 1966, Deep-water sedimentary structures, Pliocene Pico Formation, Santa Paula Creek, Ventura Basin, California: California Division of Mines and Geology, Special Report No. 89, 40 p.

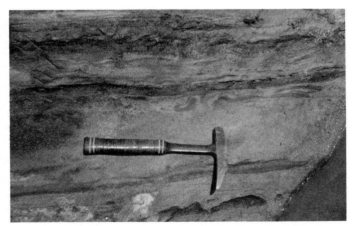

Figure 8. Pull aparts and warped clay slabs, Santa Paula Creek.

Natland, M. L., 1957, Paleoecology of West Coast Tertiary Sediments, *in* Ladd, H. S., ed., Treatise on marine ecology and paleoecology, v. 2, Paleoecology: Geological Society of America Memoir 67, p. 543–572.

Natland, M. L., and Kuenen, P. H., 1951, Sedimentary history of the Ventura Basin, California, and the action of turbidity currents: Society of Economic Paleontologists and Mineralogists Special Publication No. 2, p. 76–107.

Sedimentology of Cretaceous strata in Wheeler Gorge, Ventura County, California

T. W. Dibblee, Jr., 316 E. Mission St., Santa Barbara, California 93105

LOCATION AND ACCESS

Wheeler Gorge is the narrow part of the North Fork of Matilija Canyon that cuts through the Santa Ynez–Topatopa Range of the western Transverse Ranges, in Ventura County. This canyon is within the Los Padres National Forest.

Wheeler Gorge is accessible from California 33 which passes through it and Wheeler Springs, a small resort in this canyon 1 mi (1.6 km) west of Wheeler Gorge. Wheeler Springs is about 19 mi (30 km) north of Ventura on the coast and about 7 mi (11 km) northwest of Ojai in Ojai Valley (Fig. 1). At Wheeler Gorge the canyon is so narrow and steep-walled that three short tunnels had to be cut and three bridges built for California 33.

The North Fork of Matilija Creek and Wheeler Springs are shown on the Wheeler Springs 7½-minute Topographic Quadrangle of the U.S. Geological Survey.

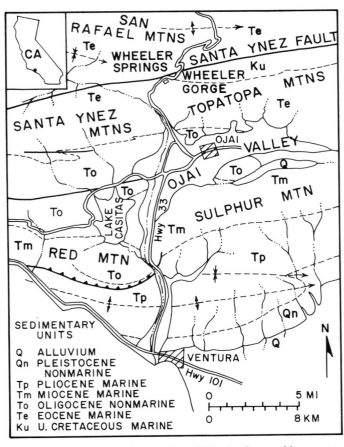

Figure 1. Index map showing location of Wheeler Gorge with respect to highways, towns, physiographic, and geologic features.

SIGNIFICANCE OF THE SITE

The western Transverse Ranges south of the Santa Ynez fault in western Ventura County expose the entire Upper Cretaceous–Cenozoic sedimentary series of the north flank of the Ventura basin, in a generally simple structure (Fig. 1). The total aggregate thickness of this varied sedimentary series, including one of the thickest Cenozoic sequences in the world, is about 42,000 ft (12,800 m).

The lower part of this enormously thick and nearly continuously deposited upended sedimentary series is a highly indurated marine clastic turbidite series of Eocene–Upper Cretaceous age that forms the east-striking Santa Ynez–Topatopa Range (Dibblee, 1982; Fig. 1). The Upper Cretaceous part of this series is superbly exposed along Wheeler Gorge. This part, exposed along the creek adjacent to California 33, has been much visited and studied by geologists. Their interpretations of the source and modes of transportation and deposition of these upper Cretaceous sedimentary rocks provide an important basis of the Late Cretaceous paleogeography of this region.

REGIONAL GEOMORPHOLOGY AND GEOLOGY

Wheeler Gorge is a deeply incised canyon eroded by the North Fork of Matilija Creek through the west end of the Topatopa Mountains. This fork joins Matilija Creek (main fork) below Wheeler Springs and drains southward via the Ventura River wash to the coast. The Santa Ynez–Topatopa mountain uplift was elevated on the south side of the 90 mi- (145 km) long Santa Ynez fault as a southward-tilted strip (Dibblee, 1982). Both forks of Matilija Creek, which drain the somewhat higher eastern San Rafael Mountains to the north, are antecedent to this uplift. Wheeler Gorge is the most narrow part of North Fork Canyon.

The stratigraphy and structure of the rock units of the Upper Cretaceous–Eocene marine turbidite sequence exposed along the North Fork of Matilija Canyon, including Wheeler Gorge, south of the Santa Ynez fault are shown on Figures 2 and 3, and the sedimentology of the Upper Cretaceous section exposed along Wheeler Gorge site is briefly described.

Unnamed Cretaceous marine strata. An unnamed marine clastic turbidite formation of Late Cretaceous age is the oldest formation exposed in the Topatopa Mountains in which about 4,600 ft (1,400 m) crops out. An unknown amount of additional lower strata of this formation are concealed. At Wheeler Gorge the uppermost 1,560 ft (475 m) of this formation crops out south of the Santa Ynez fault (Figs. 2 and 3).

At Wheeler Gorge this unnamed formation yielded a few marine fossils, including a large ammonite (*Desmoceras* sp.) and

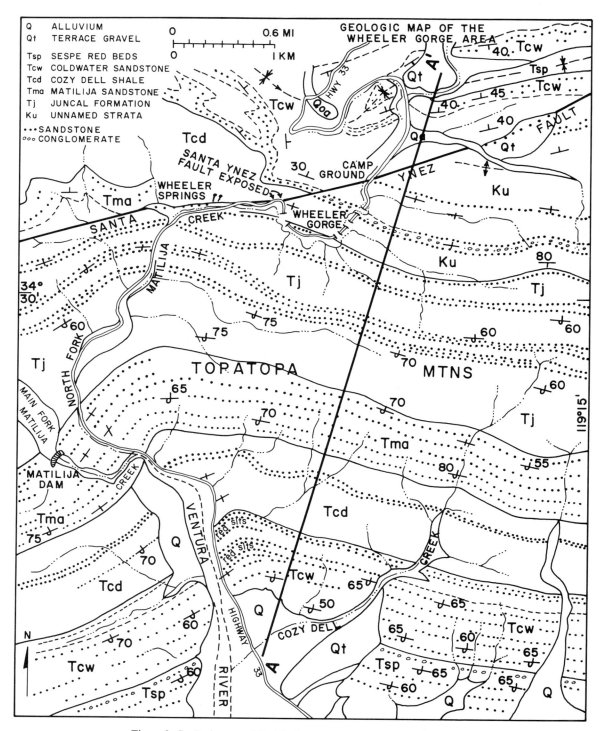

Figure 2. Geologic map of the North Fork of Matilija Creek and vicinity.

pelecypods (*Inoceramus subundatus* Meek) diagnostic of late Campanian–Maestrichtian age (Rust, 1966, p. 1390) or very late Cretaceous.

The Upper Cretaceous Formation was deposited essentially by turbidity currents probably on a shelf or fore-arc trough under a moderately deep sea of the eastern Pacific and derived from a continental basement terrane to the east.

Sedimentology of the Upper Cretaceous strata exposed at Wheeler Gorge. Along Wheeler Gorge most of the upper 1,560 ft (475 m) of the Upper Cretaceous Formation of the

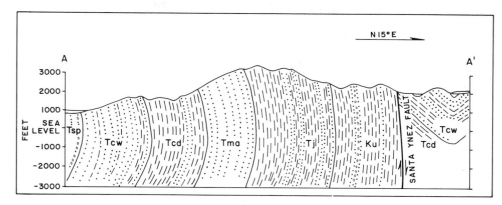

Figure 3. Cross section through the Topatopa Mountains near Wheeler Gorge. Line of section shown on Figure 2.

Topatopa Mountains is superbly exposed between the Santa Ynez fault and the overlying Juncal Formation. This section is composed of two shale sequences separated by a conglomerate member. The sedimentology of these three exposed units and their paleogeographic implications are described in detail by Rust (1966), Fisher and Mattinson (1968), and Walker (1975, 1985), from which the following paragraphs are summarized.

The shale or mudstone sequence exposed below (north of) the conglomerate member northward to the Santa Ynez fault is about 770 ft (235 m) thick. Walker (1985) divided this shale sequence into eight turbidite units. Each of these is composed of gray to black micaceous shaly mudstone that contains many very thin intercalations averaging 1 cm thick of light gray very fine-grained sandstone that grades upward into siltstone. This sequence differs from that above the conglomerate member in that the sedimentologic features generally present in the upper shale sequence are absent; the sandstone-siltstone intercalations are thinner and more sharply defined; and only the lowest exposed turbidite unit contains slumped beds.

Walker (1985) interprets the lower shale or mudstone sequence to have been deposited quietly on a generally smooth submarine basin-plain floor, probably far from any submarine channels.

The conglomerate member at Wheeler Gorge is about 350 ft (107 m) thick, and ranges from coarse cobble conglomerate to massive light gray arkosic sandstone and minor gray mudstone. The basal contact with the underlying shale sequence, well exposed at the north end of the north tunnel, is sharp and contains numerous flute-casts (photographed by Rust, 1966). Ripped-up angular mudstone fragments occur locally near the base and elsewhere in this conglomerate member. Walker (1985) divides the conglomerate of this exposure into three turbidite sequences, each of which grades upward from coarse, massive cobble conglomerate at the base through massive sandstone into fine-grained sandstone and interbedded mudstone at the top. The conglomerate is composed of smooth rounded pebbles and cobbles from 0.5 to 15 in (1 to 40 cm) in diameter in a matrix of arkosic sandstone.

Most are composed of gray quartzite, gray andesitic to dacitic porphyries, and hard granitic rocks, mostly quartz monzonite-granodiorite. Uncommon to rare clasts are of white quartz, gneiss, schist, gabbro, anorthosite, syenite, mylonite, gray chert, hard sandstone, and argillite (Rust, 1966, p. 1395).

The conglomerate member appears to be a submarine fan, deposited in a channel partly cut into the underlying shale-mudstone. The hard, durable rock types of the majority of clasts and their roundness indicates transport over great distances, possibly on land, or aggradation by wave action near shore, or both. The occurrence of this conglomerate in deep marine mudstone and its eastward pinch-out (east of Fig. 2) indicate it is not part of a deltaic fan. This condition suggests that this conglomerate may have been redeposited from a shallow near-shore deltaic fan, possibly by gravity, into its deep-sea environment, as suggested by Rust (1966), through deep-sea channels. Most paleocurrents and flute casts at Wheeler Gorge suggest the conglomerate was deposited under currents that flowed generally westward: the clast types indicate a continental basement source terrane to the east. The nearest sources of the quartzite and porphyry clasts are in the Mojave Desert region and the San Bernardino Mountains far to the east. Both of these areas are east of the San Andreas fault and may have been shifted by right slip on that fault farther southeast with respect to the conglomerate than they were in Late Cretaceous time. The rare clasts of gneiss, gabbro, anorthosite?, syenite, and mylonite reported by Rust (1966) were probably derived from the rocks now exposed in the San Gabriel Mountains east of the San Gabriel fault. Right slip on the San Gabriel fault since Late Cretaceous time has probably shifted the San Gabriel Mountains farther southeast with respect to the Wheeler Gorge area.

The shale or mudstone sequence exposed at Wheeler Gorge from above (south of) the conglomerate member to the base of the overlying Eocene Juncal Formation is about 760 ft (232 m) thick, of which about 520 ft (158 m) is exposed. It is composed of gray to black micaceous mudstone containing many thin intercalations of light gray fining-upward siltstone/very fine-grained sandstone. Walker (1985, p. 281) divides this sequence into two

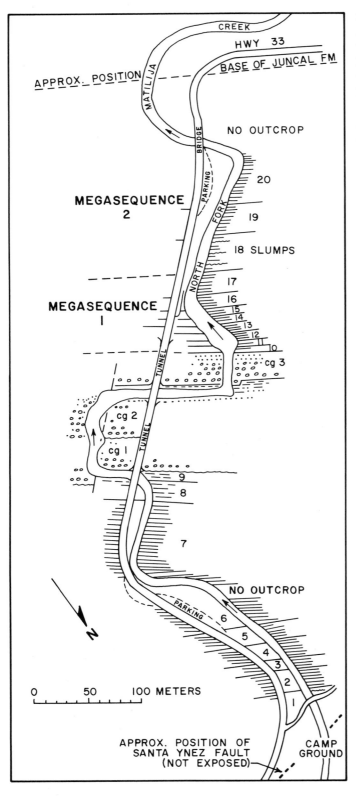

megasequences, each with a gradual upward decrease in siltstone/sandstone intercalations; he divides the lower one, 262 ft (80 m) thick, into eight turbidite units, each thinning and fining upward. Most of these contain thin Bouma-A, AC, and AB sandstone types, with ripped-up mud clasts in some A-types, convoluted laminae, scouring bed-lenticularity, and multiple slumped beds. From these sedimentology features, Nelson and others (1977) and Walker (1985) interpret the turbidite units as submarine channel-margin or levee deposits.

REFERENCES CITED

Dibblee, T. W., Jr., 1982, Geology of the Santa Ynez–Topatopa Mountains, southern California, *in* Fife, D. L., and Minch, J. A., eds., Geology and mineral wealth of the Transverse Ranges, California: South Coast Geological Society, Inc., p. 41–56.

Fisher, R. V., and Mattinson, J. M., 1968, Wheeler Gorge turbidite-conglomerate series, inverse grading: Journal of Sedimentary Petrology, v. 38, p. 1013–1023.

Nelson, C. H., Mutti, E., and Ricci-Lucchi, F., 1977, Upper Cretaceous resedimented conglomerates at Wheeler Gorge, California; Description and field guide–Discussion: Journal of Sedimentary Petrology, v. 47, p. 926–928.

Rust, B. R., 1966, Late Cretaceous paleogeography near Wheeler Gorge, Ventura County, California: American Association of Petroleum Geologists Bulletin, v. 50, p. 1389–1398.

Walker, R. G., 1975, Upper Cretaceous resedimented conglomerates at Wheeler Gorge, California; Description and field guide: Journal of Sedimentary Petrology, v. 45, p. 105–112.

——— , 1985, Mudstones and thin-bedded turbidites associated with the Upper Cretaceous Wheeler Gorge conglomerates, California; A possible channel-levee complex: Journal of Sedimentary Petrology, v. 55, p. 279–290.

Figure 4. Detailed map of the units of the unnamed Upper Cretaceous Formation exposed along Wheeler Gorge. In part modified from Walker (1985). Note orientation.

Coarse clastic facies of Miocene Ridge Basin adjacent to the San Gabriel fault, southern California

John C. Crowell, Department of Geological Sciences, University of California, Santa Barbara, California 93106

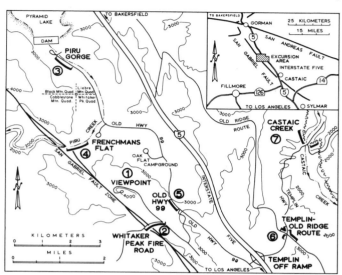

Figure 1. Location map of central Ridge Basin area, southern California, and recommended field trip stops.

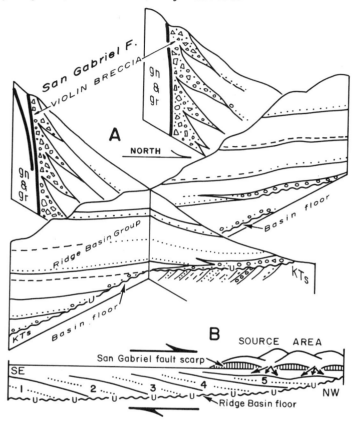

Figure 2. A, Isometric sketch showing shingling of strata within Ridge Basin and facies changes from the San Gabriel fault on the southwest, through the Violin Breccia, and into the basinal facies. Drawing is diagrammatic and not to scale. B, Sketch illustrating deposition across the San Gabriel fault scarp as the source area moves laterally with respect to the depocenter within Ridge Basin.

LOCATION AND ACCESS

The central part of Ridge Basin is easily accessible by automobile or bus from I-5, the eight-lane freeway between Los Angeles and Bakersfield, California (Fig. 1). Leave the freeway at Templin Highway and visit the localities in the order of their numbering, if time allows. If time is short, visit in this order: 4, 3, and 6. All seven localities can be visited in one full day and only locality 4 (Frenchmans Flat) requires walking, and then for only about 0.6 mi (1 km), into the narrow gorge southwestward along Piru Creek. Buses can get to all stops except 1 and 2. All are on public land administered by the U.S. Forest Service, including Oak Flat Campground, which has tables, piped water, and pit toilets. Most of the stops are on the Whitaker Peak 7½-minute Quadrangle.

DESCRIPTION AND SIGNIFICANCE

Ridge Basin, about 24 mi (40 km) long and 6 mi (10 km) wide, was filled with about 44,000 ft (13,500 m) of clastic sediments, both marine and nonmarine, during Late Miocene and Early Pliocene time. The basin formed within the San Andreas transform belt when the San Gabriel fault, bordering it on the southwest, was the principal active strand of the system. In Pliocene and Quaternary time, the region was deformed and uplifted and deeply eroded so that rapid facies changes and many types of sedimentary structures and features are now well displayed. The central part of the basin (Fig. 1) contains the best exposures, and

affords an opportunity for a visitor to appreciate sedimentation within a strike-slip regime where depocenters migrate laterally with respect to source areas. The sedimentary facies and tectonic relations are described in Crowell and Link (1982) and in maps and cross sections in Crowell and others (1982).

On the southwest the basin is bordered by the San Gabriel fault, which was continually active from about 11 to 4 Ma, as shown by the belt of coarse talus and debris-flow fanglomerates constituting the Violin Breccia. This formation has a total aggregate thickness of more than 36,000 ft (11,000 m) but extends laterally into the basin no more than 5,000 ft (1,500 m), where it interfingers with sandstone and shale along the basin trough. Near the fault, boulders and blocks up to 6.6 ft (2 m) in diameter are embedded within an earthy matrix. The clasts consist of gneisses and granitic rocks, derived from basement outcrops across the fault on the southwest, but the source area is now displaced to the

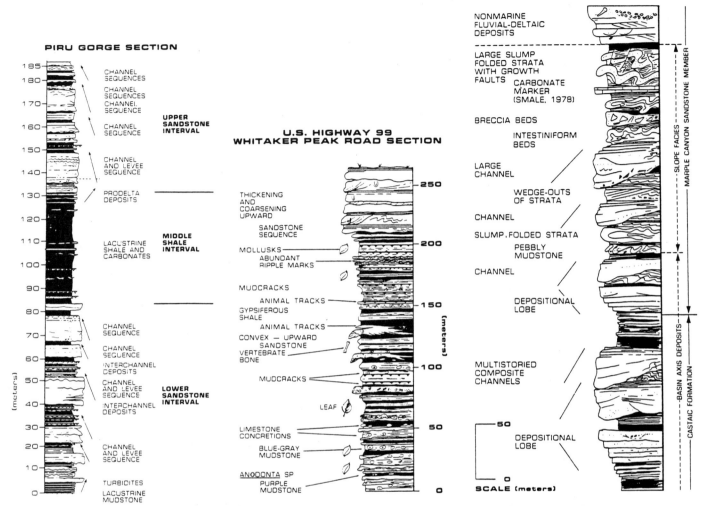

Figure 3. Stratigraphic columns at locality 3 (left), locality 5 (center), and locality 6 (right). From M. H. Link in Crowell and Link, 1982.)

northwest. This source area moved laterally alongside of the basin during the Late Miocene, dumping debris across the fault scarp into it. The result is a shingling or overlapping of younger units toward the northwest (Fig. 2). The facies changes of the coarse Violin Breccia—to debris-flow deposits and fanglomerates and on into sandstone and shale near the basin axis—are especially clear at locality 2 but are also evident at locality 4. The best accessible overall view of the basin is from the hilltop at locality 1. Dating of units comes from their stratigraphic position, from vertebrate and invertebrate fossils, from foraminifers in the marine section, and from magneto-stratigraphic studies.

Most of the sediment within Ridge Basin was derived from sources to the north and northeast, as shown by paleocurrent indicators and facies and thickness changes; only about 15 percent was derived from the southwest or Violin Breccia margin. These basinal facies are well displayed at localities 3 and 5 (Fig. 3), where they consist of lacustrine and fluviatile beds. At locality

6 the transition upward from marine beds, including shallow-water turbidites, into nonmarine fluvial-deltaic deposits is exposed. Slump structures, breccia beds, and intestiniform layers here are interpreted as perhaps caused by earthquake shocks during deposition. The unconformity at the base of the Ridge Basin Group is exposed at locality 7, where a coarse conglomerate lies on sandstone, conglomerate, and shale of the Cretaceous and Paleocene San Francisquito Formation.

REFERENCES CITED

Crowell, J. C., and Link, M. H., eds., 1982, Geologic history of Ridge Basin, southern California: Los Angeles, Pacific Section, Society of Economic Paleontologists and Mineralogists, 304 p.

Crowell, J. C., and others, 1982, Geologic map and cross sections of Ridge Basin, southern California: Los Angeles, Pacific Section, Society of Economic Paleontologists and Mineralogists, 3 sheets and explanation.

The San Andreas fault at Wallace Creek, San Luis Obispo County, California

Kerry Sieh, Division of Geological and Planetary Sciences, California Institute of Technology, Pasadena, California 91125
Robert E. Wallace, U.S. Geological Survey, 345 Middlefield Road, Menlo Park, California 94025

LOCATION

From either the coast or the interior, Wallace Creek is most easily reached by way of California 58 (Fig. 1). Precisely at the southwestern base of the Temblor Range, leave California 58 and drive southwest on an unmarked paved road about 0.2 mi (0.3 km) to a junction with an unpaved road (Fig. 2). Turn left (south) on this road and follow the San Andreas fault to Wallace Creek.

The unpaved road leading to Wallace Creek is impassable to all vehicles during and immediately following major storms. At all other times, two-wheel- or four-wheel-drive vehicles can be easily driven to within 1,300 ft (400 m) of the site. To avoid the gradual destruction of fragile tectonic landforms, park vehicles along the road near the section 33/34 boundary fence and walk the short distance north to the fault. During the dry season, be careful not to park on dry, flammable vegetation.

SIGNIFICANCE OF THE SITE

Rarely are tectonic landforms as well expressed and as well dated as they are at Wallace Creek. Along this 1.5 mi (2.5 km) length of the San Andreas fault are examples of most of the classic geomorphic features of strike-slip faults, including offset and beheaded channels, shutter ridges, and sags. The smallest features, including right-lateral offsets measuring about 33 ft (10 m), formed in association with the great Fort Tejon earthquake of 1857 (Sieh, 1978). The offsets of larger features, including a 3,700-year-old channel offset 430 ft (130 m), have grown by successive episodes of deformation (Sieh and Jahns, 1984; Wallace, 1968).

Radiometric dating of offset features at this locality demonstrates that the San Andreas fault has been slipping at an average rate of about 1.4 in/yr (34 mm/yr) for at least the past 13,000 years. This rate is only 60 percent of the total relative velocity between the North American and Pacific plates of 2.2 in/yr (56 mm/yr) determined by Minster and Jordan (1978). Hence, a significant fraction of the relative plate motion must be accommodated on other structures, and, contrary to popular belief, the plate boundary cannot be straddled by standing astride the San Andreas fault.

Division of this rate into the amount of offset accumulated during the past three large earthquakes yields recurrence interval estimates ranging between 240 and 450 years. These values are far greater than the current period of dormancy that began following the 1857 earthquake, and a great earthquake will probably not be produced by rupture involving this segment of the fault until at least 2100 A.D.

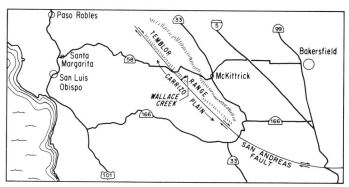

Figure 1. Location of Wallace Creek area. Wallace Creek drains southwestward out of Temblor Range and crosses San Andreas fault as it passes into Carrizo Plain.

SITE INFORMATION

The San Andreas fault is perhaps the most famous of all transform faults; certainly it is one of the most accessible. At this locality, features of the surficial trace of this major crustal structure are especially well displayed.

Tectonic Landforms

Figure 3 depicts landforms associated with the fault. Traces of the fault and mappable geologic units have been excluded from this topographic map in order to avoid obscuring the landforms. The location of the fault trace is indicated by small block triangles on the left and right margins of the map. Features discussed in the text are referenced to a 100-m grid marked from left to right on the upper and lower edge of the map.

Offset channels are one of the tectonic landforms most apparent on Figure 3. Between the 300- and 500-m marks is the channel of Wallace Creek, which is offset 430 ft (130 m) along the fault. Farther to the southeast, at the 2,250-m mark, is a younger channel, offset about 65 ft (20 m). Between the 690- and 900-m marks are four very small gullies that are offset about 30 ft (9.5 m).

At the 1,050-m mark is another small gully incised about 1.5 ft (0.5 m) into an alluvial fan. The portion of this gully upstream from the fault is no longer visible in the field. It was buried during a severe storm in February 1978. Prior to its burial, the gully displayed a measurable offset of about 30 ft (9.5 m).

Excellent examples of beheaded channels are at the 100-, 2,800-, 2,000-, and 2,100-m marks. Each of these has been displaced several hundred ft (m) from its source. The history of the

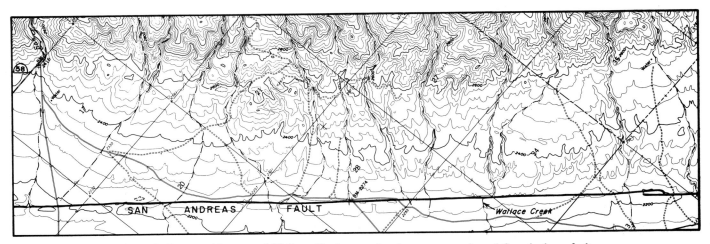

Figure 2. Topographic map of Wallace Creek area showing access roads and San Andreas fault. Location of San Andreas fault mostly from Vedder and Wallace (1970).

TABLE 1.   SMALLEST STREAM OFFSETS NEAR WALLACE CREEK AND PROPOSED INTERVALS
BETWEEN GREAT EARTHQUAKES (FROM SIEH AND JAHNS, 1984)

Stream Offsets (m)	Number of Measurements	Produced by	Slip associated with earthquake (m)	Proposed Interval Between Events (yr)
$9.5 \pm 0.5 (\pm 1\sigma)$	5	1857 event	$9.5 \pm 0.5 (\pm 1\sigma)$	
				240–320 [§]
21.8 ± 1.1	4[**]	1857 plus last prehistoric event	$12.3 + 1.2$[*]	
				300–440 [§]
32.8 or 33.5 ± 1.9	3[**] 2	1857 plus latest prehistoric events	11.0 or 11.7 ± 2.2[†]	
				240–450 [§]

[*]$21.8 - 9.5 \pm (0.5^2 + 1.1^2)^{1/2}$.
[†]$32.8 - 21.8 \pm (1.1^2 + 1.9^2)^{1/2}$ or $33.5 - 21.8 \pm (1.1^2 + 1.9^2)^{1/2}$.
[§]Slip during following earthquake in column 4 divided by average late Holocene slip rate (33.9 ± 2.9 mm/yr).
[**]Offset gullies are all between Wallace Creek and gully at 900-m mark in Figure 1.

latter two channels is discussed in more detail by Wallace (1968, p.17–18). A smaller, but very distinct, beheaded channel is visible at the 960-m mark. Very probably it was cut by streams that now flow out of the canyon at the 1,040-m mark. Much smaller beheaded gullies are present between the 570- and 870-m marks. These are barely visible in Figure 3, and some are difficult to see and interpret in the field because the channel segments near the fault are choked with debris that has been washed off the fault scarp.

These smallest beheaded and offset channels provide an important clue to the behavior of the San Andreas fault. The smallest measurable dislocations within the map area are small gullies offset about 30 ft (9.5 m). These dislocations probably formed during the great 1857 earthquake, which was produced by fault rupture along a 220-mi (360-km) or somewhat longer segment of the fault in southern California (Sieh, 1978; Agnew and Sieh,

1978). It appears that none of the 30 ft (9.5 m) has accumulated during the 20th century, because fences built across the fault in the Carrizo Plain at about the turn of the century display no offset whatsoever (Brown and Wallace, 1968). Also, the 30-ft (9.5-m) dislocations probably do not represent smaller offsets accumulated during two or more large earthquakes, because small gullies form very frequently in the Carrizo Plain. For example, a new gully formed at the 1,950-m mark in February 1978. It and many older post-1857 gullies in the Carrizo Plain display no offset at all (Sieh, 1978; Wallace, 1968). If large dislocations are produced at intervals of several decades or more, discrete populations of offsets should be observable.

In fact, larger dislocations in this area are rough multiples of the 30-ft (9.5-m) offset (Sieh and Jahns, 1984, Table 1). Several gullies are offset about 80 ft (22 m), and several more are offset about 100 ft (33 m). A logical conclusion is that the 1857 earth-

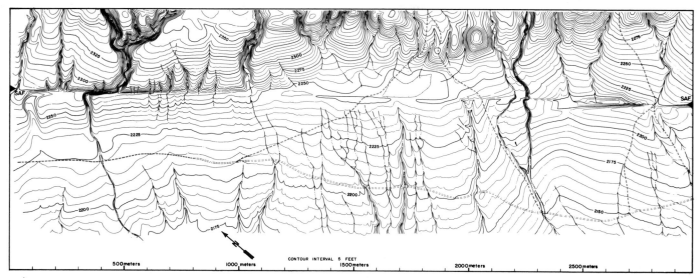

Figure 3. Topographic map of Wallace Creek area illustrates offset and beheaded channels, sag depressions, shutter ridges, and other tectonic landforms produced by recurring strike-slip movement along San Andreas fault. Elevations approximate the National Geodetic Vertical Datum (NGVD 29). Triangles (SAF) at left and right margins mark location of San Andreas fault.

quake was accompanied by about 30 ft (9.5 m) of slip and that its two predecessors were accompanied by about 40 and 36 ft (12.5 and 11 m) of slip.

Shutter ridges of various sizes are well represented along this segment of the fault. The term "shutter ridge" was originally employed by Buwalda (1936, p. 307) to describe topographic highs that had been moved across drainage courses along strike-slip faults. He envisioned these highs or ridges to have been carried by lateral fault slip into positions immediately downstream from existing drainages. The largest shutter ridge at this site can be found between the 1,100- and 1900-m marks. We are uncertain as to the origin of this shutter ridge: It may well represent a broad alluvial fan offset from a source southeast of the 2,300-m mark. Alternatively, it may be a more dynamic feature—perhaps an anticline with a southwest-plunging axial trace that has risen incrementally while being moved laterally over the past several thousand or more years. Smaller shutter ridges, composed of displaced alluvial and colluvial aprons, block the offset gullies between the 620- and 890-m marks.

Sediment is commonly "ponded" behind shutter ridges. A large region of active deposition currently exists upslope of the shutter ridge between the 1,100- and 1,900-m marks. Streams flowing into this region of topographic closure lose their carrying capacity on reaching this area of low gradient and drop their bed load. Three small alluvial fans are clearly visible on the northeastern margin of the depression. Although we have never witnessed it, water may, at times of high discharge, actually pond in this area of topographic closure behind the shutter ridge. This would enable deposition of silty and clayey suspended load as well. Eventually, alluvial fans building southwestward or quiet-water silts and clays filling the basin may be able to overtop the shutter

ridge, enabling reestablishment of a fault-crossing drainage there. This drainage would probably cross the fault at the low point in the shutter ridge at the 1,670-m mark, reoccupying the ancient beheaded drainge there. Smaller examples of "ponded" sediment occur immediately upstream from the fault in most of the small drainages between the 680- and 890-m marks. The small right-lateral jogs between the 680- and 890-m marks must result in lesser competence of these ephemeral streams immediately upstream from the fault. Perhaps this is because several meters of length, but virtually no height, are added to the long profile at the time of dislocation. Thus, dislocations add a section of shallower gradient at the fault. Alternatively, substantial losses in stream- or debris-flow velocity may occur at the fault, because the flow must abruptly bend to the right, around the small shutter ridges.

Between the 2,350- and 2700-m marks, two fault traces are arranged en echelon. Dextral slip has resulted in an increase in volume between the overlapping parts of these two faults, and a "sag" or depression has formed as the surface has dropped between the two fault planes. Spectacular evidence of this was evident immediately following a severe storm in mid-February 1978. During that storm, the three major channels that terminate in the sag delivered enough water to form an ephemeral sag pond about 12 ft (3.5 m) deep. Substantial erosion of the northwestern and southeastern channel beds also occurred at that time, and alluvial fans with deltaic fronts formed on the margins of the pond. Along the long margins of the pond, many large, elongate pits, some as much as 10 ft (3 m) deep and long, formed as the ponded water catastrophically drained into open fractures along both fault planes. Fresh pond-facing fault scarplets, up to about 6 in (15 cm) in height, formed along both long margins of the lake as near-surface debris was carried to greater depths by these waters and surficial blocks slumped in to fill the voids. Remnants of the delta fronts, collapse fissures, and fault scarps are still visible at the time of this writing (1985).

Fault scarps of several ages and degrees of activity are present within the area of Figure 3. The continuous high scarp that extends the full length of Figure 3 between zero and 300 m northeast of the fault trace is 30 ft (10 m) high and is still growing between the 490- and 650-m marks. The trace of the fault lies very near its base, and its lower slopes are much steeper than its upper slopes (about 35° versus about 5°). Sieh and Jahns (1984) demonstrated that this scarp has been growing since about 13,000 years ago, when the alluvial fan in which it formed ceased accumulating. The scarp has risen 10 ft (3 m) in the past 3,700 yrs, at an average rate of 0.03 in/yr (0.8 mm/yr). Between the 650- and 1,900-m marks, the fault scarp has either ceased growth or transforms into a broad monocline. This is indicated by several lines of evidence. First, the fault trace at the 1,900-m marks is farther away from the base of the scarp than at the 650-m mark. Second, the scarp is buried by greater volumes of debris toward the southeast (note the large alluvial fans between the 1,320- and 1,900-m marks). Third, the steepness of the lower portion of the scarp diminishes toward the southeast. Thus, this scarp appears to be growing in height northwest of and decreasing in height southeast of the 650-m mark. Thus, the block upstream from the fault appears now to be bulging upward in the northwest and subsiding in the southeast.

The scarps immediately adjacent to the fault trace alternate along strike from northeast-facing to southwest-facing: a common feature of strike-slip faults termed "scissoring." In some localities, scissoring is clearly a result of purely strike-slip offset of a non-planar ground surface. In the field, for example, scissoring of the fault scarp can be readily observed across the alluvial fan between the 960- and 1,100-m marks. On the northwestern half of the fan the fault scarp faces uphill, whereas on the southeastern half it faces downslope. This is best explained as strike-slip offset of the convex surface of the alluvial fan.

Wallace Creek

The geomorphology and stratigraphy of Wallace Creek, the large, prominent channel between the zero- and 700-m marks has been studied in detail by Sieh and Jahns (1984). By employing surficial geologic mapping, trenching, and radiocarbon dating, they were able to determine the age and evolution of Wallace Creek, and thereby determined a long-term slip-rate for the fault and made estimates of recurrence intervals for large earthquakes there. Between about 19,000 and about 13,000 years ago, the San Andreas fault at Wallace Creek traversed a broad, active alluvial fan (Fig. 4a). The fan surface was aggrading at a rate sufficient to bury fresh fault scarplets soon after their formation. About 13,000 years ago, perhaps due to climatic changes, Wallace Creek cut a channel into the fan and the fan surface ceased aggrading (Fig. 4b).

Subsequently, the fault has progressively offset the channel

several hundred ft (m). Twice the channel segment downstream from the fault has been abandoned, and a new downstream segment cut straight across the fault from the upstream segment (Fig. 4c, e). The oldest downstream segment now resides at the minus 10-m mark about 1,500 ft (475 m) away from its upstream continuation and just off the northwest edge of the map of Figure 3. Another former channel of Wallace Creek now lies beheaded at the 100-m mark. It was first cut by Wallace Creek about 10,000 years ago (Fig. 4c). Between 11,000 and 3,700 years ago this channel was offset about 250 m but was able to avoid beheading by maintaining a deep channel segment along to the fault, analogous to the segment of the presently active channel that follows the fault (Fig. 4d). During a period of merely a century or so, alluvium choked this segment of the channel. A terrace 23 ft (7 m) above the modern channel floor near the fault-crossing and between the 450- and 480-m marks is a remnant of the surface of this 3,700-year-old channel filling. This channel aggradation led to abandonment of the old channel and entrenchment of a new channel straight across the fault (Fig. 4e). In the past 3,700 years, this new channel has been offset 430 ft (130 m) to its present configuration (Fig. 4f).

SLIP RATE DURING THE LATE HOLOCENE AND ESTIMATION OF RECURRENCE INTERVALS

Knowing the date of the most recent entrenchment of Wallace Creek and the offset that accumulated since that entrenchment, one can calculate rather precisely a slip rate of 33.9 ±2.9 mm/yr (±2σ) for the San Andreas fault. The latest prehistoric dislocation of 40.6 ft (12.3 m) is estimated to have occurred between 1540 and 1630 A.D. The previous dislocation of about 36 ft (11 m) may have occurred between 1120 and 1300 A.D. If the interval between large events is indeed as long as is indicated by these estimates, the next event at Wallace Creek should not be anticipated until about 2100 A.D. or even later.

IMPLICATIONS OF THIS SITE FOR KINEMATICS OF SOUTHERN CALIFORNIA

Minster and Jordan (1978) determined from a circumglobal data set that the relative motion of the Pacific and North American plates has averaged about 2.2 in/yr (56 mm/yr) during the past 3 m.y. The geologic record at Wallace Creek shows that, at least during the past 13,000 yr, only about 1.4 in/yr (34 mm/yr) of this has been accommodated by slip along the San Andreas fault. If the 3-m.y. average is assumed to represent the Holocene average rate across the plate boundary as well, then clearly the San Andreas fault is accommodating only 60 percent of the relative plate motion. The reminder of the deformation must be accomplished elsewhere within a broader plate boundary. The San Gregorio–Hosgri fault system, which traverses the coast of central California, may have a late Pleistocene–Holocene slip rate of 0.2 to 0.5 in/yr (6 to 13 mm/yr) (Weber and Lajoie, 1977), and the Basin Ranges, to the east of the San Andreas fault, may

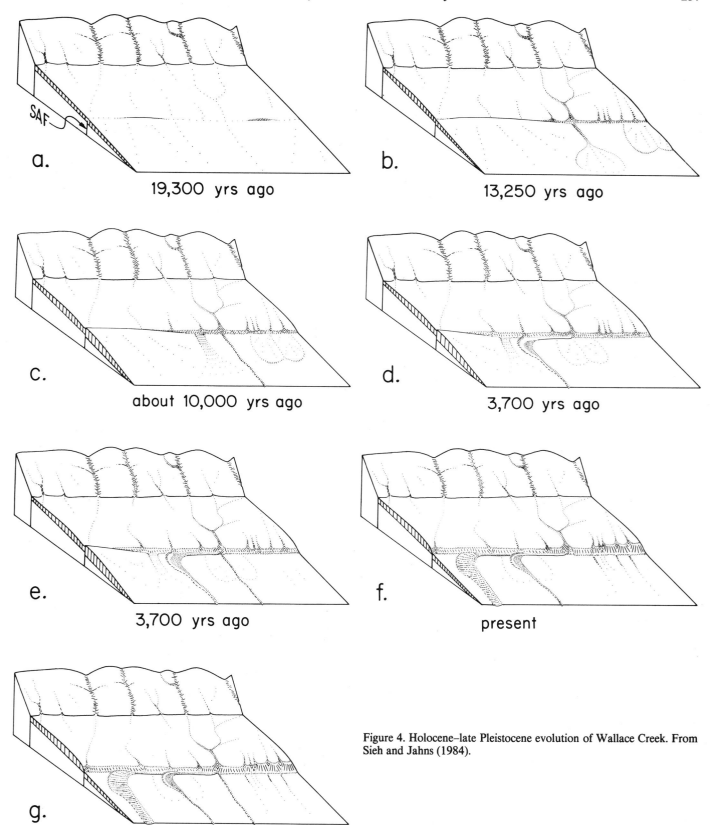

a. 19,300 yrs ago

b. 13,250 yrs ago

c. about 10,000 yrs ago

d. 3,700 yrs ago

e. 3,700 yrs ago

f. present

g. future

Figure 4. Holocene–late Pleistocene evolution of Wallace Creek. From Sieh and Jahns (1984).

be opening N.35°W. on oblique normal faults at a late Pleistocene–Holocene rate of about 0.3 in/yr (7 mm/yr) (Thompson and Burke, 1973). Most of the 2.24-in/yr (56-mm/yr) plate rate may thus be attributed to the San Andreas, San Gregorio-Hosgri, and Basin Range faults.

REFERENCES CITED

Agnew, D., and Sieh, K. E., 1978, A documentary study of the felt effects of the great California earthquake of 1857: Seismological Society of America Bulletin, v. 68, p. 1717–1729.

Brown, R. D., and Wallace, R. E., 1968, Current and historic fault movement along the San Andreas fault between Paicines and Camp Dix, California, *in* Dickinson, W. R., and Grantz, A., eds., Conference on Geologic Problems of San Andreas Fault System, Proceedings: Stanford, California, Stanford University Publications in the Geological Sciences, v. 11, p. 22–41.

Buwalda, J. P., 1936, Shutterridges, characteristic physiographic features of active faults [abs.]: Geological Society of America Proceedings 1936, p. 307.

Minster, J. B., and Jordan, T. H., 1978, Present-day plate motions: Journal of Geophysical Research, v. 83, p. 5331–5354.

Sieh, K., 1978, Slip along the San Andreas fault associated with the great 1857 earthquake: Seismological Society of America Bulletin, v. 68, p. 1731–1749.

Sieh, K. E., and Jahns, R. H., 1984, Holocene activity of the San Andreas Fault at Wallace Creek, California: Geological Society of America Bulletin, v. 95, p. 883–896.

Thompson, G. A., and Burke, D. B., 1973, Rate and direction of spreading in Dixie Valley, Basin and Range Province, Nevada: Geological Society of America Bulletin, v. 84, p. 627–632.

Wallace, R. E., 1968, Notes on steam channels offset by the San Andreas fault, southern Coast Ranges, California, *in* Dickinson, W. R., and Grantz, A., eds., Conference on Geologic Problems of the San Andreas Fault System, Proceedings: Stanford, California, Stanford University Publications in the Geological Sciences, v. 11, p. 6–21.

Weber, G. E., and Lajoie, K. R., 1977, Late Pleistocene and Holcene tectonics of the San Gregorio fault zone between Moss Beach and Point Ano Nuevo, San Mateo County, California [abs.]: Geological Society of America, Abstracts with Programs, v. 9, no. 4, p. 524.

Vedder, J. G., and Wallace, R. E., 1970, Map showing recently active breaks along the San Andreas and related faults between Cholame Valley and Tejon Pass, California: U.S. Geological Survey Miscellaneous Investigations Map I-574.

Paleocene submarine-canyon fill, Point Lobos, California

H. Edward Clifton, *U.S. Geological Survey, 345 Middlefield Road, Menlo Park, California 94025*
Gary W. Hill, *U.S. Geological Survey, 915 National Center, 12201 Sunrise Valley Drive, Reston, Virginia 22092*

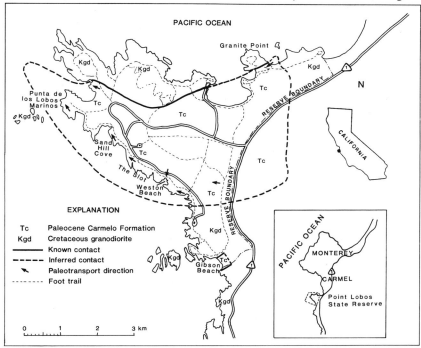

Figure 1. Distribution of the Carmelo Formation of Bowen (1965) and Cretaceous granodiorite at Point Lobos State Reserve. Geology modified from Nili-Esfahani (1965). Location names as shown in Point Lobos State Reserve literature.

LOCATION

Point Lobos, a prominent headland at the southern side of Carmel Bay on the central California coast (Fig. 1), is the site of a popular state reserve. Entrance to this reserve is from California 1, about 4 mi (6.4 km) south of the village of Carmel and 2.5 mi (4 km) southwest of the intersection of California 1 and Carmel Valley Road (County Road G16). Within the reserve, paved roads and well-maintained foot trails provide excellent access to many of the more prominent exposures (Fig. 1). Outcrops not served by foot trails are off-limits to the public; however, the geologically important exposures described herein are readily accessible.

Point Lobos State Reserve is beautifully maintained in a pristine condition by its staff, and the rules are strictly enforced. Most important, from a geologic standpoint, are strictures against collecting or disturbing any natural object within the reserve, so geological hammers are best left in vehicles. The rocks of the reserve are a striking esthetic resource and a mecca for amateur and professional photographers—they are not to be defaced.

The reserve opens in the morning (typically at 9:00) and closes before sundown. A nominal entrance fee is charged to visitors.

SIGNIFICANCE

The rocks at Point Lobos provide a beautifully exposed example of a filled part of a Paleocene submarine canyon carved into granodiorite of Cretaceous age. The fill, part of the Paleocene Carmelo Formation of Bowen (1965), consists mostly of pebble-cobble conglomerate. Interbeds of sandstone are common within the conglomerate, and a few intervals (up to 100 ft—30 m—thick) of pebble-free sandstone and mudstone exist within the fill. The rocks afford an excellent opportunity to examine evidence of the sedimentary processes that operated in such a setting, including the mechanisms of emplacement of the sand and gravel, the change of facies owing to fluctuation of sediment supply, the laterally shifting channel systems, and the trace fossils left by a prolific, exotic fauna.

SITE INFORMATION

The oldest rock exposed at Point Lobos is coarsely crystalline granodiorite. Radiometric dates indicate that this plutonic rock crystallized slightly more than 100 m.y. ago during the Cretaceous (Mattinson, 1978). Large, crudely aligned orthoclase crystals attest to a slow crystallization, probably under directed stresses.

Figure 2. Paleocene conglomerate at Punta de los Lobos Marinos, Point Lobos State Reserve. Note abutting of Paleocene strata against the lighter colored granodiorite in left background at the northern wall of the postulated submarine canyon.

Figure 3. Characteristic vertical and lateral textural variation within an organized conglomerate bed at Point Lobos.

Depositionally overlying the granodiorite, in a roughly east-west belt across the central part of the reserve, are conglomeratic sedimentary rocks—part of the Carmelo Formation of Bowen (1965). The few fossils that have been found in these rocks indicate a Paleocene age (Nili-Esfahani, 1965). The northern contact with the underlying plutonic rocks can be seen at close hand on either side of Granite Point (Fig. 1) and from a distance in its exposure east of Punta de los Lobos Marinos (Fig. 1). The coastal exposure of the contact in the southern part of the reserve is obscured by overburden; the orientation of the contact relative to stratification in the conglomerate here suggests a fault. An isolated sliver of the Carmelo Formation is downfaulted into the granodiorite at Gibson Beach near the southern boundary of the reserve (Fig. 1).

The depositional contact is steep, relative to stratification in the Carmelo Formation (Fig. 2); it is broadly sinuous (Fig. 1), and appears to have a relief of at least tens (and possibly hundreds) of meters. Pebble imbrication and ripple lamination in the sedimentary rocks indicate paleocurrents parallel to the granodiorite walls. The Carmelo Formation thus appears to fill a large valley cut into the plutonic rocks. The cove east of Granite Point (Fig. 1) contains a horizontal section across the valley floor, which slopes to the southwest.

The Carmelo Formation appears to represent the fill of a submarine canyon and probably accumulated in the middle to upper reaches of the canyon system. A gastropod found in mudstone interbedded with the conglomerate in the cove east of Granite Point suggests deposition in water depths greater than 330 to 660 ft (100 to 200 m) (Clifton, 1981), and the abundant sedimentary structures in the sandstone of the Carmelo Formation are devoid of evidence of surface wave effects. A well-developed trace-fossil assemblage suggests that deposition occurred at depths of 660 to 5,000 ft (200 to 1,500 m) (Hill, 1981).

The sediment in the canyon fill consists largely of conglomerate. Sandstone occurs as matrix to most of the conglomerate

and as interbeds within the conglomerate and, in the upper part of the succession at Point Lobos, within a few fining-upward sequences that culminate in interbedded mudstone and thin beds of sandstone. Mudstone is largely restricted to these finer intervals or to pebbly mudstone studded with dispersed pebbles and cobbles. Mudstone breccia occurs sporadically throughout the fill.

The clasts in the conglomerate are mostly well-rounded pebbles and cobbles of resistant composition, such as siliceous volcanic rocks (Nili-Esfahani, 1965). These clasts probably attest to a long and complicated depositional history in which deposition in a Paleocene submarine canyon is but the latest phase. Clasts composed of granodiorite are relatively uncommon; they tend to be anomalously large (as much as 10 ft—3 m—across) relative to other clasts and commonly are more angular. The sand-sized sediment in the fill, in contrast, tends to be feldspathic and reflective of a granodioritic source (Nili-Esfanhani, 1965).

Two fundamentally different types of conglomerate occur at Point Lobos. Disorganized conglomerate shows no internal stratification or alignment of clasts; the clasts in such conglomerate may be supported by one another, or they may be dispersed in a sandy or muddy matrix (pebbly mudstone). Such conglomerate is readily interpreted as the result of subaqueous debris flows where the dispersal of the clasts during transport is largely or totally due to the strength of the matrix material. Organized conglomerate, in contrast, is stratified and/or shows an alignment of clasts (long axes horizontal or imbricate); typically this conglomerate is clast supported. Its origins require mechanisms of transport and deposition other than simple debris flow.

The arrangement of clasts in many of the organized conglomerate beds (best seen where these beds are isolated in sandstone, but also discernible within rock composed entirely of conglomerate) suggests the nature of some of these mechanisms. A typical organized conglomeratic sedimentation unit forms a lense traceable for meters to a few tens of meters parallel to the transport direction and for a few meters transverse to the transport direction; the unit is some decimeters thick (Clifton, 1984). Well-defined textural variations exist in many of the organized conglomerate units, particularly where they are encased in sandstone (Clifton, 1984). Clasts in the conglomerate grade from fine to coarse upward within the bed (inverse grading) and in a down-transport direction toward the front of the deposit (Fig. 3). The clasts in the conglomerate commonly are aligned with long axes parallel to flow direction and inclined (imbricated) in the up-transport direction. Inversely and laterally graded conglomeratic sedimentation units are particularly evident on Punta de los

Lobos Marinos and on the point just south of Weston Beach (Fig. 1).

Conglomerate units that are encased in sandstone typically have sharp, well-defined bases and poorly defined tops. The overlying sandstone generally grades directly downward into the matrix of the conglomerate and thus appears to form the upper part of a sandstone-conglomerate couplet. Most sandstone beds show an upward-fining textural trend (normal grading). Many of the sandstone interbeds contain parallel lamination, in contrast to the conglomerate beds, which show no internal stratification. The sandstone member of a couplet commonly can be traced some meters in a down-transport direction beyond the terminus of the associated subjacent conglomerate. Such sandstone typically contains large isolated pebbles or cobbles apparently derived from the front, coarse "nose" of the conglomerate. Many of these clasts are aligned with long axes normal to transport and are imbricated. A few sandstone beds contain units about 4 in (10 cm) thick of foresets that dip steeply in a down-transport direction (as indicated by other paleocurrent indicators).

The inverse grading and flow-parallel, long-axis alignment within the conglomerate suggest transport just prior to deposition in a concentrated flow where intergranular collision among the pebbles and cobbles contributes to their dispersal as it moves. Pure "grain flow" (see Middleton and Hampton, 1976) of this type is deemed an inefficient mechanism for transport over long distances (Lowe, 1976), and accordingly the collisional sorting into inversely graded beds probably reflects either the last phase of transport from a high-density turbidity current or a "traction carpet" driven by an associated mass-flow of sand. The general absence of normal grading in the conglomerate suggests that turbulent sorting is uncommon and that the coarse clasts were rarely carried by turbulence in high-density turbidity currents. Such turbulence may have been a major factor, however, in the movement of the associated sand flows, the deposits of which typically are normally graded, and these flows may have carried the pebbles and cobbles in a colliding mass at the base of the flow. The sand flows apparently remained mobile after the gravel component "froze" into a deposit, rolling isolated coarse clasts a short distance from the down-transport nose of the conglomeratic deposit.

The floor of the canyon during intervals of predominantly gravelly sedimentation probably had an internal relief of 3.3 to 6.6 ft (1 to 2 m) produced by channeling and the small lobes left by individual gravel-bearing flows. The presence of down-canyon directed high-angle cross-bedding in associated sand beds implies that the down-canyon slope of the floor was no more than a few degrees.

Slump structures and penecontemporaneous deformational features are common within the fill. Some of these, such as the large-scale deformational features exposed in the cove in the western side of Granite Point (Fig. 1), may reflect lateral slumping from the walls of the canyon. Other features may be due to smaller-scale lateral slumping from the sides of small gullies or channels on the floor of the canyon. Two excellent examples of small rotational slumps of interbedded sandstone and mudstone lie within erosional recesses between steeply dipping resistant beds of sandstone just northwest of the base of the rock stairway that descends southward from the point on the southeastern side of Sand Hill Cove. These slumps, which locally cause nearly a meter of intercalated sandstone and mudstone strata to stand vertically between the overlying and underlying beds, clearly occurred at the sea floor and not interstratally. The sand and mud at the tops of the slumps were differentially eroded prior to (or during) emplacement of the overlying beds.

The Paleocene rocks at Point Lobos present an assemblage of trace fossils that is particularly prominent in successions of interbedded sandstone and mudstone. The faunal traces are best displayed in the exposure at Weston Beach, but they are present in the finer deposits throughout the Carmelo Formation at Point Lobos. A detailed description of the traces, which include among others *Planolites, Ophiomorpha, Chrondrites, Thalassinoides, Arenicolites,* and *Scolicia,* is presented by Hill (1981). One of the most striking traces is a complicated burrow that was previously identified as the imprint of seaweed in the rocks (Herold, 1934; Nili-Esfahani, 1965). The nature of the burrow (Fig. 4) differs depending on the arrangement of sandstone and mud interbeds relative to its main tunnel, and the trace accordingly is manifested in a surprising variety of ways depending on the location and orientation of exposure relative to the main tunnel (Hill, 1981). The nature of the burrowing organism is unknown; the trace has not been described elsewhere from the rock record.

Characteristics of the trace-fossil assemblage, such as taxonomic composition, diversity, abundance, behavioral and preservational types, and general bioturbation patterns, are useful in subdividing the rocks into specific depositional facies. In addition, the ichnoassemblage represents a mixing of "shallow-water" and "deep-water" types, leading to speculation that deposition occurred in water depths no deeper than upper to mid-bathyal (600–5,000 ft; 200–1,500 m).

Distinctive fining-upward successions are present in the upper part of the Paleocene strata exposed on the western shoreline of the reserve. Where complete, these show an upward progression from conglomerate through pebbly, then nonpebbly, sandstone into mudstone with thin sandstone interbeds. Some, as at "The Slot" midway between Sand Hill Cove and Weston Beach, are abruptly overlain by a thick succession of conglomerate.

The best developed fining-upward sequence occurs at Weston Beach; it appears to be the stratigraphically highest part of the Carmelo Formation in its seacoast exposure at Point Lobos. The section at Weston Beach (Figs. 5 and 6) is more than 100 ft (30 m) thick and is exposed on either flank of a faulted asymmetric syncline that plunges to the west. The rocks grade progressively upward through conglomerate, pebbly sandstone, thick-bedded nonpebbly sandstone, and thin-bedded sandstone, to mudstone with thin sandstone interbeds (Fig. 6).

The lower part of the sandstone section consists mostly of broadly lenticular beds of sandstone generally decimeters thick.

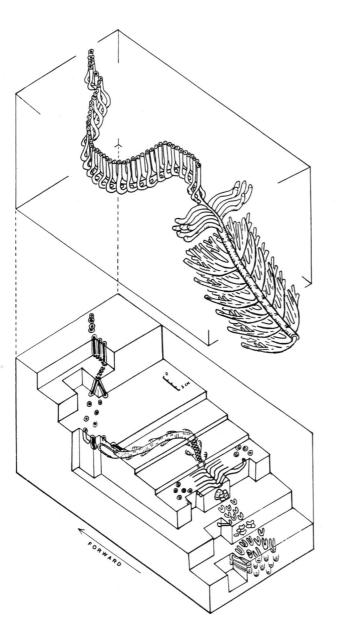

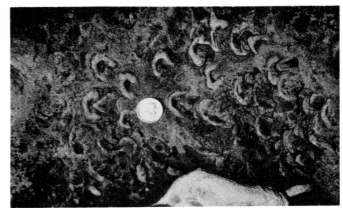

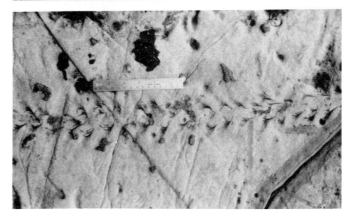

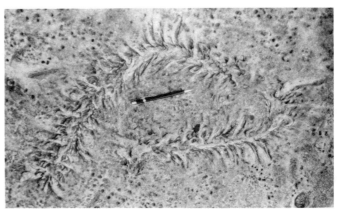

Figure 4. Trace fossils from the Carmelo Formation of Bowen (1965) at Point Lobos. Left: Schematic diagram showing various manifestations of the complex unnamed trace that is particularly abundant in the sandstone-mudstone succession at Weston Beach. Right: Examples of the different manifestations of this unnamed trace.

Figure 5. Fining-upward succession at Weston Beach. Note upward-thinning of sandstone beds.

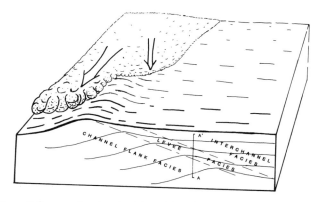

Figure 7. Interpretive sketch of the origin of the fining-upward succession at Weston Beach. Note divergence of flow where turbidity current spills over the levee and nature of vertical sequence (A-A′) produced as channel migrates laterally during sedimentation.

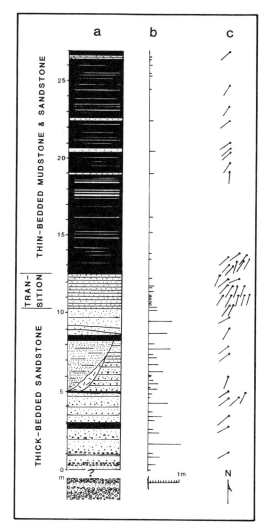

Figure 6. Sedimentary sequence as measured on the north side of Weston Beach showing (a) lithologic succession, (b) thickness and distribution of sandstone beds more than 5 cm thick, and (c) paleocurrent direction.

Most of these beds are visibly graded and show structureless to laminated (T_{a-b}) or structureless to laminated to rippled (T_{a-b-c}) Bouma sequences. A few thin beds of mudstone (or intervals of thinly interbedded mudstone and sandstone) drape over the sandstone beds. These finer-grained intervals are remarkably consistent in thickness, internal stratification, and trace-fossil assemblage over the extent of their exposure. They appear to represent episodes of relatively slow sedimentation between the flows that deposited the thicker beds of sandstone.

Paleocurrents in the thick-bedded sandstone are indicated by ripple lamination, by ripples on bedding surfaces, and—near the base of the section—by pebble imbrication. Flow was consistently toward the southwest (Fig. 6), in marked contrast to transport directions to the northwest that prevail in the conglomerate below the fining-upward sequence and in most of the other exposures along the western shore of Point Lobos (Fig. 1).

The thick-bedded sandstone is overlain by several meters of sandstone in which the beds are consistently in the range of 2 to 6 in (5 to 15 cm) thick. Most of these beds are graded and show laminated to rippled (T_{b-c}) sequences. The strata are relatively undisturbed by bioturbation. Although the contact between the two is not erosional, the thin-bedded sandstones visibly dip more steeply to the south than do the underlying strata (Fig. 5). Ripples and ripple lamination indicate paleocurrents that deviate to the south by about 30° relative to those in the section below (Fig. 6).

The thin-bedded sandstone grades up into mudstone with numerous thin beds of sandstone, nearly all of which are less than 2 in (5 cm) thick (Fig. 6). The thin sandstone interbeds are graded and/or ripple-laminated; isolated ("starved") ripples are common. Bioturbation is intense throughout this facies and in some parts of the section totally disrupts the stratification. Small, red-weathering concretions mark many horizons within the mudstone. The direction of paleocurrents in this fine-grained rock parallels that in the thick-bedded sandstone in the lower part of the section.

The upward-fining succession resembles that which would

be produced by a northerly shifting channel on the floor of the canyon (Fig. 7). If so, the thick-bedded sandstone accumulated on the north-facing channel margin. The lenticularity of the sandstone beds suggests deposition from flows other than pure turbidity currents, perhaps from a form of fluidized flow (Middleton and Hampton, 1976). Deposition of the thin-bedded sand on a south-facing levee margin (Fig. 7) would explain the divergence of attitude and paleocurrent direction in these beds relative to those below. The internal structures suggest a greater influence of tractive currents in these presumed levee deposits. The mudstone facies that caps the sequence is interpreted as an interchannel deposit. The thinness of the sandstone beds in this facies and the general intensity of bioturbation suggest that sand deposition was dominantly from relatively infrequent flows that exceeded the capacity of the channel. The direction of the paleocurrents relative to those in the subjacent conglomerate implies that the postulated channel wandered sinuously across the floor of the submarine canyon.

A slump higher in the canyon may have diverted currents enough to cut temporarily into the accretionary bank of the channel at Weston Beach. Beds in the upper part of a thick-bedded sandstone on the northern side of the cove are truncated by a steep erosional surface several meters high (Fig. 8). A mudstone-clast breccia forms talus at the base of the cut and intertongues with parallel-laminated and ripple-bedded sand away from the cut. The sand, which shows virtually no bioturbation, probably accumulated rapidly from the erosion of the graded beds at the margin of the cut. Sediment carried by the eroding currents seemingly bypassed this location. A graded and somewhat bioturbated sandstone bed of irregular thickness that caps both the fill and the cut wall represents the first down-canyon flow to be deposited in this site after the episode of erosion into the bank.

It is unclear whether the fining-upward sequences represent facies that persisted through time along the margins of channels through which gravel moved and was deposited or are facies that occupied the entire canyon floor during episodes of nondeposi-

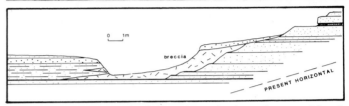

Figure 8. Top: Photograph of channel margin shown between 16 and 26 ft (5 and 8 m) in the lithologic column of Figure 6. Bottom: Sketch of lithologic relations at this channel margin. Strata shown in original horizontal position.

tion of gravel. The absence of the sequences in the superb exposures of much of the section on Punta de los Lobos Marinos and the increasing prevalence of these sequences in the upper part of the section suggest that they formed episodically, perhaps during temporary high stands of the sea when gravel deposition was suppressed.

In summary, Point Lobos is a site highly deserving of the attention of anyone interested in sedimentary geology. The reserve presents a superbly exposed array of unusual rocks in a gorgeously scenic setting. It is well worth a half or whole day of study. Bring lots of film.

REFERENCES

Bowen, O. E., 1965, Stratigraphy, structure, and oil possibilities in Monterey and Salinas Quadrangles, California, *in* Rennie, E. W., Jr., ed., Symposium of papers: Bakersfield, California, Pacific Section, American Association of Petroleum Geologists, p. 48–69.

Clifton, H. E., 1981, Submarine canyon deposits, Point Lobos, California, *in* Frizzell, V., ed., Upper Cretaceous and Paleocene turbidites, central California Coast: Pacific Section, Society of Economic Paleontologists and Mineralogists, Guide Book to Field Trip No. 6, p. 79–92.

—— , 1984, Sedimentation units in stratified deep-water conglomerate, Paleocene submarine canyon fill, Point Lobos, California, *in* Koster, E. H., and Steele, R. J., eds., Sedimentology of gravels and conglomerates: Canadian Society of Petroleum Geologists Memoir 10, p. 429–441.

Herold, C. L., 1934, Fossil markings in the Carmelo Series (Upper Cretaceous[?]), Point Lobos, California: Journal of Geology, v. 42, p. 630–640.

Hill, G. W., 1981, Ichnocoenoses of a Paleocene submarine-canyon floor, Point Lobos, California, *in* Frizzell, V., ed., Upper Cretaceous and Paleocene

turbidites, central California Coast: Pacific Section, Society of Economic Paleontologists and Mineralogists, Guide Book to Field Trip No. 6, p. 93–104.

Lowe, D. R., 1976, Grain flow and grain flow deposits: Journal of Sedimentary Petrology, v. 36, p. 188–199.

Mattinson, J. M., 1978, Age, origin, and thermal histories of some plutonic rocks from the Salinian block of California: Contributions to Mineralogy and Paleontology, v. 67, p. 233–245.

Middleton, G. V., and Hampton, M. A., 1976, Subaqueous sediment transport and deposition by sediment gravity flows, *in* Stanley, D. J., and Swift, D.J.P., eds., Marine sediment transport and environmental management: New York, John Wiley and Sons, p. 197–218.

Nili-Esfahani, A., 1965, Investigation of Paleocene strata, Point Lobos, Monterey County, California [M.A. thesis]: Los Angeles, University of California, 228 p.

Jadeitized Franciscan metamorphic rocks of the Pacheco Pass-San Luis Reservoir area, central California Coast Ranges

W. G. Ernst, Department of Earth and Space Sciences, University of California, Los Angeles, California 90024

LOCATION

The Pacheco Pass-San Luis Reservoir area is located in the central Diablo Range along the crest and eastern flank of the uplift. The site is transected by California State Highway 152, a four-lane, divided roadway near and east of the summit (elevation 1,368 ft; 410 m). The area is situated about 120 mi (200 km) southeast of San Francisco and 285 mi (450 km) northwest of Los Angeles. The location is illustrated in Figure 1. Access is provided by Highway 152, and a portion of the former road; four of the six outcrops referred to in this report are roadcuts. Because of heavy automobile and truck traffic on Highway 152, special care must be exercised when examining the rocks, and vehicles should be parked in turnouts or on the shoulder well off the pavement.

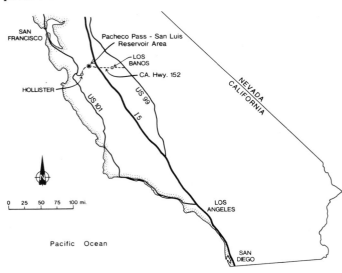

Figure 1. Index map for the Pacheco Pass-San Luis Reservoir area, showing highway access routes.

SIGNIFICANCE

Exposures of aragonite- and jadeitic pyroxene-bearing Franciscan metagraywackes and interlayered dark shales dominate a well-bedded, but folded and faulted section (McKee, 1962, a, b). Minor riebeckitic metachert layers, and lenses of both pillow lavas and gabbros, now recrystallized to omphacite ± glaucophane-crossite-bearing assemblages, add petrologic interest to the region. This is the best exposed, most accessible portion of a relatively high-pressure, low-temperature metamorphosed terrane; it is interpreted to represent part of a late Mesozoic subduction complex. The area appears to be the largest tract of jadeitic

pyroxene-bearing rocks in North America, if not the world. Blueschist facies physical conditions deduced from oxygen isotopic fractionations, experimental phase equilibria, observed mineral parageneses and phytoclast reflectance include temperatures of $150 \pm 50°C$ and pressures up to 7-8 kilobars, indicating depths of burial approaching 15-18 mi (25-30 km) (Ernst 1971a; Bostick, 1974; Maruyama and others, 1985).

SITE INFORMATION

The Diablo Range is an elongate diapiric antiform consisting of uppermost Jurassic and Cretaceous Great Valley clastic strata draped peripherally about a Franciscan core that apparently spans the same age range (Bailey and others, 1964; Ernst, 1965; Ernst and others, 1970; Cowan, 1974; Crawford, 1976). Their mutual contact is a Cenozoic high-angle reverse fault, locally known as the Ortigalita fault; this fault offsets and reactivates the so-called Coast Range thrust (Bailey and others, 1970). Structurally above this break, midfan and more nearly proximal, younger Great Valley feldspathic, turbiditic sandstones, siltstones and conglomerates (Ingersoll, 1982, 1983) stand vertically, or dip steeply away from the antiformal core, essentially parallel to the fault surface. In contrast, the Franciscan comprises locally well-bedded distal to midfan turbidites (Telleen, 1977; Jacobson, 1978; Dickinson and others, 1982), and shaly melanges (Hsü, 1968; Cloos, 1982) arranged in broadly folded, thrust-bound tectonic units. The undistinctive nature of the clastic lithologies hampers efforts to recognize stratigraphic/structural relationships in the Franciscan. Lensoid bodies of sheared, serpentinized peridotite (original rocks evidently were chiefly harzburgites) occur sparsely scattered through the Franciscan, but are most abundant along the fault contact with the tectonically higher Great Valley Sequence. Regional geologic and metamorphic relationships are illustrated in Figure 2.

According to some investigators, (e.g., Telleen, 1977; Ingersoll, 1979; Dickinson and others, 1982; Seiders and Blome, 1984), Franciscan and Great Valley clastics throughout the Coast Ranges have similar petrofacies, current directions, and inferred adjacent lithotectonic settings; hence the Franciscan is regarded as a subduction complex genetically related to the coeval Great Valley Sequence, which was deposited in the landward fore-arc basin. Other studies (e.g., Blake and Jones, 1978, 1981; Blake and others, 1982, 1984; Jayko and Blake, 1984) focus upon age, lithologic and geochemical contrasts, as well as paleomagnetic and faunal evidence for substantial northward drift of Franciscan cherts and greenstones to support the hypothesis of accidental juxtaposition of unrelated Franciscan and Great Valley strata.

A major pressure discontinuity exists across the Coast Range

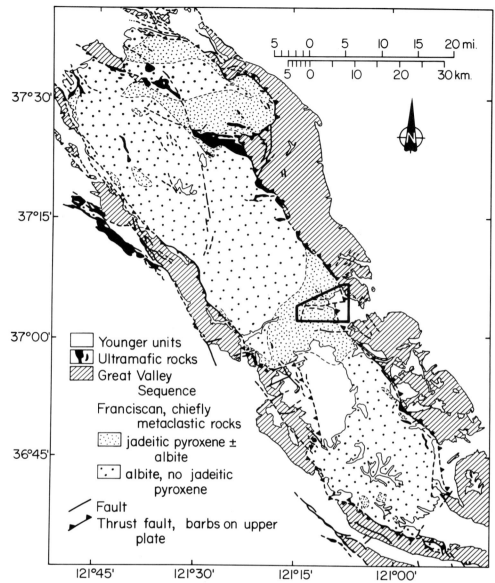

Figure 2. Regional geologic and metamorphic relationships of the Diablo Range, central California Coast Ranges (after Ernst, 1971b). The Pacheco Pass-San Luis Reservoir area shown in Figure 4 is outlined. The dashed boundary between jadeitic pyroxene-bearing and albite-bearing, sodic pyroxene-free Franciscan metagraywackes is regarded in part as an isograd, but locally may reflect faulting; Blake and others (1984) indicate that this surface represents a terrane suture.

thrust. Structurally overlying Great Valley units have been incipiently recrystallized under zeolite facies conditions, at lithostatic pressures less than 3 kilobars, whereas directly below the fault—especially on the east side of the Diablo Range—rock pressures attending blueschist facies metamorphism exceeded 7 kilobars. Elsewhere within the antiform, pumpellyite- and lawsonite-bearing Franciscan metagraywacke phase assemblages indicate that pressures ranged from 4-6 kilobars during recrystallization. The dashed-line boundary between the jadeitic pyroxene-bearing and jadeitic pyroxene-free metagraywackes is shown as an "isograd" in Figure 2, but in some places it almost certainly represents

the unrecognized faulted juxtaposition of Franciscan units of similar appearance, but recrystallized at different depths. The observed paragenetic sequence proceeding from the western, lower density albite ± calcite-bearing association eastward into the denser, aragonitic and jadeitic lithologies for the major Franciscan rock types is shown at the latitude of Pacheco Pass as Figure 3. Metamorphic pressures are postulated to have increased monotonically eastward at nearly constant temperatures. In contrast, Blake and others (1984) regard the jadeitic pyroxene "isograd" as the tectonic contact between the lower pressure, shaly, chaotic Central terrane on the west and the more coherent, higher pressure Yolla Bolly terrane on the east; hence this juncture may not represent a true metamorphic zone boundary.

A geologic map of the Pacheco Pass-San Luis Reservoir area was presented by Ernst and others (1970, Plate 3). Units dip predominantly to the east. The dominantly clastic section, which consists of monotonous, locally graded graywackes + dark shale laminae, is regarded as possessing stratal coherence, judging by the lateral continuity of interbedded cherty units. Nevertheless, pods of greenstone and minor serpentinite occur in the stratigraphically lowest part of the metaclastic sequence, member I, suggesting that this unit in part represents a melange. Mafic metaigneous rocks are present in the stratigraphically higher members chiefly as intrusives. A structural/stratigraphic interpretation is shown in Figure 4. Tectonic thickening and thinning of the units, especially member II, is evident. Six outcrops, indicated on Figure 4, will now be briefly described.

(1) Near the eastern contact with Great Valley conglomeratic sandstones, exposures of Franciscan metagraywacke + metashale near the top of member II suggest that the section is mostly right-side-up, but is overturned in places. Stratigraphic continuity is apparent, even though ptygmatically folded quartz stringers record pervasive, penetrative deformation. Very fine-grained fibrous jadeitic pyroxene pseudomorphous after detrital plagioclase (grain size ~1 mm) constitutes approximately 35 volume percent of the metagraywacke, accompanied by an equivalent percentage of quartz and more modest amounts of glaucophane + lawsonite; aragonite occurs in veins, and albite is lacking. Greenish sandy, micaceous layers appear to have been tuffaceous, but most of the metaclastics contain phases evidently derived largely from a granodioritic source. Shale chips constitute the most common rock fragments, but minor chert, trachytic volcanics and chlorite lenses sprinkled with opaques (= former serpentinite clasts) are scattered sparsely throughout.

(2) About 1.5 mi (2.4 km) to the west-southwest, Highway 152 transects layers of metachert and siliceous red metashale dipping gently to the east, and marking the base of member II. The road transects the axial trace of a small, east-dipping anticline as indicated by outcrop pattern and minor folds. These chemical and fine-grained terrigenous sediments are clearly interstratified with the dominantly metaclastic Franciscan units both here and throughout the mapped area. Beneath the siliceous horizon, pods of greenstone and metagabbro occur within the metagraywackes. Contrasting phase assemblages of these *in situ* rock types reflect the influence of bulk composition on the mineralogy of these blueschist facies lithologies. The quartzose metacherts contain a few volume percent (5 ± 5) of riebeckitic amphibole, stilpnomelane and aegirine, as well as trace amounts of lawsonite. The mafic metaigneous rocks are rich in chlorite + lawsonite and contain variable amounts of relict clinopyroxene, neoblastic omphacite and glaucophane-crossite ± aragonite. The sodic pyroxene occurs as a groundmass phase, or as a replacement of igneous titanaugite, whereas the sodic amphibole is later and occupies stringers and fracture-related patches. Albite and actinolite, minerals common to greenschists and epidote amphibolites, are con-

minerals \ zones	West of "Isograd"	East of "Isograd"
META-IGNEOUS ROCKS		
albite	– – –	
quartz	– – – – – – –	– – – – – –
lawsonite	– – – –	————
calcite	– – – – –	
aragonite		– – – – – – –
omphacite		– – – – – – –
pumpellyite	– –	
chlorite	————	————
white mica	– – – – – – –	– – – – –
stilpnomelane	– – – – – – –	– –
sphene		
glaucophane-cross.	– –	————
METACLASTIC ROCKS		
albite		– –
quartz		
lawsonite	– – – – –	
calcite	– – – – –	
aragonite		– – – – – –
jadeitic pyroxene		———————
chlorite		
white mica		
stilpnomelane		– – – –
glaucophane	– –	– –
METACHERTS		
quartz		
riebeckite	– – –	
white mica	– – – – – – –	– – – – –
stilpnomelane		
aegirine	– – – – – – –	– – – – –
lawsonite	– – – – – – –	– – – –
deerite	?	– – – – – –

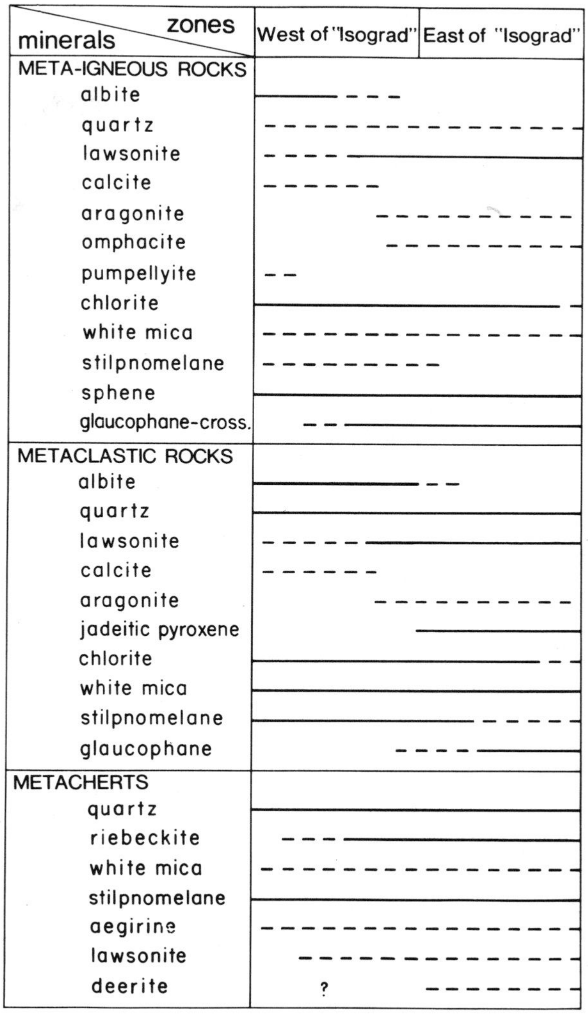

Figure 3. Presumably synchronous mineral parageneses in the principal Franciscan rock types proceeding from west to east across the Diablo Range in the vicinity of Pacheco Pass (after Seki and others, 1969; Ernst and others, 1970, Fig. 14), assuming a more or less autochthonous section.

spicuously absent. Associated metagraywackes are similar to previously described lithologies except that some specimens contain abundant relict albite grains and little or no new jadeitic pyroxene. White mica and glaucophane-crossite from a metabasaltic block directly beneath the metachert horizon just south of the highway yielded K-Ar radiometric mineral ages of 115 ± 3 and 122 ± 3 Ma, respectively (Suppe and Armstrong, 1972). These ages are 20 m.y. older than U-Pb mineral ages from metagabbroic samples obtained about 20 mi (32 km) to the southeast at Ortigalita Peak by Mattinson and Echeverria (1980).

(3) Well-bedded, stratigraphically continuous, right-side-up metagraywackes and interlayered metashales of member I crop out in a homoclinal, gently east-dipping sequence along Highway 152 approximately 0.5 mi (0.8 km) northeast of the summit. Graded bedding is distinct. Some metaclastics contain jadeitic

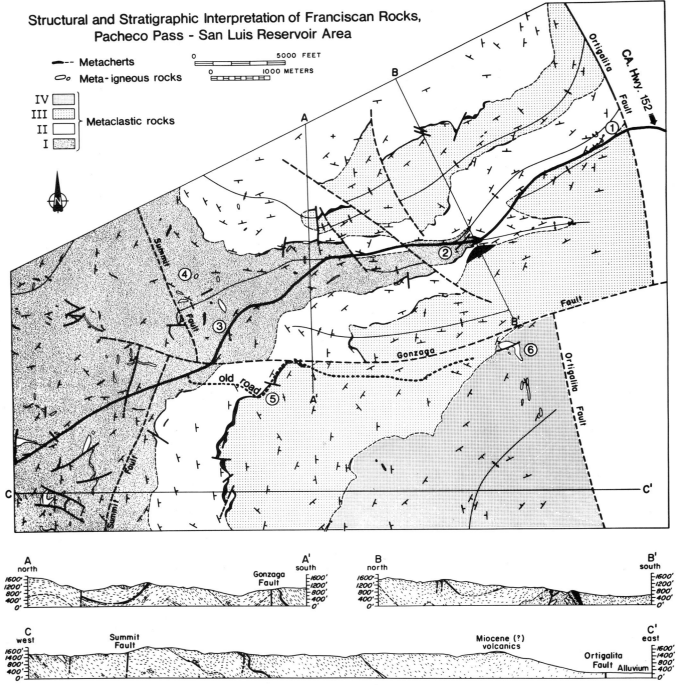

Figure 4. Interpretive geology of the Pacheco Pass-San Luis Reservoir area (after Ernst and others, 1970, Plate 3 and Figure 12). The reservoir itself is not shown. Locations described in the text are indicated. Four metaclastic members of the Franciscan Complex in this area, referred to as units I-IV, are distinguished. Pacheco Pass is located directly east of the junction of old and new highways.

pyroxene ± aragonite, whereas others carry relict albite. For three petrographically examined samples, average modes are as follows: quartz, 48; jadeitic pyroxene, 25; white mica, 11; chlorite, 5; lawsonite, 4; glaucophane, 1; and sphene, opaques, and carbonaceous material, 6. As demonstrated by x-ray diffraction techniques, interlayered metashales contain a similar phase assemblage (Ernst, 1971a; see also Cloos, 1983).

(4) A cross-country hike (boots suggested) north across the hill and descending into and along the creek bottom to a tectonic block approximately 1.0 mi (1.6 km) north of locality (3) pro-

vides a typical example of geologic problems encountered in the map area. This traverse is situated within the lowermost, in part disrupted, greenstone-bearing metaclastic unit, member I, yet some sedimentary packages exhibit significant stratal continuity. Many metagraywacke samples are albite-bearing, and probably represent phases persisting metastably due to sluggish reaction rates. Greenstone pods display sheared contacts with the surrounding metaclastic and may represent tectonic or olistolithic blocks. Landslides are numerous. The coarse-grained (2-4 mm) mafic amphibolite block is probably part of such a slumped section. It is mineralogically heterogeneous, and contains chlorite, pumpellyite, lawsonite, crossite, and locally garnet, hornblende, and/or white mica; the coarse grain size and higher grade phase associations indicate that this tectonic block is exotic to the adjacent *in situ* metamorphics. Why some metagraywackes are albitic, others jadeitic, some metaigneous rocks hornblendic ± garnetiferous amphibolite, or omphacite-bearing greenstones is perplexing, but probably reflects the tectonized nature of the melange.

(5) An exposure of relatively coarse-grained jadeitic pyroxene-bearing metagraywacke crops out along the former highway about 0.9 mi (1.4 km) east of the road junction near Pacheco Pass. An estimated 40 volume percent of the rock consists of 0.5-1.5 mm long sprays, sheaves, and prismatic clusters of Na pyroxene, which have overgrown and completely replaced albitized detrital plagioclase grains. Percentages of the accompanying phases are: quartz, 35; white mica, 10; glaucophane, 7; lawsonite, 3; stilpnomelane, 2; sphene, opaques, and carbonaceous material, 3; and a trace of $CaCO_3$. A relatively continuous 6-15 ft (2-5 m) thick chert interbed is associated with this predominantly east-dipping sequence of lowermost member III Franciscan metaclastics, but is not exposed along the road.

(6) Directly northeast of the parking area at the eastern termination of the old road, glaucophanized metagabbro, intrusive into member IV jadeitic metagraywacke, is exposed on a small hill now marked by the east portal of the Calaveras Valley irrigation tunnel. Only readily accessible at low-water stands of the San Luis Reservoir in the fall and early winter, this is an optional stop. The metagabbro contains relict titanaugite ± calcic plagioclase, now extensively overprinted and replaced by the blueschist facies assemblage lawsonite + glaucophane-crossite + chlorite + sphene ± omphacite ± aragonite. The textural and phase relations demonstrate that, similar to the Ortigalita Peak locality 20 mi (32 km) to the southeast, mafic magma intruded the Franciscan sedimentary pile prior to high-pressure recrystallization (Mattinson and Echeverria, 1980).

REFERENCES CITED

Bailey, E. H., Irwin, W. P., and Jones, D. L., 1964, Franciscan and related rocks, and their significance in the geology of western California: California Division of Mines and Geology, Bulletin 183, 171 p.

Bailey, E. H., Blake, M. C., Jr., and Jones, D. L., 1970, On-land Mesozoic oceanic crust in California Coast Ranges: U.S. Geological Survey Professional Paper 700-C, p. 70-81.

Blake, M. C., Jr., and Jones, D. L., 1978, Allochthonous terranes in northern California?—A reinterpretation, *in* Howell, D. G., and McDougall, K. A., eds., Mesozoic Paleogeography of the Western United States: Pacific Section, Society of Economic Paleontologists and Mineralogists, Pacific Coast Paleogeography Symposium 2, p. 397-400.

——1981, The Franciscan assemblage and related rocks in northern California: A reinterpretation, *in* Ernst, W. G., ed., The Geotectonic Development of California: Englewood Cliffs, New Jersey, Prentice-Hall, Inc., p. 307-328.

Blake, M. C., Jr., Howell, D. G., and Jones, D. L, 1982, Preliminary tectonostratigraphic terranes map of California: U.S. Geological Survey Open-File Report 82-593.

Blake, M. C., Jr., Howell, D. G., and Jayko, A. S., 1984, Tectonostratigraphic terranes of the San Francisco Bay region, *in* Blake, M. C., Jr., ed., Franciscan Geology of Northern California: Pacific Section, Society of Economic Paleontologists and Mineralogists, v. 43, p. 5-22.

Bostick, N. H., 1974, Phytoclasts as indicators of thermal metamorphism, Franciscan assemblage and Great Valley sequence (Upper Mesozoic), California: Geological Society of America Special Paper 153, p. 1-17.

Cloos, M., 1982, Flow melanges: Numerical modeling of geological constraints on their origin in the Franciscan subduction complex, California: Geological Society of America Bulletin, v. 93, p. 330-345.

——1983, Comparative study of melange matrix and metashales from the Franciscan subduction Complex with the basal Great Valley Sequence, California: Journal of Geology, v. 91, p. 291-306.

Cowan, D. S., 1974, Deformation and metamorphism of the Franciscan subduction zone complex northwest of Pacheco Pass, California: Geological Society of America Bulletin, v. 85, p. 1623-1634.

Crawford, K. E., 1976, Reconnaissance geologic map of the Eylar Mountain quadrangle, Santa Clara and Alameda Counties, California: U.S. Geological Survey Map MF-764.

Dickinson, W. R., Ingersoll, R. V., Cowan, D. S., Helmold, K. P., and Suczek, C. A., 1982, Provenance of Franciscan graywackes in coastal California: Geological Society of America Bulletin, v. 93, p. 95-107.

Ernst, W. G., 1965, Mineral parageneses in Franciscan metamorphic rocks, Panoche Pass, California: Geological Society of America Bulletin, v. 76, p. 879-914.

——1971a, Do mineral parageneses reflect unusually high-pressure conditions of Franciscan metamorphism? American Journal of Science, v. 270, p. 81-108.

——1971b, Petrologic reconnaissance of Franciscan metagraywackes from the Diablo Range, Central California Coast Ranges: Journal of Petrology, v. 12, p. 413-437.

Ernst, W. G., Seki, Y., Onuki, H., and Gilbert, M. C., 1970, Comparative study of low-grade metamorphism in the California Coast Ranges and the Outer Metamorphic Belt of Japan: Geological Society of America Memoir 124, 276 p.

Hsü, K. J., 1968, Principles of melanges and their bearing on the Franciscan-Knoxville Paradox: Geological Society of America Bulletin, v. 79, p. 1063-1074.

Ingersoll, R. V., 1979, Evolution of the Late Cenozoic forearc basin, northern and central California: Geological Society of America Bulletin, Part I, v. 90, p. 813-826.

——1982, Initiation and evolution of the Great Valley forearc basin of northern and central California, U.S.A: Geological Society of London, Special Publication 10, p. 459-467.

——1983, Petrofacies and provenance of Late Mesozoic forearc basin, northern and central California: American Association of Petroleum Geologists Bulletin, v. 67, p. 1125-1142.

Jacobson, M. I., 1978, Petrologic variations in Franciscan sandstone from the

Diablo Range, California, *in* Howell, D. G., and McDougall, K. A., eds., Mesozoic Paleogeography of the Western United States: Pacific Section, Society of Economic Paleontologists and Mineralogists, Pacific Coast Paleogeography Symposium 2, p. 401–417.

Jayko, A. S., and Blake, M. C., Jr., 1984, Sedimentary petrology of graywacke of the Franciscan Complex in the northern San Francisco Bay area, California, *in* Blake, M. C., Jr., ed., Franciscan Geology of Northern California: Pacific Section, Society of Economic Paleontologists and Mineralogists, v. 43, p. 121–134.

Maruyama, S., Liou, J. G., and Saskura, Y., 1985, Low-temperature recrystallization of Franciscan graywackes from Pacheco Pass, California: Mineralogical Magazine, v. 49, p. 345–355.

Mattinson, J. M., and Echeverria, L. M., 1980, Ortigalita Peak gabbro, Franciscan complex: U-Pb dates of intrusion and high-pressure-low temperature metamorphism: Geology, v. 8, p. 589–593.

McKee, B., 1962a, Widespread occurrence of jadeite, lawsonite, and glaucophane in central California: American Journal of Science, v. 260, p. 596–610.

—— 1962b, Aragonite in the Franciscan rocks of the Pacheco Pass area, California: American Mineralogist, v. 47, p. 379–387.

Seiders, V. M., and Blome, C. D., 1984, Clast compositions of upper Mesozoic conglomerates of the California Coast Ranges and their tectonic significance, *in* Blake, M. C., Jr., ed., Franciscan Geology of Northern California: Pacific Section, Society of Economic Paleontologists and Mineralogists, v. 43, p. 135–148.

Seki, Y., Ernst, W. G., and Onuki, H., 1969, Phase proportions and physical properties of minerals and rocks from the Franciscan and Sanbagawa metamorphic terranes, a supplement to Geological Society of America Memoir 124: Tokyo, Japan, Japan Society for Promotion of Science, 85 p.

Suppe, J., and Armstrong, R. L., 1972, Potassium-argon dating of Franciscan metamorphic rocks: American Journal of Science, v. 272, p. 217–233.

Telleen, K. E., 1977, Paleocurrents in part of the Franciscan Complex, California: Geology, v. 5, p. 49–51.

ACKNOWLEDGMENTS

This descriptive note was reviewed by M. C. Blake, Jr., D. S. Cowan, and Gerhard Oertel. The author's research in the California Coast Ranges has long been supported by the UCLA Research Committee and by the National Science Foundation, most recently through grant EAR83-12702.

Paleocene turbidites and modern landslides of the Point San Pedro–Devil's Slide area, San Mateo County, California

Tor H. Nilsen and James C. Yount, U.S. Geological Survey, 345 Middlefield Road, Menlo Park, California 94025

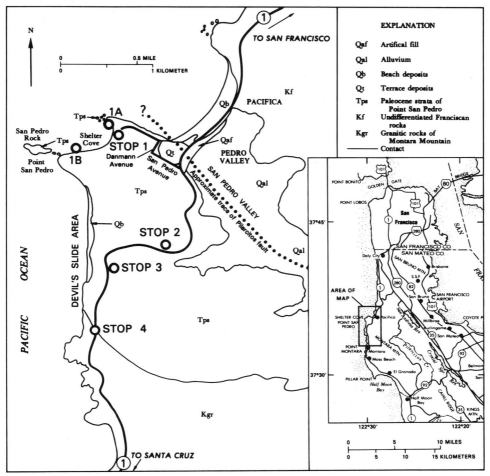

Figure 1. Index map showing general geology and field trip stop locations in the Point San Pedro area (modified from Nilsen and Yount, 1981b).

LOCATION AND ACCESSIBILITY

This site is located adjacent to the California coast in northern San Mateo County, about 15 mi (25 km) south-southwest of downtown San Francisco (Fig. 1). From San Francisco, take I-280 south to its intersection with California 1, then follow California 1 southward for approximately 9 mi (15 km) to San Pedro Avenue in Pacifica. From the central San Francisco Peninsula, take California 92 westward to its intersection with California 1 in Half Moon Bay, then proceed northward on California 1 for approximately 10 mi (16 km) to San Pedro Avenue in Pacifica. From the southern San Francisco Peninsula or Santa Cruz County, take California 1 north to San Pedro Avenue in Pacifica.

The Shelter Cove locality (Stop 1) can be reached by taking San Pedro Avenue west for three blocks from California 1, turn-

ing north on Danmann Road, and proceeding two blocks north to Shelter Cove Road. At present (July 1986), Shelter Cove Road is a privately maintained access that has sustained substantial landslide damage and is passable only on foot. The property at Shelter Cove is private, and permission to enter must be obtained from the owners (Telegan Realty, Pacifica, California). Roadcuts along California 1 are located south of San Pedro Avenue. Stop 2 is located 0.7 mi (1.1 km) to the south on the northwest side of the highway. Stop 3 in the Devil's Slide area is located 0.6 mi (1 km) farther south along California 1 and Stop 4 another 0.7 mi (1.1 km) to the south. Stops 2, 3, and 4 are subject to heavy traffic and have limited parking along a narrow, twisting, and dangerous roadway. Exposures are close to the roadway, so great caution should be exercised.

SIGNIFICANCE OF SITE

The Point San Pedro area provides excellent exposures of Paleocene submarine-fan deposits, unusual carbonate turbidites, an unconformity characterized by Paleocene conglomerate resting on Mesozoic granodiorite, and pervasive and active landsliding that has caused massive damage to California 1 in the Devil's Slide area. Many elements of a small deep-sea fan that was deposited in a restricted Paleocene continental-borderland setting within the Salinian block, an accreted tectonostratigraphic terrane underlain by granitic basement rocks, crop out in the area. Middle-fan channelized sandstone (Stop 1A, north side of Shelter Cove) and inner-fan channelized conglomerate (Stop 2) interfinger with thin-bedded fine-grained channel-margin and interchannel sandstone and shale (Stop 1B, south side of Shelter Cove, Stops 2 and 3). Although most of the Paleocene strata of Point San Pedro are fine grained and thin bedded, distribution of boulder-sized clasts of granitic material derived from nearby basement rocks indicates rapid sedimentation into a small, deep, tectonically active basin.

SITE INFORMATION

The Paleocene strata at Point San Pedro have been described by several previous workers and have been included in several previously published field trip guidebooks. Early workers believed that the strata at Point San Pedro formed the basal member of the late Mesozoic and early Cenozoic Franciscan assemblage (see summary in Morgan, 1981a). However, early Tertiary mollusks found in the strata resulted in correlation of the strata with the Martinez Formation of Paleocene age (Dickerson, 1914). Darrow (1963) subdivided the strata into an underlying Upper Cretaceous(?) unit and an unconformably overlying Paleocene unit; he placed the contact at the base of a massive conglomerate resting on thin-bedded turbidites (Stop 2). Chipping (1972) studied the sedimentology of the thin-bedded strata in the Shelter Cove area, determining that they had been deposited by northwest-flowing turbidity currents. Harbaugh (1975) described the unconformity previously suggested by Darrow (1963) in a regional field guidebook, and Howard (1979) described shallow-marine conglomerate in the area resting directly on granitic basement. In a previous guidebook (Nilsen and Yount, 1981a), and in this description, we assert that the Point San Pedro strata are entirely Paleocene in age, because the sequence above and below the conglomerate is lithologically similar, and only fossils of Paleocene age have been found.

Nilsen and Clarke (1975), in a summary of the tectonic and depositional framework of lower Tertiary deposits of central California, inferred that the strata of Point San Pedro had been deposited as a deep-sea fan in a small, restricted, borderland-type basin that deepened northwestward. Descriptions of the stratigraphy, sedimentology, structure, and petrography of the Paleocene strata of Point San Pedro by Nilsen and Yount (1981a, 1981b) and Morgan (1981a, 1981b) have helped clarify many details of the

geology of the area and form the most important references. Paleomagnetic work by Champion and others (1984) indicates that the Paleocene strata have been rotated clockwise as much as 90°, and have been transported northward about 1,250 mi (2,000 km) with the Salinian block since time of deposition.

Because there are a number of interesting geologic phenomena to be seen in the Point San Pedro area, including massive modern landslides described by Beeston and Gamble (1980), we have divided the guide into four separate stops (Fig. 1).

SITE DESCRIPTIONS

Stop 1. Shelter Cove. The strata exposed at the north point (Stop 1A), approximately 330 ft (100 m) northwest of the northernmost house, consist chiefly of thick beds of southwest-dipping, coarse-grained sandstone that we interpret to be middle-fan-channel and channel-margin deposits. The strata exposed along the north shore of Point San Pedro (Stop 1B), beginning approximately 1,000 ft (300 m) southwest of the southernmost house, consist chiefly of southwest-dipping, overturned thin-bedded turbidites. We interpret these as interchannel deposits.

The section at the north point consists of 14 fining-upward cycles that average about 10 ft (3 m) thick, with the thickest about 20 ft (6 m) and the thinnest about 3 ft (1 m) (Fig. 2). Each cycle, except for cycles 2 and 3, fines and thins upward. The basal conglomerate or conglomeratic sandstone is typically channeled into underlying shale or thin-bedded turbidites.

The basal conglomerate or conglomeratic sandstone of each cycle is typically massive or crudely parallel-stratified. Dish structure is present in some beds, especially in the upper part of the section above cycle 9. Conglomeratic clasts are less than 1 in (3 cm) long and consist of chert, quartz, volcanic rocks, or granite. Rip-up clasts of shale, mudstone, or thin-bedded turbidites, as large as 35 in (90 cm), are present in these basal units and may be deformed and intruded by the conglomeratic matrix.

Cross-stratified beds of locally conglomeratic sandstone that contain abundant small mudstone rip-up clasts overlie the basal conglomerate or conglomeratic sandstone in most cycles. These beds are as thick as 3 ft (1 m) and generally have flat bases, wavy tops, and contain rare large rip-up clasts. The internal medium- to large-scale cross-strata indicate that these beds have been deposited as sand waves by tractive currents that continually reworked and retransported the coarse sediment. The beds pinch out and abruptly change thickness laterally. In several cases, the uppermost part of the basal conglomeratic unit has been reworked into small- to medium-scale cross-strata by tractive processes. The troughs of the sand waves are locally filled with thin-bedded turbidites and sandstone containing abundant shale rip-up clasts and exhibiting injection features.

The tops of the fining-upward cycles consist of thin-bedded ripple-marked turbidites and interbedded shale. The shale is locally extensively bioturbated. The rippled thin-bedded turbidites are characterized by flat bases, wavy tops, internal cross-laminae, a general lack of size grading, and common starved ripples. The

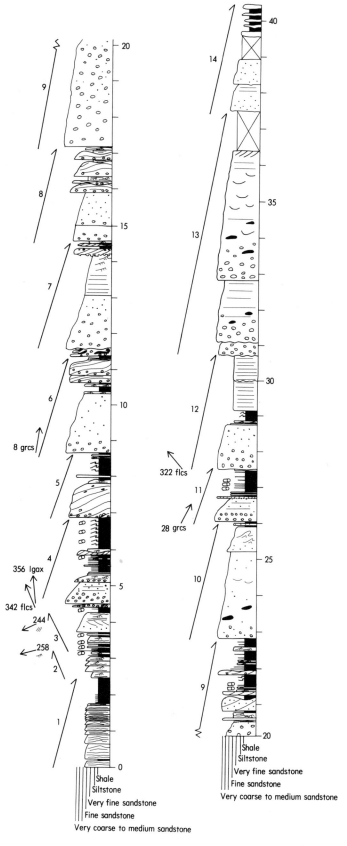

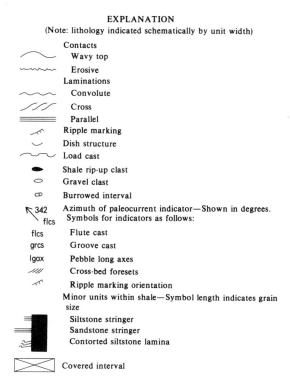

EXPLANATION
(Note: lithology indicated schematically by unit width)

Contacts
   Wavy top
   Erosive
Laminations
   Convolute
   Cross
   Parallel
   Ripple marking
   Dish structure
   Load cast
   Shale rip-up clast
   Gravel clast
   Burrowed interval
342 Azimuth of paleocurrent indicator—Shown in degrees.
flcs Symbols for indicators as follows:
   flcs Flute cast
   grcs Groove cast
   lgax Pebble long axes
   Cross-bed foresets
   Ripple marking orientation
Minor units within shale—Symbol length indicates grain size
   Siltstone stringer
   Sandstone stringer
   Contorted siltstone lamina

   Covered interval

Figure 2. Measured section of Paleocene strata of Point San Pedro from north point of Shelter Cove, Stop 1A. Numbered bars (1 through 14) at left of section indicate fining-upward or coarsening-upward cycles. Arrows at left of section indicate paleocurrent azimuths. Numbers at right indicate thickness in meters from base of section.

tops of the cycles are commonly extensively sheared; some of this shearing is synsedimentary in origin, probably related to sediment loading, because downward intrusions of sandstone and conglomerate extend into the deformed fine-grained sediment.

Thin-bedded turbidites, which presumably overlie the thick-bedded sandstones of the north point of Shelter Cove, are well exposed in the sea cliff extending along the north shore of Point San Pedro to San Pedro Rocks (Fig. 3). The best exposures are along the base of the sea cliff above the boulder-strewn beach. Poorer outcrops of the same strata can be observed along the abandoned Ocean Shore Railroad grade about 160 ft (50 m) above the beach.

The exposed sequence consists of repetitively alternating fine-grained sandstone and shale (Fig. 4A). Disruption of bedding due to synsedimentary slumping is fairly common (Fig. 4B), as are small offsets, generally less than 33 ft (10 m), due to modern landsliding and possibly postdepositional tectonism. Small-scale sedimentary structures typical of turbidites, such as graded bedding, ripple-drift lamination, parallel lamination, and convolute lamination, are common throughout the section. Small flute casts are present at the base of some of the coarser beds. The sequence is extensively bioturbated.

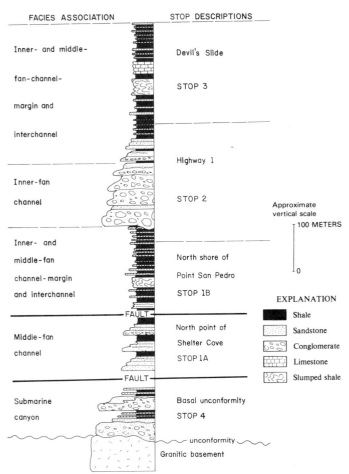

Figure 3. Schematic columnar section for Paleocene strata of Point San Pedro showing possible vertical sequence of facies associations and field trip stops (from Nilsen and Yount, 1981a, Fig. 6). Vertical scale and thicknesses of facies associations approximate.

Stop 2. Conglomerate-filled channel, California 1. This roadcut exposes conglomeratic channel deposits that rest with an erosive contact on shale-rich channel-margin and interchannel deposits (Fig. 5A). This conglomerate was originally thought to form the unconformable boundary between the Cretaceous and Paleocene parts of the Point San Pedro sequence (Darrow, 1963; Harbaugh, 1975). However, we believe that it is an intraformational erosive contact that marks the base of a major inner-fan channel complex within the wholly Paleocene sequence of Point San Pedro strata.

The sequence of thin-bedded turbidites below the conglomerate-filled channel consists of four distinct sections separated by zones or horizons of shearing (Fig. 5B). We interpret the zones of shearing to have formed at the time of sedimentation, prior to deposition of the conglomerate. The attitudes of the shear zones are subparallel to the bedding, suggesting movement of the four sections subhorizontally, probably by lateral translation or slump-

ing along an originally very gently sloping submarine topographic surface (Fig. 5C).

The thin-bedded turbidites at this stop are very typical of channel-margin to interchannel deposits. Abundant ripple-marked beds with abrupt flat bases and sharp wavy tops characterize the sections. Organization of the turbidite beds into bundles of thin cycles that both thicken and thin upward is typical of sections deposited by overbanking. The bundles reflect alternating periods of activity and inactivity related to the passing of turbidity currents down adjacent channels. Thinning-upward bundles are probably caused by successive turbidity currents of diminishing strength and thickening-upward bundles by successive currents of increasing strength. Irregular bundles result from currents of variable strength. Thick shale sections probably result from the lack of active turbidity currents, the lack of currents with sufficient strength to yield overbank deposition, or shifting of current activity to other channels that are more distant from the depositional site. Some thickening-upward bundles may be related to progradation of distal levee facies over interchannel areas.

The conglomeratic channel-fill sequence can be divided into six thinning- and fining-upward cycles marked by channelized conglomerate at their bases. In general, the percent of conglomerate, maximum clast size, and average bed thickness decrease upward and the percent of shale increases upward, suggesting an overall thinning- and fining-upward character to the channel fill. Darrow (1963) noted a maximum thickness of conglomerate of 155 ft (47 m) and a largest clast 59 in (150 cm) in length. At Stop 2, the longest clast observed in the central part of the channel fill was 54 in (138 cm) long and the maximum thickness of conglomerate is about 66 ft (20 m; Fig. 5D). On the east flank of the channel sequence, the maximum clast size above the thin-bedded turbidite sequence is about 18 in (45 cm), indicating a prominent lateral decrease in clast size away from the axis of the channel.

Darrow (1963) had previously concluded that the largest boulders in the Paleocene strata of Point San Pedro consisted of granitic rock similar to that of Montara Mountain. He also noted clasts of biotite and muscovite granite, hornblende diorite, hornblende gneiss, muscovite schist and gneiss, black slate, chert, quartzite, limestone, sandstone, and shale. There appear to be three distinct clast suites based on size and textural characteristics. The largest clasts, which are generally moderately rounded, are the granitic, dioritic, gneissic, and schistose clasts that have local sources in the Santa Cruz Mountains (Leo, 1967). The second suite consists of smaller clasts, generally less than 10 cm in length, of well-rounded chert, quartzite, and volcanic rocks; these types of clasts are abundant in Upper Cretaceous and Tertiary conglomerates throughout California, and because of their textural and compositional maturity, may have been recycled from older conglomerates. The third suite consists of sandstone, siltstone, and shale clasts that are intraformational and generated by rip-up processes, most likely along channel margins.

Stop 3. Carbonate turbidites, Devil's Slide area. Highly deformed thin-bedded turbidites at the north end of the Devil's Slide area contain interbedded bundles of thin-bedded orange-

Figure 4. Sedimentary features on south side of Shelter Cove, Stop 1B. A, thin-bedded overturned turbidites dipping to south; measuring rod in foreground is 5 ft (1.5 m) long; B, synsedimentary slump with fold axes overturned to north. Hammer for scale.

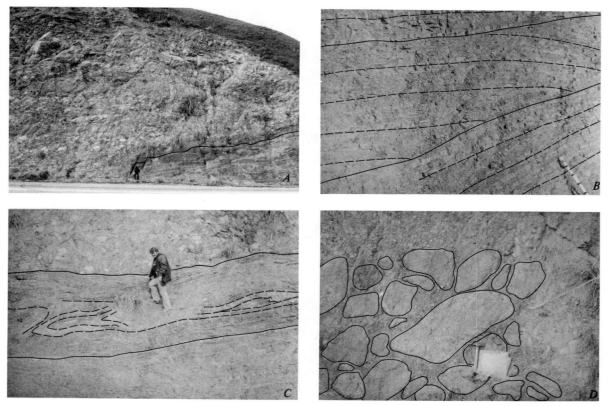

Figure 5. Sedimentary features at Stop 2, California 1. A, Conglomerate channeled into thin-bedded turbidites. Boulders in this conglomerate are as large as 6 ft (2 m) in diameter. B, Discordant relations between sections of thin-bedded turbidite sequence; measuring rod divided into 3.9 in (10-cm) divisions. C, Erosive contact between thin-bedded turbidites and overlying conglomerate. Note recumbent slump folds at left and synsedimentary boudins at right within deformed uppermost section of thin-bedded turbidites. D, Conglomerate in central part of channel showing large dimensions and moderate rounding of granitic, gneissose, and schistose clasts.

weathering carbonate turbidites. These turbidites range from very fine to medium-grained calcarenite at their base to micrite at their top. The carbonate turbidites crop out as bundles 3.3 to 10 ft (1 to 3 m) thick, with beds 0.4 to 4.7 in (1 to 12 cm) thick, within siliciclastic thin-bedded turbidites. Sedimentologic features such as cross-lamination, parallel lamination, and pervasive burrowing are present in both the carbonate beds and the adjacent siliciclastic thin-bedded turbidites. A few of the thicker carbonate turbidites contain rip-up clasts of micrite near their base.

Although some carbonate is of secondary origin, forming cement and replacing quartz and feldspar, much of the carbonate consists of detrital grains that form as much as 85 percent of the framework grains in some samples. The carbonate grains may have been derived from roof pendants of marble enclosed nearby in the granitic rocks of Montara Mountain; similar pendants are found in the Ben Lomond area of the southern Santa Cruz Mountains (Leo, 1967).

Stop 4. Landsliding and basal unconformity. Following extensive landsliding along California 1 in the Devil's Slide area during the 1979–80 rainy season, Beeston and Gamble (1980) prepared a detailed report on the engineering geology of Devil's Slide. They prepared a number of detailed geologic and engineering geologic maps that delineate the size and extent of landslide masses and show the irregular basal contact between the granitic basement and the Paleocene strata of Point San Pedro (Fig. 6). Their maps show the great extent of landsliding along this part of the coast and shearing that has developed along most geologic contacts as a result of landsliding.

The basal conglomerate is generally about 33 to 50 ft (10 to 15 m) thick where exposed along California 1 in the Devil's Slide area (Fig. 3), although sandstone and shale locally rest directly on

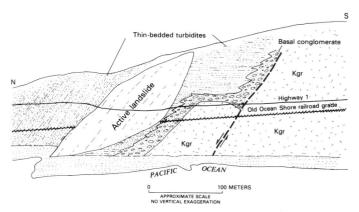

Figure 6. Relation of basal conglomerate with the granitic rocks of Montara Mountain at south end of Devil's Slide. View toward east. Oblique sketch map based in part on photograph in Beeston and Gamble (1980, Fig. 10). Symbol: Kgr, granitic rocks of Montara Mountain. Fault, dashed line; arrows indicate relative movement.

the basement. The conglomerate contains chiefly clasts of granodioritic rock, but also contains abundant well-rounded pebbles of metavolcanic and metasedimentary rocks. Adjacent to the contact, clasts of schist and gneiss are common. There is no evidence of either a nonmarine or shallow-marine origin for this basal conglomerate, although it generally doesn't crop out very well. Overlying thin-bedded turbidites appear to have been deposited in deep-marine conditions, suggesting that the basal conglomerate along the highway is also of deep-marine origin.

REFERENCES CITED

Beeston, H. E., and Gamble, J. H., 1980, Engineering geology of the Devil's Slide at San Pedro Mountain, San Mateo County, California: California Department of Transportation, Division of Construction, Office of Transportation Laboratory, Report 04-SM-1, 56 p.

Champion, D. E., Howell, D. G., and Gromme, C. S., 1984, Paleomagnetic and geologic data indicating 2,500 km of northward displacement for the Salinian and related terranes, California: Journal of Geophysical Research, v. 89, p. 7736–7752.

Chipping, D. H., 1972, Sedimentary structures and environment of some thick sandstone beds of turbidite type: Journal of Sedimentary Petrology, v. 42, p. 587–595.

Darrow, R. L., 1963, Age and structural relationships of the Franciscan Formation in the Montara Mountain Quadrangle, San Mateo County, California: California Division of Mines and Geology Special Report 78, 23 p.

Dickerson, R. E., 1914, The fauna of the Martinez Eocene of California: University of California Publications in Geological Sciences, v. 8, p. 61–180.

Harbaugh, J. W., 1975, Geology field guide to northern California: Dubuque, Iowa, Kendall/Hunt, 123 p.

Howard, A. D., 1979, Geologic history of middle California: Berkeley, University of California Press, 113 p.

Leo, W., 1967, The plutonic and metamorphic rocks of the Ben Lomond Moun-

tain area, Santa Cruz County, California, *in* Short contributions to California geology: California Division of Mines and Geology Special Report 91, p. 27–43.

Morgan, S. R., 1981a, General geology of the strata at Point San Pedro, San Mateo County, California, *in* Frizzell, V., ed., Upper Cretaceous and Paleocene turbidites, central California coast: Pacific Section, Society of Economic Paleontologists and Mineralogists, v. 20, p. 13–19.

—— , 1981b, Descriptions of field trip stops 3 and 4, *in* Frizzell, V., ed., Upper Cretaceous and Paleocene turbidites, central California coast: Pacific Section, Society of Economic Paleontologists and Mineralogists, v. 20, p. 49–51.

Nilsen, T. H., and Clarke, S. H., Jr., 1975, Sedimentation and tectonics in the early Tertiary continental borderland of central California: U.S. Geological Survey Professional Paper 925, 64 p.

Nilsen, T. H., and Yount, J. C., 1981a, Sedimentology of the Paleocene strata of Point San Pedro, California, *in* Frizzell, V., ed., Upper Cretaceous and Paleocene turbidites, central California coast: Pacific Section, Society of Economic Paleontologists and Mineralogists, v. 20, p. 21–29.

—— , 1981b, Descriptions of field trip stops 1 and 2, *in* Frizzell, V., ed., Upper Cretaceous and Paleocene turbidites, central California coast: Pacific Section, Society of Economic Paleontologists and Mineralogists, v. 20, p. 31–48.

The Merced Formation and related beds: A mile-thick succession of late Cenozoic coastal and shelf deposits in the seacliffs of San Francisco, California

H. Edward Clifton and Ralph E. Hunter, U.S. Geological Survey, 345 Middlefield Road, Menlo Park, California 94025

LOCATION

The Merced Formation and consanguineous superjacent strata are well exposed in seacliffs that extend 4.3 mi (7 km) south from Lake Merced to the trace of the San Andreas fault (Fig. 1). The exposures can be reached by walking south along the beach from oceanfront public parking areas west of Lake Merced or via several well-defined trails that lead to the beach from Fort Funston (Fig. 1), a part of the Golden Gate Recreational Area, which provides public parking. The southern part of the exposure can be reached by walking north from a public parking area at Mussel Rock (Fig. 1), which can be reached via an access road from Westline Drive in northern Pacifica (Edgemar). The central part of the exposure is presently accessible via a road to the former Thornton Beach State Park at the western end of Alemany Boulevard. This road, however, is now badly disrupted by landslides and affords uncertain future accessibility.

The character of the exposure changes as a consequence of local landsliding and seasonal variation in the level of beach sand. The deposits are best viewed following winter storms when the erosion of beach sand provides for fresh, wave-washed exposure. The observer of these deposits should continually be alert for rock falls and the possibility of being isolated by a rising tide.

SIGNIFICANCE

The seacliff exposures of the Merced Formation and its superjacent deposits permit an examination of about 5,740 ft (1750 m) of strata that, in age, probably span much of the Pleistocene and possibly part of the late Pliocene. The section appears to be structurally continuous, although the possibility of significant offset by hidden faults at the few places where the section is unexposed cannot be precluded. A range of depositional environments is represented, including shelf, nearshore, foreshore, backshore, eolian, fluvial, coastal embayment, and coastal marsh or pond. Many of the marine deposits are richly fossiliferous. The lithofacies are arranged in recurring patterns of vertical succession that reflect a combination of eustatic changes in sea level, fluctuations in sediment supply, and tectonic processes. The section provides particularly excellent examples of shallowing-upward progradational sequences and an opportunity to examine the sedimentologic and paleontologic aspects of a variety of shallow marine and coastal depositional facies.

SITE INFORMATION

The thick section of very young sedimentary deposits ex-

posed in the seacliffs southwest of San Francisco has drawn the attention of California geologists since the end of the last century (Lawson, 1893; Ashley, 1895). The seacliff exposures mark the northwestern margin of an outcrop belt that extends southeastward along the northeastern side of the San Andreas fault (Fig. 1B). The section described here extends from the northernmost exposures just west of Lake Merced southward to the vicinity of the 1906 trace of the San Andreas fault (Figs. 1A and 2).

The section includes two formally named geologic units separated by 525 ft (160 m) of strata that await formal stratigraphic designation. In our description of the strata here, we follow the approach of Hunter and others (1984) of identifying depositional sequences (most of which are unconformably bounded) by letter designation and relatively homogeneous lithologic units within the sequences by subscript numerals. Strata in the lower 5,184 ft (1,580 m) of the section compose most of the type section of the Merced Formation, a unit generally considered to be of late Pliocene and Pleistocene age (Lawson, 1893; Ashley, 1895; Glen, 1959; Hall, 1965; Addicott, 1969). Subsequent paleontological analysis (Roth, 1979; E. J. Moore, written communication, 1985) indicates that the section described here may be entirely of Pleistocene age.

The Merced Formation was divided informally into two members by Hall (1966), the boundary coinciding with the top of sequence P in this report. This subdivision was based largely on mineralogy and the inferred provenance of the sediment: the lower member is derived from the Franciscan assemblage and other local sources whereas the upper member consists of sediment generated in the drainage basins of the Sacramento and San Joaquin rivers. This change in provenance is estimated to have occurred at about 0.6 Ma (Sarna-Wojcicki and others, 1985). The two members defined by Hall also are lithologically dissimilar. His lower member is dominated by the shelf facies, which occurs but once in the section above the boundary between the members. In contrast, Hall's upper member (and the overlying strata) is dominated by eolian facies, which, at best, are poorly developed in the lower member. An ash bed in Hall's upper member (in unit S_2 of this report) is dated at 0.4 Ma (Sarna-Wojcicki and others, 1985).

The strata immediately above sequence T in this report (the "unnamed beds" of Hall, 1966) are not formally considered part of the Merced Formation, although structurally and lithologically these strata closely resemble those in the uppermost part of the Merced as presently defined (Hunter and others, 1984). Capping

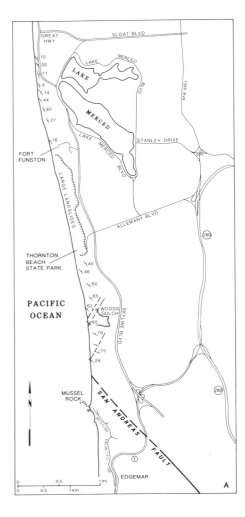

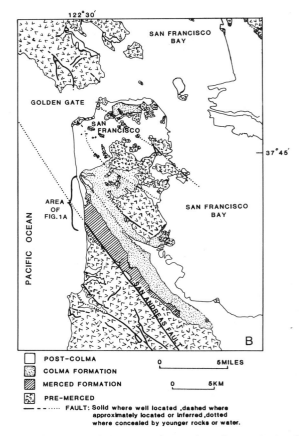

Figure 1. Index Maps. A. Map showing pertinent features in the vicinity of the type section of the Merced Formation. B. Map showing the distribution of the Merced Formation, Colma Formation, pre-Merced rocks, and post-Colma deposits on the San Francisco Peninsula.

the section is the Colma Formation, a thin unit of coastal origin that lies with angular discordance on the subjacent strata (Schlocker and others, 1958; Bonilla, 1959). The Colma Formation is probably no younger than the last interglacial highstand of the sea at 0.07 to 0.13 Ma (Hall, 1965, 1966).

The strata, with one major exception, lie in a homoclinal succession that dips to the northeast. The beds in the northern, uppermost part of the succession dip generally less than 20°; in contrast those in the lower part of the section exposed to the south dip generally in excess of 50° (Fig. 1A). Between Fort Funston and Thornton Beach, the strata strike approximately parallel to the beach and dip gently eastward. The coincidence of this structural attitude with the presence of thick, relatively incompetent muddy intervals of sequences R and Q at or near beach level may be responsible for the large landslides that occupy the area between Fort Funston and Thornton Beach (Fig. 1A).

Small faults break the succession, particularly in its lower part (Fig. 2), but the strata can be readily correlated across all observed faults. Breaks in the continuity of the succession are possible in the few places where the section is covered, as at Woods Gulch (Figs. 1A and 1B). Ashley (1895) mentions a fault

of "825 feet, down throw on the south side" at this location. This fault has not been observed by subsequent workers (Bonilla, 1959; Hall, 1965). The offset indicated by Ashley (1895) would cause slightly more than 330 ft (100 m) of the section to be repeated (Hall, 1965). Although sequences B and C and the succession of sequences ZZ-A-B-C and C-D-E-F resemble one another superficially, they differ sufficiently in detail to suggest that no significant unseen faulting occurs at Woods Gulch. Accordingly, the entire stratigraphic section exposed north of the San Andreas fault (Fig. 3) probably is structurally intact.

The timing of the deformation of the Merced is uncertain. The superjacent Colma Formation, which now lies more than 100 ft (30 m) above sea level at Thornton Beach, has clearly been elevated since its deposition. The underlying strata, in contrast, experienced a general trend of subsidence over the period of their deposition. The tilting of these strata largely predates the deposition of the Colma Formation, which overlies them in angular discordance. It is not clear whether the deformation occurred slowly and contemporaneously with the deposition of the Merced and the overlying strata—such a pattern could account for the progressive increase in deformation with age—or in a single par-

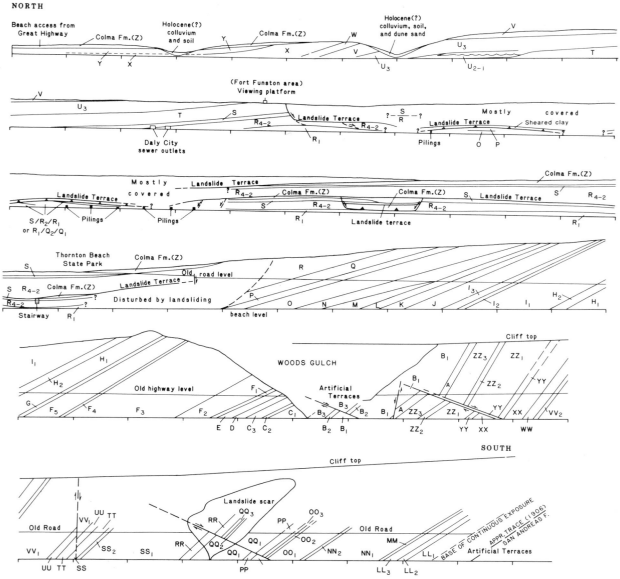

Figure 2. Generalized cross section of the Merced Formation and younger strata exposed on the seacliffs between Lake Merced and Mussel Rock, San Francisco Peninsula (southern end of section lies about 1,640 ft [500 m] north of Mussel Rock). Letters and numerical subscripts refer to sequences and lithologic units shown in stratigraphic column (Fig. 3). Sections tie together end-to-end from northernmost section (top) to southernmost section (bottom). No vertical exaggeration.

oxism shortly before deposition of the Colma Formation. In the latter case, the southward increase in degree of deformation may be due primarily to increasing proximity to the San Andreas fault at the southern (and circumstantially oldest) terminus of the section.

The depositional facies exposed in the section are described in detail by Hunter and others (1984). A brief summary of their lithologic character is presented here, with suggestions as to where they are best developed in the section or show features of special interest. Approximate locations of the sequences are shown in Figures 1A and 2.

Shelf facies ("1" on Fig. 3). This facies ranges texturally from well-sorted fine sand to sandy silt. Small pebbles may be present locally either in thin layers or isolated within the sediment. Stratification in the well-sorted sand typically is cryptic and, in some units, clearly disrupted by bioturbation. Fine lamination is common in the sandy silt, particularly near a vertical transition to well-sorted fine sand. This lamination commonly is defined by concentrations of carbonaceous detritus; it may be horizontal or show a tendency toward hummocky cross-stratification. Decimeter-thick cycles of sharp-based laminated sandy silt gradationally overlain by bioturbated, texturally

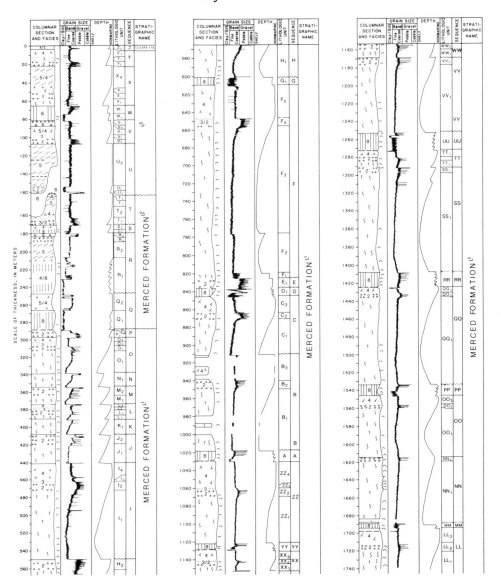

Figure 3. Stratigraphic section of the Merced Formation and younger strata exposed in the seacliffs between Lake Merced and Mussel Rock, San Francisco Peninsula. Sequences and lithologic units as introduced in Hunter and others (1984). Facies identified by numbers in the stratigraphic column: 1, shelf; 2, nearshore (may include foreshore facies in some sequences); 3, foreshore; 4, backshore; 5, eolian dune; 6, alluvial; 7, fresh-water pond/swamp/marsh; 8, embayment. Combinations of facies shown with slashed lines. Heavy dashed lines in the stratigraphic column indicate location of lignitic layers; x's mark well-developed paleosols. Occurrence of mollusks shown schematically by symbols to right of column. Grain size as estimated visually in field. Wavy lines between sequences indicate erosional surfaces. Gaps in section reflect covered intervals. Thickness in meters. [1] Denotes Hall's (1965) lower member of the Merced Formation. [2] Denotes Hall's (1965) upper member of the Merced Formation. [3] Denotes the unnamed beds of Hall (1966).

equivalent sediment probably reflect episodes of storm activity. Soft-sediment deformation in zones 3 ft (1 m) or less thick is not uncommon, and zones of intense penecontemporaneous shearing several inches or feet (several decimeters) thick are visible on surfaces freshly washed by the waves.

Mollusk remains are common in this facies as isolated disarticulated shells or shell fragments and as paired valves, some in

living position. Lenticular concentrations of shells and shell fragments attest to episodes of winnowing of the sea floor by storm waves or currents.

The shelf facies is well displayed in many of the sequences below sequence P. Sequence 0 shows particularly well the disruptive effects of large bivalves within the fine sand part of the facies. Vertical textural variations that probably reflect changes in water depth are evident in sequences OO, PP, SS, VV, B, C, F, H, and I.

Nearshore facies ("2" on Fig. 3). This facies consists of interstratified coarse- to fine-grained sand and gravel. High- and low-angle medium-scale (0.13-3 ft [4-100 cm] thick) cross-bedding is prevalent. Abraded shells and shell fragments are a common component of the gravel in many examples of this facies. Burrows, typically filled with sediment coarser than the host sediment, exist in most of the nearshore units, many of which contain the trace fossil *Macaronichnus* in their upper parts. Excellent examples of the nearshore facies exist in sequences VV, WW, ZZ, B, H, I, J. M, O, P, T, V, and Y.

Foreshore facies ("3" on Fig. 3). This facies is inferred to have been deposited in the swash zone of ancient beaches. Texturally composed of fine- to coarse-grained sand and fine gravel, the facies is typified by parallel planar stratification. Individual laminae may show inverse textural grading, but the overall deposit typically shows overall upward fining. Planar concentrations of heavy minerals may occur in this facies. In many of the sequences in the lower part of the section, the foreshore facies cannot be clearly delineated owing to the nature of exposure; in such cases, it presumably stratigraphically separates the nearshore facies and the backshore facies. The foreshore facies is well developed in sequences XX, ZZ, O, T, V, and Y.

Backshore facies ("4" on Fig. 3). This facies is inferred to have been deposited in supratidal sand flats bordering open beaches or embayments. The facies consists primarily of fine-grained sand. A few thin beds of pebbly sand exist, and thin beds of sandy clay or lignitic sand are common. Flat, gently lenticular and wavy bedding predominates. Ripple lamination is abundant in some beds and climbing-adhesion-ripple structure is fairly common. Mammal footprints occur within this facies. A spectacular example occurs in unit S_2, where bedding plane exposures protruding through the beach sand just north of the Daly City storm sewer outfall show a variety of bilobate impressions and, on occasion, well-developed canid paw prints. Some beds show load structures in their upper parts that may be due to trampling by large mammals. Root and rhyzome structures are common and provide additional evidence of subaerial exposure. A strikingly white bed of ash occurs in this facies in unit S_2; radiometric dates on the ash indicate an age of 400,000 years (Sarna-Wojcicki and others, 1985). The backshore facies is well developed in sequences throughout the section: LL, QQ, XX, ZZ, C, F, I, O, J, T, V, X, Y, and Z.

Dune facies ("5" on Fig. 3). This facies reflects deposition in coastal eolian dunes. Composed almost entirely of fine sand, the facies is typified by medium- to large-scale, medium- to high-angle crossbedding. Large-scale deformational structures are locally present; those near the bottom of unit U_3 are particularly striking owing to the concentration of dark heavy minerals in these beds. A few thin layers of mud and pebbly sand occur locally in flat-bedded parts of the facies. The few trace fossils consist of small vertical tubes in unit U_3 and possible footprint structures. Sporadic remains of terrestrial mammals have been found in the dune facies as well as in the backshore facies (Hall,

1965; 1966). Paleosols occur at several horizons within each of the dune units. The eolian dune facies is well developed in sequences (R, T, U, V, and X.

Alluvial facies ("6" on Fig. 3). Fluvial and alluvial-fan gravels and pebbly sands are relatively uncommon in this section. Where present, the facies typically contains angular to subrounded clasts of Franciscan chert and graywacke. Stratification in the gravel beds is indistinct; in alluvial sand it consists of lenticular bedding and small- and medium-scale trough crossbedding. Paleosols are common in the alluvial facies. The facies occurs in sequences R, T, U, and V.

Fresh-water pond/swamp/marsh facies ("7" on Fig. 3). This facies consists of mud, peat, or lignite. The mud is flat-bedded (laminated to thick-bedded). Burrows and/or intrastratal trails are evident in the laminated beds, and burrows or root structures commonly extend into the sand below the muddy or peaty beds. Roots and wood fragments are present within the peaty muds. Fossils in this facies include fresh-water diatoms (in unit U_2; Glenn, 1959), wing covers of nonaquatic beetles (in unit P_2; Hall, 1965, 1966) and terrestrial mammalian remains (Hall, 1965, 1966). Unit U_2 is notable for the presence of very large-scale load structures, tens of ft (m) across and several ft (m) deep, that can be seen best when much of the sand has been eroded from the beach. Thin pond or marsh deposits are common in the backshore facies and also are present in the dune facies. Peaty beds are present in sequences LL, QQ, TT, and XX, but the facies is best developed in sequences P, S, and U.

Embayment facies ("8" on Fig. 3). The embayment facies as used here represents the deposits in coastal embayments that may have ranged from brackish estuaries to tidal lagoons in which oceanic salinity was approached or exceeded. The facies accordingly is rather variable. Most of the deposits assigned to this facies contain muddy sediment, either throughout the deposit or restricted to individual beds. The embayment facies typically contains the remains of marine invertebrates, commonly in living position. Bioturbation is common and trace fossils are abundant and diverse. Sand and gravel, where present, generally show cross-stratification. A rhythmic alternation of sandy and muddy laminated intervals in sequence D suggests a tidal influence.

Some of the sequences (R, S, and W) contain vertical successions that suggest deposition in migrating tidal channels. The distinctive shell beds in unit R (the "Upper Gastropod Bed" of Ashley, 1895) appear to be lag deposits that accumulated on the floor of a tidal channel. In contrast, unit G consists of several ft (m) of more or less homogeneous shelly sand mud without an evident vertical sequence. Many of the embayment facies in the lower part of the section (in sequences PP, RR, YY, and AA) show a general upward coarsening from mud at the base to crossbedded sand and gravel at the top. The embayment facies is well developed in sequences RR, UU, YY, A, D, G, R (in beach level exposures in place below Fort Funston and in the landslide mass near its southern end at Thornton Beach), S, and W.

Facies organization. The facies described above occur in consistently repeated vertical arrays. The dominant pattern is that of systematic upward shallowing—an upward progression, where complete, of shelf to nearshore to foreshore to backshore to eolian dune facies. Only one sequence (X) displays all of these facies, and it lacks a silty component that commonly occurs in the lower part of the section, where it is attributed to deposition in deeper water. Many of the sequences, however, show a marine to nonmarine succession (LL, OO, QQ, TT, XX, ZZ, B, C, F, I, L, O, T, V, X, and Y). Even those sequences that consist only of the shelf facies (K and N) coarsen upward and thereby imply an upward shallowing. Of the 41 sequences identified in the section, 27 (accounting for 80% of the total thickness) show evidence of upward shallowing in an open-coast setting. The dominant style of sedimentation appears to be that of a shoreline prograding into an open marine basin.

Most of the progradational sequences are bounded by unconformities, above which commonly lie thin lag deposits of pebbles and shells. Widespread erosion evidently accompanied the landward translation of the shoreline during the marine transgressions that separated the progradational episodes. Such a pattern is typical of shallow marine successions elsewhere (Clifton, 1981; Ryer, 1977).

Deposition during transgressive episodes is manifested in two ways. In many of the sequences in the lower part of the section (OO, QQ, SS, FF, H, and I), the basal shelf facies is composed of well-sorted fine sand that grades upward into sandy silt, which suggests deposition in progressively deeper water.

The second manifestation of transgressive deposition lies in the embayment deposits. Several (RR, UU, YY, A, Q, S, and W) overlie a lignitic bed or alluvial unit that caps an underlying backshore or eolian unit, and upward coarsening to crossbedded sand and gravel at or near the top is common (PP, RR, YY, A, and D). It is likely that these features reflect an encroaching sea.

As sea level rose, freshwater coastal marshes (the lignitic beds) or an aggrading coastal plain or valley (the alluvial units) would develop first, to be subsequently inundated with further transgression. Mud, initially deposited in the upper reaches of an embayment, would be replaced by sand as the embayment shifted landward. Finally, crossbedded sand and gravel would be deposited as the tidal inlets migrated over the surface. This sequence of events implies the presence of a coastal barrier that was not preserved during the transgression.

Some of the embayment units may reflect the general landward limit of a transgression. In sequences Q, R, and S, the embayment facies is overlain by backshore or eolian facies. This relationship suggests renewed progradation rather than further transgression. The other examples of embayment facies are overlain by deeper water deposits (shelf facies in sequences PP, RR, UU, YY, A, G, and W), suggesting substantial subsequent inundation that probably resulted from continued transgression.

The individual effects of tectonics, eustatic sea level change, and fluctuations of sediment supply in generating the progradational and transgressive episodes are not resolved. Large-scale glacio-eustatic changes in sea level are recorded elsewhere in the world during the interval in which this section accumulated and must have been of great consequence in the development of the sequences. Given the highly active tectonic setting, however, significant rapid changes in the elevation of the land surface cannot be discounted and tectonism must be responsible for the angular unconformity at the base of sequence Z (Colma Formation). Changes in the rate of sediment supply may account for the contrast in facies above (dominantly eolian) and below (dominantly shelf) the contact between sequences P and Q, which marks the influx of sediment from the Sacramento and San Joaquin river systems (Hall, 1965; 1966).

REFERENCES CITED

Addicott, W. O., 1969, Late Pliocene molluscs from San Francisco Peninsula, California, and their paleogeographic significance: California Academy of Science Proceedings, Series 4, v. 37, no. 3, p. 57–93.

Ashley, G. H., 1895, The Neocene [sic] of the Santa Cruz Mountains, I-Stratigraphy: California Academy of Science Proceedings, Series 2, v. 5, pt. 1, p. 273–367. (Reprinted in Stanford University Publications in Geology and Paleontology, no. 1.)

Bonilla, M. G., 1959, Geologic observations in the epicentral area of the San Francisco earthquake of March 22, 1957: California Division of Mines Special Report 57, p. 25–37.

Clifton, H. E., 1981, Progradational sequences in Miocene shoreline deposits, southeastern Caliente Range, California: Journal of Sedimentary Petrology, v. 51, p. 165–184.

Glen, W., 1959, Pliocene and lower Pleistocene of the western part of the San Francisco Peninsula: University of California Publications in Geological Science, v. 36, p. 147–198.

Hall, N. T., 1965, Petrology of the type Merced Group, San Francisco Peninsula, California [M.A. thesis]: Berkeley, University of California, 126 p.

—— , 1966, Fleishhacker Zoo to Mussel Rock (Merced Formation); A Plio-Pleistocene nature walk: California Division of Mines and Geology Mineral Information Service, v. 19, no. 11, p. S22–S25.

Hunter, R. E., Clifton, H. E., Hall, N. T., Császár, G., Richmond, B. M., and Chin, J. L., 1984, Pliocene and Pleistocene coastal and shelf deposits of the Merced Formation and associated beds, northwestern San Francisco Peninsula, California: Society of Economic Paleontologists and Mineralogists Field Trip Guidebook No. 3, 1984 Midyear Meeting, San Jose, California, p. 1–29.

Lawson, A. C., 1893, The post-Pliocene diastrophism of the coast of southern California: University of California Publications Department of Geology Bulletin, v. 1, no. 4, p. 115–160.

Roth, B., 1979, Late Cenozoic marine invertebrates from northwest California and southwest Oregon [Ph.D. thesis]: University of California, 803 p.

Ryer, T. A., 1977, Patterns of Cretaceous shallow-marine sedimentation, Coalville and Rockport area, Utah: Geological Society of America Bulletin, v. 88, p. 177–188.

Sarna-Wojcicki, A. M., Meyer, C. E., Bowman, H. R., Hall, N. T., Russell, P. C., Woodward, M. J., and Slate, J. L., 1985, Correlation of the Rockland ash bed, a 400,000-year-old stratigraphic marker in northern California and western Nevada, and implications for middle Pleistocene paleogeography of central California: Quaternary Geology, v. 23, p. 235–257.

Schlocker, J., Bonilla, M. G., and Radbruch, D. M., 1958, Geology of the San Francisco North Quadrangle: U.S. Geological Survey Miscellaneous Investigations Map I-272, scale 1:24,000.

Marin Headlands, California: 100-million-year record of sea floor transport and accretion

Clyde Wahrhaftig and Benita Murchey, U.S. Geological Survey, 345 Middlefield Road, Menlo Park, California 94025

LOCATION AND ACCESSIBILITY

The Marin Headlands are the hills on the north shore of the Golden Gate opposite San Francisco (Figs. 1, 2; see Point Bonita and San Francisco North 7½-minute Quadrangles). They are entirely within the Golden Gate National Recreation Area (GGNRA), a unit of the National Park System. Collecting of rock specimens is by permit only, obtainable on adequate scientific or educational justification from the GGNRA headquarters at Fort Mason, San Francisco 94123 (415-556-0560). Application for permit should be made at least two weeks in advance of the visit.

San Francisco MUNI bus No. 76 runs on Sundays and holidays only, between San Francisco and Fort Cronkhite Visitor Center (FCVC, Fig. 1). For schedule and stops in San Francisco call the MUNI system (415-673-MUNI). The Golden Gate Transit District bus No. 10 makes daily stops at the Alexander Avenue off-ramp (Locality 1, Fig. 1); buses Nos. 20, 50, and 80 make daily stops at Spencer Avenue (Locality 2, Fig. 1); and bus No. 10 also stops at Tamalpais Valley Junction, from which a road and trail go to the Tennessee Valley trailhead (Locality 3, Fig. 1). Call 415-332-6600 for the Golden Gate Transit schedule and stops in San Francisco.

Automobile access from San Francisco is via U.S. 101 north across the Golden Gate Bridge and the second off-ramp (Alexander Avenue) on the north side of the bridge; turn left at first opportunity (about 100 m), go back underneath the freeway, and keep to right up the hill into the Golden Gate National Recreation Area (GGNRA) and Marin Headlands. Approaching from the north on U.S. 101, take the third Sausalito off-ramp (the only off-ramp between the Waldo Tunnel and the Golden Gate Bridge), turn left immediately at the stop sign, and then turn right immediately up the hill at the sign to the GGNRA. The Tennessee Valley Trailhead (Locality 3, Fig. 1) is accessible from California 1 between U.S. 101 and Tamalpais Valley Junction. All travel away from roads must be made on foot, horseback, or bicycle; motorized vehicles are prohibited off designated roads.

SIGNIFICANCE OF THE SITE

The rocks of the Marin Headlands contain the record of the 100-m.y. migration of mid-Mesozoic Pacific Ocean floor from its eruption, which was probably close to the equator, on a spreading axis, to its accretion by subduction, thousands of mi (km) to the northeast, to the North American continent. They include mid-ocean ridge basalt, with some of the best-exposed pillows to be seen on land, overlain by ribbon chert with a nearly complete radiolarian biostratigraphy from the Early Jurassic to the mid-Cretaceous (Murchey, 1984), overlain in turn by turbidites of continental or island-arc origin, presumably deposited just prior to accretion. The 10 or more thrust sheets, in which all or part of this sequence is repeated, are an unusually clear example of the imbricate stacking that took place in many subduction zones. Subsequent to accretion, the entire stack has been rotated clockwise 90° to 120° about a vertical axis, possibly because of right-lateral shear along the San Andreas system (Wahrhaftig, 1984b; Curry and others, 1984). The Headlands were among the first places along the Pacific Coast to be described geologically (Chamisso *in* Kotzebue, 1821; see Wahrhaftig, 1984b, for a review) and were the type locality for many of the formations into which Lawson (1915) divided his Franciscan Group when he first applied the name to what is now called the Franciscan assemblage.

SITE INFORMATION

Basalt

The basalt is generally weathered to depths of 15 to 30 ft (5 to 10 m) to an orange-brown mixture of clay minerals, ferric oxides, and hydroxides; natural outcrops of fresh basalt are rare except along the lower parts of sea cliffs. Few roadcuts penetrate the weathered basalt to expose unweathered dark green pillows.

The original igneous minerals as well as the basaltic glass were altered to mixtures of nontronite, chlorite, albite, epidote, pumpelleyite, and laumontite, an assemblage characteristic of the zeolite facies. From the resulting color the basalt is commonly called greenstone (Schlocker, 1974, p. 32-33). Presumably this alteration was by hot circulating sea water at the mid-ocean ridge.

The best exposures of pillows are at: (1) Locality 12, south side of Tennessee Point, currently fenced off from civilians because it is an Army explosives detonation site; (2) Locality 13 in Bonita Cove, which is accessible only during minus tides; and (3) Locality 14, the point west of Point Diablo (Fig. 3). Routes to Localities 12 and 13 are given in Wahrhaftig, 1984a, p. 42-44. Locality 14 can be reached via a fisherman's trail that follows the contour, branching east from the Bonita Cove access trail just south of the parking area. This trail is through unavoidable thickets of dense poison oak, and access to the wave-cut exposures of the pillows is particularly hazardous. Other exposures of pillows in three dimensions are at Locality 15, at the south end of Rodeo Beach, and on an inaccessible wave-washed rock visible from Point Bonita lighthouse on its south side, as well as at several points on the path to the lighthouse (Ransome, 1893). Excellent exposures of pillows in cross section are at a beach

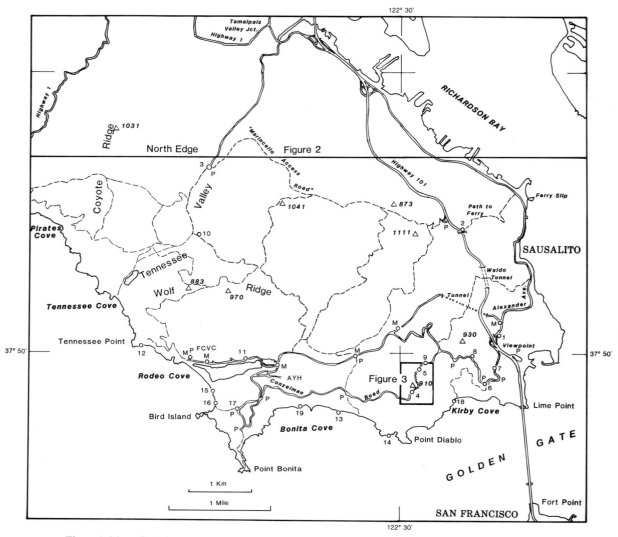

Figure 1. Map of Marin Headlands, showing routes of access and localities mentioned in text. Base from U.S. Geological Survey topographic maps of Point Bonita, San Francisco North, San Quentin, and San Rafael 7½-minute Quadrangles. M, San Francisco MUNI Route 76 bus stop; P, parking area. Numbered open circles indicate localities mentioned in text; numbered open triangles indicate summits. Numbers are altitudes in feet, solid double lines are automobile roads and highways, and single dashed lines are maintained foot trails.

outcrop (Locality 16) at the north end of the sandy beach just northeast of Bird Island. This beach can be reached by trail from a parking area at Locality 17. At this outcrop the pillow basalt is in contact on the east with medium-grained diabase. Pillows are also exposed along the access road to Kirby Cove and, during low tide, at the west end of the cove (Locality 18).

At all the above localities, a few pillows contain thin layers of basalt alternating with quartz-calcite rock that resulted from partial draining of the lava in the pillow and that are parallel to the paleo-horizontal at the time the pillows erupted (Moore and Charlton, 1984; see also Wahrhaftig, 1984b, p. 35 and Fig. 6).

Chert

The chert sequence contains abundant silt- and sand-sized radiolarian fossils, which are clearly visible with a hand lens in most chert beds. They can be extracted for study by etching the rocks with hydrofluoric acid after the method of Pessagno and Newport (1972). Seven sequential radiolarian assemblages (MH-1 through MH-7) have been identified in the chert sequence, ranging in age from Early Jurassic, Pliensbachian, to Cretaceous, late Albian, or early Cenomanian (Murchey, 1984). Scanning electron micrographs of representative fossils are shown in Figure 4.

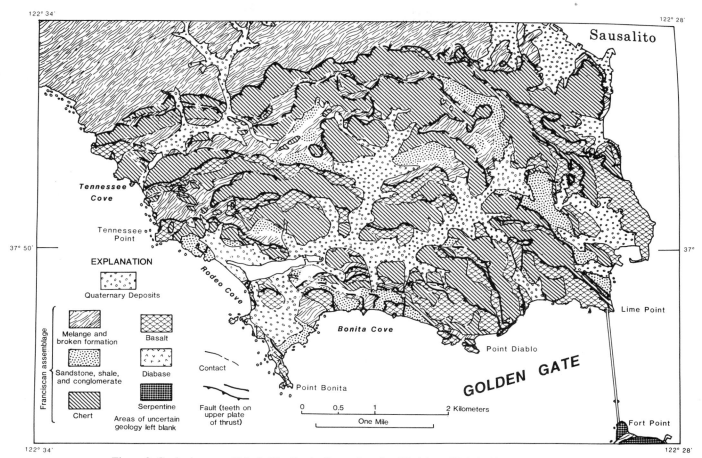

Figure 2. Geologic map of Marin Headlands. Somewhat simplified from Wahrhaftig, 1984b, Figure 13.

The reference section for the chert stratigraphy along the Alexander Avenue off-ramp and U.S. 101 (Murchey, 1984, Localities M-2a,b) is dangerous because of traffic; it is illegal to visit without a special permit from Caltrans (California Department of Transportation). An alternative section is exposed on Conzelman (Headlands) Road at and just north of Battery 129 (Localities 4 and 5, Fig. 1) and is shown in detail in Figure 4.

The oldest radiolarian assemblage (assemblage MH-1) is Pliensbachian and (or) Toarcian. Species in this assemblage are listed in Murchey (1984), and include *Canutus rockfishensis* Pessagno and Whalen, and *Trillus elkhornensis* Pessagno and Blome, shown in Figure 5. The chert containing the MH-1 assemblage rests depositionally on the basalt and at its base has lenticular conchoidally fracturing chert beds as much as 10 ft (3 m) thick. These are overlain by manganiferous maroon argillaceous chert in beds more than 4 in (10 cm) thick interbedded with maroon shale, which are overlain in turn by nonmanganiferous chert in beds less than 4 in (10 cm) thick, to a total thickness of as much as 65 ft (20 m) (see Fig. 4).

Three younger Jurassic assemblages (MH-2, MH-3, and MH-4) range in age from early Middle Jurassic (Bajocian) to Middle or early Late Jurassic. Thirty species represented in these assemblages are listed by Murchey (1984), including *Ristola turpicula,* Pessagno and Whalen (Fig. 5). These assemblages lie in a chert interval characterized by alternating beds of shale and nonmanganiferous red and green radiolarian-rich chert. This lithology can be observed at Locality 4 (see Figs. 1 and 4). At this

Figure 3. Pillow basalt at Locality 14, Figure 1 (point west of Point Diablo). Board is 25 cm wide and 35 cm long.

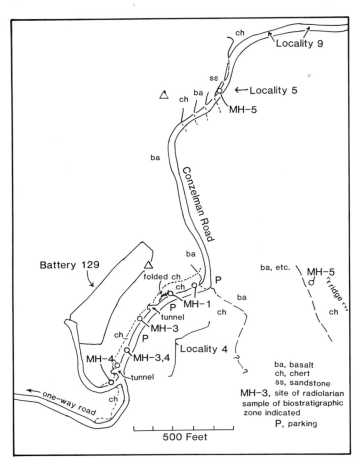

Figure 4. Sketch map of Battery 129 (Locality 4, Fig. 1). Sketched from a 1:2,400 ozalid print of a Caltrans Bay Area Transit System aerial photograph taken in 1965.

locality, assemblages MH-3 and MH-4 have been recovered.

In the Marin Headlands, Early Cretaceous assemblages are separated from Middle or early Late Jurassic assemblages by a 3- to 30-ft (1- to 10-m) recrystallized and commonly brecciated interval from which radiolarians cannot be extracted. Several Late Jurassic and Early Cretaceous radiolarian assemblages that should occur between the youngest Jurassic assemblage (MH-4) and the oldest Cretaceous assemblage (MH-5) are missing. This biostratigraphic gap represents 15 to 30 m.y., suggesting nondeposition or erosion within or above the recrystallized interval (Murchey, 1984). The recrystallized zone is well exposed along a ridge extending southeast from Locality 4. Most of the recrystallized interval consists of thinly bedded, pale green or white chert with no interbedded shale. However, orange-colored chert forming jagged crags along the ridge-crest is brecciated and contains gravity filling of darker colored chert.

Three Cretaceous radiolarian assemblages have been recognized in the less than 20 ft (6 m) of chert between the recrystallized chert unit and the depositionally overlying lithic arkose. The oldest Cretaceous radiolarian assemblage (MH-5) ranges in age

from Valanginian to Hauterivian or early Barremian. Assemblage MH-6 is a poorly preserved assemblage that occurs in partially recrystallized chert. The range of MH-6 is Hauterivian(?) to Albian(?). Directly overlying the poorly preserved MH-6 assemblage are beautifully preserved, broken radiolarians of the MH-7 assemblage, which includes *Thanarla veneta* (Squinabol), shown in Figure 5. The possible range of this assemblage is late Albian to early Cenomanian. A Cenomanian ammonite, *Mantelliceras* sp., has been recovered from float from the overlying clastic unit (Hertlein, 1956). The Cretaceous chert sequence consists of red and green ribbon chert beds characteristically less than 2 in (5 cm) thick, interbedded with very thin shales. Beds are commonly striped with red centers and green margins. The thin Cretaceous unit can be observed in a small anticline at Locality 5 on Conzelman (Headlands) Road (Fig. 4). At this locality, only MH-5 assemblage radiolarians have been recovered.

The Marin Headlands chert sequence records pelagic deposition below the calcite compensation depth over a period of approximately 100 m.y. It is one of the longest stratigraphic intervals represented in a single chert sequence in the world. Except for the uppermost part of this sequence, the Marin Headlands chert sequence contains very little volcanic or terrigenous clastic detritus. The site at which the radiolarian-rich siliceous sediments were deposited probably lay far offshore during the Early Jurassic but had moved close to the Americas by late Early Cretaceous time. Both paleontologic and paleooceanographic arguments support the hypothesis that the depositional site of the pelagic sediments was at lower latitudes relative to North America than the present locality of Marin Headlands (Murchey, 1984; Karl, 1984).

Beside Localities 4 and 5, other easily accessible chert localities are: at Locality 6, Figure 1, manganiferous chert of zone MH-1 rests on basalt. About 600 ft (200 m) east, southwesterly elongated pillow tops are exposed in the roadcut. Tightly folded chert is exposed in roadcuts at Localities 7, 8, and 9 on Conzelman Road (Fig. 6; see also Wahrhaftig, 1984a, Fig. 33 and 1984b). Other exposures of folded chert—not generally as photogenic as those just listed—are at Locality 10, just east of a horse corral in Tennessee Valley, and on the north side of the beach at Tennessee Cove. Exposures of chert without striking folding are common in cuts along roads and trails in the headlands, as well as in a few quarries, such as at Locality 11 on the north side of Rodeo Lagoon (the lagoon east of Rodeo Cove). The larger of these exposures consist of partial sections of the radiolarian biostratigraphy.

Clastic Rocks and the Turbidite Sequence

Because of their ease of weathering, natural exposures of sandstone and shale away from the sea cliffs are about nonexistent. The excellent roadcut exposures on U.S. 101 south of Waldo Tunnel cannot be examined at leisure, and most exposures in road- and trail-cuts are too deeply weathered to be informative. Most sea-cliff exposures are steep and inaccessible.

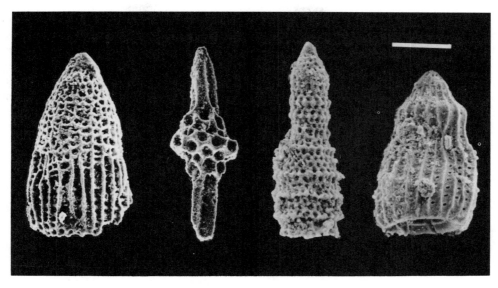

Figure 5. Electron micrographs of representative radiolarian fossils. From left to right: (1) *Canutus rockfishensis* Pessagno and Whalen, MH-1 assemblage (USGS MR 5101); scale bar, 104 μm. (2) *Trillus elkhornensis* Pessagno and Blome, MH-1 assemblage (USGS MR 5101); scale bar, 96 μm. (3) *Ristola turpicula* Pessagno and Whalen, MH-4 assemblage (USGS MR 5110); scale bar, 83 μm. (4) *Thanarla veneta* (Squinabol), MH-7 assemblage (USGS MR 5064); scale bar, 69 μm.

The cliffs at the two ends of the barrier beach of Rodeo Lagoon (the lagoon just east of Rodeo Cove) expose informative turbidite sections, the one on the north exhibiting coarse pebbly sandstone which fills channels cut in finer sandstone, and the one on the south exhibiting carbonaceous fragments on partings in the finer Bouma units. Lower parts of the turbidite sequence exposed at the north end of the beach are also exposed, but not so well, in cuts along the road to Tennessee Point.

A near-vertical cliff about 300 ft (100 m) high of massive sandstone at Locality 19 on Bonita Cove exposes at its base unusual diamict conglomerate with subangular boulders of greenstone and limestone and pebbles of red chert. To the east of this cliff, huge landslide blocks of conglomerate with well-rounded cobbles of gray quartzite and chert make a point into the cove (Seiders and Blome, 1984, Locality 62). The hazardous access to these exposures on the east side of the 300-ft (100-m) cliff is given in Wahrhaftig, 1984a, p. 45.

The best exposed and most informative turbidite section nearby is at the north end of Baker's Beach just south across the Golden Gate from the Marin Headlands (Wahrhaftig, 1984a, p. 20–22).

Other Franciscan Rocks

Coarse diabase, intrusive into basalt, and separated from sandstone on the east by a vertical fault, forms an anastomosing dike-like body trending north from Point Bonita to Locality 16. For the most part this rock is too thoroughly weathered to provide suitable material for petrographic study.

Serpentine is rare in the Headlands but is prominent in the cliffs between Baker's Beach and the Golden Gate Bridge (see southeast corner of Fig. 2, and Wahrhaftig, 1984a).

Structure

The rocks of the Marin Headlands are in an approximately 10-fold stacking of south- to southwest-dipping thrust slices consisting mainly of chert, which in many slices rests on pillow basalt and is overlain by the turbidite sequence (Fig. 2). All up-section indicators show that within individual thrust slices beds are younger upward and to the south.

Separating the thrust slices are belts of sandstone and shale commonly broken into thin phacoids of sandstone embedded in thoroughly sheared shale—a structural state called broken formation. Belts of broken formation a few m thick mark many of the faults in the Marin Headlands and are exposed, for example, at two places in the sea cliff between Rodeo Lagoon and Locality 16. More extensive exposures are between Localities 13 and 19 on Bonita Cove and at the east end of Bonita Cove beach.

Marin County north of the Headlands is a large area of mélange, which consists of a matrix of this tectonically kneaded sandstone and shale, in which are embedded large exotic blocks of basalt, chert, well-lithified sandstone, and serpentine, as well as eclogite, glaucophane schist, and other rocks of the high-pressure, low-temperature blueschist facies. The mélange is poorly exposed, but the exotic blocks, some of them schist, may be found scattered over Coyote Ridge and its spurs. Exposures of the gray clay-rich matrix are encountered along the old "Marincello access road" (now a trail) southeast of the Tennessee Valley trailhead, and also in sea cliffs around Pirate's Cove. Other exposures of

Figure 6. Tightly folded chert at Locality 9, Figure 1.

mélange are along the base of the cliffs between Baker's Beach and the south end of the Golden Gate Bridge (southeast corner of Fig. 2; see Wahrhaftig, 1984a, p. 19–25).

The minor folds in the chert have hinge lines trending south to southwest and are asymmetric, with the short limbs characteristically on the north sides of anticlinal hinges. Interpreted as the product of shear during emplacement of the thrust slices by subduction, they imply south-directed subduction, in contrast to the easterly to northeasterly direction implied by all other evidence in California. Together with paleomagnetic vectors of 105° to 115° (Curry and others, 1984), they imply a 90° to 120° clockwise rotation since docking.

Quaternary Deposits

All Quaternary deposits on Marin Headlands and along the south shore of the Golden Gate are subaerial and indicate lower stands of the sea with respect to the land than at present. All Pleistocene deposits are deeply weathered and generally have well-developed soils with thick, clay-rich B horizons that suggest that they date from the Sangamonian interval at the latest (Wahrhaftig, 1983).

The bluff behind the beach northeast of Bird Rock (accessible from Locality 17) exposes flat-lying interbedded gravel, sand, and silt in crude lenticular beds. All are poorly sorted and deeply weathered, and the gravel consists of angular chert and sandstone fragments whose imbrication indicates deposition by small streams flowing directly west. Alluvium in the valley floor passes laterally into colluvium mantling the hills on either side; this colluvium is cut off on the south by the sea cliff advancing northward from the opposite side of the point east of Bird Rock. Similar relations exist at the north end of Rodeo Beach, where they are obscured by road construction.

A complex of talus, lagoonal, and dune deposits at the north end of Baker's Beach is capped by a thick clay-rich soil (Wahrhaftig, 1984a, p. 20).

All these subaerial Quaternary deposits, facing the open ocean and extending to sea level and below, together with the complete lack of marine terraces incised into this mature fluvial landscape, demonstrate that the sea has never been higher against this topography than it is now. Hence, unlike much of the California coast, the land around the Golden Gate may be tectonically subsiding.

REFERENCES CITED

Curry, F. B., Cox, A., and Engebretson, D. C., 1984, Paleomagnetism of Franciscan rocks in the Marin Headlands, *in* Blake, M. C., Jr., ed., Franciscan Geology of Northern California: Los Angeles, California, Pacific Section, Society of Economic Paleontologists and Mineralogists, v. 43, p. 89–98.

Hertlein, L. G., 1956, Cretaceous ammonite of Franciscan group, Marin County, California: American Association of Petroleum Geologists Bulletin, v. 40, p. 1985–1988.

Karl, S. M., 1984, Sedimentologic, diagenetic, and geochemical analysis of upper Mesozoic ribbon cherts from the Franciscan Assemblage at the Marin Headlands, California, *in* Blake, M. C., Jr., ed., Franciscan Geology of Northern California: Los Angeles, California, Pacific Section, Society of Economic Paleontologists and Mineralogists, v. 43, p. 71–88.

Kotzebue, O. von, 1821, A Voyage of Discovery into the South Seas and Beering's Straits for the Purpose of Exploring a North-east Passage, Undertaken in the Years 1815–1818, translated by Lloyd, H. E.: London, Longman, Hurst, Rees, Orme, and Brown, 3 vols.

Lawson, A. C., 1915, Description of the San Francisco District: U.S. Geological Survey Folio 193, field edition.

Moore, J. G., and Charlton, D. W., 1984, Ultrathin lava layers exposed near San Luis Obispo Bay, California: Geology, v. 12, p. 542–545.

Murchey, B., 1984, Biostratigraphy and lithostratigraphy of chert in the Franciscan Complex, Marin Headlands, California, *in* Blake, M. C., Jr., Franciscan Geology of Northern California: Los Angeles, California, Pacific Section, Society of Economic Paleontologists and Mineralogists, v. 43, p. 51–70.

Pessagno, E. A., Jr., and Newport, R. L., 1972, A technique for extracting *Radiolaria* from radiolarian cherts: Micropaleontology, v. 18, p. 231–234, plate 1.

Ransome, F. L., 1893, The Eruptive Rocks of Point Bonita: University of California Publications, Bulletin of the Department of Geology, v. 1, no. 3, p. 71–114.

Schlocker, J., 1974, Geology of the San Francisco North Quadrangle, California: U.S. Geological Survey Professional Paper 782, 109 p., 3 plates.

Seiders, V. M., and Blome, C. D., 1984, Clast compositions of upper Mesozoic conglomerates of the California Coast Ranges and their tectonic significance, *in* Blake, M. C., Jr., ed., Franciscan Geology of Northern California: Los Angeles, California, Pacific Section, Society of Economic Paleontologists and Mineralogists, v. 43, p. 135–148.

Wahrhaftig, C., 1983, Stratigraphic evidence for late Quaternary relative downwarping at the Golden Gate: American Association of Petroleum Geologists, Society of Exploration Geophysicists, Society of Economic Paleontologists and Mineralogists, Pacific Sections, 58th Annual Meeting, Sacramento, California, May 18-21, 1983, Program with Abstracts, p. 144.

——, 1984a, A Streetcar to Subduction and Other Plate Tectonic Trips by Public Transport in San Francisco, revised ed.: Washington, D.C., American Geophysical Union, 72 p.

——, 1984 b, Structure of the Marin Headlands block, California; A progress report, *in* Blake, M. C., Jr., ed., Franciscan Geology of Northern California: Los Angeles, California, Pacific Section, Society of Economic Paleontologists and Mineralogists, v. 43, p. 31–50.

Table Mountain of Calaveras and Tuolumne Counties, California

Dallas D. Rhodes, *Department of Geology, Whittier College, Whittier, California 90608*

LOCATION AND ACCESSIBILITY

Table Mountain is located in the western foothills of the central Sierra Nevada Range east of Stockton, California (Figs. 1 and 2). California 108 generally follows the ridge from Knight's Ferry to Sonora. Although Table Mountain is not now the obstacle it was for the early gold seekers, only four paved roads cross it: O'Byrnes Ferry Road through Peoria Pass, Rawhide Road near Jamestown, California 49 north from Sonora, and Parrot's Ferry Road west from Columbia. From Vallecito northward, California 4 provides some views of the upstream and more discontinuous parts of the flow, as does California 108 north to Sonora Pass. Most of the lower part of Table Mountain is privately owned and permission should be sought from local land holders. The higher areas are in national forests. Poison oak grows in abundance along the flanks of Table Mountain and caution is advisable.

SIGNIFICANCE OF SITE

The Stanislaus (or Tuolumne) Table Mountain is one of the best examples of inverted topography to be found anywhere. This ancestral valley of the Stanislaus River exists today as a meandering, discontinuous flat-topped ridge capped by a lava flow. The Table Mountain flow extends for more than 60 mi (100 km) from the crest of the Sierra Nevada, near Sonora Pass, to Knight's Ferry, where the flow is hidden beneath the sediments of the Mehrten Formation. Table Mountain also provides an excellent record of Tertiary fault activity in the Sierra foothills. In addition, the Table Mountain flow is the type locality of latite, a term first applied by Ransome (1898). Published studies of the area include ones by Lindgren (1911), Jenkins (1948, 1970), Taliaferro and Solari (1948), and Eric and others (1955).

REGIONAL SETTING

The Table Mountain channel is one of several Tertiary river channels preserved in the western Sierra (Fig. 2). The gravel deposits in these channels were the object of much of the placer mining during the California gold rush. The Tertiary channel deposits are of two distinct ages, termed by Lindgren (1911) "prevolcanic" and "intervolcanic." The Table Mountain or "cataract channel" is intervolcanic.

By the beginning of the Cenozoic, the "subadjacent" igneous and metamorphic rocks of the Sierra Nevada were deeply eroded in the vicinity of Table Mountain. Local relief was as great as 3,300 ft (1,000 m; Bateman and Wahrhaftig, 1966). In the western foothills, the relatively fine-grained sediments of the mid-Eocene Ione Formation were deposited on this surface. At higher elevations, the auriferous prevolcanic stream gravels were laid down.

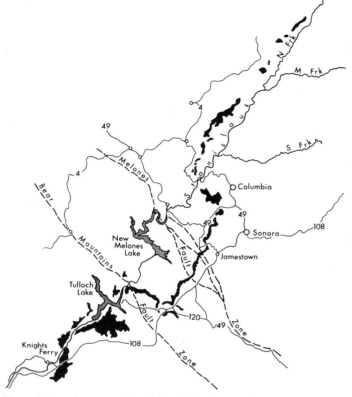

Figure 1. Location map of Stanislaus Table Mountain and surrounding area.

In the present Stanislaus drainage, the Oligocene to Miocene Valley Springs rhyolite tuffs blanketed the pre-existing topography about 33.4 Ma (Slemmons, 1966), burying the prevolcanic channels and requiring the establishment of new drainage lines (Fig. 3). The "cataract channel," the Pliocene course of the Stanislaus River, was one of the valleys cut during this "intervolcanic" period. This new river system captured drainage from the South Fork basin of the Tertiary Calaveras River (Fig. 2).

About 9 Ma, volcanic activity in the vicinity of Sonora Pass produced an eruption of latite lava that flowed both east and west from the crest of the range (Dalrymple, 1964). The Table Mountain Latite of the Stanislaus Formation (Slemmons, 1966) or Group (Noble and others, 1974) is the only stratigraphic marker connecting the Sierra Nevada to both the Basin and Range and the Great Valley regions. The westward flow entered the "cataract channel" and moved downslope for more than 60 mi (100 km), filling the Table Mountain valley with 49 to 295 ft (15 to 90 m) of lava. The base of the flow and its relation to the underlying sediments are well exposed just below Pulpit Rock (Fig. 4).

In late Miocene and early Pliocene time, andesitic material of the Relief Peak Formation was deposited on the "intervol-

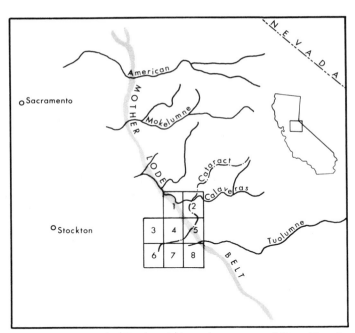

Figure 2. Map showing Tertiary drainage in west-central Sierra Nevada and topographic maps covering the lower part of Table Mountain. Key to 7½-minute Quadrangles: (1) Angels Camp, (2) Columbia, (3) Copperopolis, (4) Melones Dam, (5) Sonora, (6) Knights Ferry, (7) Keystone, (8) Chinese Camp.

canic" surface, preserving the consequent drainage developed by erosion of the rhyolitic materials. Plio-Pleistocene uplift of the Sierra rejuvenated the rivers draining the range.

The modern canyons of the Stanislaus and Tuolumne rivers were cut during the Pleistocene and Holocene, and most of the Tertiary volcanic cover was stripped from the area, exposing the underlying bedrock series (Fig. 3). Because the latite cap of the Table Mountain channel was far more resistant to erosion than the surrounding pyroclastic debris, the topography was inverted. The former Table Mountain channel, originally a topographic low, now stands above the countryside as a ridge. The downvalley portion of the Table Mountain flow has undergone relatively little erosion. Its form and features preserve many of the characteristics of the original Pliocene valley and can be used for paleogeomorphological interpretation. Displacement of the flow by several structures provides a record of some of the subsequent history of the western Sierra Nevada.

GEOLOGY AND GEOMORPHOLOGY OF TABLE MOUNTAIN

At its source near Sonora Pass, the Table Mountain Latite consists of up to 40 flows of similar composition with an aggregate thickness of 1,500 ft (460 m; Slemmons, 1966). Near Sonora, the unit appears to be a single flow with a maximum thickness of about 200 ft (60 m). The rock is an olivine-augite latite containing abundant phenocrysts of glassy plagioclase,

which are aligned with the flow direction in some places. Locally, both the top and the bottom of the flow are vesicular, and many vesicles are notably elongated in the flow direction. Columnar fracturing of the middle of the flow is common. In the old railroad cut along the Rawhide Flat road northwest of Jamestown, the columnar structure is particularly well displayed. The surface of the flow is often quite rough and covered with small rounded latite boulders.

Relatively little gold was produced from the Table Mountain channel, although it is located in the heart of the Mother Lode Belt (Fig. 2). Most of the drainage of the ancestral Stanislaus River was across the volcanic terrain, not the gold-bearing bedrock. The gold found in the Table Mountain gravels may have been derived from pre-volcanic auriferous gravels, particularly those of the Tertiary Calaveras River where the Table Mountain channel crosses it, and from channel deposits near Columbia (Eric and others, 1955; Fig. 3). The most profitable mining was done from "pre-volcanic" gravels buried beneath the flow.

Unlike other "table mountains" in the Sierra foothills, the lower part of the Stanislaus Table Mountain flow has not been severely dissected by erosion. In a part of the Tertiary valley just west of Columbia that has not been inverted, the flow averages 500 to 600 ft (150 to 180 m) in width. Just downvalley, where topography has been inverted, the flow has a similar dimension, indicating that little of its original width has been destroyed. Tributary junctions preserved as spurs on both sides of the flow also attest to the lack of significant erosion of the flanks of the ridge. Vesicular lava fragments on the top of Table Mountain indicate that little of the original surface of the flow has been removed. The flow surface may have been protected for some time by sediment that continued to move down the valley. Rounded white vein-quartz cobbles and pebbles scattered across the surface of the flow may be remnants of an alluvial cover.

Table Mountain preserves the geometry of the "cataract channel." It is meandering and has a rather low sinuosity, about 1.3 (Fig. 4). The meanders have radii of curvature that vary from about 500 to 1,500 ft (150 to 460 m). Meander wave lengths are also variable, ranging from approximately 1,000 to 4,200 ft (320 to 1280 m). The channel width, radii of curvature of bends, and meander wave lengths of rivers are related to their discharge. Based on these characteristics, the mean-annual flood of the Table Mountain channel was estimated (Rhodes, 1980). Although all the measured parameters are variable, they indicate a

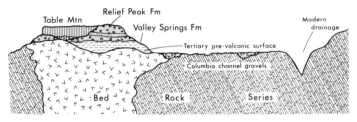

Figure 3. Diagrammatic geological cross section near Columbia (after Jenkins, 1970).

Figure 4. Aerial photograph of Table Mountain.

relatively narrow range of discharges, on the order of 1,100 to 5,000 ft³/sec (30 to 140 m³/sec).

Two pieces of evidence indicate that relief in the Table Mountain drainage was considerable at the time of the flow. Spurs projecting from the lava flow are preserved tributary junctions to the Table Mountain channel. At least 16 junctions are present along the 9-mi (15-km) length of the channel that is least eroded (Fig. 4). Nearly all of the paleo-tributaries enter the main channel at high angles (i.e., greater than 45°). High junction

angles are associated with tributaries that have significantly steeper slopes and are of much lower stream order than the main channel. Both are characteristic of basins with substantial relief. There is also direct evidence that the "cataract channel" valley was relatively deep. Even where the topography is not inverted, the flow has steep margins, indicating that it did not spill out of the channel onto a floodplain. Furthermore, because the valley was steep sided, the base of the flow varies in elevation by as much as 200 ft (60 m) over short distances (Woodward-Clyde, 1978).

The downvalley slopes of the flow surface vary from about 60 to 300 ft/mi (12 to 57 m/km; Fig. 5). These slopes are larger than the gradient of the present Stanislaus River in the same area, where values of 26 ft/mi (5 m/km) are typical. Part of the Table Mountain slope is a result of Plio-Pleistocene uplift and tilting of the Sierra (Christensen, 1966). Between Knight's Ferry and Columbia, Table Mountain crosses two major fault zones of the Sierra Nevada foothills–the Bear Mountains zone and the Melones fault zone (Figs. 2, 4, and 5). In this 17-mi (27-km) long portion of Table Mountain, the lava flow has been displaced by a total of 250 ft (76 m) due to warping and faulting produced by Cenozoic tectonic movements along these Mesozoic structures. Near Rawhide Flat, two normal faults cut across Table Mountain (Figs. 4 and 5). On both faults, the recorded movement is down-to-the-east. The east Rawhide Flat scarp is 56 ft (17 m) high, and the western scarp is 89 ft (27 m) in height. Woodward-Clyde Consultants (1978) estimated that these scarps were produced by tectonic activity during the last 10,000 to 100,000 years. These substantial scarps do not persist in the easily erodible Mesozoic rocks on either side of the ridge. Many other linear fractures that cross the flow are also apparently tectonic. Approximately 71% of the fractures are aligned between 10° and 40° west of north,

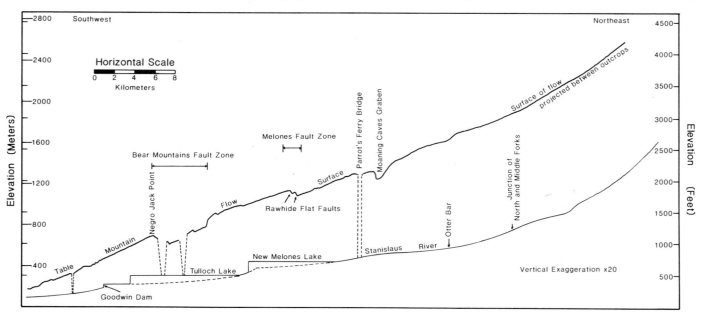

Figure 5. Longitudinal profiles of Table Mountain and modern Stanislaus River projected to N45°E.
Modified after Woodward-Clyde (1978) Figures C.4-15 a and b.

subparallel to the major fault systems. Because these structures bear no consistent relationship to the orientation of the flow, they are probably unrelated to the cooling of the lava.

The Table Mountain flow supplies a means of assessing both the nature of uplift of the central Sierra and the rate of downcutting of the modern river. In general, the difference in elevation between the Stanislaus River and the top of Table Mountain increases from west to east, toward higher altitudes. For example, just below Goodwin Dam the flow stands 460 ft (140 m) above the river. At Otter Bar, about 25 mi (40 km) to the northeast, the modern river is 1,870 ft (570 m) below the lava cap. Irregularities in this trend occur because of local deformation. At Negro Jack Point (about 6 mi [10 km] above Goodwin Dam), the flow is offset by a fault, placing the up-raised western side 740 ft (225 m) above the river. This exception, and similar ones, do not obscure the basic trend, which indicates that major tilting of the Sierran block toward the west has occurred since 9 Ma. The present-day canyon of the Stanislaus River, which reaches well below the prevolcanic bedrock surface, has also been cut during that period. Vertical incision rates, averaged over the time since the Table Mountain flow, vary from about 0.6 in (1.5 cm)/1,000 yr at Goodwin Dam to 2.4 in (6 cm)/1,000 yr at Otter Bar. Uplift and tilting of the central Sierra appear to be continuing today, and the rate may actually be accelerating (Huber, 1981).

CONCLUDING STATEMENT

Our understanding of the Sierra Nevada's Cenozoic history is by no means clear. As Bateman and Wahrhaftig (1966, p. 145) stated: "There is uncertainty and controversy over the time of uplift and tilting, the total amount of uplift, and the amount of internal deformation." Table Mountain is a rich source of information on the geologic history of the Sierra Nevada. This part of the Mother Lode Belt also played an important role in American literary history as the setting for numerous stories by Mark Twain and Bret Harte. In one of Harte's (1883) lesser known poems, he chronicles the events that lead to the demise of "The Society upon the Stanislaus." One participant in the meeting claimed that some recently discovered bones belonged to a prehistoric creature; another was sure that they were the remains of one of his mules. The lesson in how not to conduct a scientific debate may bear repeating:

Now I hold it is not decent for a scientific gent
To say another is an ass,—at least, to all intent;
Nor should the individual who happens to be meant
Reply by heaving rocks at him, to any great extent.

Then Abner Dean of Angel's raised a point of order, when
A chunk of old red sandstone took him in the abdomen,
And he smiled a kind of sickly smile, and curled up on the floor,
And the subsequent proceedings interested him no more.

For, in less time than I write it, every member did engage
In a warfare with the remnants of a palaeozoic age;
And the way they heaved those fossils in their anger was a sin,
Till the skull of an old mammoth caved the head of Thompson in.

And this is all I have to say of these improper games,
For I live at Table Mountain, and my name is Truthful James;
And I've told in simple language what I knew about the row
That broke up our Society upon the Stanislow.

REFERENCES CITED

Bateman, P. C., and Wahrhaftig, C., 1966, Geology of the Sierra Nevada, in Bailey, E. H., ed., Geology of northern California: California Division of Mines and Geology Bulletin 190, p. 107–172.

Christensen, M. N., 1966, Late Cenozoic crustal movements in the Sierra Nevada of California: Geological Society of America Bulletin, v. 77, p. 163–182.

Dalrymple, G. B., 1963, Potassium-argon dates of some Cenozoic volcanic rocks of the Sierra Nevada, California: Geological Society of America Bulletin, v. 74, p. 379–390.

Eric, J. H., Stromquist, A. A., and Swiney, C. M., 1955, Geology and mineral deposits of the Angel's Camp and Sonora Quadrangles, Calaveras and Tuolumne counties, California: California Division of Mines and Geology Special Report 41, 55 p.

Harte, B., 1883, The poetical works of Bret Harte: Boston, Houghton, Mifflin and Company, 324 p.

Huber, N. K., 1981, Amount and timing of late Cenozoic uplift and tilt of the central Sierra Nevada, California; Evidence from the upper San Joaquin River basin: U.S. Geological Survey Professional Paper 1197, 28 p.

Jenkins, O. P., 1948, Geologic history of the Sierran gold belt, in Geologic guidebook along Highway 49—Sierran gold belt; The Mother Lode Country (centennial edition): California Division of Mines and Geology Bulletin 141, p. 23–30.

Jenkins, O. P., 1970, Geology of placer deposits: California Division of Mines and Geology Special Publication 34, 27 p.

Lindgren, W., 1911, The Tertiary gravels of the Sierra Nevada, California: U.S. Geological Survey Professional Paper 73, 226 p.

Noble, D. C., Slemmons, D. B., Korringa, M. K., Dickinson, W. R., Al-Rawi, Y., and McKee, E. H., 1974, Eureka Valley Tuff, east-central California and adjacent Nevada: Geology, v. 2, p. 139–142.

Ransome, F. L., 1898, Some lava flows of the western slope of the Sierra Nevada, California: U.S. Geological Survey Bulletin 89, 74 p.

Rhodes, D. D., 1980, Exhumed topography—a case study of the Stanislaus Table Mountain, California: National Aeronautics and Space Administration Technical Memorandum 82385, p. 397–399.

Slemmons, D. B., 1966, Cenozoic volcanism of the central Sierra Nevada, California, in Bailey, E. H., ed., Geology of northern California: California Division of Mines and Geology Bulletin 190, p. 199–208.

Taliaferro, N. L., and Solari, A. J., 1948, Geology of the Copperopolis Quadrangle, California: California Division of Mines and Geology Bulletin 145, 2 plates, scale 1:62,500.

Woodward-Clyde Consultants, 1978, Stanislaus Nuclear Project preliminary safety analysis report [unpublished manuscript].

ACKNOWLEDGMENTS

My initial study of Table Mountain was supported by a NASA Summer Faculty Fellowship at the Jet Propulsion Laboratory.

Multiple deformation in the Bennettville area of the Saddlebag Lake pendant, central Sierra Nevada, California

D. D. Trent, *Citrus College, Glendora, California 91740*

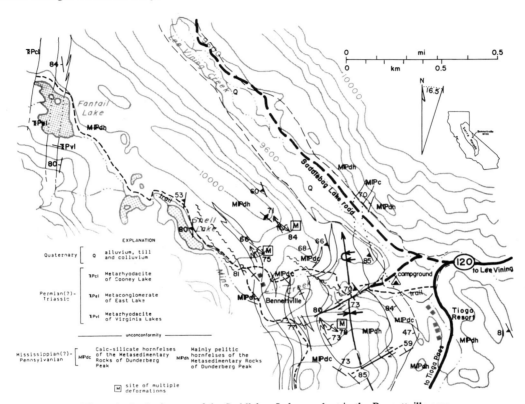

Figure 1. Geologic map of the Saddlebag Lake pendant in the Bennettville area.

LOCATION AND ACCESS

The Bennettville area of the Saddlebag Lake pendant (Fig. 1) straddles the boundary of the Tuolumne Meadows and Mono Craters quadrangles. Bennettville is located in the SE¼, NW¼,Sec.19,T.1N.,R.25E. Park vehicles on California 120 at the road junction to Saddlebag Lake, 9.6 mi (19 km) west of Lee Vining and 2.4 mi (4 km) east of Tioga Pass. Bennettville is reached by walking west through the campground to a well-defined fisherman's trail that begins at the far western edge of the campground on the north side of Mine Creek. The trail roughly parallels Mine Creek to Bennettville, an abandoned gold mining camp, where two buildings remain in arrested decay.

SIGNIFICANCE OF SITE

Evidence of multiple deformation of a Sierra Nevada roof pendant, perhaps representing episodic and regional deformations of the Triassic Sonoma orogeny and a middle Cretaceous orogeny are observable in the Bennettville area of the Saddlebag Lake

roof pendant (Brook, 1977; Brook and others, 1979). The pendant is typical of a number of roof pendants in the east-central Sierra Nevada that reveal multiple deformations and which are surrounded by and intruded by a complex of Mesozoic granitic plutons (Nokleberg and Kistler, 1980). The easy access, excellent exposures, well-developed structures, and exquisite alpine scenery make this an attractive locality for examination of these controversial rocks and their structures.

PETROLOGY AND STRATIGRAPHY

Metasedimentary rocks underlie most of the Bennettville area. The metasediments consist of massive bedded quartzite interbedded with argillite, dark-gray quartzite, quartz sericite schist, light-tan quartzite, light-tan calc-silicate hornfels, quartzo-feldspathic hornfels, brown pelitic hornfels, minor marble, thin interbeds of light metasiltstone, and minor metaconglomerate near the top of the unit (Brook, 1977; Seitz, 1983; this report). Relict sedimentary structures are rare because of extreme deformation, but graded bedding and cross bedding exist locally in the quartzo-feldspathic hornfels and may be used to determine bedding tops.

a

b

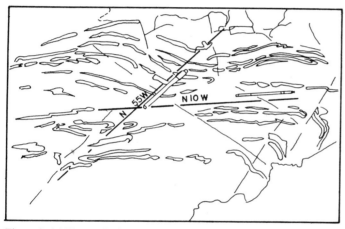

Figure 2. (a) Photo of minor F_1 isoclinal folds and a superposed F_2 open fold in interbedded siliceous and pelitic hornfels of the metasedimentary rocks of Dunderberg Peak. (b) The F_1 axial plane labeled N10W belongs to the set of axial planes with an average strike of N25°W. The F_2 axial plane labeled N55W belongs to the set of axial planes with an average strike of N48°W.

The age of the metasediments has been determined from Pennsylvanian conodonts (Nokleberg, 1983), Mississippian (?) crinoid fragments and corals (Brook and others, 1979; Kistler and Nokleberg, 1979), and by stratigraphic correlation with other Sierran pendants.

Schweickert (1986, personal communication) believes the fossils are from float and that these rocks are equivalent to the Ordovician and Silurian Palmetto Formation of west-central Nevada. Stratigraphic thickness of the metasedimentary sequence is uncertain due to the intense deformation, but Seitz (1983) estimated it to be approximately 3,000 ft (915 m) in the vicinity of Saddlebag Lake. Seitz correlated the sequence with the Dunderberg Peak rocks to the north on the basis of stratigraphic position and similar lithologies. The metasedimentary rocks of the Bennettville area are part of the Lewis sequence (Kistler and

Swanson, 1981) and are included in Nokleberg's (1983) Owens terrane.

Permian (?) and Triassic metavolcanic rocks and pelitic hornfels unconformably overlie the metasedimentary sequence and are exposed west of Fantail Lake. These rocks were metamorphosed from rhyodacitic tuff, volcanic breccia, marl, volcaniclastics, basalt flows, and conglomerate (Brook, 1977; Seitz, 1983). Immediately above the unconformity with the underlying metasedimentary sequence is a 200-ft-thick (60 m) unit of light-gray, medium-grained metarhyodacite tuff correlated by Seitz (1983) with the metarhyodacite tuff of Virginia Lakes to the north. Above the metatuff is the most distinctive and consistently reliable stratigraphic marker horizon in the Saddlebag pendant, a sequence of metaconglomerate, quartzite, hornfels, and metavolcanic rocks. At Fantail Lake, about 0.8 mi (1.4 km) northwest of Bennettville, the unit is about 400 ft (120 m) thick and contains more than 20 separate conglomerate beds, the pebbles of which are stretched parallel to the foliation. Seitz named this unit "the metaconglomerate of East Lake" and interpreted it as the basal conglomerate of the overlying metavolcanic rocks. Immediately above the conglomerate is metarhyodacite tuff, which Seitz named "the metarhyodacite of Cooney Lake."

Eighteen specimens from the metavolcanic rocks yield a whole rock isochron age of 237 ± 11 Ma. If the ages of 17 other specimens from the Matterhorn Peak area are combined with the 18 specimens mentioned above, the resulting isochron age is 224 ± 14 Ma (Kistler and Swanson, 1981; Seitz, 1983). The metavolcanics are therefore Permian (?) and Triassic in age. These rocks lie within the lower Koip sequence of Kistler and Swanson (1981), and within the lower High Sierra terrane of Nokleberg (1983). Schweickert and Lahren (1986) consider these rocks to be equivalent to the Diablo and Candelaria Formations of the Miller Mountain and Candelaria Hills area of Nevada.

STRUCTURE

About 330 ft (100 m) north of Bennettville, east of Mine Creek along the trail to Shell Lake, are glacially smoothed expanses of Mississippian(?) and Pennsylvanian interbedded siliceous and pelitic hornfels. These outcrops reveal numerous minor first-generation (F_1) isoclinal folds on the limb of a large, north-plunging F_1 syncline (Fig. 1). These folds in turn are refolded by minor second-generation (F_2) open folds (Fig. 2). Average strikes of the axial planes and average plunges of the fold axes are shown on stereo projections (Fig. 3; D. D. Trent and T. L. Henyey, 1981, unpublished data). The F_1 folds are interpreted as Triassic (Sonoman) and the F_2 folds as middle Cretaceous (Brook, 1977; this study). The timing of deformation is based on comparison with similar structural trends in other Sierran pendants, and by stratigraphic correlation with other Sierran pendants (Brook, 1977; Brook and others, 1979; Nokleberg and Kistler, 1980; Russell and Nokleberg, 1977). An earlier deformation is recognized in the older Paleozoic rocks in the eastern part of the Saddlebag Lake pendant, and in other pendants to the south

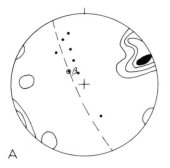
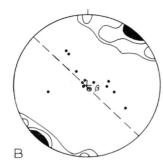

A   B

Figure 3. Structure diagrams of minor folds in the Bennettville area. A, F_1 data; 19 poles to axial surfaces and axial plane cleavage contoured at 5, 11, 16, and 21 percent per one percent area. B, F_2 data; 24 poles to axial surfaces and axial plane cleavage contoured at 4, 16, and 21 percent per one percent area. Dots represent orientations of fold axes. Tick marks indicate true north.

(Nokleberg and Kister, 1980; Nokleberg, 1983). This earlier deformation is interpreted as forming during the Devonian–Mississippian Antler orogeny (Brook, 1977, Brook and others, 1979; Nokleberg and Kistler, 1980; Russell and Nokleberg, 1977). The earlier folds formed in the older event have been deformed subsequently by the same Triassic Sonoman and middle Cretaceous deformations observed in the Bennettville area. The Permian (?) and Triassic sequence of metavolcanics and hornfels to the west of Mine Creek are in an area of more difficult access. These rocks exhibit similar isoclinal F_1 and open F_2 folds. A summary of the structural elements for the two deformations in the Bennettville area of the Saddlebag Lake pendant is given in Table 1.

TABLE. 1. AGE, TYPE, AND STRUCTURAL TRENDS OF THE MAJOR AND MINOR FOLDS NEAR BENNETTVILLE

Age of Deformation	Folds		
	Type	Average strike of axial planes	Average plunge of axes
Mid-Cretaceous	Second-generation (F_2) open and asymmetric	N48°W	84°SE
Triassic (Sonoman)	First-generation (F_1) isoclinal	N25°W	71°NW

TECTONIC INTERPRETATION

Structural studies in metamorphic pendants in the Sierra Nevada by Brook (1977), Brook and others (1974), Kistler (1966), and Bateman and Wahrhaftig (1966) concluded that isoclinal deformation and regional metamorphism occurred following the deposition of the youngest Paleozoic rocks and before the eruption of the Mesozoic volcanic rocks. Thus the two deformations in the Bennettville area described in this report appear to have occurred after the Devonian–Mississippian Antler deformation. Schweickert and others (1984) and Schweickert and Lahren (1986) questioned the interpretation of Brook, Nokleberg, Kistler, and Bateman, and presented evidence suggesting four episodes of deformation—Devonian–Mississippian Antler, Trias-

sic Sonoman, Late Jurassic Nevadan, and middle Cretaceous—and that the rocks of the Saddlebag Lake pendant are part of the Roberts Mountains and Golconda allochthons. Hamilton (1978) questioned all of the above-mentioned timing of the deformations of the pendant because such timing requires that the major deformation and much of the metamorphism is unrelated to Mesozoic plutonism. Hamilton proposed that the refolded folds do not represent differing structural histories, but are due to differing responses to a similar history experienced by both the Paleozoic and Mesozoic rocks because of contrasting mecanical behavior. "The marbles do indeed display exceedingly complex folding—tight folds superimposed on sheared-apart isoclinal folds—whereas the metavolcanic rocks display laminar shear and tight and isoclinal folds which although of simpler aspect have axes of similar orientation. This contrasted behavior is typical of most metacarbonate and metavolcanic rocks anywhere" (Hamilton, 1978, p. 48). Part of the basis for the differences between their interpretations is that Brook, Nokleberg, Kistler, and Bateman regarded the folds as produced by compression. This requires that each refolding requires a new and differently oriented stress system. Hamilton, to the contrary, regarded the folds as representing discontinuities in shear velocities, "the folds having migrated through the rocks like advancing caterpillar tractor treads: coaxial tight and isoclinal folds form simultaneously, not sequentially" (Hamilton, 1978, p. 48). Hamilton agreed that the youngest metavolcanic rocks must have had a simpler history than the metasedimentary and older metavolcanic rocks, but he contended that studies have yet to prove that all of the younger metavolcanic rocks have had simpler structural histories than all of the older rocks.

A recent refined interpretation is that the Saddlebag Lake pendant is part of a collage of tectonostratigraphic terranes that were accreted to the North American plate in an Alaskan-type arc environment similar to that of southern Alaska in Cenozoic time (Nokleberg, 1983). Brook and others (1979) and Nokleberg (1983) interpreted the rocks in the Bennettville area as belonging to the High Sierra tectonostratigraphic terrane, which appears to have been accreted to and thrust over older rocks of the Owens terrane to the east. This accretion occurred in the Triassic along the Owens Valley fault which lies east of the Bennettville area.

Orientations and deformational styles of major and minor folds similar to those in the Saddlebag Lake pendant are present in other pendants throughout the central Sierra Nevada. The regional extent of these structures and stratigraphic correlations between the pendants in the central Sierra are the bases for establishing the veracity and timing of multiple deformations and the possible accretion tectonics of this region.

TRAILSIDE GUIDE

To visit the best examples of the multiply deformed minor folds requires about a 0.6-mi (1-km) hike. Stout hiking shoes are recommended. Begin at the parking area at the junction of California 120 and the road to Saddlebag Lake and walk west along the graded road through the U.S. Forest Service campground. The road crosses a meadow underlain by Quaternary glacial and Holocene stream deposits. A well-used fisherman's trail begins at the far-western edge of the campground where exposures of brown Mississippian (?) and Pennsylvanian calcsilicate hornfels are first encountered.

At this point, Mine Creek discharges into the meadow from a small, steep-sided canyon. The course of the creek for about 330 ft (100 m) into the canyon follows a small fault that has offset the east limb of the large F_1 syncline in a right-lateral sense. From here the trail climbs more steeply and traverses the syncline. Occasional minor folds may be observed along the trail in the interbedded light-colored quartzo-feldspathic and brown pelitic hornfels. The trail crosses the axial region of the syncline in the area of a small meadow.

Chattermarks, striae, and glacial polish are common on the outcrops of the hornfels along the trail, this area having been covered by some 1,000 ft (300 m) of glacial ice several times during the Quaternary. The glaciers here were part of the great ice fields that capped the Sierra Nevada; the outflow lobes in this region spilled down Lee Vining canyon to the shore of the ancestral Mono Lake (Bateman and Wahrhaftig, 1966, p. 158; Putnam, 1950; Russel, 1889).

About 0.5 mi (0.9) km from the parking area is the site of Bennettville. Here are two buildings and several leveled areas marking the sites of former buildings dating from the 1880s that are relics of the halcyon days of western mining speculation.

Bennettville was an unsuccessful mining camp that was active briefly from 1881 to 1884. Machinery for the mine was hauled on sleds by cable and windlass from Lundy during the winter of 1883. The same year marked the completion of the Great Sierra Wagon Road to Bennettville from a rail connection to the west. The road was later to become an important part of the modern Tioga Pass Highway. By the summer of 1884 a 1,716-ft (520-m) adit had been drilled. However, litigation among the stockholders closed down the operation when the adit was only a few tens of meters short of reaching the silver-bearing quartz lode. Some futile attempts were made to reopen the mine, the most recent being in 1933 (Hubbard, 1958; Chalfant, 1947). The mine adit and waste-rock pile are located west of Bennettville across Mine Creek, and a few prospect pits exist 150 ft (50 m) north of the old buildings.

Minor folds occur at numerous localities in the Bennettville area, but those noted on Figure 1 from 350 to 650 ft (100 to 200 m) north of the old buildings are the most obvious and easily located. The more adventurous visitor may wish to examine the unconformity between the Mississippian(?) and Pennsylvanian metasediments and the overlying Permian and Triassic metavolcanics. This contact is about 0.8 mi (1.4 km) northwest of Bennettville at Fantail Lake and is reached by following the fisherman's trail along the east side of Shell Lake to the north end of Fantail Lake. The base of the Permian (?) and Triassic section is marked by the light-gray metarhyodacite of Virginia Lakes that is overlain by the distinctive metaconglomerate of East Lake.

REFERENCES CITED

Bateman, P. C., and Wahrhaftig, C., 1966, Geology of the Sierra Nevada, *in* Bailey, E. H., ed., Geology of northern California: California Division of Mines and Geology Bulletin 190, p. 107–172.

Brook, C. A., 1977, Stratigraphy and structure of the Saddlebag Lake roof pendant, Sierra Nevada, California: Geological Society of America Bulletin, v. 88, p. 321–334.

Brook, C. A., Nokleberg, W. J., and Kistler, R. W., 1974, Nature of the angular unconformity between the Paleozoic metasedimentary rocks and the Mesozoic metavolcanic rocks in the eastern Sierra Nevada, California: Geological Society of America Bulletin, v. 87, p. 571–576.

Brook, C. A., Gordon, M., Jr., MacKey, M. J., and Chetelat, G. F., 1979, Fossiliferous upper Paleozoic rocks and their structural setting in the Ritter Range and Saddlebag Lake roof pendants, central Sierra Nevada, California: Geological Society of America Abstracts with Programs, v. 11, no. 3, p. 71.

Chalfant, W. A., 1947, Gold, guns and ghost towns: Stanford, California, Stanford University Press, 175 p.

Hamilton, W., 1978, Mesozoic tectonics of the western United States, *in* Howell, D. G., and McDougall, K. A., eds., Mesozoic paleogeography of the western United States: Los Angeles, California, Society of Economic Paleontologists and Mineralogists, Pacific Section, Pacific Coast Paleogeography Symposium 2, p. 33–70.

Hubbard, D. H., 1958, Ghost mines of Yosemite: Fresno, California, Awani Press, 36 p.

Kistler, R. W., 1966, Structure and metamorphism in the Mono Craters Quadrangle, Sierra Nevada, California: U.S. Geological Survey Bulletin 1221-E, 53 p.

Kistler, R. W., and Nokleberg, W. J., 1979, Carboniferous rocks of the eastern Sierra Nevada, *in* Saul, R. B., and others, eds., The Mississippian and Pennsylvanian (Carboniferous) systems in the United States–California, Oregon,

and Washington: U.S. Geological Survey Professional Paper 1110-CC, 26 p.

Kistler, R. W., and Swanson, S. E., 1981, Petrology and geochronology of metamorphosed volcanic rocks and a middle-Cretaceous volcanic rock in the east-central Sierra Nevada, California: Journal of Geophysical Research, v. 86, no. B11, p. 10489–10501.

Nokleberg, W. J., 1983, Wallrocks of the central Sierra Nevada batholith, California: A collage of accreted tectono-stratigraphic terranes: U.S. Geological Survey Professional Paper 1255, 24 p.

Nokleberg, W. J., and Kistler, R. W., 1980, Paleozoic and Mesozoic deformations in the central Sierra Nevada, California: U.S. Geological Survey Professional Paper 1145, 28 p.

Putnam, W. C., 1950, Moraine and shoreline relationships at Mono Lake, California: Geological Society of America Bulletin, v. 61, p. 115–122.

Russel, I. C., 1889, The Quaternary history of Mono Valley, California: U.S. Geological Survey Annual Report, v. 8, p. 261–394.

Russell, S., and Nokleberg, W. J., 1977, Superposition and timing of deformations in the Mount Morrison roof pendant and in the Sierra Nevada, California: Geological Society of America Bulletin, v. 88, p. 335–345.

Schweickert, R. A., and Lahren, M. M., 1986, Mesozoic thrust belt along the eastern edge of Yosemite National Park (YNP), California: Geological Society of America Abstracts with Programs, v. 18, no. 2, p. 181.

Schweickert, R. A., Fisher, G. R., and Lahren, M. M., 1984, Structural and tectonic history of the Saddlebag Lake pendant (SLP), eastern Sierra Nevada, California: Geological Society of America Abstracts with Programs, v. 16, no. 5, p. 293.

Seitz, J. F., 1983, Geologic map of the Tioga Lake, Hall Natural Area, Log Cabin-Saddlebag, and Horse Meadows roadless areas, Mono County, California: U.S. Geological Survey Miscellaneous Field Studies Map MF-1453-A, scale 1:62,500.

Great Valley Group, west side of Sacramento Valley, California

Raymond V. Ingersoll, *Department of Earth and Space Sciences, University of California, Los Angeles, California 90024*

LOCATION

The best exposures of the Upper Cretaceous part of the Great Valley Group (formerly sequence) occur along Cache Creek northwest of Guinda and Putah Creek west of Winters (Fig. 1). The Cache Creek exposures are reached by driving along California Route 16, and the Putah Creek exposures occur along California Route 128. All exposures are in road cuts or river cuts on public property, and may be reached by automobile or bus, with minor walking over easy terrane.

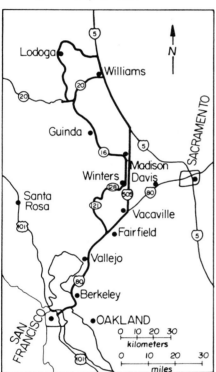

Figure 1. Location map for Cache Creek section along California Route 16, northwest of Guinda, and Putah Creek section along California Route 128, west of Winters. (From Ingersoll and Dickinson, 1981.)

be sampled at these localities (Ingersoll, 1978a, 1983). Most of the turbidites were derived from the coeval Sierra Nevada magmatic arc to the east and were deposited at lower-bathyal to abyssal depths (Ingersoll, 1979, 1983).

SITE INFORMATION

Figure 2 shows schematic stratigraphic sections exposed along Putah Creek (a) and Cache Creek (b). Figure 3 is a view of the retrogradational sequence exposed at Monticello Dam along Putah Creek, and Figure 4 shows typical outer-fan lobe deposits of the progradational sequence exposed along Cache Creek. Detailed site information, as well as regional summaries and discussions, may be found in the following guidebooks: Bertucci and Ingersoll (1983), Graham (1981), Ingersoll and Dickinson

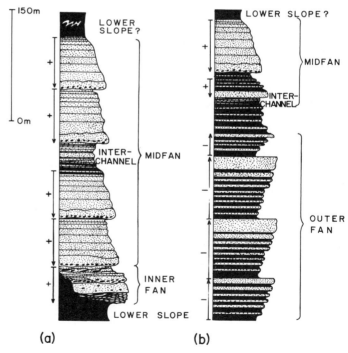

(a)   (b)

Figure 2. (a) Retrogradational suite (lower slope, inner fan, midfan, lower slope (?), outer fan [not shown]) exposed at Monticello Dam (Putah Creek). This suite was deposited contemporaneously with a transgression eastward that resulted in the eastward migration of fan facies associations (also, see Fig. 3). (b) Progradational suite (basin plain [not shown], outer fan, midfan, lower slope [?]) exposed along Cache Creek. This suite was deposited near the basin axis, where fan facies associations prograded into the basin as it filled (also, see Fig. 4). (From Ingersoll, 1978b.)

SIGNIFICANCE

The Great Valley Group is the sedimentary fill of one of the best studied and best understood forearc basins in the world (Dickinson and Seely, 1979; Ingersoll, 1978a, b, c, 1979, 1982, 1983; Ingersoll and Dickinson, 1981). The Cache Creek and Putah Creek localities contain some of the best exposures of submarine-fan turbidite facies in California (Ingersoll, 1978b). In addition, some of the petrofacies of the Great Valley Group may

Figure 3. View looking south at Monticello Dam (Putah Creek), showing retrogradational submarine-fan sequence of the Venado Member of the Cortina Formation. Lower-slope mudrock of the Boxer Formation is exposed at the west end (right) of the outcrop. Inner-fan-channel conglomerate, pebbly mudstone, and pebbly sandstone make up the lowermost resistant unit. These are overlain by several upward-thinning cycles of sandstone and pebbly sandstone deposited in midfan channels. Just to the southeast of the water intake are levee (overbank) deposits, also formed in a midfan environment. At the east end of the outcrop are folds that may be syndepositional slump features, formed either along a channel margin or in a lower-slope environment. The section shown is approximately 1,485 ft (450 m) thick (see Fig. 2).

Figure 4. Typical outer-fan depositional-lobe deposits exposed along Cache Creek (looking north from California Route 16) within the Sites Sandstone Member of the Cortina Formation. Thickest bed is approximately 6.6 ft (2 m) thick. This section is near the middle of the schematic section shown in Figure 2b.

(1981), Ingersoll and others (1977), and Ingersoll and others (1984). The reader is urged to acquire some or all of these guidebooks before proceeding to the outcrops. Ingersoll and Dickinson (1981) is the best guidebook with which to start.

REFERENCES CITED

Bertucci, P. F., and Ingersoll, R. V., eds., 1983, Guidebook to the Stony Creek Formation, Great Valley Group, Sacramento Valley, California (Annual Meeting Pacific Section SEPM): Pacific Section, Society of Economic Paleontologists and Mineralogists, 25 p.

Dickinson, W. R., and Seely, D. R., 1979, Structure and stratigraphy of forearc regions: American Association of Petroleum Geologists Bulletin, v. 63, p. 2–31.

Graham, S. A., ed., 1981, Field guide to the Mesozoic-Cenozoic convergent margin of northern California: Pacific Section, American Association of Petroleum Geologists, 118 p.

Ingersoll, R. V., 1978a, Petrofacies and petrologic evolution of the Late Cretaceous fore-arc basin, northern and central California: Journal of Geology, v. 86, p. 335–352.

——1978b, Submarine fan facies of the Upper Cretaceous Great Valley Sequence, northern and central California: Sedimentary Geology, v. 21, p. 205–230.

——1978c, Paleogeography and paleotectonics of the late Mesozoic forearc basin of northern and central California, *in* Howell, D. G., and McDougall, K. A., eds., Mesozoic paleogeography of the western United States: Pacific Section, Society of Economic Paleontologists and Mineralogists Pacific Coast Paleogeography Symposium 2, p. 471–482.

——1979, Evolution of the Late Cretaceous forearc basin, northern and central California: Geological Society of America Bulletin, v. 90, Part I, p. 813–826.

——1982, Initiation and evolution of the Great Valley forearc basin of northern and central California, *in* Leggett, J. K., ed., Trench-forearc geology: Sedimentation and tectonics on modern and ancient active plate margins: Geological Society of London Special Publication 10, p. 459–467.

——1983, Petrofacies and provenance of late Mesozoic forearc basin, northern and central California: American Association of Petroleum Geologists Bulletin, v. 67, p. 1125–1142.

Ingersoll, R. V., and Dickinson, W. R., 1981, Great Valley Group (sequence), Sacramento Valley, California, *in* Frizzell, V., ed., Upper Mesozoic Franciscan rocks and Great Valley sequence, central Coast ranges, California (Annual Meeting Pacific Section SEPM Field Trips 1 and 4): Pacific Section, Society of Economic Paleontologists and Mineralogists, p. 1–33.

Ingersoll, R. V., Rich, E. I., and Dickinson, W. R., 1977, Great Valley sequence, Sacramento Valley: Cordilleran Section, Geological Society of America, Fieldtrip Guidebook, 72 p.

Ingersoll, R. V., Schweickert, R. A., Kleist, J. R., Graham, S. A., and Cowan, D. S., 1984, Field guide to the Mesozoic-Cenozoic convergent margin of northern California: Revised roadlogs (field trip 16), *in* Lintz, J., Jr., ed., Western geological excursions, Volume 4: Geological Society of America and Department of Geological Sciences, Mackay School of Mines, University of Nevada, Reno, p. 304–353.

Unconformity between Coast Range ophiolite and part of the lower Great Valley sequence, South Fork of Elder Creek, Tehama County, California

M. C. Blake, Jr., A. S. Jayko, D. L. Jones, and B. W. Rogers, U.S. Geological Survey, 345 Middlefield Road, Menlo Park, California 94025

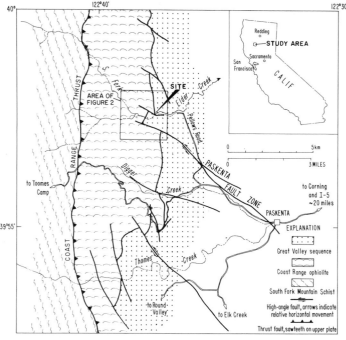

Figure 1. Small-scale geologic map showing location of Paskenta area, California.

LOCATION AND ACCESSIBILITY

The South Fork of Elder Creek is located on private property about 6 mi (10 km) northwest of Paskenta, Calif. (Fig. 1). To visit this locality it is necessary to call Mr. Les Sutfin (916-824-4628) and arrange to pick up the key to the gate at his home in Corning. From the Paskenta Store, drive 3.3 mi (5.3 km) north on the Toomes Camp road to the locked gate on the north side of the road. From here, take the Pellows Road (four-wheel drive vehicle recommended) 3.6 mi (5 km) north to the end of the road, then walk west along the trail parallel to the South Fork of Elder Creek for approximately 0.5 mi (0.8 km) to the unconformity (Fig. 2).

SIGNIFICANCE OF SITE

Large masses of ultramafic rock that lie between metamorphic rocks of the Franciscan assemblage to the west and the Great Valley sequence to the east were interpreted by Irwin (1964) and Brown (1964) to have been emplaced as a cold intrusion along a major fault zone. However, in June 1969, Edgar Bailey and

David Jones discovered a well-exposed unconformity between Jurassic strata of the Great Valley sequence and underlying mafic and ultramafic rocks along the South Fork of Elder Creek in Tehama County, northern California. Subsequent field work determined that the mafic and ultramafic rocks were part of an ophiolite sequence interpreted to be the oceanic crust upon which the Great Valley sequence was deposited. Farther west, blue schist–grade Franciscan rocks (South Fork Mountain Schist) structurally underlie the ophiolite along the Coast Range thrust. These relations led to the proposal that the Franciscan rocks were subducted beneath the Coast Range ophiolite and Great Valley sequence (Bailey and others, 1970).

Since 1970, numerous papers have been written on various aspects of the Coast Range ophiolite and the Great Valley sequence, and numerous new ophiolite localities have been described (Hopson and others, 1981). Nevertheless, nowhere are the basal contacts as well exposed as they are along the South Fork of Elder Creek.

SITE INFORMATION

The basal contact is best seen on the north bank of the South Fork of Elder Creek (Fig. 2), where mudstone of the Upper Jurassic Great Valley sequence (Knoxville Formation of earlier workers) rests on ophiolitic breccia consisting of angular to subrounded clasts of mafic and ultramafic rock derived from the underlying Coast Range ophoilite (Fig. 3), with interstratified thin units of pillow breccia. The breccia at the site along the South Fork of Elder Creek is only about 330 ft (100 m) thick; however, along Digger Creek to the south, the thickness of about 3,300 ft (1,000 m) suggests the paleo sea floor had substantial relief. The breccia, in turn, unconformably overlies layered wehrlite and clinopyroxenite (Fig. 4). At this contact, the underlying ultramafic rocks are intruded by a diorite dike at 155 Ma (Lanphere, 1971, and oral communication, 1985). This dike places a lower limit on the age of the breccia. The upper limit is constrained by the age of the basal strata of the Great Valley sequence that contain *Buchia rugosa* of late Kimmeridgian age (~152 Ma) (Jones, 1975).

These geologic relations suggest the following sequence of events. During the Middle Jurassic (~165 Ma), the Coast Range ophiolite was formed, probably in a back-arc basin to the west of the west-facing Sierran-Klamath arc (see bibliography for arguments based on geochemistry of ophiolite). During the Late Jurassic (~155–150 Ma), the ophiolite was faulted, uplifted, and locally eroded, removing the imbricated upper levels of pillow

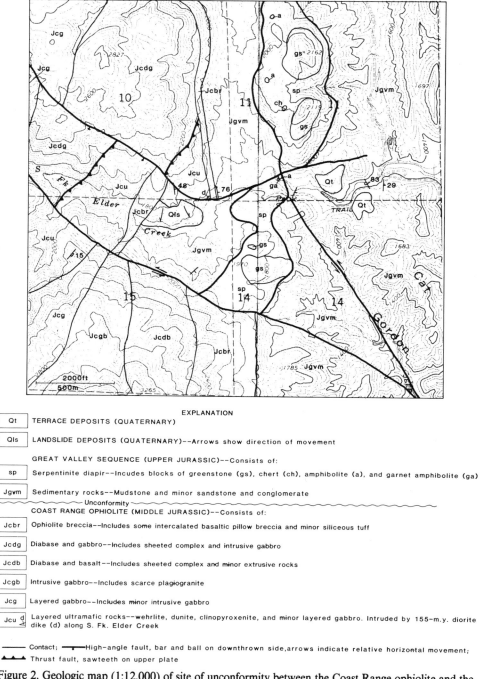

EXPLANATION

| Qt | TERRACE DEPOSITS (QUATERNARY) |
| Qls | LANDSLIDE DEPOSITS (QUATERNARY)--Arrows show direction of movement |

GREAT VALLEY SEQUENCE (UPPER JURASSIC)--Consists of:

| sp | Serpentinite diapir--Incudes blocks of greenstone (gs), chert (ch), amphibolite (a), and garnet amphibolite (ga) |
| Jgvm | Sedimentary rocks--Mudstone and minor sandstone and conglomerate |

〜〜〜〜〜〜 Unconformity 〜〜〜〜〜〜

COAST RANGE OPHIOLITE (MIDDLE JURASSIC)--Consists of:

Jcbr	Ophiolite breccia--Includes some intercalated basaltic pillow breccia and minor siliceous tuff
Jcdg	Diabase and gabbro--Includes sheeted complex and intrusive gabbro
Jcdb	Diabase and basalt--Includes sheeted complex and minor extrusive rocks
Jcgb	Intrusive gabbro--Includes scarce plagiogranite
Jcg	Layered gabbro--Includes minor intrusive gabbro
Jcu d	Layered ultramafic rocks--wehrlite, dunite, clinopyroxenite, and minor layered gabbro. Intruded by 155-m.y. diorite dike (d) along S. Fk. Elder Creek

——— Contact; ——┬—— High-angle fault, bar and ball on downthrown side, arrows indicate relative horizontal movement;

▲▲▲ Thrust fault, sawteeth on upper plate

Figure 2. Geologic map (1:12,000) of site of unconformity between the Coast Range ophiolite and the Great Valley sequence, California.

lavas, sheeted dikes and sills, intrusive gabbro, and layered gabbros. The breccia, which contains pieces of all of these elements, was deposited directly on the disrupted surface of the oceanic crust at this locality (see Fig. 2). Deformation of the Coast Range ophiolite, which coincides in time with the Nevadan orogeny of the Sierran-Klamath province, was closely followed by deposition of the basal strata of the Great Valley sequence; scarce detri-

tal quartz, potassium feldspar, and carbonaceous material in these basal strata were probably derived from the continental margin to the east.

Upstream from the site, a high-angle normal fault drops the basal sedimentary strata against the ultramafic rocks (Fig. 2). Those wishing to see more of the ophiolite should walk up the South Fork of Elder Creek an additional 3 mi (4.8 km). Also of

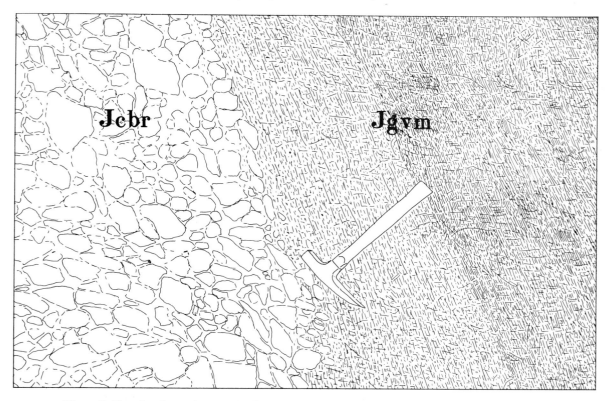

Figure 3. Drawing from photograph of depositional contact of Great Valley sequence (Jgvm) on ophiolite breccia (Jcbr), South Fork of Elder Creek.

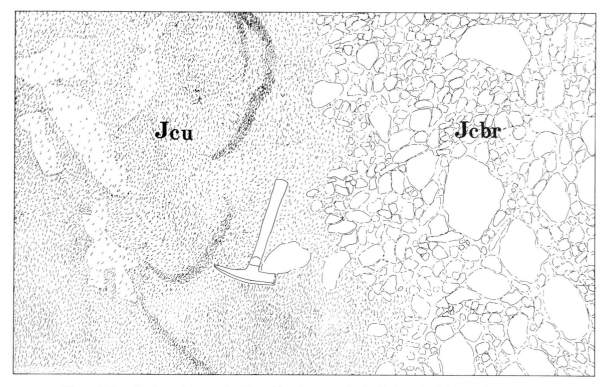

Figure 4. Drawing from photograph of depositional contact of ophiolite breccia (Jcbr) on clinopyroxenite (Jcu), South Fork of Elder Creek.

interest is a serpentinite diapir downstream from the site (Fig. 2). In addition to blocks of greenstone and chert, this diapir contains blocks of amphibolite and garnet amphibolite, believed to be derived from beneath the Coast Range ophiolite. The diapir probably formed within the Paskenta fault zone (Fig. 1), one of a series of large left-lateral structures confined to the upper plate of the Coast Range thrust (Jones and Irwin, 1971).

SELECTED BIBLIOGRAPHY

Bailey, E. H., Blake, M. C., Jr., and Jones, D. L., 1970, On-land Mesozoic oceanic crust in California Coast Ranges, *in* Geological Survey Research, 1970: U.S. Geological Survey Professional Paper 700-C, p. C70–C81.

Bailey, E. H., and Blake, M. C., Jr., 1974, Major chemical characteristics of Mesozoic Coast Range ophiolite in California: U.S. Geological Survey Journal of Research, v. 2, p. 637–656.

Brown, R. D., Jr., 1964, Geologic map of the Stonyford quadrangle, Glenn, Colusa, and Lake Counties, California: U.S. Geological Survey Mineral Investigations Field Studies Map MF-279.

Hopson, C. A., Mattinson, J. M., and Pessagno, E. A., 1981, Coast Range ophiolite, western California, *in* Ernst, W. G., ed., The geotectonic development of California: Englewood Cliffs, New Jersey, Prentice-Hall, p. 419–450.

Irwin, W. P., 1964, Late Mesozoic orogenies in the ultramafic belts of northwestern California and southwestern Oregon, *in* Geological Survey Research, 1964: U.S. Geological Survey Professional Paper 501-C, p. C1–C9.

Jones, D. L., 1975, Discovery of *Buchia rugosa* of Kimmeridgian age from the base of the Great Valley sequence: Geological Society of America Abstracts with Programs, v. 7, no. 3, p. 330.

Jones, D. L., and Irwin, W. P., 1971, Structural implications of an offset early Cretaceous shoreline in northern California: Geological Society of America Bulletin, v. 82, p. 815–822.

Lagabrielle, Y., Roure, F., Coutelle, A., Maury, R., and Thonon, P., 1986, The Coast Range ophiolite (northern California), possible arc and marginal basin remnants, their relations with the Nevadan orogeny: Bulletin of the Geological Society of France (in press).

Lanphere, M. A., 1971, Age of the Mesozoic oceanic crust in the California Coast Ranges: Geological Society of America Bulletin, v. 82, p. 3209–3212.

Shervais, J. W., and Kimbrough, D. L., 1985, Geochemical evidence for the tectonic setting of the Coast Range ophiolite; A composite island arc-oceanic crust terrane in western California: Geology, v. 13, p. 35–38.

ACKNOWLEDGMENTS

We dedicate this site to Edgar Bailey, who died in 1983, in memory of his discovery and realization of the importance of the geologic relations at the South Fork of Elder Creek.

The ultramafic rocks at Eunice Bluff, Trinity Peridotite, Klamath Mountains, California

Bruce A. Carter, *Department of Physical Sciences (Geology), Pasadena City College, Pasadena, California 91106*
James E. Quick, *Department of Earth and Space Sciences, University of California at Los Angeles, Los Angeles, California 90024*

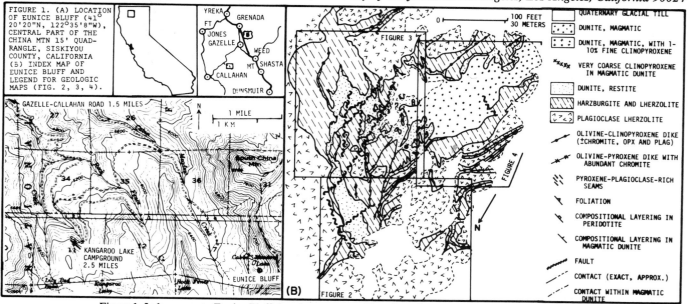

Figure 1. Index maps to Eunice Bluff showing access, and location of more detailed maps in Figures 2-4.

LOCATION AND ACCESS

Eunice Bluff is located in the Klamath Mountains of northern California, approximately 30 mi (48 km) south of the Oregon border and 10 mi (16 km) west of I-5 (Fig. 1A). It is easily reached by conventional automobile on paved and graded roads, followed by a short walk through open forest. From Gazelle, take the Gazelle-Callahan road west 15.5 mi (24.8 km) and turn left (south) on the road to Kangaroo Lake. Drive 3.7 mi (5.9 km) and turn left (east) on a U.S. Forest Service road with a large steel gate. Drive 2.5 mi (4 km) to Cabin Meadow Creek, turn right (south) and drive 1 mi (1.6 km) up the creek. At this point the road forks; cross the creek and continue an additional 1.2 mi (1.9 km) up the east side of the creek to the dead end of the road, avoiding the minor logging roads that turn off to the left. From the dead end, walk up the logging skid-road 0.25 mi (0.4 km) east past a small meadow, and then walk 0.25 mi (0.4 km) south-southeast up the creek to Cabin Meadow Lakes. Eunice Bluff is the prominent light-colored outcrop above the south end of the upper lake. The beginning of the trip, survey point A, is on the large ledge that is about 400 ft (122 m) south of and 100 vertical ft (30.5 m) above the upper lake. The outcrop is usually accessible from May through November but may be covered with snow at times during the rest of the year.

SIGNIFICANCE

The Trinity peridotite—an enormous, lower Paleozoic,

eastward-dipping sheet of ultramafic rocks (Lipman, 1965; Irwin, 1966)—contains a more diverse assemblage of ultramafic lithologies than any other North American alpine-type peridotite (Gaullaud, 1975; Quick, 1981a). These rocks are well exposed on many large glaciated outcrops, where they are relatively unaffected by crustal deformation and serpentinization. The mineralogy of the rocks is easily identified, and the spatial relations between different lithologies are clear.

Lithologies and structures are particularly well represented at Eunice Bluff, where it can be seen that the upper mantle is not homogeneous even within a small area. Primary plagioclase lherzolite is intimately associated with lherzolite, harzburgite, dunite, clinopyroxenite, and websterite. Although field relations are complex, many analogous features have been previously described in mantle xenoliths (Wilshire and Pike, 1975), and detailed mapping demonstrates the consistent repetition of petrogenetically significant relationships. The outcrop is a "snapshot" of the evolving upper mantle, in which a protolith of plagioclase lherzolite was invaded by ascending basaltic magmas. Reaction between these transient magmas and the plagioclase lherzolite wall rocks produced "restite" lherzolite, harzburgite, and dunite by leaching of plagioclase and pyroxene from a plagioclase lherzolite protolith—a process that may be considered to be a form of mantle metasomatism.

Conduits through which the melts passed are marked by

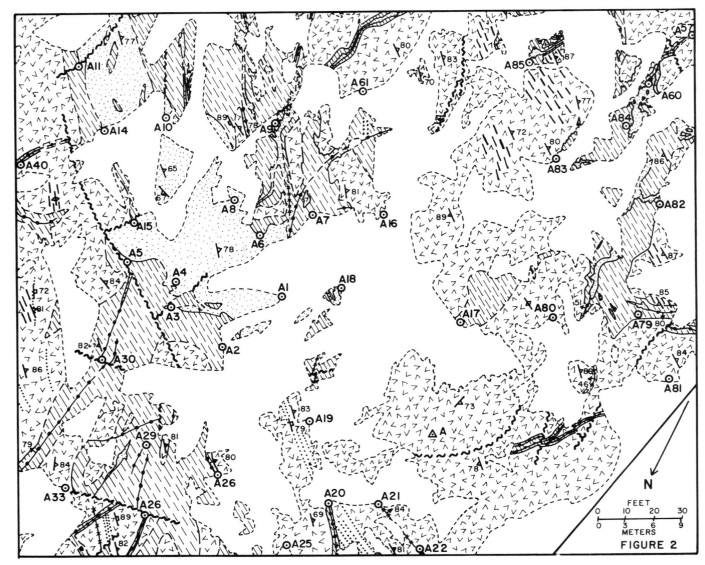

Figure 2. Northern part of Eunice Bluff. Circled locations marked by steel stakes. For legend, see Figure 1.

elongate dunite bodies, and fractional crystallization of magma in these conduits formed chromite segregations and olivine-clinopyroxene dikes. A mass of dunite and wehrlite 100 by 200 ft (30 × 60 m) formed where the magma pooled and crystallized olivine and clinopyroxene in a small upper mantle magma chamber. In order to accurately portray all significant features, Eunice Bluff was mapped in detail by plane table at a scale of 1:240. More than 100 stations are recorded on the map and keyed to permanent stainless steel markers on the outcrop. Equipped with this map, a geoscientist with no previous experience in petrology of ultramafic rocks may explore a unique natural laboratory that preserves the effects of upper mantle petrogenetic processes such as partial melting, melt migration, fractional crystallization, and metasomatism.

SITE INFORMATION

The geology of Eunice Bluff is shown in detail on Figures 2, 3, and 4 (locations in Figure 1B). Significant features of these Figures are described in the order they would be encountered by a visitor walking through Eunice Bluff from the base to the top of the outcrop. Figure 2 illustrates plagioclase lherzolite wall rock and dike relationships. Figure 3 illustrates features produced by metasomatic processes in an area where numerous magma-filled conduits have reacted with the primary wall rock. Figure 4 illustrates a small magma chamber in which olivine and minor pyroxene crystallized from the magma.

The primary mineralogy of the rocks at Eunice Bluff is easily determined in the field. In most rocks, light-brown to buff weathering olivine is the most abundant mineral, forming a matrix

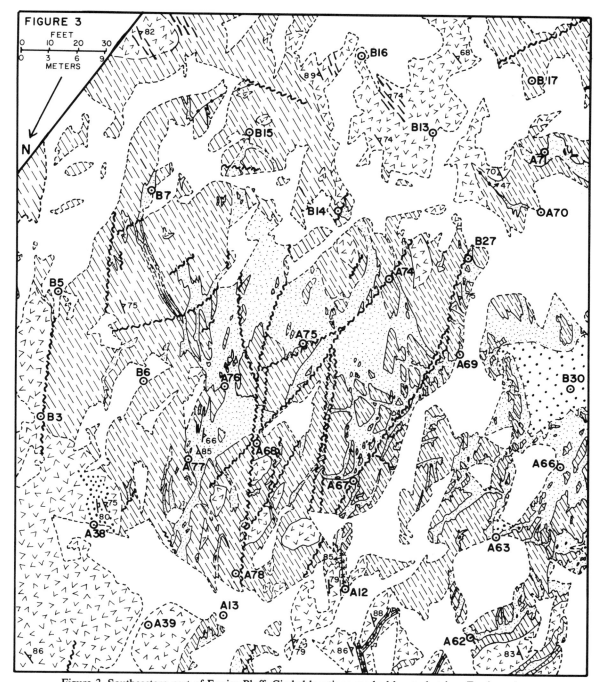

Figure 3. Southeastern part of Eunice Bluff. Circled locations marked by steel stakes. For legend, see Figure 1.

through which the other minerals are dispersed. Clinopyroxene weathers dark-green to white in local zones of intense serpentinization. Orthopyroxene weathers medium-brown to greenish-brown. Chromite is black and metallic. Plagioclase, mostly pseudomorphed by clinozoisite, is chalky white. Pyroxenes and lenses of plagioclase and chromite (±pyroxene) tend to weather out above less-resistant olivine.

Plagioclase lherzolite (70-80 percent olivine, 15 to 20 percent orthopyroxene, 2 to 10 percent clinopyroxene, 1 to 2 percent chromite, 2 to 10 percent plagioclase) is the most abundant lithology at the north and south sides of Eunice Bluff. This rock type is similar in composition to models for undepleted upper mantle (e.g., Jaques and Green, 1980), and is the primary lithology at Eunice Bluff. Dunite (90 to 99 percent olivine, 1 to 2 percent

Figure 4. Southwestern part of Eunice Bluff. Circled locations marked by steel stakes. For legend, see Figure 1.

chromite, ± minor clinopyroxene, ± trace orthopyroxene) crops out as a large body near the center of Eunice Bluff and as many small, irregular to dike-like bodies.

As described below, structural and textural evidence suggests that some dunite is formed by direct crystallization of olivine from a basaltic magma, but also that other dunite is produced by the metasomatic interaction of these transient melts with the wall

rocks. Lherzolite (75 to 85 percent olivine, 15 to 20 percent orthopyroxene, 2 to 10 percent clinopyroxene, 1 to 2 percent chromite) and harzburgite (80 to 90 percent olivine, 10 to 20 percent orthopyroxene, 1 to 2 percent chromite, trace percentages of clinopyroxene) are particularly abundant in envelopes around dunite bodies. Structural evidence suggests that these two lithologies were produced from a plagioclase lherzolite protolith by metasomatic interaction with the transient melts. Wehrlite (40 to 90 percent olivine, 10 to 60 percent clinopyroxene, trace percentages of orthopyroxene, trace percentages of chromite) occurs as segregations within the large dunite body. Wehrlite, clinopyroxenite (more than 90 percent clinopyroxene), minor percentages of olivine, trace percentages of chromite) and websterite (less than 20 percent olivine, 50 to 80 percent clinopyroxene, less than 10 percent orthopyroxene, trace percentages of chromite) are mapped as olivine-clinopyroxene dikes and cut all of the above lithologies except dunite; textural evidence suggests that these lithologies formed by direct crystallization from transient basaltic melts.

Northern Part of Eunice Bluff (Fig. 2)

The characteristics of the plagioclase lherzolite are well exposed near point A. The rock is porphyroclastic and has pyroxene and plagioclase concentrated in small (5 to 30 mm) lenses that define a pronounced metamorphic foliation produced by plastic flow in the mantle. This foliation is also defined by flattened chromite grains and trails of chromite grains, and is commonly referred to as a tectonite fabric (e.g., Boudier, 1978). Primary compositional layering is well exposed between points A19 and A20, which consists of subparallel, pyroxene-rich layers that are cross-cut by the foliation. Similar layering is typically the oldest structure in many alpine-type peridotites and is thought to have formed by pressure solution creep (Dick, 1977) or possibly as cumulate layering or as dike swarms (Quick, 1981a). Near points A85 and A83 there are numerous small 0.5 to 15 in-long (1 to 40 cm) lenses of plagioclase-enriched peridotite that are cross-cut by the foliation. Similar features in the Lanzo peridotite are interpreted to represent small segregations of basaltic melt produced by incipient melting of the plagioclase lherzolite during plastic deformation (Boudier, 1978).

Some typical olivine-clinopyroxene dikes and dunite bodies are also well displayed. Near point A20 a thin olivine-clinopyroxene dike cross-cuts compositional layering and is bordered by thin reaction rims of lherzolite. The dike consists of large pyroxene crystals that are slightly elongate perpendicular to the dike, with variable amounts of interstitial olivine. The mineralogy of these dikes is consistent with crystallization from a basaltic magma (Quick, 1981a). Nd/Sm and Rb/Sr isotopic data (Jacobsen and others, 1984) from nearby outcrops demonstrates that the pyroxenes in these dikes are in isotopic disequilibrium with the plagioclase lherzolite. This suggests that the melts that crystallized the olivine-clinopyroxene dikes were not derived from the local plagioclase lherzolite.

Near point A82, an elongate dunite body cutting lherzolite contains small segregations of chromite near its core. Near point A61, another dunite dike cuts plagioclase lherzolite but is bordered by a reaction rim of harzburgite/lherzolite. Examination of Figures 2, 3, and 4 demonstrates that dunite is always separated from plagioclase lherzolite by a zone of lherzolite/harzburgite, implicating a reaction relationship between dunite and plagio-

clase lherzolite. This suggests that the dunite, harzburgite, and lherzolite were formed in situ by the extraction of plagioclase and pyroxene from a plagioclase lherzolite protolith. These features are interpreted to have formed during movement of basaltic magma through discrete channels; pyroxene and chromite appear to have crystallized from the melt while dunite, harzburgite, and lherzolite were produced by the metasomatic reaction of the melts with the host plagioclase lherzolite (Quick, 1981a, b; Noller and Carter, 1986).

Southeastern Part of Eunice Bluff (Fig. 3)

Additional evidence for metasomatism is visible in the area covered by Figure 3. Near point A77, a dunite dike with distinctive chromite segregations cuts lherzolite. Although the chromite probably crystallized from a transient magma, most of the dunite appears to have formed by the removal of pyroxene from a lherzolite protolith. Note that the foliation (defined by small, elongate chromite grains or trains of grains) is traceable from the lherzolite into most of the dunite in that area, although the dunite cross-cuts both foliation and primary layering in the lherzolite. This suggests that the foliation in the dunite was inherited from the protolith. Between A77 and A76 note that the irregular dunite contact is controlled by the primary layering, such that dunite appears to be invading along pyroxene-rich layers. Also in this area there are numerous lensoidal relicts of lherzolite surrounded by dunite. Between points A67 and A74 numerous dunite bodies coalesce, isolating relict blocks of lherzolite and producing a "pseudo-breccia." Note that the foliation in each block is parallel to the foliation in the rest of the outcrop. These relationships reinforce the hypothesis that most of the dunite in these areas is a "restite," formed by removal of pyroxene from preexisting lherzolite.

Southeastern part of Eunice Bluff (Fig. 4).

This area is underlain by a dunite with different characteristics than the restite dunite. Between points B30 and B29, note that the dunite is devoid of foliation, chromite is rare or absent, and a small percentage of large, poikilitic clinopyroxene is present. Between points A55 and A42, note the presence of abundant, small (5–10 mm) interstitial clinopyroxene grains; in this area, the rock is more properly termed wehrlitic dunite or wehrlite. Locally, large (1 to 8 in) (2 to 20 cm) clinopyroxene crystals are concentrated into poorly defined layers (e.g., near points B29 and A55) that are transverse to the trend of the dunite body. In a few places (e.g., near points A55 and A46), large orthopyroxene crystals overgrow clinopyroxene, and at least two large crystals of pargasitic hornblende are observed surrounding pyroxene cores. The absence of foliation and the presence of poikilitic clinopyroxene suggest that this dunite formed by precipitation of olivine followed by clinopyroxene from a melt. This dunite body may have formed by fractional crystallization of a magma in a small mantle magma chamber. The presence of orthopyroxene indicates that the magma ultimately reached the olivine-clinopy-roxene-orthopyroxene cotectic and the presence of amphibole suggests that the magma was "wet."

Near points A51 and A73 (Fig. 4) several olivine-clinopyroxene dikes run southwestward from the margin of the large dunite body. These have approximately the same orientation as many of the dikes in Figures 2 and 3. If the large dunite body did form in a magma chamber, it is possible that the dikes at points A51 and A73 represent conduits through which magma flowed upward from the chamber, and that the dikes in Figures 2 and 3 were feeders to the chamber.

REFERENCES CITED

Boudier, F., 1978, Structure and petrology of the Lanzo peridoite massif (Piedmont Alps): Geological Society of America Bulletin, v. 89, p. 1574–1591.

Dick, H.J.B., 1977, Partial melting in the Josephine peridotite I; The effect on mineral composition and its consequence for geobarometry and geothermometry: American Journal of Science, v. 277, p. 801–832.

Gaullaud, L., 1975, Structure and petrology of the Trinity mafic-ultramafic complex, Klamath Mountains, Northern California, *in* Lindsley-Griffin, N., and Kramer, J. C., eds., Geology of the Klamath Mountains, Northern California: 73rd Annual Meeting, Cordilleran Section, Geological Society of America, p. 112–133.

Irwin, W. P., 1966, Geology of the Klamath Mountains Province: California Division of Mines and Geology Bulletin, v. 190, p. 19–38.

Jacobsen, S. B., Quick, J. E., and Wasserburg, G. W., 1984, A Nd and Sr isotopic study of the Trinity peridotite; Implications for mantle evolution: Earth and Planetary Science Letters, v. 68, p. 361–378.

Jaques, A. L. and Green, D. H., 1980, Anhydrous melting of peridotite at 0-15 Kb pressure and the genesis of tholeiitic basalts: Contributions to Mineralogy and Petrology, Beitraege zur Mineralogie und Petrologie, v. 73, p. 287–310.

Lipman, P. W., 1965, Structure and origin of an ultramafic pluton in the Klamath Mountains, California: American Journal of Science, v. 262, p. 199–222.

Noller, J. S., and Carter, B., 1986, The origin of various types of chromite schlieren in the Trinity peridotite, Klamath Mountains, California, *in* Augustithis, S. S., ed., Metallogeny of Basic and Ultrabasic Rocks: Athens, Theophrastus Publishing (in press).

Quick, J. E., 1981a, Petrology and petrogenesis of the Trinity peridotite, an upper mantle diapir in the eastern Klamath Mountains, northern California: Journal of Geophysical Research, v. 86, p. 11837–11863.

—— , 1981b, The origin and significance of large, tabular dunite bodies in the Trinity peridotite, northern California: Contributions to Mineralogy and Petrology, v. 78, p. 413–422.

Wilshire, H. G., and Pike, J. E., 1975, Upper-mantle diapirism; Evidence from analogous features in alpine peridotite and ultramafic inclusions in basalt: Geology, v. 3, p. 467–470.

Medicine Lake Volcano and Lava Beds National Monument, California

Julie M. Donnelly-Nolan, U.S. Geological Survey, 345 Middlefield Road, Menlo Park, California 94025

LOCATION

Medicine Lake Volcano is located in the Modoc Plateau physiographic province in northeastern California, about 30 mi (50 km) northeast of Mt. Shasta. It is a Pleistocene and Holocene shield volcano whose products cover about 900 mi^2 (2,500 km^2); volume is estimated to be about 130 mi^3 (600 km^3). Lava Beds National Monument is located on the northern flank of the volcano. The monument encompasses mostly basaltic and some andesitic lavas. Higher on the volcano, basaltic lavas are mostly absent, andesite dominates, and high-silica lavas are present including the spectacular late Holocene rhyolites and dacites of Glass Mountain, Little Glass Mountain, and the Medicine dacite flow (Anderson, 1941). A wide variety of volcanic and tectonic phenomena can be seen at Medicine Lake Volcano. Many features are young and very well exposed, making it an ideal place for a field trip to see the range of volcanic activity from basaltic to rhyolitic (Fig. 1).

Medicine Lake sits at an elevation of 6,676 ft (2,005 m) within a 4.5-by 7.5-mi (7- by 12-km) caldera; the highest point on the caldera rim is 7,913 ft (2,398 m), and the plateau surrounding the volcano is at about 4,000 ft (about 1,200 m). The volcano is traversed by numerous roads, and access is good although most roads are unpaved. Four-wheel-drive vehicles are unnecessary for this trip and for access to most of the volcano, but high clearance is recommended. For weather and road conditions, it is wise to call ahead to one or more of the national forests whose lands cover most of the volcano: Modoc National Forest (Doublehead Ranger District, Tulelake, CA) for the eastern half including the caldera and the campgrounds at Medicine Lake; Shasta-Trinity National Forest (headquarters in Redding, CA), the southwestern part; Klamath National Forest (headquarters in Yreka, CA) for the northwestern part.

Campgrounds are also available at Lava Beds National Monument. Be forewarned that campgrounds can fill up on holiday weekends. Depending on the weather and on the depth of winter snowfalls, it may not be possible to drive across the caldera in early summer. It is recommended that this field trip guide be followed between the Fourth of July and early October. Roads may be open and clear in June and into early November, but be sure to contact the local Forest Service or Park Service office to find out. Lava Beds National Monument can be visited year-round.

The weather is unpredictable from day to day at high eleva-

tions. Temperatures commonly drop to freezing even during mid-summer nights at Medicine Lake. A typical summer day will reach 90°F (32°C) at lower elevations, 70-80°F (21-27°C) higher on the volcano, with a buildup of clouds in the afternoon perhaps resulting in isolated thunderstorms. Food and gas are not available along the route of the trip; water is available only at Medicine Lake and at the Visitors Center and campground in Lava Beds National Monument.

A few other words of advice. Wear gloves and eye protection when sampling and climbing on young, glassy, high-silica flows. You will need lights (and something warm to wear) for stop 8 and any other caves you might visit; flashlights can be borrowed if necessary at the Visitors Center in Lava Beds National Monument. Remember not to collect samples in the monument. Watch out for rattlesnakes at lower elevations, particularly in the monument. And be aware that new roads are being built all the time and old ones are destroyed. Do obtain the Forest Service and topographic maps for the area (Bartle, Hambone, Medicine Lake, Mt. Dome, and Timber Mountain 15′ quadrangles), but remember that the Forest Service maps available as of this writing (early 1985) are only approximations to reality and that the topographic maps are about 30 years old.

I have written this as a one-day trip. It is designed to show a variety or rock types and morphologic features. If each stop takes half an hour, the whole trip should take about 10 hours. Many other stops are possible, and I have listed a few below if you wish to make it a multiple-day trip. A few other stops are written up in Donnelly-Nolan and others (1981). The classic geologic reference is Anderson (1941); this paper contains a very useful, although generalized map. Some petrologic references include Condie and Hayslip (1975), Mertzman (1977a, 1977b), and Grove and Baker (1984).

The turnoff to Medicine Lake is on Highway 89 about 16.5 mi (26.5 km) east of McCloud and about 20 mi (32 km) east of Interstate Highway 5. Driving east, the turnoff is just past the Bartle store; there should be a small sign on your left. Turn north onto a good paved road that goes through the place called Bartle on the Bartle 15′ quadrangle. Continue north about 4 mi (about 6.5 km) to a junction with another paved road on the right; again, there should be a sign pointing toward Medicine Lake. Turn right (note that if you take the left-hand road, it continues north as a good paved road past Little Glass Mountain and across the northwest flank of Medicine Lake Volcano, then turns west and

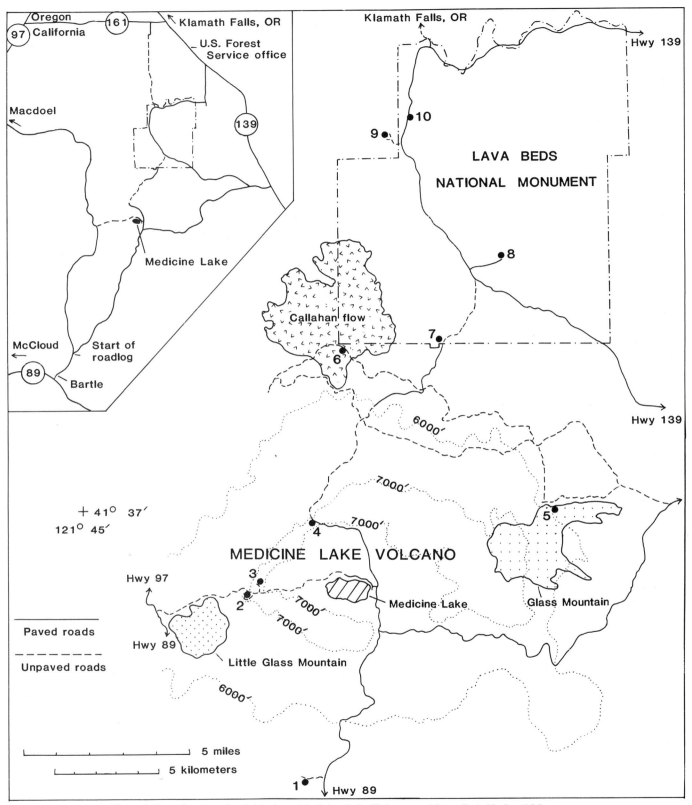

Figure 1. Location map for field trip to Medicine Lake Volcano and Lava Beds National Monument. Inset map shows locations of major highways surrounding the area. Trip goes from south to north; north points to top of map.

connects with Highway 97 at Macdoel). Trip mileage begins at this intersection.

ROADLOG

Odometer reading in miles (kilometers in parentheses).

0.0 Junction of paved roads 4 mi (6.5 km) north of Highway 89 and Bartle. Turn right and head northeast across Pliocene high-alumina basalt known as the Warner Basalt (see Anderson, 1941, p. 353-354). Note thick red soil, better developed than any soil you will see on the younger lavas of Medicine Lake Volcano. After about 9 mi (14.5 km), the road drops over an east-facing fault scarp onto young Medicine Lake basaltic lavas. Road crosses lava tube that begins at Giant Crater and can be traced for at least 14 mi (23 km), making it one of the longest known lava tubes (Greeley and Baer, 1971). Continue driving north.

17.6 (28.2) Sign and parking area on left for Jot Dean ice cave. As you continue across this Holocene basalt that erupted from Double Hole Crater (Baer, 1970), note the deflation that took place after the flow formed a "high-water mark," crusted over, and the crust was later let down as lava flowed away via one or more tubes.

18.4 (29.4) Turn left onto good gravel road. Road sign should point to Grasshopper Flat. Follow road west for 0.5 mi (0.8 km) to large spatter cone on your left. Stop and park off the road.

18.9 (30.2) Stop 1. Double Hole Crater Walk south toward crater across thin pahoehoe high-alumina basalt. Note small surface tubes and pahoehoe toes. Climb up on this relatively large spatter vent and look at welded spatter. This stop gives you an opportunity to see up close one of the several types of vents for the many basaltic flows of Medicine Lake Volcano. Other vent types include much smaller spatter cones aligned to form spatter ramparts, cinder cones, and pit craters (stop 7). With the exception of the Lake Basalt (Anderson, 1941) in the caldera, which has an unusual texture and may represent some form of cumulate, you will not cross basalt again until you reach the north flank of the volcano. The basalt of Double Hole Crater is one of a set of early Holocene basalts that erupted from several nearby vents including Giant and Chimney Craters. The initial eruptions were contaminated with crustal material (Grove and others, 1984), followed by as much as 2.2 mi^3 (10 km^3) of high-alumina basalt that was transported south about 25 mi (40 km) via lava tubes. Turn vehicles around and return to paved road.

19.4 (31.0) Turn left and continue north on paved road toward Medicine Lake.

25.6 (41.0) Junction with good paved road on your right. Continue straight ahead to the north, but note that the road to the right goes down the east side of the volcano and joins Highway 139.

27.2 (43.5) Turn left on paved road. Sign should point to Medicine Lake.

27.5 (44.0) Turn right. Drive around north side of lake past campgrounds. Pavement ends at west entrance to Medicine campground; continue west.

29.2 (46.7) Forest Service ranger station. Continue straight ahead. About half a mile (about 1 km) further where you have a choice between equally good roads, keep right. The road is rough and slow.

31.5 (50.4) Base of Little Mount Hoffman cinder cone. Road forks. Take the left fork (straight ahead), put your vehicle in low gear, and drive to top of cinder cone. If it is early in the summer and there are snowdrifts around the base of the cone, there may be snowdrifts across the road to the top. In this case, you may wish to walk to the top rather than having to back down this rather steep and exposed road. The view at the top is worth it.

32.1 (51.4) Stop 2. Little Mount Hoffman lookout. The view from the top of this cinder cone is spectacular. Below you is the late Holocene rhyolite of Little Glass Mountain. Behind the glass flow, Mt. Shasta dominates the western skyline. To the south on a clear day, you can see Lassen Peak; to the north are Mt. McLoughlin and, still farther away, Mt. Scott on the eastern rim of Crater Lake caldera. To the east, you can see the dome of Glass Mountain on the eastern rim of Medicine Lake caldera. Drive back down to the base of Little Mount Hoffman.

32.7 (52.3) At road junction, go straight ahead (east), back toward Medicine Lake.

32.9 (52.6) Turn left. Follow narrow dirt road about 0.3 mi (0.5 km) to first wide spot on left where several cars can be parked.

33.2 (53.1) Stop 3. Cracks. Cross the road and walk about N 30°E into logged area. You will begin to see large cracks in the ground. Continue walking northeast about 100 yd (100 m) to the largest crack, about 30 ft (10 m) deep and 30 ft (10 m) wide. This set of cracks developed over a rhyolite dike (Fink and Pollard, 1983) during emplacement of Little Glass Mountain and several domes of identical composition farther to the northeast. Note that the ragged edges of the cracks could be fit back together again in an east-west sense. The walls of the cracks expose the interior of a fountain-fed andesite flow. You can easily see the agglutinated texture and angular lithic fragments typical of this type of flow. The andesite was erupted from the glaciated cinder cone just south of the parking spot. Return to vehicles and retrace route 0.3 mi (0.5 km) to road junction.

33.5 (53.6) Turn left and retrace route to Medicine Lake. I recommend stopping at one of the campgrounds or at the picnic area at the east end of the lake for lunch.

37.3 (59.7) Turn left on paved road away from Medicine Lake.

37.6 (60.2) Turn left on main paved road across caldera. Continue north across the caldera floor around the east and north sides of the late Holocene Medicine dacite flow. Drive west up northern caldera wall. Stop at wide spot in road just before road turns north over the caldera rim.

40.6 (65.0) Stop 4. North rim of caldera. Below you to the south is the Medicine dacite flow and beyond that, Medicine Lake. The muted topography of the caldera is partially a result of glaciation which has stripped the tops of lava flows and rounded off cinder cones. Medicine Lake appears to sit in a bed of glacial clay that prevents the water from draining away into the permeable volcanic rocks. The caldera does not appear to be the result of ash-flow eruptions; rather, Anderson's 1941 suggestion that it resulted from voluminous eruption of andesite lavas around the rim of the caldera appears to be the best explanation. The roadcut here exposes one of these andesites; another was exposed in the cracks at stop 3. Follow the road over the north rim.

40.9 (65.4). Pavement ends at intersection. Continue straight ahead.

45.1 (72.2) Intersection of major dirt roads. Turn right on Cougar Butte Road. Continue east on main gravel road to first major road on right.

51.4 (82.2) Turn right and proceed south. Good dirt road becomes narrow track in pumice. When you arrive in the pumice mining area, drive toward the front of the big rhyolite flow you can see sparkling ahead of you. The roads in the mining area are continually changing; in 1985 as you continue south you should come to a major haul road. Turn left onto this road, then turn right very soon onto a narrow dirt track that goes around the front of the big rhyolite lobe you can see above you. Park at the edge of the flow where a big pile of obsidian blocks a bulldozer road that goes up onto the flow.

53.4 (85.4) Stop 5. Glass Mountain. Walk up the bulldozer cut and examine the rhyolite. Pieces vary from black aphyric obsidian to pumiceous, lighter-colored samples, and samples with interesting textures that show evidence of breakage, oxidation, and flowage while still hot. On your left, under the rhyolite lobe, is the dacite that comprises the first part of the Glass Mountain flow (Anderson, 1933; Eichelberger, 1975). Go out onto the dacite and notice the abundant basaltic andesite inclusions. Two mixing events are suggested: first, a basaltic component is mixed into rhyolite magma and homogenization to dacite takes place; second, basaltic andesite magma is injected into the dacite, forming blobs that chill against the host silicic magma and perhaps cause the eruption to begin. For further discussion of this stop, see Donnelly-Nolan and others (1981) and Eichelberger (1981).

Glass Mountain lava erupted from a N 30° W-trending dike along which are 13 vents. According to Eichelberger (1975), three of the vents contributed to form the Glass Mountain flow. The other 10 vents produced small domes and flows, nine of which are north and one south of the main flow. The Glass Mountain eruption took place sometime between 200 and 1,300 years ago based on carbon-14 dates. There are large trees growing in pumiceous zones on the flow, suggesting that 200 years is too young. See Heiken (1978) for further discussion of this topic. Heiken discusses both Glass Mountain and Little Glass Mountain tephra deposits, concluding that the former feature is the younger. Return to vehicles and retrace your route north through the pumice mining area onto the road by which you came. Continue north to T-intersection with Cougar Butte Road.

55.4 (88.6) Turn left onto Cougar Butte Road.

55.7 (89.1) Turn right at first good road on your right. This road goes generally northwest down the north flank of the volcano. Continue on it until you intersect a paved road.

61.2 (97.9) Drive directly across paved road onto narrow dirt road. (Note that if you turn left at this intersection you will return

directly to stop 4 and to Medicine Lake; if you turn right, you will go directly to stop 7 and Lava Beds National Monument.) Drive straight ahead, staying on this old railroad grade for about 3 mi (about 5 km), driving directly into a late Holocene basaltic andesite flow shown on Anderson's 1941 map as the Callahan flow. The road actually cuts through the base of the flow's vent, Cinder Butte, where it was quarried for roadbed material. Slow down and look for the first parking area on your left that can hold several vehicles.

64.3 (102.9). Stop 6. Callahan flow. Rather than traversing the surface, the road cuts into the flow giving you a three-dimensional view of the sometimes blocky, sometimes smooth-surfaced flow morphology. This particular location allows you to look into the center of a small flow lobe. Because the flow appears to have cooled in place here, paleomagnetists have chosen this location for sampling; note the one-inch-diameter core holes. The flow has a carbon-14 age of 1,110 ± 60 years (Donnelly-Nolan and Champion, in press).

There is a very nice view to the north and northeast from here across the sparsely vegetated flow. To the north are several normal faults, down to the east and trending north toward Klamath Falls. To the northeast you can see some of the cinder cones and flows of Lava Beds National Monument; at this location you are just outside the southwest corner of the monument. Turn your vehicles around here and return to the paved road.

67.4 (107.8). Turn left onto the paved road. Drive north just past where the paved road turns into dirt at the southern boundary of Lava Beds National Monument. On your left is the parking area for Mammoth Crater.

69.0 (110.4) Stop 7. Mammoth Crater. Follow the short paved path west to a view of Mammoth Crater. This pit crater vented a large volume of basalt and basaltic andesite. This is one of several vents including Modoc Crater and Bearpaw Butte that produced the late Pleistocene unit referred to by Donnelly-Nolan and Champion (in press) as the basalt of Mammoth Crater. This unit covers about two-thirds of the monument and extends beyond monument boundaries to the east and west; volume is estimated to be about 2.2 mi³ (10 km³). This unit is analogous to basalt erupted at Giant Crater and related vents on the south side of the volcano (see description of stop 1). The earliest eruptions appear to have produced basaltic andesite lavas that built up around the vents; later, more fluid basalt was transported via lava tubes as far as 15 mi (25 km) away. You will see one of these tubes at stop 8. Continue driving north on dirt road to intersection with paved road.

71.6 (114. 6). Turn left on main road through Lava Beds National Monument. Proceed north to first paved road on right. Sign should point to Skull Cave.

72.0 (115.2). Turn right and drive east to Skull Cave.

73.1 (117.0). Stop 8. Skull Cave. You will need flashlights and/or Coleman lanterns to explore the cave; flashlights can be borrowed at the Visitors Center, 1.5 mi (2.4 km) south on the main monument road. Even if you don't have lights or the time to obtain them (and note that you can return to this stop after dark if you are running late), just walking into the entrance of the cave is an impressive experience. This large tube transported lava to the northeastern corner of Lava Beds National Monument. Proceeding into the cave, stairs take you down to ice at the bottom, two levels below. Thus, at least three lava tubes are stacked one on top of the other at this location. This is one place where the law of superposition fails because tubes are often used more than once and the inner lava is youngest. Thus, younger lava can be under older lava. Return to the main monument road.

74.2 (118.7). Turn right and drive north about 4 mi (6.5 km) where you will go northwest across a Holocene aa basalt flow referred to as the basalt of Devils Homestead. Ahead of you is the fault scarp called Gillem Bluff. The road turns north up the face of the fault. Look for a road on your left. The sign indicates that the road is not maintained for public use.

78.8 (126.1) Turn left. The road soon becomes dirt. Follow it west, then north to first right-hand turn.

79.2 (126.7). Turn right and drive about 0.2 mi (0.3 km) north, keeping left at the next intersection, and parking next to a shallow gully on your left.

79.4 (127.0) Stop 9. Andesitic ash-flow tuff. Walk down into the gully. Brownish-red andesitic ash-flow tuff forms low outcrops on both sides. Estimated thickness here prior to some quarrying was about 5 or 6 ft (less than 2 m). This small patch of partially welded andesitic tuff is one of many on the north and west sides of Medicine Lake Volcano. It is also present in a few small patches on the east side. The spatial distribution of the tuff indicates that it was erupted at or near the center of the volcano. The largest pumice and lithic fragments are found in the single outcrop in the caldera, suggesting a source in the caldera. The tuff is younger than one of the andesite flows on the northwest rim of Medicine Lake Volcano, but that andesite flow is not broken by any caldera-forming faults, implying that eruption of the tuff was not responsible for creating the caldera.

The complete absence of this unit anywhere on the caldera rim, together with indications of hydrothermal alteration at the caldera outcrop, suggests that the tuff erupted through an ice cap on the volcano and deposited in the caldera on the only exposed ground surface where a fumarole had melted the ice (Donnelly-Nolan, 1983). Coarse gravels and dry channels cut in the tuff on the northwest flank of the volcano indicate that meltwater from the interaction of the ash-flow and the ice formed a catastrophic flood (Nolan and Donnelly-Nolan, 1984). The tuff has not been dated directly, but the indirect argument about the presence of an ice cap, together with evidence from younger lava flows, some of which have been glaciated, point to a late Pleistocene age predating the latest glaciation. Stratigraphically, the tuff is of major importance as the volcano's only marker bed. Anderson (1941)

recognized the andesitic ash-flow tuff but interpreted its typical exposure in the bottoms of gullies to indicate that it was one of the volcano's oldest units. Recent mapping (Donnelly-Nolan, unpublished) shows that the tuff is younger than about 90 percent by volume of Medicine Lake Volcano. Retrace your route to the main paved road through Lava Beds National Monument.

80.0 (128.0). Turn left. Drive along Gillem Bluff and turn right into parking area.

80.8 (129.3). Stop 10. Devils Homestead Overlook. Directly below you is the basalt of Devils Homestead, erupted from spatter vents at Fleener Chimneys further south along the Gillem Bluff fault scarp. To the south is a panoramic view of Lava Beds National Monument and the shield shape of Medicine Lake Volcano, the north flank dotted with cinder cones. To the north, the fault scarp of Gillem Bluff continues directly north forming the western margin of Tule Lake basin.

This trip can easily be expanded to two or more days. In Lava Beds National Monument alone, many more stops can be made. I suggest Captain Jacks Stronghold for an interesting example of geo-history (see Waters, 1981), also Fleener Chimneys, Black Craters, a climb to the top of Schonchin Butte, and Valentine Cave for a nice, clean cave with lots of interesting flow features. At the monument Visitors Center, you'll find publications for sale along with interesting displays, and you can obtain directions to the nearly 300 caves in the monument. In the middle of the Visitors Center parking lot is the entrance to Mushpot Cave where movies are shown, including one that describes lava tube formation.

Medicine dacite flow and Little Glass Mountain contain interesting suites of inclusions ranging from gabbro to basalt, andesite, and granite (Mertzman, 1981; Grove and Donnelly-Nolan, 1983). Little Glass Mountain also displays some very interesting flow features that have been described by Fink (1981); directions are in Donnelly-Nolan and others (1981). A drive down the east side of the volcano (see directions to the paved road between stops 1 and 2 of this roadlog) offers a Vista Point where you can look south over the Burnt Lava flow and High Hole Crater (southern analogs of the Callahan flow and Cinder Butte) to Lassen Peak in the distance. Further east are roads that will take you to the south side of the Glass Mountain flow (these are usually indicated by signs). Once you have descended the volcano, you can stop at the railroad tracks to view the Warner Basalt. Park at the railroad crossing and walk a short distance north along the tracks to the first good exposure. East of the railroad tracks and west of Highway 139 is a turnoff to Lava Beds National Monument if you wish to enter the monument from the southeast.

REFERENCES CITED

Anderson, C. A., 1933, Volcanic history of Glass Mountain, northern California: American Journal of Science, v. 26, p. 485–506.

——1941, Volcanoes of the Medicine Lake Highland, California: University of California Publications, Bulletin of the Department of Geological Sciences, v. 25, no. 7, p. 347–422.

Baer, R. L., 1970, Petrology of Quaternary lavas and geomorphology of lava tubes, south flank of Medicine Lake Highland, California [Master's thesis]: University of New Mexico, 120 p.

Condie, K. C., and Hayslip, D. L., 1975, Young bimodal volcanism at Medicine Lake volcanic center, northern California: Geochimica et Cosmochimica Acta, v. 39, p. 1165–1178.

Donnelly-Nolan, J. M., 1983, Two ash-flow tuffs and the stratigraphy of Medicine Lake Volcano, northern California Cascades: Geological Society of America Abstracts with Programs, v. 15, no. 5, p. 330.

Donnelly-Nolan, J. M., and Champion, D. E., Geologic map of Lava Beds National Monument: U.S. Geological Survey Map I-1804, 1:24,000 scale (in press).

Donnelly-Nolan, J. M., Ciancanelli, E. V., Eichelberger, J. C., Fink, J. H., and Heiken, Grant, 1981, Roadlog for field trip to Medicine Lake Highland, *in* Johnston, D. A., and Donnelly-Nolan, J. M., eds., Guides to some volcanic terranes in Washington, Idaho, Oregon, and northern California: U.S. Geological Survey Circular 838, p. 141–149.

Eichelberger, J. C., 1975, Origin of andesite and dacite: Evidence of mixing at Glass Mountain in California and at other circum-Pacific volcanoes: Geological Society of America Bulletin, v. 86, p. 1381–1391.

——1981, Mechanism of magma mixing at Glass Mountain, Medicine Lake Highland Volcano, California, *in* Johnston, D. A., and Donnelly-Nolan, J. M., eds., Guides to some volcanic terranes in Washington, Idaho, Oregon, and northern California: U.S. Geological Survey Circular 838, p. 183–189.

Fink, J. H., 1981, Surface structure of Little Glass Mountain, *in* Johnston, D. A., and Donnelly-Nolan, J. M., eds., Guides to some volcanic terranes in Washington, Idaho, Oregon, and northern California: U.S. Geological Survey Circular 838, p. 171–176.

Fink, J. H., and Pollard, D. D., 1983, Structural evidence for dikes beneath silicic domes, Medicine Lake Highland Volcano, California: Geology, v. 11, p. 458–461.

Greeley, Ronald, and Baer, Roger, 1971, Hambone, California and its magnificent lava tubes—Preliminary report: Geological Society of America Abstracts with Programs, v. 3, no. 2, p. 128.

Grove, T. L., and Baker, M. B., 1984, Phase equilibrium controls on the tholeiitic versus calc-alkaline differentiation trends: Journal of Geophysical Research, v. 89, p. 3253–3274.

Grove, T. L., Baker, M. B., Champion, D. E., and Donnelly-Nolan, J., 1984, The role of assimilation and fractional crystallization in the compositionally zoned Giant and Chimney Craters eruption, Medicine Lake Volcano, California (abstract): EOS, American Geophysical Union Transactions, v. 65, no. 45, p. 1153.

Grove, T. L., and Donnelly-Nolan, J., 1983, Role of amphibole in the differentiation history of Medicine Lake Highland lavas (abstract): EOS, American Geophysical Union Transactions, v. 64, no. 45, p. 900.

Heiken, Grant, 1978, Plinian-type eruptions in the Medicine Lake Highland, California, and the nature of the underlying magma: Journal of Volcanology and Geothermal Research, v. 4, p. 375–402.

Mertzman, S. A., Jr., 1977a, The petrology and geochemistry of the Medicine Lake Volcano, California: Contributions to Mineralogy and Petrology, v. 62, p. 221–247.

——1977b, Recent volcanism at Schonchin and Cinder Buttes, northern California: Contributions to Mineralogy and Petrology, v. 61, p. 231–243.

——1981, Genesis of Recent silicic magmatism in the Medicine Lake Highland, California: Evidence from cognate inclusions found at Little Glass Mountain: Geochimica et Cosmochimica Acta, v. 45, p. 1463–1478.

Nolan, K. M., and Donnelly-Nolan, J. M., 1984, Catastrophic flooding related to a sub-glacial eruption at Medicine Lake Volcano, northeastern California (abstract): EOS, American Geophysical Union Transactions, v. 65, no. 45, p. 893.

Waters, A. C., 1981, Captain Jack's Stronghold (The geologic events that created a natural fortress), *in* Johnston, D. A., and Donnelly-Nolan, J. M., Guides to some volcanic terranes in Washington, Idaho, Oregon, and northern California: U.S. Geological Survey Circular 838, p. 151–161.

Shelf and deep-marine deposits of Late Cretaceous age, Cape Sebastian area, southwest Oregon

Ralph E. Hunter and H. Edward Clifton, U.S. Geological Survey, Menlo Park, California 94025

LOCATION

The outcrops described in this report are in the vicinity of Cape Sebastian, Curry County, Oregon (Fig. 1). The principal outcrop (Point A, Fig. 2) is in the Cape Sebastian Sandstone as restricted by Bourgeois (1980), at the south tip of the Cape Sebastian headland. Visitors can most easily reach the outcrop by a 1.2-mi-long (2 km) foot trail that leads down to the south tip of the headland from the parking area at the south end of the road through Cape Sebastian State Park (Point B, Fig. 2). The entrance to the state park is on U.S. 101, 6 mi (10 km) south of Gold Beach, Oregon. The outcrops of the Hunters Cove Formation of Dott (1971) are located along the shore of Hunters Cove and can most easily be reached by walking northward along the beach from the parking area alongside U.S. 101 just south of the cove (Point C, Fig. 2).

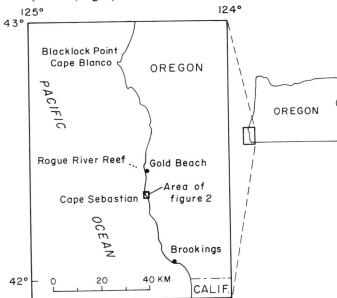

Figure 1. Index map of southern Oregon coast.

SIGNIFICANCE OF LOCALITY

The Cape Sebastian Sandstone as restricted by Bourgeois (1980) is a shallow-marine sandstone of Late Cretaceous age that contains well-exposed cyclic deposits interpreted to have formed during alternating storms and periods of relative calm. A complete cycle (average thickness 5.2 ft; 1.6 m) consists of a lower, storm-deposited sandstone unit characterized by hummocky cross-stratification, an upper bioturbated sandstone unit that formed during relatively calm conditions, and an intermediate unit composed of planar- to ripple-bedded sandstone and shale that formed during the waning phase of the storm and at the beginning of the following calm period. Changes in grain size and sedimentary structures vertically through the section suggest that the formation accumulated in gradually deepening water.

The Cape Sebastian Sandstone unconformably overlies thick-bedded turbidite sandstones and pebbly sandstones of the Late Cretaceous Houstenaden Creek Formation of Bourgeois and Dott (1985) and underlies, probably conformably, thin-bedded turbidite sandstones and shales of the Late Cretaceous Hunters Cove Formation of Dott (1971). The Hunters Cove Formation includes a submarine-slump breccia and several intervals of turbidite sandstone beds without interbedded shale.

GEOLOGIC SETTING

The small area of Upper Cretaceous rocks in the vicinity of Cape Sebastian, Oregon, is remarkable for the range of depositional environments represented. Although the beds have been moderately folded and faulted, as shown by the detailed mapping of Howard and Dott (1961), the structural complexity is much less than that of the unconformably underlying Otter Point Formation, of Late Jurassic age (Koch, 1966). The Upper Cretaceous rocks were divided by Dott (1971) into the Cape Sebastian Sandstone and the overlying Hunters Cove Formation, consisting of sandstone and shale. The Cape Sebastian Sandstone was subsequently restricted by Bourgeois (1980) to beds that overlie an angular unconformity recognized by Hunter and others (1970) within the formation as defined by Dott (1971). Bourgeois and Dott (1985) have named the sequence that underlies the unconformity the Houstenaden Creek Formation.

Apart from their occurrence in the Cape Sebastian area, Upper Cretaceous rocks are of very limited occurrence in coastal southwestern Oregon. They are known from the Rogue River Reef, from Blacklock Point, and less certainly from a few other localities (Fig. 1; Hunter and others, 1970; Dott, 1971; Bourgeois and Dott, 1985). The paleogeography of the depositional basin is unknown, as is the source of the light-colored, quartz- and feldspar-rich sand and of many of the pebble types (Bourgeois and Dott, 1985). Fossils indicate that the Upper Cretaceous beds in the Cape Sebastian area range from Campanian (or possibly as old as Albian) to possibly Maestrichtian in age (Dott, 1971; Bourgeois, 1980; Bourgeois and Dott, 1985).

HOUSTENADEN CREEK FORMATION

About 660 ft (200 m) of the Houstenaden Creek Formation of Bourgeois and Dott (1985) is exposed on the west face of the

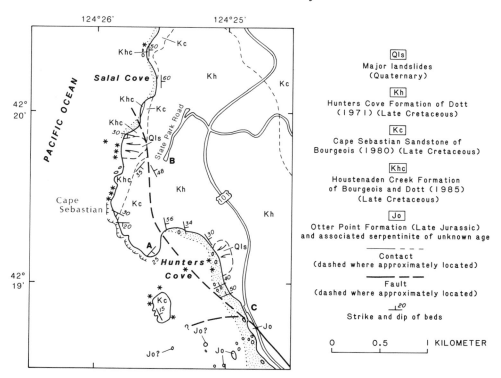

Figure 2. Geologic map of Cape Sebastian area. Base modified from Gold Beach 15-minute Quadrangle. Geology modified from Howard and Dott (1961), Hunter and others (1970), Dott (1971), Bourgeois (1980), Hunter and Clifton (1982), and Bourgeois and Dott (1985). Principal outcrop described in this report is labeled A; parking areas are labeled B and C.

Cape Sebastian headland (Fig. 3), and the upper few ft (m) can be seen at the tips of two small points that bound the cove, informally named Salal Cove by Bourgeois (1980), just north of the headland (Fig. 2). All these localities are difficult to reach. The formation in the Cape Sebastian area consists of sandstone, pebbly sandstone, and conglomerate in thick, normally graded turbidite beds (Hunter and others, 1970). Bourgeois and Dott (1985) consider this facies to overlie a less conglomeratic, more shaly facies of the formation; this more shaly facies is exposed at the type locality of the formation near the mouth of Houstenaden Creek, 7.4 mi (12 km) south of Cape Sebastian. The exclusively sandy and gravelly nature of the turbidites in the Cape Sebastian area suggest an origin in the proximal (inner) part of a submarine fan or fan channel.

The Houstenaden Creek Formation was deformed and partly eroded before deposition of the Cape Sebastian Sandstone. At a site on the west face of the Cape Sebastian headland that is probably not accessible by foot, the Cape Sebastian Sandstone overlies the Houstenaden Creek Formation with angular unconformity (Fig. 4). Near the mouth of Myers Creek, 1.9 mi (3 km) southeast of Cape Sebastian, the Houstenaden Creek Formation is apparently absent, for beds assigned to the Cape Sebastian Sandstone unconformably overlie the Otter Point Formation at that locality (Bourgeois, 1980).

CAPE SEBASTIAN SANDSTONE

The Cape Sebastian Sandstone as restricted by Bourgeois (1980) is well exposed on the Cape Sebastian headland and in Salal Cove, just north of the headland (Fig. 2). In this area the formation is about 660 ft (200 m) thick and has been divided into four facies (Bourgeois, 1980; Fig. 3). The formation is not richly fossiliferous, but enough fossils have been found to indicate a marine origin for all facies.

Conglomeratic Facies

The basal 33 ft (10 m) of the Cape Sebastian Sandstone consists of conglomerate and crossbedded to planar-bedded pebbly sandstone. This facies is exposed on the west face of the Cape Sebastian headland (Fig. 4) and in Salal Cove, but it is difficult to reach at both places. The facies is also found at several points outside the area shown in Figure 2 (Bourgeois, 1980). The basal conglomerate bed contains sandstone boulders that apparently are concretions eroded from the underlying Houstenaden Creek Formation of Bourgeois and Dott (1985). The crossbedding is mostly high-angle (original dip angles 15°–30°) and of medium scale (set thicknesses 8 to 12 in; 20–30 cm). Bivalve shell fragments and trace fossils are common.

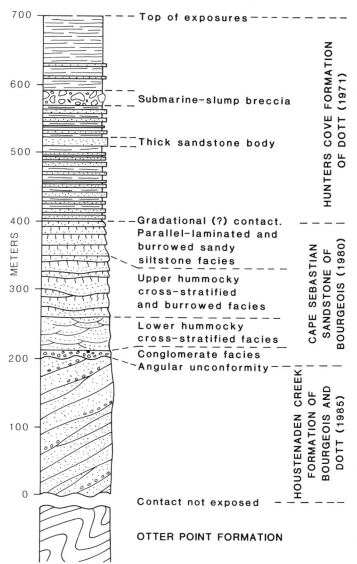

Figure 3. Stratigraphic section of rocks exposed in Cape Sebastian area. Modified from Bourgeois (1980) and Bourgeois and Dott (1985).

Figure 4. Angular unconformity (marked by arrows) between conglomeratic facies of Cape Sebastian Sandstone and underlying, more steeply dipping beds of Houstenaden Creek Formation, west side of Cape Sebastian headland.

shore extending north (downsection) from the south tip of the Cape Sebastian headland.

The cross-stratification in this and the overlying facies (Figs. 5 and 6) has most of the features listed by Harms and others (1975) as typical of hummocky cross-stratification (Bourgeois, 1980). The sets of cross-strata have low-angle (original dip angles generally less than 15°) erosional bounding surfaces. Crosslaminae immediately above the lower bounding surfaces are approximately parallel to these surfaces. Dip angles of the crosslaminae tend to decrease upward within a set, because of a gradual filling of the scour depressions defined by the wavy lower bounding surface. Dip directions of the crosslaminae are widely scattered and show little or no preferred orientation. The cross-stratification in this facies differs somewhat from typical hummocky cross-stratification in having few convex-up segments of bounding surfaces or of the crosslaminae, evidently because such segments were preferentially eroded (Bourgeois, 1980; Dott and Bourgeois, 1982, 1983).

Although the details of origin are still controversial, hummocky cross-stratification probably forms in relatively shallow water during storms, when oscillatory wave-induced currents are combined with a superimposed net current (Harms and others, 1975, 1982; Bourgeois, 1980; Hunter and Clifton, 1982; Dott and Bourgeois, 1982, 1983; Walker and others, 1983). In this facies any nonstorm deposits were completely eroded during subsequent storms, leaving a sequence of amalgamated storm deposits. The water depth is inferred, on the basis of the sandy texture and the absence of high-angle crossbedding, to have been somewhat greater than that in which the conglomeratic facies was deposited.

Upper Hummocky Cross-Stratified and Burrowed Facies

Above the lower hummocky cross-stratified facies is an approximately 300-ft-thick (90 m) interval containing cyclic alter-

The coarse texture and high-angle crossbedding of the conglomeratic facies suggest deposition in a very shallow shoreface environment. Wave-generated nearshore currents probably formed the megaripples whose migration was responsible for the formation of the crossbedding.

Lower Hummocky Cross-Stratified Facies

The conglomeratic facies is overlain by about 165 ft (50 m) of medium- to fine-grained sandstone and pebbly sandstone in which hummocky cross-stratification is almost the only sedimentary structure. Shells and shell fragments are present but not common. This hummocky cross-stratified facies is well exposed and accessible along a 1,300 ft-long (400 m) stretch of rocky

Figure 5. Hummocky cross-stratification in Cape Sebastian Sandstone near south tip of Cape Sebastian headland. The hummocky cross-stratified sandstone overlies burrowed sandstone.

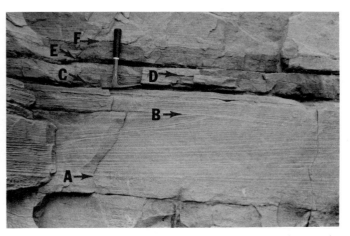

Figure 6. Complete cycle of the type occurring in the upper hummocky cross-stratified and burrowed facies of the Cape Sebastian Sandstone near south tip of Cape Sebastian headland. Hummocky cross-stratified sandstone sharply overlies burrowed sandstone of underlying cycle (contact labeled A) and is overlain (contact B) by planar- and rippled-bedded sandstone containing a thin shale bed (C). Burrowed sandstone gradationally overlies planar-bedded sandstone (contact D); thin shale bed (E) within burrowed interval is atypical. Top of cycle is erosional contact (F) between burrowed sandstone and hummocky cross-stratified sandstone of overlying cycle.

nations of hummocky cross-stratified sandstone and intensely bioturbated sandstone (Bourgeois, 1980; Hunter and Clifton, 1982). Shells are rare, but trace fossils are common. The facies is well exposed at the south tip of the Cape Sebastian headland.

The cycles have erosional bounding surfaces and average 5 ft (1.6 m) in thickness. A complete cycle contains three intervals (Fig. 6): (1) a lower interval of fine-grained sandstone characterized by hummocky cross-stratification; (2) a middle interval of fine- to very fine-grained sandstone characterized by planar- and ripple-bedding and with a thin shale bed in its middle part; and (3) an upper interval of fine- to very fine-grained silty sandstone in which almost all of the original stratification was destroyed by burrowing. The shale bed in the middle interval occurs in only a small fraction of the cycles. The hummocky cross-stratification in this facies differs from that in the underlying facies in having better preservation of convex-up segments of bounding surfaces and of crosslaminae.

The cycles in the upper hummocky cross-stratified and burrowed facies resemble cycles in hummocky cross-stratified deposits elsewhere (Dott and Bourgeois, 1982). The cycles are interpreted to have originated by the alternation of storm and relatively calm conditions (Hunter and Clifton, 1982). Erosion during the intensifying phase of a storm produced the lower bounding surface of a cycle, according to this interpretation. The lower, stratified, fining-upward part of a cycle, up to the top of the shale bed in the middle interval of the cycle, was deposited during the gradually waning phase of a storm. The part of the planar- and ripple-bedded sandstone above the shale bed was probably deposited during relatively calm conditions after the storm but before reestablishment of a normal benthic fauna. The bioturbated sandstone interval was deposited during periods of relative calm, which may have included minor storms. A somewhat deeper environment for this facies than for the underlying facies is suggested by the somewhat finer texture and by the preservation of bioturbated deposits.

Parallel-Laminated and Burrowed Sandy Siltstone Facies

The upper approximately 165 ft (50 m) of the Cape Sebas-

tian Sandstone consists of alternating parallel-laminated very fine-grained sandstone and intensely bioturbated sandy siltstone rich in organic matter (Bourgeois, 1980). As in the underlying facies, cyclic bedding is apparent, particularly because the bases of the sandstone beds are sharp and slightly erosional whereas the tops are gradational into the overlying siltstone beds. Trace fossils are common, and some shells are present. This uppermost facies of the formation is exposed in Salal Cove but is difficult to reach.

The cycles in the uppermost facies of the Cape Sebastian Sandstone are interpreted to have originated, like those in the underlying facies, by deposition during alternating storms and periods of relative calm. The decrease in average grain size from the underlying facies to this facies, together with the disappearance of hummocky cross-stratification, suggests that this facies was deposited in deeper water than the underlying facies. Outer-shelf depths may have been reached (Bourgeois, 1980).

HUNTERS COVE FORMATION

The Hunters Cove Formation of Dott (1971) is well exposed along the shore of Hunters Cove. It consists largely of thinly and evenly interbedded sandstone, siltstone, and olive gray shale (Fig. 7), and is at least 990 ft (300 m) thick (Fig. 3). The contact between the Cape Sebastian Sandstone and the Hunters Cove Formation is not exposed in Hunters Cove, where the two formations are in fault contact, but is thought to be gradational on the basis of poor exposures in Salal Cove (Dott, 1971; Bourgeois, 1980). The formation is overlain locally by Pleistocene terrace deposits and modern sediments.

Figure 7. Thinly and evenly interbedded sandstone turbidites and shale beds of the Hunters Cove Formation at Hunters Cove.

Figure 8. Submarine-slump breccia in the Hunters Cove Formation at Hunters Cove. Note indistinct bedding.

The sandstone beds of the Hunters Cove Formation are typically less than 12 in (30 cm) thick and are characterized by sedimentary structures that indicate deposition by turbidity currents (Phillips and Clifton, 1974; Bourgeois and Dott, 1985). Among such structures are sole marks, normal grain-size grading within beds, and Bouma sequences (i.e., the vertical sequence of sedimentary structures within a bed recognized by Bouma [1962] to be characteristic of turbidites). Most of the Bouma sequences are incomplete, being without the lower unlaminated interval (T_a) of a complete sequence; most commonly, the lowermost interval of the Bouma sequence represented in the Hunters Cove turbidites is the interval characterized by small-scale crosslamination (T_c). Normal grading within the turbidites is common, but the overall upward decrease in grain size within a typical bed is superimposed on the alternations in grain size that define the lamination and crosslamination.

In addition to the thinly interbedded sandstone and shale, the Hunters Cove Formation contains a thick breccia interpreted to be a submarine-slump deposit (Phillips and Clifton, 1974; Bourgeois and Dott, 1985). The breccia crops out near both the north and south ends of Hunters Cove, but near the south end recent landsliding has obscured the original character of the slump deposit and has produced modern breccia difficult to distinguish from the original breccia. Nevertheless, the presence of breccia within the Hunters Cove Formation is indicated by the intact sequence near the north end of the cove and by features such as sand injection structures and calcite-filled veinlets cutting across the breccia. The breccia is composed largely of intraformational clasts of sandstone and mudstone, some of boulder size, derived from the Hunters Cove Formation (Fig. 8). Also present in the breccia are large blocks apparently derived from the boulder conglomerate at the base of the Cape Sebastian Sandstone; these blocks form sea stacks on the present beach. The breccia suggests the presence of steep scarps, probably fault scarps, during the deposition of the Hunters Cove Formation.

Another rock type in the Hunters Cove Formation is well-sorted, light-colored sandstone in intervals several feet (m) thick without interbedded shale. One of these sandstone intervals is well exposed near the south end of Hunters Cove, and similar sandstone bodies occur in and near Burnt Hill cove, 6 mi (10 km) south of Cape Sebastian. The sandstone is characterized by alternating intervals of planar lamination and climbing current ripples, convolute lamination (Dott and Howard, 1962), fluid-escape structures, and load structures at the contact of the sandstone and underlying shale (Fig. 9). Each sandstone interval seems to represent the deposits of several separate turbidity currents.

The Hunters Cove Formation is interpreted to be a submarine-fan deposit (Bourgeois and Dott, 1985). The thinness of the turbidites might be taken to suggest deposition on the distal part of a fan, but a more proximal depositional site is possible if the turbidity currents were of small size and low velocity. The slump breccia and the relatively thick sandstone bodies are more compatible with a proximal depositional site than with a distal one.

CONCLUSIONS

The Upper Cretaceous rocks in the Cape Sebastian area were deposited in a tectonially active area. Deformation and uplift occurred between the time of deposition of the Houstenaden Creek Formation and that of the Cape Sebastian Sand-

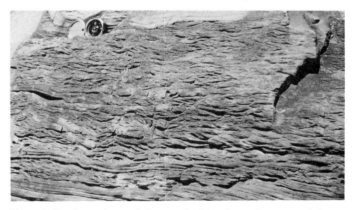

Figure 9. Sandstone interval characterized by alternating zones of planar lamination and climbing-ripple structures. Hunters Cove Formation at Hunters Cove.

stone, and faulting probably took place while the Hunters Cove Formation was being deposited. During the deposition of the Cape Sebastian Sandstone the water depth is inferred to have increased gradually, probably due in part to subsidence and in part to eustatic sea level rise (Bourgeois, 1980; Bourgeois and Dott, 1985). The inferred gradualness of the deepening suggests a close approach to long-term balance between the rate of deposition and the rate of relative rise of sea level. On a shorter time scale (probably from a few decades to a few centuries), the rate of deposition was extremely variable due to cycles of major storms and relatively calm conditions.

REFERENCES CITED

Bouma, A. H., 1962, Sedimentology of some Flysch deposits: Amsterdam, Elsevier Publishing Company, 168 p.

Bourgeois, J., 1980, A transgressive shelf sequence exhibiting hummocky stratification; The Cape Sebastian Sandstone (Upper Cretaceous), southwestern Oregon: Journal of Sedimentary Petrology, v. 50, p. 681–702.

Bourgeois, J., and Dott, R. H., Jr., 1985, Stratigraphy and sedimentology of Upper Cretaceous rocks in coastal southwest Oregon; Evidence for wrench-fault tectonics in a postulated accretionary terrane: Geological Society of America Bulletin, v. 96, p. 1007–1019.

Dott, R. H., Jr., 1971, Geology of the southwestern Oregon coast west of the 124th meridian: Oregon Department of Geology and Mineral Industries Bulletin 69, 63 p.

Dott, R. H., Jr., and Bourgeois, J., 1982, Hummocky stratification; Significance of its variable bedding sequences: Geological Society of America Bulletin, v. 93, p. 663–680.

——, 1983, Hummocky stratification: Significance of its variable bedding sequences: Reply: Geological Society of America Bulletin, v. 94, p. 1249–1251.

Dott, R. H., and Howard, J. K., 1962, Convolute lamination in non-graded sequences: Journal of Geology, v. 70, p. 114–121.

Harms, J. C., Southard, J. B., Spearing, D. R., and Walker, R. G., 1975, Depositional environments as interpreted from primary sedimentary structures and stratification sequences: Society of Economic Paleontologists and Mineralogists, Short Course No. 2 Lecture Notes, 161 p.

Harms, J. C., Southard, J. B., and Walker, R. G., 1982, Structures and sequences in clastic rocks: Society of Economic Paleontologists and Mineralogists, Short Course No. 9, Lecture Notes, variously paged.

Howard, J. K., and Dott, R. H., Jr., 1961, Geology of Cape Sebastian State Park and its regional relationships: Ore Bin, v. 23, p. 75–81.

Hunter, R. E., and Clifton, H. E., 1982, Cyclic deposits and hummocky cross-stratification of probable storm origin in Upper Cretaceous rocks of the Cape Sebastian area, southwestern Oregon: Journal of Sedimentary Petrology, v. 52, p. 127–143.

Hunter, R. E., Clifton, H. E., and Phillips, R. L., 1970, Geology of the stacks and reefs off the southern Oregon coast: Ore Bin, v. 32, p. 185–201.

Koch, J. G., 1966, Late Mesozoic stratigraphy and tectonic history, Port Orford--Gold Beach area, southwestern Oregon coast: American Association of Petroleum Geologists Bulletin, v. 50, p. 25–71.

Phillips, R. L., and Clifton, H. E., 1974, Late Cretaceous submarine slump-breccia, southwest Oregon: Geological Society of America Abstracts with Programs, v. 6, p. 235.

Walker, R. G., Duke, W. L., and Leckie, D. A., 1983, Hummocky stratification; Significance of its variable bedding sequences: Discussion: Geological Society of America Bulletin, v. 94, p. 1245–1249.

Mount Mazama and Crater Lake caldera, Oregon

Charles R. Bacon, U.S. Geological Survey, 345 Middlefield Road, Menlo Park, California 94025

LOCATION AND ACCESS

Most of Mount Mazama lies within Crater Lake National Park in southern Oregon; its lower flanks are within Rogue River and Winema National Forests. Access to the rim of Crater Lake caldera is excellent, via paved roads from Oregon 62 and 230, but Mazama's flanks within the park are generally only reachable by hiking trails. This guide serves for a long one-day excursion through the park, concentrating on features near the caldera rim road, with additional points of interest that could be visited on a second day. New U.S. Geological Survey 7½-minute topographic maps covering Mount Mazama and adjacent areas were published in 1985 at a scale of 1:24,000. The existing 1:62,500 scale topographic map (Crater Lake National Park and Vicinity, 1956) was not made by modern methods, and the information on roads is not current. This map is in revision, and a modern version is expected shortly. Up-to-date information on access, accommodations, and visitor services can be obtained from Crater Lake National Park. The brochure entitled "Crater Lake," which contains an adequate road map of the park at approximately 1:85,000 scale, is available at the park entrance booth, visitor center (Rim Village), or park headquarters. July and August are the best months to visit the park. In most years, snow limits access to the caldera rim through June, and early fall storms may interfere with a trip in September. It is wise to contact the park for information on early summer or fall snow conditions and to be prepared for thunderstorms or cold weather at any time.

SIGNIFICANCE

Crater Lake partially fills the caldera within Mount Mazama, Crater Lake National Park, southern Oregon. Mount Mazama is an andesitic stratovolcano cluster that collapsed 6845 ± 50 B.P. during a catastrophic, compositionally zoned pyroclastic eruption of about 50 km³ of magma. Prior to its climactic eruption, Mount Mazama was one of the major Quaternary volcanoes of the Cascade Range. Mazama's eruptive history has been delineated by recent mapping of its caldera walls and flanks. Products of the climactic eruption partially fill surrounding valleys and mantle the slopes and caldera rim. The first detailed account of the geology of the park was U.S. Geological Survey Professional Paper 3, by Diller and Patton, published in 1902. Howel Williams, in his classic monograph (1942), showed that the caldera formed by collapse and suggested that the compositional zonation reflected sequential tapping of a layered magma chamber. Other authors have reported on the zoned ignimbrite, most recently Ritchey (1980). A detailed account of the petrology of the entire climactic eruption and related precursory lavas is in preparation (C. R. Bacon and T. H. Druitt, unpublished manuscripts, 1986). The climactic eruption took place in two stages: single-vent and ring-vent phases; the ring-vent phase coincided with caldera collapse (Bacon, 1983). The single-vent phase produced the widespread airfall (climactic pumice fall) from a Plinian eruption column, followed by the Wineglass Welded Tuff, an ignimbrite deposited by ash flows that formed when the column collapsed. Deposits of the ring-vent phase range from pumiceous ignimbrite in the valleys around Mazama to coignimbrite lithic breccia (lag breccia) near the caldera rim (Druitt and Bacon, 1986).

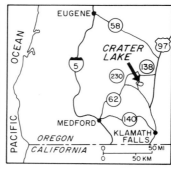

Figure 1. Index map.

SITE DESCRIPTIONS

A summary of the eruptive history of Mount Mazama and descriptions of many of the localities noted in this guide can be found in recent publications by Bacon (1983, 1985, and 1986). I refer to these papers, and make corrections that result from improved knowledge of the volcano, throughout the guide. Ages of units are K-Ar dates by M. A. Lanphere (unpublished data, 1985) unless otherwise indicated. Collecting is prohibited in the park, although permits may be obtained at park headquarters, and climbing or walking on the caldera walls is not allowed except on the Cleetwood trail. The trip starts at the junction of the road to Crater Lake and Oregon 62.

0.0 mi (0.0 km). Junction of Oregon 62 and road to Crater Lake. Road cut west of junction exposes three Pleistocene olivine andesite flows in a normal fault scarp. Arrant Point 0.8 mi (1.3 km) to the south is a subvolcanic intrusion of olivine andesite cut by the same fault. Drive south (toward Klamath Falls) on Oregon 62.

1.6 (2.6) Stop 1. Turnout with View of Annie Creek Canyon. Excellent exposure of valley-filling medial-facies pumiceous ignimbrite of ring-vent phase of climactic eruption (Bacon, 1983, figs. 6c, 6d, 7, and 9; Druitt and Bacon, 1986, fig. 12). Lower half of section is silicic ignimbrite dominated by rhyodacite pumice clasts with 70.4 percent SiO_2 (all SiO_2 values are recalculated volatile-free). Most of upper half is mixed ignimbrite containing 20 to 80 percent silicic pumice, the rest being andesitic to basaltic scoria of 61 to 51 percent SiO_2. The uppermost approximately 10 m are mafic ignimbrite with less than 20 percent silicic pumice. Scoriae in valley-filling ignimbrite generally are hornblende and plagioclase phyric; olivine- and clinopyroxene-rich clasts occur near the top of the section. Most scoriae apparently are cumulates. The color change from buff below to gray above reflects increasing emplacement temperature of the matrix of the deposit upward, as much as its composition. Note the bleached zone of fumarolic alteration beneath the uppermost 1 m of fine ash. Erosion resistant pinnacles and coarse columnar joints (near falls) result from vapor-phase induration. Entire section was deposited without significant breaks, as no sharp grain size breaks are evident. Note, however, rusty stained lithic-rich horizon near lower part of deposit. Good exposures of mixed ignimbrite are accessible just beyond the north end of the long turnout. Garfield and Applegate peaks on the caldera rim are visible on the skyline to the north. Lithic breccia high on the slopes of these peaks and in the adjacent valleys forms the proximal facies of the same eruptive unit exposed here at Annie Creek. Return to junction of Oregon 62 and road to Crater Lake.

0.0 (0.0). Junction of Oregon 62 and road to Crater Lake. Turn right on road to Crater Lake and stop at Crater Lake National Park entrance station. Proceed to Stop 2.

4.0 (6.4) Stop 2. Park Headquarters and Lithic Breccia Locality. Information, literature, and restrooms are available at park headquarters between 8 a.m. and 5 p.m. Walk 0.1 mi (0.2 km) to first cut on right north of headquarters on road to caldera rim. Cut exposes coarse lithic breccia of proximal facies of ring-vent-phase pyroclastic-flow deposits of the climactic eruption (Bacon, 1983; Druitt and Bacon, 1986, fig. 3). Similar material underlies floor of Munson Valley in this vicinity. Lag breccia also occurs on summits of all nearby peaks, both in the lee of or upslope from obstacles, and locally on the upper slopes of Mount Mazama (Druitt and Bacon, 1986, fig. 2). Most clasts here are variably altered andesite from conduit or vent walls. Systematic variation in clast lithology around the caldera is evidence for multiple vents during this part of the eruption. Note rounding of these clasts due to thermally induced spalling of corners. Some clasts are highly altered friable andesite; others are partially fused granitoids. Deposit is clast supported and subtly imbricated. Matrix here contains hornblende andesite pumice. At some localities juvenile material zones from rhyodacite pumice at the base upward to mafic scoria. Road cut is in a bedform near the east side of Munson Valley. Longitudinal and transverse bedforms are common in the lag breccia.

4.5 (7.2). Second cut beyond park headquarters on road to caldera rim exposes dacitic debris-flow deposit (Stop 3), overlain by about 1 m of airfall pumice believed to have been erupted from the Llao Rock center (Stop 4), overlain by lag breccia.

5.3 (8.5) Stop 3. Dacitic Fragmental Deposits. Long straight cut with turnout on south (left) side immediately after hairpin turn from northeast to southwest exposes two subunits of approximately 40,000 B.P. dacitic fragmental rocks (Bacon, 1983, p. 77–79). Eastern subunit is monolithologic debris-flow or avalanche deposit that apparently records collapse of dome(s) southwest of Mazama's summit. Similar material is present in the tree-covered swale high on Garfield Peak visible to the east. Western subunit is the younger and contains the same dacite, altered dacite, and dense prismatically-jointed dacite and pumiceous dacite blocks believed to represent the final pulse of magma responsible for destruction of the dome(s) (Bacon, 1985, fig. 7). The western subunit is thought to be a lithic-pyroclastic-flow deposit. The dacite contains chilled inclusions of andesite magma (Bacon, 1986, figs. 7 and 8, table 3).

6.8 (10.9) Stop 4. Rim Village and View of Caldera. Morning light is good for viewing north and west caldera walls (Bacon, 1983, fig. 4), afternoon for east walls (Stop 12; Bacon, 1983, fig. 3). The locus of andesitic volcanism moved from east to west during construction of Mount Mazama. Imposing cliff on north caldera wall is Llao Rock, the vent-filling rhyodacite flow and associated pumice fall deposit dated at 7015 ± 45 B.P. by radiocarbon (Bacon, 1983). This compositionally zoned silicic eruption (72–70 percent SiO_2) preceded the climactic eruption by about 140 years (Bacon, 1983, p. 105). The walls below Llao Rock consist of andesite and dacite flows and minor dacite tephra. As many as five eroded surfaces are present. Lava at lake level here has been dated at about 190,000 B.P. Sheetlike thin flows here and elsewhere in the walls consist of agglutinated spatter topped by rubble. Such fountain-fed lavas grade from bomb beds and agglutinate near source, through streaky lava flows with thick rubbly tops, to distal homogeneous lava with relatively thinner rubble zones. Eye-shaped cliff to east of Llao rock marks the rhyodacite of Steel Bay (71.6 percent SiO_2), believed to be one of the earliest products of the climactic magma chamber, emplaced about 30,000 B.P. The ages of this flow and its cousins, Grouse Hill and Redcloud Cliff, are constrained to within the interval of about 20,000 to 30,000 B.P. by field relations and measurements of paleomagnetic secular variation (D. E. Champion, unpublished data, 1985). East of Llao Rock are Pumice Point and Cleetwood Cove. Cliffs at rim at Cleetwood Cove are in rhyodacite Cleetwood Flow, identical in composition (volatile-free) to climactic pumice and still hot when caldera collapsed. Dark lava descending to water level there is the "backflow," which oozed into the caldera during collapse (Bacon, 1983, p. 100–101, fig. 11). West of Llao Rock are Devils Backbone dike, Hillman Peak, and The Watchman. Gray vegetated slope just below rim south of Devils Backbone marks lithic-pyroclastic-flow deposit correlated with that of Stop 3. Hillman Peak consists of pyroxene andesite, overlain by hornblende andesite (forms gray scree slopes on near-vent fall deposits), overlain by pyroxene andesite with prominent augite and blocky plagioclase. All are about 70,000 years old. The Watchman flow is about 50,000 years old. The dike feeding the dacite flow that forms The Watchman can be seen on the caldera wall below the saddle between The Watchman and Hillman Peak. In the southwest wall are lavas older than Hillman Peak, an eroded surface overlain by dacitic ignimbrite that weathers orange, and andesite flows as young as about 50,000 B.P. (Bacon, 1985, figs. 6 and 7). Wizard Island is the tip of a thick pile of postcaldera andesite that apparently was erupted when the lake level was about 100 m lower. Return to Rim Drive and turn right (north).

8.9 (14.3) Stop 5. Andesite Flows. Convenient turnout on west side of road across from cliff in andesite lava. Younger flow shows glassy brecciated base grading into subtle columns, then upward into platy joints that parallel flow base. Platy zone grades up into coarsely blocky-jointed interior. Original rubbly top has been removed by glaciation. In some flows a second set of platy joints forms steeply dipping sheets high in the flow where this set dominates. Plagioclase phenocrysts in this flow have a distinctive seriate texture. Older flow is more typical of Mazama andesites with plagioclase, augite (green), hypersthene (brown), and titanomagnetite phenocrysts, common glomerocrystic clots, and sparse magmatic inclusions. Olivine (yellow) is a common minor constituent of many Mazama andesites.

9.3 (15.0). Cut on northeast of road exposes crudely bedded lag breccia lying on glaciated surface of dacite lava similar to the Watchman flow (Druitt and Bacon, 1986, fig. 6). Fumarolic alteration shows breccia was deposited hot. Note platy jointing in the Watchman flow between here and Stop 6.

9.8 (15.8) Stop 6. The Watchman Flow. Turnout where Rim Drive heads north across the Watchman dacite flow gives good view to south to Mounts McLaughlin (38 mi; 61 km) and Shasta (125 mi; 200 km). Union Peak is the eroded core of a cinder cone atop a basaltic shield in the middle distance. In the road cut a remnant of the pumiceous blocky carapace of the dacite flow can be examined and seen to grade down through glassy dacite into devitrified rock. This lava has 67.5 percent SiO_2 and contains rare andesitic magmatic inclusions. Its porphyritic texture and two-pyroxene, plagioclase, and Fe-Ti oxide phenocrysts are typical of Mazama dacites.

11.2 (18.0) Stop 7. View to North and Williams Crater. Large turnout on northwest side of road gives view of Red Cone (north) and Bald Crater (north-northwest), basaltic cinder cones. In the middle distance, left to right, are Mount Bailey, Diamond Peak (beyond Diamond Lake), and Mount Thielsen. Three Sisters volcanoes may be seen in the far distance on clear days. To reach Williams Crater (formerly Forgotten Crater, Bacon, 1983, p. 80–82), walk south along Rim Drive. In the first cut, note lag breccia bleached by fumarolic alteration lying on basaltic lapilli from Williams Crater, which in turn rest on andesite breccia of the youngest unit of the Hillman Peak center. The andesite lava has distinctive blocky plagioclase phenocrysts, hypersthene, and prominent augite. Just past this cut, turn west and walk about 1300 ft (400 m) to Williams Crater, noting the small dacite dome immediately south of where you left the road. This dome is the easternmost lava of the Williams Crater complex, which consists of a basalt flow, cinder cone, and three more flows of commingled andesite and dacite with basalt inclusions. Before reaching Williams Crater, note the arcuate bedform of lag breccia banked against it.

Williams Crater contains bombs cored with angular fragments of commingled andesite-dacite lava, themselves bearing inclusions of the basalt. Walk to the south side of the cinder cone and visit the roughly 10-m high cliff in commingled lava to see a variety of proportions of andesite and dacite. A basaltic dike evidently intersected the margin of a silicic magma reservoir, venting first basalt, but forcing convective mixing in the reservoir. Basalt thoroughly mixed with dacite to form hybrid andesite, which then mingled with the dacite, and blobs of basalt became incorporated in both andesite and dacite. Fragments of the commingled magma were thrown out in the basaltic eruption. Finally, viscous commingled lava oozed from vents on a trend radial to the present caldera. That all of these events took place rapidly is indicated by paleomagnetic results obtained by D. E. Champion, which show the same direction of thermoremanent magnetization for all units of Williams Crater. The dacite is chemically most similar to that of the fragmental deposits in Munson Valley (Stop 3); tephra from Williams Crater lies on the deposit south of Devils Backbone noted at Stop 4.

12.4 (20.0 km) Stop 8. Hornblende Andesite, Lag Breccia, and Cumulate Blocks. Below turnout on east side of road is a glacially striated outcrop of hornblende andesite of Hillman Peak (middle unit). Similar rock crops out about 100 ft (30 m) west of the road. Nearby are mafic blocks in the final veneer deposit of the climactic eruption. These are plagioclase-pyroxene cumulates with variable amounts of olivine or poikilitic hornblende. The veneer locally mantles lag breccia, which to the west occurs in bedforms with clasts to 5 m (Druitt and Bacon, 1986, fig. 10). Llao Rock can be seen well from this point. The bedded climactic pumice fall present near its summit was eroded from its flanks by ring-vent-phase pyroclastic flows. The dark streak truncating airfall beds on Llao Rock is scoria-bearing ignimbrite veneer.

12.7 (20.4). Turn right at junction of Rim Drive and north entrance road. At caldera rim, lag breccia grades down into pumiceous ignimbrite, which grades down into a small outcrop of the Wineglass Welded Tuff (Stop 12) that lies on bedded fall deposits.

13.0 (20.9) Stop 9. Llao Rock Obsidian and Proximal Ignimbrite. Park in large turnout on west side of road past big cut in obsidian. Ignimbrite veneer lies on lag breccia at north end of cut. About 20 cm of pink vitric ash, the ash-cloud deposit caps the veneer. The climactic pumice fall and most of the original blocky pumiceous carapace of the Llao Rock flow have been removed by the pyroclastic flows. Pumiceous rhyodacite grades down into obsidian, which becomes lithophysal and spherulitic deeper into the flow south along the cut. Note abundant andesitic magmatic inclusions (Bacon, 1986, figs. 5 and 6, table 2).

14.4, (23.3), 14.8 (23.8). Views northeast to Grouse Hill (70–71 percent SiO_2), previously thought to be Holocene (Bacon, 1983, 1985) and now believed to be Pleistocene (c. 20,000–30,000 B.P.). Grouse Hill is chemically and mineralogically similar to the rhyodacites of Steel Bay and Redcloud Cliff.

14.9 (24.0). Cut in the rhyodacite of Steel Bay (hill 7352′ of Bacon, 1983, 1985, fig. 8). Rock contains plagioclase, hornblende, augite, hypersthene, titanomagnetite, and ilmenite phenocrysts, rare andesitic inclusions. Cuts between here and Stop 10 are in andesite erupted about 50,000 B.P.

16.0 (25.7) Stop 10. Pumice Point. Park at long turnout and walk about 1000 ft (300 m) along old road to rim at Pumice Point. Visible on the steep slope below are the following deposits, from base to top: (1) Faintly orange airfall pumice erupted from a vent near the east wall of the caldera at Pumice Castle (Stop 16) and resting on glaciated andesite; (2) a glaciated andesite lava flow dated at about 50,000 B.P. and filling an erosional depression in the pumice of (1); (3) a few feet of light-gray Llao Rock pumice fall forming a gentle slope above the lava; (4) climactic pumice fall, pinkish, in well-defined beds with nonwelded pyroclastic-flow interbeds, supporting the steep slope; (5) orange incipiently welded Wineglass Welded Tuff grading upward into (6) lag breccia; (7) ignimbrite veneer; (8) ash-cloud deposit, as at Stop 9; (9) crystal and lithic ash, presumably re-ejected from the caldera by phreatic explosions following collapse; and (10) a few feet of wind-reworked material forming a small dune at the caldera rim. Near the base of (4) is a meter-thick, reddish, incipiently welded bed of climactic airfall pumice that contains rare hornblende andesite scoria and partially fused granitoid blocks.

17.2 (27.7). Parking lot for trail to lake and tour boats; beginning of one-way road. Lake tour takes 2 hours, with optional stop at Wizard Island. Highly recommended for a second day. Inquire at park headquarters or Crater Lake Lodge for current schedule and fee. Allow 20 minutes to descend and 30 minutes to hike out of caldera. Trail has good exposures of partly welded Wineglass Welded Tuff and cuts showing climactic airfall, lithic-rich Cleetwood airfall, and Llao Rock airfall lying on till. Lava flow below till is silicic andesite dated at about 100,000 B.P. Underlying material may be an avalanche deposit of related blocks that preceded emplacement of the lava flow.

17.5 (28.2) Stop 11. Cleetwood Cove. Park at turnout across road from brick-red pumice. Below are cliffs in Cleetwood rhyodacite flow and tongue of lava (backflow) that oozed down caldera wall during ring-vent phase of climatic eruption (Bacon, 1983, fig. 11). Cliffs east of cove show internal flow structure of lava. Road cut exposes climactic pumice fall lying on Cleetwood flow. Near their mutual contact these units are highly oxidized. Nowhere but on top of this lava flow is the pumice fall oxidized and sintered. The pumice, which was hot enough when it landed to support steep slopes, fell as a hot blanket on the cooling lava flow, trapping and heating air. This trapped air, and possibly also degassing of the lava, resulted in the oxidation. Fumarolic alteration cuts the airfall and overlying lag breccia, showing that the entire climactic eruption took place before the Cleetwood flow cooled completely (Bacon, 1983, fig. 8). About 600 ft (180 m) north of here is the first of a series of small rift valleys transverse to the direction of flow, formed as the brittle upper surface of the flow broke up while its fluid interior flowed north (presumably in response to seismicity associated with caldera collapse) and south into the caldera. Faults bounding the valleys cut climactic pumice fall and thin Wineglass Welded Tuff (Stop 13). Lithic breccia is banked against the walls of these valleys, showing that caldera collapse began before the ring-vent phase ended (Bacon, 1983, fig. 12).

19.0 (30.6) Stop 12. Wineglass Welded Tuff, Palisade flow, and view of caldera walls. Park at the west end of the turnout and walk to the caldera rim. You are standing on the Wineglass Welded Tuff, the top of which has been scoured away by the ring-vent-phase pyroclastic flows. Note how the Wineglass is thickest in the topographic depression to the east, thins to the west, and is absent from the high point (Roundtop) to the east. The Wineglass consists of up to four flow units of rhyodacitic ignimbrite that form a single cooling unit. It was deposited by heat-conserving, valley-hugging ash flows that emanated from the collapsed eruption column of the single-vent phase (Bacon, 1983, p. 92–93, fig. 9). Here the top of the ignimbrite shows gash fractures parallel to the caldera rim, indicating that, where sufficiently thick, the tuff slumped toward the caldera while still hot and plastic. This is compelling evidence for the rapidity of events during the climactic eruption. The large pumice slope on the wall south of Skell Head south-southeast of here shows a fine section of Llao Rock and climactic airfalls overlain by the Wineglass Welded Tuff capped by lag breccia (Druitt and Bacon, 1986, fig. 5). There the Wineglass shows four distinct flow units, the top one being a vitrophyre. The lag breccia appears crudely bedded into five units. The lower two contain rhyodacite pumice, the next two hornblende andesite pumice, and the top one mafic scoria, showing that the breccia mimics the stratigraphic order of the zoned valley-filling ignimbrite. The craggy top of the Palisade flow (dacite) that forms the cliffs to the west has been glaciated, and the flow occupies an old glacial valley carved against the older silicic andesite of Roundtop to the east.

This is also a good place to see features on the east and south caldera walls. The cliff-forming flows below the rim at Grotto Cove southeast of here are inclusion-bearing silicic andesites, the younger of which is the subject of Stop 14. The Redcloud Cliff rhyodacite flow (70–71 percent SiO_2) forms the prominent cliff on the east wall, filling its vent crater and spilling to the north over earlier dacite (Bacon, 1985, fig. 8). It is compositionally similar to the rhyodacites of Grouse Hill and Steel Bay and is late Pleistocene in age (c. 20,000–30,000 B.P.). Immediately south and stratigraphically below Redcloud Cliff are Pumice Castle and related dacite flows described at Stop 16. Andesites below the dacites fall into at least two groups between about 220,000 and 340,000 B.P. Between Pumice Castle and Kerr Notch, the lower of two beheaded glacial valleys, is Sentinel Rock (best seen from the west in afternoon light), where thick intracanyon low-silica dacite flows (about 300,000 B.P.) lie on glaciated older andesite (about 350,000 B.P.). Between Kerr and Sun notches is Dutton Cliff. The oldest rocks exposed on the caldera wall, about 400,000 B.P., are at water level below Dutton Cliff and comprise the Phantom Cone (Williams, 1942). Stripes on the wall with moderate initial dips highlight pervasively altered rubbly tops of agglutinated andesite flows; their dense interiors, having been less permeable, show little effect of hydrothermal alteration from this distance. Phantom Ship is the small island and is partly composed of dikes related to the adjacent Phantom Cone. From Dutton Cliff to water level beneath Sun Notch three sets of andesite flows are exposed. The top is fresh olivine-bearing pyroxene andesite, the middle altered andesite, and the lowest altered andesite with abundant magmatic inclusions. Applegate Peak and Garfield Peak form the summits of the south wall, west of Sun Notch. All are pyroxene andesite and low-silica dacite flows except the top two flows of Garfield Peak, which are hornblende andesite (about 220,000 B.P.). The altered flows near water level below the big talus slopes between Applegate and Garfield peaks comprise Chaski slide, a block of caldera wall that failed to completely slide into the caldera. West of Garfield Peak is the head of Munson Valley, site of Crater Lake Lodge and Rim Village, which are built on the dacitic debris-flow and lithic-pyroclastic-flow deposits of Stop 3.

20.0 (32.2) Stop 13. Wineglass. Park at small unpaved turnout on right where Rim Drive meets the caldera rim after traversing the andesite flow of Roundtop. Wineglass refers to the shape of a scree chute below here. From the southeast side of the valley, excellent exposures of the Wineglass Welded Tuff lying on climactic pumice fall can be seen. The uppermost 1 m of the fall deposit is oxidized because emplacement of the Wineglass on warm airfall trapped and heated air. Lag breccia lies on the Wineglass. This is the locality where hot fiamme oozed over the surface of an open fracture (a larger version of those seen from above at Stop 12) before the tuff completely cooled; the block on the caldera side of the fracture has since fallen away, revealing tongues of devitrified glass hanging over the inclined fracture surface. Because the caldera must have been present at the time of fracturing, caldera collapse took place shortly after emplacement of the Wineglass; i.e., during the ring-vent phase of the climactic eruption (Bacon, 1983, p. 102). Beneath the scree below, climactic pumice fall lies on Cleetwood airfall pumice, which rests on Llao Rock pumice. Below the airfall units is till containing glassy andesite derived from the flow underneath. This lava flow displays spectacular radiating columnar joints that grade into shattered glassy blocks upward into the till. The flow evidently encountered ice. Similar features can be seen in many other less obvious examples of chilled lava around the caldera.

21.0 (33.8) Stop 14. Inclusions in Andesite. Park on the east side of Rim Drive in an unpaved turnout where the road bends from southeast to south just past a cut in lava on the southwest side of the road. Walk back to this cut and observe abundant magmatic inclusions of andesite (60 percent SiO_2) in silicic andesite (62 percent SiO_2). The host is the younger of two compositionally identical flows that are about 70,000 years old (Bacon, 1985, fig. 6; 1986, figs. 9–15, Table 4). Inclusions larger than about 12 cm have exceptionally porous (diktytaxitic) cores with denser rinds a few inches thick. Because of a relatively small initial thermal contrast between porphyritic host and nearly aphyric inclusion magmas, incomplete undercooled crystallization of inclusions resulted in a large fraction of vapor-saturated residual liquid (75 percent SiO_2). Most of this liquid was expelled from cores of inclusions, which were by then rigid crystal meshes, by gas filter-pressing owing to differential vapor pressure between host lava and cores of large inclusions. Consequently, cores are more mafic (56 percent SiO_2) than the original inclusion bulk composition. Although the rind texture is not common, andesitic inclusions are abundant in many silicic andesite and dacite flows of Mount Mazama, particularly in the relatively old rocks of Mount Scott and the east and south caldera

walls. The cut at this stop is high in the holocrystalline interior of the flow; the upper glassy zone of the flows is preserved locally near their distal ends, where glass-lined segregation vesicles can be seen in inclusions. Inclusions are most abundant near the tops of flows, apparently because tops of flows were erupted later and advanced more rapidly than the underlying lava. Inclusions may have been incorporated into the host from a deeper level in the magma reservoir by forced convection during flow of magma up the eruption conduit. Return to vehicle and note small gully on west (right) that exposes climactic pumice fall lying on Llao Rock pumice separated by about 2 cm of fine ash. Directly to the south is a glaciated exposure of obsidian at the north end of the Red-cloud Cliff rhyodacite flow. Pyroclastic flows of the ring-vent phase evidently abraded this surface.

23.6 (38.0). Parking area for Mount Scott trail and view of Mount Scott. Mount Scott consists of the oldest dated lavas of Mount Mazama, about 420,000 B.P. This cone is made up of sheets of agglutinated low-silica dacite. Andesitic inclusions are abundant. Glaciation has exposed the core of the volcano. Hydrothermal alteration variably affects these rocks, and these features can be seen readily on the trail to the summit. The summit ridge is capped by climactic pumice fall and lag breccia.

23.8 (38.3). Turn right on Cloudcap road.

24.6 (39.6) Stop 15. Rhyodacite lava and pumice, welded dacite airfall pumice, and view of northeast flank of Mount Mazama. Cut on south (left) exposes four airfall pumice deposits and a steeply dipping rib of lava. From east to west, in stratigraphic order, the units area: (1) Buff to orange dacite pumice with blocky plagioclase and pyroxene, similar to the tephra of Pumice Castle; (2) white hornblende rhyodacite pumice with abundant coarse lithic blocks, the proximal airfall for the Red-cloud Cliff vent; (3) rhyodacite lava grading from pumiceous blocks up through obsidian into lithophysal and spherulitic rhyodacite, believed to be a glaciated "bathtub ring" of lava left after most of an oversteepened dome slid north to form the Red-cloud Cliff flow; (4) Llao Rock airfall poorly exposed beneath scree; (5) climactic pumice fall supporting a small cliff; and (6) lag breccia of the ring-vent phase of the climactic eruption. Walk around the west end of this exposure and up over the lava to a brick-red 2-m cliff east of the cut. This is welded airfall pumice similar to that of Pumice Castle. In the distance are Mount Thielsen and tree-covered lateral moraines extending to the east, the Gibraltar-like west-facing scarp of Walker Mountain 31 mi (50 km) northeast near the town of Chemult (approximate limit of ring-vent-phase ignimbrite), the toad-back silhouette of Newberry Volcano 68 mi (109 km) northeast, and the shieldlike form of Yamsay Mountain 34 mi (55 km) east. In the middle distance are the late Pleistocene shield and cone of Timber Crater to the north and ignimbrite-floored flats of Klamath Marsh to the east. Closer features are the higher domes of the Sharp Peak group of at least 12 late Pleistocene rhyodacite domes thought to have been erupted from the climactic chamber (Bacon, 1985, fig. 8), several pre-Mazama rhyodacite domes surmounting thick flows (Bear Butte, Lookout Butte, Scout Hill, approximately

600,000 to 700,000 B.P.; climactic lag breccia occurs on the summit of Lookout Butte!), cliffs to the north in the silicic andesite flow of Stop 14, cliffs in the Redcloud Cliff flow, and the dacite flows of Scott Bluffs.

24.9 (40.1). Cloudcap overlook. Return to Rim Drive.

26.0 (41.8). Turn right on Rim Drive.

27.1 (43.6) Stop 16. Pumice Castle Overlook. Large turnout where Rim Drive bends to the south. The view of the northeast wall of the caldera includes the Cleetwood flow, Palisade flow, Roundtop flow, and Wineglass. Redcloud Cliff dominates the east wall in the foreground. Cliff at rim south of Redcloud Cliff is in Pleistocene dacite flow lying on tephra of Pumice Castle. The latter consists of dacite pumice that is nonwelded at the south end of the exposure but becomes progressively more densely welded to the north. Ages of lava flows above and below the tephra of Pumice Castle suggest that this dacite erupted about 70,000 B.P. Just south of Redcloud Cliff, this deposit includes several vitrophyric layers and a stubby lava flow. The top of the airfall has been "fused" by the overlying lava flow. Pumice Castle is the prominent brick red–orange set of towers with resistant welded layers. Below the pumice are sheets of mafic andesite (about 220,000 B.P.) lying on altered andesite lavas (as old as 340,000 B.P.) and debris-avalanche deposits. Numerous dikes cut the pre– Pumice Castle rocks. In the trees just below the caldera rim immediately east of the turnout is an exposure of partly welded Wineglass Welded Tuff, the southeasternmost such outcrop at the rim. Confinement of the Wineglass to depressions from Merriam Point clockwise to here indicates the vent for the single-vent phase of the climactic eruption was northeast of the summit(s) of Mount Mazama. Large lithic blocks in the climactic pumice fall above Pumice Castle are consistent with proximity to the vent, which may have been located between here and Cleetwood Cove. Here, as elsewhere, lag breccia lies on the Wineglass. Thick lag breccia also mantles airfall above Redcloud Cliff and the dacite flow above Pumice Castle. The next two turnouts provide good views of the south caldera walls.

29.6 (47.6). Turn left on road to Pinnacles (two-way).

34.2 (55.0). Small unpaved turnout on left provides good view into Sand Creek where ring-vent-phase ignimbrite is columnar jointed due to incipient welding and (or) vapor-phase crystallization. Such jointing is common in narrow canyons in medial ignimbrite (e.g., Castle Creek and lower Annie Creek). Light gray color of columnar jointed tuff indicates relatively high emplacement temperature of mixed and mafic ignimbrite. Rhyodacite pumice in these occurrences is purplish gray.

35.7 (57.4) Stop 17. The Pinnacles. The Pinnacles are resistant to erosion because of vapor-phase crystallization adjacent to gas-escape pipes and sheets within medial-facies ignimbrite. Note lithic concentrations and intense fumarolic alteration within the pipes. The best view of this oft-photographed scene is obtained from the west end of the exposure, where a safe descent can be made on the tree-covered slope. Buff silicic and mixed ignimbrite form the lower part of the exposure and gray mixed and mafic ignimbrite the upper (Druitt and Bacon, 1986, fig. 11).

The color change reflects increasing emplacement temperature and crystal content of the matrix upward. Andesitic scoria is common in the upper part of the buff zone and rare near its base. The base of the ignimbrite is not exposed. Meter-scale stratification does not correspond to sharp grain-size breaks. Note red oxidized top of the sheet and fine ash-cloud deposit. Wavy ledges near top are not primary depositional features. This locality is spectacular, particularly in late afternoon light, but is atypical of medial-facies ignimbrite in its sharp color change and stratified appearance. Compare with Stop 1 or the upper part of Castle Creek. Hills to northeast and east are mostly pre-Mazama rhyodacite. Return to Rim Drive.

41.6 (66.9). Turn left on Rim Drive. Note lag breccia on right.

43.5 (70.0). View to south of Klamath Graben, upper Klamath Lake (30 mi; 48 km), and Medicine Lake Volcano (100 mi; 160 km), the broad shieldlike form beyond upper Klamath Lake. Snow-capped Mount Shasta (110 mi; 177 km) may be visible.

44.5 (71.6). View northwest to Applegate Peak. Silicic andesite and low-silica dacite flows grade from agglutinated lava upward into bomb beds in each "flow" at the caldera rim and south down the flank into streaky lava flows with rubbly tops. Lavas in cuts here on Rim Drive are strongly flow-banded olivine-bearing silicic andesites of Dutton Cliff except the last cut before Sun Notch, which exposes altered, older andesite.

45.5 (73.2) Sun Notch. A short walk to the caldera rim yields a fine view of Phantom Ship and Dutton Cliff.

47.5 (76.4) Crater Peak Trailhead. Columnar base and platy interior of silicic andesite flow well exposed in cut where Rim Drive leaves valley of Sun Creek and crosses Vidae Ridge immediately east of trailhead.

49.0 (78.8). Moraine on east wall of Munson Valley.

49.9 (80.3) Park Headquarters. End of road log.

REFERENCES CITED

Bacon, C. R., 1983, Eruptive history of Mount Mazama and Crater Lake caldera, Cascade Range, U.S.A.: Journal of Volcanology and Geothermal Research, v. 18, p. 57–118.

——— , 1985, Implications of silicic vent patterns for the presence of large crustal magma chambers: Journal of Geophysical Research, v. 90, p. 11243–11252.

——— , 1986, Magmatic inclusions in silicic and intermediate volcanic rocks: Journal of Geophysical Research, v. 91, p. 6091–6112.

Diller, J. S., and Patton, H. B., 1902, The geology and petrography of Crater Lake National Park: U.S. Geological Survey Professional Paper 3, 167 p.

Druitt, T. H., and Bacon, C. R., 1986, Lithic breccia and ignimbrite erupted during the collapse of Crater Lake caldera, Oregon: Journal of Volcanology and Geothermal Research, v. 29, p. 1–32.

Ritchey, J. L., 1980, Divergent magmas at Crater Lake, Oregon; Products of fractional crystallization and vertical zoning in a shallow, water-under-saturated chamber: Journal of Volcanology and Geothermal Research, v. 7, p. 373–386.

Williams, H., 1942, The geology of Crater Lake National Park, Oregon: Carnegie Institution of Washington Publication 540, 162 p.

Depoe Bay, Oregon

Parke D. Snavely, Jr., U.S. Geological Survey, 345 Middlefield Road, Menlo Park, California 94025

LOCATION AND ACCESSIBILITY

The Depoe Bay field site is located on the central Oregon coast, adjacent to U.S. Highway 101, opposite the fishing village of Depoe Bay, Lincoln County, Oregon (Fig. 1). The pillow lavas of the Depoe Bay Basalt are well exposed in sea cliffs just beyond the low seawall along the west side of the highway. The Sandstone of Whale Cove and breccias of the Cape Foulweather Basalt can be reached by steep trails that lead down to the base of the sea cliffs along the north side of the outer bay. Here, and elsewhere along the Oregon coast, extreme caution should be exercised not to work too close to sea level, as unexpectedly large waves can sweep a person off the rocks. A one-hour excursion aboard a fishing boat that leaves from the inner bay is highly recommended, as this trip provides an opportunity to view the geology along this scenic coastline.

SIGNIFICANCE OF SITE OR LOCALITY

Exceptional sea-cliff exposures at Depoe Bay afford a unique opportunity to study the stratigraphic relationship between two petrochemically different basalt units that are interbedded in middle Miocene shallow-water strata. The lower basalt unit, the Depoe Bay Basalt, overlies fossiliferous sandstone and siltstone beds of the Astoria Formation which is exposed in the inner bay. Particularly well-developed pillow lavas, with a variety of shapes and internal cooling structures, are well exposed along the entire east margin of the outer bay. The Sandstone of Whale Cove, which overlies the Depoe Bay Basalt, occurs in several prow-shaped channels that cut and deform a thin-bedded siltstone estuarine sequence. The internal structures in these massive concretionary sandstone channels exhibit large-scale penecontemporaneous deformational features such as convolute bedding, load casts, and slump structures. Peperite dikes that cut these sandstone deposits are feeders to the overlying tuff-breccia of the Cape Foulweather Basalt.

Rapid deposition and attendant loading of the Whale Cove Sandstone by the basalt breccia has produced clastic dikes that extend upward for several meters into the Cape Foulweather Basalt. The Cape Foulweather Basalt, which contains fragments of carbonized wood, forms the jaws of the outer bay and can be studied on the wave-cut platform that extends north and south of the bay. Late Pleistocene marine terrace deposits, which in places are offset by small faults, cap the Tertiary rocks on both sides of the outer bay.

SITE INFORMATION

The scenic Oregon coast between Cape Foulweather and Government Point is a unique wedding of ancient and recent marine environments. The prominent headlands at Cape Foulweather and the rugged coastline north of the cape to Government Point are composed of basalt flows and breccia of middle Miocene age (Fig. 1). These volcanic rocks, which overlie and are interbedded with fossiliferous nearshore middle Miocene sandstone and siltstone, extend seaward onto the continental shelf (Snavely and others, 1980; Snavely and Wells, 1984). Two basalt units with distinctive petrochemistry have been mapped along the northern Oregon coast; the oldest was named the Depoe Bay Basalt and the younger, the Cape Foulweather Basalt (Fig. 1; Snavely and others, 1973, 1976a, 1976b). The older unit is virtually identical in chemical composition to the Grande Ronde Basalt and the younger to the Frenchman Springs Member of the Wanapum Basalt, both part of the Columbia River Basalt Group (Swanson and others, 1979). This striking petrochemical similarity to flows of the Columbia River Basalt Group has led some workers (Beeson and others, 1979) to suggest that the middle Miocene basalts along coastal Oregon are invasive tongues of basalt flows erupted from fissures on the Columbia Plateau. The writer, however, does not accept this interpretation, because, in the Oregon Coast Range, the Depoe Bay Basalt intrudes rocks as old as the oceanic basalt of early Eocene age as well as Paleogene sedimentary and volcanic rock that overlies these basement rocks. Therefore, the writer concludes that the coastal basalts along the central Oregon coast were erupted from local vents and fissures located near the present coast (Snavely and others, 1973).

At the Depoe Bay field site, the stratigraphic relationship between the Depoe Bay Basalt and the Cape Foulweather Basalt is clearly shown in the outer bay (Fig. 2). There, isolated-pillow breccia of the Depoe Bay Basalt forms the east margin of the outer bay. Excellent exposures of this unit can be seen directly below the rock wall bordering the west side of U.S. Highway 101. The Depoe Bay Basalt unconformably overlies marine sandstone of the Astoria Formation, which is exposed in the inner bay and contains a shallow-water megafauna. The basalt is overlain by the Sandstone of Whale Cove that crops out along both the south and north ends of the outer bay. The Cape Foulweather Basalt forms the projecting headlands of the outer bay; it unconformably overlies the Sandstone of Whale Cove.

The breccia matrix that surrounds the isolated-pillows of Depoe Bay Basalt is composed of glassy or very fine-grained basalt that is palagonitized on weathered surfaces. Some breccia fragments are broken pillow rims. The pillows have ropy rims, and some have multiple chilled margins and drained-out cores (Fig. 3). Finely comminuted basaltic glass or calcareous sand and silt fill some of the drained pillows (Fig. 3). Clusters of hot-dog-shaped pillows and very elongate pillows are exposed in several places within the isolated-pillow breccia and appear to have resulted from more rapid extrusion of lava into the marine environment. Small quarries 0.5 mi (0.8 km) south of Depoe Bay

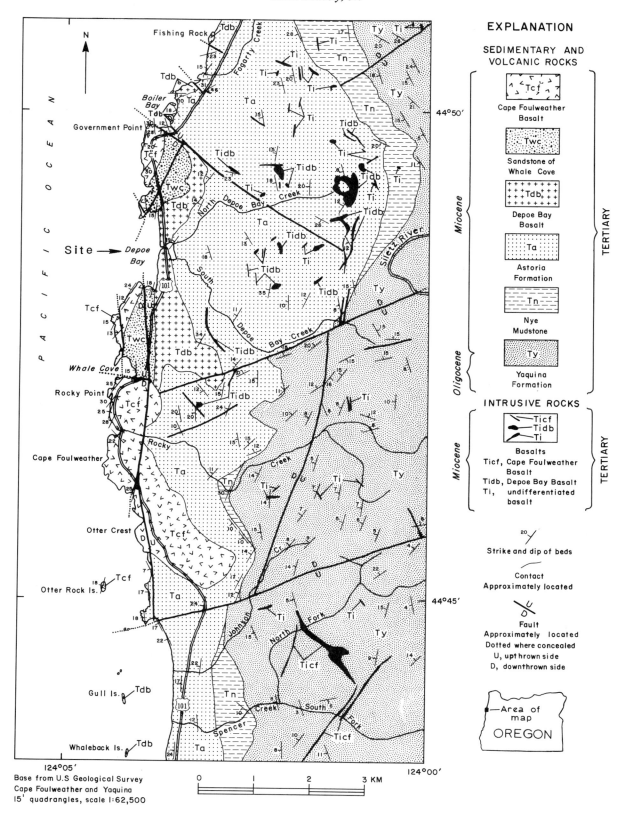

Figure 1. Generalized geologic map in the vicinity of Depoe Bay, Oregon, based upon mapping by Snavely and others, 1976a and 1976b.

Figure 2. View north across Depoe Bay, Oregon, type locality of the Depoe Bay Basalt. The Depoe Bay Basalt is well exposed along the east side of the bay (adjacent to U.S. Highway 101) and is overlain by the Sandstone of Whale Cove. The Cape Foulweather Basalt overlies the sandstone and forms the jagged coastline.

Figure 3. Isolated-pillow breccia of the Depoe Bay Basalt at Depoe Bay, Oregon. Note the discontinuous chilled zone developed inward from the chilled margin of the pillow in foreground and the sandstone-filled tension cracks and sandstone-filled core of the originally hollow pillow in center (under hammer.

expose a thick, poorly jointed subaerial flow of Depoe Bay Basalt. Immediately north and south of these quarries, the lava apparently flowed into the Miocene sea and formed the isolated-pillow breccia. Numerous dikes, sills, and irregular bodies of Depoe Bay Basalt intrude sandstone and siltstone of the middle Miocene Astoria Formation immediately east of Depoe Bay (Fig. 1), and some probably were feeders to the extrusive basalt at Depoe Bay. A 50-ft- (15 m-) thick flow of Depoe Bay Basalt was penetrated in the Standard-Union Nautilus well at a depth of about 3,730 ft (1,130 m), 10.8 mi (18 km) seaward from the coast. A sill of Depoe Bay Basalt also was encountered in this well at a depth of about 8,250 ft (2,500 m) (Snavely and others, 1980).

The Depoe Bay Basalt is aphanitic to fine grained, relatively uniform in composition, and is characterized by a high content of both SiO_2 and alkalies; it is quartz-normative.

The Cape Foulweather Basalt extends along the coast from Otter Crest, south to Cape Foulweather to Government Point (Fig. 1), but does not extend seaward as far as the Nautilus well. The Cape Foulweather Basalt can readily be distinguished from the Depoe Bay Basalt on the basis of petrochemistry, and it can be identified in the field by its sparse, but ubiquitous labradorite phenocrysts, which do not occur in the Depoe Bay Basalt. The Cape Foulweather Basalt is characterized by a lower content of SiO_2 and higher TiO_2, P_2O_5, and total iron oxide; it is quartz-normative. At Cape Foulweather, most of the extrusive tuff-

breccia is probably of subaerial origin. The water-laid, well-bedded lapilli tuff at Government Point apparently formed part of a fringing marine apron around the main vent at Cape Foulweather. Grading in many individual beds suggests that they were deposited by density currents; more massive units probably represent breccia transported by submarine landslides. The Cape Foulweather flows and breccia are restricted to inner shelf areas (Snavely and Wells, 1984).

On the north side of Boiler Bay, just north of Government Point, peperite and basalt dikes of the Depoe Bay Basalt intrude sandstone and siltstone of the Astoria Formation (Fig. 1). Cascade Head, the prominent headland 15 mi (25 km) to the north, is underlain by upper Eocene subaerial alkalic basalt flows. Cape Lookout, the seaward-projecting low headland about 39 mi (56 km) to the north, is composed of a beautifully exposed sequence of pillow basalts and subaerial flows of the Depoe Bay Basalt.

REFERENCES CITED

Beeson, M. H., Perttu, Rauno, and Perttu, Janice, 1979, The origin of the Miocene basalts of coastal Oregon and Washington: An alternative hypothesis: Oregon Geology, v. 41, no. 10, p. 159–166.

Snavely, P. D., Jr., MacLeod, N. S., and Wagner, H. C., 1973, Miocene tholeiitic basalts of coastal Oregon and Washington and their relations to coeval basalts of the Columbia Plateau: Geological Society of America Bulletin, v. 84, no. 2, p. 387–424.

Snavely, P. D., Jr., MacLeod, N. S., Wagner, H. C., and Rau, W. W., 1976a, Geologic map of the Yaquina and Toledo Quadrangles, Lincoln County, Oregon: U.S. Geological Survey Miscellaneous Investigations Series Map I-867.

—— 1976b, Geologic map of Cape Foulweather and Euchre Mountain Quadrangles, Lincoln County, Oregon: U.S. Geological Survey Miscellaneous Investigations Series Map I-868.

Snavely, P. D., Jr., Wagner, H. C., and Lander, D. L., 1980, Geological cross section of the central Oregon continental margin: Geological Society of America Map and Chart Series MC-28J, scale 1:250,000.

Snavely, P. D., Jr., and Wells, R. E., 1984, Tertiary volcanic and intrusive rocks on the Oregon and Washington Continental Shelf: U.S. Geological Survey Open-File Report 84-282, 17 p.

Swanson, D. A., Wright, T. L., Hooper, P. R., and Bentley, R. P., 1979, Revisions in stratigraphic nomenclature of the Columbia River Basalt Group: U.S. Geological Survey Bulletin 1457-G, 59 p.

Late High Cascade volcanism from summit of McKenzie Pass, Oregon: Pleistocene composite cones on platform of shield volcanoes: Holocene eruptive centers and lava fields

Edward M. Taylor, Department of Geology, Oregon State University, Corvallis, Oregon 97331

LOCATION AND ACCESS

The observation platform at Dee Wright viewpoint facility on the summit of McKenzie Pass, Oregon State Highway 242, can be reached by driving east from Eugene, Oregon, 55 mi (89 km) on U.S. 126 then 22 mi (35 km) on Oregon 242 or by driving west from Sisters, Oregon, 15 mi (24 km) on Oregon 242 (Fig. 1). The highway is paved but narrow with sharp curves. The site elevation is 5,300 ft (1,617 m); it is generally closed by snow November through July.

GEOLOGIC SIGNIFICANCE OF SITE

The central High Cascade Range, as seen from McKenzie Pass, is distinct from the western Cascades and from the eastern Cascade foothills with respect to its younger constructional volcanic landforms modified by glaciation. It is a broad platform composed of overlapping mafic shield volcanoes dominated by the composite cones of North Sister and Mount Washington. Silicic volcanism also contributed to this platform but was restricted to the vicinity of Mount Jefferson and the Three Sisters. McKenzie Pass is central to an area in which Holocene eruptions have produced cones, domes, and extensive lava fields of basaltic, andesitic, dacitic, and rhyodacitic composition.

SITE INFORMATION

From the observatory roof, the following landmarks are seen proceeding clockwise from true north in azimuthal degrees. An orientation device is mounted on a central pedestal.

0° - Mount Jefferson: Composite cone of basaltic andesite capped by andesite. Dacite domes on east slope are associated with latest Pleistocene eruption of tephra, pyroclastic flows, and associated debris flows.

7° - Cache Mountain: Glaciated basaltic andesite volcano, 0.9 m.y. KAr, capped by two late Pleistocene basaltic cones and lavas bearing quartz xenocrysts and abnormal concentration of incompatible trace elements.

11° - Bald Peter (far horizon): Basaltic andesite shield volcano, 2.1 m.y. KAr, cut nearly in half during Pleistocene glaciation.

20° - Dugout Butte (forested foreground): Composed of glaciated diktytaxitic olivine basalt flows similar to bulk of High Cascade platform.

30° - Green Ridge (far horizon): A north-south fault block ridge, 20 mi (32 km) long with down-dropped block on west side. Lavas on crest 5.1 m.y. KAr; main episode of faulting occurred approximately 4.5 m.y. ago.

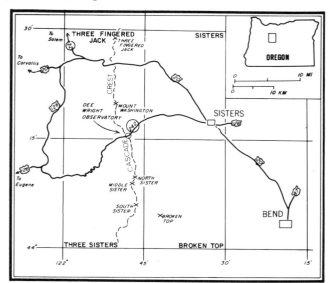

Figure 1. Index map to McKenzie Pass area, central High Cascade Range, Oregon.

40° - Black Butte (symmetrical cone in background): Basalt and basaltic andesite cone built over fault trace at south end of Green Ridge. Older than most High Cascade volcanoes; 1.4 m.y. KAr. Preservation due to lack of glaciation in rain shadow east of Cascades.

38° - Bluegrass Butte (foreground): Glaciated basaltic andesite cinder cone. Ridge extending east is a lateral moraine.

82° - Black Crater (fills most of eastern sector): Late Pleistocene basaltic shield volcano. Lower west flank has been glaciated; summit "crater" is a cirque open to the northeast.

105° to 155° - Unnamed Cascade crestline ridge composed of glaciated basaltic cinders, bombs, and lavas which issued from a 5-mi (8 km) long chain of cones, probably as spectacular lava fountains which produced broad flows choked with large bombs. Main flow moved 9 mi (14 km) east.

168° - North Sister (elevation 10,085 ft, 3,076 m): Basaltic andesite composite cone resting on a broad shield. Central plug and radial dikes exposed by glacial erosion. Oldest of Three Sisters. Yapoah Cone (left) and Collier Cone (right) can be seen at the base of North Sister.

174° - Middle Sister (elevation 10,045 ft, 3,064 m): Composite volcano supporting Collier Glacier. Older than South Sister which is obscured from view. Predominantly olivine basalt porphyry but also contains flows of basaltic andesite, andesite, dacite, and rhyodacite.

178° - Summit of Little Brother and ridge sloping west: Glaciated remnant of basaltic composite cone with exposed central plug and radial dikes. Older than North Sister.

188° - Four-in-One Cinder Cone (below skyline): A northeast-southwest andesitic ridge cone breached in four places by flows that moved northwest 2,600 years ago.

197° - The Husband volcano (on skyline, partly obscured): Although of late Pleistocene age, it is one of the oldest and most extensively dissected volcanoes in this region. The exposed plug of basaltic andesite is 0.25 mi (0.4 km) in diameter.

218° Condon Butte: Late Pleistocene cinder cone surrounded by glaciated lobes of basaltic andesite. Nested summit craters. Small knob visible at left base is an unnamed glaciated dome of rhyodacitic obsidian.

235° Horsepasture Mountain (far horizon): High-standing ridge crest in western Cascades.

256° - Scott Mountain: Summit cone on a broad basaltic shield. Cone and higher parts of shield were glaciated, but lavas at lower positions were not.

282° - South Belknap Cone: Cone was formed and breached 1,800 years ago, then surrounded by basaltic andesite lavas from a vent which became active about 1,500 years ago at south base of Belknap Crater.

285° and 306° Unnamed twin steptoes (in foreground lava field) are glaciated basaltic andesite cones surrounded by lava from Little Belknap volcano.

309° - Belknap Crater (summit cone on skyline): Focal point of a long-continued and complex episode of Holocene basalt and basaltic andesite eruptions. The broad shield which fills the northwest view is 5 mi (8 km) in diameter; it is estimated to be 1,700 ft (520 m) at maximum thickness and 1.3 cubic mi (5.4 cubic km) in volume. The volcano probably contains a core of cinders which interfingers with peripheral lavas and whose surface expression is the summit cone. Basaltic andesite issued from vents at the north and south bases of the cone approximately 1,500

years ago. Lava poured 12 mi (19 km) to the west and ash was ejected from a northern summit crater. The main bulk of Belknap ash, which has been traced over an area exceeding 100 square mi (259 square km), was ejected earlier from a larger south crater. Still earlier lavas were basaltic and moved eastward 7 mi (11 km) from their vents.

321° - Little Belknap volcano: A subsidiary shield, built 2,900 years ago on the east flank of the larger Belknap shield.

340° - Mount Washington (elevation 7,795 ft, 2,377 m): Glaciated remnant of a large basaltic andesite composite cone on a broad shield whose base is approximately 6 mi (10 km) in diameter. Central plug is flanked by a north-south swarm of dikes. Sharp prominence on northeast slope is a thick deposit of palagonitic tuff, penetrated and made resistant by dikes.

"Lava River Trail" east of the observatory provides a short pathway across the broken surface of a Holocene lava flow. The observatory is located on a blocky lobe of lava from Yapoah Cone which erupted basaltic andesite during the interval 2,600 to 2,900 years ago. The first lobe moved over this area toward the northwest, but as it advanced onto the opposing slope of Belknap volcano, its movement was checked, arcuate pressure ridges appeared on its surface, and subsequent lava was deflected to the northeast. The area just east of the observatory then became the main channel. The trail leads across the center of this channel where the roof of a small lava tube collapsed and a host of vertical cracks were produced in the lava by thermal contraction and loss of gases just prior to final consolidation. On the east side of the main channel is an imposing levee. The visitor might imagine the channel in full flood, rising at least as high as the levee crest and spilling over the eastern flank. When the supply of lava diminished, the molten interior drained away and the channel surface subsided. Lacking support, large segments of the levee tipped toward the central channel, causing deep, irregular tension cracks to open in the levee crest. The trail affords a view of one of these cracks and of several lava tongues along the east base of the levee.

REFERENCES CITED

Armstrong, R. L., Taylor, E. M., Hales, P. O., and Parker, D. J., 1975, K-Ar dates for volcanic rocks, central Cascade Range of Oregon: Isochron/West, no. 13, p. 5–10.

Luedke, R. C., and Smith, R. L., 1982, Map showing distribution, composition and age of Late Cenozoic volcanic centers in Oregon and Washington: U.S. Geological Survey Miscellaneous Investigations Map I–1091–D, scale 1:1,000,000.

Taylor, E. M., 1981, Central High Cascade roadside geology—Bend, Sisters, McKenzie Pass, Santiam Pass, Oregon, *in* Johnson, D. A., and Donnelly-Nolan, J., eds., Guides to some volcanic terranes in Washington, Idaho, Oregon, and northern California: U.S. Geological Survey Circular 838, p. 55–83.

Williams, H., 1944, Volcanoes of the Three Sisters region, Oregon Cascades: University of California Publications in Geological Sciences, vol. 27, p. 37–83.

Record of early High Cascade volcanism at Cove Palisades, Oregon: Deschutes Formation volcanic and sedimentary rocks

Edward M. Taylor and Gary A. Smith, Department of Geology, Oregon State University, Corvallis, Oregon 97331

LOCATION AND ACCESSIBILITY

This site is a series of scenic viewpoints on the east rim of Crooked River canyon above Cove Palisades State Park, near the boundary between sections 2 and 11, T.12S., R.12E., U.S. Geological Survey 7.5-minute quadrangle Round Butte Dam, Oregon. From downtown Madras, Oregon, follow signs marked "Cove" or "Cove State Park" on paved highways westward to the park entrance. From Redmond, Oregon, travel north on U.S. 97, turn northwestward at the exit to Culver, then follow signs marked "Cove" on paved highways to the park entrance (Fig. 1). Highways to the overlook and park are open all year. Just east of the park entrance, take the paved road north ("To Viewpoints") along the east edge of the canyon rim to any of three viewpoint parking areas.

GEOLOGIC SIGNIFICANCE OF SITE

At the confluence of the Crooked, Deschutes, and Metolius rivers, deep canyons have been cut into rimrock lavas and the underlying Deschutes Formation (Fig. 2), exposing a detailed record of late Miocene-early Pliocene volcanism in the adjacent Cascades, sedimentation in braided streams, and volcanism-induced high-sediment-load floods and debris flows.

SITE INFORMATION

From viewpoints on the east rim of the Crooked River canyon overlooking The Cove, one can observe in the pattern of "arms" of Lake Billy Chinook reservoir, the course of the Crooked River to the south, the Deschutes River to the southwest, and a 1.9-mi-long (3 km) septum of basaltic lava between them, known as The Island. The Metolius canyon, downstream from The Island, holds a western arm of the reservoir (Fig. 1). Snow-clad Cascade volcanoes, from the Three Sisters on the southwest to Mount Hood on the northwest, are visible on a clear day. Mount Jefferson is the prominent peak almost directly west of the park. The forested skyline in front of Mount Jefferson

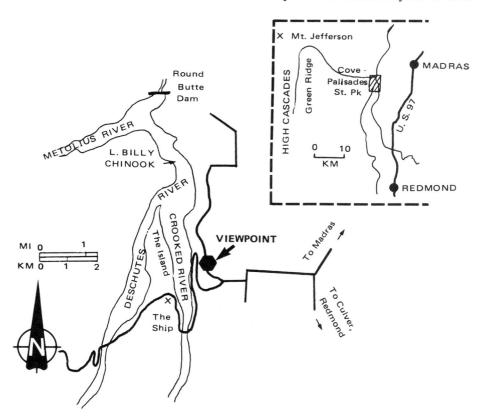

Figure 1. Location of field trip stop at Cove Palisades State Park, central Oregon.

Figure 2. View looking southward from overlook above a marina at Cove Palisades State Park. Deschutes Formation crops out prominently in roadcuts along Crooked River arm of Lake Billy Chinook and on The Ship. Intracanyon basalts form The Island and the bench in the Crooked River canyon.

follows the crest of 19-mi-long (30 km) Green Ridge whose west face is an abrupt down-to-the-west fault escarpment on which are exposed volcanic rocks correlative with those in the canyon walls of The Cove (Fig. 3).

Intracanyon lavas. Isolated benches of basaltic lava can be seen along the sides of Crooked River and Deschutes River canyons (Fig. 2). The canyon benches and The Island are erosional remnants of a tongue of basalt that extended 63 mi (100 km) north from source vents on the lower flank of the Pleistocene-Holocene Newberry volcano, south of Bend. More than 700,000 years ago, these lavas entered the Crooked River canyon northeast of Redmond, followed the canyon to The Cove, moved up the Deschutes River for 1.9 mi (3 km), and poured downstream beyond the position of Round Butte Dam. Younger lavas from Newberry moved directly down the Deschutes River canyon as far as The Cove. Thick V-shaped wedges of basalt partly filled the canyons before they were reexcavated by the modern rivers.

The Deschutes Formation. Coarse-grained volcaniclastic sediment interbedded with pumiceous lapilli and ash, sheets of basalt, and ash-flow tuffs, crop out extensively on the canyon walls and are the chief constituents of the Deschutes Formation. Lavas near the base and top of the formation have yielded radiometric ages of approximately 7.6 Ma and 4.5 Ma, respectively (Smith and Snee, 1984; Armstrong and others, 1975). Near The Cove, the Deschutes Formation is about 825 ft (250 m) thick and dominated by epiclastic and pyroclastic rocks. The formation thickens westward until at Green Ridge it is more than 2,310 ft (700 m) thick and is dominated by lavas and coarse-grained ash-flow tuffs.

Some of the widespread sheets of rimrock-forming basalt and similar interbedded lavas were produced from cinder cones within the Deschutes valley. However, the volcaniclastic deposits were derived from an ancestral High Cascade Range to the west. More than 70 distinct ash-flow tuff units have been recognized

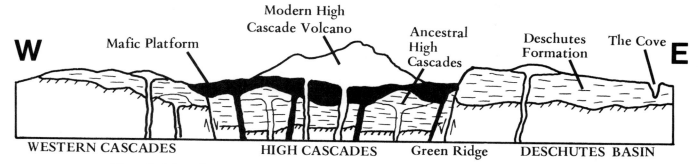

Figure 3. Schematic cross section of the central Oregon Cascade Range (after Taylor, 1981). The Deschutes Formation, and equivalent rocks capping western Cascade ridges, were derived from an ancestral High Cascade volcanic chain. The older volcanoes subsided into an intra-arc graben in the early Pliocene and subsequently were buried by a platform of basalts and basaltic andesites upon which the modern stratovolcanoes have been constructed.

toward western sources where they become thicker, coarser grained, and more densely welded. Deposits of accretionary lapilli and air-fall pumice occur throughout the Deschutes Formation. They were carried eastward by prevailing winds from eruptive centers in the High Cascades.

Volcanoes that produced Deschutes Formation rocks. The volcanic source terrain that gave rise to the lavas and tuffs of the Deschutes Formation occupied the position of the modern central High Cascade Range. However, the composite cones of that province are all less than 1 Ma and most of the underlying platform of mafic shield volcanoes is less than 2 Ma. The Deschutes Formation volcanic source terrain subsided into a north-south-trending graben structure approximately 4.5 Ma (Taylor, 1981). The Deschutes volcanoes are no longer exposed at the surface because they have been buried beneath the High Cascade platform (Fig. 3). The Deschutes Formation provides the best exposed, most lithologically diverse, and most accessible record of this earlier episode of High Cascade volcanism.

Points of interest. It is suggested that the visitor follow the paved highway through Cove Palisades State Park, over the bridges spanning the Crooked and Deschutes arms of the reservoir, and ascend the grade up the west wall of the Deschutes River canyon (Fig. 1). Items of special interest include:

1) Highway roadcuts on the descent from the canyon rim to the marina expose coarse-grained, volcaniclastic sediments and ash-flow tuffs of the Deschutes Formation. The prominent white, rhyodacitic ash-flow tuff is also exposed on the promontory called The Ship, just south of The Island, and farther west in the Deschutes River canyon. The overlying orange ash-flow tuff is also of rhyodacitic composition. Both of these massive units contain pumice lapilli, bombs, and lithic fragments up to 1 m across, supported in a matrix of ash and crystals as is typical of the deposits of ash flows. This texture contrasts with the better-sorted, stratified, reworked pyroclastic sediment composing the thick white deposit at the top of the grade. Air-fall pumice lapilli from a late Pleistocene eruption of Mount Jefferson occurs as a light-colored exposure above the road, just before the switchback at the marina.

2) From the marina to the Crooked Arm bridge, exposure alternates between Deschutes Formation volcaniclastics and intracanyon basalt lavas with spectacular cooling-joint patterns. A distinctive, pink ash-flow tuff is exposed at road level along this stretch and is noteworthy for the abundance of rounded cobbles near its base which were entrained from a river bed by the moving ash flow. Upstream from the bridge, the Crooked River emerges from an inner canyon incised through the intracanyon basalt flows (Fig. 2).

3) Between the Crooked and Deschutes rivers, the road is constructed primarily on landslide debris. Deschutes Formation sediments and ash-flow tuffs are spectacularly exposed at The Ship. A parking area at The Ship is adjacent to a petroglyph-bearing boulder recovered from the Crooked River before impoundment of the reservoir. The view upstream from the Deschutes Arm bridge includes erosional remnants of the Crooked River intracanyon flows which backed up the Deschutes River and, just above water level at the bend in the river, the distal end of the younger basalts which flowed directly down the Deschutes.

4) The roadcuts along the grade west of the Deschutes River provide one of the best sections through the Deschutes Formation. The lowest 312 ft (95 m) of the section (to the second switchback) contains gravel and interbedded sand and silt representing channel and overbank deposition, respectively, by a northeast-flowing river. These sediments are interbedded with two ash-flow tuffs, an olivine basalt flow, and several debris-flow and flood deposits. The predominantly fluvial facies are overlain abruptly, at the horizon of the white ash-flow tuff, by a 82-ft-thick (25 m) sequence of poorly sorted, debris-flow and high-sediment-load flood deposits whose deposition was probably induced, directly or indirectly, by explosive volcanism in the ancestral High Cascades. The succeeding 115 ft (35 m) of poorly exposed section is mostly plane-bedded sheetflood deposits and paleosols. The uppermost 50 ft (15 m) is composed of interbedded massive sand and tephra with two flows of diktytaxitic olivine basalt. Abundant pedogenic features (e.g., burrows, root traces, oxidation to tan colors, blocky jointing) indicate that these sands are immature paleosols. The low sedimentation rates implied by this paleosol-dominated sequence suggest an abrupt change from the depositional record exposed below. The change probably documents the subsidence of the High Cascade graben along faults near Green Ridge which isolated the Deschutes basin from the volcanic source areas. Ash-flow tuffs, which one expects would be emplaced contemporaneously with the thick air-fall tephras exposed here, are not found within the basin and presumably were impounded within the graben by west-facing escarpments, along with eruption-induced sedimentary deposits.

REFERENCES CITED

Armstrong, R. L., Taylor, E. M., Hales, P. O., and Parker, D. J., 1975, K-Ar dates for volcanic rocks, central Cascade Range of Oregon: Isochron/West, no. 13, p. 5–10.

Peterson, N. V., and Groh, E. H., 1970, Geologic tour of Cove Palisades State Park near Madras Oregon: The Ore Bin, v. 32, p. 141–168.

Smith, G. A., and Priest, G. R., 1983, A field trip guide to the central Oregon Cascades, first day: Mount Hood—Deschutes basin: Oregon Geology, v. 45, p. 119–126.

Smith, G. A., and Snee, L. W., 1984, Revised stratigraphy of the Deschutes basin, Oregon: Implications for the Neogene development of the central Oregon Cascades [abs.]: EOS, Transactions of the American Geophysical Union, v. 65, p. 330.

Taylor, E. M., 1981, Central High Cascade roadside geology—Bend, Sisters, McKenzie Pass, Santiam Pass, Oregon, *in* Johnston, D. A., and Donnelly-Nolan, J., eds., Guides to some volcanic terranes in Washington, Idaho, Oregon, and northern California: U.S. Geological Survey Circular 838, p. 55–83.

John Day Fossil Beds National Monument, Oregon: Painted Hills unit

Paul T. Robinson, *Centre for Marine Geology, Dalhousie University, Halifax, Nova Scotia, Canada*

LOCATION

The Painted Hills unit, formerly Painted Hills State Park, lies along Bridge Creek about 9.5 mi (16 km) northwest of Mitchell, Wheeler County, Oregon, near the intersection of T.10 and 11S. and R.20 and 21E. It can be reached by taking the Bridge Creek road north from U.S. 26, which runs between Prineville and Mitchell (Fig. 1). The turnoff is located 4 mi (6.5 km) west of Mitchell and is marked by a sign directing travelers to the Painted Hills. The Bridge Creek road is paved and is passable in all weather by passenger cars.

SIGNIFICANCE OF THE SITE

The Painted Hills area provides excellent exposures of the lower John Day Formation and upper Clarno Formation. A deep weathered profile marks the top of the Clarno Formation and several resurrected Oligocene hills can be observed in the area. The varicolored tuffs and tuffaceous sediments of the lower John Day Formation illustrate the effects of diagenesis and incorporation of the red Clarno saprolite. Excellent exposures of the Picture Gorge ignimbrite are also present. Just north of the Painted Hills proper, across Bridge Creek, are good exposures of the upper John Day Formation, which is unconformably overlain by flows of the Columbia River Basalt Group. The type locality of the Bridge Creek flora is located in the lower John Day Formation.

SITE INFORMATION

The Painted Hills consist of varicolored tuffs and tuffaceous claystones of the lower John Day Formation capped by the Picture Gorge ignimbrite. Underlying the John Day Formation is the Clarno Formation, a sequence of andesitic to rhyolitic lava flows, tuffs, and tuffaceous sediments. Above the Picture Gorge ignimbrite is the upper John Day Formation, composed of gray to buff tuffs and tuffaceous sediments, which in turn are capped by thick lava flows of the Columbia River Basalt Group. The entire sequence is folded into a broad, open, northeast-trending syncline (Fig. 2).

Clarno Formation

The Clarno Formation is a heterogeneous unit consisting largely of andesitic and basaltic lava flows and breccias, with lesser amounts of volcaniclastic sedimentary rocks, tuffs, and silicic lava flows and domes. Individual lithologic units are discontinuous and stratigraphic markers are generally absent. Fossil flora and fauna suggest an Eocene to early Oligocene age; K-Ar

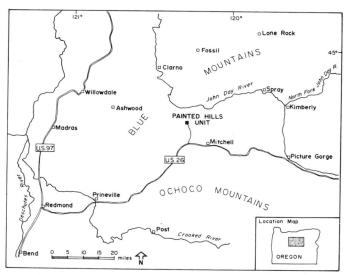

Figure 1. Map of north-central Oregon showing location of John Day Fossil Beds National Monument, Painted Hills unit.

dates generally support this interpretation (Fiebelkorn and others, 1982). A few dates suggest that locally Clarno volcanism continued into John Day time.

Only the upper Clarno Formation is exposed in the Painted Hills area. In this locality the formation consists chiefly of rhyolitic and andesitic lavas, rhyolitic vitric tuffs, and varicolored claystones and conglomerates (Hay, 1963).

Hydrothermal alteration, which is common in the lower parts of the formation, is lacking, although vitric material is replaced by clinoptilolite, montmorillonite, and opal. Two K-Ar dates of 36 and 37 m.y. from samples in the Painted Hills area (Hay, 1962a) are considered to be slightly too young (Swanson and Robinson, 1968), since dates from an ashflow tuff at the base of the John Day Formation give ages of 37.1 ± 1.0 and 37.4 ± 1.1 m.y. (Swanson and Robinson, 1968; Fiebelkorn and others, 1982).

At the top of the Clarno Formation is a deep red saprolite developed on a hilly surface with up to 300 ft (90 m) of relief (Waters, 1954; Hay, 1962a). Within the Painted Hills area, several resurrected pre–John Day hills crop out in Sections 35 and 36, T.10S., R.20E. (Fig. 2). Here, the saprolite is 10 to 16.5 ft (3 to 5 m) thick and is developed on porphyritic pyroxene andesite. Overlying the saprolite is a discontinuous colluvial layer also composed largely of kaolinite, and less commonly, montmorillonite (Hay, 1962a) derived from the weathered profile. This deposit is generally 10 to 20 ft (3 to 6 m) thick, but varies significantly, thickening in depressions and thinning against hills.

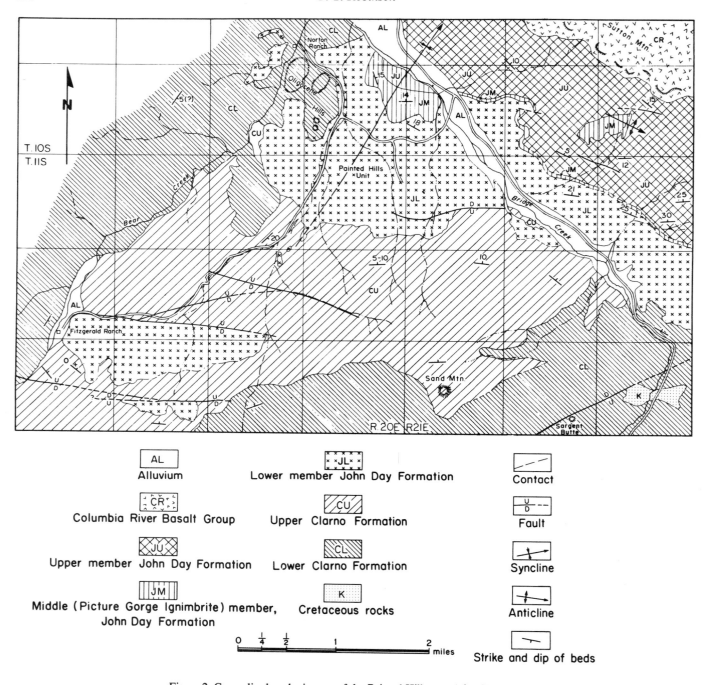

Figure 2. Generalized geologic map of the Painted Hills area (after Hay, 1963).

John Day Formation

In its type locality east of the Blue Mountains the John Day Formation consists largely of airfall tuffs and tuffaceous sediments, with a prominent ash flow sheet near the middle of the unit (Hay, 1962a; 1963; Fisher, 1966). To the west of the Blue Mountains the formation becomes thicker and coarser grained, containing numerous intercalated ash flow sheets and lava flows (Waters, 1954; Peck, 1964; Robinson, 1975).

In the eastern area, the formation has a maximum aggregate thickness of approximately 2,500 ft (750 m) and has been divided into four members by Fisher and Rensberger (1972). From oldest to youngest, these are the Big Basin member, composed of deep red claystones; the Turtle Cove member, composed of green, buff, and red zeolitized tuff; the Kimberly member, consisting of light gray to buff unzeolitized tuff; and the Haystack Valley member, a sequence of unzeolitized, gray, reworked tuff. This subdivision is based largely on diagenetic features and therefore has limited stratigraphic significance. Hay (1963) divided the formation into lower, middle, and upper members, based on the presence of the widespread ash flow sheet near the middle of the unit (Picture Gorge ignimbrite of Fisher, 1966).

In the Painted Hills, the lower member consists of about 825 to 1,200 ft (250 to 350 m) of varicolored tuffaceous claystones and vitric tuff (Hay, 1963) (Fig. 3). The claystones form massive beds of pale brown, olive gray, yellow, or green montmorillonite that weathers to form smooth, rounded hills (Fig. 4). The thin, prominent red layers consist chiefly of kaolinite, presumably derived from the saprolite at the top of the Clarno Formation (Hay, 1962a). All of the claystones formed by diagenetic alteration of pyroclastic material, most of which was andesitic to dacitic based on relict crystals of andesine, pyroxene, hornblende, biotite, and ilmenite.

Intercalated with the claystones are numerous beds of white-to cream-colored vitric tuff that form resistant ledges a few inches (cm) to 10 ft (3 m) thick. Most beds contain sanidine or sodic plagioclase crystals, and a few contain quartz, suggesting a rhyodacitic to rhyolitic composition. One bed contains distinctive soda-rich sanidine crystals with myrmekitic rims similar in composition and texture to sanidine from the ash flow tuff of member G of the western facies. This bed is considered to be the airfall equivalent of member G (Hay, 1962b; Woodburne and Robinson, 1977).

A distinctive crystal-rich ash flow tuff crops out at the base of the formation approximately 9 mi (15 km) northeast of the Painted Hills in Section 15 T.8S.,R.21E. This moderately to weakly welded tuff has been correlated with the basal ash flow sheet of the western facies on the basis of its phenocryst assemblage and distinctive feldspar composition (Hay, 1963; Woodburne and Robinson, 1977).

Locally, in the Painted Hills area, thin flows of olivine basalt crop out in the lower 250 ft (80 m) of the formation. Individual flows rarely exceed 50 ft (15 m) in thickness and are laterally discontinuous. These are somewhat alkaline, high-titania basalts similar to those found elsewhere in the formation (Robinson, 1969). The discontinuous nature of the flows and the presence of several dikes suggest that the lavas were erupted locally.

The Picture Gorge ignimbrite, which comprises the middle member of the formation, is a compound ash flow sheet composed of two cooling units and a thin, basal airfall tuff, with an aggregate thickness of approximately 100 ft (30 m) in the Painted Hills area. Fisher (1966) estimated that the ignimbrite originally covered an area of approximately 2,000 mi^2 (5,000 km^2) being thickest in the Mitchell area and thinning gradually to the northeast.

Both cooling units are composed of a coarse-grained, eutaxitic, sparsely phyric, rhyolitic tuff containing flattened pumice fragments up to 10 in (25 cm) long and numerous irregular fragments of black glass up to 2 in (5 cm) across. Crystals of sodic plagioclase, sanidine, quartz, and clinopyroxene average about 1 percent of the rock.

In the Painted Hills area, the upper member is approxi-

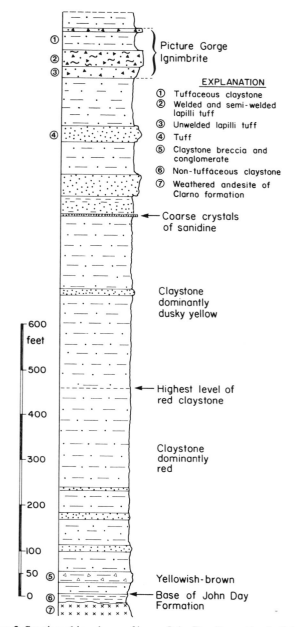

Figure 3. Stratigraphic column of lower John Day Formation in Painted Hills unit (after Hay, 1963).

mately 825 ft (250 m) thick and is composed chiefly of tuffaceous claystone and vitric tuff, compositionally similar to that of the lower member. The principal differences are diagenetic, the upper member being lighter in color and less altered than the lower. The buff-colored, unzeolitized upper portion corresponds to the Kimberly member, as defined by Fisher and Rensberger (1972). Airfall tuffs are most abundant in the lower part of the upper member. One distinctive air fall tuff (Deep Creek Tuff of Fisher, 1962; 1963) lies about 10 to 200 ft (3 to 60 m) above the Picture Gorge ignimbrite and ranges up to 30 ft (10 m) thick. It is

Figure 4. View of Painted Hills unit looking north. In foreground are low, rounded hills of varicolored tuffaceous claystones. White bands are layers of airfall tuff. Hill in middle distance is capped by Picture Gorge ignimbrite; on skyline are flows of Columbia River Basalt Group overlying upper John Day Formation.

composed of cream-colored, coarse-grained ash in which fresh black shards are commonly preserved. Bits of coarse-grained reworked tuff and conglomerate are present locally, particularly in the upper part of the member.

Unconformably overlying the John Day Formation in the Painted Hills area are flows of the Columbia River Basalt Group. These are thick massive flows of the Picture Gorge type. They are generally medium- to coarse-grained aphyric lavas, some of which have well-developed ophitic textures.

The entire section is folded into a broad, open syncline, the axial plane of which passes directly through the Painted Hills (Hay, 1963). Dips in the Picture Gorge basalts generally are less than 10°, whereas those of the underlying John Day and Clarno Formations are significantly steeper. On a regional basis the John Day Formation thins toward the axis of the Blue Mountain uplift and is overlapped by flows of the Columbia River Basalt Group (Fisher, 1967; Robinson, 1975).

REFERENCES CITED

Fiebelkorn, R. B., Walker, G. W., MacLeod, N. S., McKee, E. H., and Smith, J. G., 1982, Index to K-Ar age determinations for the state of Oregon: U.S. Geological Survey Open-File Report 82-596, 40 p., index map, scale 1:1,000,000.

Fisher, R. V., 1962, Clinoptilolite tuff from the John Day Formation, eastern Oregon: Ore Bin, v. 24, p. 197–203.

——— , 1963, Zeolite-rich beds of the John Day Formation, Grant and Wheeler Counties, Oregon: Ore Bin, v. 25, p. 185–197.

——— , 1966, The geology of a Miocene ignimbrite layer, John Day Formation, eastern Oregon: University of California Publications in Geological Sciences, v. 67, 73 p.

——— , 1967, Early Tertiary deformation in north-central Oregon: American Association of Petroleum Geologists Bulletin, v. 51, p. 111–123.

Fisher, R. V., and Rensberger, J. M., 1972, Physical stratigraphy of the John Day Formation, central Oregon: University of California Publications in Geological Sciences, v. 101, 45 p.

Hay, R. L., 1962a, Origin and diagenetic alteration of the lower part of the John Day Formation near Mitchell, Oregon: in Petrologic Studies: Geological Society of America, Buddington Volume, p. 191–216.

——— , 1962b, Soda-rich sanidine of pyroclastic origin from the John Day Formation of Oregon: American Mineralogist, v. 47, p. 968–971.

——— 1963, Stratigraphy and zeolite diagenesis of the John Day Formation of Oregon: University of California Publications in Geological Sciences, v. 42, p. 199–262.

Peck, D. L., 1964, Geological reconnaissance of Antelope-Ashwood area, north-central Oregon: U.S. Geological Survey Bulletin 1161-D, p. D1–D26.

Robinson, P. T., 1969, High titania alkali-olivine basalts of north-central Oregon, U.S.A.: Contributions to Mineralogy and Petrology, v. 22, p. 349–360.

——— , 1975, Reconnaissance geologic map of the John Day Formation in the southwestern part of the Blue Mountains and adjacent areas, north-central Oregon: U.S. Geological Survey Miscellaneous Geologic Investigations Map I-872, scale 1:125,000.

Swanson, D. A., and Robinson, P. T., 1968, Base of the John Day Formation in and near the Horse Heaven mining district, north-central Oregon: U.S. Geological Survey Professional Paper 600D, p. D154–D161.

Waters, A. C., 1954, John Day formation west of its type locality [abs.]: Geological Society of America Bulletin, v. 65, p. 1320.

Woodburne, M. O., and Robinson, P. T., 1977, A new late Hemingfordian mammal fauna from the John Day Formation, Oregon and its stratigraphic implications: Journal of Paleontology, v. 51, p. 750–757.

Columbia River Gorge: The geologic evolution of the Columbia River in northwestern Oregon and southwestern Washington

Marvin H. Beeson, *Department of Geology, Portland State University, P.O. Box 751, Portland, Oregon 97207*
Terry L. Tolan, *Geosciences Group, Rockwell International, P.O. Box 800, Richland, Washington 99352*

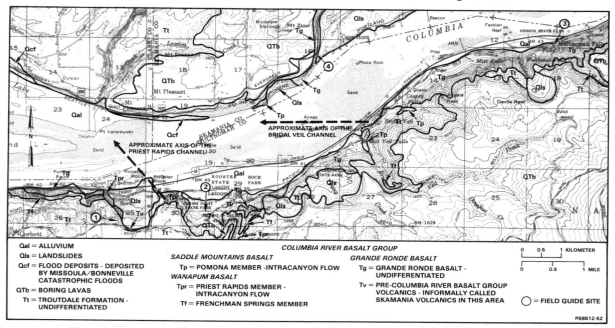

Figure 1. Map showing the geology and location of field guide sites described in the western portion of the Columbia River Gorge.

LOCATION

The Columbia River Gorge is located astride the Cascade Range along a 72 mi (120 km) stretch of the Oregon/Washington border (Fig. 1). The western half of the gorge is covered by the Camas and Bridal Veil 15-minute quadrangles. Primary access to the gorge is by two major highways which traverse its length—the two-lane State Highway 14 along the Washington side and the four-lane Interstate 84 freeway along the Oregon side. Also along the Oregon side are segments of the old Columbia River Scenic Highway, which when completed in 1917 was the first paved highway to extend the length of the gorge. All sites to be visited on the Oregon side are along the old Columbia River Scenic Highway (Fig. 1).

SIGNIFICANCE OF SITE

The Columbia River Gorge has interested geologists for more than 100 years because it provides a natural cross section through the northern Oregon/southern Washington Cascade Range. Of equal interest to many geologists is the geologic history of the Columbia River itself and the evolution of the spectacular present-day gorge.

Over the years, geologists have not agreed on the locations of the paths of the ancestral Columbia River in relation to the present-day course. These differences of opinion have been extreme. Recent advances in our understanding of the Neogene units exposed within the gorge (Fig. 2) have provided new insights into the overall history of the ancestral Columbia River (Anderson, 1980; Tolan and Beeson, 1984; Tolan and others, 1984a, b).

The history of the ancestral Columbia River which is emerging from this new work is a 16 m.y. record of dynamic processes and cataclysmic events which included obliteration of former ancestral Columbia River canyons by flood basalt flows and inundation of another channel by lava flows and debris created by Cascadian volcanism. The sites described in this guide have been chosen to show important features and relationships that have helped us understand the key events which have shaped the Neogene evolution of the Columbia River in the gorge area.

NEOGENE HISTORY OF THE COLUMBIA RIVER— A SUMMARY

The following short summary of the Neogene history of the Columbia River is presented to give the reader a better perspective and understanding of what will be seen at the sites described below. For a more complete description and discussion of the

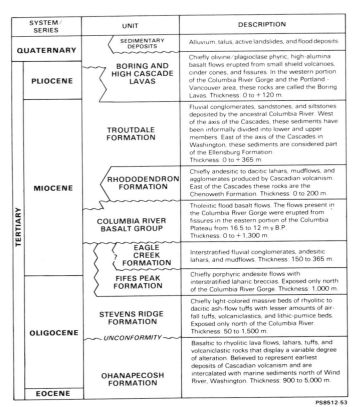

SYSTEM/SERIES		UNIT	DESCRIPTION
QUATERNARY		SEDIMENTARY DEPOSITS	Alluvium, talus, active landslides, and flood deposits.
TERTIARY	PLIOCENE	BORING AND HIGH CASCADE LAVAS	Chiefly olivine-/plagioclase phyric, high-alumina basalt flows erupted from small shield volcanoes, cinder cones, and fissures. In the western portion of the Columbia River Gorge and the Portland - Vancouver area, these rocks are called the Boring Lavas. Thickness: 0 to + 120 m.
	MIOCENE	TROUTDALE FORMATION	Fluvial conglomerates, sandstones, and siltstones deposited by the ancestral Columbia River. West of the axis of the Cascades, these sediments have been informally divided into lower and upper members. East of the axis of the Cascades in Washington, these sediments are considered part of the Ellensburg Formation. Thickness: 0 to + 365 m.
		RHODODENDRON FORMATION	Chiefly andesitic to dacitic lahars, mudflows, and agglomerates produced by Cascadian volcanism. East of the Cascades these rocks are the Chenoweth Formation. Thickness: 0 to 200 m.
		COLUMBIA RIVER BASALT GROUP	Tholeiitic flood basalt flows. The flows present in the Columbia River Gorge were erupted from fissures in the eastern portion of the Columbia Plateau from 16.5 to 12 m.y. B.P. Thickness: 0 to + 1,300 m.
		EAGLE CREEK FORMATION	Interstratified fluvial conglomerates, andesitic lahars, and mudflows. Thickness: 150 to 365 m.
	OLIGOCENE	FIFES PEAK FORMATION	Chiefly porphyric andesite flows with interstratified laharic breccias. Exposed only north of the Columbia River Gorge. Thickness: 1,000 m.
		STEVENS RIDGE FORMATION	Chiefly light-colored massive beds of rhyolitic to dacitic ash-flow tuffs with lesser amounts of air-fall tuffs, volcaniclastics, and lithic-pumice beds. Exposed only north of the Columbia River. Thickness: 50 to 1,500 m.
		— UNCONFORMITY —	
		OHANAPECOSH FORMATION	Basaltic to rhyolitic lava flows, lahars, tuffs, and volcaniclastic rocks that display a variable degree of alteration. Believed to represent earliest deposits of Cascadian volcanism and are intercalated with marine sediments north of Wind River, Washington. Thickness: 900 to 5,000 m.
	EOCENE		

PS8512-53

Figure 2. Generalized stratigraphy of the Columbia River Gorge area. Modified from Tolan and others (1984a).

stratigraphy and Neogene paleodrainage history of the gorge region, the reader is referred to Tolan and Beeson (1984) and Tolan and others (1984a, b).

The discernible story of the ancestral Columbia River in the gorge region began about 15.6 Ma, during a hiatus in Columbia River basalt volcanism. Prior to this time, the gorge area had been inundated and buried by 330 to 3,300 ft (100 to 1,000 m) of Grande Ronde Basalt (Fig. 3) in approximately 1 m.y. The Grande Ronde flows (see Reidel and others, this volume) were able to cross the Miocene Cascade Range through a 37-mi-wide (60 km) lowland which stretched from the present-day gorge south into Oregon. Preexisting stream systems that must have drained the Columbia Plateau and extended through this area were completely overwhelmed by the Grande Ronde flows and are now hidden from our direct view. It wasn't until the long quiescent period (perhaps as long as several 100,000 years) between the end of the Grande Ronde and the onset of Frenchman Springs Member volcanism (Fig. 3) that the first known ancestral Columbia River channel came into being.

The path of this early Columbia River through the Miocene Cascades lay within a developing synclinal trough (part of the Yakima Fold Belt) that extended southwestward from The Dalles, Oregon, to under the site of present-day Mount Hood (Fig. 4). These Yakima Fold structures continued to grow throughout Columbia River basalt time (see Caron and others,

this volume) and played a major role in controlling the position of future Columbia River channels. This early course of the ancestral Columbia River was inundated by flows of the Frenchman Springs Member (Fig. 3). This, coupled with active Miocene Cascadian volcanoes shedding debris (Rhododendron Formation, Fig. 2) into the trough, succeeded in closing off this route and forced the waters draining off the Columbia Plateau northward into the next synclinal trough.

In the gorge region, another short hiatus followed after Frenchman Springs volcanism had ceased. This was caused by the failure of the Roza Member flows (Fig. 3) to cross the Cascade Range. This brief hiatus allowed a new Columbia River channel to develop through the Miocene Cascade Range and eastward to near the present-day town of Mosier, Oregon, by about 14.5 Ma (Fig. 4). This channel of the Columbia River was short-lived, as it was totally destroyed by an advancing Priest Rapids Member flow (Figs. 3, 5a) forcing the ancestral Columbia River northward to a new path once again.

The third channel that the ancestral Columbia River established, the Bridal Veil channel (Tolan and Beeson, 1984) was longer lived than its predecessors in that the river remained in it for about 10 m.y. One of the primary reasons for the longevity of the Bridal Veil channel was that initially there was more than 1 m.y. for the river to incise a deep canyon through the Miocene Cascades before a Columbia River basalt flow (Pomona Member, Figs. 3, 5b) reached it. Thus the Bridal Veil channel was capable of containing the Pomona flow and was not destroyed. This allowed the Columbia River to remain in the Bridal Veil channel.

For the next 6 m.y. following the emplacement of the Pomona flow, the ancestral Columbia River deposited sands and gravels (lower member Troutdale Formation, Fig. 2) within the confines of the Bridal Veil channel, interrupted only by an occa-

SERIES	GROUP	SUB-GROUP	FORMATION	MEMBER	K-Ar DATE (m.y.)	MAGNETIC POLARITY
MIOCENE / MIDDLE	COLUMBIA RIVER BASALT GROUP	YAKIMA BASALT SUBGROUP	SADDLE MOUNTAINS BASALT	POMONA	12	R
				EROSIONAL UNCONFORMITY		
			WANAPUM BASALT	PRIEST RAPIDS	14.5	R
				LOCAL EROSIONAL UNCONFORMITY		
				ROZA		T
				FRENCHMAN SPRINGS		N,E
				LOCAL EROSIONAL UNCONFORMITY		
MIOCENE / LOWER			GRANDE RONDE BASALT		15.5	N₂
						R₂
						N₁
					16.5	R₁

E = EXCURSIONAL MAGNETIC POLARITY
N = NORMAL MAGNETIC POLARITY
R = REVERSED MAGNETIC POLARITY
T = TRANSITIONAL MAGNETIC POLARITY

PS8512-54

Figure 3. Generalized stratigraphy of the Columbia River Basalt Group in the Columbia River Gorge region. N = normal magnetic polarity, R = reversed magnetic polarity, T = transitional magnetic polarity, E = excursional magnetic polarity, K-Ar dates in millions of years.

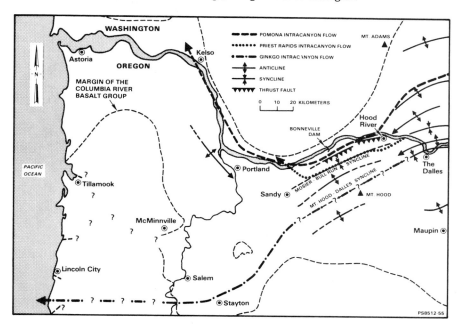

Figure 4. Sketch map showing the positions of Columbia River basalt intracanyon flows in western Oregon and Washington. The northeast-trending folds depicted on the map represent the extension of the Yakima Fold Belt through the Cascade Range and into western Oregon (Beeson and others, 1985). The synclinal troughs played a major role in controlling the various courses of the ancestral Columbia River through the Cascade Range. Note that the earliest intracanyon flow followed an ancestral Columbia River course which crossed the Miocene Oregon Coast Range far to the south of the present-day Columbia River. The existence of this southern pathway across the Miocene Coast Range removes the need for a local, coastal vent source for these flows and clearly shows that they have a Columbia Plateau origin as previously suggested by Beeson and others (1979).

sional Rhododendron lahar (Fig. 2). This period ended between 6 and 4 Ma with the onset of basaltic volcanism (Boring/High Cascade Lavas, Fig. 2) within the Cascades. High-alumina basalt flows from widely scattered vents produced during this episode of Cascadian volcanism repeatedly reached the Bridal Veil channel. Phreatic brecciation of these lavas flowing into the ancestral Columbia River produced tremendous amounts of hyaloclastic and clastic debris that now characterizes the upper member of the Troutdale Formation (Tolan and Beeson, 1984; Fig. 2). This repeated influx of debris caused rapid aggradation of the Bridal Veil channel that eventually allowed the river to escape the confines of the canyon. Continued local basaltic volcanism forced the Columbia River to the north of the Bridal Veil channel and finally capped this former course.

The incision of the present-day Columbia River Gorge, which began with the onset of uplift related to the present-day High Cascades, marked the end of Troutdale deposition. Field relationships tentatively suggest that the onset of rapid Cascadian uplift may have begun as late as 2 Ma (Tolan and Beeson, 1984).

West of the rising High Cascades, the Columbia River began to cut its present-day canyon near where the more resistant Columbia River basalt laps out against older, less resistant rocks. The Columbia River was prevented from reoccupying and incising a new canyon in its former site (the Bridal Veil channel) by

the high-alumina basalt flows that capped this pathway. East of the rising Cascades, the Columbia River appears to have begun to incise its canyon along the northern margin of the Bridal Veil channel. Thus, the present-day course of the Columbia River was established.

The Columbia River Gorge was widened in Pleistocene time by catastrophic floods that resulted from the failure of glacial dams which had created ice marginal lakes northeast of the Columbia Plateau (see Carson and others, this volume). Other than widening and sculpting the gorge, these flood events had no lasting impact on the course of the Columbia River through the Cascade Range.

SITES

Four sites within the western half of the Columbia River Gorge are described to illustrate some of the evidence for the evolutionary history of the Columbia River summarized above. More detailed descriptions of these and other sites can be found in Tolan and others (1984a, b).

Site 1. Women's Forum State Park/Chanticleer Point, Oregon (Figs. 1 and 6). Women's Forum State Park on the old Columbia River Scenic Highway can be reached by taking the Corbett exit (No. 22) off Interstate 84. Proceed up the hill to the town of Corbett (Fig. 1) and the junction with the old Columbia

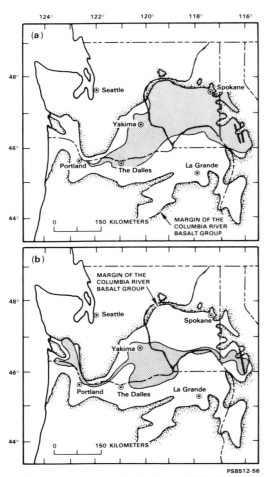

Figure 5. Maps showing the distribution of (A) the Priest Rapids Member of the Wanapum Basalt and (B) the Pomona Member of the Saddle Mountains Basalt. Known source dikes shown schematically. From Tolan and Beeson (1984).

graphc high that the Columbia River basalt flows failed to cover.

Below and to the west of Crown Point is Rooster Rock slide block (B, Fig. 6) and the Crown Point landslide (C, Fig. 6). The slide plane of the Crown Point landslide is probably the unconformable contact between the younger Priest Rapids intracanyon flow and earlier Columbia River basalt flows or the older volcanic rocks.

A remnant of the 12 Ma Pomona intracanyon flow (D, Fig. 6), along with the Grande Ronde flows which formed the canyon wall (E, Fig. 6), marks the northern margin of the east-west-trending Bridal Veil channel on the Washington side of the river. The eastern projection of the Bridal Veil channel into Oregon (I, Fig. 6) is not well seen from this location. However, an excellent view of the Oregon portion of the Bridal Veil channel will be seen at site 4. West of the Pomona exposure, post-Pomona lower member sandstones and conglomerates of the Troutdale Formation can be seen in a cliff face (F, Fig. 6).

Mount Zion (G, Fig. 6) is a Boring Lavas volcano that postdates the Troutdale Formation. A small basaltic-andesite intracanyon flow (H, Fig. 6) filling a channel cut into the top of the Troutdale Formation can be seen emanating from Mount Zion. Similar vents in the Portland, Oregon, area are inferred to be less than 730,000 years old.

Far up the gorge is Beacon Rock (J, Fig. 6), a post-Columbia River basalt volcanic neck.

Site 2. View Towards the East Side of Crown Point (Figs. 1 and 7).
To reach site 2 from site 1, proceed east on the old Columbia River Scenic Highway for 0.5 mi (0.8 km) to the Crown Point/Larch Mountain junction. Stay left on the old Columbia River Scenic Highway and travel 2.8 mi (4.5 km) to the Latourell Road junction. Turn left off the old Columbia River Scenic Highway onto Latourell Road and proceed about 0.3 mi (0.5 km) to the junction with Latourell Henry and Park streets. Turn left onto Latourell Henry Street and drive 0.3 mi (0.5 km) to the end of the street and park in the area before the private driveway.

From this vantage point, you can see a longitudinal cross section through the Priest Rapids intracanyon flow and the underlying bedded Priest Rapids hyaloclastite (Fig. 7). This relationship of an intracanyon flow overlying a thick accumulation of bedded hyaloclastite of like composition occurs along the entire known length of the ancestral canyon (Fig. 4). This relationship is unique among Columbia River basalt intracanyon flows and therefore requires some special circumstances for its origin (Tolan and Beeson, 1984, p. 464–470).

Other Columbia River basalt intracanyon flows that flowed down the river channel, from the headwaters towards the mouth, encountered little water in the channel except where tributary streams were dammed up by the advancing flow front. Subsequent flowage of lava into these tributaries encountered newly ponded water where pillow lavas and hyaloclastites were formed. In the case of the Priest Rapids intracanyon flow, the hyaloclastite displays crude bedding and other features (Tolan and Beeson, 1984, p. 464-467; Tolan and others, 1984b, p. 104) suggestive of

River Scenic Highway. Turn left (east) at this junction onto the old Columbia River Scenic Highway and travel 1.6 mi (2.6 km) to Women's Forum State Park (Fig. 1).

The view up the Columbia River from Women's Forum State Park (Fig. 6) shows two former paths of the ancestral Columbia River which are now intersected by the present-day river.

Crown Point (A, Fig. 6) is a portion of the Priest Rapids intracanyon flow (Fig. 4), which totally destroyed the second ancestral Columbia River channel through the Cascade Range about 14.5 m.y. ago. The single Priest Rapids flow which filled and overflowed the ancestral Columbia River canyon at Crown Point consists of 508 ft (155 m) of hackly jointed basalt which in turn overlies more than 198 ft (60 m) of bedded Priest Rapids hyaloclastite (see discussion at site 2). Here the pre-Priest Rapids river had cut its northwest-trending canyon at the contact between earlier Columbia River basalt flows and older Cascadian volcanic rocks. In the Crown Point area, these older rocks probably represented a volcanic edifice which formed a paleotopo-

Figure 6. View from site 1 - Women's Forum State Park. See text for explanation of labeled features.

debris flow-like processes indicating that the hyaloclastite was generated upstream from here and transported in a water mixture downstream. Also important to note is the lack of pillow lavas at the base of the Priest Rapids intracanyon flow indicating it did not encounter standing water in the canyon.

East of the head of the ancestral Columbia River canyon in the western Columbia Plateau (Fig. 4), this Priest Rapids flow has an extensive pillow complex at its base which overlies a thin lacustrine interbed. These features indicate the existence of a shallow lake just east of the head of the canyon. The relative positions of the lake, head of the river canyon, and the advancing lava flow front appear to have created a situation where pent-up lake waters were forced to flow across the lava flow front (Tolan and Beeson, 1984, p. 469–470). The escaping lake waters chilled the molten lava causing it to violently break up (phreatic brecciation) creating sand- to cobble-size fragments of glassy lava (hyaloclastite) that were carried in slurry-like surges into and down the ancestral Columbia River canyon where it accumulated. We estimate at least 2 mi^3 (8 km^3) of hyaloclastic debris was generated and flushed into the canyon in advance of the Priest Rapids lava flow drastically reducing the capacity of the canyon to contain the oncoming lava flow which resulted in the total obliteration of this course of the ancestral Columbia River.

Site 3. Multnomah Falls (Fig. 1). To reach Multnomah Falls from site 2, return to the Latourell Road/old Columbia River Scenic Highway junction. Turn left (east) onto the old

Columbia River Scenic Highway and travel approximately 6.6 mi (10.6 km).

Multnomah Falls is the best known and probably the most popular tourist stop in the gorge. It is the highest falls in the gorge, with a total drop of 620 ft (189 m) for the two falls. The cliff exposures here provide an excellent opportunity to examine many of the features commonly seen in Columbia River basalt flows.

Eleven low MgO Grande Ronde Basalt flows are exposed along Multnomah Creek, including one N_1 flow, five R_2 flows, and five N_2 flows. Six of these flows crop out from the river to the top of the upper falls; the rest are exposed upstream above the falls.

The Grande Ronde Basalt was erupted in a relatively rapid series of flows from fissures in the eastern portion of the Columbia Plateau between 16.5 to 15.6 Ma (see Reidel and others, this volume). Although pillow basalts occur at the base of some of the flows here, no low MgO Grande Ronde flows are known to occur in this area as intracanyon flows. This is probably the result of insufficient time between eruptions to allow the development of a stream canyon this far from the sea. However, at Latourell Falls, an N_2 high MgO Grande Ronde Basalt flow filled a small canyon eroded at the contact between the earlier Columbia River basalt flows and the older volcanic rocks (Tolan and others, 1984b, p. 106).

Another feature of interest here is the variety of cooling joint

Figure 7. View of the east side of Crown Point from site 2. The Priest Rapids intracanyon flow consists of hackly jointed entablature (A) and a thin basal colonnade (B) which directly overlie a thick, bedded, Priest Rapids hyaloclastite (C). Note the lack of a pillow basalt complex at the contact between the flow and the hyaloclastite. See text for discussion of these features.

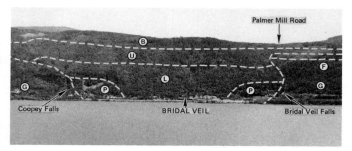

Figure 8. View of the natural cross section through the Bridal Veil channel as seen from site 4. Here, the ancestral Columbia River incised a canyon into Frenchman Springs (F) and Grande Ronde (G) flows that was more than 800 ft (244 m) deep. Remnants of the Pomona Member intracanyon flow (P) indicate that it only partly filled the Bridal Veil channel about 12 Ma. In post-Pomona time, the ancestral Columbia River began again to deposit lower member Troutdale (L) sands and gravels. The onset of Boring/High Cascade volcanism 6 to 4 Ma. is reflected in the composition of the upper member Troutdale (U) sediments. The contact between the upper and lower members is gradational and shown in a generalized manner. The total thickness of the Troutdale sediments exposed here exceeds 1,100 ft (335 m). The Boring/High Cascade flows (B) also capped this channel and prevented the ancestral Columbia River from reoccupying this former course when the Cascades were uplifted.

patterns displayed within the flows as well as the steep-walled topography that results from erosion controlled by these joints. The presence of prominent vertical cooling joints in most of these flows, combined with the weak interflow zones (e.g., vesicular flow tops) result in steep cliffs and abundant waterfalls. One may imagine that the ancestral canyons of the Columbia River appeared much like this except that the uplands were relatively flat and wide prior to the numerous eruptions of Cascadian volcanoes whose products presently dominate the higher topography.

Observations of waterfalls occurring over Columbia River basalt flows have shown that falls often occur where flows are flat lying or dipping upstream. This condition allows blocks produced by vertical cooling joints to be stable until support is withdrawn by erosion of the weaker interflow material at the base of the flows. The rate of erosion of interflow material probably largely controls the rate of retreat of the falls. The two falls are produced here because of a more easily eroded zone at the base of the upper falls. Furthermore, the amphitheater-shaped valley common to many of the falls within the gorge is due to freeze-thaw action of water from splash mist that penetrates the joints.

The catastrophic floods, the last of which roared down the gorge about 13,000 years ago, must have scoured most of the loose soils and talus from the sides of the gorge as well as eroded some bedrock. The extent to which these floods produced the hanging valleys and waterfalls is debatable. We think the waterfalls would exist here even if no catastrophic floods had occurred.

Site 4. View of the Natural Cross Section Through the Bridal Veil channel (Figs. 1 and 8). To reach site 4 from Multnomah Falls, continue east on the old Columbia River Scenic Highway and rejoin Interstate 84 eastbound. Continue east on Interstate 84 for 8.3 mi (13.3 km) and take the Cascade Locks Exit. Cross over the Columbia River on the toll bridge. Turn left (west) onto State Highway 14. It is approximately 16.6 mi (26.6 km) from the bridge junction to site 4. Stop at the narrow turnout on the left side of the highway. Watch out for traffic!

The modern Columbia River has dissected a path oblique to the axis of the westerly trending Bridal Veil channel of the ancestral Columbia River. Figure 8 delineates the various features of the Bridal Veil channel that can be seen from this vantage point.

REFERENCES CITED

Anderson, J. L., 1980, Pomona Member of the Columbia River Basalt Group—An intracanyon flow in the Columbia River Gorge, Oregon: Oregon Geology, v. 42, no. 12, p. 195–199.

Beeson, M. H., Fecht, K. R., Reidel, S. P., and Tolan, T. L, 1985, Regional correlations within the Frenchman Springs Member of the Columbia River Basalt Group—New insights into the middle Miocene tectonics of northwestern Oregon: Oregon Geology, v. 47, no. 8, p. 87–96.

Beeson, M. H., Perttu, R., and Perttu, J., 1979, The origin of the Miocene basalts of coastal Oregon and Washington—An alternative hypothesis: Oregon Geology, v. 41, no. 10, p. 159–166.

Tolan, T. L., and Beeson, M. H., 1984, Intracanyon flows of the Columbia River Basalt Group in the lower Columbia River Gorge and their relationship to the Troutdale Formation: Geological Society of America Bulletin, v. 95, no. 4, p. 463–477.

Tolan, T. L., Beeson, M. H., and Vogt, B. F., 1984a, Exploring the Neogene history of the Columbia River—Discussion and geologic field guide to the Columbia River Gorge, Part I—Discussion: Oregon Geology, v. 46, no. 8, p. 87–97.

——1984b, Exploring the Neogene history of the Columbia River—Discussion and geologic field guide to the Columbia River Gorge, Part II—Road log and comments: Oregon Geology, v. 46, no. 9, p. 103–112.

The Wallowa Mountains, northeast Oregon

William H. Taubeneck, Department of Geology, Oregon State University, Corvallis, Oregon 97331-5506

LOCATION

The Wallowa Mountains are mostly within the Eagle Cap Wilderness Area, which is accessible only on foot or horseback. The only road that penetrates the mountains extends south from Lostine (Fig. 1) along the Lostine River and is continuously within 1,300 ft (400 m) of nearly all of the east contact of unit one (Fig. 1) of the Wallowa batholith; this road ends at location H in the east part of unit two.

INTRODUCTION

Pre-Tertiary rocks of the Wallowa Mountains are exposed within an elongated uplift of Miocene Columbia River Basalt in which arching preceded major northwest faulting along opposite sides of the mountains (Taubeneck, 1963). The Wallowa fault, on the northeast side of the uplift, has a maximum displacement of 6,600 to 7,900 ft (2,000 to 2,400 m). The uplift is bounded on the southwest by the Eagle fault. Arching is shown on the northwest by a gentle flexure of about 5° in Columbia River Basalt and on the east (southeast of Wallowa Lake) by dips in basalt of as much as 25°. Structural relations involving arching and faulting along the northern front of the Wallowa Mountains are visible from Oregon 82 and roads east of Joseph (Fig. 1).

Prebatholithic rocks of greenschist-facies regional metamorphism are the Late Triassic Clover Creek Formation, the overlying Martin Bridge Formation, and the Late Triassic to Early Jurassic Hurwal Formation. The Clover Creek Formation consists of basic volcanic flows, conglomerate, sandstone, and mudstone. The overlying Martin Bridge Formation is a reef-associated carbonate complex. The Hurwal Formation is mostly mudstone with minor limestone and sedimentary breccia (Nolf, 1966).

The Wallowa batholith is a Jurassic composite intrusion emplaced in a mafic-to-felsic sequence that commenced with many small gabbroic bodies, continued with four major units (Fig. 1) of zoned tonalite-granodiorite, and terminated with at least 22 small felsic masses, mostly of less than 0.4 mi² (1 km²) each. The batholith is well exposed in the east, but is overlain by Columbia River Basalt in the west. The composite Cornucopia stock (Taubeneck, 1964, 1967) in the southeast Wallowa Mountains (Fig. 1) is the most significant satellite. K-Ar and Rb-Sr dates and Sr isotopic compositions (Armstrong and others, 1977) of rocks from the batholith suggest an age of emplacement of about 160 Ma in a volcanic arc environment.

This text describes exposures of the Wallowa batholith (site 1) and dikes of the Chief Joseph swarm (Taubeneck, 1970) that were feeders for the Columbia River Basalt.

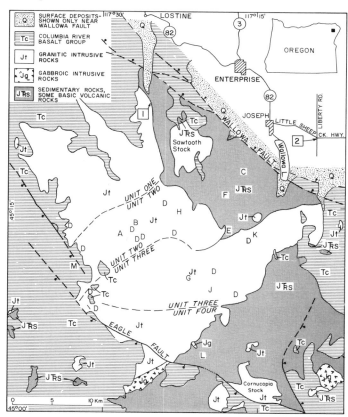

Figure 1. Generalized geologic map of the Wallowa Mountains. See text for discussion of map symbols, and for locations of sites 1 and 2 (in boxes) and lettered localities.

SITE ONE: WALLOWA BATHOLITH

Site 1 (Fig. 2) is the deep road cuts on either side of the Pole Bridge over the Lostine River, about 7.5 mi (12 km) south by road from the small town of Lostine (Fig. 1). The road is accessible to all vehicles. Pole Bridge is in Sec.15,T.2S.,R.42E., Enterprise Quadrangle, Wallowa County, Oregon.

SIGNIFICANCE

Granitic rocks along much of the eastern border of unit one (Fig. 1) of the Wallowa batholith are texturally and mineralogically distinctive among the zoned granitic intrusions of northeast Oregon. These border rocks are gneissic granodiorite with local mylonite gneiss and indicate intense forceful emplacement for this part of the batholith. Elsewhere in the Wallowa Mountains and other parts of northeast Oregon, border rocks of zoned intrusions are tonalite with planar structures that are less pronounced. Planar structure is absent in the interior of each zoned intrusion.

SITE INFORMATION

The granodiorite at Site 1 is gneissic with mafic minerals deflected around augen of plagioclase. Conspicuous deformational features characterize all thin sections. Twinning planes in many plagioclase are bent 20° or more. Recrystallization of plagioclase into a fine-grained mosaic is common along many crystal margins where plagioclase crystals impinge against one another at pressure points. Annealing occurred wherever either single fracturing or wholesale crushing accompanied plastic deformation. A few larger hornblendes are recrystallized into aggregates of 20 to 35 much smaller crystals, all within a lens-shaped margin of the original hornblende. Fine-grained biotite from 0.1 to 0.4 mm in length winds around "eyes" of plagioclase and hornblende in contrast to the much larger crystals of biotite that normally occur in border rocks of the batholith. An average modal analysis for five specimens (two thin sections each) from roadcuts on either side of the river is 10.0 percent orthoclase, 19.7 percent quartz, 47.1 percent plagioclase, 10.4 percent biotite, 12.0 percent hornblende, 0.2 percent augite cores in hornblende, and 0.6 percent accessories.

The road south of Lostine penetrates about 3 mi (5 km) into the northeast part of zoned unit two (Fig. 1). Accordingly, differences between chemical compositions and modal analyses of border rocks of quartz-poor tonalite and core rocks of granodiorite in unit two are recorded as examples of the variations within one of the four major zoned intrusions of the batholith. Border rocks chemically contain less than 1.6 percent K_2O and between 58 and 61 percent SiO_2, whereas core rocks (vicinity of locations A and B, Fig. 1) contain 2.3–2.6 percent K_2O and 66–67 percent SiO_2. Three border rocks collected within 164 ft (50 m) of the contact contain an average of 0.3 percent orthoclase, 17.5 percent quartz, 59.4 percent plagioclase, 10.4 percent biotite, 11.8 percent hornblende, and 0.6 percent accessories, whereas seven core rocks contain an average of 10.1 percent orthoclase, 25.7 percent quartz, 50.2 percent plagioclase, 8.0 percent biotite, 5.3 percent hornblende, and 0.7 percent accessories.

Emplacement of the Batholith

Structural features in country rocks also indicate that the batholith was forcefully emplaced. Relations are best documented in the northern Wallowa Mountains where the major structure is a strongly asymmetric syncline that trends north-northwest. The syncline was a comparatively open fold prior to emplacement of units one and two of the batholith. The east limb of the syncline dips westward at about 25° in the general vicinity of Chief Joseph Mountain (Fig. 1, location C). The west limb of

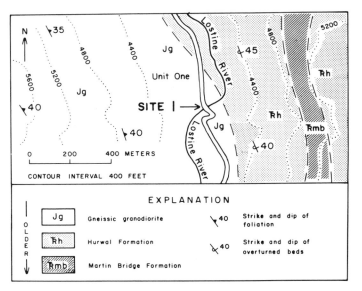

Figure 2. Geologic map of area near Pole Bridge.

the syncline is deformed into a tight series of isoclinal folds against units one and two of the batholith (Nolf, 1966).

Isoclinally folded sedimentary rocks were deflected about 1.9 mi (3 km) to the east along a cross fold during the forceful emplacement of unit two of the batholith (Nolf, 1966). Much eastward deflection is exposed in the canyon wall on the east side of the Lostine River and can be viewed from the Bowman Trail from 160 to 230 ft (50 to 70 m) before reaching the third switchback (elevation 6,000 ft; 1,800 m) from the well-marked trailhead with parking space. The trailhead is alongside the road about 14 mi (23 km) south of Lostine. Looking east from the Bowman Trail, marble of the Martin Bridge Formation is exposed through a vertical distance of 3,300 ft (1,000 m) from where it abruptly swings eastward at low elevations in the canyon to where it disappears eastward beyond the crest of the canyon wall.

Forceful emplacement of units two and three is dramatically demonstrated by the tremendous thickness of marble between locations E and F (Fig. 1). More than 6 mi^2 (15 km^2) of marble occur in this area. The marble accumulated plastically by eastward deformation of the Martin Bridge Formation during emplacement of unit two, followed by northward concentration of marble during emplacement of unit three. Undeformed sections of the Martin Bridge Formation are about 1,150 ft (350 m) thick (Nolf, 1966).

Dismembered screens of metasedimentary rocks that may include a few small gabbroic bodies extend for as much as several kilometers into the batholith between and along the contacts of the four major units. Border rocks of the four units are so similar that no intrusive contacts between the four units were detected more than 2.2 mi (3.5 km) into the batholith. The approximate location of the three contacts to the west and southwest of the last sizeable (about 130 ft; 40 m long) metasedimentary inclusions

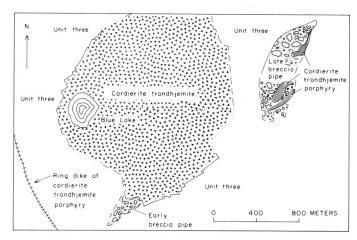

Figure 3. Geologic map of area near Blue Lake.

was determined (Taubeneck, 1985) by (1) higher mafic index of border tonalite, (2) stronger planar structure, (3) small gabbroic masses in the contact zone, (4) inclusions of comb-layered rocks (Moore and Lockwood, 1973) that originally were part of the gabbroic intrusions, and (5) occasional small xenoliths of recognizable country rocks.

Hundreds to thousands of metasedimentary xenoliths in 14 dikes of Columbia River Basalt within the batholith suggest that much of the batholith is underlain by metasedimentary rocks. Most xenoliths are well-bedded and many are fine-grained. The xenoliths are not gneissic or migmatitic, nor are any micas or amphiboles present to define or enhance a foliation. The mafic mineral in almost all assemblages is a clinopyroxene that varies from diopside to hedenbergite. Orthopyroxene occurs locally. Wollastonite is the other most distinctive mineral. Tight isoclinal folds in some xenoliths are very similar to folds of hand-specimen size that occur in metasediments of the Hurwal Formation near the contact of the batholith and in the dismembered screens between major units of the batholith. Locations of 12 dikes are denoted on Figure 1 by the letter D. The other two dikes are farther west within a small "window" of granitic rocks surrounded by Columbia River Basalt.

Cordierite in Granitic Rocks

Granitic rocks of the Wallowa Mountains include a suite of unique cordierite trondhjemite with associated comb-layered rocks, explosion breccias, porphyries, and local mineralization. Magmatic cordierite occurs in trondhjemite, trondhjemite porphyry, pegmatite, and quartz veins with or without sulphides and magnetite.

Early cordierite trondhjemites in the Cornucopia stock (Taubeneck, 1964) evolved into later ones within the batholith such as the intrusion at Blue Lake (Fig. 1, location G). This lake is about 8 mi (13 km) by trail from the campground (Fig. 1, location H) at the end of the road south of Lostine. A pegmatitic

contact distinguishes the Blue Lake cordierite trondhjemite from other cordierite-bearing intrusions and nearly all epizonal plutons (Buddington, 1959). The contact (Fig. 3) is exposed almost continuously southeast of Blue Lake where it is commonly accentuated by a brownish stain from the oxidation of local concentrations of magnetite. An average modal analysis for 13 rocks (4 thin sections per specimen) from the Blue Lake cordierite trondhjemite is 34.3 percent quartz, 54.0 percent plagioclase, 4.6 percent orthoclase, 3.5 percent muscovite, 1.9 percent biotite, 1.1 percent cordierite, and 0.6 percent accessories.

Euhedral cordierite crystals are hexagonal in shape and are commonly surrounded by a rind of biotite. Cruciform twins are rare. Most crystals of cordierite are pseudomorphs of muscovite and lesser amounts of green biotite. Cordierite varies in length from less than 1 mm to as much as 5 cm. Size, crystal habit, and dark color make cordierite the most conspicuous mineral in the leucocratic rocks. Cordierite crystallized very early and 16 crystals in the 52 thin sections are enclosed within plagioclase; some are near the center of plagioclase.

Quartz also crystallized early, and about half of the 52 thin sections contain one or more bipyramidal crystals from 5 to 8 mm across. Most large crystals are somewhat rounded or egg shaped rather than bounded by perfect pyramidal faces. Early crystallization of quartz is substantiated by many excellent bipyramidal crystals in chemically similar intrusions of cordierite trondhjemite porphyry (Fig. 3) about 1 mi (1.7 km) east of Blue Lake.

The small intrusions of cordierite trondhjemite porphyry in the breccia pipe (Fig. 3) east of Blue Lake contain visible pyrite and more than five times as much Cu as the average of 20 ppm for the Blue Lake cordierite trondhjemite. Open spaces occur in the upper half of the breccia pipe and are lined with crystals of epidote, quartz, and pyrite. Fragments in the pipe are angular, unlike some fragments in the explosion breccia (Fig. 3) south of Blue Lake, which were rounded to pebble and cobble shapes during fluidization (Reynolds, 1954).

A concentrated area of cordierite trondhjemite, cordierite trondhjemite porphyry, and explosion breccia about 2 mi (3.5 km) southeast of Blue Lake is centered on the ridge (Fig. 1, location J) midway between Hidden Lake and Frazier Pass. The explosion breccia is without open spaces and includes fragments of contact schists of the hornblende-hornfels facies from beneath the batholith. The various trondhjemites are encicled by inward-dipping quartz veins that contain cordierite and rare andalusite. The veins are as much as 3 ft (1 m) wide and occur as cone sheets in a circular zone with a maximum diameter of about 0.6 mi (1 km). Well within the zone of quartz veins, and cutting the cordierite trondhjemite, is a ring dike of massive quartz as much as 80 ft (25 m) wide. Near the outer margin of the zone of quartz veins is an incomplete ring dike up to 2 ft (0.6 m) wide of quartz, pyrite, and molybdenite partly defined by prospect pits and dumps that date from the early 1900s. All rocks within the batholith that contain cordierite are enclosed by a ring dike of trondhjemite porphyry, which also contains cordierite (Fig. 3).

Cordierite is cited by White and others (1986) as the most diagnostic mineralogical indicator of S-type granitic rocks, but the cordierite trondhjemites of the Wallowa Mountains have many fundamental features of I-type rocks (Chappel and White, 1974). The peraluminous character of two rocks of the Blue Lake intrusion is indicated by normative corundum of 1.62 and 1.89 percent and A/CNK ratios of 1.18 and 1.16. These values are consistent with those of S-type rocks. However, low K_2O/Na_2O ratios of 0.28 and 0.36 are those of I-type rocks, in contrast to high ratios of more than one for S-type rocks (White and Chappell, 1983). Fe_2O_3/FeO ratios of 1.00 and 0.98 are even higher than in most I-type rocks and in marked contrast to the low ratios of S-type rocks (White and Chappell, 1983). The opaque oxide in the Wallowa rocks is magnetite rather than ilmenite, which characterizes S-type rocks (White and others, 1986). The initial $^{87}Sr/^{86}Sr$ ratio is low (R. L. Armstrong, written communication, 1985) in contrast to the high values of S-type rocks (White and Chappell, 1983). In summary, not all peraluminous granitic rocks containing cordierite have the other characteristics of S-type rocks.

Comb-Layered Rocks of Cordierite and Quartz

The most remarkable igneous rocks of the Wallowa Mountains are sequences of comb-layered rocks near the contacts of the Crater Lake cordierite trondhjemite (Taubeneck, 1964) of the Cornucopia stock (Fig. 1). Comb-layering in igneous rocks (Taubeneck and Poldervaart, 1960; Moore and Lockwood, 1973) is characterized by the orientation of highly elongated crystals that are roughly perpendicular to alternating layers almost always of plagioclase and of one or more mafic minerals. The crystals commonly have curved, branching, and feathery morphologies, suggesting growth into a layer of super-cooled melt ahead of the solidification front (Lofgren and Donaldson, 1975). The growth direction as determined by the widening and branching of crystals is invariably toward the interior of the intrusion. Comb-layered rocks occur either at or near (generally within 330 ft; 100 m) intrusive contacts. The proximity of comb-layered sequences to contacts is a useful guide in unraveling complicated intrusive histories involving closely related magma pulses (Taubeneck, 1964; Shannon and others, 1982).

Quartz and cordierite were the high temperature phases in the silica-rich peraluminous magmas from which cordierite trondhjemite crystallized. Therefore, these two minerals comprise most of the comb-layered sequences. Comb-layered rocks associated with cordierite trondhjemite are markedly different in mineralogy from layered sequences in intrusions that crystallized from magmas of nearly all other compositions because quartz crystallized instead of plagioclase, and cordierite instead of one or more of the mafic minerals in Bowen's reaction series. Monomineralic layers of either quartz or cordierite occur, but both minerals crystallized simultaneously during formation of most layers. Much cordierite is unaltered. The greenish to bluish gray colors of the cordierite make the layers readily visible.

Comb-layered rocks in Crater Lake cordierite trondhjemite are best observed near contacts with Cornucopia tonalite west of Pine Lakes (Taubeneck, 1964, Fig. 2). The layered sequences surround all or part of some inclusions of tonalite. Most layered sequences are from 2 to 28 in (5 to 70 cm) wide but nearly 6 ft (2 m) of layered rocks occur at the contact with tonalite about 1 mi (1.6 km) west of Pine Lakes. At this locality, in the last few centimeters of the sequence, plagioclase rather than cordierite crystallized with quartz. Intrusion and brecciation of the layered rocks by cordierite trondhjemite commonly prevents tracing any sequence in the intrusion for distances of more than 130 ft (40 m). Autoliths are common and widely dispersed.

The average grain size in most layers is from 0.2 mm to 5.0 mm; variations between adjacent layers may be extreme. Cordierite in a few layers is as much as 12 mm long and reaches lengths of 18 mm in one layer. Quartz generally is not more than 3 mm long, but elongation ratios commonly are 4 to 7 and a few crystals have ratios of as much as 12. Crystals of cordierite in monomineralic layers are highly elongated but rarely show growth direction. Most cordierite associated with quartz, however, shows directional features such as (1) gradual widening of crystals from base toward extremities, (2) tree-like branching in which as many as four branches rather commonly extend upward and outward, and (3) rapid flaring outward in cone-shaped forms.

Significant recent contributions pertaining to the crystallization of comb-layered rocks include those of Donaldson (1977), Lofgren (1980), and Brigham (1983).

SITE TWO: DIKES OF COLUMBIA RIVER BASALT

Site 2 (Fig. 1) is at the intersection of Little Sheep Creek Highway and Liberty Road about 4 mi (6.4 km) east from Oregon 82 in the town of Joseph (Fig. 1). Site 2 is on the section line between Sec.35,T.2S.,R.45E. and Sec.2,T.3S.,R.45E., Joseph Quadrangle, Wallowa County, Oregon.

Significance

Dikes of Columbia River Basalt "are the most striking visible feature of the Wallowa range" (Swartley, 1914). The estimated 1,800 dikes in the Wallowa Mountains are part of the huge Chief Joseph swarm (Taubeneck, 1970), which extends northward for more than 165 mi (265 km) from the Snake River Plain in westernmost Idaho. At least 40 percent of the dikes in this swarm are in the Wallowa Mountains. Dikes of this swarm were the major feeders for Columbia River Basalt.

Site Information

The most conspicuous dikes visible (morning hours only) from Site 2 are those near the summit of Craig Mountain (Fig. 1, location K) at a distance of about 9.5 mi (15.3 km) to the S31°W. Observers can locate the summit of Craig Mountain by looking south-southwest into the mountains along the only deep canyon that appears on the skyline. The summit is the isolated peak on the skyline on the west side of the slot created by the canyon. Dikes of basalt are the nearly vertical dark stripes that cut the light-colored granitic rocks of the peak. Closer observation of the dikes is possible from Oregon 82 near the south end of Wallowa Lake.

Dikes are much more numerous in pre-Tertiary rocks than in Columbia River Basalt, which indicates that thousands of dikes (most of the feeders) of the Chief Joseph swarm are concealed in the many areas north and south of the Wallowa Mountains where the plateau basalt is relatively undissected. Dike density in pre-Tertiary rocks in the Wallowa Mountains commonly is 2 to 7 dikes per mi^2 (up to 3 per km^2), but some areas contain only about one dike per mi^2 (less than 1 per km^2) whereas others contain more than 10. There is no single axis of dike eruption in any part of the mountains such as near Cornucopia (Waters, 1961).

Many Cenozoic structural features of Washington and Oregon, including the orientation of dike swarms of Columbia River Basalt, were attributed to north-south compression by Wise (1963). Only the extensional component of compression has been recognized in dike patterns in Columbia River Basalt in the Pacific Northwest, but the shear components are also represented in patterns within the Wallowa batholith and Cornucopia stock. The apparent absence of dikes in Columbia River Basalt along shear directions may reflect the difference in behavior in a stress field of a superincumbent stack of flows in contrast to massive basement rocks.

Most dikes in the Wallowa Mountains and elsewhere in the Chief Joseph swarm trend either north or north-northwest. A notable feature of dikes within the Wallowa Mountains is a group that trends about N55°W across the south part of the batholith. These dikes extend from country rocks near the batholith in the vicinity of location L to Columbia River Basalt near location M. Some dikes of this group are older than intersecting north-trending dikes, whereas others are younger. Dikes that trend N55°W are approximately parallel to the Wallowa fault, Eagle fault, and also the Brothers fault zone (Walker, 1969, 1974) that crosses much of central Oregon. Dikes with west-northwest trends strengthen the possibility of a tectonic relationship between the Wallowa fault block, western Snake River Plain (Mabey, 1976) and Brothers fault zone. Perhaps the relationship is somewhat as visualized by Lawrence (1976).

Crustal Assimilation in Basalt Petrogenesis

Isotopic studies of flows of Columbia River Basalt from widely scattered localities throughout the Columbia Plateau have suggested to some workers (McDougall, 1976; Carlson and others, 1981; Carlson, 1984) that crustal assimilation was a significant factor in the petrogenesis of the flood basalts. Melting and size reduction of inclusions of pre-Tertiary rocks in dikes within the Wallowa batholith indicate that relatively few xenoliths survive in flows. Dispersed inclusions which do survive are small and inconspicuous. Consequently, direct studies of the role of crustal assimilation in basalt petrogenesis will be mostly confined to the study of contaminated dikes.

Visible "cooking" of granitic wall rock along some dikes is attested by paralleling zones of altered rock as much as 6 ft (2 m) thick. Thin sections show as much as 50 percent melting in specimens of wall rock. Common xenocrysts of quartz and plagioclase within the outer 3 cm of some dikes verify wall rock contamination.

Many dikes in the batholith contain inclusions of granitic rocks, but they rarely exceed 5 percent by volume of the rock within any part of a dike. Contacts of many granitic xenoliths are irregular and some are even moth-eaten. Thin sections of one indistinct xenolith contain about 55 percent plagioclase and quartz of a medium-grained rock in an interlocking network of basalt. Disintegration of xenoliths apparently is a common process in contaminated dikes because many xenocrysts of quartz and plagioclase occur in specimens of basalt from areas that contain numerous inclusions. Rounded and embayed crystals of quartz include amoebalike shapes. Plagioclase xenocrysts are much less irregular but some have cuspate borders.

Metasedimentary inclusions from beneath the batholith comprise from 1 to 35 percent of either all or part of 14 dikes. Thousands of xenoliths occur along the lengths (up to 1 mi; 1.6 km) of three dikes. About 95 percent of the xenoliths are less than 8 in (19 cm) in greatest dimension; many are less than 1 in (3 cm). Most xenoliths are elongated with length to width ratios between 3 and 10. Megascopic features of the inclusions reveal extreme size reduction along and across bedding planes. Thin sections show fingers of basalt as narrow as 0.25 mm extending along bedding planes. Specimens of basalt from areas characterized by thousands of megascopic xenoliths contain microscopic lenses and slivers of metasediments that show chemical reaction with the magma.

Vugs that generally are lined with zeolites are a notable feature of dikes containing many inclusions. The vugs resemble the miarolitic cavities in some epizonal granitic rocks. The volatile constituents responsible for the vugs were concentrated in the basaltic magma during dehydration reactions associated with the conversion of unstable hydrous minerals of the xenoliths such as biotite and hornblende into stable anhydrous phases such as magnetite, hypersthene, and clinopyroxene. For example, the transformation of biotite into hundreds of very small cubes of magnetite, and hornblende into hypersthene and spongy clinopyroxene, are typical in granitic inclusions.

Field and petrographic observations indicate that crustal assimilation has modified the chemistry of some dikes within the Wallowa batholith. However, the relative number of contaminated to uncontaminated dikes in the Wallowa Mountains cannot be gauged until dikes are carefully examined in the manner of

Delaney and Pollard (1981) and Ross (1986). Dikes are not as simple as is suggested by nearly all maps. For example, dikes conventionally are mapped as straight lines (Pollard and Muller, 1976) regardless of whether the dike is segmented, slightly en echelon, or curved. Routine mapping procedures in which dikes are not walked from end to end ignore significant structural features of many dikes and do not necessitate the examination of common localities for the occurrence of inclusions, such as the vicinity of directional changes and dike terminations. In summary, not many dikes have been carefully examined either in the Wallowa Mountains or elsewhere in the Pacific Northwest.

The general location of nearly all dikes of Columbia River Basalt in Idaho, Washington, and Oregon has been determined in recent years through the efforts of many workers. Rewarding fields for future research include a focus on both structural and chemical features of the dikes including detailed studies of dikes that contain xenoliths.

REFERENCES CITED

Armstrong, R. L., Taubeneck, W. H., and Hales, P. O., 1977, Rb-Sr and K-Ar geochronometry of Mesozoic granitic rocks and their Sr isotopic composition, Oregon, Washington and Idaho: Geological Society of America Bulletin, v. 88, p. 397–411.

Brigham, R. H., 1983, A fluid dynamic appraisal of a model for the origin of comb layering and orbicular structure: Journal of Geology, v. 91, p. 720–724.

Buddington, A. F., 1959, Granite emplacement with special reference to North America: Geological Society of America Bulletin, v. 70, p. 671–747.

Carlson, R. W., 1984, Isotopic constraints on Columbia River flood basalt genesis and the nature of the subcontinental mantle: Geochimica et Cosmochimica Acta, v. 48, p. 2357–2372.

Carlson, R. W., Lugmair, G. W., and MacDougall, J. D., 1981, Columbia River volcanism; The question of mantle heterogeneity or crustal contamination: Geochimica et Cosmochimica Acta, v. 45, p. 2483–2499.

Chappell, B. W., and White, A.J.R., 1974, Two contrasting granite types: Pacific Geology, v. 8, p. 173–174.

Delaney, P. T., and Pollard, D. D., 1981, Deformation of host rocks and flow of magma during growth of minette dikes and breccia-bearing intrusions near Ship Rock, New Mexico: U.S. Geological Survey Professional Paper 1202, 61 p.

Donaldson, C. H., 1977, Laboratory duplication of comb layering in the Rhum pluton: Mineralogical Magazine, v. 41, p. 323–336.

Lawrence, R. D., 1976, Strike-slip faulting terminates the Basin and Range Province in Oregon: Geological Society of America Bulletin, v. 87, p. 846–850.

Lofgren, G. E., 1980, Experimental studies on the dynamic crystallization of silicate melts, *in* Hargraves, R. B., ed., Physics of magmatic processes: Princeton, New Jersey, Princeton University Press, p. 487–551.

Lofgren, G. E., and Donaldson, C. H., 1975, Curved branching crystals and differentiation in comb-layered rocks: Contributions to Mineralogy and Petrology, v. 49, p. 309–319.

Mabey, D. R., 1976, Interpretation of a gravity profile across the western Snake River Plain, Idaho: Geology, v. 4, p. 53–55.

McDougall, I., 1976, Geochemistry and origin of basalt of the Columbia River Group, Oregon and Washington: Geological Society of America Bulletin, v. 87, p. 777–792.

Moore, J. G., and Lockwood, J. P., 1973, Origin of comb layering and orbicular structure, Sierra Nevada batholith, California: Geological Society of America Bulletin, v. 84, p. 1–20.

Nolf, B. O., 1966, Structure and stratigraphy of part of the northern Wallowa Mountains, Oregon [Ph.D. thesis]: Princeton, New Jersey, Princeton University, 135 p.

Pollard, D. D., and Muller, O. H., 1976, The effect of gradients in regional stress and magma pressure on the form of sheet intrusions in cross section: Journal of Geophysical Research, v. 81, p. 975–984.

Reynolds, D. L., 1954, Fluidization as a geological process and its bearing on the problem of intrusive granites: American Journal of Science, v. 252, p. 577–614.

Ross, M. E., 1986, Flow differentiation, phenocryst alignment, and compositional trends within a dolerite dike at Rockport, Massachusetts: Geological Society of America Bulletin, v. 97, p. 232–240.

Shannon, J. R., Walker, B. M., Carten, R. B., and Geraghty, E. P., 1982, Unidirectional solidification textures and their significance in determining relative ages of intrusions at the Henderson Mine, Colorado: Geology, v. 10, p. 293–297.

Swartley, A. M., 1914, Ore deposits of northeast Oregon: Oregon Bureau of Mines and Geology, v. 1, no. 8, 229 p.

Taubeneck, W. H., 1963, Wallowa Mountain uplift, northeastern Oregon [abs.]: Geological Society America Special Paper 73, p. 69.

——, 1964, Cornucopia stock, Wallowa Mountains, northeastern Oregon; Field Relationships: Geological Society of America Bulletin, v. 75, p. 1093–1116.

——, 1967, Petrology of Cornucopia tonalite unit, Cornucopia stock, Wallowa Mountains, northeastern Oregon: Geological Society of America Special Paper 91, 56 p.

——, 1970, Dikes of Columbia River Basalt in northeastern Oregon, western Idaho, and southeastern Washington, *in* Columbia River Basalt Symposium, 2nd, Cheney Washington, 1969, Proceedings: Cheney, Eastern Washington State College Press, p. 73–96.

——, 1985, Elusive contacts within the Wallowa batholith, northeast Oregon: Geological Society of America Abstracts with Program, v. 17, p. 412.

Taubeneck, W. H., and Poldervaart, A., 1960, Geology of the Elkhorn Mountains, northeastern Oregon; Part 2, Willow Lake intrusion: Geological Society of America Bulletin, v. 71, p. 1295–1322.

Walker, G. W., 1969, Geology of the High Lava Plains Province, *in* Mineral and Water Resources of Oregon: Oregon Department Geology and Mineral Industries Bulletin 64, p. 77–79.

——, 1974, Some implications of late Cenozoic volcanism to geothermal potential in the high lava plains of south-central Oregon: Ore Bin, v. 36, p. 109–119.

Waters, A. C., 1961, Stratigraphic and lithologic variations in the Columbia River Basalt: American Journal of Science, v. 259, p. 583–611.

White, A.J.R., and Chappell, B. W., 1983, Granitoid types and their distribution in the Lachlan Fold Belt, southeastern Australia *in* Roddick, J. A., ed., Circum-Pacific plutonic terranes: Geological Society of America Memoir 159, p. 21–34.

White, A.J.R., Clemens, J. D., Holloway, J. R., Silver, L. T., Chappell, B. W., and Wall, V. J., 1986, S-type granites and their probable absence in southwestern North America: Geology, v. 14, p. 115–118.

Wise, D. U., 1963, An outrageous hypothesis for the tectonic pattern of the North American Cordillera: Geological Society of America Bulletin, v. 74, p. 357–362.

Mount St. Helens, Washington, with emphasis on 1980-85 eruptive activity as viewed from Windy Ridge

Michael P. Doukas and Donald A. Swanson, U.S. Geological Survey, Cascades Volcano Observatory, 5400 MacArthur Boulevard, Vancouver, Washington 98661

LOCATION AND ACCESSIBILITY

Mount St. Helens, in the Cascade Range of southwest Washington, is the centerpiece of the Mount St. Helens National Volcanic Monument. Windy Ridge, in the monument about 4.2 mi (7 km) northeast of the volcano, offers the only readily accessible (in 1986) viewpoint for observing many of the effects of the 1980-85 eruption. A series of paved and graveled roads, narrow and curvy but suitable for buses, leads to the ridge (Fig. 1). The area west of Windy Ridge is closed to public access in 1986; for details, contact monument headquarters (U.S. Forest Service, Rt. 1, Box 369, Amboy, WA 98601, telephone 206-247-5473). Roads between Randle and Swift Reservoir are usually closed by snow between early November and early June. Construction of an overlook on Johnston Ridge, 4.8 mi (8 km) northwest of the volcano, is planned for the late 1980s; this overlook, which offers a better view of the crater than does Windy Ridge, will be accessible by road from Castle Rock but probably not from Windy Ridge.

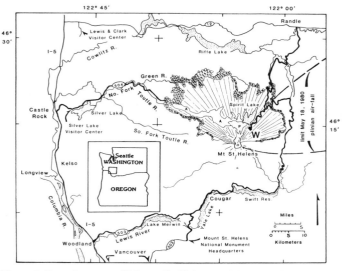

Figure 1. Location map of Mount St. Helens and devastated area. Area of radiating lines, inner devastated zone where virtually every large tree was uprooted or broken off, branches stripped, and trunks abraded and left alined in direction of local blast motion; heavy stippled area, scorch zone of devastated area where vegetation was left standing but killed by heating.

SIGNIFICANCE

The May 18, 1980, eruption of Mount St. Helens had great impact on the region. The landscape near the volcano, a play-ground for thousands, was vastly changed. Tephra fallout affected tens of thousands of downwind residents. People were re-awakened to the volcanic origin of the Cascade Range and reminded—or taught—that eruptions are commonplace.

The eruption had great impact on volcanology as well. Recognition of the consequences of sudden depressurization, caused by landslide, of a shallow magmatic-hydrothermal system spawned reinterpretation and .discovery at many volcanoes worldwide. Many pioneering studies have investigated the complex processes that generated the May 18 landslide and debris avalanche, lateral blast, plinian column, pyroclastic flows, and lahars. Other studies deal with later small explosions and with the growth of a dacite dome in the new crater. Most of the events were observed in unprecedented detail; consequently, interpretative studies are unusually well constrained. The devastated area presents unmatched opportunities for long-range geomorphic and hydrologic investigations as well as biologic studies related to floral and faunal repopulation.

SITE INFORMATION

Past Activity

Mount St. Helens is the youngest and most studied Cascade volcano. The cone formed mostly within the last 1,000 years but caps an older volcanic center active more than 40,000 years to 2,500 years ago (Hoblitt and others, 1980; Mullineaux and Crandell, 1981). Before 2,500 years ago, the volcano repeatedly erupted dacitic pyroclastic flows, tephra, and domes (Table 1). Since then, more mafic lava including basalt has also been erupted. Lava flows extend 9.6 mi (16 km) from the volcano, pyroclastic flows 11.4 mi (19 km), and lahars 33.6 mi (56 km). Ash from past eruptions is widespread; tephra set Y occurs more than 580 mi (965 km) away in Alberta.

During the last 4,500 years, Mount St. Helens has been more active and explosive than any other volcano in the conterminous United States (Crandell and Mullineaux, 1978). Many dated deposits (Crandell and others, 1981) document the high frequency of eruptions. Before the current activity, Crandell and others (1975) suggested from past history that the volcano could erupt soon, possibly before the end of the century.

Eruption of 1980-1985

The 1980 activity of Mount St. Helens, described in U.S. Geological Survey Professional Papers 1249 (Foxworthy and Hill, 1982) and 1250 (Lipman and Mullineaux, 1981), began on

TABLE 1. SUMMARY OF ERUPTIVE HISTORY OF
MOUNT ST. HELENS
(After Mullineaux and Crandell, 1981; Crandell and others, 1981).

Eruptive Period	Approximate Age (years before present)	Eruptive Products[a] Tephra Sets[e] (air-laid [t][b]; soils [s])
Modern		Dome (dac),pyroclastic flows, avalanche, and a lateral blast. 1980 tephra
Goat Rocks	[c]180-123	Dome (dac) lava flows. T(A.D. 1800) tephra
Dormant interval of about 200 yr*		
Kalama	[d]500-350	Dome (dac), pyroclastic flows, lava flows (and) X tephra W (A.D. 1482) tephra
Dormant interval of about 650 yr [s]		
Sugar Bowl	1,150	Dome (dac), pyroclastic flows, deposits of a lateral blast. Sugar Bowl tephra
Dormant interval of about 550 yr		
Castle Creek	>2,200-1,700	Lava flows (bas, and) pyroclastic flows. B tephra
Dormant interval of about 300 yr [t]		
Pine Creek	3,000-2,500	Domes, (dac), pyroclastic flows. P tephra
Dormant interval of about 300 yr [s] [t]		
Smith Creek	4,000-3,300	Pyroclastic flows. Y tephra
Dormant interval of about 4000 yr [s]		
Swift Creek	13,000-8,000	Domes (dac), pyroclastic flows. J tephra; S tephra
Dormant interval of about 5000 yr [s] [t]		
Cougar	21,000?-18,000?	Domes (dac), pyroclastic flows and a few lava flows (dac, and) K tephra; M tephra [s] [t]
Dormant interval of about 15,000 yr [s] [t]		
Ape Canyon	50,000-36,000?	Pyroclastic flows. C tephra [s] [t]

Note: dac=dacite; and=andesite; bas=basalt;
    > = greater than.
[a]Lahars formed during many eruptive periods but are not listed here.
[b]Fine-grained deposits [t] interbedded with tephra have undetermined origin.
[c]Years before 1980, based on tree-ring dates (Yamaguchi, 1983) and historical records.
[d]Years before 1980, based on tree-ring dates (Yamaguchi, 1983) and 14C dates.
[e]Tephra sets W and Y subdivided in text into separate east and north subsets (We, Wn and Ye, Yn).
*Dormant intervals are periods during which no eruptive products from Mount St. Helens have been documented.

March 15 with small earthquakes beneath the volcano. Seismicity picked up markedly on March 20. The first phreatic eruption occurred on March 27 coincident with strong seismicity. A summit crater formed and continued to enlarge for two months as phreatic activity continued. All tephra from this time was derived from old rock. However, viscous magma was intruding high into the cone, forming a cryptodome whose surface manifestation was the infamous "bulge" of the north flank. The bulge grew outward at a maximum rate of 8.25 ft/d (2.5 m/d) with no acceleration or other recognized change until the climactic eruption.

The May 18 eruption at 0832 local time was apparently triggered by a magnitude 5.1 earthquake that dislodged the unstable north flank into three great landslides. The slides evolved into a debris avalanche that sped 15 mi (25 km) in about 10 min down the valley of the North Fork Toutle River. Unloading of the volcano by the slides relieved confining pressure on the cryptodome and its associated hydrothermal system; the depressurized gases rapidly expanded, generating a north-directed lateral explosion or blast. A pyroclastic surge or flow fed by the blast fanned outward, felling trees and killing most wildlife in a 220 mi^2 (550 km^2) area. Two columns convectively rose from the devastated area and by 0900 joined to reach a height of 15 mi (25 km).

The landslides and blast removed the upper 1,320 ft (400 m) of the cone and left a crater 2,060 ft (625 m) deep, 1.6 mi (2.7 km) long, and 1.2 mi (2.0 km) wide. About 30 min after the blast, debris falling from the crater wall and lesser vesiculating dacitic magma from the roots of the cryptodome were explosively ejected in an eruption column that varied between 8.4 and 9.6 mi (14 and 16 km) high throughout the morning. Dark gray ash, largely lithic debris from this column, fell more than 900 mi (1,500 km) away. The column became lighter in color and more energetic at about noon, possibly as a fresh supply of gas-rich magma reached the surface, and pumiceous pyroclastic flows spilled northward from the crater and covered part of the debris avalanche, forming the pumice plain. Light gray magmatic ash from this 8.4-11.4 mi (14-19 km) column fell on older dark gray lithic ash in eastern Washington and northern Idaho. The activity declined and ended that night.

Many mudflows formed on May 18, mostly by melting of snow and glacier ice. The largest, down the North Fork Toutle River, formed as the debris avalanche dewatered (Janda and others, 1981; Voight and others, 1983). This flow destroyed or heavily damaged 200 homes and deposited more than 94 million yd^3 (72 million m^3) of sediment in the Cowlitz and Columbia Rivers, where clogged shipping channels required costly dredging (Schuster, 1981).

Five explosive eruptions between May 18 and October 18, 1980, produced pyroclastic flows and tephra falls, and three ended with emplacement of a small dacite lava dome. From October 18 until at least June 1985, a composite dome grew in an episodic but predictable way (Table 2; Swanson and others, 1983).

Route to Windy Ridge

The route to Windy Ridge provides good exposures of older deposits of Mount St. Helens and lahars, tephra falls, and blast deposits from May 18, 1980. It starts in Cougar, a 30-min drive east of Woodland (Fig. 2). Cougar is the last chance for fuel and food for 165 mi (260 km). The trip requires a full day.

Tumuli of the 1,900 yr Cave Basalt (Castle Creek eruptive

TABLE 2.   ERUPTIVE ACTIVITY OF MOUNT ST. HELENS
DURING 1980-1985

Date	Dominant Activity	Other Products
1980 March	Explosive	Lithic ash
April	Explosive	Lithic ash
May	Explosive	Avalanche; lateral blast; Lahar; Pyroclastic flow May 18 tephra
May	Explosive	Pyroclastic flow; May 25 tephra
June	Explosive Dome growth	Pyroclastic flow; June 12 tephra
July	Explosive	Pyroclastic flow; July 22 tephra
August	Explosive Dome growth	Pyroclastic flow; August 7 tephra
October	Explosive Dome growth	Pyroclastic flow October 17 tephra
December	Dome growth; minor explosive	
1981 February	Dome growth	
April	Dome growth	
June	Dome growth	
September	Dome growth	
October	Dome growth	
1982 March	Dome growth; minor explosive	Lahar; March 19 tephra
April	Dome growth	
May	Dome growth	
August	Dome growth	
1983 February through December	Dome growth (continuous); minor explosive	Lahar
1984 January	Dome growth (end of continuous growth)	
February	Dome growth	
March	Dome growth	
May	Minor explosive	Lahar
June	Dome growth	
September	Dome growth	
1985 May	Dome growth	

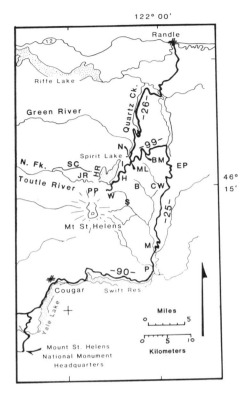

Figure 2. Field trip route from Cougar to Randle via Windy Ridge. Route follows USFS roads 90, 25, 99, and 26. P, Pine Creek bridge; M, Muddy River bridge; CW, Clearwater Creek basin; EP, Elk Pass; BM, Bear Meadows; ML, Meta Lake; N, Norway Pass trailhead; I, Independence Pass; B, Bean Creek; H, Harmony basin; S, Smith Creek; W, Windy Ridge parking lot; PP, Pumice Plain; JR, Johnston Ridge; HR, Harry's Ridge; SC, South Fork Coldwater Creek.

period; Table 1) occur along the road 3 mi (4.8 km) east of Cougar. The Cave Basalt is tube-fed pahoehoe erupted from vents on the south flank of Mount St. Helens, 7.8 mi (13 km) away (Greeley and Hyde, 1972).

The road crosses a power canal 1 mi (1.6 km) beyond the tumuli and climbs through Cougar-age (Table 1) volcaniclastic deposits that once blocked the Lewis River. In early Cougar time, a debris avalanche similar to that of 1980 may have come from Mount St. Helens (Newhall, 1982). Its deposit forms reddish-pink outcrops along the road just west of the Swift Dam overlook.

The route follows the north shore of Swift Reservoir on Forest Service (USFS) road 90, passing cuts in pyroclastic flows of Cougar age, basalt and andesite flows of Marble Mountain (a Pleistocene shield volcano centered 3 mi (5 km) north of the road), and middle Tertiary volcanic rocks. At the east end of the lake is a fan of debris carried by the Muddy River and Pine Creek lahars of May 18, 1980, and modified by subsequent high flows.

Just beyond the USFS Information Center, continue straight on USFS road 25 to Pine Creek, which heads on the southeast flank of Mount St. Helens. A major lahar down Pine Creek on May 18 took approximately 28 min to travel 14.4 mi (24 km) from the cone. The lahar was 33 ft (10 m) deep at the bridge (Janda and others, 1981); a mudline can still be seen on trees upstream from the bridge.

At Muddy River, about 3.6 mi (5.8 km) beyond Pine Creek, shifting channels and other fluvial processes have destroyed much of the May 18 lahar deposit. Large boulders, mudlines on trees, and downstream-pointing "bayonet" trees remain upstream and downstream from the bridge. Gullies cut into the parking-lot fill expose a May 18 lahar deposit of clast-supported gravel.

Many exposures of airfall tephra (1980 and older layers) can be seen for the next 20 mi (32 km) along USFS road 25. About 1.2 mi (2 km) from the Muddy River bridge are the southern margin of tephra fallout from the July 22, 1980, eruption and the 8-in (20-cm) isopach for both the We and Ye tephra layers

(Table 1). In another 3.7 mi (6 km) is the southern margin of the tephra deposit of May 18, 1980, barely visible because of erosion and vegetation.

In 5 mi (8 km), good views above Clearwater Creek show a very small part of the area devastated on May 18. On the east side of the valley is the limit of tree blowdown, about 9.6 mi (16 km) from the volcano. Much of the fallen or damaged timber has been salvaged.

At Elk Pass, about 7 mi (11 km) farther, is the approximate thickness axis of the May 18 tephra fall. The deposit, here about 6 in (15 cm) thick, can best be seen by digging anywhere west of the road. Turn left on USFS road 99 about 3.9 mi (6.2 km) north of Elk Pass.

About 4.6 mi (7.4 km) farther, at Bear Meadow, Gary Rosenquist and Keith Ronholm each took a remarkable set of photographs of the early stage of the May 18 eruption. The following description is from eyewitness accounts in Rosenbaum and Waitt (1981):

One observer watching with binoculars saw the north side of the volcano get "fuzzy, like there was dust being thrown down the side." Several seconds later the north face began to slide. The "bulge was moving . . . the whole north side was sliding down." The first cloud appeared to form at the base of a "cirque-like" wall from which the bulge had moved. In about 20 sec the landslide was out of view behind a ridge.

At this time, two distinct clouds seemed to issue from separate vents. An extremely dark cloud grew vertically from the summit. A lighter cloud (the blast cloud), which seemed to come from the area vacated by the landslide, expanded uniformly except for a large "arm" that shot out to the north in the direction of the avalanche.

As the blast cloud grew, what appeared to be a shock wave similar to that associated with a nuclear explosion moved ahead of it. About 1½ min after the blast, a noise resembling a thunder clap was heard at Bear Meadow, accompanying a sudden pressure change. This noise was followed by continuous rumbling "like a freight train."

The blast cloud moved toward Bear Meadow, reaching the ridge nearest the mountain 25-30 sec after the start of the landslide. When the cloud hit the ridge, it rose and "boiled" upward. Just before the top of the mountain became obscured, observers saw the south side of the summit crumble into the hole formed by the landslide.

Seven or eight minutes after the eruption's start, rocks began falling on a witness 2-3 mi (3.2-4.8 km) northeast of Bear Meadow. He collected two "golfball-sized" rocks (determined later to be cryptodome dacite) from the fall, which continued for roughly 30 sec. As he drove, this material was replaced by "mud drops" that flattened on impact. The largest observed flattened mud drop was 19 mm across. Gradually mudfall abated, but fine ash fell more heavily until the witness could no longer see to drive. Ashfall then abated slowly.

The northeast edge of the sear or scorch zone occurs 2 mi (3.2 km) beyond Bear Meadow. The luxuriant forest contrasts strikingly with the blast-devastated landscape, although some of the area was already denuded by clearcutting before 1980. The road from here to Windy Ridge is continuously inside the devastated area. At the junction of roads 26 and 99 is a car destroyed on the morning of May 18; it belonged to a family of three killed at their small mine cabin on the ridge 1.5 mi (2.4 km) west of here.

Continue straight on USFS road 99 for 0.2 mi (0.8 km) to the trail to Meta Lake, an interesting 10-min walk through downed timber to a lake only moderately disturbed by the blast. Small trees and trout survived because of snow and ice cover. One blowndown tree near the start of the trail shows tightly clustered rings indicative of stressed growth for a few years following 1800 A.D., when tephra set T was erupted (Yamaguchi, 1983).

Thick tephra layers are exposed along road 99 from its junction with road 94 to the Windy Ridge parking lot (Fig. 3). Airfall tephra and blast deposit from 1980 overlie forest duff and tephra sets T (dacite pumice), X (thin andesite ash), and Wn (dacite pumice) (Table 1). A possible blast deposit related to emplacement of Sugar Bowl dome low on the northeast flank of the volcano underlies Wn. Tephra set B (andesite and basalt ash with one thin dacite layer) lies beneath the Sugar Bowl deposit and overlies set P (dacite ash), which in turn overlies layer Yn (yellow-brown dacite pumice) and deposits from set J.

Windy Ridge

The Independence Pass trail, 1.5 mi (2.4 km) beyond the junction of roads 99 and 94, is a rewarding 1,300-ft (400-m) walk through downed trees. At the end of the trail is a fine view of Mount St. Helens and Spirit Lake. The top of the lava dome in the crater can be seen. The Spirit Lake basin was covered by virgin coniferous forest before its denudation. A tack in a cut log along the trail marks tightly clustered annual rings of 1801-1805 A.D. that reflect stressed growth for a few years after pumice of set T fell. The blast deposit can be recognized by its gray color, poor sorting, and content of shredded wood. The hike provides a chance to examine diverse effects of the blast on trees. Small trees buried under snow were undamaged. Very few trees were transported or reoriented after falling. Standing trunks on and behind the ridge to the north indicate that the surge lofted but still snapped off their crowns.

Along the road, the first gap in the ridge beyond the Independence Pass trail overlooks Harmony Falls basin and the north end of Spirit Lake. Evidence that the avalanche displaced water from Spirit Lake can be seen across the lake. Logs are sparse in a zone extending far above lake level; those present commonly point downslope in contrast to those higher up, which were felled by the blast and point uniformly away from the volcano. Many logs washed from the slope now float in the lake; in 1980 much of the lake surface was covered by logs and woody debris. Waitt (1984) infers the following sequence of events:

(1) The surge rushed northeastward, leveling forest and depositing layer Al (Fig. 3). Patterns of downed timber indicate that topography partly channeled movement. Most of Spirit Lake basin was swept by a northeast-moving current, but the head of Harmony basin was engulfed by a current rushing northward out of Smith Creek and over the divide into the head of Harmony basin. This northward current then turned west into the basin.

(2) The first landslide, moving more than 150 mi/hr (250 km/hr), entered Spirit Lake. It created a catastrophic wave that

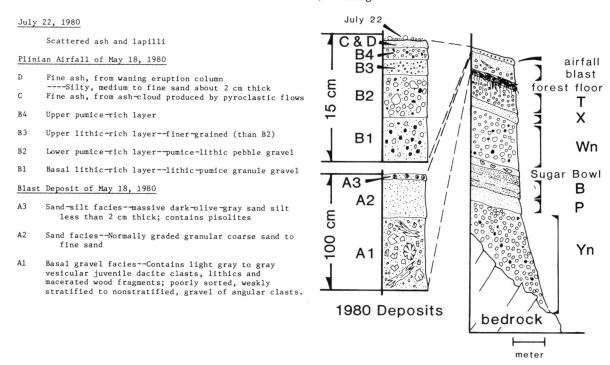

July 22, 1980

    Scattered ash and lapilli

Plinian Airfall of May 18, 1980

D   Fine ash, from waning eruption column
        ----Silty, medium to fine sand about 2 cm thick

C   Fine ash, from ash-cloud produced by pyroclastic flows

B4   Upper pumice-rich layer

B3   Upper lithic-rich layer--finer-grained (than B2)

B2   Lower pumice-rich layer--pumice-lithic pebble gravel

B1   Basal lithic-rich layer--lithic-pumice granule gravel

Blast Deposit of May 18, 1980

A3   Sand-silt facies--massive dark-olive-gray sand silt
        less than 2 cm thick; contains pisolites

A2   Sand facies--Normally graded granular coarse sand to
        fine sand

A1   Basal gravel facies--Contains light gray to gray
        vesicular juvenile dacite clasts, lithics and
        macerated wood fragments; poorly sorted, weakly
        stratified to nonstratified, gravel of angular clasts.

Figure 3. Schematic stratigraphic column of pyroclastic deposits from Mount St. Helens along USFS road 99 in devastated area. Airfall and blast deposit of 1980 in left column (Waitt, 1981; Waitt and Dzurisin, 1981) with detailed description of subunits. Relation of 1980 deposits to prehistoric deposits in right column.

surged up Harmony basin and rinsed slopes clean of timber and layer A1 to a level marked by the sharp trimline on the north valley side.

(3) Losing momentum, this water descended to the basin floor and poured back into the lake, carrying most logs with it but leaving some grounded in "rafts."

(4) The gradually waning pyroclastic current continued, depositing layer A2.

(5) Layer A3 (blast fallout) and later airfalls accumulated.

The next four gaps provide outstanding views of Spirit Lake basin, upper Smith Creek (east of Windy Ridge), and the volcano.

The Windy Ridge parking lot is at the end of the public road, about 4.2 mi (7 km) from the crater. The blast deposit here is about 3.3 ft (1 m) thick and veneered by 1980 airfall deposits; it is recognized from a distance by its distinctive gray color. It contains gravel- to sand-sized rock fragments from the edifice of Mount St. Helens, pieces of slightly vesicular gray dacite from the cryptodome, and fragments of shredded wood.

The best viewpoint is reached by a short climb to the top of the hill just north of the parking lot. The debris-avalanche deposit dams Spirit Lake and forms the hummocky topography in the distance (Fig. 4). On the south end of Harry's Ridge, the first ridge west of Spirit Lake, all trees and soil were removed down to bedrock by the avalanche and blast. Light gray pyroclastic-flow and ash-cloud deposits, mainly from May 18 but also from June

12 and July 22, 1980, cover and lap against the debris avalanche south of Spirit Lake. Many of these deposits have been stripped or buried by lahars generated by rapid snowmelt during explosions. Several fumaroles apparently rooted in the pyroclastic flows still emitted steam in 1985, visible best on humid days. Much of the northern part of the crater can be seen from here, but in April 1985, only the very top of the dome, 760 ft (230 m) high, 2,640 ft (800 m) wide, and located in the southern part of the crater, shows over the shoulder of the Sugar Bowl dome.

Water level in Spirit Lake is maintained by a gravity-fed tunnel cut through Harry's Ridge to South Coldwater Creek, which drains into the North Fork Toutle River. The portal is visible about halfway along the western shore of the lake. This spillway prevents overtopping and possible catastrophic erosion of the unstable debris-avalanche blockage. Between November 1982 and April 1985, Corps of Engineers contractors maintained lake level using barge-mounted pumps that withdrew water and sent it down a channel on the pumice plain.

A tiny shed high on Harry's Ridge houses surveying and radio equipment used by volcanologists of the U.S. Geological Survey to monitor volcanic activity. Most current monitoring is conducted within the crater and on the dome. The Geological Survey notifies the USFS and other governmental agencies when monitoring indicates significant changes in activity. All decisions regarding access and land use in Mount St. Helens National Volcanic Monument are made by the USFS.

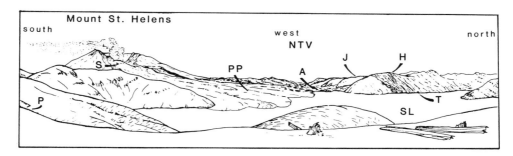

Figure 4. Panoramic view looking west from hill north of Windy Ridge parking lot. Volcanic gas and water vapor always rise from dome. During humid weather, water vapor condenses to form visible plumes from dome and from fumaroles in pyroclastic flows of Pumice Plain. P, Windy Ridge parking lot; J, Johnston Ridge; S, Sugar Bowl dome; H, Harry's Ridge; PP, Pumice Plain; SL, Spirit Lake; NTV, North Fork Toutle River Valley; T, Spirit Lake outlet tunnel; A, Debris-avalanche (Spirit Lake blockage). Drawing by Bobbie Myers.

Route North to Randle

Leave Windy Ridge and return to the junction of roads 99 and 26 at the damaged car. From here, either return to Cougar via the inbound route or turn northward on road 26 into a scenic area of blowdown. It is long but possible to drive to Cougar via roads 26 and 25, but it is easier to continue on road 26 to the intersection with U.S. Route 12 in Randle.

Meta Lake is visible southwest of road 26, 0.25 mi (0.4 km) beyond the junction. Good examples of downed timber and pre-1980 tephra are on the north side of the road. One mile (1.6 km) farther is the Norway Pass trailhead and a spectacular view of tree blowdown in the upper Green River basin. The downed timber is from a mature forest; some trees were more than 500 yr old. The blowdown here will be left to decay naturally.

At the junction of USFS roads 26 and 2612, turn left and drive a short distance to Ryan Lake. Deposits of 1980 tephra are present but thinner than at Windy Ridge. The blast deposit here is also thin, but nonetheless campers and horses at the lake died from asphyxiation caused by ingestion of fine ash accompanying the blast. Return to road 26 and turn left.

Road 26 continues down Quartz Creek, providing excellent views of the margin of the seared zone and effects of the blast on alder high on the east side of the valley. The transition to undamaged forest is remarkably abrupt where the road leaves the devastated area, about 13.5 mi (22.5 km) from the volcano. From here continue to Randle, about 24 mi (40 km) from the volcano, where burned branches and tephra fell on May 18. To return to Cougar, turn right at the intersection with road 25, 9.3 mi (14.9 km) from where road 26 leaves the zone of devastation.

REFERENCES CITED

Crandell, D. R., and Mullineaux, D. R., 1978, Potential hazards from future eruptions of Mount St. Helens volcano, Washington: U.S. Geological Survey Bulletin 1383–C, 26 p.

Crandell, D. R., Mullineaux, D. R., and Rubin, Meyer, 1975, Mount St. Helens volcano; recent and future behavior: Science, v. 187, p. 438–441.

Crandell, D. R., Mullineaux, D. R., Rubin, Meyer, Spiker, Elliott, and Kelley, M. L., 1981, Radiocarbon dates from volcanic deposits at Mount St. Helens, Washington: U.S. Geological Survey Open-File Report 81-844, 15 p.

Foxworthy, B. L., and Hill, Mary, 1982, Volcanic eruptions of 1980 at Mount St. Helens—The first 100 days: U.S. Geological Survey Professional Paper 1249, 125 p.

Greeley, Ronald, and Hyde, J. H., 1972, Lava tubes of the Cave Basalt, Mount St. Helens, Washington: Geological Society of America Bulletin, v. 83, p. 2397–2418.

Hoblitt, R. P., Crandell, D. R., and Mullineaux, D. R., 1980, Mount St. Helens eruptive behavior during the past 1,500 yr: Geology, v. 8, p. 555–559.

Janda, R. J., Scott, K. M., Nolan, K. M., and Martinson, H. A., 1981, Lahar movement, effects, and deposits, in U.S. Geological Survey Professional Paper 1250, p. 461–478.

Lipman, P. W., and Mullineaux, D. R., eds., 1981, The 1980 eruptions of Mount St. Helens, Washington: U.S. Geological Survey Professional Paper 1250, 844 p.

Mullineaux, D. R., and Crandell, D. R., 1981, The eruptive history of Mount St. Helens, in U.S. Geological Survey Professional Paper 1250, p. 3–15.

Newhall, C. G., 1982, A prehistoric debris avalanche from Mount St. Helens: EOS, v. 63, p. 1141.

Rosenbaum, J. G., and Waitt, R. B., Jr., 1981, Summary of eyewitness accounts of the May 18 eruption, in U.S. Geological Survey Professional Paper 1250, p. 53–67.

Schuster, R. L., 1981, Effects of the eruptions on civil works and operations in the Pacific Northwest, in U.S. Geological Survey Professional Paper 1250, p. 701–718.

Swanson, D. A., Casadevall, T. J., Dzurisin, Daniel, Malone, S. D., Newhall, C. G., and Weaver, C. S., 1983, Predicting eruptions at Mount St. Helens, June 1980 through December 1982: Science, v. 221, no. 4618, p. 1369–1376.

Voight, B., Janda, R. J., Glicken, Harry, and Douglass, P. M., 1983, Nature and mechanics of the Mount St. Helens rockslide-avalanche of 18 May 1980: Geotechnique, v. 33, p. 243–273.

Waitt, R. B., 1981, Devastating pyroclastic density flow and attendant air fall of May 18—stratigraphy and sedimentology of deposits, in U.S. Geological Survey Professional Paper 1250, p. 439–458.

—— 1984, Deposits and effects of devastating lithic pyroclastic density current from Mount St. Helens on 18 May 1980—Field guide for northeast radial: U.S. Geological Survey Open-File Report 84-839, 10 p.

Waitt, R. B., and Dzurisin, D., 1981, Proximal air-fall deposits from the May 18 eruption–stratigraphy and field sedimentology, in U.S. Geological Survey Professional Paper 1250, p. 601–628.

Yamaguchi, D. K., 1983, New tree-ring dates for recent eruptions of Mount St. Helens: Quaternary Research, v. 20, p. 246–250.

Lone Butte and Crazy Hills: Subglacial volcanic complexes, Cascade Range, Washington

Paul E. Hammond, *Department of Geology, Portland State University, Portland, Oregon 97207*

LOCATION AND ACCESS

Lone Butte and Crazy Hills, about 3,900 to 4,750 ft (1,200 to 1,450 m) in elevation, are located approximately 25 mi (40 km) north of Stevenson and 15 mi (25 km) southeast of Mount St. Helens, in Skamania County, Washington, about midway between Mount St. Helens and Mount Adams (Fig. 1). They are perched on the eastern edge of the Lewis River canyon and along the western margin of the Indian Heaven basaltic volcanic field. The area lies entirely within the Gifford Pinchot National Forest.

They can be reached by vehicle along Washington 504 and U.S. Forest Service roads 90 (the Lewis River Road) and 3211 about 60 mi (100 km) east of Woodland, and along the Wind River Highway and U.S. Forest Service roads 30 and 3211 about 30 mi (50 km) north of Carson and the Columbia River (Fig. 1). Well-graded, graveled U.S. Forest Service roads provide local access (Fig. 2). Road 32 travels the eastern side of Crazy Hills, 3211 runs along the southern side, and 3220 transects the hills north-south. Most interconnecting roads are shown in the Burnt Peak and Lone Butte 7½-minute quadrangles, which cover the area. Cautious driving is recommended because logging truck traffic is occasionally heavy. The area is generally snowbound and inaccessible to wheeled vehicles between the months of November and May. Although the area has been extensively logged in recent years, it is covered with low shrubbery, and the terrain is mantled by Mount St. Helens tephra and soil. Exposures of bedrock are limited primarily to road, stream, and isolated hillside cuts.

SIGNIFICANCE

Lone Butte and Crazy Hills, in the southern Washington Cascade Range, are subdued landforms, composed chiefly of palagonitized basaltic hyaloclastites and pillow lavas. They are remnants of subglacial volcanic complexes, which erupted during the Hayden Creek Glaciation, about 130,000 to 190,000 yrs ago. They are among the largest well-exposed and easily accessible of the few recognized occurrences of subglacial volcanic landforms in the Cascade Range within the conterminous United States.

SITE INFORMATION

Lone Butte and Crazy Hills are built atop the northwestern lava flows of the Indian Heaven basaltic volcanic field (Fig. 2). They rest chiefly upon the 3.75 ma basalt of Thomas Lake (Table 1). Lone Butte probably overlaps lava flows of the younger basalt of Rush Creek (Fig. 3). Northward, the Crazy Hills overlap strata of the Western Cascade group (Hammond, 1980). The lava flows of basalt of Lake Comcomly encircle the Crazy Hills along their southern side; the lava flows of basalt of Indian Heaven overlap the southern flank of the hills.

Drift of the Hayden Creek Glaciation, which was deposited

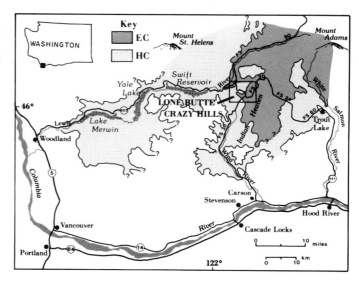

Figure 1. Map of southwestern Cascade Range in Washington, showing location of Lone Butte–Crazy Hills subglacial volcanic complexes. Maximum extent of glacier advance: HC, Hayden Creek Glaciation; EC, Evans Creek Glaciation. Area west of Swift Reservoir after Mundorff, 1984.

during formation of the complex (Table 1), is only locally preserved, having been covered by lava flows and eroded or covered by drift of the Evans Creek Glaciation. Hayden Creek drift is exposed between Curly Creek and road FS 51 where it separates basalt of Thomas Lake from basalt of Lake Comcomly (Fig. 2). Fluvial gravel deposits that could have been deposited by a jökulhlaup (a glacial outburst causing a flood) have not been found marginal to the lower part of the complex above the Lewis River. This area was erosionally modified by the Lewis River glacier during the younger Evans Creek Glaciation.

Drift of the Evans Creek Glaciation (Table 1) overlaps the subglacial volcanic complex on its eastern side (Stops 4, 5, and 6) and covers the lava flows of the basalts of Lake Comcomly and Indian Heaven (Fig. 2). Meltwater during the glaciation partly eroded the northern side of the Crazy Hills and fluvially deposited well-stratified beds of reworked hyaloclastite rich in lithic clasts (Stop 8).

Layers of Mount St. Helens J-, Y-, and W-tephra (Mullineaux, 1986; Table 1) are preserved atop drift of the Evans Creek Glaciation (Stops 7 and 8); whereas to the west atop higher parts of Crazy Hills, rodent-disturbed layers of S- and J- as well as Y- and W-tephra overlie less than 3 ft (1 m) of hyaloclastitic till of Hayden Creek Glaciation.

Lone Butte and Crazy Hills are good examples of the two

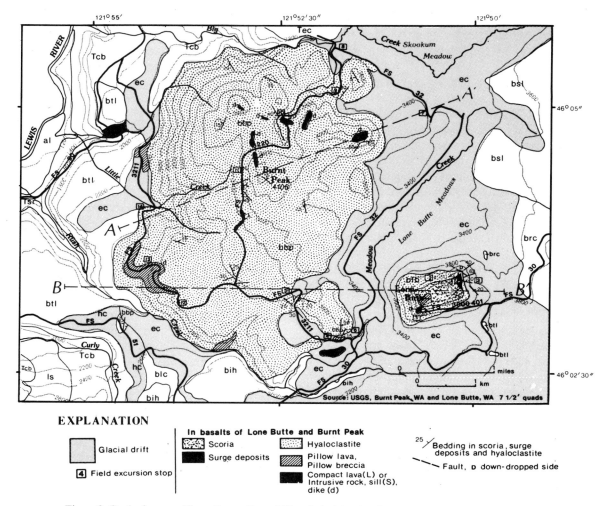

Figure 2. Geologic map of Lone Butte–Crazy Hills subglacial volcanic complex in southern Washington Cascade Range. For key to stratigraphic units see Table 1.

	In basalts of Lone Butte and Burnt Peak	
Glacial drift	Scoria	Hyaloclastite
4 Field excursion stop	Surge deposits	Pillow lava, Pillow breccia
		Compact lava(L) or Intrusive rock, sill(S), dike (d)

25 / Bedding in scoria, surge deposits and hyaloclastite

— Fault, □ down-dropped side

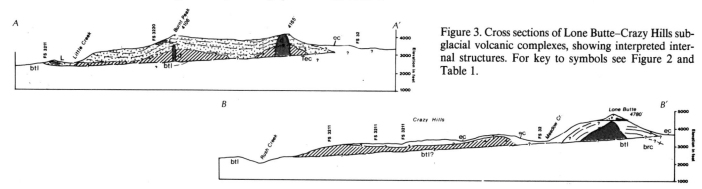

Figure 3. Cross sections of Lone Butte–Crazy Hills subglacial volcanic complexes, showing interpreted internal structures. For key to symbols see Figure 2 and Table 1.

common subglacial volcanic landforms, tuya and moberg hills, respectively. Both landforms are composed of pillow lava and hyaloclastite in varying proportion. A tuya is a volcano that erupts initially beneath a glacier, melts through the ice, and develops an upper, subaerial part, which commonly consists of a flat-topped form capped by a lava flow (Jones, 1969). Moberg is the Icelandic term for palagonite rock (Kjartansson, 1959). A moberg hill is a rounded landform composed chiefly of palagoniti-

zed hyaloclastite. Hyaloclastite is very fine-grained to coarsely fragmented rock, consisting of a high percentage of glass relative to crystalline rock fragments, generally of basaltic composition, and commonly poorly sorted. It is produced where lava flows or intrudes into water, ice, or water-rich sediment. The fragmentation is caused by one or several processes: (1) spalling of the glassy rinds of the pillows, (2) autobrecciation of the pillows by the continued flowage of unchilled lava, (3) multiple steam

TABLE 1. STRATIGRAPHIC UNITS IN THE LONE BUTTE-CRAZY HILLS AREA

Age (yr B.P.)	Symbol	Unit	Thickness (m)	Description
	al	alluvium	1-10	Unconsolidated fluvial and lacustrine deposits and drift in Lewis River valley
	ls	landslide	10-30	Unconsolidated, unstratified, unsorted debris
400		W-tephra of Mount St. Helens*	0.05-0.10	Light-gray, sand and fine pumice ash
3300-3500		Y-tephra of Mount St. Helens*	0.05-0.30	Light-gray yellow to orange-yellow pumice ash
10,700		J-tephra of Mount St. Helens*	0.10-0.15	Reddish-yellow pumice lapilli
12,910		S-tephra of Mount St. Helens*	0.10-0.15	Light-yellow orange fine pumice ash
ca 15,000-25,000[a]	ec	Drift of Evans Creek Glaciation	2-35	Poorly sorted and stratified detritus consisting of chiefly rounded clasts of diverse basalts in a purple-gray silt matrix, with an upper yellowish-brown oxidation zone up to 1 m thick
29,000[b]	bih	Basalt of Indian Heaven	2-28	Light-gray abundantly phyric olivine-plagioclase basalt
	blc	Basalt of Lake Comcomly	2-30	Dark-gray sparsely phyric plagioclase-olivine basalt with scattered glomerocrysts with olivine cores and plagioclase coronas
ca 130,000-	hc	Drift of Hayden Creek Glaciation	2-35	Poorly sorted and stratified detritus consisting of chiefly basalt and Western Cascade lithologies in a silt matrix oxidized to a yellowish brown
	bbp	Basalt of Burnt Peak	122-366	Medium-gray moderately phyric olivine-plagioclase basalt
-ca 190,000	blb	Basalt of Lone Butte	400	Dark-gray, dense, sparsely phyric olivine basalt
891,000 ±50,000[c]	bsl	Basalt of Surprise Lakes	60-80	Light-gray, dense, moderately phyric plagioclase-olivine basalt
	brc	Basalt of Rush Creek	>20	Medium-gray, dense, moderately phyric augite-olivine basalt
3,750,000 ±500,000[c]	btl	Basalt of Thomas Lake	35-150	Light-gray abundantly phyric augite-olivine-plagioclase basalt with randomly oriented platelets of plagioclase, 1-6 mm in size

--- MAJOR UNCONFORMITY ---

Western Cascade Group (Tec, Tcb, Tsr), volcanics and volcaniclastics >15 Ma.

---

* Descriptions and ages after Mullineaux (1986).
[a] Ages after Coleman and Pierce (1981).
[b] ^{14}C date provided by Meyer Rubin, USGS, Reston (written communication, April 1984).
[c] K-Ar date provided by E. H. McKee, USGS, Menlo Park (written communication, May 1984). Age of bsl is probably between 910,000 and 970,000 yrs. and age of btl is between 3,250,000-3,400,000 and 4,250,000-3,870,000, because lava flows have normal remnant magnetic polarity.

(phreatic) explosions caused by water trapped within the chilled lava, or (4) disruption and redistribution of the fragments of pillow lavas by subglacial streams. Much of the hyaloclastites of the subvolcanic complexes here have been generated and deposited by the latter process. Fresh hyaloclastite is rich in sideromelane (basalt glass) and is gray to black. Palagonite is hydrated hyaloclastite and is composed of abundant clay minerals; its colors range from brown through red and orange to yellow.

Lone Butte is a conical tuya, rising 1,400 ft (430 m) to an elevation of 4,780 ft (1,457 m), lying east of and isolated from the Crazy Hills (Fig. 2). It covers more than 1 mi² (2.5 km²) and contains 0.07 mi³ (0.3 km³). Prior to the end of Hayden Creek Glaciation, it was nearly twice its present area and volume. It contains a 770-ft-thick (235 m) base of palagonitized hyaloclastites, which are overlain by 130 ft (40 m) of hyaloclastitic surge beds at its eastern end, and foreset-bedded pillow breccia more than 130 ft (40 m) thick at its western end. These units are overlain by a 475-ft-thick (145 m) remnant of a scoria cone. The western flank of the scoria cone is capped by about 200 ft (65 m) of subaerial lava flows. Four dikes of basalt, 30 to 65 ft (10 to 20 m) wide, cut hyaloclastites on the flanks of the butte. The dikes are interpreted to be vertical offshoots from an east-west intrusive zone within the core of the butte (Fig. 3).

In contrast to Lone Butte, the Crazy Hills are a complex of rounded to flat-topped moberg hills and ridges, covering 8.2 mi² (21.3 km²), rising 400 to 1,200 ft (122 to 366 m) in height, and containing 0.4 mi³ (1.7 km³; Fig. 2). Burnt Peak, 4,106 ft (1,252 m), is among the highest points. The hills are composed chiefly of palagonitized hyaloclastites, occurring in massive, thick to thin laminated beds, with clasts ranging from silt to boulder in size. Most exposures near the bases of the hills consist of pillow lavas, suggesting that the entire complex is underlain by pillow lavas and subglacial to intraglacial lava flows (Fig. 3). The hills at higher elevations, located in the northeast, are upheld by discontinuous masses, elliptical to ball in shape, of glassy to vesicular to densely crystalline dikes, sills, and lava flows. The name Crazy Hills was applied by early surveyors who were perplexed by the irregular valley system. Hyaloclastites, even where palagonitized, are permeable and support few streams. These subglacial deposits were first recognized by Pedersen (1973).

The basalts that compose Lone Butte and the Crazy Hills differ geochemically and petrographically; therefore, they are given separate stratigraphic names. The basalt of Lone Butte (blb) is calcalkaline; the basalt of Burnt Peak (bbp), which composes all of the Crazy Hills, is classified as a tholeiite. The former, blb, is dense and contains abundant olivine phenocrysts; whereas bbp has a generally vesicular to diktytaxitic texture and contains abundant plagioclase and olivine phenocrysts.

Lone Butte and Crazy Hills are interpreted to have formed during the Hayden Creek Glaciation, between about 190,000 and 130,000 yrs ago (Crandell and Miller, 1974; Coleman and Pierce, 1981). This glaciation also produced the Amboy Drift of Mundorff (1984), when the Lewis River glacier last advanced almost the total length of the river from its origin at Mount Adams (Fig. 1). The glacier filled the Lewis River canyon opposite the site of Crazy Hills and overlapped eastward several kilometers onto the Indian Heaven volcanic field. During the same glaciation a local ice cap spread westward from the northern end of Indian Heaven. The two glaciers merged in the area of Lone Butte and Crazy Hills.

No evidences have been found at the subglacial complexes of an earlier extensive glaciation, such as the Wingate Hill Glaciation (Crandell and Miller, 1974), at about 0.80 ma (Coleman and Pierce, 1981). Furthermore, Mount Adams and its andesitic predecessors, the source area for major glaciers, are no older than 0.47 ma (Hildreth and Fierstein, 1983).

Five stages can be identified in the development and modification of the subglacial complexes. An important criterion in relating the development of the two complexes is the presence of the hyaloclastitic surge beds at Lone Butte. These deposits indicate that the volcano built up to the surface of the covering glacier during the early part of the Hayden Creek Glaciation when only the glacier of Indian Heaven covered the area and before the Lewis River glacier had advanced significantly. The Crazy Hills were not formed until the Hayden Creek Glaciation was well advanced.

Stage 1. During the early part of the Hayden Creek Glaciation, Lone Butte erupted; its basal, massive hyaloclastite beds were deposited beneath the 770-ft-thick (235 m) Indian Heaven glacier. The Crazy Hills volcanic complex had not yet erupted. Stage 1 ended when the Lone Butte volcano broke the surface of the Indian Heaven glacier (Fig. 4A).

Stage 2. Still during the early part of the Hayden Creek Glaciation, Lone Butte erupted and grew above the level of the Indian Heaven glacier. Its eruption first produced surge deposits that formed a tuff ring (Stop 3) at least 130 ft (40 m) high, within a lake at the surface of the glacier. When water was no longer able to gain access to the volcanic vent, continuing eruptions built a scoria cone (Stop 2) another 475 ft (145 m) high. Lava flows erupted from this cone and, on its western side, flowed into the lake to form pillow breccia (Stop 1). The growth rate of the volcano exceeded the rate of advance of the glaciation. Eventually, with the advance of the Lewis River glacier, its height rose over the top of Lone Butte (Fig. 4B).

Stage 3. During the later part of the Hayden Creek Glaciation, Lone Butte had ceased eruption and was extensively eroded by the covering glacier (Fig. 4B). The northern part of the tuff ring and much of the scoria cone were removed, exposing agglutinate and welded spatter in the core of the cone (Stop 2) and several dikes. During this same stage the Crazy Hills volcanic complex erupted from one or more vents located at or north and east of Burnt Peak (Figs. 2 and 3) and extruded lava beneath the glacier. Here, closer to the eruptive sources, up to 1,180 ft (360 m) of hyaloclastites accumulated beneath the Lewis River glacier. Any evidence that the hills broke through the surface of the glacier has been removed by subsequent glacial erosion during stage 4. In the southwestern part of the Crazy Hills the deposits average less than 650 ft (200 m) in thickness and form a topographic bench (Fig. 3). The thinness and uniform surface of these deposits is attributed to the initial southwestward spreading of pillows and lava flows beneath the glacier (Stops 4 and 14), under the influence of the direction of glacier flow and topographic gradient. Much of the bedding in the hyaloclastites is inclined in this direction.

Stage 4. Still during the later part of the Hayden Creek Glaciation, the Lewis River glacier covered and eroded the Crazy Hills and Lone Butte. Their eruptive activity had ceased. After retreat of the glaciers of the Hayden Creek Glaciation, the subvolcanic landforms were not substantially modified until Evans Creek Glaciation.

Stage 5. During the Evans Creek Glaciation, Lone Butte was further eroded along its margins by a small glacier from the Indian Heaven volcanoes, which was thinner than the height of the butte (Fig. 4C). The glacier streamlined the butte to an elongated east-west shape, and advanced onto the eastern side of the Crazy Hills and eroded their southern slope, leaving a mantle of drift (Fig. 2). A small tongue of the glacier overtopped a saddle in the southeastern part of the hills (Stop 5). A larger tongue advanced down Rush Creek to the southwestern corner of the hills. The Lewis River glacier advanced only as far as Curly Creek and trimmed only the western base of the hills (Fig. 4C).

SPECIFIC SITES

Geology at fourteen stops is described in the Lone Butte and Crazy Hills subglacial volcanic complex (Fig. 2). The tour begins from a parking area at the junction of roads FS 30 and 32. It first visits the southwestern side of Lone Butte, then returns to the southeastern part of the Crazy Hills. From there the tour proceeds counterclockwise along road FS 32 to the north and southward on road FS 3220 to its junction with 3200. Then it goes west and north on road FS 3211. The visitor can either retrace the route to the junction of roads 30 and 32 or continue north on FS 3211 to road 90 and south on 90 to road 51, and return east to road 30 (Fig. 1). The total distance of the tour, including the return to 30 via 51, is 23 mi (36 km). The number in the parentheses is the estimated time to spend visiting each stop.

Stop 1. (30 min) Southwest quarry of Lone Butte (Fig. 2). The quarry wall, 130 ft (40 m) high, exposes partly palagonitized pillow breccia of basalt of Lone Butte. Large pillow lobes and slabs of breccia are inclined about 35° toward the quarry floor. Walk north about 200 ft (60 m); the west dike of the butte can be seen about 200 ft (60 m) above the base of the talus.

Stop 2. (2 hr) Summit of Lone Butte tuya (Fig. 2). Climb the slope above the quarry along a game trail just south of the

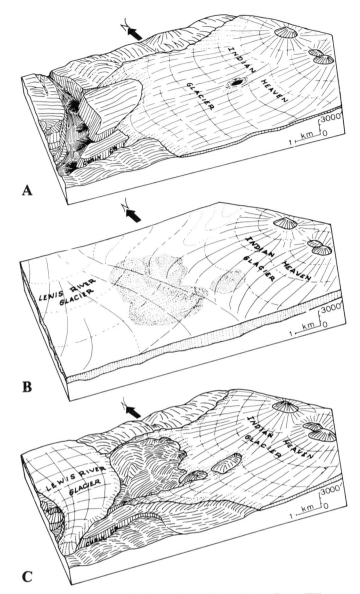

Figure 4. A. Sketch of oblique view of Lone Butte–Crazy Hills area, showing initial subglacial growth of Lone Butte volcano beneath Indian Heaven glacier during early part of Hayden Creek Glaciation (stage 1). B. Shows subglacial development of Crazy Hills and erosion of Lone Butte during later part of Hayden Creek Glaciation (stage 3). C. Shows glacial extent with respect to subglacial volcanoes during maximum Evans Creek Glaciation (stage 5).

quarry wall, gaining 400 ft (120 m) in elevation, and passing upward through the 220-ft-thick (67 m) capping lava flow and eastward along the ridge top to the summit, through 90 ft (27 m) of agglutinate and welded scoria of the eroded summit cone. From the summit area continue northeastward.

Stop 3. (1 hr) Northeast dike and palagonitized hyaloclastitic surge beds of Lone Butte (Fig. 2). Climb down 260 ft (80 m) along the northeast ridge from the summit. The dike is in sharp contact with the beds to the southeast. The beds are uniformly stratified, 1 to 15 cm in thickness, and show both normal and reverse grading. Lithic fragments vary between beds from blocky glassy to scoriaceous basalt. Fluvially rounded cobbles of basalt of Lone Butte are common, having been derived from the subglacially eroded base of the butte. The beds southeast of the dike lie at about the rim of the tuff ring. North of the dike the beds form the inner crater slope of the ring, dipping 43° to 62° to the northwest. The contact between the surge beds and the overlying scoriaceous beds occurs to the north just above the base of the dike. Retrace route back to quarry. Return to junction roads FS 30 and 32; proceed north on 30 to 3211 and turn west (Fig. 2).

Stop 4. (20 min) Southeast quarry of Crazy Hills (Fig. 2). The wall exposes 40 ft (12 m) of well-formed pillows within a sparse (pillow-supported) matrix of thinly layered, fine-grained hyaloclastite. Pillows grade upward 3 to 7 ft (1 to 2 m) into palagonitized, hyaloclastitic, coarse-grained to pebbly sandstone, about 25 to 35 ft (8 to 10 m) thick. About 7 ft (2 m) of drift of Evans Creek Glaciation overlies the hyaloclastite. Pinnacles, 7 to 10 ft (2 to 3 m) high, of palagonitized hyaloclastite are probable remnants left by erosion of a stream marginal to the Indian Heaven glacier during the Evans Creek Glaciation (Fig. 4C). The vertical distance between the surface of the glacier during the Hayden Creek Glaciation to the top of the pillow lava may have been the volatile fragmentation depth, about 1,000 ft (300 m); Fig. 3), for erupting basalt of Crazy Hills. The volatile fragmentation depth (Fisher and Schmincke, 1984) is that depth at which the volatile pressure in erupting lava exceeds the hydrostatic pressure of overlying water or a glacier. In most subglacial deposits of Iceland that depth is about 660 ft (200 m; Allen, 1980). Here the bedding and current structures in the overlying hyaloclastites indicate that they were likely deposited fluvially by a subglacial stream. On the east side of the quarry the hyaloclastite dips steeply and is cut by normal faults of small displacement. From the east side, walk south of the road 500 ft (150 m) to the 100-ft-high (30 m) crest of a till-mantled ridge underlain by glassy lava of basalt of Burnt Peak. The Indian Heaven glacier, during the Evans Creek Glaciation, eroded and removed much of the hyaloclastites from this ridge and the southeastern end of the Crazy Hills to expose the pillows and lava flow. Continue west on FS 3211 (Fig. 2), passing through drift of the Evans Creek Glaciation, slumped and contorted hyaloclastite beds, and various facies of hyaloclastites.

Stop 5. (15 min) In the cut on the west, palagonitized till of the Evans Creek Glaciation fills a channel, about 25 ft (8 m) deep, eroded into more than 30 ft (10 m) of thick-bedded, palagonitized, lithic hyaloclastite. The till contains abundant, rounded, varied basaltic clasts from Indian Heaven. The Indian Heaven glacier breached a saddle in the hills to northeast of the stop (Fig. 2). Retrace route to east. At junction with road FS 32, turn north.

Stop 6. (10 min) The cut on the west exposes more than 10 ft (3 m) of till of the Evans Creek Glaciation (Fig. 2). About 3 ft (1 m) of oxidation is exposed at the top. Many clasts here are of dense, abundantly olivine phyric basalt of Lone Butte. Continue north on FS 32 to just past the junction with spur 3200 121 (Fig. 2).

Stop 7. (10 min) To the west, about 1.5 ft (0.5 m) of rodent-disturbed Mount St. Helens J-, Y-, and W-tephra overlie more than 13 ft (4 m) of boulder-rich till of the Evans Creek Glaciation. The till has about 1.6 ft (0.5 m) of oxidation at its top. Continue westward to the junction with FS 3220 (Fig. 2).

Stop 8. (20 min) The road cut on the south exposes, beneath forest duff, about 3 ft (1 m) of Mount St. Helens J-, Y-, and W-tephra, overlying about 1.5 (0.5 m) of soil developed atop fluvially deposited beds of lithic hyaloclastite (outwash deposit), which, in turn, sharply overlie more than 6 ft (2 m) of till of the Evans Creek Glaciation. Continue south on FS 3220 (Fig. 2).

Stop 9. (20 min) Road cuts on both sides expose varied facies in near horizontally bedded lithic-poor hyaloclastite and a steep-walled channel-fill of lithic-rich hyaloclastite. Walk 300 ft (100 m) south to examine lithic-rich hyaloclastite, which was derived by fluvial disruption of pillow lava. Continue south on FS 3220.

Stop 10. (30 min) A quarry on the south side, above the road, exposes two sills, 6 to 13 ft (2 to 4 m) thick, which intrude obliquely, well-stratified, thin-bedded coarse- to fine-grained hyaloclastic sandstone. Strata above the higher sill are offset by small faults. Note glassy, irregular margins of the sills. Lower sill has margins that have been locally fragmented to form a peperite. A peperite is a breccia formed where an intrusion, commonly of basalt, enters wet sediment at a shallow depth, and is disrupted by steam explosions. A major dike underlies the flat-topped ridge to the east. Continue southward to west side of Burnt Peak (Fig. 2). Where FS 3220 begins its descent to the west, spur 3200 061 to the northwest leads to a good section of Mount St. Helens tephra.

Stop 11. (30 min) Along the east side of FS 32, below Burnt Peak (Fig. 2), extensive cuts on the east side expose fragments of lava flows, pillow lava, and various facies of hyaloclastites near the top of the subglacial complex. Deposits between 3,600 and 3,700 ft (1,097 and 1,128 m) elevation are diamictites, and were deposited by the Lewis River glacier during the late part of the Hayden Creek Glaciation. Angular boulders of basalt were stripped from the dike along the crest of the ridge to the north of Burnt Peak (Fig. 2). The topographically lower bench to the southwest of Burnt Peak was formed by subglacial flowage of lava from one or more vents at Burnt Peak and at points to the northeast. Pillow lavas below the bench were eventually partly eroded and covered with hyaloclastites by subglacial streams. Continue southward to the junction with road FS 3211. Turn west on 3211 (Fig. 2).

Stop 12. (20 min) Here, along about 300 ft (100 m) of roadcut, occurs well-stratified, opal-coated, coarse-grained hyaloclastic sandstone. In the center is a 30- to 40-ft-wide (10 to 12 m) pod of lava. Continue westward on road FS 3211 (Fig. 2).

Stop 13. (30 min) An excellent exposure, to the north, of 13 to 45 ft (4 to 14 m) of pillow lava with abundant (matrix-supported) finely layered, fine-grained hyaloclastite. The pillow lava is overlain along an irregular contact by 0 to 16 ft (0 to 5 m)

of similar finely layered, fine-grained to cross-bedded coarse-grained sandstone. The cut is topped by 3 to 10 ft (1 to 3 m) of lithic-rich hyaloclastite, which is locally palagonitized. The lithic hyaloclastite is the product of fluvial reworking of pillow lava. Continue northward on road FS 3211 (Fig. 2).

Stop 14. (30 min) An intraglacial lava flow of the basalt of Burnt Peak is exposed in west-facing cut. The principal unit in the high cut is 25 to 35 ft (8 to 10 m) thick, radially jointed lava flow, partly enclosed in pillow lava. The top of the cut was eroded by the Lewis River glacier during late part of the Hayden Creek Glaciation. It is overlain by thin layers of till and Mount St. Helens tephra. To the south, massive, lithic hyaloclastite underlies the pillow lavas; to the north, bedded hyaloclastic sandstone underlies the lithic hyaloclastite. Retrace route to the junction of FS 32 and 30 (Fig. 2), or continue northward to the junction with road FS 90 (Fig. 1). Turn south (left) onto 90 and continue to FS 51. Take 51 eastward to FS 30, the Wind River Road.

ACKNOWLEDGMENTS

I am most grateful to John Eliot Allen and Rodney D. Swanson for greatly improving the manuscript, and Ellen Bartsch, Cartographic Center, Portland State University, for completing the cartography. Portland State University Environmental Sciences and Resources Publication no. 204.

REFERENCES CITED

Allen, C. C., 1980, Icelandic subglacial volcanism; Thermal and physical studies: Journal of Geology, v. 88, p. 108–117.

Coleman, S. M., and Pierce, K. L., 1981, Weathering rinds on andesitic and basaltic stones as a Quaternary age indicator, western United States: U.S. Geological Survey Professional Paper 1210, 56 p.

Crandell, D. R., and Miller, R. D., 1974, Quaternary stratigraphy and extent of glaciation in the Mount Rainier region, Washington: U.S. Geological Survey Professional Paper 847, 59 p.

Fisher, R. V., and Schmincke, H.-U., 1984, Pyroclastic rocks: Berlin, Springer-Verlag, 472 p.

Hammond, P. E., 1980, Reconnaissance geologic map and cross sections of the southern Washington Cascade Range, lat 45°30′–47°15′N, long 120°45′–122°22.5′W: Portland, Oregon, Portland State University, Publications of the Department of Geology, scale 1:125,000.

Hildreth, W., and Fierstein, J., 1983, Mount Adams volcano and its 30 "parasites": Geological Society of America Abstracts with Programs, v. 15, no. 5, p. 331.

Jones, J. G., 1969, Intraglacial volcanoes of the Laugarvatn region, southwest Iceland—I: Quarterly Journal of the Geological Society of London, v. 124, pt. 3, p. 197–211.

Kjartansson, G., 1959, The Moberg Formation, in Thorarinsoon, S., Einarsson, T., and Kjartansson, G., eds., On the geology and geomorphology of Iceland: Geografisk Annaler, v. 41, p. 135–169.

Mullineaux, D. R., 1986, Summary of the pre-1980 tephra-fall deposits erupted from Mount St. Helens, Washington State, USA: Bulletin of Volcanology, v. 48, no. 1, p. 17–26.

Mundorff, M. J., 1984, Glaciation in the lower Lewis River basin, southwestern Cascade Range, Washington: Northwest Science, v. 58, no. 4, p. 269–281.

Pedersen, S. A., 1973, Intraglacial volcanoes of the Crazy Hills area, northern Skamania County, Washington: Geological Society of America Abstracts with Programs, v. 5, no. 1, p. 89.

77

Evidence for dozens of stupendous floods from Glacial Lake Missoula in eastern Washington, Idaho, and Montana

Richard B. Waitt, U.S. Geological Survey, 5400 MacArthur Boulevard, Vancouver, Washington 98661

LOCATION AND ACCESS

Burlingame Canyon (Fig. 1). Walla Walla valley, southeast Washington about 2.5 mi (4 km) south of Lowden. Lat. 46°01′ N.; Long. 118°36′ W. SW¼SW¼Sec.5,T.6N.,R.34E. (SW¼ of Lowden, Washington 7½-minute Quadrangle). Drive on U.S. 12 to Lowden, 11 mi (17.7 km) west from Walla Walla, 18 mi (29 km) east from Wallula Junction at U.S. 395. From Lowden, drive south for 0.25 mi (0.4 km) to intersection; turn south (right) on road to Gardena, and drive south another 2.4 mi (3.9 km) to house on right just beyond crossing of irrigation ditch. Burlingame 'canyon' is another 0.1 mi (0.16 km) along road and 200 ft (60 m) west of road. (Please do not try to park or walk in the clear, vaguely fenced area between road and head of canyon; this is man-made in-ground nesting for pollinating bees.) PLEASE OBTAIN PERMISSION FROM DITCHMASTER at house on west side of road just south of irrigation ditch. Burlingame 'canyon' is on private land; the area is hazardous to the unwary, for vertical 100-ft-high (30 m) walls are unprotected at top by fence and are formed in collapsible silt: BE CAREFUL. Access is a slope at head (northeast end) of canyon.

Manila Creek, Sanpoil valley (Fig. 1). Section located in Manila Creek valley, western tributary of lower Sanpoil River (arm of Lake Franklin D. Roosevelt behind Grand Coulee Dam), north-central Washington. NE¼SW¼Sec.18,T.29N.,R.33E. (near southwest corner of USGS Keller 15-minute Quadrangle). On Washington 21, drive about 4.8 mi (7.7 km) north from ferry landing or Lake FDR or 5.5 mi (8.8 km) south from town of Keller to intersection of *old* Manila Creek Road (not new road, which joins Washington 21 about 0.3 mi (0.5 km) farther north). Turn west on old Manila Creek Road and drive about 1 mi (1.6 km) to prominent tall exposure on conspicuously bedded deposits on north (right) side of road (Arrieta Locality; see Fig. 4). Park here; higher parts of stratigraphic section are exposed 0.2 mi (0.3 km) up road (Switchback Locality, Fig. 4).

SIGNIFICANCE OF SITES

Glacial Lake Missoula, Montana, was dammed by the Cordilleran ice sheet during each of several Pleistocene glaciations, most recently 15,500 to 13,000 yr B.P. during the late Wisconsin. The lake, whose volume was as much as 600 mi³ (2,500 km³), discharged through the ice dam as stupendous floods, which carved the Channeled Scabland of Washington and followed the Columbia River valley west to the Pacific Ocean (Fig. 1). Between 1923 and 1932, J Harlen Bretz published a series of imaginative papers on the Channeled Scabland that describe large-scale erosional and depositional landforms wholly alien to rivers of ordinary discharge. Bretz's unorthodox hypothesis of

gigantic floods through Washington seemed outrageous to many and was contested for decades (e.g., Flint, 1938). But at length, after unarguable field evidence—such as giant current dunes—became known and the source of water identified, stupendous flood became an acceptable explanation (Bretz and others, 1956). Later the idea received quantitative legitimacy (Baker, 1973), and the colossal scale of landforms was made plain by space-age imagery (Baker and Nummedal, 1978).

Rhythmically bedded deposits in southern Washington have been variously attributed to: (1) fluctuations within ordinary lacustrine and fluvial environments (Flint, 1938; Lupher, 1944); (2) fluctuating currents within a transient lake during only one or a few great floods (Bretz and others, 1956; Baker, 1973; Mullineaux and others, 1978); and (3) several dozen floods, each of which deposited one graded bed (Waitt, 1980, 1985). By the last hypothesis, floodwater backed up dead-end valleys off the main Scabland floodways to form transient ponds in which suspended load settled. Because the side valleys were protected from violent currents, flood-laid strata were not eroded by later floods but became buried and preserved.

Varved lake sediment separates successive flood-laid beds at many localities in northern Washington. The numbers of varves indicate durations of 6 decades to a few years—generally becoming fewer upsection—between successive floods (Waitt, 1984, 1985; Atwater, 1984, 1986). The bottom sediment of glacial Lake Missoula is also varved; it constitutes dozens of fining-upward sequences, each the record of a gradually deepening then swiftly emptying lake (Chambers, 1971; Waitt, 1980). Figure 2 shows the inferred relation of Lake Missoula's bottom deposits to the interbedded lake and catastrophic-flood deposits in northern Idaho and Washington and to the flood-laid beds in southern Washington. The behavior of repeated discharge every few decades or years suggests that glacial Lake Missoula emptied due to the hydraulic instability that causes glacier-outburst floods (jökulhlaups) from present-day glaciers in Iceland and elsewhere (Waitt, 1985).

The Lake Missoula floods swept some 500 mi (800 km) from the ice dam to the sea; the lake itself was 186 mi (300 km) long (Fig. 1). No one locality gives all the needed data to prove the floods were numerous. The sites below are among the most informative of hundreds scattered about a vast area between western Montana and southern Washington.

This guide contains brief descriptions of four sites in the area of the Cordilleran Section and two sites in the area of the Rocky Mountain Section that have been more completely described in publication. Each of these gives different information, though three are more important than others. The essence of the data and its interpretation can be had from the sections at Burlingame

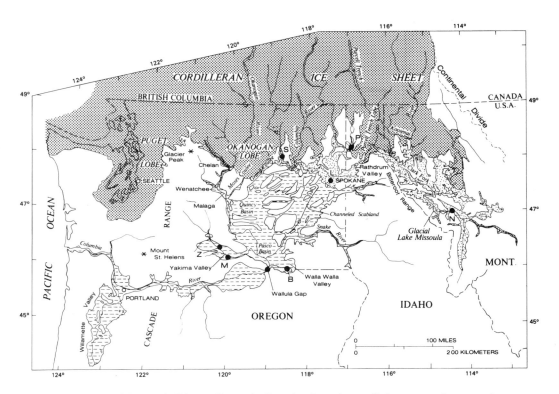

Figure 1. Map of Columbia River valley and tributaries. Irregular small-dot pattern shows maximum area of glacial Lake Missoula east of Purcell Trench ice lobe and maximum extent of glacial Lake Columbia east of Okanogan lobe. Dashed-lined pattern shows area that, in addition to these lakes, was swept by the Missoula floods. Late Wisconsin Cordilleran ice sheet margin (heavy-dot pattern) is from Waitt and Thorson (1983, Fig. 3-1). Large dots indicate sites of bedded flood sediment mentioned in text or figures: B, Burlingame canyon; L, Latah Creek; M, Mabton; N, Ninemile Creek; P, Priest valley; S, Sanpoil valley; Z, Zillah. From Waitt (1985, fig. 1).

Canyon (southern Washington), Manila Creek (Sanpoil Valley, northern Washington), and Ninemile Creek (northwestern Montana; Fig. 1).

SITE INFORMATION

Burlingame Canyon. (Waitt, 1980, 1985). Burlingame 'canyon', cut in the 1930s by wastewater from the nearby irrigation ditch, exposes southern Washington's most complete stratigraphic section of slackwater deposits of the late Wisconsin floods from Lake Missoula. Burlingame canyon is the keystone exposure from which the one-graded-bed-per-flood hypothesis was erected (Waitt, 1980). (But to the contrary, these same beds have been used to advocate that rhythmic deposits in southern Washington are deposits of one or a few pulsating floods [Bjornstad, 1980].) The canyons exposes a rhythmic stratigraphic section of 39 normally graded beds. Most beds have an upward sequence of sedimentary structures—conspicuous plane laminae, to ripple-drift laminae, to drapes, to obscure plane laminae, to ripple-drift laminae, to drapes, to obscure plane laminae or

massive—crudely comparable to that of distal turbidites (Fig. 3). Sparse crystalline, quartzite, and metasedimentary-rock erratics show that the depositing water invaded from the Columbia River valley, for the Walla Walla valley is formed entirely within the Columbia River Basalt Group; ripple-drift laminae in the rhythmites indicate that paleocurrents generally flowed eastward (*up*valley).

The visitor to Burlingame canyon may debate whether many successive beds were deposited by fluctuating currents during one flood (Bjornstad's view) or whether instead each bed represents a separate flood (Waitt's view). If on the one hand the many rhythmites accumulated during one flood, certain conditions would have controlled sedimentation and the character of the deposits: the water would have remained ponded as much as 650 ft (200 m) deep above the top of exposure, and the accumulating sediment would have had to remain loose and saturated. If on the other hand terrestrial environments intervened for decades between successive floods, other dominating conditions would have influenced the character of deposits: the sediment would have become dewatered, and animals would have repopulated

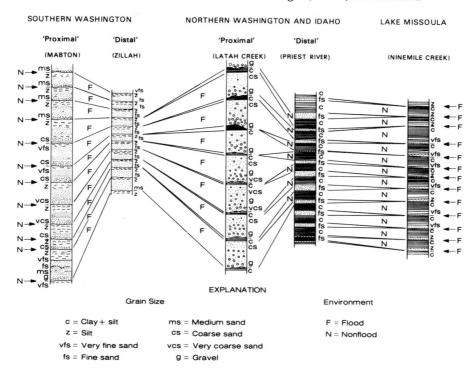

Figure 2. Inferred relations between rhythmites in southern Washington, northern Washington and Idaho, and western Montana (Lake Missoula). From Waitt (1985, Fig. 17). See Figure 1 for section locations.

the area. Much of the diagnostic evidence lies at contacts between graded beds.

Loess between any two beds indicates that a terrestrial, eolian environment intervened between the floods that deposited the two beds. But the occurrence of loess has been debated, for the water-laid tops of most graded beds are texturally nearly identical to loess, from which the flood-laid sediment was indeed derived.

Channels between rhythmites indicate that an erosional process of some sort interrupted the accumulation of flood sediment. One conspicuous channel exposed near the canyon bottom has near-vertical sides: is this likely if the sediment had remained continuously saturated during rapid accumulation?

Slopewash (inferred) partly infills some of the channels. This material is finer and darker (organic coloring?) than is the flood-laid sediment.

Volcanic ash overlies the eleventh rhythmite below top of section (Fig. 3). This characteristic ash couplet is identified as "set S" from Mount St. Helens, dated at about 13,000 yr B.P. (Mullineaux and others, 1978; Waitt, 1980). Both ash layers are structureless and nearly uncontaminated. Is this possible had the ash settled through deep, turbid water during a flood episode? Or does it instead suggest that the ash settled in a terrestrial interflood environment?

Rip-up clasts of the fine material that forms or overlies rhythmite tops may be found low in the section. Rip-up clasts imply that the sediments had dewatered and become coherent before they were reworked by a succeeding current into a new bed.

Rodent burrows filled with reworked flood-laid sediment may be found throughout the entire 100 ft (30 m) of section. Because rodents burrow less than 6 ft (2 m) below the ground surface, the filled burrows imply that rodents repopulated the surface numerous times during the accumulation of this 100 ft (30 m) of sediment.

The striking similarity of graded beds to each other suggests a common origin, not a mixture of two or more different origins. Thus if any one horizon—such as that containing the ash couplet—demands a terrestrial environment, then all other nearly identical horizons in the sequence suggest, if not demand, a similar origin. But it is difficult to *prove* a terrestrial episode at the top of each and every rhythmite.

A general upsection thinning and fining of rhythmites is apparent, especially in the upper third of the section. This regional characteristic is attributed to the ice dam becoming thinner, and therefore glacial Lake Missoula and floods therefrom becoming smaller, during deglaciation.

The bases of the upper 10 or so beds are relatively fine. These beds are similar to those at Zillah used as examples of 'distal' flood beds on Figure 2.

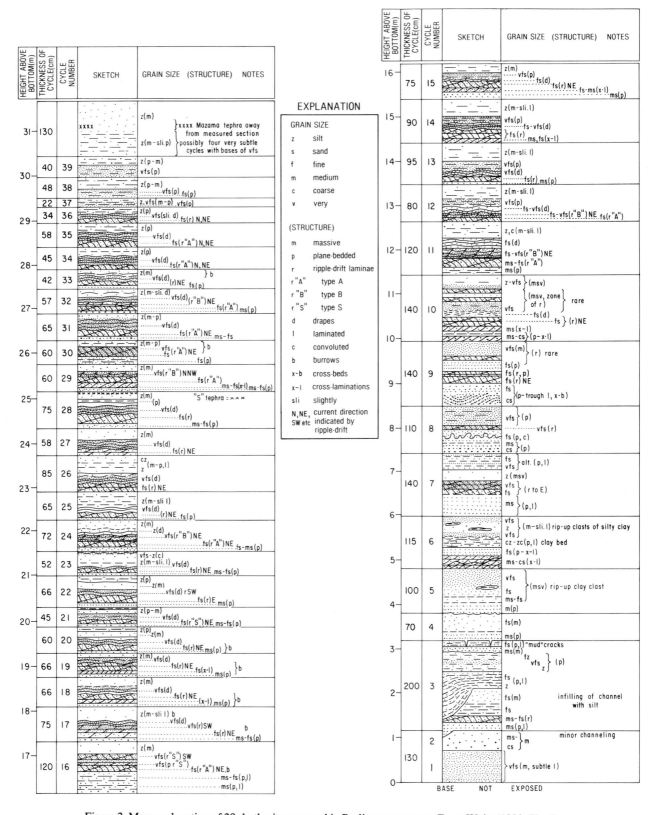

Figure 3. Measured section of 39 rhythmites counted in Burlingame canyon. From Waitt (1980, Fig. 5). See Figure 1 for section locations.

The 100-ft (30-m) section of rhythmic graded beds is capped by about 3 ft (1 m) of loess. The loess is Holocene age, as can be seen near the southwest end of the canyon where the loess encloses Mazama (Crater Lake) ash whose radiocarbon age is about 6,850 yr B.P. (Bacon, 1983).

Mabton. (Waitt, 1980, 1985). Section located in the lower Yakima valley, south-central Washington. NW¼NW¼Sec.31, T.9N.,R.23E. and adjacent NE¼NE¼Sec.36,R.22E. (near northwest corner of USGS Prosser 15-minute Quadrangle); about 1 mi (1.6 km) north of Mabton, Washington, along main road between Mabton and Sunnyside. Conspicuous exposure is continuous along bluffs defining south side of Yakima River valley for 300 ft (90 m) on both sides of the road. West of road please obtain permission from landowner (through gate and up driveway to top of bluff).

Many features in the succession of graded beds here are similar to those at Burlingame canyon, including upsection thinning and fining of rhythmites and the fact that exactly 11 rhythmites overlie the conspicuous ash couplet. Differences include: (1) Paleocurrent indicators are west directed (but that is upvalley here); (2) coarse bedload deposit of locally derived basalt forms base of many rhythmites, especially those low in section; (3) The Mount St. Helens "set S" ash couplet is much thicker, Mabton being about halfway between the volcano and Burlingame canyon; bases of both ash layers are uncontaminated; (4) Two additional thin ash laminae lie at top of the rhythmites that overlie and underlie the rhythmite capped by the prominent ash couplet; (5) Dunes at the base of several rhythmites consist half of freshwater shells; the shells must have been concentrated in an adjacent pond—accumulated there over years or decades—before being swept up by an incoming flood; and (6) shells from the base of second rhythmite below the ash couplet give radiocarbon ages of 14,060 ± 450 yr B.P. (USGS-684) and 13,130 ± 350 yr B.P. (W-2983).

Manila Creek (Sanpoil valley). (Atwater, 1986). The composite section in Manila Creek contains about 2,000 to 2,500 varves. These varves are punctuated by 89 graded coarser beds, each attributed to a Missoula flood (Atwater, 1986). The flood beds become generally thinner and finer while varves become thicker toward the top of the composite section (Fig. 4)—to the point that flood-laid beds become difficult to distinguish. Wood fragments from a compound varve in midsection yield a radiocarbon age of 14,490 ± 290 yr B.P. (USGS-1860). Interpretation is that glacial Lake Columbia dammed by the Okanogan lobe (Fig. 1) existed for 2 to 3 millennia and was invaded by scores of separate floods from glacial Lake Missoula. This composite section with 89 inferred flood beds is the most complete in the region.

Varves between flood beds low in the composite section generally number 30 to 50; they decrease to only 1 or 2 in upper part of section; an estimated 200 to 400 varves overlie the highest recognized flood-laid bed. These varve counts confirm what can be qualitatively inferred from the upsection thinning and fining of flood beds at this and other sections: the late Wisconsin glacial

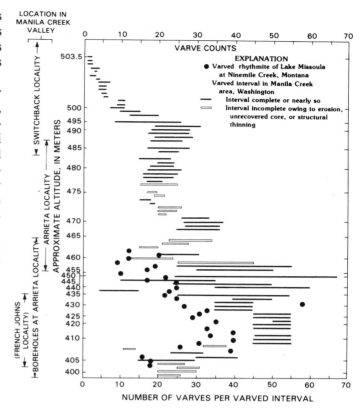

Figure 4. Data on varves interbedded between dozens of flood-laid beds at composite section at Manila Creek, tributary of Sanpoil valley, northern Washington (from Atwater, 1986, Fig. 17).

lake Missoula floods became smaller and more frequent as the controlling ice dam (Purcell Trench lobe) thinned during deglaciation. From this and other evidence in the Sanpoil valley, Atwater infers relative fluctuations of several lobes of the Cordilleran ice sheet (Figs. 1, 4).

Latah (Hangman) Creek. (Waitt, 1984, 1985). Section located in Latah (Hangman) Creek valley in northeastern Washington about 5 mi (8 km) south of Spokane, Washington, in SW¼SE¼Sec.31,T.25N.,R.43E. (USGS Spokane SW 7½-minute Quadrangle). From intersection of I-90 and U.S. 195 just west of Spokane, proceed south on U.S. 195 for 2.5 mi (4 km). Site is the tall, conspicuous, south-facing exposure 300 to 1,200 ft (90 to 365 m) east of highway. Park on shoulder and well off pavement of *north*bound lanes of highway. Traffic is fast and sometimes heavy: BE CAREFUL. From highway, walk through grassland and ford the creek.

Lower Latah Creek valley is part of the principal floodway from glacial Lake Missoula to the Channeled Scabland; it is far from the ice-sheet margins that dammed glacial Lake Columbia, an arm of which flooded this area (Fig. 1). Compared with the

R. B. Waitt

Priest valley section, the flood beds here are therefore coarse and thick ('proximal' on Fig. 2) and varves are thin and fine.

The lower two-thirds of section below a conspicuous gravel-lined disconformity is a succession of 16 gravel-sand beds 3 to 10 ft (1 to 3 m) thick punctuated by inconspicuous silty-clay varved beds only 0.4 to 8 in (1 to 20 cm) thick. The relatively coherent varved beds form benches; yet they are so thin and locally discontinuous that the visitor may have to search to find them. The flood beds give evidence of high-energy emplacement: they are coarse and carry rare clasts as large as 3 ft (1 m); foreset beds dip at low angle (<10°) and upvalley; some beds are channeled into underlying beds; some beds were invasive into and beneath varved beds; some varved beds are locally lifted and dismembered, and fragments of dismembered varved beds occur as rip-up clasts in flood beds. As many as 51 varves have been counted in an individual bed; but because the varved beds are eroded, such figures must be taken to represent a minimum number of years between floods.

Above the gravel-lined disconformity is a succession of about 12 lenticular beds, also with low-angle upvalley–dipping foreset beds, but without separating varve beds. These beds are inferred to be flood beds shed up the valley by giant floods after this arm of the glacial lake had been filled with sediment or after the lake had subsided to below the level of this section.

Priest valley, Idaho. (Waitt, 1984, 1985). Section located in Priest River valley in northwestern Idaho 'panhandle' in SW¼SE¼Sec.1,T.56N.,R.5E. (USGS Priest River, Idaho 7½-minute Quadrangle). From town of Priest River on U.S. 2, turn north on Idaho 57 toward Priest Lake. Proceed 3.5 mi (5.6 km) to intersection of Peninsula Road. Turn sharp right onto Peninsula Road and continue 0.2 mi (0.3 km) nearly to top of long hill. Exposure is conspicuous on east (right) side of road.

This site is situated behind a moraine and former ice tongue that blocked the valley mouth and dammed a lake in Priest valley. Only the largest Missoula floods were able to overtop the moraine to reach this site. This site, being close to an ice margin,

has relatively thick varves; the floods having been channeled roundaboutly, the flood beds are relatively thin and fine ('distal' on Fig. 2).

The exposure shows about 50 ft (15 m) of silt-to-clay varved lake beds punctuated by beds of sand with upvalley–directed ripple-drift laminae (Fig. 2). Each of the sand beds, which are 4 to 5 grain-size (ϕ) intervals coarser than the varved beds, represents a vigorous current invading the valley. Most of the sand beds show an upward succession of plane beds to ripple drift to drapes to plane beds as at Burlingame canyon. The varves between flood beds number 20 to 50: the period between floods was 2 to 5 decades.

Ninemile Creek, Montana. (Chambers, 1971, appendix III; Chambers, 1984; Waitt, 1980, 1985; Waitt and Thorson, 1983). Located in northwestern Montana in lower Ninemile Creek near its confluence with the Clark Fork River. Conspicuous tall roadcut on both sides of I-90 just east of bridge over Clark Fork River 6 mi (9.6 km) west of Frenchtown, Montana, which is about 15 mi (24 km) west of Missoula (south center of USGS Alberton, Montana 15-minute Quadrangle). Park far off pavement and BE CAREFUL of fast traffic including large trucks.

This long exposure shows a succession of nearly 40 graded beds. Each bed grades up from cross-laminated sand or silt, through silty compound thick varves, to simple clayey thin varves. This regular motif evidently represents a gradually deepening lake whose shoreline progressively migrated north and thus isolated the main source of sediment (ice-sheet margin) from the center of the lake. The abrupt upward transition from thin clayey varves to cross-laminated sand represents an abrupt change from a deep to a shallow lake. This section therefore records about 40 gradual deepenings and swift lowerings of glacial Lake Missoula. There being no notable unconformities or soils within the section, this history must all be late Wisconsin. The upsection thinning and fining of rhythmites is probably a result of northward retreat of the Cordilleran ice sheet, which thus gradually removed the main sediment source during deglaciation.

REFERENCES CITED

Atwater, B. F., 1984, Periodic floods from glacial Lake Missoula into the Sanpoil arm of glacial Lake Columbia, northeastern Washington: Geology, v. 12, p. 464–467.

—— , 1986, Pleistocene glacial-lake deposits of the Sanpoil River valley, northeastern Washington: U.S. Geological Survey Bulletin 1661, 39 p.

Bacon, C. R., 1983, Eruptive history of Mount Mazama and Crater Lake caldera, Cascade Range, U.S.A.: Journal of Volcanology and Geothermal Research, v. 18, p. 57–115.

Baker, V. R., 1973, Paleohydrology and sedimentology of Lake Missoula flooding in eastern Washington: Geological Society of America Special Paper 144, 79 p.

Baker, V. R., and Nummedal, D., 1978, The Channeled Scabland: Washington, D.C., National Aeronautics and Space Administration. 186 p.

Bretz, J H., Smith, H.T.U., and Neff, G. E., 1956, Channeled Scabland of Washington; New data and interpretations: Geological Society of America Bulletin, v. 67, p. 957–1049.

Chambers, R. L., 1971, Sedimentation in glacial Lake Missoula [M.S. thesis]: Missoula, University of Montana, 100 p.

Chambers, R. L., 1984, Sedimentary evidence for multiple glacial Lakes Missoula: Montana Geological Society, Field Conference Northwestern Montana, p. 189–199.

Flint, R. F., 1938, Origin of the Cheney-Palouse scabland tract, Washington: Geological Society of America Bulletin, v. 49, p. 461–524.

Lupher, R. L., 1944, Clastic dikes of the Columbia Basin region, Washington and Idaho: Geological Society of America Bulletin, v. 55, p. 1431–1462.

Mullineaux, D. R., Wilcox, R. E., Ebaugh, W. F., Fryxell, R., and Rubin, M., 1978, Age of the last major scabland flood of the Columbia Plateau in eastern Washington: Quaternary Research, v. 10, p. 171–80.

Waitt, R. B., 1980, About forty last-glacial Lake Missoula jökulhlaups through southern Washington: Journal of Geology, v. 88, p. 653–679.

—— , 1984, Periodic jökulhlaups from Pleistocene glacial Lake Missoula; New evidence from varved sediment in northern Idaho and Washington: Quaternary Research, v. 22, p. 46–58.

—— , 1985, Case for periodic, colossal jökulhlaups from Pleistocene glacial Lake Missoula: Geological Society of America Bulletin, v. 96, p. 1271–1286.

Waitt, R. B., and Thorson, R. M., 1983, The Cordilleran ice sheet in Washington, Idaho, and Montana, in Wright, H. E., Jr., ed., Late-Quaternary Environments of The United States, Vol. 1: Minneapolis, University of Minnesota Press, p. 53–70.

Columbia River Basalt Group, Joseph and Grande Ronde canyons, Washington

S. P. Reidel, *Geosciences Group, Rockwell International, P.O. Box 800, Richland, Washington 99352*
P. R. Hooper, *Department of Geology, Washington State University, Pullman, Washington 99164*
S. M. Price, *Geosciences Group, Rockwell International, P.O. Box 800, Richland, Washington 99352*

LOCATION AND ACCESSIBILITY

The Grande Ronde-Joseph Canyon area is located in southeast Washington near the common boundaries of Washington, Oregon, and Idaho (Fig. 1). The locality lies within the canyon country which forms the southeast part of the Columbia Plateau Province near the eastern termination of the Blue Mountains and the southern boundary of the Lewiston Basin.

The site can be reached by following Washington SR 129 south from Clarkston to Asotin, Washington, and then by following the Snake River road (1st street in Asotin) south from Asotin for 25 mi (40 km) to the confluence of the Snake and Grande Ronde rivers. The Snake River road is paved for the first 13 mi (21 km) but then becomes a well-maintained gravel road that serves the area. All sites are easily reached from this road. Land within the area either is owned by the Washington Department of Game, Fish, and Wildlife Service or is private. Information about current ownership can be obtained from the headquarters of the Chief Joseph Fish and Wildlife area in Joseph Canyon (H, Fig. 2). Permission must be obtained to visit private land, but landowners understand the significance of the area and have always granted access graciously.

SIGNIFICANCE OF SITE

The Columbia Plateau consists of a large accumulation of layered plateau basalt flows and is an excellent example of continental flood basalts. The Columbia River Basalt Group has an estimated volume of about 40,940 mi^3 (170,649 km^3) and covers about 63,213 mi^2 (163,721 km^2) (Tolan and Reidel, unpublished data, 1985). It is one of the youngest known and best preserved flood basalt provinces, having formed between the early and late Miocene (17.0 and 6.0 Ma).

The Grande Ronde and Joseph canyons formed as tributaries to the Snake River incised into the basalt. The result is one of the most dramatic exposures of the basalt stratigraphy on the Columbia Plateau. Between 45 and 50 flows are present in more than 3,000 ft (900 m) of relief. At the base, the oldest formation, the Imnaha Basalt, rests on the Permo-Triassic Seven Devils Volcanic Group. These are overlain by the most voluminous formation, the Grande Ronde Basalt. Overlying these are flows of the youngest formation, the Saddle Mountains Basalt. The excellent exposure permits examination of the physical characteristics and contact relationships of each flow. In addition, dikes of the Grande Ronde dike swarm pass through the area. It was through these fissures that the Columbia River basalt reached the surface. Numerous dikes can be observed cross cutting flows on the canyon walls. On the floors of the canyons, sedimentary deposits of the Pleistocene Bonneville and Missoula catastrophic floods can be observed. In addition to the geologic significance, the area has played an important role in the Nez Percé Indian history of the Pacific Northwest. Joseph Canyon was the birthplace of Chief Joseph, an important leader of the Nez Percé during the mid-1800s.

STRATIGRAPHY AND STRUCTURE

The Columbia River Basalt Group consists of five formations (Swanson and others, 1979; Fig. 3), three of which, the Imnaha, Grande Ronde, and Saddle Mountains, are exposed in the area.

The oldest formation is the Imnaha Basalt (about 17.0 Ma) which is confined to the eastern part of the plateau. It forms about 7 percent of the volume of the Columbia River basalt and is distinguished from the overlying Grande Ronde Basalt by its coarse porphyritic texture, drusy weathering, and significantly different major, minor, and trace element composition. Most flows have normal paleomagnetic polarity, but the oldest flows are reversed. Hooper and others (1984) have subdivided the Imnaha Basalt into 11 chemical types based upon field appearance, petrographic criteria, and chemical composition. The 11 chemical types fall into two broad subgroups, the older American Bar and

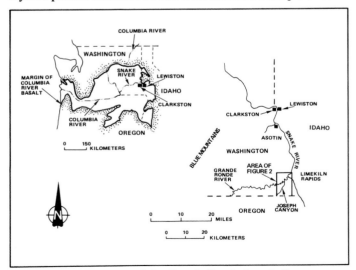

Figure 1. Location map of the Grande Ronde-Joseph Canyon area, southeast Washington.

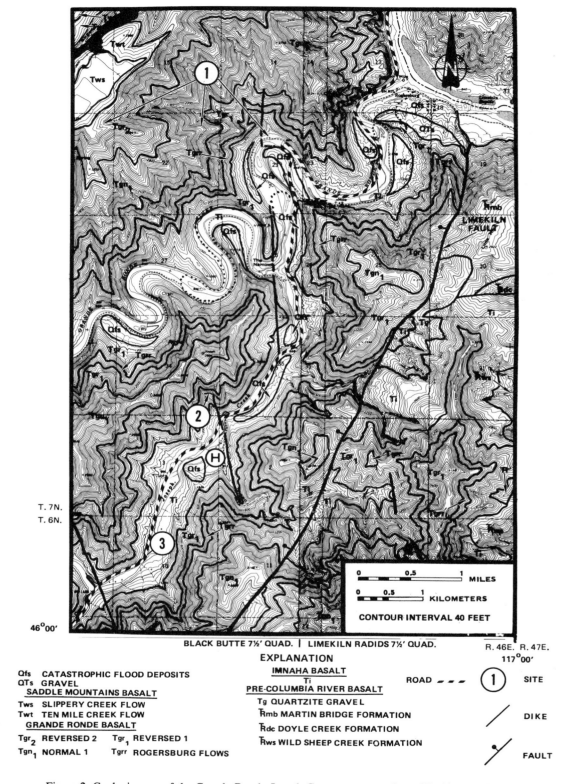

Figure 2. Geologic map of the Grande Ronde-Joseph Canyon area, southeast Washington. Geology from Reidel (in Reidel and others, 1986).

the younger Rock Creek. The Rock Creek predominates but both types interfinger. The mineralogy is dominated by plagioclase, olivine, and augite, with the Rock Creek subgroup having larger phenocrysts and a coarser-grained matrix.

The Grande Ronde Basalt forms about 85 percent of the Columbia River basalt. It is the most extensive formation of the group, covering most of the Columbia Plateau but also having spread to the Pacific Ocean in western Washington and Oregon (see Beeson and Tolan, this volume) and east into Idaho. The thickest known section (2.0 mi; 3.2 km) has been encountered in boreholes in the Pasco Basin (east-central Washington) but the Grande Ronde area has one of the thickest natural exposures. The Grande Ronde Basalt was erupted in a short time interval (17.0-15.5 Ma) with approximately one flow every 5,000 years; this rate was less than the present Hawaiian rate but of greater volume (Shaw and Swanson, 1970, p. 297). The Grande Ronde Basalt has been regionally subdivided into four magnetostratigraphic units which have been mapped across the Columbia Plateau. The flows are typically aphyric but small phenocrysts of plagioclase, olivine, pigeonite, and rare resorbed orthopyroxene can be found. Compositionally the Grande Ronde Basalt is a quartz tholeiite. Within this area, there is a cyclical compositional trend through the stratigraphic section that is dominated by a Ca-rich clinopyroxene composition (Reidel, 1983). This results in a relatively characteristic composition for each of the magnetostratigraphic units in the area.

The Saddle Mountains Basalt is the youngest formation (14.5-6.0 Ma) and comprises about 1 percent of the Columbia River basalt. The Saddle Mountains flows have diverse compositions, petrography, and paleomagnetic polarity, making each flow relatively unique and easier to identify in the field.

The physical features of basalt flows are well displayed in the area. A typical flow consists of three general parts: a colonnade at the base, an entablature in the middle, and a flow top. The colonnade and entablature are distinguished principally by the size and regularity of jointing of the columns; the colonnade tends to have regular, vertical polagonal jointed columns. The entablature has a less regular joint pattern (curvicolumnar). Flow tops are typically vesicular and are often oxidized in the Grande Ronde Basalt. The contact between the colonnade and entablature is sharp, but these features may vary over great distances. The horizontal joint pattern of dikes allows them to be easily distinguished from basalt flows. Other primary features such as flow-top breccias and pillowed-hyaloclastic zones are uncommon in this area but are found to the west.

Despite the amount of detailed data available, a consensus on the origin of the basalts and their evolution is lacking. The homogeneity of most flows points to well-mixed and often replenished reservoirs with no significant change in bulk composition between the reservoir and the surface (Reidel, 1983; Hooper and others, 1984; Hooper, 1984). Evidence also suggests that fractionation has occurred in both the Imnaha and Grande Ronde basalts at the crust-mantle boundary (21 mi; 35 km). Large grabens, such as the Lewiston Basin, may have resulted from subsidence related to magma withdrawal. The role of contamination by assimilation of the crust and the nature of the mantle are some of the currently debated petrogenetic topics.

SERIES	GROUP	SUB-GROUP	FORMATION
MIOCENE	COLUMBIA RIVER BASALT GROUP	YAKIMA BASALT SUBGROUP	SADDLE MOUNTAINS BASALT (14.5 - 6 Ma)
			WANAPUM BASALT (15.5 - 14.5 Ma)
			GRANDE RONDE BASALT (17.0 - 15.5 Ma)
			PICTURE GORGE BASALT
			IMNAHA BASALT (17.5 - 17.0 Ma)

Figure 3. Stratigraphy of the Columbia River Basalt Group. Modified from Swanson and others (1979).

During the eruption of the basalt, the Columbia Plateau was deforming under north-south-oriented maximum horizontal compression. East-west extensional features related to this compression dominate the area. The generally north-south-oriented dikes are interpreted to be a principal response to east-west extension. The area is located at the southern boundary (Limekiln Fault, Fig. 2) of the Lewiston Basin, a large graben in an area that was deforming by at least late Grande Ronde time (Reidel, 1982) and continued to deform throughout the eruption of the Columbia River basalt (Hooper and Camp, 1981). More than 1,650 ft (500 m) of post-basalt movement has occurred on the Limekiln Fault. Most faults within the area have normal displacements although some faults also show a component of subhorizontal movement.

SITE INFORMATION

Three sites within the locality are described as an introduction to the geology of this area. Each site is readily accessible, and together they demonstrate the geologic characteristics of continental flood basalts from the Columbia Plateau. For general references on the stratigraphy of the Columbia River basalt, see Waters (1961) and Swanson and others (1979). For references on the stratigraphy of the Columbia River basalt of this area, see Reidel (1982, 1983) and Hooper and others (1984). For references on the general structure of the area, see Hooper and Camp

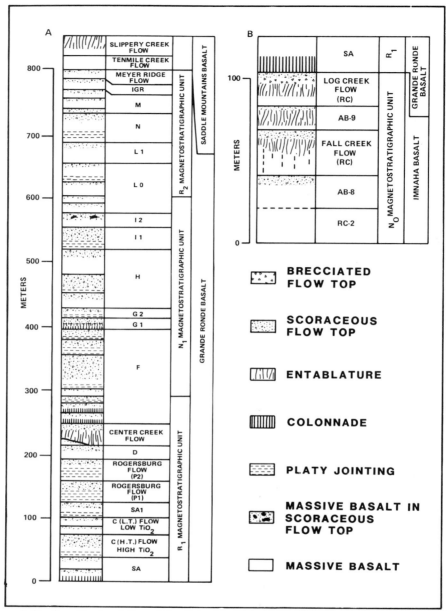

Figure 4. Stratigraphic description of the Columbia River basalt from sites 1 and 3. A-stratigraphic description of the Grande Ronde Basalt from site 1, the Grande Ronde River section. B-stratigraphic description of the Imnaha Basalt from site 3, the Joseph Creek section.

(1981), Camp and Hooper (1981), Camp and others (1982), and Reidel (1982, 1983). Swanson and others (1980) provide a reconnaissance geologic map (1:250,000) of southeast Washington; a detailed geologic map of the locality is provided in Reidel and others (1986), a portion of which is reproduced in Figure 2. A geologic map by Hooper and others (1985) includes the area along the Snake River south from Clarkston to just north of the locality.

Site 1. Grande Ronde Basalt Type Section. From the confluence of the Snake and Grande Ronde rivers, the gravel road follows the north side of the Grande Ronde River for 2.6 mi (4.2 km) to a bridge with public parking on the west side (Fig. 2). A short walk (0.1 mi; 0.16 km) along a dirt road that continues up the north side of the river leads to the base of the type section for the Grande Ronde Basalt. Beginning with the first flow along the river, the section contains about 3,000 ft (892 m) of nearly flat-lying Columbia River basalt that consists of Grande Ronde Basalt capped by two flows of the Saddle Mountains Basalt (Fig. 4a). Only the R_1, N_1, and R_2 Grande Ronde magnetostratigraphic units are present. Here the Grande Ronde Basalt consists of 35

Figure 5. Photograph showing the field relationship at site 2 between Grande Ronde Basalt flow C (H.T.) (see Fig. 4) and the dike that fed it.

flows or flow lobes which form 20 chemical subgroups that can be traced throughout the area (Reidel, 1983). The flows of the Grande Ronde Basalt form prominant ledges throughout the canyon. At 1,140 ft (349.5 m) elevation, a dike (Fig. 2) cuts the section. This dike has a chemical composition similar to the first set of flows of the N_1 magnetostratigraphic unit (Group F, Reidel, 1983). Two very distinct plagioclase-phyric flows of the R_1 magnetostratigraphic unit, the Rogersburg flows (P_1 and P_2), occur from elevations 1,230-1,330 ft (374.9-405.4 m) and can be easily traced throughout the area. Two flows above the Rogersburg flows is the Center Creek flow; this flow is principally entablature jointed with two flow lobes, one of which is seen to pinch out here. In flows higher up in the section, well-developed colonnades, zones of platy jointing, and thick zones of flow top scoria (vesicular basalt) can be observed. The contact with the overlying Saddle Mountains Basalt is not well exposed here but is marked by a spring line. Elsewhere it consists of a thick soil zone that represents a hiatus in volcanic activity when the N_2 Grande Ronde Basalt, Wanapum Basalt, and some older flows of the Saddle Mountains Basalt were deposited farther west and north.

Site 2. Basalt Flow-Dike Connection. From site 1, the road crosses the Grande Ronde River and enters Joseph Canyon after traversing one of the spectacular incised meanders that inspired the river's name. Site 2 (Fig. 2) is 3.4 mi (5.6 km) from site 1 and at the bridge that crosses Joseph Creek. The bridge on the Limekiln Rapids quadrangle differs in location from that of the present bridge; Figure 2 shows the new bridge location. Many of the gravel deposits on the floors of the canyons (Fig. 2) are from the Pleistocene Missoula and Bonneville catastropic floods.

Webster and others (1982) have described at least 10 rhythmites of Touchet beds in the road cut at about the junction of the Grande Ronde River and Joseph Creek canyons.

Physical connections between basalt flows and the dikes that fed them are rare, although many dikes can be observed standing out on canyon walls. Mapping by Reidel (in Reidel and others, 1986) has shown a direct connection between a dike (well exposed on the other side of the canyon along the road that leads to the Fish and Wildlife Headquarters) and Grande Ronde Basalt flow C (H. T.) (Fig. 5) of Reidel (1983). A short walk up the north side of the canyon wall, beginning from the south side of the old bridge ramp and through the gate in the fence along the old road, allows one to examine the nature of flow-dike connections. The dike is exposed along the north side of a small stream valley; it can be easily recognized by the horizontal jointing and aphanitic texture that stand out in contrast to the coarse porphyritic texture of the Imnaha Basalt through which it passes. Note the glassy selvages that represent the chilled dike margins; these are especially evident where it crosses resistant layers of basalt flows. Near the Grande Ronde-Imnaha Basalt contact, the dike changes trend abruptly from northwest to north. At this point the dike becomes very vesicular (presumably due to degassing of the magma as it neared the surface) and is easily eroded. The dike remains vesicular up to the flow-dike contact, but more important, the dike contains blocks of cooled lava and breccia. These blocks represent lava that cooled at the surface but flowed back into the fissure at the end of the eruption. Primary flow banding of the lava blocks and vent material have various orientations but are found only to a depth of about 200 ft (61 m). At the flow-dike contact, the flow is very different from that observed in the Grande Ronde section (site 1). No columnar jointing is present, only glassy basalt with a highly oxidized surface, prominent flow banding, and some primary breccia. Chilled glassy banding and spatter are very apparent along the length of exposure.

Site 3. Imnaha Basalt. A well-studied stratigraphic section (Fig. 4b) of Imnaha Basalt (Hooper and others, 1984) occurs 1.8 mi (2.7 km) up Joseph Creek along the gravel road (just beyond state land). Five Imnaha flows are exposed below the Grande Ronde Basalt. The Log Creek flow, an exceptionally coarse-grained flow of the Rock Creek subgroup, occurs at the top of the section beneath the Grande Ronde Basalt. The Fall Creek flow, a very plagioclase-phyric flow also of the Rock Creek subgroup, is separated from the Log Creek flow above and from still another Rock Creek subgroup flow below by flows of the American Bar subgroup. Unlike the Grande Ronde Basalt, these flows do not have resistant flow tops but form the prominent glassy slopes throughout the canyon. Where the flows form resistant ledges, the main body of the flow rather than the flow top is responsible. Excellent specimens of the zeolites: natrolite, stilbite, and analcime occur in the flow along the road.

REFERENCES CITED

Camp, V. E., and Hooper, P. R., 1981, Geological studies of the Columbia Plateau: I. Late Cenozoic evolution of the southeast part of the Columbia River basalt province: Geological Society of America Bulletin, v. 92, Part I, p. 659–668.

Camp, V. E., Hooper, P. R., Swanson, D. A., and Wright, T. L., 1982, Columbia River basalt in Idaho: Physical and chemical characteristics, flow distribution, and tectonic implications, *in* Bonnichsen, Bill, and Breckenridge, R. M., eds., Cenozoic Geology of Idaho: Idaho Bureau of Mines and Geology Bulletin 26, p. 55–75.

Hooper, P. R., 1984, Physical and chemical constraints on the evolution of the Columbia River basalt: Geology, v. 12, p. 495–499.

Hooper, P. R., and Camp, V. E., 1981, Deformation of the southeast part of the Columbia Plateau: Geology, v. 9, p. 323–328.

Hooper, P. R., Kleck, W. D., Knowles, C. R., Reidel, S. P., and Thiessen, R. L., 1984, The Imnaha Basalt, Columbia River Basalt Group: Journal of Petrology, v. 25, p. 473–500.

Hooper, P. R., Webster, G. D., and Camp, V. E., 1985, Geologic map of the Clarkston 15 minute Quadrangle, Washington and Idaho: Washington Division of Geology and Earth Resources Geologic Map GM–31, scale 1:48,000.

Reidel, S. P., 1982, Stratigraphy of the Grande Ronde Basalt, Columbia River Basalt Group, from the lower Salmon and northern Hell's Canyon area, Idaho, Washington and Oregon, *in* Bonnichsen, Bill, and Breckenridge, R. M., Cenozoic Geology of Idaho: Idaho Bureau of Mines and Geology Bulletin 26, p. 77–101.

——— 1983, Stratigraphy and petrogenesis of the Grande Ronde Basalt from the deep canyon country of Washington, Oregon, and Idaho: Geological Society of America Bulletin, v. 94, p. 519–542.

Reidel, S. P., Hooper, P. R., Webster, G. D., and Camp, V. E., 1986, Geologic map of the Anatone and Limekiln Rapids 15 minute Quadrangles, Washington: Washington Division of Geology and Earth Resources Geologic Map GM Series, scale 1:48,000 (in press).

Shaw, H. R., and Swanson, D. A., 1970, Eruption and flow rates of flood basalts, *in* Gilmour, E. H., and Stradling, D., eds., Proceedings of the Second Columbia River Basalt Symposium, Cheney, Washington, March 1969: Cheney, Washington, Eastern Washington State College Press, p. 271–300.

Swanson, D. A., Wright, T. L., Camp, V. E., Gardner, J. N., Helz, R. T., Price, S. M., Reidel, S. P., and Ross, M. E., 1980, Reconnaissance geologic map of the Columbia River Basalt Group, Pullman and Walla Walla quadrangles, southeast Washington and adjacent Idaho: U.S. Geological Survey Miscellaneous Geological Investigations Map I–1139, scale 1:250,000.

Swanson, D. A., Wright, T. L., Hooper, P. R., and Bentley, R. D., 1979, Revisions in stratigraphic nomenclature of the Columbia River Basalt Group: U.S. Geological Survey Bulletin 1457-G, 59 p.

Waters, A. C., 1961, Stratigraphic and lithologic variations in the Columbia River Basalt: American Journal of Science, v. 259, p. 583–611.

Webster, G. D., Kuhns, M.J.P., and Waggoner, G. L., 1982, Late Cenozoic gravels in Hells Canyon and the Lewiston Basin, Washington and Idaho, *in* Bonnichsen, Bill, and Breckenridge, R. M., eds., Cenozoic Geology of Idaho: Idaho Bureau of Mines and Geology Bulletin 26, p. 669–683.

Geology of the Vantage area, south-central Washington: An introduction to the Miocene flood basalts, Yakima Fold Belt, and the Channeled Scabland

Robert J. Carson, Department of Geology, Whitman College, Walla Walla, Washington 99362
Terry L. Tolan and Stephen P. Reidel, Geosciences Group, Rockwell International, Richland, Washington 99352

LOCATION

The Vantage area is located along the Columbia River in the central portion of Washington State at the boundary between Grant and Kittitas counties (Fig. 1). The area lies entirely within the eastern portion of the Yakima Fold Belt of the Columbia Plateau. The area is covered by the Beverly, Evergreen Ridge, Ginkgo, and Vantage 7.5-minute quadrangles. The Vantage area can be reached via Interstate 90 from the east or west or State Route 243 from the south (Fig. 1). All sites within the Vantage area are along paved roads.

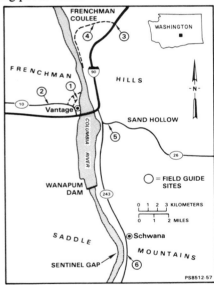

Figure 1. Index map showing the locations of field guide sites (denoted by numbered circles) and major highways that serve this area. See Figures 3, 4, and 5 for more specific information on the location of each site.

SIGNIFICANCE OF SITE

The Vantage area affords an excellent opportunity to see in proximity a number of geologic features and relationships which are important to our understanding of the Neogene history of this region. These features include: 1) lava flows of the Columbia River Basalt Group which are part of the youngest continental flood basalt plateau in the world, 2) anticlinal ridges of the Yakima Fold Belt through which water gaps provide natural cross sections, 3) petrified wood that was part of a 15.5-m.y.-old fossil forest with the greatest known species diversity in the world (Beck, 1945), and 4) evidence for catastrophic floods from glacial

Lake Missoula which swept across the Columbia Plateau and created the Channeled Scabland. The geological investigation of this region has intensified within the past decade or so, particularly because it has been suggested that the Columbia River basalt flows might be suited as a repository for high-level nuclear waste. The site under consideration is the U.S. Department of Energy's Hanford Reservation which lies approximately 30 mi (50 km) southeast of the Vantage area. Also since 1980, major petroleum companies have been exploring this region for its potential to produce commercial quantities of natural gas. This activity has included drilling several deep wells which have penetrated through the Columbia River basalt and into the underlying rocks.

GEOLOGIC SETTING

The purpose of this section is to provide a brief overview of the main geologic features and processes that have shaped this region and are the focus of this guide. The topics to be covered are: 1) the stratigraphy of the Vantage area, 2) Yakima folds, and 3) the catastrophic floods. More specific and detailed information on these topics can be found in the references cited in each section.

Stratigraphy

The oldest exposed rocks in the Vantage area are the tholeiitic flood basalt flows of the Miocene Columbia River Basalt Group (CRBG) (Waters, 1962). It has been estimated that more than 60,000 mi^3 (250,000 km^3) of CRBG lava was erupted from fissures and vents in eastern Washington (Tolan and Reidel, 1985, unpublished data), western Idaho, and northeastern Oregon from about 17 to 6 m.y. ago. The combination of large volume, low viscosity, and high temperature of the erupting lavas enabled them to cover more than 62,200 mi^2 (161,175 km^2) in Washington, Idaho, and Oregon (Tolan and Reidel, 1985, unpublished data). More than 99 percent by volume of the CRBG lavas were erupted over a very short period—from about 17.5 to 14.5 m.y. ago. Changes and variations in the geochemical compositions, paleomagnetic polarity, and physical characteristics of the CRBG flows have allowed for their division into five formations and 14 members (Swanson and others, 1979). The Vantage area is underlain by 5,000 ft to more than 10,000 ft (1,500 to more than 3,000 m) of Columbia River basalt which can be divided into Grande Ronde Basalt, Wanapum Basalt, and Saddle Mountains Basalt.

Grande Ronde Basalt. The Grande Ronde Basalt was erupted from about 17 to 15.5 m.y. ago and is the most volum-

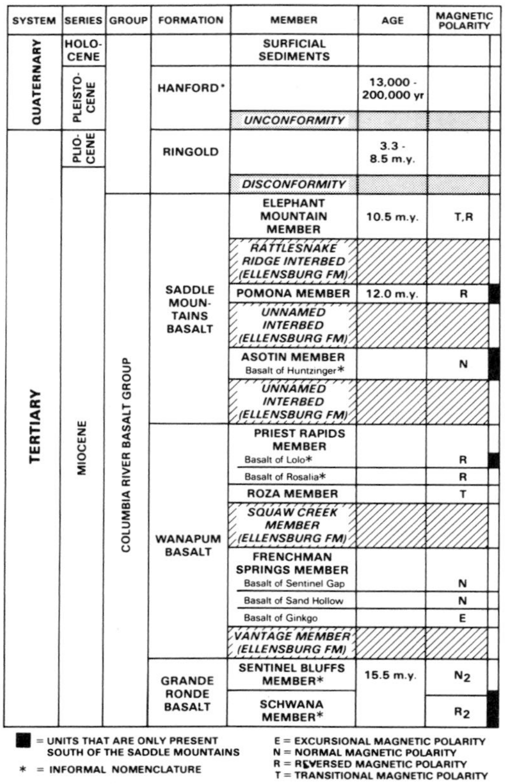

SYSTEM	SERIES	GROUP	FORMATION	MEMBER	AGE	MAGNETIC POLARITY
QUATERNARY	HOLOCENE			SURFICIAL SEDIMENTS		
QUATERNARY	PLEISTOCENE		HANFORD*		13,000 - 200,000 yr	
QUATERNARY	PLEISTOCENE			UNCONFORMITY		
TERTIARY	PLIOCENE		RINGOLD		3.3 - 8.5 m.y.	
TERTIARY				DISCONFORMITY		
TERTIARY	MIOCENE	COLUMBIA RIVER BASALT GROUP	SADDLE MOUNTAINS BASALT	ELEPHANT MOUNTAIN MEMBER	10.5 m.y.	T,R
TERTIARY	MIOCENE	COLUMBIA RIVER BASALT GROUP	SADDLE MOUNTAINS BASALT	RATTLESNAKE RIDGE INTERBED (ELLENSBURG FM)		
TERTIARY	MIOCENE	COLUMBIA RIVER BASALT GROUP	SADDLE MOUNTAINS BASALT	POMONA MEMBER	12.0 m.y.	R ■
TERTIARY	MIOCENE	COLUMBIA RIVER BASALT GROUP	SADDLE MOUNTAINS BASALT	UNNAMED INTERBED (ELLENSBURG FM)		
TERTIARY	MIOCENE	COLUMBIA RIVER BASALT GROUP	SADDLE MOUNTAINS BASALT	ASOTIN MEMBER Basalt of Huntzinger*		N ■
TERTIARY	MIOCENE	COLUMBIA RIVER BASALT GROUP	WANAPUM BASALT	UNNAMED INTERBED (ELLENSBURG FM)		
TERTIARY	MIOCENE	COLUMBIA RIVER BASALT GROUP	WANAPUM BASALT	PRIEST RAPIDS MEMBER Basalt of Lolo*		R ■
TERTIARY	MIOCENE	COLUMBIA RIVER BASALT GROUP	WANAPUM BASALT	Basalt of Rosalia*		R
TERTIARY	MIOCENE	COLUMBIA RIVER BASALT GROUP	WANAPUM BASALT	ROZA MEMBER		T
TERTIARY	MIOCENE	COLUMBIA RIVER BASALT GROUP	WANAPUM BASALT	SQUAW CREEK MEMBER (ELLENSBURG FM)		
TERTIARY	MIOCENE	COLUMBIA RIVER BASALT GROUP	WANAPUM BASALT	FRENCHMAN SPRINGS MEMBER Basalt of Sentinel Gap		N
TERTIARY	MIOCENE	COLUMBIA RIVER BASALT GROUP	WANAPUM BASALT	Basalt of Sand Hollow		N
TERTIARY	MIOCENE	COLUMBIA RIVER BASALT GROUP	WANAPUM BASALT	Basalt of Ginkgo		E
TERTIARY	MIOCENE	COLUMBIA RIVER BASALT GROUP	GRANDE RONDE BASALT	VANTAGE MEMBER (ELLENSBURG FM)		
TERTIARY	MIOCENE	COLUMBIA RIVER BASALT GROUP	GRANDE RONDE BASALT	SENTINEL BLUFFS MEMBER*	15.5 m.y.	N2
TERTIARY	MIOCENE	COLUMBIA RIVER BASALT GROUP	GRANDE RONDE BASALT	SCHWANA MEMBER*		R2 ■

■ = UNITS THAT ARE ONLY PRESENT SOUTH OF THE SADDLE MOUNTAINS
* = INFORMAL NOMENCLATURE

E = EXCURSIONAL MAGNETIC POLARITY
N = NORMAL MAGNETIC POLARITY
R = REVERSED MAGNETIC POLARITY
T = TRANSITIONAL MAGNETIC POLARITY

Figure 2. Generalized stratigraphy of the Vantage area. Modified after Reidel (1984). The black bar in the right-hand column denotes units that are only present south of the Saddle Mountains. N = normal magnetic polarity, R = reversed magnetic polarity, T = transitional magnetic polarity, E = excursional magnetic polarity.

inous of the CRBG formations (greater than 85% by volume; Reidel, 1983). The Grande Ronde Basalt has been formally subdivided into four magnetostratigraphic units (Swanson and others, 1979) with only the two youngest of these (R2, N2) exposed in the Vantage area (Fig. 2). These magnetostratigraphic units consist of a series of flows that have the same natural paleomagnetic polarity.

Vantage Member—Ellensburg Formation. The end of Grande Ronde volcanism marked the beginning of a hiatus over much of the Columbia Plateau which probably lasted for perhaps as long as 100,000 to 200,000 years before the onset of Wanapum volcanism. Sediments deposited during this hiatus in the western portion of the plateau are defined as belonging to the Vantage Member (Fig. 2) of the Ellensburg Formation (Mackin, 1961; Swanson and others, 1979). These epiclastic and volcaniclastic sediments of the Vantage Member, in most cases, represent deposits associated with the reestablishment of an integrated drainage system on and across the plateau. Prior to Vantage time, the repeated eruptions of voluminous Grande Ronde flows (average one every 10,000 to 20,000 years) prevented the reestablishment and development of an integrated drainage system on the plateau.

Wanapum Basalt. The Wanapum Basalt constitutes only about 3 percent by volume of the CRBG and in the Vantage area

consists of three members (Fig. 2) which were erupted between about 15.5 and 14.5 m.y. ago.

The oldest member present in this area is the Frenchman Springs Member which has been further subdivided into three units (Mackin, 1961; Fig. 2). Two of these units, the basalt of Ginkgo and the basalt of Sentinel Gap, have thick, foreset-bedded, basal pillow-basalt complexes indicating that they flowed into lakes. These lakes had been created by the damming of the paleodrainage system that had developed during Vantage time by the earliest Frenchman Springs flows. It is in the basalt of Ginkgo that most of the petrified wood is found.

Overlying the Frenchman Springs Member in the Vantage area is the Roza Member (Fig. 2). The Roza Member consists of one to two flows or flow units that are characterized by abundant plagioclase phenocrysts.

The youngest Wanapum member, the Priest Rapids Member, consists of two units (Fig. 2). The youngest of the two flows, the basalt of Lolo, is only found south of the Sentinel Gap area (Reidel, 1984).

Squaw Creek Member—Ellensburg Formation. The Squaw Creek member consists chiefly of diatomite, siltstone, and shales (Mackin, 1961) that lie between the Frenchman Springs Member and the Roza Member (Fig. 2). This interbed is found mainly in the northeast portion of the Vantage area. These sediments were deposited in a lake that is thought to have been impounded by the northern flow-edge of the Sentinel Gap flow (Mackin, 1961).

The Saddle Mountains Basalt. The Saddle Mountains Basalt consists of up to 10 compositionally diverse members and is the youngest formation within the CRBG, making up about 1 percent of the total volume. Members of this formation were erupted intermittently from about 14.0 to 6.0 m.y. ago. In this locality, the Saddle Mountains Basalt is represented by the Asotin, Pomona, and Elephant Mountain members (Reidel, 1984).

Post-Wanapum Ellensburg Formation Interbeds. Intercalated with the Saddle Mountains Basalt flows are sedimentary deposits of the Ellensburg Formation (Fig. 2). These interbedded sediments are primarily fluvial deposits and are named on the basis of confining basalt flows, not lithology. Since not all Saddle Mountains flows are present everywhere (e.g., the Vantage area), problems as to how to properly name an interbed occur. As a result, several interbeds are left unnamed (Fig. 2) instead of stretching existing definitions or creating new names.

Ringold Formation. The Ringold Formation (Fig. 2) overlies the CRBG in the eastern and southern portion of the Vantage area and consists of fine sand, silt, conglomerate, and fanglomerate (Grolier and Bingham, 1978).

Hanford Formation. Remnant Pleistocene catastrophic flood deposits, here informally called the Hanford formation (Fig. 2), are found intermittently throughout the Vantage area below the 1,200 ft (365 m) elevation. These glaciofluvial sediments were deposited as two facies, the Pasco Gravels and the Touchet Beds (Flint, 1938). The gravels were considered to be related to the last major floods from glacial Lake Missoula in Montana but thorium-uranium age dates suggest some gravels may be pre-Wisconsinan. The Touchet Beds are the rhythmically bedded, fine-grained facies of the Hanford Formation and are interpreted as slackwater deposits resulting from ponding of flood waters behind anticlinal ridges.

Yakima Fold Belt

The Yakima Fold Belt consists of a series of generally east-west-trending anticlinal ridges and synclinal valleys that were produced under north-south regional compression. These folds extend from the western half of the Columbia Plateau into and through the Cascade Range (Beeson and Tolan, this volume).

The anticlinal folds are typically asymmetric, with a thrust or high-angle reverse fault(s) along the steeper limb. The fold geometry can vary from open to tight and the folds are typically segmented. The narrow anticlinal ridges are separated by synclinal valleys that are in many cases very broad, flat basins.

Detailed studies of several Yakima folds on the Columbia Plateau (e.g., Reidel, 1984) have revealed that they have had a long, complex deformational history. These studies have shown that flows of the CRBG, sedimentary interbeds of the Ellensburg Formation, and suprabasalt sediments thin over the anticlinal ridges and thicken into the synclines. The variations in thickness demonstrate that these folds were developing since at least Grande Ronde time (16.5 to 15.5 m.y. B.P.). Rate of deformation studies indicate that the uplift rate was probably greatest during Grande Ronde time (between 1,968 to 820 ft/m.y.; 600 to 250 m/m.y.) and slowed by the end of the Miocene (less than 131 ft/m.y.; 40 m/m.y.). Reidel (1984) points out that there is an interesting correlation between the growth rates of the folds and the estimated supply and eruption rates of the CRBG suggesting a connection between volcanism and deformation.

In the past 10 years, many different models have been advanced to explain the origin and evolution of the Yakima folds. Major differences exist between the models including: 1) extent of sub-CRBG basement involvement (e.g., thin-skinned decollements vs. basement faulting), and 2) the relationship between folding and faulting. For more information on this topic, see Reidel and others (1984, p. 267-269).

Catastrophic Floods

During the Quaternary, the Columbia Plateau experienced a series of catastrophic floods which resulted from the repeated failures of glacial ice dams that released tremendous quantities of impounded water. It was these flood waters from glacial Lake Missoula which eroded the enormous system of proglacial channels across the Columbia Plateau that collectively form the Channeled Scabland (Bretz and others, 1956; Bretz, 1969). Although the origin of the Channeled Scabland is now well known and accepted, this was not always the case. It was Dr. J Harlen Bretz who, through a series of papers from 1923 to 1932, first advanced the "outrageous hypothesis" that portions of the Columbia Plateau has been sculpted by a great flood (the Spokane Flood). These papers sparked a heated controversy and debate which raged for more than 40 years until the mid 1960s when Bretz's hypothesis was vindicated. The story of Bretz's defense and the eventual vindication of this hypothesis against the anti-catastrophist opposition (Baker, 1978) is one of the most fascinating episodes in the history of Northwest geology.

The focus of controversy has now shifted from was there a flood to how many floods have there been? Most investigators agree that there have been more than several catastrophic floods; at least one flood event per advance and retreat of the Cordilleran Ice Sheet. However, Waitt (1980) interprets rhythmically bedded slackwater flood deposits (the Touchet Beds) to suggest about 40 flood events. Mount St. Helens' "set S" ash is found between Touchet Beds rhythmites in many places and has been used to date the last floods from glacial Lake Missoula at about 13,000 years ago (Waitt, 1980). For a more detailed discussion of the catastrophic floods, the reader is referred to Baker (1973, 1983), Baker and Nummedal (1978), and Atwater (1984).

SITES

Sites 1 and 2—Ginkgo Petrified Forest State Park

The Ginkgo Petrified Forest State Park is an area of excellently preserved Miocene (15.5 m.y. old) petrified wood that notably occurs in a basal pillow complex of the Ginkgo flow (Fig. 2) of the Frenchman Springs Member. The two sites described here on the state park grounds provide excellent opportunities to examine the petrified wood both in its original geologic setting and in museum exhibits; Site 1 is the park museum and surroundings and Site 2 is the Interpretative Trail Center where the petrified wood can be seen in its original geologic setting.

Site 1—Museum (Fig. 3). Exit from Interstate 90 at the town of Vantage (Exit 136). Turn left (north) onto the old Vantage Highway (State Route 10) and drive 0.8 mi (1.28 km) north through the town of Vantage to the intersection with the State Park road. Turn right onto this road and follow the signs to the museum.

Within the museum are excellent displays of the various genera of petrified wood collected from the Vantage area as well as interesting displays on the natural history of this area. Most of the petrified logs around the museum are from the pillowed base of the Ginkgo flow and are excellently preserved.

Outside and behind the museum is a scenic overlook from which one can see most of the area covered by this field guide. Far to the north on the east side of the Columbia River is the mouth of Frenchman Springs Coulee (Site 4). Frenchman Springs Coulee is a cataract which was carved out by catastrophic flood waters as they poured into the Columbia River channel from the east. Directly across the river from here can be seen Babcock Bench which is a stripped structural surface created by the catastrophic floods. Below the bench are N_2 Grande Ronde Basalt flows and above the bench are flows of the Wanapum Basalt. The break in the cliffs to the southeast is the mouth of Sand Hollow (Site 5). Far to the south, beyond Wanapum Dam,

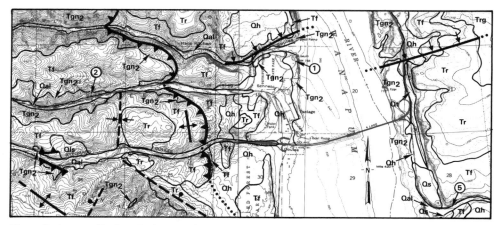

Figure 3. A generalized geologic map of the Vantage, Washington area (Sites 1, 2, and 5). See Figure 5 for explanation of map symbols and for map scale. Geology by T. L. Tolan.

is Sentinel Gap (Site 6)—a water gap through the Saddle Mountains.

At the south end of the museum is a short trail leading to Indian petroglyphs that were carved on basalt columns. These columns were removed from nearby areas prior to the filling of the Wanapum reservoir.

Site 2—The Interpretative Trail (Fig. 3). The interpretative trails can be reached from the museum area by returning to the old Vantage Highway (State Route 10), turning right onto the old Vantage Highway, and then driving approximately 3 miles (4.8 km).

The interpretative trails (Fig. 3) consist of a 0.6 mi (1 km) long trail system which mainly stays within the basal pillow complex of the Ginkgo flow (Fig. 2) of the Frenchman Springs Member. In the parking lot area at the start of the trail, the top of the Grande Ronde Basalt is exposed. Also along the lower portion of the trail, one may see exposures of gray-white volcaniclastic sediment of the Vantage Member of the Ellensburg Formation (Fig. 2).

Within the Ginkgo pillow complex are found petrified woods representing a diverse assemblage of conifers and deciduous trees; more than 50 genera may be present within the Ginkgo pillow complex (Beck, 1945), and detailed paleobotanical studies have been done on 24 species from here by Prakash and Barghoorn (1961a, b) and Prakash (1968). Along the interpretative trail system the petrified logs have been left in their original position within the Ginkgo pillow complex, and in most cases the logs have been identified by genus.

Site 3—Roza Peperite (Fig. 4). To reach Site 3 from Site 2, return to the town of Vantage via the old Vantage Highway (State Route 10). At the west end of town, take the east-bound Interstate 90 ramp. Proceed approximately 7.4 mi (11.8 km) east on Interstate 90 and take the Champs deBrionne Winery Road exit (Exit 143). Park on the shoulder of the off ramp near the intersection at the bottom of the ramp.

At this location, road cuts provide excellent exposures of a Roza peperite. A peperite is a basalt/sediment breccia which

formed when advancing lava invaded and plowed through water-saturated sediments causing violent phreatic brecciation of the lava and mixing with the sediment (Schmincke, 1967). Here the sediment invaded by the Roza flow was a diatomite of the Squaw Creek Member of the Ellensburg Formation (Fig. 2). Not only did the Roza lava mix with this sediment, but the resulting peperite was lifted and rafted along on the top of the Roza flow, thus making the peperite appear to be stratigraphically younger than the underlying Roza flow (Swanson and others, 1979). Invasion and rafting of sediments by CRBG flows on the Columbia Plateau was not an unusual occurrence. The flow overlying the Roza peperite belongs to the Priest Rapids Member (Fig. 2).

Site 4—Frenchman Springs Cataract (Fig. 4). To reach Site 4 from Site 3, turn left at the intersection onto Champs deBrionne Winery Road and drive 0.8 mi (1.3 km) to the next road junction. Turn left at this junction and proceed approximately 1.2 mi (1.9 km). Park at the small turnout on the left side of the road.

From here one has an excellent view of the northern alcove

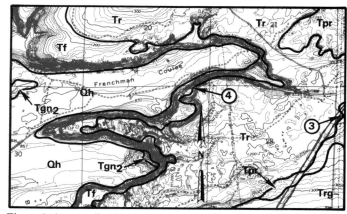

Figure 4. A generalized geologic map of the Frenchman Springs Coulee area (Sites 3 and 4). See Figure 5 for explanation of map symbols and for map scale. Geology modified from Grolier and Bingham (1971) by T. L. Tolan.

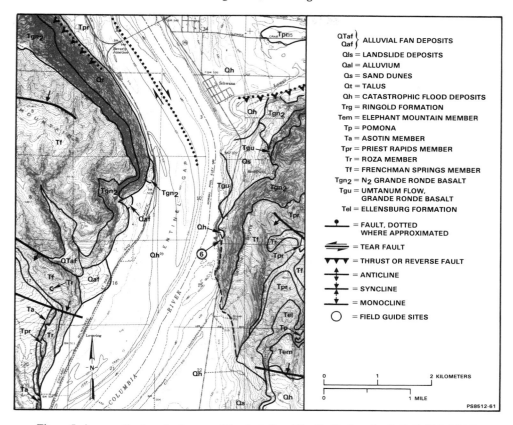

Figure 5. A generalized geologic map of Sentinel Gap (Site 6). Geology by S. P. Reidel (1984).

of Frenchman Springs Cataract that was created by the catastrophic floods. As flood waters from the Quincy Basin to the east reached this point, they encountered a drop of more than 660 ft (200 m) to the Columbia River. Tremendous waterfalls were created near the Columbia River channel; the erosional action caused these waterfalls to retreat eastward. The alcove immediately south of here has a preserved plunge pool more than 165 ft (50 m) deep at the base of its "dry falls."

The tremendous erosive work of the flood waters has provided beautiful exposures of the basalt flows that underlie this area. The road you are on is built on the Roza/Frenchman Springs (Sand Hollow flow; Fig. 2) contact and provides an excellent opportunity to examine a flow contact. Just east from the turnout is a spiracle at the base of the Roza Member. The spiracle once had the shape of an inverted cone which was probably partly, or entirely filled with vesicular breccia. This feature was created by a steam blast probably derived from superheating a local pocket of water-saturated sediments (Mackin, 1961). Walking east back along the road you can see the vesicular and brecciated flow top of the Sand Hollow flow. Note the buckling and folding displayed by slabs of crust within the flow top. In the walls of the coulee can be seen excellent examples of colonnade and entablature jointing (Reidel and others, this volume). Note how these jointing patterns change laterally within the same flow.

On the horizon to the north are white piles of sediment from diatomite mines. In the decade from 1948 to 1957, seven mines were active within a few kilometers of here, but in 1984 only one company was mining diatomite from two pits.

Site 5—Ginkgo Pillow Complex at Sand Hollow (Fig. 3). To reach Site 5 from Frenchman Springs Coulee, return to Interstate 90 and proceed southwest toward Vantage. Leave Interstate 90 at Exit 137 before crossing the Columbia River. Follow State Route 26 south for 1.8 mi (2.9 km) to the mouth of Sand Hollow. Park on the turnout on the right side of the highway.

Exposed in the road cut is a portion of a lava delta consisting of pillowed basalt and hyaloclastic debris which formed when the Ginkgo flow of the Frenchman Springs Member (Fig. 2) entered a lake that covered the Vantage area. The foreset-bedding displayed by the pillow complex is a good indicator of flow direction; at this locality the Ginkgo flow was moving in a westward direction. This exposure also provides an opportunity to examine features typical of pillow lavas and pillow complexes. Some of the pillow lava features to note here are: 1) the radial jointing of individual pillows, 2) the glassy rind on the outer margin of the pillows, 3) hollow interiors of some of the pillows which formed when still fluid lava flowed out of the center after the margins of the pillow had cooled; in a few cases the pillows have even collapsed, and 4) the bulbous elongation of the pillows in the down-slope direction. Also very common here are long stringers of basalt that are conformable to the primary dip of the complex. These stringers of Ginkgo basalt are essentially subaqueous lava tubes. The hyaloclastic debris between the pillows was created by the phreatic brecciation of the Ginkgo lava. The glassy debris was once black in color, but alteration of the glass to palagonite has imparted an orange color to this debris. Also within this complex are several tree molds. Overlying the pillow complex is columnar, subaerial Ginkgo basalt which advanced across the lava delta.

At the west end of this road cut is a mixture of coarse colluvium and fine-grained slackwater sediments associated with the catastrophic floods. These slackwater sediments exhibit crude rhythmic bedding and clastic dikes.

Site 6—Sentinel Gap/Saddle Mountains (Fig. 5). To reach Site 6, return to the intersection of State Route 26 and State Route 243 and turn left onto State route 243. Follow State Route 243 south for 10.5 mi (16.8 km) into Sentinel Gap. Park on the turnout on the right side of the highway.

Sentinel Gap is a water gap through the Saddle Mountains uplift that provides an excellent cross section through a Yakima fold. The uplift is an asymmetrical, north vergent anticlinal ridge with a faulted north limb (Fig. 5). The fold axis on the east side of the gap is about 1.6 mi (2.6 km) south of the axis on the west side, suggesting a tear fault and structural control of the river (Reidel, 1984). On the southwest side of the gap, the Huntzinger flow of Mackin (1961; Asotin Member on Fig. 2) fills a valley cut

into the Priest Rapids Member preserving a portion of the ancestral course of the Columbia River at approximately 13 m.y. ago.

The Saddle Mountains fault is a high-angle reverse to thrust fault that is best exposed on the northwest side of Sentinel Gap (Fig. 5) opposite Crab Creek. Here the fault zone consists of one major fault and several smaller faults that together make up a zone more than 164 ft (50 m) thick. Uplift on the Saddle Mountains fault has exposed flows of the Grande Ronde Basalt, Wanapum Basalt, and Saddle Mountains Basalt. These flows thin and in some cases pinch out onto the uplift suggesting that the structure began growing during the Miocene and continued to grow after the basalt eruptions had ceased. The Vantage Member of the Ellensburg Formation (Fig. 2), the light-colored interbed exposed on the cliffs on the west side of the gap, also thins and pinches out onto the uplift; this is especially evident on the east side of the gap.

Evidence for the catastrophic floods abounds at this site. The vast flood plain of the Columbia River in the gap is composed of catastrophic flood gravels as is the small bar exposed along the road several hundred meters north of the parking area. A prominent boulder train can be observed south of the gap along the road. Volcanic ash from the 6,600-year-old eruption of Mt. Mazama (now Crater Lake) is exposed in colluvium at the parking area.

REFERENCES CITED

Atwater, B. F., 1984, Periodic floods from glacial Lake Missoula into the Sanpoil arm of glacial Lake Columbia, northeastern Washington: Geology, v. 12, p. 464–467.

Baker, V. R., 1973, Paleohydrology and sedimentology of Lake Missoula flooding in eastern Washington: Geological Society of America Special Paper 144, 79 p.

——1978, The Spokane Flood controversy, *in* Baker, V. R., and Nummedal, D., eds., The Channeled Scabland—A guide to the geomorphology of the Columbia Basin, Washington, prepared for the Comparative Planetary Geology Field Conference held in the Columbia Basin, June 5-8, 1978: Washington, D.C., National Aeronautics and Space Administration, p. 3–15.

——1983, Late-Pleistocene fluvial systems, *in* Porter, S. C., ed., The Late Pleistocene-Late Quaternary environments of the United States, volume 1: Minneapolis, University of Minnesota Press, p. 115–129.

Baker, V. R., and Nummedal, D., eds., 1978, The Channeled Scabland—A guide to the geomorphology of the Columbia Basin, Washington, prepared for the Comparative Planetary Geology Field Conference held in the Columbia Basin, June 5-8, 1978: Washington, D.C., National Aeronautics and Space Administration, 186 p.

Beck, G. F., 1945, Ancient forest trees of the sagebrush area in central Washington: Journal of Forestry, v. 43, no. 5, p. 334–338.

Bretz, J H., 1969, The Lake Missoula floods and the Channeled Scabland: Journal of Geology, v. 77, p. 505–543.

Bretz, J H., Smith, H.T.U., and Neff, G. E., 1956, Channeled scablands of Washington: New data and interpretations: Geological Society of America Bulletin, v. 67, p. 957–1049.

Flint, R. F., 1938, Origin of the Cheney-Palouse scabland tract, Washington: Geological Society of America Bulletin, v. 49, p. 461–523.

Grolier, M. J., and Bingham, J. W., 1971, Geologic map and sections of parts of Grant, Adams and Franklin Counties, Washington: U.S. Geological Survey Miscellaneous Geologic Investigations Series Map I-589, scale 1:62,500.

——1978, Geology of parts of Grant, Adams and Franklin Counties, east-central Washington: Washington Division of Geology and Earth Resources Bulletin 71, 91 p.

Mackin, J. H., 1961, A stratigraphic section in the Yakima Basalt and Ellensburg Formations in south-central Washington: Washington Division of Mines and Geology Report of Investigations 19, 45 p.

Prakash, U., 1968, Miocene fossil woods from the Columbia basalts of central Washington, III: Paleontographica, Ser. B, v. 122, nos. 4–6, p. 183–200.

Prakash, U., and Barghoorn, E. S., 1961a, Miocene fossil woods from the Columbia basalts of central Washington, II: Arnold Arboretum Journal, v. 42, p. 165–203.

——1961b, Miocene fossil woods from the Columbia basalts of central Washington: Arnold Arboretum Journal, v. 42, p. 347–362.

Reidel, S. P., 1983, Stratigraphy and petrogenesis of the Grande Ronde Basalt from the deep canyon country of Washington, Oregon and Idaho: Geological Society of America Bulletin, v. 94, p. 519–542.

——1984, The Saddle Mountains—the evolution of an anticline in the Yakima Fold Belt: American Journal of Science, v. 284, p. 942–978.

Reidel, S. P., Scott, G. R., Bazard, D. R., Cross, R. W., and Dick, B., 1984, Post-12 million year clockwise rotation in the central Columbia Plateau, Washington: Tectonics, v. 3, no. 2, p. 251–273.

Schmincke, H-U., 1967, Fused tuff and peperites in south-central Washington: Geological Society of America Bulletin, v. 78, p. 319–330.

Swanson, D. A., Wright, T. L., Hooper, P. R., and Bentley, R. D., 1979, Revisions in stratigraphic nomenclature of the Columbia River Basalt Group: U.S. Geological Survey Bulletin 1457-G, 59 p.

Waitt, R. B., Jr., 1980, About forty last-glacial Lake Missoula jokulhlaups through southern Washington: Journal of Geology, v. 88, p. 653–679.

Waters, A. C., 1962, Basalt magma types and their tectonic associations-Pacific Northwest of the United States, *in* The crust of the Pacific Basin: American Geophysical Union Geophysical Monograph 6, p. 158–170.

Ingalls Tectonic Complex and Swauk Formation, Ruby Creek area, Cascade Mountains, Washington

Samuel Y. Johnson, U.S. Geological Survey, Box 25046, MS 916, Denver Federal Center, Denver, Colorado 80225
Robert B. Miller, Department of Geology, San Jose State University, San Jose, California 95192-0102

LOCATION

The Ruby Creek area encompasses several outcrops of the Late Jurassic or Early Cretaceous Ingalls Tectonic Complex and the Eocene Swauk Formation in the central Cascades of Washington (Fig. 1, Stop 1). To get to the outcrops on Ruby Creek, either: (1) drive south on U.S. 97 (Swauk Pass Highway) from its junction with U.S. 2 for 8.9 mi (14.3 km), or (2) drive north on U.S. 97 from its junction with U.S. 131 for 26.6 mi (42.8 km). Turn east off U.S. 97 onto the well-marked Ruby Creek Road (logging road 2322) and park. Walk up the road for several hundred meters across a fault contact between the Ingalls and the Swauk. Locations of other outcrops are shown on Figure 1, and their accesses from the Ruby Creek exposures are described later in the text.

SIGNIFICANCE

The Ingalls Tectonic Complex (Pratt, 1958; Southwick, 1974; Tabor and others, 1982; Miller, 1985) is the largest and most complete of several Mesozoic ophiolites in northwestern Washington. It consists predominantly of ultramafic tectonites, with less abundant altered gabbros, diabases, basalts, cherts, and fine-grained graywackes. The internal structure of the Ingalls Tectonic Complex is dominated by two east-west–striking fault zones characterized by steeply dipping serpentinite melange. Miller (1985) has suggested that the Ingalls formed in a large ocean or large marginal basin and was deformed in an oceanic fracture zone. The Ingalls was then thrust onto the metamorphic core of the North Cascades and intruded by the Late Cretaceous (86–93 Ma) Mount Stuart batholith. The Swauk Formation comprises more than 23,000 ft (7,000 m) of sandstone, mudstone, and conglomerate, and is one of the thickest nonmarine sequences in North America. The Swauk is one of several thick Eocene sequences in Washington that formed within a regional network of Late Cretaceous and early Tertiary, north-northwest–striking, dextral strike-slip faults (Johnson, 1985). The Ingalls served partly as both basement and source terrane for the Swauk. Diverse facies of the Ingalls and the Swauk Formation in the Ruby Creek area illustrate the complex history of each unit and provide insight into the Mesozoic and early Cenozoic tectonic history of the Washington Cascades.

INGALLS TECTONIC COMPLEX

The contact between the Ingalls Tectonic Complex and the Swauk Formation at Ruby Creek is a north-south–striking, high-

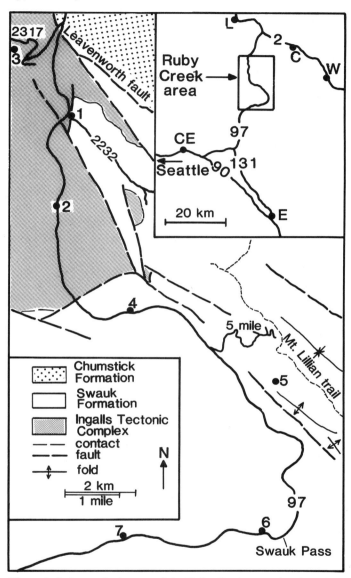

Figure 1. Index geologic map of the Ruby Creek area, showing numbered field locality sites. Geology is from Tabor and others (1982). Inset highway map abbreviations are as follows: C, Cashmere; CE, Cle Elum; E, Ellensburg; L, Leavenworth; W, Wenatchee.

angle fault (Fig. 1; Tabor and others, 1982). Directly west of the fault are strongly sheared and slickensided serpentinites of the Ingalls. Farther west, the serpentinites are separated from a heterogeneous assemblage of Ingalls mafic rocks by a steep, southeast-dipping fault. Altered gabbros, diabases, and basalts are exposed in an outcrop about 82 ft (25 m) in length. These rocks may represent the shallow levels of an intrusive complex. They have undergone static greenschist-facies metamorphism characterized by uralitization of clinopyroxene and sassuritization of plagioclase. Clinozoisite veins and patches are also abundant in the gabbros and diabases. Some of these rocks have been tectonically brecciated and cataclastic microstructures are common.

On the west side of U.S. 97, across from the Ruby Creek Road, are argillites and lithic wackes that form part of the largest (3.7 mi^2; 9.5 km^2) tectonic block of sedimentary rocks in the Ingalls. These rocks were originally referred to as the Peshastin Formation (Smith, 1904) and are more than 3,300 ft (1,000 m) thick (Southwick, 1974). Argillite and less common fine-grained lithic wacke predominate, and ribbon chert occurs locally. Radiolarians in a chert near Ruby Creek are Late Jurassic (Pessagno, E. A., in Tabor and others, 1982).

Sedimentary structures are poorly developed in the clastic rocks. Bedding is uncommon and rhythmic and graded bedding are rare. The argillites are extensively fractured, but only locally are penetratively deformed. Metamorphism is low-grade and primary sedimentary structures are generally preserved.

The fine-grained lithic wackes consist of subangular to angular clasts set in a clay-rich matrix. Felsic to mafic volcanic fragments, albitized plagioclase, quartz, chert, and shale are the dominant clast types. The abundance of volcanic clasts and their wide compositional range suggest derivation from a calc-alkaline volcanic terrane (Southwick, 1974; Miller, 1985). Deposition in an open-ocean setting or a large marginal basin is implied by the typical fine-grained texture of the clastic rocks and the presence of radiolarian chert.

SWAUK FORMATION

East of the fault, the Swauk Formation forms an east-dipping, well-exposed section more than 105 ft (32 m) thick (Fig. 2). This section consists of greenish-gray conglomerate, conglomeratic sandstone, and medium- to coarse-grained sandstone (Fig. 3), and is cut by numerous small faults of minor displacement. Conglomerate beds are as thick as 22 ft (6.7 m) and are mostly massive. Primary stratification is suggested by crude horizontal alignment of clasts in some beds. Clasts are poorly sorted, typically subrounded to rounded, as large as 18 in (45 cm) in maximum dimension, and are either framework supported or are interspersed in a matrix of coarse-grained sandstone. Two paleocurrent measurements on rare imbricate clasts indicate sediment transport to the east. Sandstone and conglomeratic sandstone beds are as thick as 15.6 ft (4.7 m), lenticular, and are either massive or characterized by crude horizontal or low-angle bedding. Contacts between beds are planar, or less commonly erosional.

Based on a count of 306 pebbles, conglomerate clasts have the following lithologies: greenstone, 51.3%; gabbro to granodiorite, 27.4%; graywacke and argillite, 10.5%; chert, 4.6%; quartz and quartzite, 3.6%; rhyolite to andesite, 2.0%; serpentinite, 0.6%. Sandstones are rich in lithic fragments.

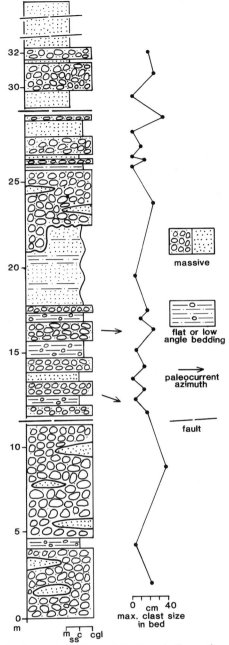

Figure 2. Stratigraphic section of the Swauk Formation at the Ruby Creek outcrop.

The Swauk section adjacent to the fault is capped by highly faulted sandstones that were not measured, and by a covered interval. For the next several hundred meters to the southeast on the Ruby Creek Road, there are sporadic exposures of trough-crossbedded to flat-bedded fine- to coarse-grained sandstone, massive to crudely stratified pebble conglomerate, and massive mudstone. These strata differ from the Swauk adjacent to the fault in that they are much more fine-grained, and the sandstones are arkosic and contain a much smaller proportion of lithic frag-

Figure 3. Massive to crudely stratified conglomerate and sandstone of the Swauk Formation at the Ruby Creek outcrop, interpreted as proximal alluvial fan deposits.

ments. Six paleocurrent measurements on pebble imbrications and trough crossbed axes have a vector mean of 170° ± 17°, indicating sediment transport to the south.

Swauk conglomerate and sandstone adjacent to the fault contact at Ruby Creek are interpreted as proximal alluvial fan deposits on the basis of coarse grain size, poor sorting, and thick, lenticular bedding. The typical lack of internal structure in conglomerates suggests that mass-flow depositional processes were dominant. Framework-supported conglomerates might represent partially winnowed debris flows. However, the rare presence of erosional surfaces and horizontally aligned and imbricate clasts suggests that stream processes were also active. Most clast types occur as bedrock to the west in the Ingalls, the apparent sediment source. Granodiorite clasts were probably derived from a more distal source in the Mount Stuart batholith. It seems likely that the alluvial fan formed along a scarp related to offset on the fault at Ruby Creek.

The abrupt change in grain size in the Swauk outcrops a few hundred yards (m) to the southeast of the fault at Ruby Creek suggests that the inferred alluvial fan was quite small. Paleocurrent and lithologic contrasts suggest that this small fan interfingered to the east with a fluvial system carrying arkosic debris that was probably derived from a northern crystalline source.

OTHER INGALLS TECTONIC COMPLEX OUTCROPS OF THE RUBY CREEK AREA

Stop 2: Drive south from the Ruby Creek turnoff on U.S. 97 for 2.1 mi (3.4 km) and pull off on the east (left) side of the road in a large parking area. This is the site of the old gold-mining town of Blewett. The remains of an arrastre, built in 1861, are preserved here and the history of the site is described on a large marker. Walk south from the parking area along the east side of the road for about 131 ft (40 m) to a prominent outcrop of strongly sheared and slickensided serpentinite. Serpentinite forms

the matrix of the tectonic melange that characterizes the internal structure of the Ingalls Tectonic Complex (Miller, 1985). Original minerals and textures of the parent peridotite have been mostly destroyed in this outcrop. Shear zones and an associated anastomosing foliation strike east-west and dip steeply to the south. Foliated zones wrap around lenses of massive serpentinite, which are particularly abundant in the southern portion of the outcrop. These zones may have developed in an oceanic fracture zone prior to obduction of the Ingalls.

Several small bodies of rodingitized mafic rocks of the Ingalls and a clinopyroxenite dikelet occur within the serpentinite. Rodingitization (Ca^{+2} metasomatism) accompanied serpentinization, resulting in the replacement of plagioclase by hydrogrossular. Some of the rodingites apparently are thin (<10 in [25 cm] thick) dikes, whereas others are tectonic pods ranging up to 13 ft (4 m) in length. Relict primary textures indicate that at least some of the pods are altered gabbros.

A dike of middle Eocene basalt of the Teanaway Formation intrudes the serpentinite and is sub-parallel to the shear foliation in the latter lithology. The dike is slickensided and weakly deformed near its contacts, perhaps as a result of movement along the eastern border fault of the Ingalls Tectonic Complex (see Ruby Creek stop). Slickenside fibers generally plunge moderately to steeply to either north or south.

Stop 3: Drive north from Ruby Creek on U.S. 97 for 1.9 mi (3.1 km) and turn southwest (left) on the road marked Ingalls Creek (Figs. 1, 4). Several stores and a gas station are on the east side of U.S. 97 at the turnoff. Drive southwest for 1.1 mi (1.8 km) and turn right (northeast) on logging road 2317. This road is steep, but passable. Drive up road 2317 for 0.4 mi (0.7 km) to a hairpin curve, beyond which is a gate that is generally locked. Walk or drive another 0.6 mi (1.0 km) up the road (Fig. 4). From here, walk southward to the flat segment of the ridge between Ingalls Creek and Hansel Creek. This segment can be reached most directly by walking up a short, steep slope (about 200 ft; 60 m of elevation gain). However, this route is brushy and one may prefer to traverse southeastward to the ridge and then climb along the ridge to the flat segment. A forest fire burned this portion of the ridge in 1976. Upon reaching the ridge crest, examine exposures of mafic rocks of the Ingalls Tectonic Complex along the flat segment of the ridge and eastward down the ridge. J. M. Mattinson (1981, personal communication) obtained a 156 Ma U-Pb date on gabbro from this ridge.

Numerous outcrops display spectacular examples of heterogeneous deformation and metamorphism of mafic rocks. Diabases and gabbros with typical igneous textures are cut by zones of strongly foliated and lineated mylonitic amphibolites that range from an inch (a few cm) to greater than 3 ft (1 m) in thickness. Some zones show sigmoidal fabrics typical of ductile shear zones (Ramsay and Graham, 1970), whereas others consist of nearly uniform tectonites. In a few places, weakly foliated metagabbros are cut by anastomosing ductile shear zones that rotate their earlier fabric. Marked grain-size reduction characterizes all of these shear zones.

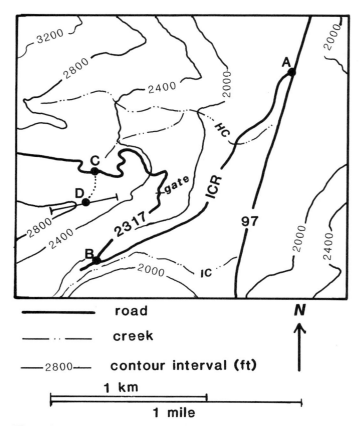

Figure 5. Sandstone, conglomerate, and mudstone of the Swauk Formation interpreted as distal alluvial fan deposits. Stop 4. Person (lower left) for scale.

Figure 4. Map showing access to Stop 3 on the divide between Hansel Creek (HC) and Ingalls Creek (IC) (base from the Liberty 15-minute quadrangle). Turn off U.S. 97 onto Ingalls Creek Road (ICR) at A. Turn right on to road 2317 at B. Drive and(or) walk to C. Walk to outcrops at D along ridge crest.

they were cooling. According to Miller (1985), the mafic rocks were intruded at a spreading ridge, and then deformed shortly thereafter in an active segment of an oceanic fracture zone. Dynamothermal metamorphism occurred where the rocks were sheared, but only static metamorphism ensued where the mafic rocks were spared from shearing.

Strongly tectonized tonalitic sills and dikes intrude both the massive and foliated mafic rocks at several places on the ridge. These intrusive bodies are now biotitic gneisses and have a strong mineral lineation and moderately developed foliation. Large relict plagioclases suggest an original porphyritic texture. These rocks may be silicic differentiates of the Ingalls or are possibly related to the nearby Late Cretaceous Mount Stuart batholith.

OTHER OUTCROPS OF THE SWAUK FORMATION OF THE RUBY CREEK AREA

Stop 4: Drive south from the Ruby Creek turnoff on U.S. 97 for 5.5 mi (8.9 km) and pull off on the south (right) side of the road. Road cuts on both sides of the highway expose east-dipping interbedded conglomerate, fine- to coarse-grained sandstone, and mudstone (Fig. 5). Conglomerate beds are massive to crudely stratified, poorly sorted, and contain clasts as large as 10 in (25 cm) in maximum dimension. Clast lithologies are similar to those at the Ruby Creek outcrop and had their main source in the Ingalls Tectonic Complex. Sandstones are crossbedded, massive, and flat-bedded. Mudstones are both massive and laminated. Locally sandstones and mudstones are extensively burrowed and mottled.

Conglomerates at Stop 4 are texturally and compositionally similar to conglomeratic alluvial fan deposits at Ruby Creek and in the Swauk Pass area (Fraser, 1985), and have an inferred distal alluvial fan origin. Abundant crossbedding suggests stream processes were dominant. Correlative strata to the southeast grade laterally into lake beds (Roberts, 1985). Locally abundant bioturbation might have occurred along the margins of this lake.

Foliation is defined by alternating layers of hornblende and plagioclase, and lineation by aligned fine- to medium-grained amphibole. Foliation generally strikes east-west and dips moderately to steeply to the south. Lineation typically plunges about 35°–40° to the east.

A variety of primary features are preserved in the weakly deformed fine-grained to pegmatitic mafic rocks. Gabbros probably predominate, but diabasic textures are also common. Some diabases occur as dikes and irregular bodies that intrude the gabbros. Basaltic dikelets, approximately 4 in (10 cm) in thickness, also cut gabbro locally.

Metamorphism of the mafic rocks is markedly heterogeneous. Massive rocks show greenschist-facies assemblages and static textures compatible with hydrothermal, ocean-floor-type metamorphism. The amphibolitic zones are characterized by the syn-kinematic assemblage of hornblende-plagioclase. The upper-amphibolite-facies assemblage of diopside-calcic plagioclase (An_{68}) occurs locally (Miller, 1980).

The variety of mineral assemblages and fabrics probably resulted from heterogeneous shearing of the mafic rocks while

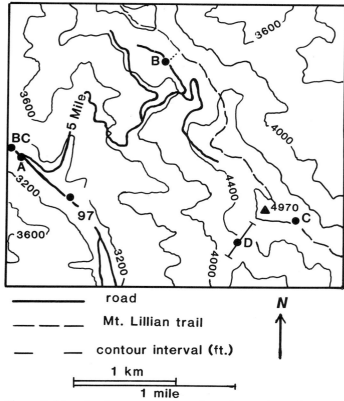

A

B

Figure 6. Map showing access to Stop 5 on the southwest flanks of Tronsen Ridge (base from the Liberty 15-minute Quadrangle). Turn off U.S. 97 at A near Bonanza Campground (BC) on to Five Mile Road and drive to B. Walk to the crest of Tronsen Ridge and walk south on the Mount Lilliam trail to C. Walk off the trail down the southwest-facing slope of Tronsen Ridge to examine the outcrops at D.

Figure 7. A. View to the northwest of the fine-grained strata at stop 5, interpreted as lacustrine deposits. B. Close up of the Swauk Formation on Tronsen Ridge. Resistant sandstone ledges form the upper parts of coarsening-upward cycles interpreted as delta-lobe deposits.

On the north side of the road, steeply-dipping Swauk beds are cut by a prominent vertical basaltic dike of the 47 Ma Teanaway dike swarm (Tabor and others, 1984). This intrusive relationship indicates that folding of the Swauk Formation occurred before 47 Ma.

Stop 5: Drive north from the Ruby Creek outcrops for 7.0 mi (11.3 km) and turn left on Five Mile Road (log road 7224), approximately 0.1 mi (0.1 km) past the entrance to the Bonanza Campground (Figs. 1, 6). This road is maintained by logging companies, and its quality is dependent on recent logging activity. Most of the time it can be easily traveled by passenger car. Drive up this road for approximately 3.3 mi (5.3 km) and park in the saddle just below the crest of Tronsen Ridge. Walk south on the ridge crest on the Mount Lillian trail for about 1.2 mi (2 km). Turn west off the trail and walk to the summit of the peak 4970 ft (1515 m) up. Walk southwest down the southwest-facing slope of Tronsen Ridge through several hundred yards of well-exposed Swauk Formation (Fig. 7).

These strata consist dominantly of repetitive coarsening-upward cycles consisting from base to top of (1) massive black mudstone, (2) ripple- and flat-laminated mudstone and very fine-grained sandstone, and (3) flat- and ripple-laminated fine- and medium-grained sandstone. Cycle thickness varies considerably but is typically about 33 ft (10 m; Roberts, 1985). Crossbedded sandstones fill erosional channels in the tops of some cycles. Roberts (1985) has interpreted these strata as mainly lacustrine delta deposits, with coarsening-upward cycles reflecting progradation and abandonment of delta lobes. Channeled sandstones represent fluvial facies of the delta plain. Paleocurrent directions from the fluvial facies suggest sediment transport to the east (Roberts, 1985). These lake beds interfinger with and overlie the conglomeratic alluvial fan deposits viewed at Ruby Creek and stop 4.

Stop 6: Drive south from the Ruby Creek outcrops for approximately 12.5 mi (20.1 km) on U.S. 97. Pull off on the south (left) side of the road opposite bold outcrops of the Swauk Formation, about 1,300 ft (400 m) south of Swauk Pass (Fig. 8).

Figure 8. Interbedded sandstone and mudstone of the Swauk Formation near Swauk Pass.

This outcrop consists of more than 60 ft (20 m) of interbedded sandstone and mudstone. Sandstones form thick (>6 ft; 2 m) trough crossbedded units and thin (<6 ft; 2 m) ripple- and flat-laminated sheet-like bodies. Sheet sandstones are commonly mottled and rooted. Mudstones are typically massive, flat-laminated, or ripple-laminated. Abundant organic debris and plant fossils are commonly preserved on laminae surfaces. Fraser (1985) interpreted these strata as meandering river deposits. Crossbedded sandstones, sheet sandstones, and mudstones are interpreted as channel, crevasse splay, and overbank deposits, respectively.

These deposits characterize a >7,200-ft-thick (>2,200 m) stratigraphic unit in the Swauk Pass area that underlies the alluvial fan and lacustrine deposits seen at Ruby Creek and at Stops 4 and 5. Based on paleocurrent data, Fraser (1985) and Taylor (1985) have inferred sediment sources to the north and east.

Stop 7: Drive north on U.S. 97 from the Ruby Creek out-

crops for 14.6 mi (23.5 km). Pull off and park on the left (south) side of the road. Walk into the quarry north of the road to examine the Silver Pass Volcanic Member (Tabor and others, 1982, 1984) of the Swauk Formation, which occurs as an interbed in the unit of fluvial deposits viewed at Stop 6. Here, the Silver Pass consists of about 50 ft (15 m) of volcanic breccia, lapilli tuff, tuff, and tuffaceous sandstone (Fraser, 1985). Breccia and tuff are massive and show varying degrees of mixing with arkosic debris. Lapilli are moderately sorted and commonly rounded. Tuffaceous sandstone is massive or crudely stratified. Based on (1) the presence of nonvolcanic detritus, (2) crude stratification in the sandstone, (3) rounded lapilli, and (4) gradational contacts with underlying and overlying fluvial deposits, Fraser (1985) interpreted these rocks as air fall deposits that were slightly reworked by fluvial processes. Tabor and others (1982, 1984) obtained fission track dates on zircon of 48.4 ± 2.3 Ma and 50.5 ± 1.2 Ma from this Silver Pass interbed.

HISTORY OF THE SWAUK FORMATION IN THE RUBY CREEK AREA

The Swauk Formation in the Ruby Creek area records a dynamic interaction between tectonics and sedimentation. The 50 to 52 Ma Silver Pass Volcanic Member (Stop 7) is the oldest part of the Swauk Formation viewed in the Ruby Creek area, and is overlain by about 1,970 ft (600 m) of fluvial deposits (Stop 6) derived from northern and eastern sources. Subsequently, Swauk drainage patterns in the Ruby Creek area were reorganized and more than 9,850 ft (3,000 m) of alluvial fan (Stops 1 and 4), lacustrine (Stop 5), and fluvial sediments at least partly derived from western sources were deposited. This thick alluvial sequence was folded prior to the 47 Ma intrusion of Teanaway dikes (Stop 4). This dramatic history of basin margin faulting, extremely rapid subsidence, reorganizations in drainage patterns, and intrabasinal volcanism is characteristic of Eocene nonmarine basins in Washington (Johnson, 1985).

REFERENCES CITED

Fraser, G. T., 1985, Stratigraphy, sedimentology, and structure of the Swauk Formation in the Swauk Pass area, central Cascades, Washington [M.S. thesis]: Pullman, Washington State University, 218 p.

Johnson, S. Y., 1985, Eocene strike-slip faulting and basin formation in Washington, in Biddle, K. T., and Christie-Blick, N., eds., Strike-slip deformation, basin formation, and sedimentation: Society of Economic Paleontologists and Mineralogists, Special Publication 37, p. 283–302.

Miller, R. B., 1980, Structure, petrology, and emplacement of the ophiolitic Ingalls Complex, north-central Cascade Mountains, Washington [Ph.D. thesis]: Seattle, University of Washington, 422 p.

——— , 1985, The ophiolitic Ingalls Complex, north-central Cascades Mountains, Washington: Geological Society of America Bulletin, v. 96, p. 27–42.

Pratt, R. M., 1958, Geology of the Mount Stuart area, Washington [Ph.D. thesis]: Seattle, University of Washington, 229 p.

Ramsay, J. G., and Graham, R. H., 1970, Strain variations in shear belts: Canadian Journal of Earth Sciences, v. 7, p. 786–813.

Roberts, J. W., 1985, Stratigraphy, sedimentology, and structure of the Swauk Formation along Tronsen Ridge, central Cascades, Washington [M.S. thesis]:

Pullman, Washington State University, 186 p.

Smith, G. O., 1904, Description of the Mount Stuart Quadrangle: U.S. Geological Survey Atlas, Mount Stuart folio, no. 106, 10 p.

Southwick, D. L., 1974, Geology of the alpine-type ultramafic complex near Mount Stuart, Washington: Geological Society of America Bulletin, v. 85, p. 391–402.

Tabor, R. W., Waitt, R. B., Jr., Frizzell, V. A., Jr., Swanson, D. A., Byerly, G. R., and Bentley, R. D., 1982, Geologic map of the Wenatchee 1:100,000 Quadrangle, Washington: U.S. Geological Survey Miscellaneous Investigations Map MI-1311.

Tabor, R. W., Frizzell, V. A., Jr., Vance, J. A., and Naeser, C. W., 1984, Age and stratigraphy of lower and middle Tertiary volcanic rocks of the central Cascades, Washington; Application to the tectonic history of the Straight Creek fault: Geological Society of America Bulletin, v. 95, p. 26–44.

Taylor, S. B., 1985, Stratigraphy, sedimentology, and paleogeography of the Swauk Formation in the Liberty area, central Cascades, Washington [M.S. thesis]: Pullman, Washington State University, 202 p.

Dry Falls of the Channeled Scabland, Washington

Victor R. Baker, Department of Geosciences, University of Arizona, Tucson, Arizona 85721

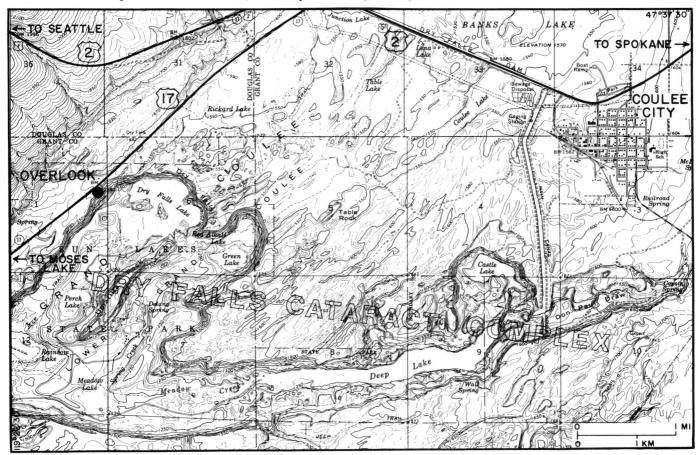

Figure 1. Part of Coulee City 7½-minute Quadrangle; contour interval, 10 ft. (3 m). All but the upper left (northwest) and lower left (southwest) corners of this quadrangle were swept by cataclysmic Missoula flood water. The Dry Falls cataract complex is located southwest of Coulee City, extending to the west margin of the quadrangle.

LOCATION

Dry Falls is located on the Coulee City 7½-minute Quadrangle (Fig. 1) at Lat. 47°36′N, Long. 119°21′W; sec. 6, T.24N., R28E, in Grant County, Washington. It can be reached by passenger car or bus on Washington 17, 18.5 mi (29.6 km) north of Soap Lake or 1½ mi (2.4 km) south of the junction of U.S. 2 and Washington 17 near Coulee City, Washington. The falls are easily viewed from a public overlook at Dry Falls State Park.

SIGNIFICANCE

Dry Falls (Fig. 2) is a great cataract 3.3 mi (5.5 km) wide and 396 ft (120 m) high, that formed during cataclysmic late Pleistocene floods that emanated from glacial Lake Missoula in Montana. Between 16,000 and 12,000 years ago, glacial ice impounded as much as 70×10^{12} ft^3 (480 mi^3) [2×10^{12} m^3 (2,000 km^3)] of water in western Montana. The glacial dam, located in northern Idaho, failed repeatedly, releasing immense discharges of water into the Columbia River system (Fig. 3). Much of this water was diverted across the Columbia Plateau, where it formed the region that was named the Channeled Scabland by the late J Harlen Bretz of the University of Chicago. Hydraulic calculations for the flood peak (Baker, 1973) indicate that the flows achieved a peak discharge of more than 700×10^6 ft^3/sec (20×10^6 m^3/sec), and mean flow velocities reached 33 to 100 ft/sec (10 to 30 m/sec). Dry Falls formed as this high-velocity flood water eroded the Miocene basalt layers of the Grand Coulee region.

SITE INFORMATION

Dry Falls consists of a series of horseshoe-shaped headcuts with plunge pool lakes in alcoves at the base of cliffs 396 ft (120 m) high (Fig. 2). The cataract occurs at the northern end of the

369

Figure 2. Oblique aerial view of Dry Falls, showing the Grand Coulee in the background (with Banks Lake covering its floor). Note the longitudinal grooves on the basalt surface above the cataract lip.

Lenore Canyon area, now containing Park Lake and Lenore Lake. The whole complex was excavated as cataclysmic flood water incised the zone of fractured basalt along the Coulee Monocline (Bretz, 1932, 1969). The cataracts head an inner channel (Baker, 1981) that receded headward into the Hartline Basin, near Coulee City, Washington.

The basalt surface north of the Falls is marked by a prominent series of longitudinal grooves, aligned with the paleoflow direction of the cataclysmic flood and cut across structural trends (Baker, 1981). The grooves, which lead right up the lips of the cataract, measure up to 10 ft (3 m) in height and are spaced 99 to 198 ft (30 to 60 m) apart (Fig. 2). Plunge pool lakes (Dry Falls Lake, Green Lake, Deep Lake, and Castle Lake) occur in alcoves at the base of the cataract. Large accumulations of boulders can be observed immediately downstream of the plunge pools.

Dry Falls is one of numerous areas where the cataclysmic Pleistocene floods produced spectacular erosion forms in basalt bedrock. Planes of weakness within the basalt bedrock were an important influence on fluvial erosional forms produced by Missoula flooding. The individual basalt flows average 82.5 to 198 ft (25-60 m) thick. They are characterized by a variety of depositional and cooling features, which provided variable resistance to

the flood erosion (Fig. 4) that occurred as a result of the turbulent plucking action of the flood water.

Figure 5 shows a series of hypothetical cross sections that illustrate how Missoula flood water eroded the Columbia Plateau materials. The first flood water encountered a plateau capped by loess hills (top diagram in Fig. 5). The phase (I) was then followed by a series of changes as the underlying basalt was exposed and eroded (lower diagram in Fig. 5). First, the high-velocity water quickly exposed the underlying basalt, leaving an occasional streamlined loess hill as a remnant of the former cover (Phase II). The entablature of the uppermost basalt flow was then encountered. This probably yielded to groove development, possibly associated with longitudinal roller vortices. The first exposure of well-developed columnar jointing, perhaps at the top of a flaring colonnade along the irregular cooling surface, introduced a very different style of erosion (Phase III). Large sections of columns could now be removed at this site with the simultaneous development of vertical vortices (kolks). With the enlargement and coalescence of the resultant potholes, the surface assumed the bizarre butte-and-basin topography that characterizes much of the Channeled Scabland (Phave IV). The eventual topographic form was the development of a prominent inner channel (Phase

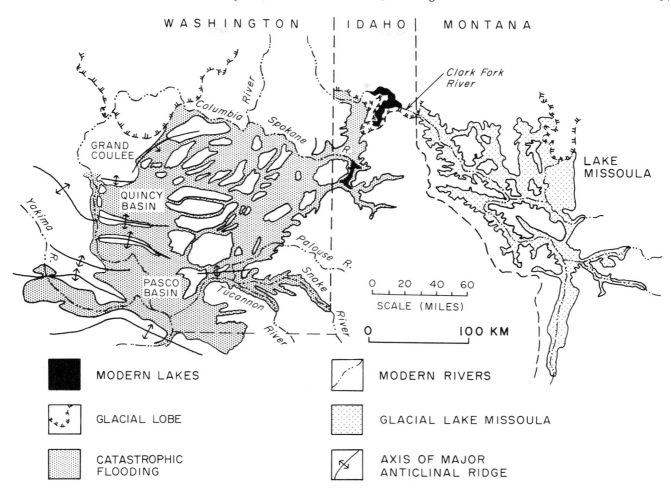

Figure 3. Generalized map showing the relationship of glacial Lake Missoula to the Channeled Scabland region in eastern Washington.

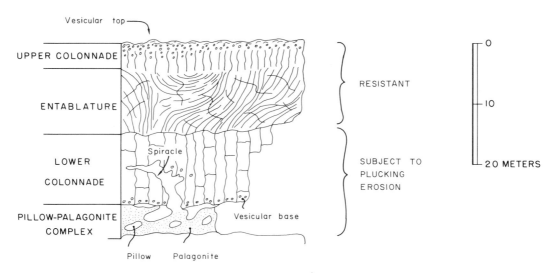

Figure 4. Cross section of an idealized Yakima Basalt flow showing structural features important to flood erosion processes. Note that large columns and pillow-palagonite layers were preferentially eroded, undermining the more resistant layers.

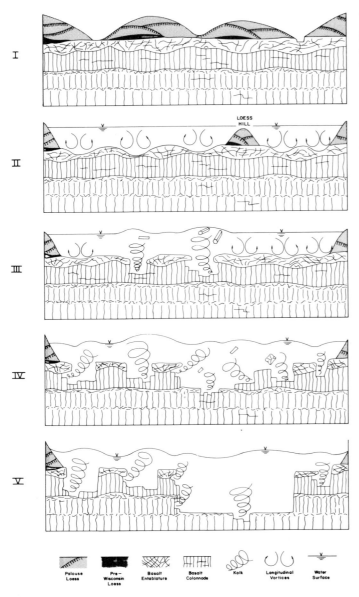

Figure 5. Hypothetical sequence of flood erosion (from top to bottom) showing the action of turbulent cataclysmic flows. Longitudinal vortices and kolks (vertical vortices) eroded the basalt by plucking action. The lateral enlargement of inner channels probably proceeded by the under-cutting of resistant entablature, as columns were plucked out by kolks.

V). Such inner channels may have been initiated downstream at structural steps in the basalt, and then migrated headward by cataract recession. The Dry Falls area is an excellent example of this type of scabland erosion.

The origin of the Channeled Scabland was the subject of one of the most famous controversies in the history of geology (Baker, 1978). When Bretz formulated the cataclysmic flood hypothesis for this region in 1923, his idea was rejected by much of the geological community. In 1973, after 50 years of controversy, the quantitative consequences of the hypothesis were shown to be valid. Even more exciting has been the demonstration that scabland-type erosion processes occurred on the planet Mars (Baker, 1982).

REFERENCES CITED

Baker, V. R., 1973, Paleohydrology and sedimentology of Lake Missoula flooding in eastern Washington: Geological Society of America Special Paper 144, 79 p.

——, 1978, The Spokane Flood controversy and the origin of the Channeled Scabland: Science, v. 202, p. 1249–1256.

——, 1981, Catastrophic Flooding; The origin of the Channeled Scabland: Stroudsburg, Pennsylvania, Dowden, Hutchinson, and Ross, 360 p.

——, 1982, The channels of Mars: Austin, Texas, The University of Texas Press, 198 p.

Bretz, J H., 1928, The Channeled Scablands of eastern Washington: Geographical Review, v. 18, p. 446–477.

——, 1932, The Grand Coulee: American Geographical Society Special Publication 15, 89 p.

——, 1969, The Lake Missoula floods and the Channeled Scabland: Journal of Geology, v. 77, p. 505–543.

Melange rocks of Washington's Olympic coast

Weldon W. Rau, *Washington State Department of Natural Resources, Division of Geology and Earth Resources, Olympia, Washington 98504*

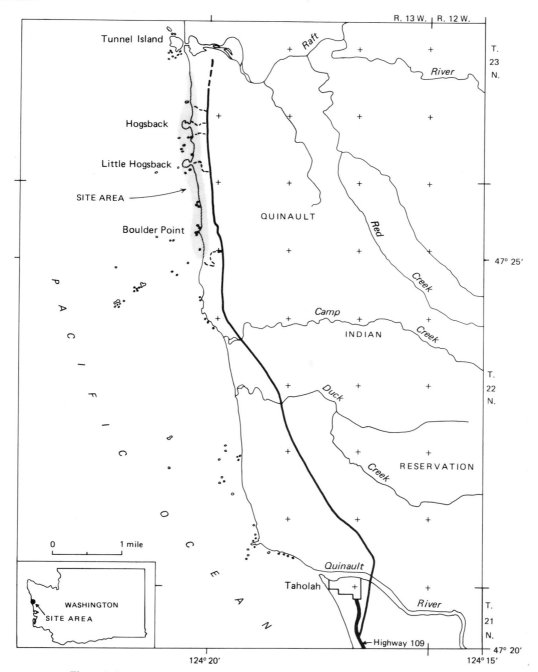

Figure 1. Index map showing site location of melange rocks of the Washington coast.

INTRODUCTION

One of the most impressive outcrops of melange rocks along the Washington coast is exposed continuously for 2.5 mi (4 km) in sea cliffs immediately south of the mouth of Raft River on the Quinault Indian Reservation, Grays Harbor County (Taholah 15-minute quadrangle) (Fig. 1). This site affords a rare opportunity for viewing a rock type that has resulted largely from tectonism and is particularly significant to an understanding of the general geology and tectonic framework of the Olympic Moun-

Figure 2. Hoh melange rocks exposed in the Hogsbacks area.

tains and adjacent coastal-offshore region. Locally the site is referred to as the Hogsbacks area after two major promontories, Little Hogsback to the south, and Hogsback to the north.

Access to the area is from the south and from the Indian community of Taholah where State Highway 109 ends. The site

is 5 mi (8 km) north of Taholah and is reached by a gravel road. Because it is a beach locality on the Quinault Indian Reservation, permission to walk on the beaches must be obtained from the Quinault Tribal Office in Taholah. Instructions will also be given there for the best route from the road to the beach. Several trails are less than 0.25 mi (0.4 km) in length.

THE SITE

The sea cliffs and adjacent beach outcrops of this area are composed of a melange of rocks consisting of chaotically arranged blocks of siltstone, mudstone, bedded siltstone and sandstone, and graywacke sandstone, together with exotic blocks of altered volcanic materials and metamorphic rocks set in a mudlike, relatively soft matrix of clays and siltstone fragments (Fig. 2). The materials of this and other similar outcrops along the coast have been informally referred to as "Hoh breccia" after a local Indian name. Native Americans and early settlers called them "smell muds" due to the petroliferous odor discernible, particularly from freshly broken blocks of siltstone. Resistant blocks vary greatly in size. The largest, as erosional remnants, form the major headlands and offshore rocks of this coastal area (Fig. 3). Phacoidal shapes are common among smaller resistant blocks, and surfaces are commonly slickensided. Secondary clay minerals are common throughout and consist largely of kaolinite, illite-montmorillonite, and chlorite (Rau and Grocock, 1974). The expandable nature of these clay minerals is a major factor in the structural weakness of melange outcrops, particularly along the coast where high tides together with storm waves frequently moisten the outcrops. Calcite veins are common in many of the resistant blocks.

RELATIONS

Zones of melange rocks have been mapped in detail along

Figure 3. Looking northward at the Hogsbacks area where several mi (km) of continuous outcrop expose a major zone of Hoh melange.

Figure 4. Contact between the dark-colored Hoh melange and overlying lighter colored siltstone of the Quinault Formation, exposed at the south end of the site.

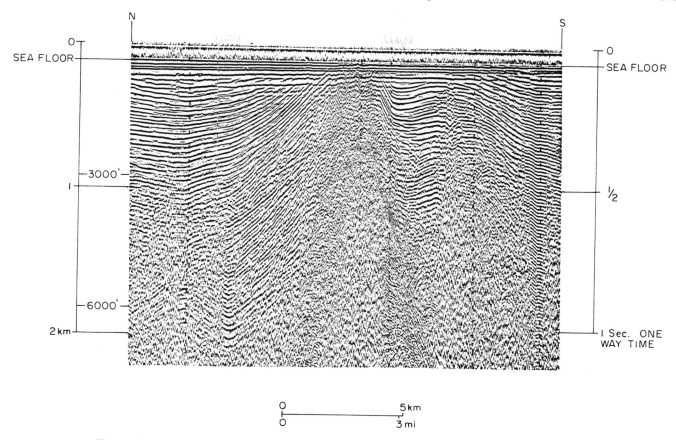

Figure 5. A diapiric structure as interpreted from an offshore seismic profile near Point Grenville. Photo courtesy of Parke D. Snavely, U.S. Geological Survey.

the coastal area for some 45 mi (72 km) between Point Grenville and LaPush and inland some 8 to 10 mi (13 to 16 km) (Rau, 1975, 1979). Similar rocks are known to extend northward to at least The Point of Arches (Snavely and others, 1980a). To the south, outcrops are rare but subsurface data indicate such melange materials are at depth at least as far south as the Ocean City area (Rau and McFarland, 1982). Offshore data from a limited number of wells and interpretations from seismic records indicate that materials comparable to those of this site are extensive in places, largely at depth beneath the Continental Shelf (Snavely and Wagner, 1981, 1982).

GEOLOGIC AGE

Foraminiferal faunas from five locations scattered throughout this site, as well as others nearby, all indicate an early to middle Miocene age (Saucesian-Relizian Stages) (Rau, 1975). However, older faunas ranging from middle Eocene to Oligocene in age (Ulatisian, Narizian, Refugian, and Zemorrian stages) are known from localities both to the north and south of this immediate coastal area (Rau, 1975, 1979; Snavely, personal communication, 1985).

Younger strata of the Quinault Formation of Pliocene age overlie melange rocks at the southern end of this melange out-

crop. Depending upon conditions of the sea cliffs due to almost continuous slumping, the precise contact can sometimes be seen gently dipping southward (Fig. 4). The dark gray color of melange rocks is in marked contrast with the light gray color of the Quinault Formation. Alluvium conceals the northern contact, but the Quinault Formation is again exposed about 0.25 mi (0.4 km) to the north at the mouth of Raft River.

Melange rocks of this coastal region of Washington are regarded as one of two major groups of rocks that make up the Hoh rock assemblage (Rau, 1973). The second group of Hoh rocks is strata believed to be largely turbidite in origin, consisting of tightly folded, steeply dipping but coherent sequences of siltstone, graywacke sandstone, and conglomerates. Geologic mapping between Point Grenville and LaPush reveals that zones of melange Hoh rocks similar to those exposed at this site separate large packets of Hoh sedimentary sequences measuring several miles in length and breadth. These strata contain foraminiferal faunas similar in age to those of the melange rocks of this site and other nearby outcrops.

ORIGIN

The origin of melange rocks of this and other outcrops along the west coast of the Olympic Peninsula has long been a topic of discussion among geologists. In recent years it has been concluded that major tectonic forces directly related to the convergence of the Pacific and North American plates, subduction, and the accretion of rocks to the North American plate have generated these zones of melange. Several variations on the specific mechanism of origin have been presented. Stewart (1971) stressed that major faulting or differential movements between large segments of the earth's crust have created shear zones and thus formed a chaotic arrangement of various rock types. Weissenborn and Snavely (1968) suggested that a combination of gravity thrust and intraformational gravity sliding may have been the major process of origin. Rau (1979), based on mapping between Point Grenville and LaPush, also interpreted melange zones as major thrust zones between plate segments of sedimentary sequences. Large blocks of sedimentary sequences appear to be separated by a combination of major north-trending thrust zones of melange rocks and other, largely strike-slip, northeast-trending faults.

Recent studies of data from offshore wells together with interpretations of seismic data support the general concept that, as the result of plate convergence, melange materials, like these seen along the Washington coast, were generated in zones of shearing primarily along major thrusts (Snavely, personal communication, 1985). During middle late Eocene and late middle Miocene times, two principal accretionary terranes of melange and broken formations are believed to have been generated on the continental margins of Washington and Oregon (Snavely and others, 1980b; Snavely and Wagner, 1982). Onshore exposures of these tectonostratigraphic units along the west side of the Olympic Peninsula were apparently transplanted eastward along fault-bounded "mini-plates" (Snavely and Wells, 1984). Snavely (written communication, 1985) refers to melange terranes formed in middle late Eocene time and exposed largely along the north part of the coast as the "Ozette assemblage" and the upper middle Miocene melange terrane of this coastal area as the Hoh rock assemblage of Rau (1973).

In addition to a tectonic framework of thrusting and subduction, perhaps combined with major gravity sliding, it is also believed that many melange rocks have further moved upward as diapirs. The apparent mobility of materials of this and other outcrops together with offshore seismic records (Fig. 5) provides substantial evidence that such materials do move upward through other Hoh and similar rocks as well as, in some places, into overlying younger rocks of the Quinault Formation and equivalent offshore strata (Rau and Grocock, 1974). Furthermore, seismic-reflection profiles on the continental shelf of both Oregon and Washington indicate that some diapirs are presently growing as they deform and breach Holocene sediments (Snavely and Wagner, 1982).

REFERENCES CITED

Rau, W. W., 1973, Geology of the Washington coast between Point Grenville and the Hoh River: Washington Division of Geology and Earth Resources Bulletin 66, 58 p.
—— 1975, Geologic map of the Destruction Island and Taholah quadrangles, Washington: Washington Division of Geology and Earth Resources Geologic Map GM–13, scale 1:62,500.
—— 1979, Geologic map in the vicinity of the lower Bogachiel and Hoh River valleys, and the Washington coast: Washington Division of Geology and Earth Resources Geologic Map GM–24, scale 1:62,500.
Rau, W. W., Grocock, G. R., 1974, Piercement structure outcrops along the Washington coast: Washington Division of Geology and Earth Resources Information Circular 51, 7 p.
Rau, W. W., McFarland, C. R., 1982, Coastal wells of Washington: Washington Division of Geology and Earth Resources Report of Investigations 26, 4 sheets.
Snavely, P. D., Jr., Niem, A. R., MacLeod, N. S., Pearl, J. E., Rau, W. W., 1980a, Makah Formation—A deep marginal basin sedimentary sequence of late Eocene and Oligocene age in the northwestern Olympic Peninsula, Wash-

ington: U.S. Geological Survey Professional Paper 1162-B, 28 p.
Snavely, P. D., Jr., and Wagner, H. C., 1981, Geologic cross section across the continental margin off Cape Flattery, Washington, and Vancouver Island, British Columbia: U.S. Geological Survey Open-File Report 81–978, 6 p.
—— 1982, Geologic cross section across the continental margin of southwestern Washington: U.S. Geological Survey Open-File Report 82–459, 10 p.
Snavely, P. D., Jr., Wagner, H. C., and Lander, D. L., 1980b, Geological cross section of the central Oregon continental margin: Geological Society of America Map and Chart Series MC–28J, scale 1:250,000.
Snavely, P. D., Jr., and Wells, R. E., 1984, Tertiary volcanic and intrusive rocks on the Oregon and Washington Continental Shelf: U.S. Geological Survey Open-File Report 84–282, 17 p.
Stewart, R. J., 1971, Structural framework of the western Olympic Peninsula, Washington: Geological Society of America Abstracts with Programs, v. 3, no. 2, p. 201.
Weissenborn, A. E., and Snavely, P. D., Jr., 1968, Summary report on the geology and mineral resources of Flattery Rocks, Quillaute Needles, and Copalis National Wildlife Refuges, Washington: U.S. Geological Survey Bulletin 1260–F, p. F1–F16.

A Tertiary accreted terrane: Oceanic basalt and sedimentary rocks in the Olympic Mountains, Washington

R. W. Tabor, U.S. Geological Survey, 345 Middlefield Road, Menlo Park, California 94025

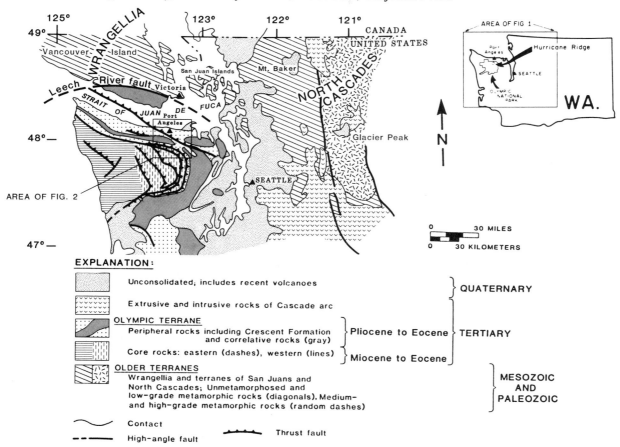

Figure 1. Generalized geologic map of northwestern Washington showing structural setting of the Olympic terrane. Data from King and Beikman (1974), Tabor and Cady (1978a), and Snavely (in Muller and others, 1983, p. 11).

INTRODUCTION AND GEOLOGIC SIGNIFICANCE

The rocks of the Olympic Peninsula and their counterparts to the south in the Coast Ranges of southern Washington and Oregon comprise one or more Tertiary accreted terranes which are part of the collage of terranes that make up most of northwestern Washington (Coney and others, 1980).

The Olympic Mountains in Olympic National Park are one of the best examples of an accreted subduction complex of Tertiary age found in the conterminous United States. The Hurricane Ridge Road, south of Port Angeles, crosses parts of the complex where the observer can see spectacular outcrops of pillow basalt, pelagic interbeds, and submarine fan deposits, as well as the complex structures engendered during accretion of these rocks to the continental margin of North America.

Terrane (or tectonostratigraphic terrane) as used in this

guide refers to a discrete structural block with stratigraphic and/or structural coherence; it is separated from its neighboring terranes by faults. Moreover, a large terrane that was coherent during its later geologic history, can be subdivided into smaller terranes whose earlier geologic histories differ.

LOCATION

Hurricane Ridge can be reached via Port Angeles (Fig. 1), an industrial town perched on a terrace below the mountain escarpment overlooking the Strait of Juan de Fuca. The town is about two hours by automobile and ferry, and about 45 minutes by air, from Seattle.

In Port Angeles (Fig. 2), the Hurricane Ridge Road branches west from 8th Street a short distance south of the Olympic National Park Visitors Center. The Visitors Center is 1.1

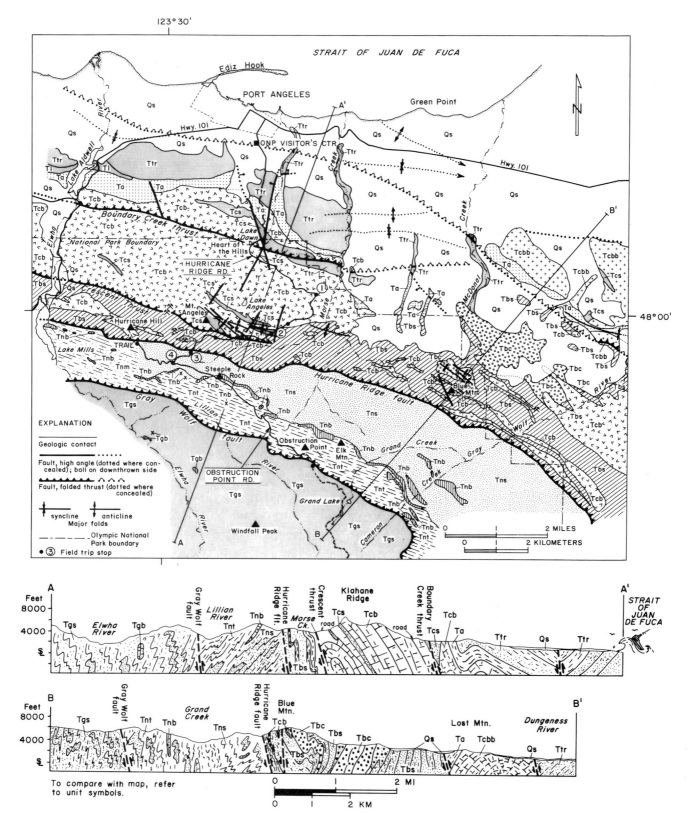

Figure 2. Simplified geologic map of the Hurricane Ridge area modified from Tabor (in Muller and others, 1983, p. 32-34).

mi (1.8 km) south of U.S. Highway 101 in Port Angeles. The road proceeds for 17.4 mi (28.0 km) up Hurricane Ridge to the Hurricane Ridge Lodge at an altitude of about 5,000 ft (1,525 m). Except for sightseeing buses in operation during the summer, there is no public transportation to Hurricane Ridge. However, car-rental services are available in Port Angeles. The best time to visit Hurricane Ridge is in July or August, the months that offer the best weather. The weather also may be fair in June, September, and October, but with less certainty. The Hurricane Ridge Road is kept open on most weekends during the winter months, but snow may obscure almost all outcrops.

The area described in this guide is entirely within Olympic National Park, a place designated for the preservation of the natural world; collecting of samples is not allowed without special permit. To prevent misunderstandings with other visitors and

park rangers, the geologist would be wise to keep his hammer out of sight.

GEOLOGIC OVERVIEW

The Olympic Mountains comprise two major geologic terranes: (1) a peripheral belt of lower and middle Eocene oceanic basaltic rocks of the Crescent Formation and overlying middle Eocene to Miocene and minor Pliocene marine sedimentary rocks that wraps partly around (2) a core of marine sedimentary rocks that are approximately coeval with the peripheral rocks but everywhere in fault contact with them (Fig. 1). Peripheral rocks are fossiliferous, faulted, and folded, but generally stratigraphically continuous. The major basal unit of the peripheral rocks, the Crescent Formation, is a very thick sequence of tholeiitic pillow basalts, breccias, rare columnar basalt near the top of the section, and interbedded clastic and pelagic sedimentary rocks. The Crescent Formation forms a horseshoe-shaped outcrop pattern, open to the west.

Within the arms of the basaltic horseshoe, the western part of the core is mostly Eocene to Miocene marine sedimentary rocks. It is nonslaty and locally includes coherent packets of rocks that are stratigraphically continuous. Complex folds and faults are common, and some areas are so totally disrupted that the rocks have the aspect of melange (Stewart, 1970; Rau, 1973, 1980; Tabor and Cady, 1978a, 1978b).

In the eastern core, the Eocene to early Oligocene rocks have a pronounced slaty fabric. They are pervasively sheared and best characterized as broken formations. The eastern core rocks are mostly shale, siltstone, and sandstone with minor amounts of conglomerate, basalt, basaltic volcaniclastic rocks, diabase, and gabbro. Sandstones are feldspathic to volcanic subquartzose. Graded beds, sole marks, and rhythmic bedding suggest that the sandstones are turbidites. The sedimentary rocks have been variously metamorphosed to slate and semischist in the prehnite-pumpellyite and the greenschist facies of regional metamorphism. Basaltic rocks have become greenstone and greenschist.

The Olympic core rocks lack blocks of recognizable exotic material such as blueschist or eclogite. Although a few small feldspathic peridotite dikes have been found, no mantle material has been recognized.

Core units are recognized by gross lithology. They form long, irregular, curved packets, roughly concave westward (Fig. 1). They vary from relatively intact interbedded sandstone and slate to completely disrupted broken formations of foliate sandstone or semischist in a matrix of slate or phyllite. Many beds are overturned to the east. Units are separated locally by wide zones of intensely sheared rocks, some of which are locally mylonitic. Eastern core rocks are separated from the peripheral rocks by a folded thrust fault, the Hurricane Ridge fault. Western core rocks are also in fault contact with peripheral rocks (Fig. 1); to the north the original thrust has become strike-slip.

Due to overlying Quaternary deposits, the nature of the contact between the peripheral rocks of the Olympic Peninsula

(Figure 2, continued)

EXPLANATION:

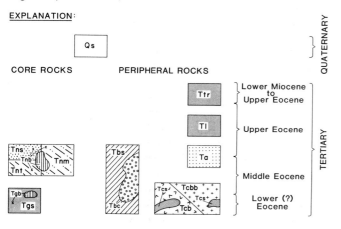

DESCRIPTION OF MAP UNITS

Qs   Surficial deposits—Predominantly glacial drift

CORE ROCKS

NEEDLES-GRAY WOLF LITHIC ASSEMBLAGE
Tns   Sandstone and slate—Gray and greenish brown, fine to medium-grained lithic to feldspathic sandstone with 5-7 percent detrital muscovite and biotite. Foliated sandstone common. Black weathering to silver gray or light brown, slate is micaceous
Tnb   Basaltic rocks—Basalt contains phenocrysts of altered plagioclase and titanaugite, rarely oxyhornblende in intersertal matrix, commonly altered to chlorite, sphene, and calcite. Pillows abundant, some breccia, minor diabase and gabbro. Interbedded tuffs and volcanic-rich sedimentary rocks include rare gray or red limestone
Tnt   Thin-bedded slate and siltstone, with less than 30 percent micaceous sandstone
Tnm   Micaceous sandstone and slate undifferentiated

GRAND VALLEY LITHIC ASSEMBLAGE
Tgs   Sandstone, foliated sandstone, and semischist with 40-70 percent siltstone, slate, and minor phyllite
Tgb   Pillow basalt, greenstone, and minor red argillaceous limestone

PERIPHERAL ROCKS
Ttr   TWIN RIVER GROUP—Sandstone, siltstone, shale, and conglomerate
Tl   LYRE FORMATION—Conglomerate and sandstone
Ta   ALDWELL FORMATION—Siltstone, minor sandstone, and conglomerate

CRESENT FORMATION—Divided into:
Tcbb   Basalt flows and mudflow breccias—Flows are characterized by closely spaced, random joints and locally are columnar jointed, or more rarely pillowed
Tcb   Basalt, massive flows, pillows, and breccia; minor diabase and gabbro—Basalt contains clinopyroxene (rare titanaugite) and soda to calcic plagioclase microphenocrysts in a matrix of chlorite and or montmorillinoid minerals and rare brown glass
Tcs   Volcanic sedimentary rocks—Variegated red to green basaltic breccia, tuff, and volcanic conglomerate. Beds of maroon to green siliceous or limy argillite, thin to thick beds of volcanic sandstone, red foraminiferal limestone, and chert

BLUE MOUNTAIN UNIT—Divided into:
Tbs   Sandstone and argillite—Gray or black, weathering to brown, very fine to medium-grained lithic sandstone, volcanic rich, rarely micaceous; fair to poorly sorted and angular with thin to rarely thick beds. Unit includes rare conglomerate, tuffaceous argillite, and red limestone
Tbc   Conglomerate and pebbly sandstone—Predominantly pebbles and cobbles of volcanic rocks and chert in greenish-gray, very coarse-grained feldspatholithic sandstone

and the Mesozoic and Paleozoic terranes of the adjoining North American continent to the east is unknown (Fig. 1). Most workers consider this unexposed contact a suture between the accreted oceanic basalt and the continent to the east (Glassley, 1974, p. 792; Cady 1975, p. 579), but Johnson (1984) suggests modification of the suture by strike-slip faulting. However, on Vancouver Island, the apparently left lateral Leech River fault separates the Olympic terrane from the much older Wrangellia terrane making up the rest of Vancouver Island and coastal lands beyond (Fairchild and Cowan, 1982, p. 1830–1833; Jones and others, 1977). Recent deep seismic reflection profiles indicate the Leech River fault flattens with increasing depth to the north, suggesting that rocks of the Olympic terrane have been thrust under the older rocks (Yorath and others, 1985, p. 551–552).

The Crescent Formation in Washington and its counterparts in the Coast Ranges of Oregon form a long irregular belt. Paleomagnetic studies of the basalt in southern Washington and Oregon offer dramatic evidence of clockwise rotation of the basaltic belt into its present position along the coast of North America (Simpson and Cox, 1977, p. 585).

Although several versions of rotational history have been proposed, Wells and others (1984, p. 280) summarize the view that most of the rotation took place after accretion and was due to asymmetric backarc spreading with a contribution from the right-lateral shear between the oceanic and North American plates. The previously accreted belt of basalt pivoted on the Olympic Peninsula as the southern end (in Oregon) swung westward.

Somewhat equivocal data suggest that in the Olympic Mountains the horseshoe pattern is due to oroclinal bending as the accreting basalt wedge was pushed into the reentrant between Vancouver Island (that is, Wrangellia) and the North Cascades on the mainland (Livingston and Tobin, 1971; Beck and Engebretson, 1982; see also Tabor, 1975, p. 36).

HURRICANE RIDGE ROAD

The geologic section and structures exposed along the Hurricane Ridge Road can best be viewed at the four stops (Fig. 2) described below.

1. Overview of the structural setting. At 9 mi (14.5 km) from the Visitors Center, an overlook and parking lot (just before the first tunnel cut through pillow basalt of the Crescent Formation) provide a grand view of northern Puget Sound and its major geologic terranes. A National Park Service display points out the major geographic features.

Immediately to the north, the prominent rounded double hill (partly cleared of trees for a subdivision on the southeast side) is composed of Eocene marine sandstone and shale of the Aldwell Formation and overlying Twin River Group of the peripheral rocks (Fig. 2). The lowland of the Olympic Peninsula is carved from the folded rocks of the Twin River Group overlain by thick glacial drift from the Cordilleran ice sheet. The distant mountains on the west across the Strait of Juan de Fuca are underlain by the Metchosin Formation, the Canadian equivalent

of the Crescent Formation. The Leech River fault runs from west to east behind the mountains and swings southward into the Strait of Juan de Fuca south of Victoria and the low hilly islands to the northeast (Fig. 1). These hilly islands are the San Juan Islands, a probable thrust stack of Mesozoic and Paleozoic oceanic rocks, some intact, some sheared to melange (Brandon and others, 1983). Farther east is the western portion of the North Cascade Range, carved from Mesozoic and Paleozoic thrust slices of oceanic rocks. The rocks of the San Juan Islands and the North Cascades were probably in place when the first of the Olympic terranes (that is, the Crescent Formation) arrived, although there may have been some later displacement of terranes due to strike-slip faulting (Johnson, 1984).

On a clear day, the glacier-covered volcanic cones of Mount Baker and Glacier Peak rise white and shining to the northeast and east above the darker mountains in the distance.

2. Accreted oceanic basalt. After rounding the edge of Klahhane Ridge above the valley of Morse Creek (10.3 mi; 16.6 km), the Hurricane Ridge Road passes extraordinary exposures of pillow basalt, basalt breccia, and diabase of the Crescent Formation. The beds dip steeply northeast, and some interbeds of red argillaceous coccolithic limestone can be seen (see Garrison, 1973). Park vehicles carefully on the widened south shoulder of the highway at 12.3 mi (19.8 km) to observe interbeds of basaltic sandstone in the volcanic rocks. Graded bedding in the dark-red and green sandstone at this location indicates that the rocks are on the southern limb of a southward-overturned anticline (Fig. 2, section AA'). Foraminifers from these interbeds are middle Eocene and indicate a warm-water, inner-shelf environment (W. W. Rau as reported by P. Snavely, written communication, 1983).

The geochemistry of the low-K_2O tholeiitic lavas has been examined by numerous workers, but the exact mode of origin is not yet agreed upon. On the basis of geochemistry, the Crescent Formation has been considered to be a typical ridge basalt, part of an island arc, Hawaiian-type seamounts, or combinations of ridge and seamounts (MacLeod and Snavely, 1973; Snavely and others, 1968; Glassley, 1974; Cady, 1975; Glassley and others, 1976; Muller, 1980).

Based on radiometric dating, Duncan (1982) proposes that the Crescent Formation originated near the north end of a symmetrical chain of seamounts generated by a hotspot on a spreading ridge and was subsequently rotated into its present position. Employing a rigorous analysis of rotational history and plate motion, Wells and others (1984, p. 282) also favor the hotspot origin, but stipulate several alternate scenarios, including leaky transform faults, rifting, and reorganization of northeast Pacific plates prior to accretion and differential rotation of Olympic-Coast Range basalt packets.

3. Hurricane Ridge Fault: A Major Subduction Structure. The Crescent Formation is underlain by dark volcanic-rich sandstone and argillite and rare cherty conglomerate of the Blue Mountain unit (Tabor and Cady, 1978a). These rocks crop out along the Hurricane Ridge Road as it winds along the side of Klahhane Ridge. At 14.9 mi (24 km), where there is good parking

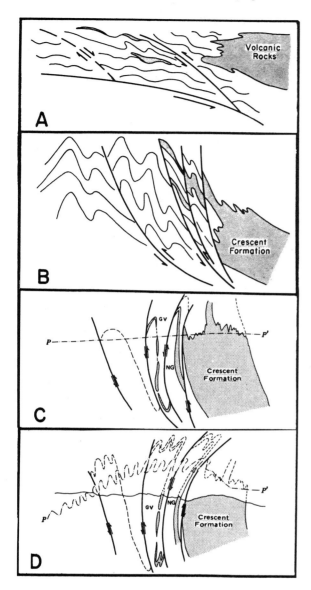

Figure 3. Schematic diagram to show development of Olympic core structure. Modified from Tabor and Cady (1978b). A and B. The thick sequence of lower Tertiary sandstone and shales with thin interbeds of pillow basalt was folded along roughly northwest-southeast-trending axes, imbricated, and underthrust eastward. Major folding as portrayed by Tabor and Cady (1978b, Fig. 27) would also produce the pattern seen today, but faulting and imbrication would be more consistent with incremental slicing off of ocean floor sediments as the oceanic plate descended beneath the North American plate. An alternative to the subduction process would be a stacking up of overthrusts emplaced from a western welt formed by the impingement of a Humbolt-type triple junction against the continent (Fox, 1983, p. 33–35). C. The major units of the eastern core have been established. NG and GV are Needles-Gray Wolf and Grand Valley lithic assemblages, respectively. P-P′ is a form line to show deformation in D. In map plan (Fig. 1), continued deformation under roughly east-west compression pressed the core rocks into an arc of basaltic rocks that was forming in response to deformation. The easternmost core rocks were bent into the horseshoe fold to the north, but were mostly sheared off below the basalt to the south and southeast. D. Deformation continued as younger material accreted to the west, and as the pile of sedimentary rocks, now thoroughly lithified, became even more constrained by the basaltic horseshoe (in map plan), the core rocks yielded upward and outward by shear folding. An overall mushroomlike dome, highly asymmetric to the east and northeast, developed with a fan of cleavage. Conspicuous pencil structures formed where the new cleavage intersected older, deformed cleavage and bedding. In much of the eastern core, the sense of shear in this last stage of deformation was opposite that of the earlier stages. Many of the east-dipping shear planes in the eastern core that had undergone eastward underthrusting in the early stage of deformation were now back-folded eastward, and movement continued as eastward reverse faulting.

thick accumulations of oceanic basalt should not stratigraphically overlie thick sedimentary deposits.

Although the structural relationship of the Crescent Formation and Blue Mountain unit is unresolved, the latter has stratigraphic and lithologic affinity to the Crescent (Tabor and others, 1972), and the important structural break, the Hurricane Ridge fault, between the peripheral rocks and the core rocks, can be viewed at 16.0 mi (25.7 km) at the western head of a tributary to Morse Creek. Limited parking is available here, but the fault is best appreciated by walking from relatively unsheared, thin-bedded sandstone and argillite exposed in outcrops to the northeast (down road) through highly sheared rocks that become more slaty to the southwest (uproad). In the southwesternmost outcrops (south of small creeks), lenses of micaceous sandstone, some with sheared contacts, are prominent in the slaty argillite. The Hurricane Ridge fault, although broad and diffuse here, separates rocks with typical eastern-core structures and lithologies from typical peripheral rocks.

4. Overview of Olympic Core Rocks. Continuing up Hurricane Ridge, the highway cuts through the main crest of the ridge and swings west providing a grand view of the mountainous core. Park at 17.4 mi (28.0 km) to see outcrops of core rocks and typical pencil slates or continue on to the main parking lot (17.5 mi; 28.6 km) near the Hurricane Ridge Lodge. The panorama of ridges and peaks seen here are carved from rocks of the Needles-Gray Wolf lithic assemblage on the southeast to Mount Olympus

for the Mount Angeles-Klahhane Ridge Trail, look up from outcrops of overturned sandstone and argillite rhythmites of the Blue Mountain unit to see the ribbed cliffs of basalt and breccia on Mount Angeles (Fig. 2). Whether the Blue Mountain rocks underlie the Crescent Formation stratigraphically or have been thrust under it is not known. West of Hurricane Ridge, similar shallow-water sedimentary rocks (containing middle and early late Eocene foraminifers) are considered by Snavely (in Muller and others, 1983, p. 10-11) to be in thrust contact with the lower part of the Crescent Formation, which at this location contains deep-water early Eocene nannofossils. Based on divergent structures in the Crescent and Blue Mountain rocks along the road, it can be argued that the contact here is a fault (see Tabor in Muller and others, 1983, p. 38); farther to the east and southeast, evidence of faulting is less convincing (see Tabor and Cady, 1978a; Tabor and others, 1972; Cady and others, 1972). Theoretically, on the west, carved from thick-bedded sandstone on the edge of the eastern core. In the immediate foreground is the canyon of the Lillian River; beyond it lies the main drainage of the Elwha River. Between the Lillian and the Elwha are sedimentary and metasedimentary rocks of the Grand Valley lithic assemblage.

To see more outcrops of core rocks, hike out the Hurricane Hill nature trail or drive out the Obstruction Point Road.

Figure 3 summarizes the development of Olympic core rocks during their accretion to the North American plate. Because the Juan de Fuca plate had a strong northward component to its motion (relative to the North American plate) during the Tertiary, each wedge of sediment in the accreted mass may have approached the continent tangentially before it was scraped off the subducted plate and added to the accretionary prism. Structural data from the eastern core do not indicate that the deformational process was either continuous or sporadic, but the sedimentary record in peripheral rocks of the northwestern Olympic Peninsula strongly suggests episodic deformation (Snavely in Muller and others, 1983, p. 23).

The thickened pile of folded and accreted sedimentary rocks indicates that the final uplift of the Olympic Mountains could well have been isostatic. This uplift would increase the overturning of structures on the east, (Fig. 3D), especially if there was considerable differential uplift in the center of the range as suggested by rocks of the highest metamorphic grade uplifted to higher elevations (see Tabor, 1972, p. 1811).

REFERENCES CITED

Beck, M. E., Jr., and Engebretson, D. C., 1982, Paleomagnetism of small basalt exposures in the west Puget Sound Area, Washington, and speculations on the accretionary origin of the Olympic Mountains, *in* Special issue on accretion tectonics: Journal of Geophysical Research, v. 87, no. 5, p. 3755–3760.

Brandon, M. T., Cowan, D. S., Muller, J. E., and Vance, J. A., 1983, Pre-Tertiary geology of San Juan Islands, Washington and southern Vancouver Island, British Columbia: Geological Association of Canada, Mineralogical Association of Canada and Canadian Geophysical Union, Field Trip Guidebook, Trip 5, annual meeting, Victoria, 65 p.

Cady, W. M., 1975, Tectonic setting of the Tertiary volcanic rocks of the Olympic Peninsula, Washington: U.S. Geological Survey Journal of Research, v. 3, no. 5, p. 573–582.

Cady, W. M., Tabor, R. W., MacLeod, N. S., and Sorensen, M. L., 1972, Geologic map of the Tyler Peak quadrangle, Washington: U.S. Geological Survey Geological Quadrangle Map GQ–970.

Coney, P. J., Jones, D. L., and Monger, J.W.H., 1980, Cordilleran suspect terranes: Nature, v. 288, p. 329–333.

Duncan, R. A., 1982, A captured island chain in the Coast Range, Oregon and Washington: Journal of Geophysical Research, v. 87, p. 827–837.

Fairchild, L. H., and Cowan, D. S., 1982, Structure, petrology, and tectonic history of the Leech River complex northwest of Victoria, Vancouver Island: Canadian Journal of Earth Sciences, v. 19, p. 1817–1835.

Fox, K. F., Jr., 1983, Melanges and their bearing on late Mesozoic and Tertiary subduction and interplate translation at the west edge of the North American plate: U.S. Geological Survey Professional Paper 1198.

Garrison, R. E., 1973, Space-time relations of pelagic limestones and volcanic rocks, Olympic Peninsula, Washington: Geological Society of America Bulletin, v. 84, p. 583–593.

Glassley, W. E., 1974, Geochemistry and tectonics of the Crescent volcanic rocks, Olympic Peninsula, Washington: Geological Society of America Bulletin, v. 85, p. 785–794.

Glassley, W. E., Lyttle, N. A., and Clarke, D. B., 1976, New analyses of Eocene basalt from Olympic Peninsula, Washington—Discussion and reply: Geological Society of America Bulletin, v. 87, p. 1200–1204.

Johnson, S. Y., 1984, Evidence for a margin-truncating fault (pre-late Eocene) in western Washington: Geology, v. 12, p. 538–541.

Jones, D. L., Silberling, N. J., and Hillhouse, John, 1977, Wrangellia—A displaced continental block in northwestern North America: Canadian Journal of Earth Sciences, v. 14, no. 11, p. 2565–2577.

King, P. B., and Beikman, H. M., 1974, Geologic map of the United States, scale 1:2,500,000 U.S. Geological Survey.

Livingston, J. L., and Tobin, D. G., 1971, Paleomagnetic evidence of tectonic rotation in the Olympic Peninsula, Washington (abstract): EOS, Transactions of the American Geophysical Union, v. 52, p. 921.

MacLeod, N. S., and Snavely, P. D., Jr., 1973, Volcanic and intrusive rocks of the central part of the Oregon Coast Range: Oregon Department of Geology and Mineral Industries Bulletin 77, p. 47–74.

Muller, J. E., 1980, Chemistry and origin of the Eocene Metchosin volcanics, Vancouver Island, British Columbia: Canadian Journal of Earth Sciences, v. 17, no. 2, p. 199–209.

Muller, J. E., Snavely, P. D., Jr., and Tabor, R. W., 1983, The Tertiary Olympic terrane, southwest Vancouver Island and northwest Washington: Geological Association of Canada, Mineralogical Association of Canada and Canadian Geophysical Union, Field Trip Guidebook, Trip 12, annual meeting, Victoria, 57 p.

Rau, W. W., 1973, Geology of the Washington coast between Point Greenville and the Hoh River: Washington Department of Natural Resources, Division of Geology and Earth Resources, Bulletin 66, 58 p.

—— 1980, Washington coastal geology between the Hoh and Quillayute Rivers: Washington Department of Natural Resources, Division of Geology and Earth Resources, Bulletin 72, 57 p.

Simpson, R. W., and Cox, A., 1977, Paleomagnetic evidence for tectonic rotation of the Oregon Coast Range: Geology, v. 5, p. 585–589.

Snavely, P. D., Jr., MacLeod, N. S., and Wagner, H. C., 1968, Tholeiitic and alkalic basalts of the Eocene Siletz River volcanics, Oregon Coast Range: American Journal of Science, v. 266, p. 454–481.

Stewart, R. J., 1970, Petrology, metamorphism, and structural relations of graywackes in the western Olympic Peninsula, Washington [Ph.D. thesis]: Stanford University, 129 p.

Tabor, R. W., 1972, Age of the Olympic metamorphism, Washington—K-Ar dating of low-grade metamorphic rocks: Geological Society of America Bulletin, v. 83, p. 1805–1816.

—— 1975, Guide to the geology of Olympic National Park: University of Washington Press, Seattle, 144 p.

Tabor, R. W., and Cady, W. M., 1978a, Geologic map of the Olympic Peninsula: U.S. Geological Survey Miscellaneous Investigations Series Map I-994, scale 1:125,000.

—— 1978b, The structure of the Olympic Mountains, Washington—analysis of a subduction zone: U.S. Geological Survey Professional Paper 1033, 38 p.

Tabor, R. W., Yeats, R. S., and Sorensen, M. L., 1972, Geologic map of the Mount Angeles quadrangle, Washington: U.S. Geological Survey Geologic Quadrangle Map GQ-958, 1:62,500.

Wells, R. E., Engebretson, D. C., Snavely, P. D., Jr., and Coe, R. S., 1984, Cenozoic plate motions and the volcano-tectonic evolution of western Oregon and Washington: Tectonics, v. 3, no. 2, p. 275–294.

Yorath, C. J., Clowes, R. M., Green, A. G., Sutherland-Brown, A., Brandon, M. T., Massey, N.W.D., Spencer, C., Kanusewich, E. R., and Hyndman, P. D., 1985, LITHOPROBE-Phase 1: Southern Vancouver Island: Preliminary analysis of reflection seismic profiles and surface geological studies: Current Research, Part A, Geological Survey of Canada Paper 85-1A, p. 543–554.

The Late Cretaceous San Juan thrust system, Washington: Nappes related to the arrival of Wrangellia

Mark T. Brandon, *Department of Geology and Geophysics, Yale University, Box 6666, New Haven, Connecticut 06511*
Darrel S. Cowan, *Department of Geological Sciences, University of Washington, Seattle, Washington 98195*

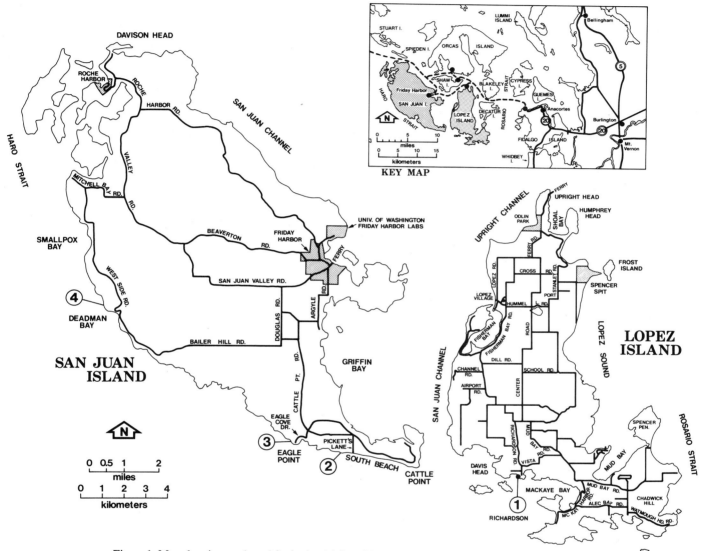

Figure 1. Map showing roads and ferries in vicinity of San Juan Islands. Numbers refer to stops in this guide.

LOCATION AND ACCESSIBILITY

The San Juan Islands are located at the north end of Puget Sound in northwestern Washington State. This field guide includes four localities: one on Lopez Island and three on San Juan Island. To reach the islands (Fig. 1), take I-5 to Exit 230, then drive west on Washington 20 to the Anacortes ferry terminal. Washington State ferries provide frequent daily service for cars and passengers to Lopez and San Juan Islands. Long waits should be expected on summer weekends.

SIGNIFICANCE

A variety of rock units ranging in age from early Paleozoic to Late Cretaceous are exposed in the San Juan archipelago. Mid-Cretaceous and older rocks occur in several thrust sheets or nappes, which are separated by a set of thrust faults and fault zones. These thrusts are broadly folded around axes gently plunging to the southeast (Fig. 2). The San Juan thrusts and nappes are part of a broader Late Cretaceous thrust system that extends 48 mi (80 km) eastward into the North Cascade Mountains. To the

SAN JUAN THRUST-AND-NAPPE SYSTEM
ROCK UNITS IN THE THRUST SYSTEM,
SHOWN IN ASCENDING STRUCTURAL
ORDER

 Decatur Terrane. A coherent terrane consisting of M. to U. Jurassic ophiolitic and arc-volcanic rocks of the Figalgo Complex (cross-hatch pattern), and U. Jurassic to Lw. Cretaceous sandstone, mudstone and conglomerate of the overlying Lummi Formation (stipple pattern).

 Lopez Structural Complex. Imbricated slices of Jurassic to mid-Cretaceous sandstone, pebbly mudstone, pillow lava and chert. Also contains rare slices of Turtleback quartz diorite.

 Constitution Formation. Massive volcaniclastic sandstone, with interbedded sequences of mudstone, chert, pillow lava and green tuff. Jurassic to Lw. Cretaceous.

 Orcas Formation. Ribbon chert and minor pillow basalt. Triassic and Lower Jurassic. Locally imbricated with slices of Turtleback Complex and Garrison Schist.

Deadman Bay Formation. Pillow basalt with minor chert and Asiatic-fusulinid limestone. Lw. Permian to Triassic.

 Turtleback Complex and East Sound Group. An undifferentiated stratigraphic sequence consisting of lower Paleozoic plutonic rocks of the Turtleback Complex, and upper Paleozoic arc-volcanic rocks and limestone of the East Sound Group.

EXTERNAL UNITS
ROCK UNITS FORWARD OF, AND BELOW, THE
SAN JUAN THRUST SYSTEM

Chuckanut Formation. Non-marine sandstone and conglomerate. Lower Tertiary.

Nanaimo Group. Marine and non-marine sandstone, conglomerate and shale. Upper Cretaceous.

Spieden Group. Sandstone and conglomerate with arc-volcanic clasts, U. Jurassic and Lw. Cretaceous.

Haro Formation. Sandstone and conglomerate with arc-volcanic clasts. Also contains minor shelly interbeds. U. Triassic.

Figure 2. Generalized geologic map of San Juan Islands (modified from Brandon and others, 1983, and in prep.). Major faults are approximately located. Quaternary sediments are not shown. "R" denotes the village of Richardson. "A" denotes Tethyan fusulinid localities in the Deadman Bay Volcanics. Garrison Schist, which is restricted to a narrow zone beneath the Rosario thrust, is not included on this map.

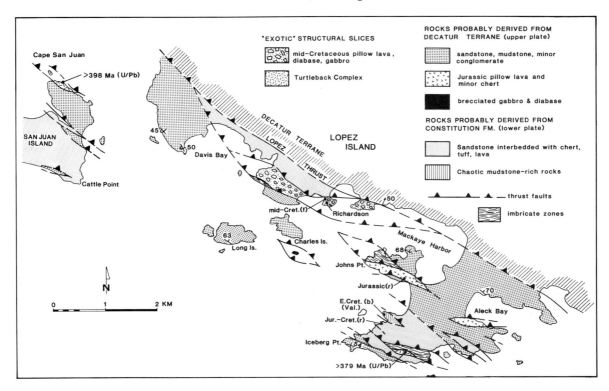

Figure 3. Geologic map of Lopez Structural Complex for Stop 1 at Richardson, Lopez Island (from Brandon and others, in prep., based on mapping by Cowan). Localities for U/Pb zircon dates are labeled U/Pb. Fossil localities are labeled as follows: r, radiolaria from chert; b, *Buchia* from sandstone; f, foraminifera from mudstone.

northwest of the San Juan nappes lies the Wrangellia terrane (Jones and others, 1977; Muller, 1977) exposed on Vancouver Island and adjacent smaller islands. We interpret the San Juan–Cascades thrust-and-nappe system as having been emplaced westward onto the Wrangellia terrane as Wrangellia was driven eastward and beneath the continental margin of North America during the Late Cretaceous (Brandon and Cowan, 1985). The field guide localities were chosen for several reasons: (1) they provide important constraints on the timing of deformation, (2) they illustrate styles of thrust-related deformation, and (3) they epitomize important stratigraphic units within the nappes. More detailed stratigraphic and structural information is in Vance (1975, 1977), Whetten and others (1978), Brown and others (1979), and Brandon and others (in prep.). The geology of the North Cascades is summarized in Misch (1966, 1977). Muller (in Brandon and others, 1983) provides a brief sketch of the geology of southern Vancouver Island nearest the San Juan Islands. Other more comprehensive field guides to the San Juan Islands and surrounding areas are Brandon and others (1983), Brown (1977), Vance (1977), Cowan and Whetten (1977), and Misch (1977).

FIELD GUIDE LOCALITIES

Stop 1: Richardson, Lopez Island. An exotic slice of mid-Cretaceous basalts in the Lopez Structural Complex.

From the ferry terminal on Lopez Island, drive south to the hamlet of Richardson on the south shore of the island (Fig. 1). Park at the Richardson store where the road ends in a cul-de-sac. The outcrops to be examined are on the east side of the road and on the coast extending 250 ft (75 m) north of the store.

The brownish-red mudstones in the 10-ft (3-m)-high road cut opposite the store are very important because they are the youngest dated rocks in the San Juan thrust system. Foraminifera from these mudstones, first discovered by Danner (1966), have been dated as latest Albian (mid-Cretaceous, about 100 Ma) by W. Sliter (personal communication, 1986). These mudstones occur as an interbed in a small fault-bounded basaltic unit within the Lopez Structural Complex (Figs. 2, 3), a 1.8-mi (3-km)-thick imbricate fault zone that separates two relatively coherent Mesozoic units: the structurally lower Constitution Formation and the overlying Decatur terrane. The most prominent structure visible in this outcrop is a northeast-dipping slaty cleavage that postdates imbrication within the Lopez Complex.

More of this volcanic unit is exposed in the steep 15-ft (5-m)-high seacliff immediately north of the store. Note that the mudstone is part of a northeast-dipping stratigraphic sequence including (from bottom to top) pillow basalt, red and black mudstone, pillow breccia, and, finally, more pillow basalt at the north end of the exposure. The pillows indicate that the sequence is upright. Based on their trace-element composition (high TiO_2,

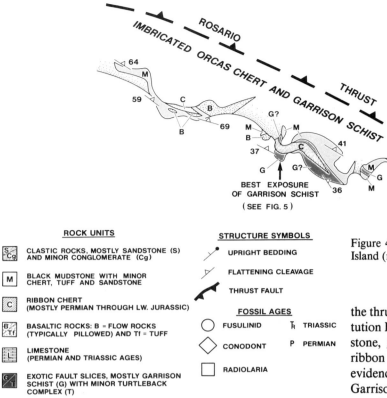

ROCK UNITS

S / Cg	CLASTIC ROCKS, MOSTLY SANDSTONE (S) AND MINOR CONGLOMERATE (Cg)
M	BLACK MUDSTONE WITH MINOR CHERT, TUFF AND SANDSTONE
C	RIBBON CHERT (MOSTLY PERMIAN THROUGH LW. JURASSIC)
B / Tf	BASALTIC ROCKS: B = FLOW ROCKS (TYPICALLY PILLOWED) AND Tf = TUFF
L	LIMESTONE (PERMIAN AND TRIASSIC AGES)
G / T	EXOTIC FAULT SLICES, MOSTLY GARRISON SCHIST (G) WITH MINOR TURTLEBACK COMPLEX (T)

STRUCTURE SYMBOLS

- UPRIGHT BEDDING
- FLATTENING CLEAVAGE
- THRUST FAULT

FOSSIL AGES

- FUSULINID — Tr TRIASSIC
- CONODONT — P PERMIAN
- RADIOLARIA

Figure 4. Outcrop map for Stop 2 in the South Beach area, San Juan Island (modified from Brandon, 1980).

light rare-earth element enriched; Brandon and others, in prep.), these pillowed basalts probably erupted in an "oceanic island" setting, and therefore might represent a mid-Cretaceous seamount. The mudstone contains small lenses of sand-sized volcanic quartz and feldspar, indicating that an intermediate arc volcanic terrane was nearby.

Amygdules and veins of metamorphic aragonite and pumpellyite are present in the Richardson basalts. Sandstone-rich units, present elsewhere in the Lopez Complex, contain lawsonite-aragonite metamorphic assemblages. These minerals formed during a very low-temperature, high-pressure regional metamorphism that affected most of the rocks in the San Juan thrust system. This metamorphic event occurred in the interval 100-83 Ma, and was related to rapid tectonic burial of nappes within the thrust system (Brandon and others, in prep.).

Stop 2: South Beach, San Juan Island. Exotic slices of Permo-Triassic metamorphic rock within the Rosario thrust zone.

From the ferry terminal at Friday Harbor, San Juan Island, drive through town, and then head south toward Cattle Point (Fig. 1). Take Pickett's Lane down to South Beach, and then follow the dirt road west about 0.2 mi (0.4 km) to the outcrops at the west end of the beach. The area to be examined on foot starts at these first outcrops, shown as the most easterly outcrops in Figure 4, and continues about 2,000 to 2,600 ft (600 to 800 m) to the west along the coast.

This area lies at the southwestern end of the Rosario thrust zone, which strikes offshore beyond this point. At this location,

the thrust dips northeast, placing massive sandstone of the Constitution Formation over a structurally complex assortment of mudstone, green volcanic rocks (locally pillowed), green tuff and ribbon chert, which are assigned to the Orcas Chert. The best evidence for large thrust displacements is the exotic slices of Garrison Schist scattered through this fault zone. The Garrison experienced a greenschist- to amphibolite-grade, high-pressure metamorphism (barrositic amphiboles) during the Permo-Triassic, prior to its tectonic imbrication with sub-greenschist-grade Orcas Chert during the Late Cretaceous (Brandon and others, in prep.).

At the east side of the map (Fig. 4), the first outcrops belong to the Constitution Formation, and consist of a northeast-dipping depositional sequence including a thin horizon of mudstone, green tuff, and ribbon chert, overlain by a thick massive sandstone unit. Of particular interest is the clear interbedding of clastic rocks with radiolarian ribbon chert.

Farther west along the coast is a highly imbricated sequence of Orcas Chert with exotic slices of Garrison Schist. These rocks lie beneath the Rosario thrust, which is the highest recognized

Figure 5. Photograph (looking east) of a slice of Garrison Schist (g) in Rosario thrust zone. The slice is surrounded by mudstone and chert of the Orcas Chert (m); see Figure 4 for location.

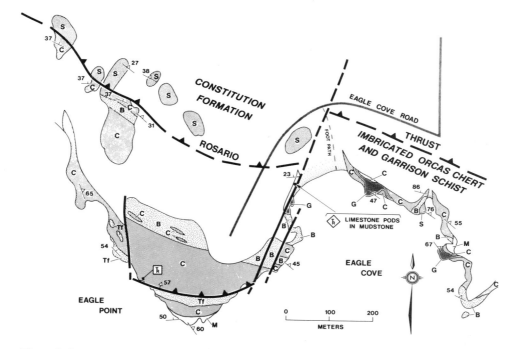

Figure 6. Outcrop map for Stop 3 in the Eagle Point and Eagle Cove area, San Juan Island (modified from Brandon, 1980); see Figure 4 for legend.

thrust within this imbricated fault zone. Here and elsewhere in the San Juan Islands, the slices of dark- to light-green Garrison Schist are localized within a 300- to 600-ft (100- to 200-m)-thick zone beneath the Rosario thrust. The Garrison in the South Beach area is a brecciated, fine-grained mafic schist consisting of chlorite+actinolite+epidote+plagioclase. Brecciation and cataclasis are attributed to tectonic emplacement of these fault slices. Structural relationships are best exposed at the location indicated on the map (Fig. 4) where a large, tabular slice 6 × 20 × 20 ft; (2 × 6 × 6 m) of Garrison is surrounded by disrupted black mudstone, ribbon chert, and minor unfoliated basalt of the Orcas (see Fig. 5). Note the small imbricate fault zone developed beneath the south side of this Garrison slice. These exotic slices were emplaced into the Rosario fault zone prior to the high-pressure regional metamorphism (lawsonite-prehnite-aragonite), as indicated by thin-section textures that show the cataclastic fabric of the schist cut by undeformed veins of aragonite.

Stop 3: Eagle Point and Eagle Cove, San Juan Island. Constitution, Orcas and Garrison Formations at the Rosario thrust.

Drive from Stop 2 toward Friday Harbor along Cattle Point road. Turn south (left) on Eagle Cove Drive, and park near Eagle Point at the end of the road. Walk southwest to seacliffs around the Triassic fossil locality labeled in Figure 6.

The Rosario thrust continues from South Beach northwestward to Eagle Cove (Figs. 2 and 6), where it also is a northeast-dipping fault zone containing exotic slices of Garrison Schist. Exposures of Constitution sandstone are restricted to the low outcrops north and northwest of Eagle Point. The Rosario thrust

is mapped at the highest occurrence of chert and mudstone of the Orcas Formation (Fig. 6); the overlying Constitution Formation was relatively unaffected by faulting within the Rosario zone.

Radiolarian ribbon chert and volcanic flows of the Orcas Formation occur as disrupted fault slices in this area and are best exposed on Eagle Point. Radiolaria from cherts at this locality have been dated as Triassic (D. L. Jones, personal communication, 1980). Note the mesoscale tight folds in the ribbon chert outcrops between the parking area and the point.

Slices of Garrison Schist are exposed in outcrops around Eagle Cove (Fig. 6). An entrained sequence of limestone pods in black mudstone is exposed on the west side of the cove. These pods, which have yielded Late Triassic conodonts (Savage, 1984), probably represent small olistoliths or slide blocks.

Stop 4: Deadman Bay, San Juan Island. Tethyan-fusulinid limestones in the Deadman Bay Volcanics.

Return to Cattle Point Road and drive north toward Friday Harbor. Turn west and connect with Bailer Hill Road, which turns into West Side Road where it meets the coast. Deadman Bay is a small cove located just south of where the road makes two sharp switchback turns (Fig. 7); park on the short road off West Side Road just before the first switchback. Walk southwest to the coast.

Deadman Bay is located at the southern end of a 2.4-mi (4-km) long fault slice of Deadman Bay Volcanics (Fig. 7). Gray and green ribbon chert exposed at the south side of the bay belongs to the Orcas Chert, which structurally overlies the Deadman Bay Volcanics along a northeast-dipping thrust fault. The trace of this fault follows the West Side Road to the north.

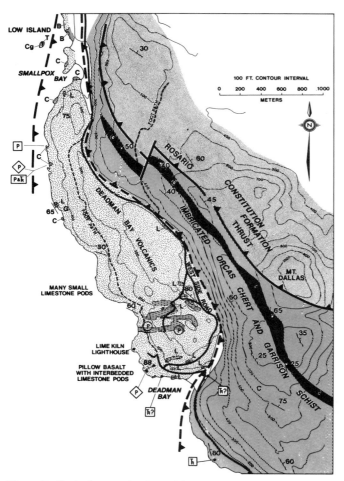

Figure 7. Geologic map for Stop 4 in Deadman Bay area, San Juan Island (modified from Danner, 1966; Vance, 1975; and unpublished mapping by Brandon and Cowan); see Figure 4 for legend.

The Deadman Bay Volcanics are well exposed along the coast on the rocky headlands between the bay and the lighthouse to the north. The unit is dominated by red and green pillow basalt, breccia, and tuff with subordinate interbedded limestone. It is commonly disrupted by faults, but generally has a persistant easterly strike, and in this area a near-vertical dip; geopetal structures indicate younging to the north. Limestones in the unit are massive and gray, and contain small amounts of intercalated green tuff. Where present in the limestones, bedding is typically contorted and appears to have been affected by soft-sediment slumping. Carbonate material occurs interstitially in the pillowed flows, and might have been sucked into the pillowed framework by rapidly convecting currents generated by the cooling submarine flows, or by churning as the lavas flowed across carbonate accumulations. The limestones were converted to aragonite marble during Late Cretaceous high-pressure metamorphism.

Crinoid debris and fragments of other fossils can be found in many limestone pods. A thin limestone bed, clearly interbedded in the pillow basalt sequence at Deadman Bay, has yielded late(?)

Leonardian (late Early Permian) conodonts (Fig. 7; M. J. Orchard, written communication, 1985). Danner (1966) has identified early Guadalupian (early Late Permian fusulinids) from limestone in the quarry to the north (Fig. 7). These fusulinids belong to the Tethyan or Asiatic fusulinid province, which suggests that the Deadman Bay Volcanics are exotic to North America. Trace-element geochemistry (high TiO_2 and light rare-earth element enrichment; Brandon and others, in prep.) indicates that the volcanics were probably erupted in an "oceanic island" setting.

REFERENCES CITED

Brandon, M. T., 1980, Structural geology of Middle Cretaceous thrust faulting on southern San Juan Island, Washington [M.S. thesis]: Seattle, University of Washington, 123 p.

Brandon, M. T., and Cowan, D. S., 1985, The Late Cretaceous San Juan Islands–Northwestern Cascades thrust system: Geological Society of America Abstracts with Programs, v. 17, p. 343.

Brandon, M. T., Cowan, D. S., Muller, J. E., and Vance, J. A., 1983, Pre-Tertiary geology of San Juan Islands, Washington and southeast Vancouver Island, British Columbia, Field Trip Guidebook: Geological Association of Canada, Victoria Section, 65 p.

Brown, E. H., 1977, The Fidalgo ophiolite, in Brown, E. H., and Ellis, R. C., eds., Geological Excursions in the Pacific Northwest: Bellingham, Western Washington University, p. 309–320.

Brown, E. H., Bradshaw, J. Y., and Mustoe, G. E., 1979, Plagiogranite and keratophyre in ophiolite on Fidalgo Island, Washington: Geological Society of America Bulletin, Part I, v. 90, p. 493–507.

Cowan, D. S., and Whetten, J. T., 1977, Geology of Lopez and San Juan Islands, in Brown, E. H., and Ellis, R. C., eds., Geological Excursions in the Pacific Northwest: Bellingham, Western Washington University, p. 321–338.

Danner, W. R., 1966, Limestone resources of western Washington: Washington Division of Mines and Geology Bulletin 52, 474 p.

Jones, D. L., Silberling, N. J., and Hillhouse, J., 1977: Wrangellia; A displaced terrane in northwestern North America: Canadian Journal of Earth Sciences, v. 14, p. 2565–2577.

Misch, P., 1966, Tectonic evolution of the Northern Cascades of Washington State; A west-Cordilleran case history: Canadian Institute of Mining and Metallurgy Special Volume 8, p. 101–148.

Misch, P., 1977, Bedrock geology of the North Cascades, in Brown, E. H., and Ellis, R. C., eds., Geological Excursions in the Pacific Northwest: Bellingham, Western Washington University, p. 1–62.

Muller, J. E., 1977, Evolution of the Pacific Margin, Vancouver Island, and adjacent regions: Canadian Journal of Earth Sciences, v. 14, p. 2062–2085.

Savage, N. M., 1984, Late Triassic (Karnian) conodonts from Eagle Cove, southern San Juan Island, Washington: Journal of Paleontology, v. 58, p. 1535–1537.

Vance, J. A., 1975, Bedrock geology of San Juan County, in Russell, R. H., ed., Geology and Water Resources of the San Juan Islands: Washington Department of Ecology Water Supply Bulletin 46, p. 3–19.

——, 1977, The stratigraphy and structure of Orcas Island, San Juan Islands, in Brown, E. H., and Ellis, R. C., eds., Geological Excursions in the Pacific Northwest: Bellingham, Western Washington University, p. 170–203.

Whetten, J. T., Jones, D. L., Cowan, D. S., and Zartman, R. E., 1978, Ages of Mesozoic terranes in the San Juan Islands, Washington, in Howell, D. G., and McDougall, K. A., eds., Mesozoic Paleogeography of the Western United States: Pacific Section, Society of Economic Paleontologists and Mineralogists, p. 117–132.

ACKNOWLEDGMENTS

D. L. Jones and W. Sliter of the U.S. Geological Society and M. J. Orchard of the Geological Survey of Canada kindly provided unpublished paleontological data cited here. G.S.C. contribution 24986.

The Fidalgo ophiolite, Washington

Daryl Gusey, U.S. Forest Service, Naches, Washington 98937
E. H. Brown, Department of Geology, Western Washington University, Bellingham, Washington 98225

LOCATION AND ACCESS

The Fidalgo ophiolite is located on scenic Fidalgo Island, approximately 60 mi (96 km) north of Seattle, Washington. Stop 1 is at Washington Park in Anacortes. To get there from I-5, take Washington 20 (exit 230) at Burlington west to Anacortes. In Anacortes, follow signs to the ferry terminal, that is, Commercial Avenue north to 12th Street, then left onto 12th Street (Washington 20 spur). Stay in the left lane at the ferry terminal junction and proceed straight ahead to Washington Park. Field trip mileage begins at the entrance to Washington Park. Follow the loop road. Miles and kilometers are shown in parentheses.

SIGNIFICANCE

The Fidalgo ophiolite in northwestern Washington is a well-preserved sequence of Jurassic rocks that, from the base upward, consists of serpentinite, layered gabbro, a dike complex of plagiogranite and keratophyre and spilite, coarse sedimentary breccia, pelagic argillite, tuffaceous argillite, and lithic graywacke (Figs. 1, 2). This complex is part of the Decatur Terrane in the San Juan Islands identified by Whetten and others (1978). The silicic nature of the volcanic rocks and field relationships between the volcanic rocks and the various sedimentary units suggests that the Fidalgo ophiolite formed in an island arc setting (Brown, 1977a; Gusey, 1978). In terms of regional geology, the significance of an arc origin for the Fidalgo ophiolite is not clear, as the tectonic evolution of northwestern Washington and southern British Columbia has yet to be well explained. Other western Washington Mesozoic ophiolites, however, have also been interpreted to have formed in an oceanic island arc system or within a small oceanic basin (Vance and others, 1980; Miller, 1985). In age and lithology, the Fidalgo ophiolite is similar to the California Coast Range ophiolite.

DESCRIPTIONS OF ROCKS

(0.4 mi; 0.6 km) Stop 1. Washington Park—Serpentinite.
Stop near the leaning tree. Tidal exposures in this vicinity and to the west for several hundred ft (m) display the ultramafic rock of the Fidalgo ophiolite. Look for chromite layers, relict pyroxenes, and original dunite-peridotite contacts in the serpentinite.

The serpentinite exposed here is part of an ultramafic belt (the Fidalgo Formation of McClellan, 1927) that is also exposed on nearby Burrows and Cypress Islands. Detailed analysis of the ultramafic rock on Cypress Island has shown that the unserpentinized parts of the mass are largely harzburgite with a tectonite fabric, but possible relict igneous textures are visible (Raleigh, 1965).

The serpentinite is thought to be the base of the Fidalgo ophiolite. Nowhere on Fidalgo Island, or elsewhere in the San Juan Islands, has a primary, unfaulted contact between the serpentinite and layered gabbro (Stop 2) been observed. However, the close spatial relationship of gabbroic and ultramafic rock, and the indication from graded bedding in the gabbro suggests that the serpentinite lies down-section from the gabbro. Also, a small (99 × 165 ft; 30 × 50 m) lens of serpentinized peridotite occurs within the plagiogranite dike complex near the southern end of Fidalgo Island.

(1.8 mi; 2.9 km) Havekost Monument. From here, Stop 2 is visible to the south across Burrows Bay.

(2.4 mi; 3.8 km) At the exit to Washington Park head east on Sunset Avenue.

(3.0 mi; 4.8 km) Turn right onto Anaco Beach Road.

(5.0 mi; 8.0 km) Turn right onto Delmar Drive.

(5.2 mi; 8.3 km) Stop 2. Alexander Beach—layered gabbro.
Park in the parking area near the rest rooms. This is Alexander Beach—private property. Permission has been granted by Mr. Don Taylor, president of the Del Mar Community, for geologists to visit the outcrops along the shoreline. Please respect private property—**no rock hammers**. Hike north to the second, most prominent rocky point (about 0.3 mi; 0.5 km). A visit to this locality is best made at other than high tide.

At this locality, cumulus gabbro is the host to dikes of foliated hornblende gabbro, plagiogranite, diabase, and basalt (Fig. 3). Graded bedding in the gabbro indicates tops to the northeast toward the volcanic and sedimentary parts of the ophiolite, and away from the serpentinite exposed at Washington Park and Burrows Island. Features of interest are the strong flow foliation in the hornblende gabbro, and the composite dikes of hornblende gabbro and plagiogranite. The intrusive relationships seen here are complex.

The occurrence of hornblende instead of pyroxene in the gabbroic dikes and the hydrothermal alteration of the cumulus gabbro near the dikes indicate a high water content of the intrusive magmas. Chemical trends and sharp cross-cutting field relations of the plagiogranite dikes to the cumulus gabbro suggest that they are not comagnetic. See Brown and others (1979) for a discussion of the geochemistry of the rocks exposed at Alexander Beach and elsewhere on the island. From the parking lot, go back up the hill.

(5.5 mi; 8.8 km) Turn right onto Marine Drive.

(6.4 mi; 10.2 km) Turn left onto Havekost Road.

(7.3 mi; 11.7 km) Turn right into the Marine Asphalt Company quarry.

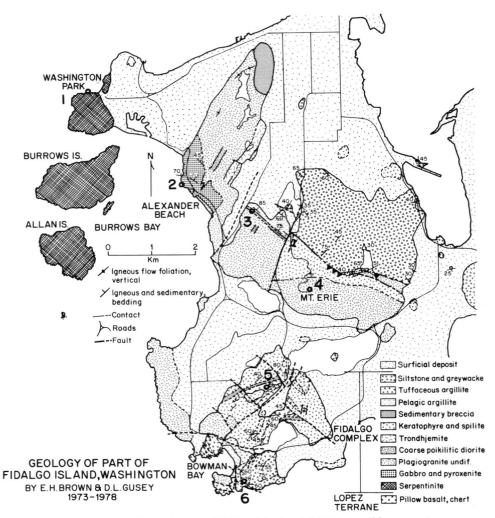

Figure 1. Geologic map of Fidalgo Island and field trip localities.

(7.7 mi; 12.3 km) Stop 3. Marine Asphalt Company quarry—plagiogranite, sedimentary breccia, pelagic argillite. Stop at the office to ask for permission to go into the quarry.

Here, plagiogranite exposed in the south quarry wall is overlain along an irregular contact by coarse sedimentary breccia containing clasts of plagiogranite, keratophyre, and spilite. Exposed on the west quarry wall, depositionally overlying the breccia, is steeply dipping pelagic argillite.

Elsewhere on Fidalgo Island, sedimentary breccias apparently occupying the same stratigraphic position, i.e., beneath pelagic argillite, contain clasts of serpentinite, cumulate gabbro, and pegmatitic diorite, as well as plagiogranite and volcanic rocks. Clast size ranges from 0.4 to 12 in (1 to 30 cm). These breccias may represent talus accumulations at the base of a submarine fault.

Generally, the pelagic argillite unit is a radiolarian-bearing, metal-enriched, carbonate-free chloritic argillite with local minor tuffaceous debris. Radiolaria from an argillite bed approximately 180 ft (55 m) from the base of the unit have been dated as Late Kimmeridgian to Early Tithonian (Gusey, 1978). The concentration of metals (Ba, Cu, Ni, Co, and Mn) in the pelagic argillite is similar to that in present-day Pacific pelagic sediments (Brown, 1977b).

Of special interest within the pelagic argillite unit are thin (~2 in; 5 cm) interbeds of a chloritic sandstone. This sandstone consists of sand-size chlorite grains in a micritic matrix. Diffractograms of the clay-size fraction indicate the presence of a well-crystallized, expandable chlorite. In thin section, mesh textures and cleavage traces are common in the chlorite. Also, chlorite has been observed replacing pyroxene. Accessory minerals include chromite, quartz, plagioclase, clinopyroxene, garnet, amphibole, and volcanic and sedimentary rock fragments. These beds are interpreted to have been deposited by turbidity currents from a nearby ultramafic source and subsequently chloritized (Gusey, 1978).

(8.1 mi; 13.0 km) At the quarry entrance, turn left onto Havekost Road.

(9.9 mi; 16.0 km) Turn left onto Rosario Road.

(10.9 mi; 17.4 km) Turn left at Lake Erie Grocery onto Heart Lake Road.

(12.1 mi; 19.4 km) Turn right at "v" in the road.

(12.2 mi; 19.5 km) Turn right onto the steep and winding Mt. Erie Road. Proceed to the top of Mt. Erie.

(13.9 mi; 22.2 km) Stop 4. Mt. Erie—plagiogranite and keratophyre. Follow the footpath next to the radio tower to the viewpoint overlooking southern Fidalgo Island. Here, dioritic rock is intruded by dikes of keratophyre. Radiometric ages of the plagiogranite here and at Alexander Beach range from 155 ± 5 Ma to 170 ± 10 Ma (Brown, 1977a; Whetten and others, 1978).

Plagiogranite, as used in this report, refers to diorite, quartz diorite, trondhjemite, and albite granite. In the plagiogranite, the igneous minerals preceding the low-grade metamorphism were quartz, plagioclase, hornblende, Fe-Ti oxide, and rarely, minor biotite (Brown and others, 1979). Relict igneous textures are clearly seen through the imprint of low-grade metamorphism. There is no indication of a metasomatic event that might have added SiO_2, removed K-feldspar, or otherwise substantially altered the composition of the original igneous rock. The range of chemical compositions exhibited by the plagiogranite and the keratophyre and spilite is essentially the same (see Table 2 of Brown and others, 1979).

Down the road 0.2 mi (0.3 km) from the top of Mt. Erie, coarse diorite and hornblendite are well exposed on the uphill side of the road. Keratophyre and spilite are exposed in road cuts at 0.6 to 1.2 mi (1.0 to 1.9 km) from the top of Mt. Erie.

Keratophyre and spilite occur mainly as flows and flow breccias stratigraphically above the plagiogranite. This rock unit ranges from felsic (60 to 65 percent SiO_2) to mafic (48 percent SiO_2) rock. In thin section, the igneous minerals are clinopyroxene, plagioclase, quartz, and opaques. Plagioclase and pyroxene occur as phenocrysts, commonly in synneusis clumps. The groundmass is composed of plagioclase microlites and quartz, locally showing trachytic texture. Samples taken at close intervals across the flows show little variation in major element chemistry except for Na enrichment and K depletion at the flow tops (Brown and others, 1979). Proceed back down the Mt. Erie Road.

(15.6 mi; 25.0 km) Turn left onto Heart Lake Road.

(16.9 mi; 27.0 km) Turn right at Lake Erie Grocery onto Rosario Road.

(17.9 mi; 28.6 km) Stay to the left on Rosario Road at its junction with Marine Drive.

(19.3 mi; 30.9 km) Turn left onto Sharpe Road.

(19.6 mi; 31.4 km) Straight onto Ginnet Road.

(20.3 mi; 32.5 km) The road narrows—stay to the left and park just past the mobile home.

(20.5 mi; 32.8 km) Stop 5. The Red Rock Quarry—tuffaceous argillite. This area is within Deception Pass State Park. Hike about 1,000 ft (300 m) beyond the gate along the road to the Red Rock Quarry at the end of the road. At the quarry, red and green radiolarian-bearing tuffaceous argillite is exposed. Dates obtained for radiolaria from the tuffaceous argillite unit

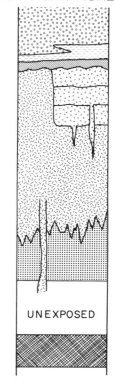

Figure 2. Generalized stratigraphic sections for Fidalgo Island. Same legend as for Figure 1.

range from lower Kimmeridgian to upper Valanginian (Gusey, 1978).

Tuffaceous argillite occurs as interbeds that are widespread throughout the keratophyre and spilite unit south of Mt. Erie. The argillite is typically associated with sedimentary and pyroclastic breccias. These interbeds range up to 165 ft (50 m) in thickness and appear to occur as lens-shaped bodies.

Sand-sized and larger pyroclastic debris, including euhedral plagioclase, altered shards of volcanic glass, volcanic rock fragments, and angular quartz grains are common within the tuffaceous argillite interbeds. Sedimentary breccia consisting of subangular to subrounded clasts (to 4 in; 10 cm) of volcanic detritus in an argillaceous matrix is also commonly associated with this unit.

The tuffaceous argillite interbeds probably represent sediment ponding on an irregular volcanic terrane. The coarse nature of the pyroclastic debris and its textural immaturity suggests a nearby volcanic center as its source.

(21.3 mi; 34 km) Junction of Sharpe and Ginnet Roads. Go straight to Rosario Road.

(21.6 mi; 34.6 km) Turn left onto Rosario Road.

(23.5 mi; 37.6 km) Turn right onto Washington 20.

(24.0 mi; 38.4 km) Stop 6. Deception Pass—graywacke. The best exposure of the graywacke of the Fidalgo ophiolite is here at

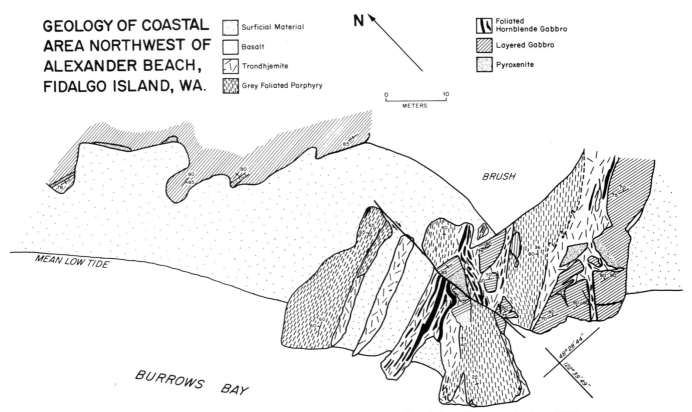

Figure 3. Geologic map of the coastal area north of Alexander Beach, from Brown and others (1979). Field trip Stop 2.

the north end of Deception Pass. At this locality, the graywackes are folded and faulted, making measurement of the section difficult. However, a cross section of an unfaulted(?) section northeast of here suggests that the unit exceeds 1,650 ft (500 m) in thickness.

At several outcrops elsewhere in the southern part of Fidalgo Island, graywacke is intercalated with volcanic flows, tuffs, and debris flows.

Within the graywacke unit, both massive and graded bedding can be observed. Individual beds vary in thickness from a few centimeters to several meters. Mineralogically, the graywacke consists of subangular grains of volcanic rock fragments and euhedral to subhedral plagioclase crystals. Quartz is present in amounts of less than 5 percent. No macrofossils have been found, but a few radiolaria have been seen in thin section (Gusey, 1978).

The abundance of volcanic rock fragments and euhedral to subhedral plagioclase, the paucity of quartz, and field relationships suggest that the source area was the adjacent volcanic terrain. Depositional features suggest deposition by turbidity flow.

A volcanic arc origin of the Fidalgo ophiolite is suggested not only by the abundance of silicic igneous rocks, but also by the large component of pyroclastic/volcaniclastic debris in the complex as seen here and in the Red Rock Quarry.

REFERENCES CITED

Brown, E. H., 1977a, Ophiolite on Fidalgo Island, Washington: *in* Coleman, R. G., and Irwin, W. P., eds., North American Ophiolites: Oregon Department of Geology and Mineral Industries Bulletin, no. 95, p. 67–73.

—— , 1977b, The Fidalgo Ophiolite: Field Trip No. 11, *in* Brown, E. H., and Ellis, R. C., eds., Geological Excursions in the Pacific Northwest, Geological Society of America 1977 Annual Meeting Guidebook: Bellingham, Western Washington University, p. 309–320.

Brown, E. H., Bradshaw, J. Y., and Mustoe, G. E., 1979, Plagiogranite and keratophyre in ophiolite on Fidalgo Island, Washington: Geological Society of America Bulletin, v. 90, p. 493–507.

Gusey, D. L., 1978, The geology of southwestern Fidalgo Island [M.S. thesis]: Bellingham, Western Washington University, 85 p.

McClellan, R. D., 1927, Geology of the San Juan Islands [Ph.D. thesis]: Seattle, University of Washington, 185 p.

Miller, R. B., 1985, The ophiolitic Ingalls Complex, north-central Cascade Mountains, Washington: Geological Society of America Bulletin, v. 96, p. 27–42.

Raleigh, C. B., 1965, Structure and petrology of an alpine peridotite on Cypress Island, Washington, U.S.A.: Beitrage zur Mineralogue und Petrographie, v. 11, p. 719–741.

Vance, J. A., Dungan, M. A., Blanchard, D. P., and Rhodes, J. M., 1980, Tectonic setting and trace element geochemistry in Mesozoic ophiolitic rocks in western Washington: American Journal of Science, v. 280-A, p. 359–388.

Whetten, J. T., Jones, D. L., Cowan, D. S., and Zartman, R. E., 1978, Ages of Mesozoic terranes in the San Juan Islands, Washington: Pacific Coast Paleogeography Symposium 2, Mesozoic Paleogeography of the Western United States, Society of Economic Paleontologists and Mineralogists, 573 p.

The type section of the Skagit Gneiss, North Cascades, Washington

Peter Misch, Department of Geological Sciences, University of Washington, Seattle, Washington 98195

LOCATION AND SIGNIFICANCE

Figure 1 shows access to, and location of, the Skagit Gneiss field sites. These sites are shown on the geologic map, Figure 2. The 12.4-mi (20-km) type section is now accessible by Washington 20. A detailed road log was published by Misch (1977, p. 35–46). Twelve recommended stops are indicated on the map by mileages (calibrated by mile markers). For a quick trip, the best stops are at mile 125.1 and 125.8.

The Skagit Gneiss provides a spectacular display of migmatitic regional metamorphism in the core of a young mobile belt. It is in pristine condition like the Nanga Parbat migmatites but is readily accessible.

SITE INFORMATION

The Skagit Gneiss was named after the transverse canyon of the antecedent Skagit River (Misch, 1966, 1968; for earlier reference, see those papers). Except for orthogneiss tracts, the unit has a heterogeneous, migmatitic character. It usually contains essentially isochemical remnants of nongranitic rocks that, as a rule, are more mafic than the surrounding gneiss and in almost all cases are of supracrustal origin. ("Granitic" is used here for all quartz-bearing rocks of plutonic aspect.) Though subordinate in volume at most places, such remnants (paleosomes) are exceedingly numerous. They range in thickness from less than an inch to several yards and more, in places exceeding 330 ft (100 m). Their shapes are tabular to elongate-lenticular parallel to s, or ribbon-like where b is dominant.

The most common nongranitic lithology is metasedimentary *biotite-quartz-plagioclase schist (paragneiss),* henceforth referred to as "biotite schist." (Minerals used in rock names are listed in order of increasing abundance.) Most it of it derived from psammite, predominantly graywacke that chemically resembles immature, plagioclase-rich metagraywacke in low-grade Cascade River Schist (Fig. 2). Highly quartzose varieties are minor. Al-excess, pelitic varieties are rare; some are impure-quartzitic. About half of the biotite schists contain porphyroblastic almandine, and about one-quarter contain subordinate amphiboles.

Amphibolites rank second among the nongranitic remnants. They comprise metaigneous and metasedimentary varieties. The former include both metavolcanics and some metagabbros, while the latter in places are associated with calc-silicate rocks. Relatively few of the amphibolites are garnetiferous. Many contain biotite, subordinate except where wholesale biotitization has occurred locally. The amphibolitic group of essentially isochemical remnants also includes leucoamphibolites (color index <40) and plagioclase-rich leucocratic hornblende schists, both predominantly metasedimentary. In the amphibolites *s. str.,* as well as in the more leucocratic members of the group, hornblende may be accompanied by cummingtonite, which exceptionally is dominant. Subordinate diopside occurs sporadically.

Very minor, but widely distributed *calc-silicate rocks and*

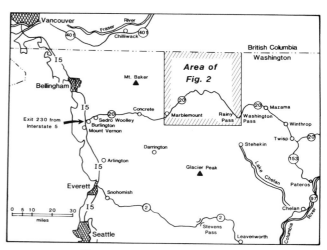

Figure 1. Location and road access.

marbles form thin intercalations in the common metasedimentary rock types. Other minor lithologies are impure quartzites that grade into biotite schists, para-amphibolites and calc-silicate rocks; and very rare quartz-pebble and other *metaconglomerates.* All of the supracrustal lithologies found within the Skagit Gneiss have counterparts in the isochemical Cascade River Schist (Misch, 1966). At least partial stratigraphic equivalence is indicated. The probable depositional age of the Cascade River Schist is Triassic.

Synkinematic regional Skagit metamorphism is of Barrovian type, encompassing chlorite to sillimanite zone. The Skagit Gneiss ranges from middle kyanite-staurolite to sillimanite grade; only the latter is encountered in the Skagit gorge. The main-stage migmatitic metamorphism was dated at 90 to 60 Ma by Mattinson (1972); some orthogneisses are as young as Eocene (see below).

The remnants of biotite schists, amphibolites, and so on show all transitions into *migmatitic gneisses.* Contacts vary from gradational to abrupt (Fig. 3). Some sharp contacts between biotite schist and leucosome are accentuated by biotite-enriched selvages (Fig. 3b). Such selvages mostly remain very thin (a cm or so); exceptionally, they compensate for *all* of an adjacent layer of leucosome. Such selvages commonly are absent even at sharp contacts, not to mention those contacts where essentially isochemical paleosome directly grades into leucocratic gneiss. The lack of systematic, wholesale basification of paleosomes seems to exclude simple closed-system interpretations of the Skagit migmatites—most obviously where amphibolite remnants are associated with large quantities of leucosome. Viewing the migmatites as a whole, the ratio leucosome/paleosome ranges from a few percent to more than 90 percent.

Both in biotite schists and amphibolites, the change to leu-

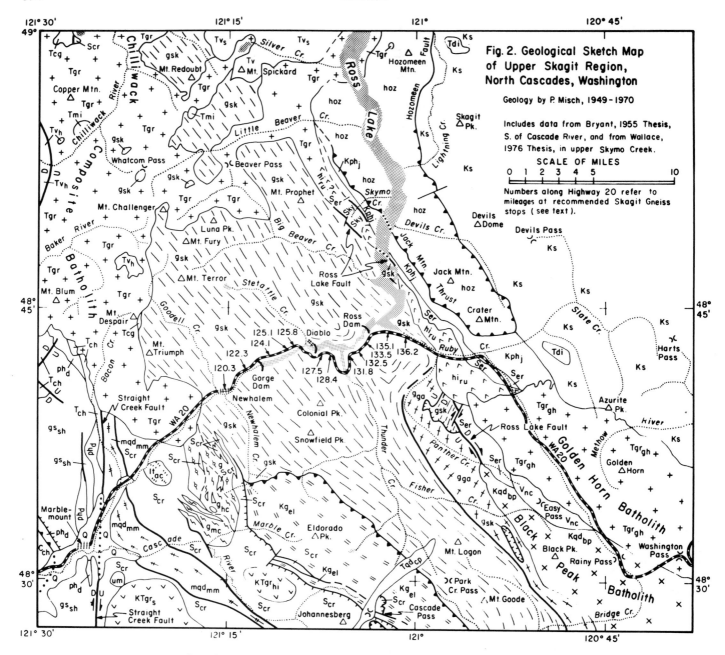

Figure 2 (this and facing page). Geological sketch map of upper Skagit region.

cocratic gneisses commonly is accomplished by an increase in number and size of plagioclase porphyroblasts (Fig. 3a, c, e, j). First sprouting as individuals—irregularly scattered or in trains along preferred s planes—such porphyroblasts next coalesce into swelling and pinching stringers and, finally, into thicker layers of usually also quartz-enriched plagioclase-porphyroblastic leucosome, giving rise to *banded gneisses* (Fig. 3c, e–h, j). Where the porphyroblasts grow very large, the resulting leucotrondhjemitic gneiss is pegmatitic (Fig. 3c, e–g, j). There also are banded gneisses that lack conspicuous porphyroblast development

(Fig. 3b, d, h [part], i). In places, paleosome-leucosome contacts have been sharpened by late cataclastic shear.

The *trondhjemitic biotite-quartz-plagioclase leucosomes* in the banded gneisses usually lack K-feldspar, whether paleosomes are biotite schists or amphibolites. Where a leucosome contains mafic species other than biotite (garnet, amphiboles), these invariably are identical with species present in the associated paleosome. However, mafic species of a paleosome are not necessarily present in the leucosome, owing to widespread biotitization. Plagioclase compositions in leucosomes are closely allied to

LEGEND FOR FIG. 2

CENOZOIC STRATA:

(all continental):

Q Quaternary (small areas omitted)

Tv Tertiary volcanics with associated clastics, including:

Tv$_s$ Skagit volcanics, mid-Tertiary

Tv$_h$ Hannegan Volcanics, Late Pliocene

Tcg Early Teriary clastics rich in conglomerate

T$_{ch}$ Chuckanut Formation, Eocene clastics

NON-METAMORPHIC LATE CRETACEOUS AND TERTIARY PLUTONIC ROCKS:

Tertiary granitic rocks, including:

Tqd$_{cp}$ Cascade Pass quartz diorite, Miocene

Tgr Chilliwack Composite Batholith, quartz-dioritic to quartz-monzonitic, (Eocene) Oligocene to Miocene, including E. extension near Ross Lake (Perry Creek Pluton, Oligocene to Miocene)

Tgr$_{gh}$ Golden Horn Batholith, Eocene, alkalic leucogranite to quartz monzonite

Tdi Diorite stocks E. of Ross Lake, Early Tert.

Tmi mafic intrusions, stocks within Chilliwack Composite Batholith, Eocene in part

KTgr$_s$ Snow King Massif } Late Cretaceous and/or
KTgr$_{hi}$ Hidden Lake Stock } Tertiary granitic rocks

hi$_{ru}$ Ruby Creek Heterogeneous Plutonic Belt, gabbroic to granitic, largely Late Cretaceous (?), with rafts of S$_{er}$

Kqd$_{bp}$ Black Peak Batholith, predominantly quartz diorite, early Late Cretaceous

W. OF STRAIGHT CREEK FAULT:

C$_{ch}$ Chilliwack Group, Late Paleozoic

SHUKSAN METAMORPHIC SUITE:

gs$_{sh}$ Shuksan Greenschist, actinolitic greenschist with blueschist intercalations

ph$_d$ Darrington Phyllite

━━ overthrust

━━ decollement thrust

SKAGIT CRYSTALLINE CORE:

SKAGIT METAMORPHIC SUITE-SENSU STRICTO:

S$_{cr}$ Cascade River Schist

gsk Skagit Gneiss (numerous small orthogneiss bodies in the dominantly migmatitic gneiss are not shown separately, nor are minor zones of essentially non-migmatized schist and locally amphibolite

MAJOR ORTHOGNEISS BELTS ASSOCIATED WITH SKAGIT SUITE:

g$_{ga}$ Gabriel Peak Orthgneiss, grades eastward into Kqd$_{bp}$, lower grade than g$_{sk}$ and commonly blastomylonitic

Kg$_{el}$ Eldorado Orthogneiss, early Late Cretaceous quartz diorite subjected to Skagit Metamorphism

mqd$_{mm}$ Marblemount Meta Quartz Diorite, mostly gneissose, Triassic intrusive subjected to Skagit Metamorphism

GRANITIC INTRUSIVES ASSOCIATED WITH CASCADE RIVER SCHIST:

lt$_{ac}$ Alma Creek Leucotrondhjemite, faintly gneissose, intruded at end of Skagit Metamorphism, presumed Eocene

g$_{hc}$ Haystack Creek Leucotrondhjemitic Orthogneiss, intruded during late stage of Skagit Metamorphism, probably Early Eocene

g$_{mc}$ Marble Creek Trondhjemitic Orthogneiss, intruded before or during Skagit Metamorphism, with quartz-dioritic and granodioritic varieties

g undifferentiated orthogneisses in S$_{cr}$

um metaperidotite in S$_{cr}$ (numerous other um bodies in S$_{cr}$ and g$_{sk}$, all subjected to Skagit Metam., are too small to show)

E. OF ROSS LAKE FAULT:

Ks Cretaceous sedimentary formations of Methow-Pasayten Belt, undifferentiated

Kph$_j$ Jack Mountain Phyllite derived from Lower Cretaceous (+ Upper Jurassic?) strata

V$_{nc}$ North Creek Volcanics, andesitic, with interbedded clastics, pre-late Jurassic, presumed Mesozoic

S$_{er}$ Elijah Ridge Schist derived from V$_{nc}$ and presumed northern equivalents

hoz Hozomeen Group, Permian and Triassic

sky Skymo Creek granulite complex, pre-g$_{sk}$(?); map belt includes later intrusives correlated with hi$_{ru}$

---

those in associated paleosomes (Misch, 1968). K$_2$O content, which commonly is around 2.3 to 2.5 weight percent in the biotite schists, is systematically lower in the leucosomes and as low as 1 to 0.5 percent in leucotrondhjemitic end products. Na$_2$O, which is between 3 and 3.5 percent in a majority of the schists, is consistently higher in the leucosomes, commonly being around 4.5 percent but tending to exceed 5 percent in highly leucocratic end products. SiO$_2$ is up to 10 percent higher in the leucosomes than in the schists, whereas Al$_2$O$_3$ is nearly constant. Predictably, FeO, MgO, TiO$_2$, and MnO are lower in the leucosomes than in the biotite schists. Amphibolitic paleosomes and associated leucosomes show similar, but on the whole larger, compositional differences; K$_2$O is initially low; it increases in intermediate products as amphibole is biotitized, and systematically decreases in leucocratic end products. Details on the chemistry of the main-stage migmatites were given by Misch (1968) and Babcock (1970).

Dikes and pods of *late, discordant pegmatites* (Fig. 3d and i) have formed by replacement, by segregation, by local intrusive mobilization, and in places by injection from more distant sources. Crosscutting, small mobilized dikes (Fig. 3d, f, g) are

widespread and encompass the entire compositional range of the parent banded gneisses, from paleosome to leucosome. Controls are by local structure rather than by composition. Some mobilized dikelets are characterized by reduced grain size, the result of "cataclastic-plastic flow" (annealed as a rule).

Relatively small bodies of metaplutonic rocks occur throughout the Skagit Gneiss. In the type section, quartz-dioritic and subordinate dioritic, concordant *orthogneisses* account for an estimated 25 percent of the rock volume. They differ from the migmatites by relative homogeneity, by a lack of supracrustal remnants, and by relict gross plutonic texture. The orthogneisses were intruded at early to late stages of the metamorphic cycle; some of them were tectonically dismembered during that cycle. At the southern border of the area of Figure 2, orthogneisses become predominant within the Skagit Gneiss and include varieties with major K-feldspar. One of these was dated at 50 Ma by Hoppe (1985).

Metagabbro stocks range from pre-metamorphic (now migmatized amphibolite) to late-metamorphic (postdating migmatization, but still deformed and biotitized). Numerous small *metaperidotite pods* have fully participated in the regional meta-

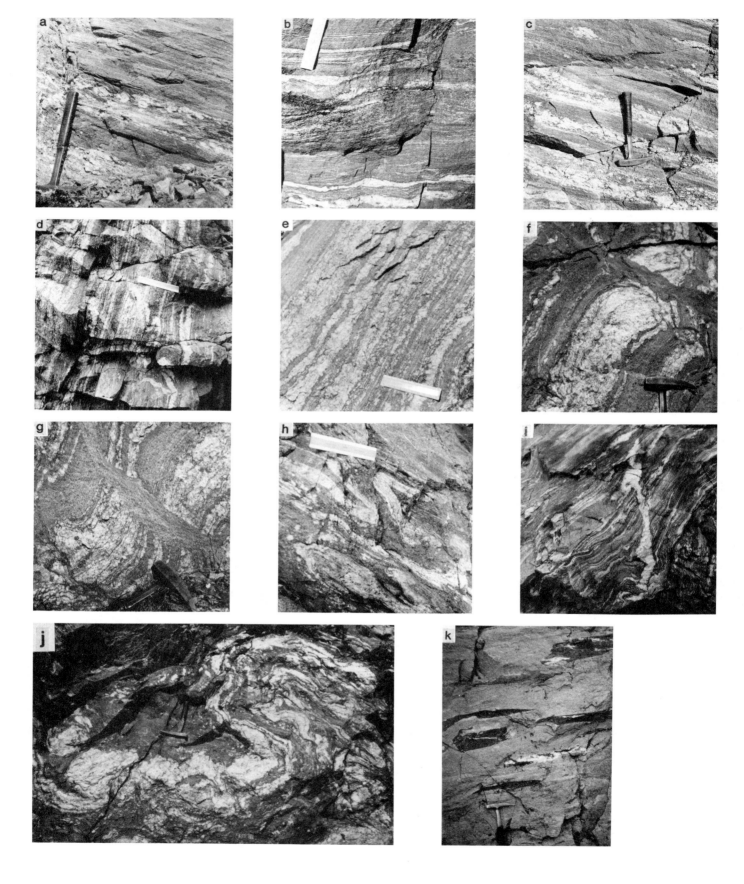

morphism. They display moderate to extreme metasomatic changes. Although on the whole strongly tectonized, some of these pods have maintained massive cores. These alpine peridotites are small mantle fragments, tectonically introduced into the Skagit protolith before, or during an early stage of, the metamorphic cycle. Commonly, their contacts show signs of disharmonic further movement during that cycle and, in places, of further tectonic dismemberment.

Youngest in the complex are *leucocratic orthogneiss dikes* intruded at the end of the metamorphic cycle. They are very widespread, though minor in volume. They discordantly cut the folded migmatites and in places contain xenoliths of those migmatites (Fig. 3k). They participated in the latest stage of deformation, largely in terms of b lineation. Usually they are homogeneous and display relict igneous textures, metamorphic recrystallization remaining somewhat incomplete. In contrast to migmatite leucosomes, they characteristically carry major K-feldspar; most compositions are granodioritic to granitic (quartz-monzonitic). These dikes were dated at 45 Ma by R. L. Armstrong (Babcock and others, 1985). Dike complexes of dioritic to silicic rocks, which were injected after the end of the metamorphic cycle, occur in some places.

The *structure of the Skagit Gneiss* is regionally characterized by intense, paracrystalline, compressional deformation of considerable complexity in patterns and in history; the history may vary in details from place to place. Thorough penetrative deformation is represented by the foliation, s, which parallels any preserved bedding (transposed) and the metamorphically gener-

---◄---

Figure 3. Skagit Gneiss exposures along Washington 20: Main-stage, synkinematic migmatites; late, crosscutting dikes. Paleosomes are biotite schists (paragneisses) except for (e)–(g). Locations given by road mileages. Scale: Hammer is 12.6 in (32 cm) long; its head measures 6.7 in (17 cm); ruler is 7.9 in (20 cm) long. (a) Incipient stage of migmatization by growth of plagioclase porphyroblasts. Advanced leucosome on lower left. Mile 132.8. (b) Early stage of development of banded migmatite. Some contacts between schist and leucosome accentuated by thin biotite-enriched selvages. Mile 122.3. (c) Well-developed plagioclase-porphyroblastic, banded migmatite. Note lack of basification in schist layers. Mile 132.8. (d) Advanced development of banded migmatite, porphyroblastic in part. Note lack of any significant basification in schist remnants. Late, discordant leucosome vein locally is still folded. Mile 127.9. (e)–(g) Amphibolite-associated, plagioclase-porphyroblastic, banded migmatites at mile 125.8. (e) Typical pattern. All layers are less mafic than essentially isochemical amphibolite remnants at this locality. (f) Mafic layers still are relatively close to amphibolite; identical in composition with discordant mobilized dike. Controlling fracture lacks displacement. (g) Discordant mobilized dike is of intermediate composition. Controlling fracture shows fault displacement. (h) Tight paracrystalline folds in banded migmatite. Some areas are plagioclase-porphyroblastic. Note lack of basification of schist. Mile 125.1. (i) Banded migmatite discordantly cut by dike of pegmatitic leucosome. Mile 127.6. (j) Plastic paracrystalline folds in plagioclase-porphyroblastic, banded migmatite. Some refolded remnants of earlier isoclinal folds. Note lack of basification of schist. Mile 125.1. (k) Discordant orthogneiss dike intruded at end of metamorphic cycle. Elongate xenoliths of wall-rock migmatite have been rotated into parallelism with flow foliation. Mile 127.5.

ated banding, and by the usually strong b lineation, which parallels fold axes and is the direction of any stretching normal to that of shortening. Attitudes of b are regionally consistent; it predominantly trends northwest. Paracrystalline folding of s (Fig. 3h and j) shows a variety of plastic, commonly complex patterns; ranges from open to closed and locally fan-shaped; includes refolding during the same cycle; and commonly is disharmonic. The last characteristic, in conjunction with synchronous fracturing, may lead to small-scale intrusive mobilization.

Recommended Stops Along Washington 20 (identified by mileages)

120.3, in Newhalem. Prominent cliff in quartz diorite orthogneiss, homogeneous except for minor pegmatitic veins; fully recrystallized.

122.3–4, West of First Road Tunnel. Graywacke-derived biotite schists (Fig. 3b) grade into plagioclase-porphyroblastic and banded migmatitic gneisses with trondhjemitic leucosomes (see general description of main-stage migmatites). About 200 ft (60 m) west of the tunnel, folded banded gneiss is cut by a mobilized dike of fully annealed brecciated migmatite, encompassing the same compositional range from paleosome to leucosome as the host. At the tunnel, quartz-dioritic orthogneiss is cut by pegmatite controlled by late shears; cataclasis is largely annealed.

124.1, West of Second Tunnel. Graywacke-derived biotite schists, mostly garnetiferous, some with hornblende. All gradations into plagioclase-porphyroblastic and banded migmatites; trondhjemitic leucosomes range from coarsely porphyroblastic to nonpegmatitic. A large cliff face, nearly normal to b, displays intricate folds and some late crosscutting elements, including pegmatite and mobilized dikelets that encompass the entire wall-rock compositional range. East side of cliff parallels b and shows almost perfect structural parallelism.

125.1–2. Spectacular display of migmatization and structure of graywacke-derived garnet-biotite schists. Some schists carry cummingtonite (rarely plus anthophyllite); a thin diopsidic bed (dolomitic graywacke) has hornblendic selvages. Minor sillimanite occurs locally; rare armored relics of kyanite and staurolite in garnet attest to history of progression. Gradations into plagioclase-porphyroblastic and banded synkinematic migmatites (Fig. 3h and j); end-product leucosomes are leucotrondhjemitic. Paracrystalline fold patterns (Fig. 3h and j) include refolding.

125.8, West of Gorge Lake Bridge. Slablike face is excellent example of advanced synkinematic migmatization of orthoamphibolite. Subordinate well-preserved amphibolite remnants are conspicuous in core of synform above the base of the slab and farther east below the slab. Much of the amphibolite is garnetiferous; in part of it, cummingtonite joins hornblende; there is some late biotitization of amphiboles. There is no evidence for wholesale basification of amphibolite during leucosome production. The only basification seen is restricted to a 6-in (15-cm), late-metamorphic shear zone in garnet amphibolite east of the synform's core; here, most plagioclase has been removed by pressure solution; hornblende and cummingtonite show major biotitization (addition of K along the shear).

Migmatization (Figure 3e) displays the same plagioclase-porphyroblastic and banded patterns as seen in the biotite schists, but intermediate products differ compositionally: The amphibolite changes first to dioritic and then to quartz-dioritic gneiss, commonly with hornblende± cummingtonite still present aside from largely new biotite. The final products of *both* starting materials are leucotrondhjemitic, commonly pegmatitic gneisses; some subtle geochemical differences may survive.

East of the synform's core, banded migmatite in the base of the slab is cut by a fairly mafic dike identical in composition with the concordant layer from which it was mobilized under fracture control (Fig. 3f). In the face above, east-dipping small faults, which offset the migmatitic banding, are annealed by thin mobilized dikelets; most are leucocratic, but some are more mafic (Fig. 3g). Obviously, small-scale intrusive mobilization was structurally rather than compositionally controlled.

127.5, 0.1 Mile East of Side Road to Diablo Dam, Turnout on Left. Biotite schist-derived migmatite (Fig. 3i) is discordantly cut by leucogranitic orthogneiss dikes intruded at the end of the metamorphic cycle (see above). One dike displays angular xenoliths of migmatite (Fig. 3k); elongate fragments were rotated into parallelism with the dike, which shows some flow foliation.

128.4, 0.1 Mile West of End of Long Road Cut, Turnout on Left. White marble dike has intruded migmatitic gneiss. The diopside- and grossularite-bearing marble is an annealed mylonite and contains cataclastic xenoliths of gray silicic gneiss. Later, the whole was engulfed in a dioritic-granitic igneous dike complex, producing minor contact-metamorphic wollastonite.

131.8, Scenic View Point (with Exhibit of North Cascades Rocks). The panorama almost is entirely in Skagit Gneiss. Road cut shows an injection complex; gneissose leucotrondhjemitic pegmatite dikes represent a late-kinematic stage of the Skagit cycle.

132.5, Turnout on Right. Pod-shaped metaperidotite within gneiss has fully recrystallized during Skagit metamorphism but remains massive except near contacts. It consists of anthophyllite and tremolite (to Mg-pargasite), commonly joined by forsterite and less commonly by enstatite; minor Mg-Al-chlorite persists, as does rare talc (texturally distinct from retrogressive talc and serpentine); minor phlogopite occurs sporadically. Some metasomatism is indicated. Metasomatism becomes thorough near contacts, where schistose phlogopite "blackwall" is developed (±Mg-Al-chlorite, ± anthophyllite), followed on the inside by an anthophyllite zone, locally with coarse cross-fiber replacing fine-grained isotropic texture.

133.5, Horsetail Creek Bridge. Below, and northeast of the bridge, are spectacular exposures of biotite schist-derived banded gneisses. Most leucosomes are plagioclase-porphyroblastic and converge toward leucotrondhjemitic gneissose pegmatite. Intricate mobile fold patterns include local refolding, are commonly disharmonic, and are associated with structurally con-trolled, small-scale mobilization; locally, mobilized dikelets have subsequently been folded. There are a number of small, usually concordant bodies of tectonized, metasomatized metaperidotite. At the cat track south of the bridge, a thin ultramafic schist on the creek's west bank contains these zones: core of tan anthophyllite with silvery talc and emerald-green tremolite ± light-brown phlogopite; anthophyllite-tremolite-phlogopite zone; tremolite-phlogopite zone; outward, the tremolite changes to darker-green actinolitic hornblende, and the phlogopite, to darker-brown biotite; rind of biotite-bearing hornblendite with some introduced plagioclase. Across the creek, another thin tremolite-phlogopite schist grades outward into impure hornblendite. The high road cut northeast of the bridge displays a number of intercalations of such ultramafic schists (five in the interval of 150 to 200 ft [45 to 65 m] from the bridge); they pinch and swell. but maximum thickness rarely exceeds 1.6 ft (0.5 m), whereas the minimum becomes zero where boudinage is complete. A thicker pod of metaperidotite-derived impure hornblendite occurs 330 ft (100 m) from the bridge. At its southwestern contact, the normal metasomatic zonation is reversed (related to pegmatitic injection?); another example of reversed zonation is seen 150 ft (45 m) from the bridge.

135.1, Second Turnout for View up Ross Lake. Highest point on left is Mount Prophet, in Skagit Gneiss.

136.2, Lillian Creek (No Parking). Metamorphosed uniform volcanic breccia from 136.1 to 136.4. Light-green angular fragments consist of grandite + salite + calcic plagioclase, locally with garnet-enriched pink cores. The matrix is dark-green amphibolite. Annealed deformation is dominated by lineation and ranges from weak to intense. Where strongly tectonized, the rock has a thinly banded or striped appearance. This lithology has no known counterpart in the Skagit Gneiss and lacks any migmatization. White veinlets consist of albite + prehnite ± epidote. Perhaps this rock is a small fault block of Elijah Ridge Schist (contacts not exposed).

REFERENCES CITED

Babcock, R. S., 1970, Geochemistry of the main-stage migmatitic gneisses in the Skagit Gneiss complex [Ph.D. thesis]: Seattle, University of Washington, 147 p.

Babcock, R. S., Armstrong, R. L., and Misch, P., 1985, Isotopic constraints on the age and origin of the Skagit Metamorphic Suite and related rocks: Geological Society of America Abstracts with Programs, v. 17, p. 339.

Hoppe, W. J., 1985, Origin and age of the Gabriel Peak Orthogneiss, North Cascades, Washington [M.Sc. thesis]: Lawrence, University of Kansas, 79 p.

Mattinson, J. M., 1972, Ages of zircons from the Northern Cascade Mountains, Washington: Geological Society of America Bulletin, v. 83, p. 3769–3784.

Misch, P., 1966, Tectonic evolution of the Northern Cascades of Washington State; A West Cordilleran case history, *in* Gunning, H. C., ed., A symposium on the tectonic history and mineral deposits of the western Cordillera: Canadian Institute of Mining and Metallurgy Special Volume 8, p. 101–148.

—— , 1968, Plagioclase compositions and non-anatectic origin of migmatitic gneisses in Northern Cascade Mountains of Washington State: Contributions to Mineralogy and Petrology, v. 17, p. 1–70.

—— , 1977, Bedrock geology of the North Cascades, *in* Brown, E. H., and Ellis, R. C., eds., Geological excursions in the Pacific Northwest, Geological Society of America 1977 Annual Meeting guidebook: Bellingham, Western Washington University, p. 1–62.

Republic graben, Washington

Falma J. Moye, Department of Geology, Idaho State University, Pocatello, Idaho 83209

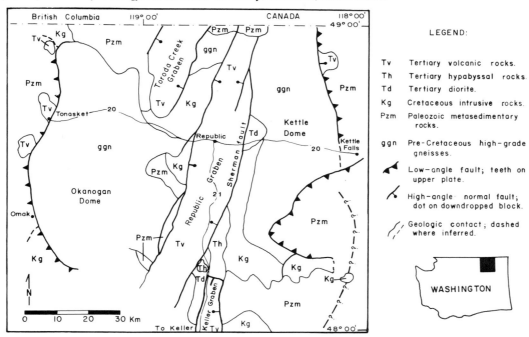

Figure 1. Location and general geology of the Republic graben and adjacent areas in northeast Washington.

LOCATION AND ACCESS

The Republic graben, located in northeastern Washington, extends from the Canadian border south approximately 48 mi (80 km) to the Columbia River. It can be reached on hard surface road from the east and west along Washington 20 from Kettle Falls or Tonasket, respectively, or from the south along Washington 21 from Wilbur (Fig. 1). The town of Republic, where lodging, food, and gasoline can be obtained, is located about 6 mi (10 km) north of the geographic center of the graben.

All of the stops shown in Figure 2 are accessible by two-wheel drive vehicle. Most of the area described in this report is on land owned by the Colville Indian Reservation, which has headquarters in Nespelem, Washington, 40 mi (64 km) south of Republic; Figure 2 shows the northern boundary at the reservation. North of Thirteenmile Creek, land is controlled by the National Forest Service or is privately owned. Permission should be obtained from private land owners and the Colville Tribal Headquarters before entering their land from the road right-of-way.

REGIONAL SIGNIFICANCE OF THE REPUBLIC GRABEN

Detailed stratigraphic and structural relations of Eocene volcanic and plutonic rocks in this area have contributed to an understanding of the Eocene extensional tectonics and concomitant magmatism in the Pacific Northwest. The Republic graben is one of several volcano-tectonic structures in northeast Washington that preserve remnants of widespread Eocene volcanism, plutonism, and sedimentation (Figs. 1 and 2). This volcanic field is part of an Eocene magmatic arc which extends from southern British Columbia across northeast Washington and central Idaho and into Montana and Wyoming; it is characterized by voluminous calc-alkalic and alkalic lava flows, ash flows, and consanguineous intrusive rocks of intermediate composition.

STRATIGRAPHY AND PETROLOGY

The volcanic stratigraphy defined by Muessig (1962) in the Republic graben was found to be regionally correlatable with other Eocene volcanic rocks in northeast Washington (Pearson and Obradovich, 1977). This widespread stratigraphic correlation implies that an extensive sheet of volcanic and volcaniclastic rocks covered the region and that remnants were preserved by graben faulting. Evidence where the volcanic cover is preserved suggests that numerous eruptive centers across the region were associated with the volcano-tectonic structures. The volcanic sequence as defined by Muessig (1962) is, in ascending order:

O'Brien Creek Formation. This Formation is composed of light-colored volcaniclastic and epiclastic rocks up to 1 mi (1.3 km) thick in the Republic graben (Muessig, 1967). These unconformably overlie pre-Tertiary rocks and were shed from adjacent highlands into the ancestral Republic grabens. At the southern end of the graben, extensive light-colored rhyodacite

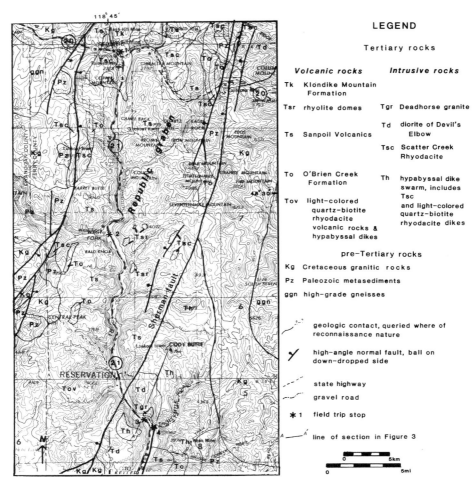

Figure 2. Geologic map and location of field-trip stops at the southern end of the Republic graben.

and rhyolite lava flows, ash flows, and intrusive rocks are the principle types of volcanic rocks. A K-Ar age of 53 ± 1.5 Ma has been determined for this unit (Pearson and Obradovich, 1977).

Sanpoil Volcanics. The Sanpoil Volcanics consist primarily of dark-colored, porphyritic lava flows and ash flows, ranging in composition from acid andesite to rhyodacite. Individual flows range from less than 66 ft (20 m) to more than 990 ft (300 m) in thickness; maximum thickness of the section reaches 1.5 mi (2.4 km) in the Toroda Creek graben (Pearson, 1967) and 1.3 mi (2.2 km) at the type locality in the Republican graben (Fox and Beck, 1985). The greatest thickness of individual flows is reached in the West Fork area (Fig. 2), where flows ponded over a source area in the center of the graben. Small rhyolite domes and dikes that intrude the Sanpoil Volcanics are confined to the area near the center of the graben (Fig. 2). Radiometric age of the Sanpoil Volcanics range from 49.6 ±3.0 to 53.8 ± 1.8 Ma (Pearson and Obradovich, 1977). The rhyolite domes yield an age of 49.53 ± .3 Ma (Moye, 1984).

Intrusive equivalents of the Sanpoil include Scatter Creek Rhyodacite and diorite of Devil's Elbow. The Scatter Creek is exposed along the length of the graben within and outside the structure (Fig. 2). The diorite is an elongate northeast-trending subvolcanic intrusion exposed along the east margin of the graben and is intruded by the granite of Deadhorse Creek which is temporally and chemically correlatable with the rhyolite domes in the center of the graben (Fig. 2). Radiometric ages for these units are 50.28 ± 0.34 Ma for Scatter Creek, 49.92 ± 0.33 Ma for diorite of Devil's Elbow, and 49.63 ± 0.31 Ma for granite of Deadhorse Creek (Atwater and Rinehart, 1984; Moye, 1984).

Klondike Mountain Formation. The basal Tom Thumb Tuff Member consists of lahar and volcanic conglomerate overlain by tuffaceous lake bed sediments. The upper section of the Klondike Mountain Formation consists of aphyric lithoidal and glassy lava flows that are rhyodacite and quartz latite in composition. This unit is confined to an area at the northern end of the Republic graben and the Toroda Creek graben. These rocks were deposited on an erosional surface cut on Sanpoil Volcanics. The Klondike Mountain is about 3,000 ft (900 m) thick in the Republic graben; in the Toroda Creek graben to the west, it is estimated to be 0.7 mi (1.2 km) thick (Pearson and Obradovich, 1977). The radiometric age of the Klondike Mountain is imprecise but is given as 46 Ma by Pearson and Obradovich (1977).

Structural evolution of the Republic graben was concomitant with volcanism. Muessig (1967) suggested that graben faulting began before volcanism and that the greatest amount of subsidence coincided with extrusion of large volumes Sanpoil lavas over their source area. A gravity survey across the graben south of West Fork suggests depth to basement is more than 1 mi (2 km) (Moye, 1984). Geologic mapping at the southern end of the Republic graben has shown that an extensive dike swarm parallel to and most abundant near the graben border faults was comagmatic with the O'Brien Creek and Sanpoil Volcanics (Pearson and Obradovich, 1977; Moye, 1984).

The eastern border fault of the graben, the Sherman fault, is a high-angle normal fault; numerous step faults parallel this structure in both the downdropped and upthrown blocks, and many of these are filled by Scatter Creek Rhyodacite dikes. The western border structure is not so clearly defined; evidence suggests that it may be a low-angle normal fault. The geometry of the border faults suggests that the graben is asymmetrical with the greatest amount of displacement on the eastern border faults.

SITE DESCRIPTIONS

The following recommended field trip stops (Fig. 2) have been selected to acquaint the visitor with the stratigraphy in the graben and with the principle intrusive rocks which, at the southern end of the graben, seem to be comagmatic with the volcanic rocks.

Stop 1. Intersection of Washington 21 and 20 at the southern outskirts of Republic. The Tom Thumb Tuff Member of the Klondike Mountain Formation at this locality consists of fine-grained tuffaceous sediments deposited in lakes which were formed along faults or behind dams of volcanic rock or landslide (Muessig, 1967). At this locality, plant fossils originally regarded as Oligocene indicate a subtropical climate. Species include: *Ginkgo adiantoides* (Unger) Heer; *Metasequoia occidentalis* (Newberry) Chaney; *Pinus* spp.; *Alnus* spp; *Pseudolarix americana* Brown; and *Thuites* sp.

From Stop 1 drive south on Washington 21 toward Keller. For about the next 13 mi (21 km), the road parallels the Sanpoil River which has cut through lava flows and volcanic breccias and has exposed the type section of the Sanpoil Volcanics. Any of the exposures in the canyon which starts about 11 mi (18 km) south of Stop 1 are representative of typical Sanpoil. The lava flows are generally dark-gray to black or brown with abundant phenocrysts of hornblende, plagioclase, pyroxene, and biotite in variable relative percentages set in a stony groundmass.

Stop 2. West Fork. The canyon opens southward into a broad valley at the confluence of the West Fork and the Sanpoil River. Park at the intersection of West Fork Road with Washington 21.

West Fork affords a view of the great thickness of the Sanpoil lava flows in the center of the graben. I regard this area as the site where greatest subsidence coincided with evacuation of a magma reservoir at depth. The massive lava flows ponded over the vent area which may correspond to a hydrothermally altered Scatter Creek intrusive 0.3 mi (0.5 km) to the northwest.

The cliff on the east side of the valley is mineralogically and chemically zoned from the base to the top. The lower lava flows are primarily hornblende rhyodacites with coarse hornblende phenocrysts up to several centimeters in size. Toward the top of the cliff, there is an abrupt transition to a phenocryst assemblage which consists of augite, biotite, and plagioclase. This mineralogical change does not correspond to interruption of a more or less continual eruption of lava, but rather may reflect compositional variation in the magma chamber owing to concentration of volatiles toward the top. Major-element geochemical variation is most obvious in SiO_2 content, which ranges from 62.54 wt.% at the base of 60.51 wt.% at the top. Similar mineralogical and compositional zonation occurs in thick sections of Sanpoil Volcanics farther south along the Sanpoil River (Moye, 1984).

From Stop 2, continue to drive south on Washington 21. You will see that the morphology of the Sanpoil lava flows changes from thick uninterrupted flows to thinner discrete flow units bounded by flow top and basal breccias. Ash-flow tuffs and tuff breccias, which do not occur in the West Fork area, are intercalated in the sequence. The Sherman fault crosses the highway south of the intersection of Washington 21 with South Nanamkin Creek Road (Fig. 2).

Stop 3

Park at the intersection of Thirtymile Creek Road and Washington 21. From this point, walk back along the highway approximately 0.25 mi (0.4 km) to the outcrop on the east side of the road.

Two options are possible at this site: the stout of heart may consider a traverse up through the steep section on the west side of the highway. Those not so inclined may visit the outcrop along the highway which is representative of the diorite of Devil's Elbow, which is temporally and compositionally similar to the Sanpoil Volcanics and Scatter Creek Rhyodacite. The diorite intrudes its roof, which consists of a hypabyssal dike swarm comprised of Scatter Creek Rhyodacite and light-colored rhyodacite dikes, the intrusive equivalent of the O'Brien Creek Formation. Locally, the diorite appears to intrude the Sherman Creek fault; however, contact relations are too obscure to be conclusive (Moye, 1984). Figure 3 schematically illustrates the relation of the diorite to the hypabyssal roof rocks. The diorite is compositionally and texturally zoned from the base to the top: it grades upward from a medium-grained, equigranular biotite-hornblende monzodiorite at the base through biotite-hornblende quartz monzodiorite to fine-grained, subporphyritic, hornblende-biotite granodiorite at the top. Locally, the upper part of the Devil's Elbow is indistinguishable from Scatter Creek Rhyodacite. These relations can be seen on the traverse to the top of the hill on the west side of the river. The traverse takes at least one-half day. Beware of rattlesnakes.

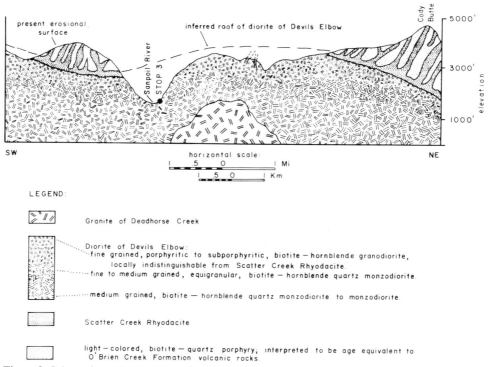

Figure 3. Schematic geologic cross section showing compositional variation of the diorite of Devil's Elbow and relation to hypabyssal roof rocks. Line of section shown on Figure 2.

Stop 4

Turn east onto Thirtymile Creek Road from Stop 3, drive 0.9 mi (1.5 km), and park at the steep talus slope on the north side of the road.

At this location, the diorite of Devil's Elbow exposed along the road is a medium-grained biotite-hornblende monzodiorite. A traverse up the talus slope to the conspicuous outcrops about 400 ft (122 m) above the road gives the opportunity to see cross-cutting dike relations in the hypabyssal dike swarm, which forms the roof of the diorite. Unfortunately, the contact relations of the diorite and the hypabyssal dike swarm are obscured by the talus.

Three distinctive rock units are exposed in outcrops above the road. From oldest to youngest, these are coarse-grained granite of the Cretaceous granite complex, light-colored quartz-biotite rhyodacite dikes, and dark-colored, biotite-hornblende Scatter Creek Rhyodacite dikes. The granitic rocks comprise less than 10% of the rock volume here, owing to extreme extension that accompanied formation of the graben and emplacement of the dike swarm. The light-colored dikes that chilled against granitic rocks are compositionally and petrographically similar to lava flows of O'Brien Creek Formation. Dark-colored rhyodacite dikes, which have distinctive chill margins against both of the other units, are Scatter Creek Rhyodacite and are genetically and temporally related to Sanpoil Volcanics and diorite of Devil's Elbow (Moye, 1984).

REFERENCES CITED AND ADDITIONAL USEFUL PUBLICATIONS

Atwater, B. F., and Rinehart, C. D., 1984, Preliminary geologic map of the Colville Indian Reservation, Ferry and Okanogan counties, Washington: U.S. Geological Survey Open-File Report 84-389, scale 1.

Fox, K.F., and Beck, M.E., 1985, Paleomagnetic results for Eocene volcanic rocks from northeastern Washington and the Tertiary tectonics of the Pacific Northwest: Tectonics, v. 4, no. 3, p. 323–341.

Moye, F. J., 1984, Geology and petrochemistry of Tertiary igneous rocks in the western half of the Seventeenmile Mountain 15-minute quadrangle, Ferry County, Washington [Ph.D. thesis]: Moscow, University of Idaho, 242 p.

Moye, F. J., Carlson, D. H., and Utterback, B., 1982, Volcanism and plutonism in a portion of northeastern Washington: Geological Society of America Abstracts with Programs, v. 14, p. 218–219.

Muessig, S., 1962, Tertiary volcanic and related rocks of the Republic area, Ferry County, Washington: U.S. Geological Survey Professional Paper 450-D., p. D56–D58.

Muessig, S. F., 1967, Geology of the Republic and a part of the Aeneas quadrangle, Ferry County, Washington: U.S. Geological Survey Bulletin 1216, 135 p.

Parker, R. L., and Calkins, J. A., Geology of the Curlew quadrangle, Ferry County, Washington: U.S. Geological Survey Bulletin 1169, 95 p.

Pearson, R. C., 1967, Geologic map of the Bodie Mountain quadrangle: U.S. Geological Survey Map GQ 636, scale 1:62,500.

——, 1977, Preliminary geologic map of the Togo Mountain quadrangle, Ferry County, Washington: U.S. Geological Survey Open-File Report 77-371, scale 1:48,000.

Pearson, R. C., and Obradovich, J. D., 1977, Eocene rocks in northeast Washington—Radiometric ages and correlation: U.S. Geological Bulletin 1433, 41 p.

Staatz, M. H., 1964, Geology of the Bald Knob quadrangle, Ferry and Okanogan counties, Washington: U.S. Geological Survey Bulletin 1161-F, 79 p.

Garibaldi area, southwestern British Columbia; volcanoes versus glacier ice

W. H. Mathews, Geology Department, University of British Columbia, Vancouver, British Columbia V6T 2B4, Canada

SIGNIFICANCE AND LOCATION

One of the few areas in the world where volcanoes have erupted under, into, through, or onto glacier ice, and the only place in North America that is reasonably accessible, is some 50 mi (80 km) north of the city of Vancouver in and near Garibaldi Provincial Park in southwestern British Columbia. Distinctive land forms recording the contest between hot lava and cold ice have been found at a score of sites scattered over a mountainous area 12 mi (20 km) from north to south and 6 mi (10 km) from east to west. Only the western edge of this area is accessible by road (Fig. 1); one site (at Brandywine Falls) can be reached directly by car. Others are visible at a distance from the road; several can be reached by a more or less strenuous hike on a well-marked foot trail, but the more distant sites require overnight camping or can be enjoyed with field glasses as distant views. All the sites are in provincial parks or along public roads.

SITES ACCESSIBLE BY ROAD

Take Highway 99 north from Vancouver past Horseshoe Bay (Fig. 1) (divergence of Highway 99 from Highway 401 = mi 0 = km 0) and along the east side of the fiord called Howe Sound. At mi 19.4 (km 31.0) is Britannia Beach, site of a major abandoned copper mine and mill, now home of the British Columbia Museum of Mining. Continue to Garibaldi lookout at mi 22.0 (km 35.0).

The view from this point (Fig. 2) on a clear day is dominated by Mount Garibaldi, 15 mi (25 km) to the northeast, a Pelean-type volcano (summit elevation 8,787 ft [2,678 m]) which erupted about 13,000 years ago. At that time the Late Wisconsin ice sheet, which had earlier covered all land up to about 6,600 ft (2,000 m), had shrunk to about the 4,400 ft (1,350 m) level. A cone of glowing-avalanche debris, built mostly on the ice, collapsed as the ice sheet melted away, leaving intact only a small segment of the original cone visible to the right of the summit (Fig. 2). The pyramidal peak in the center of the mass marks the site of the former dome and its intrusive neck.

Continuing north on Highway 99, one can get a rather different glimpse of Mount Garibaldi from the road at and immediately south of Cheekye Creek (mi. 33.8; km 54.5). A landslide scar is seen up Cheekye Gorge (to the east); it exposes a 2,500 ft (750-m) stratigraphic thickness of Pelean breccia. As the late Wisconsin ice melted away, much of the breccia fell away and flowed, or was washed, through Cheekye Gorge to be deposited in kame terraces against the dwindling ice sheet and, in postglacial time, onto a growing alluvial fan which extends 1.8 mi (3 km) downstream from Highway 99.

Some 14 mi (22.5 km) farther north, Highway 99 crosses

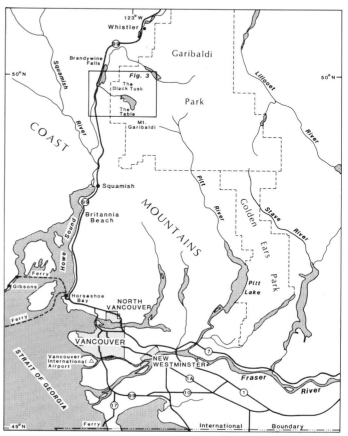

Figure 1. Index map of southwestern British Columbia showing location of the Mount Garibaldi area.

Rubble Creek (mi. 47.8; km 77) (Fig. 3); toward the southeast, up Rubble Creek Valley, there is a view of a spectacular raw cliff, 'The Barrier,' at the valley's head. This cliff, at the lower end of an ice-dammed lava flow, has been the source of multiple rockslides that have swept as much as 2.75 mi (4.6 km) down Rubble Creek to its junction with Cheakamus River, which it dammed to form Daisy Lake. A second precipitous ice-dammed lava lobe, which has not collapsed, lies to the right (west) of The Barrier. A tour to the base of this cliff by car and foot is outlined below.

Continue north on Highway 99 past the Cheakamus Dam and Daisy Lake picnic ground (mi. 49; km 79). The island in Daisy Lake, the reservoir above Cheakamus Dam, and the peninsula to the south of it are parts of a basaltic flow curiously restricted to a narrow linear belt.

Another flow with a similar character is to be seen at the

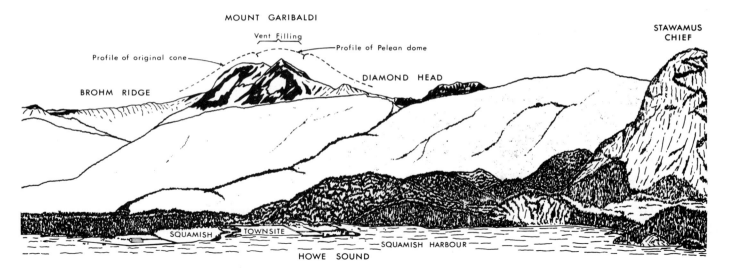

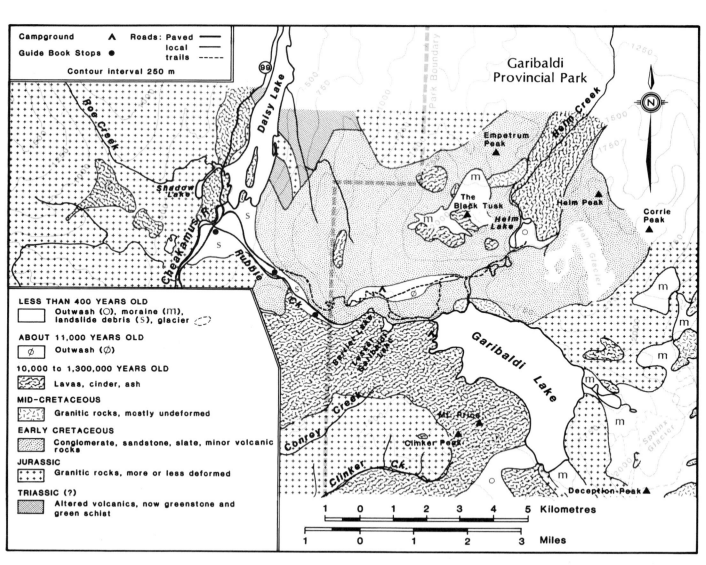

next site, Brandywine Falls Park (mi. 52.6; km 85). Walk south from the parking lot to the intersection of the British Columbia Railway and Highway 99, then east to the adjacent railway cut. This cut reveals the core of a narrow basaltic lava flow and, with a little digging, the base of the flow can be seen (above track level) on top of granulated basalt. The columnar joint pattern within the lava (Fig. 4) is consistent with cooling in a setting similar to present; little, if any, of the lava has been eroded away. This flow is interpreted as having flowed and congealed within a tunnel, or trench, in the Wisconsin ice sheet, built, like an esker, on and above the adjacent glacial deposits and older flows, but confined by icy walls to a narrow and sinuous course.

SITES AND VIEWPOINTS ACCESSIBLE ONLY BY FOOT

Upper Rubble Creek (Fig. 3) has two sites revealing the character of proglacial lavas that can be approached only by foot. A paved road, branching from Highway 99 at mi 48.0 (km 77.2), immediately north of the Rubble Creek bridge, leads 1.5 mi (2.5 km) up Rubble Creek valley to the start of the Black Tusk trail. This begins at the upvalley end of the parking lot at the road's end, and after a short switchback and a flight of stairs, climbs gradually southeastward for about 1.2 mi (2 km). At 1.4 mi (2.4 km) from the beginning, a branch trail leads down about 165 ft (50 m) to the floor of Rubble Creek valley and breaks out of mature forest into brushy second growth. Follow the trail to the first opening that gives a view of the cliff on the opposite wall of the valley. (It is pointless to go much farther, as (1) an unfordable stream blocks access to the cliff, (2) the cliff itslf could be scaled only with considerable danger, and (3) significant structures are better appreciated from a distant perspective—helped, if possible, with binoculars.)

The facing cliff is approximately 1,600 ft (500 m) high and is the terminus of a lava flow emanating from Clinker Peak, 3 mi (5 km) to the southeast and out of sight. Here the lava is a black, glass-rich dacite with excellent columnar jointing. Radiating clusters of columns are common and low dips predominate, as if cooling was inward from an initial steep but irregular surface. The cooling surfaces related to individual clusters of columns simulate, on a scale of tens of meters, the dribbles of wax that congeal on the side of a lighted candle; indeed, this candle analogy has been suggested for their origin. The glassy texture of the lava matrix may, moreover, record quick cooling of the lava not as a single solid mass, on the scale of the cliff, but rather as successive accretions on an already steep face. At the top of the cliff, 1,650 ft (500 m) above the valley floor, the form changes to that of a normal lava flow; there are coarse aa surface textures, marginal levees, and transverse wrinkles. This change in character at about the 4,500 ft (1,370 m) level is considered to mark the upper limit of the dwindling Wisconsin ice sheet at the time of the eruption from Clinker Peak. Lava from the vent, after flowing across the upper ice-free slopes, is presumed to have reached the edge of the ice which still filled Cheakamus Valley and extended

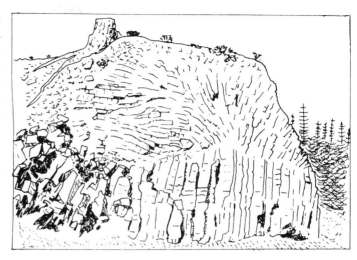

Figure 4. Columnar jointing pattern in north wall of railroad cut at Brandywine Falls. From a 1946 photo.

up Rubble Creek valley to this point. Initial contact with the ice would retard the advance of the flow front and encourage the development of a dam of solidified lava which, in turn, would pond hot fluid lava behind it. The heat from the lava dam would promote melting of the ice and produce a gap between lava front and ice—a gap either water-filled or empty, depending on opportunity for drainage through the ice. Periodic overflows of fluid lava (like hot wax in the candle) could spill down the front of the dam into the gap, congeal, and produce accretions on an already steep face. The face as we see it now, probably little changed from the original, stands as mute testimony to a phenomenon that can rarely be seen today and then only in remote areas where glaciers and volcanic vents coexist.

The Barrier (see above), another rather different part of the lava front just described, is visible only with difficulty from the brushy flats of Rubble Creek valley and cannot be safely approached from below because of repeated rock falls onto and across its talus apron, particularly in warm, dry weather. A better view can be obtained from a higher branch of The Black Tusk trail diverging to the southeast at 6.2 km, some 3.2 km beyond and almost 500 m above the branch just described. The upper branch, which leads to Barrier, Lesser Garibaldi, and Garibaldi lakes, has a short spur about 200 m from its start; this takes the hiker out to open rocks (green Cretaceous arenites) and The Barrier viewpoint.

The Barrier, as seen from this site, is a spectacular, near-vertical cross section of part of the Clinker Peak flow. The top 200 ft (60 m) is a red flow breccia, the source of many of the rock falls. Below this, making up three quarters of the 800 ft (250 m) thickness of a single flow, is a dacite made up of alternating black and gray bands, some 4 to 12 in (10 to 30 cm) thick, that become reddish upward. The presee of large joint columns near the far end of The Barrier indicates that the banded mass cooled as a

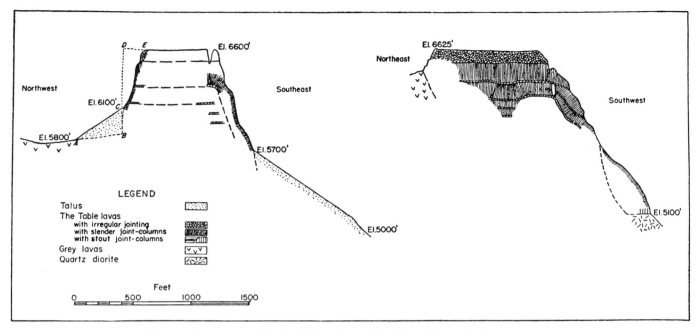

Figure 5. Transverse and longitudinal sections of The Table, Mount Garibaldi map-area.

unit, and the near-vertical attitude of the prisms reveals that the lava in their vicinity cooled through the upper, rather than the front, surface. Major rockslides, the latest in 1855–56, have removed the former front of the flow, presumably a part that matched the structure and origin of the lava cliff seen across the valley from the last site. But here we look at the heart of the flow, the reservoir of once hot and fluid lava, ponded behind the lava dam developed at the glacier margin.

For the enthusiastic geologist with spare time, daylight, and energy, who wishes to see more of the products of proglacial vulcanism—albeit from a distance—there is The Table, a volcano built through the ice. Take any one of several trails leading to The Black Tusk Meadows, cross these to the slopes of either Panorama Ridge or The Black Tusk, and ascend to or beyond the 1,900 m contour. Some 6 to 8 km to the southeast, across Garibaldi Lake and in front of Mount Garibaldi, is The Table, a peculiar flat-topped mountain that looks like an inverted coffee cup, with a handle (complete with a hole) at its western end (Fig. 5).

The Table was initially more nearly a vertical prism (like a giant tabular iceberg, elliptical in plan) than it is today after the recession of the upper slopes of some tens of meters. The internal structure revealed by this recession is that of a series of nearly horizontal cooling units, each with a layer of very short, stout columns below and a much thicker layer of slender, vertical columns above (Fig. 5). One of these units, followed westward, becomes a pair of units which extend down the flank of The Table to its western base; there, the lava rests on a completely unweathered till. The overlying flow is seen on the south face of the mountain, where it extends downward as a plaster on the precipitous face.

The Table is interpreted to be the filling of a cylindrical vent thawed through the Wisconsin ice sheet at some time close to its climax. Lava covering the bottom of this vent would spread until making contact with the icy walls, then chill to form an annular ring within which the remaining hot lava would cool. Before the first layer was completely cooled, a second was built on the first, also spreading outward to the same ice contact. A gap may have developed between congealed lava and the ice, down which hot lava could spill to form the "icing on the flanks of a layer cake."

Closer inspection of the mountain could be rewarding but is discouraged by the difficulty of access and the extreme hazard caused by loose and falling rock. A "volcano fly-by" (Souther, 1977) is another means of viewing the strange forms and structures associated with ice-contact volcanism but should be undertaken only in favorable weather.

REFERENCES CITED

Mathews, W. H., 1951, The Table, a flat-topped volcano in southern British Columbia: American Journal of Science, v. 249, p. 830–841.
——— , 1952a, Mount Garibaldi, a supraglacial Pleistocene volcano in southwestern British Columbia: American Journal of Science, v. 250, p. 81–103.
——— , 1952b, Ice-dammed lavas from Clinker Mountain, southwestern British Columbia: American Journal of Science, v. 250, p. 553–565.
——— , 1958, Geology of the Mount Garibaldi map area, southwestern British Columbia, Canada; Part II, Geomorphology and Quaternary volcanic rocks: Geological Society of America Bulletin, v. 69, p. 179–198.
——— , 1975, Garibaldi geology, a popular guide to the geology of the Garibaldi Lake area: Cordilleran section, Geological Association of Canada, 48 p.
Roddick, J. A., Mathews, W. H., and Woodsworth, J. G., 1977, Southern end of the Coast Plutonic Complex; Fieldtrip guidebook #9: Cordilleran Section, Geological Association of Canada, 33 p.
Souther, J., 1977, Volcano fly-by; Fieldtrip guidebook #16: Cordilleran Section, Geological Association of Canada, 15 p.

Cache Creek–Ashcroft area, western Intermontane Belt, British Columbia

William B. Travers, Department of Geological Sciences, Cornell University, Ithaca, New York 14853

LOCATION AND SIGNIFICANCE

Outcrops in and near the settlements of Cache Creek and Ashcroft, southern British Columbia (Fig. 1), provide an opportunity to view some of the structures and rocks that help to explain the evolution and amalgamation of the allochthonous terranes of the southern Canadian Cordillera. Sediments and igneous rocks of the Cache Creek and Nicola Groups formed in late Paleozoic to late Triassic times in oceanic and island arc settings. The sediments of the Ashcroft Formation were deposited from Early through Middle Jurassic time in an arc-related basin. In Late Jurassic or possibly earliest Cretaceous time, thrust faulting juxtaposed Cache Creek-Nicola rocks over Ashcroft strata. This event is thought to have sutured these terranes to Quesnellia of which the Ashcroft Formation is generally considered to be a part. This faulting may have been related to the early (Late Jurassic) deformation in the Omineca Crystalline Belt when east-directed thrust faults joined Quesnellia to that crystalline terrane.

SITE DESCRIPTION

The field guide begins at the traffic light in Cache Creek at the intersection of the Trans Canada 1 and British Columbia 97 (Caribou Highway to Clinton and 100 Mile House). Proceed west from the light 0.1 mi (0.16 km) across the Bonaparte River to Stop 1. Turn left (south).

Stop 1, Cache Creek—cumulative distance: 300 ft (100 m). Excellent exposures of the type Cache Creek are seen in and above the road cuts that run down the west edge of Cache Creek village. This road will rejoin Trans Canada 1 about 0.4 mi (0.65 km) south. The Cache Creek Group was named from these exposures by Selwyn in 1872. This, the Eastern Belt (Fig. 2) of the Cache Creek Group, consists of a mélange of gypsiferous argillite, cherty argillite, ribbon chert and phyllite, which composes the matrix of the Cache Creek mélange. In the hills around Cache Creek village are numerous bold outcrops or "knockers" of intensely folded ribbon chert, basalt, greenstone, chloritic schists, serpentinite, graywacke and limestone. These bold outcrops are allochthonous blocks, usually lenticular in outline, that are surrounded by a matrix of intensely folded and sheared rock of mostly gypsiferous argillite and cherty argillite. Beds may be traced for short distances within individual blocks, but bedding is thoroughly disrupted within the argillitic matrix. Small folds and faults within a particular block may have similar orientations but are different from block to block. The proportion of argillite to ribbon chert in the matrix varies considerably, as does the distribution of various rock types that compose the blocks. These variations allow the Cache Creek Group to be divided into three belts of outcrops in southern British Columbia (Monger, 1981). The Eastern Belt, seen at Stop 1, has blocks of predominantly

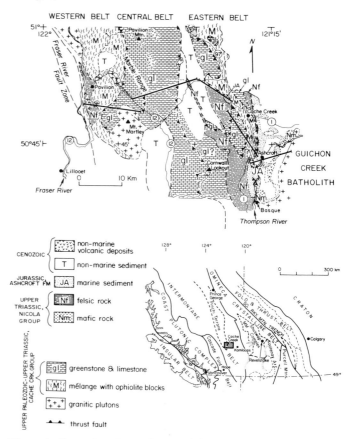

Figure 1. General geology of part of the western Intermontane Belt showing the location of the general cross section, Figure 4. (From Travers, 1978 and 1982, and Monger, 1981). See insert for location.

radiolarian chert, greenstone, limestone, serpentine, and basalt, in order of abundance, and appears to be the most deformed. The Central Belt has small to very large (1-2 km) blocks of mostly light gray, partially recrystallized limestone with a lesser number of chert and basalt blocks. Verbeeckinid fusulinids found in limestone are said to be of the Tethyan realm and hence are exotic to late Paleozoic North America (Monger and Ross, 1971). A huge limestone block is exposed a few miles west in the Marble Canyon along British Columbia 12 (to Pavilion and Lillooet). The Western Belt is mostly chert blocks with fewer volcanic, sandstone and limestone blocks. The Eastern and Western Belts are similar in composition and style of deformation. However, they were not necessarily formed together or at the same time.

In the Eastern Belt, the limestones yield fusulinids of Upper Pennsylvanian and Lower Permian stages (Sada and Danner, 1974). Locally, the Cache Creek cherts have Upper Triassic

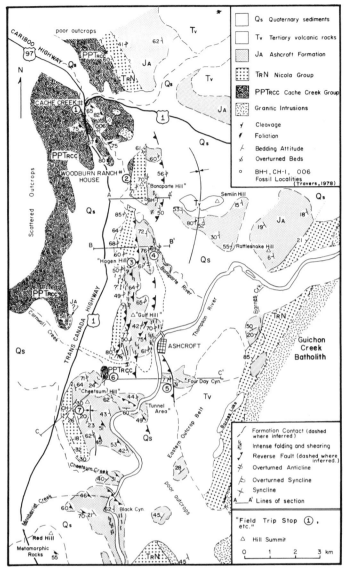

Figure 2. Detailed geology map of the Ashcroft area showing locations of detailed cross sections A–A', B–B', and C–C' (Figure 3). Field trip stops are the circled numbers (From Travers, 1978).

(Middle Norian) radiolaria (Travers, 1978) and Triassic conodonts (Orchard, 1984). The Central Belt limestones contain Middle and Late Permian fusulinids (Monger and Ross, 1971) and Early and Late Triassic conodonts (Orchard, 1984). The Western Belt contains poorly preserved fusulinids of Permian age (Trettin, 1961 and 1980).

The style of deformation and rock types seen along the road at Stop 1 are typical of subduction zone complexes as seen in many parts of the world. The Franciscan of California (Hsu, 1973), the Argille scagliose of the Northern Apennines, mélanges on the islands in the Western Pacific region (Karig and Sharman, 1975), as well as the type mélange in Wales (Greenley, 1919; Wood, 1973), are similar and have usually been interpreted as

subduction zone complexes. However, it is seldom clear whether the deformation is due to gravity sliding, to underthrusting during subduction, or both. Mud diapirism and gravity sliding into trench and forearc basins and later (post-subduction) tectonism can all contribute to mélange deformation (Cowan, 1985). Much of the deformation seen here in Cache Creek argillite occurred under conditions of ductile flow. Tight flow folds lack brittle fractures in hinges, while broad zones of ductile faulting are characterized by a crude, irregular flow fabric. This may be interpreted as prelithification deformation when the water content of muds was still quite high. Early formation of gypsum during diagenesis would also lower shear strength in the argillaceous sediments. Eocene shales of the Argille scagliose, Italy, deform in modern subaerial debris flows, as do parts of the Cretaceous Francisco Group, California. Thus, soft sediment deformation can occur long after sedimentation.

Return to Trans Canada 1 at the traffic light (intersection of British Columbia 97). Turn right (south). Proceed south from Cache Creek village 0.8 mi (1.2 km). Turn left (east) on the gravel road that leads into the Woodburn Ranch. Drive 0.15 mi (0.25 km) to a fork in the road. Continue straight to the modern ranch house to ask permission to view the geology on the east bank of the Bonaparte River. Return to the road fork, turn right and drive down the hill onto the modern flood plain of the Bonaparte River. Turn left at the bottom of the hill and park near a small wooden bridge over the river 0.2 mi (0.3 km) from fork. Walk across the bridge, turn right, walk about 300 ft (100 m) east and then turn south across badlands terrane above the low bluff, following the east bank of the river for about 1,500 ft (500 m). Walk 600 ft (200 m) up slope (east) across serpentinite to the base of the steep part of the hill to Stop 2.

Stop 2, Woodburn Road—cumulative distance: 1.1 mi (1.75 km). At the top of the bare serpentinite slope is the contact between the Cache Creek Group and debris flows of the Nicola group (cross section A–A', Fig. 3). The contact zone is only a few centimeters wide and is well exposed. The contact is a near vertical gouge zone. This is the best exposure of the Cache Creek–Nicola contact in southern British Columbia. Several unmetamorphosed basalt dikes, presumably of Tertiary age, are intruded into the serpentinite (Note that the largest basalt dike has "pillow" structures on both sides!). The basalt dike nearest the Cache Creek–Nicola contact is deformed into a single fold with near vertical plunge and with an axial plane trending northeast. This fold may be a drag fold due to displacement along the (fault) contact. If so, it would demonstrate right lateral strike-slip displacement of Tertiary age. However, several closely spaced, unaltered basalt dikes are found a few meters south in the Nicola and they terminate at the Cache Creek contact. If any one of those dikes was part of the dike seen in the serpentinite, the strike-slip offset is only a few tens of feet.

The basal Nicola debris flow contains black, gray, and red chert clasts that could have come from Cache Creek cherts. No radiolaria have been found in these clasts. The Nicola here is about 1,300 ft (400 m) thick. Farther up the hill (east) are steep,

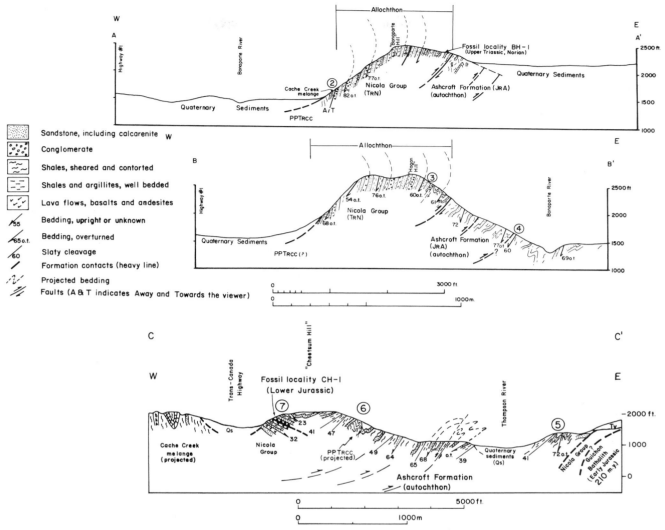

Figure 3. Detailed cross sections A–A', B–B', and C–C' showing field trip stops and fossil localities (see Figure 2 for location). See Travers, 1978 for discussion of fossils. Bonaparte Hill is a klippe of Nicola and lower Ashcroft strata resting upon middle and upper Ashcroft strata (section A–A') while Hagen Hill (section B–B') is a klippe of Nicola rocks on Ashcroft Formation (modified from Travers, 1978). Cheetsum Hill is a klippe of Nicola and lower Ashcroft strata resting on middle and upper Ashcroft strata and a slice of Cache Creek rock, PP ʈ cc (modified from Travers, 1978).

west dipping beds of argillite and graded sandstone with abundant chert, greenstone and limestone grains. Some sandstones are graded and finely crossbedded calcarenites which indicate that stratigraphic up is to the east; thus Nicola strata here are overturned and verge east (cross section A–A', Fig. 3).

The Cache Creek/Nicola contact here on the Woodburn Ranch may be interpreted in several ways: (1) that Cache Creek rocks may have been placed against the Nicola Group by strike-slip faulting—fault(s) may have a large component of normal faulting, probably down on the west; (2) that the Nicola may have been deposited on top of the Cache Creek mélange (if so, the basal Nicola debris was eroded from the exposed Cache Creek rocks); (3) that the Cache Creek may have been thrust over

Nicola strata possibly in late Triassic time—the Nicola debris flow could be derived from the advancing Cache Creek allochthon, or it may have been derived from older parts of the Nicola island arc (in this view, minor Tertiary faulting may have done little to disturb the contact); or (4) that the Cache Creek may have been juxtaposed against Nicola rocks by debris flows and/or thrust faulting in Late Jurassic time. Then, probably still in Late Jurassic time (post Callovian), the Cache Creek and Nicola rocks may have been thrust over Ashcroft strata, as seen in Stop 3. Presumably, this post-Callovian thrusting rotated Nicola strata to their present near-vertical position, and may have also rotated the Cache Creek–Nicola contact as seen here. If so, the Cache Creek–Nicola contact was originally near-horizontal.

Consequently, it may have been a low-angle gravity fault that dipped gently east or a low-angle thrust fault that dipped gently west. In either case, the hill to the east (Bonaparte Hill) is a klippe that rests on highly deformed Ashcroft strata similar to those seen at Stop 4.

Directions to Stop 3. Return to Trans Canada 1, turn left and proceed south 1.2 mi (1.9 km) to the turn onto the highway to Ashcroft. Turn left (southeast). The large rounded hill directly in front is Hagen Hill (sometimes called Elephant Hill). It is a klippe of overturned Nicola strata resting on tightly folded and faulted Ashcroft strata. The fault contact dips west about 45 degrees. This contact is best seen near the top of Hagen Hill (Stop 3).

From Trans Canada 1, travel southeast toward Ashcroft 0.2 mi (0.35 km) to the Hagen house on the left in order to ask permission to take their ranch road to the radio tower at the top of Hagen Hill. The road to the tower is suitable for four-wheel drive vehicles, pick-up trucks, or other high-center vehicles, but not for ordinary sedans. If you lack a suitable vehicle, skip Stop 3 (and the stop at the Hagen ranch house) and continue along the highway to Ashcroft. At 0.65 mi (1.0 km) on the right above the road are outcrops of Nicola argillite and volcanics. These are the most westerly Nicola outcrops on Hagen Hill, which lacks outcrops of the Cache Creek Group. Continue along the highway, until you are 0.9 mi (1.5 km) from the Trans Canada 1 to Stop 4.

If you are able to drive to the top of Hagen Hill from the Hagen ranch house, go back along the road toward Trans Canada 1 about 300 ft (100 m) and turn left (south) onto the ranch road. Travel south past haystacks about 1.6 mi (2 km). Keep straight ahead at first road fork, then angle left at the second road fork (about 500 ft; 150 m beyond the first fork). Continue on up the hill about 0.5 mi (0.8 km) to two sharp switchback turns near the top. The second switchback turns north. Outcrops in small road cuts just beyond this turn are in Ashcroft sandstone and shale. No fossils have been found here, but the beds are probably Lower Jurassic (Sinemurian or Pliensbachian), based on presumed correlation to basal Ashcroft beds elsewhere in the Ashcroft Valley. For approximately the next 0.2 mi (0.3 km) the road travels north-northwest and crosses the faulted Nicola/Ashcroft contact which trends about due north. The last 300 ft (100 m) of the road swings to the west to reach the radio tower at the top of the hill. Low cliffs on the right (north) are Nicola argillites, sandstones, and volcanics.

Stop 3, Hagen Hill (Elephant Hill)—cumulative distance: 4.5 mi (7.3 km). The fault contact trends along the east side of the hill at the base of the small cliffs (Figs. 2, 3). Park at the radio tower and walk back along the road about 300 ft (100 m). Leave the road and walk along the base of the cliff for about 300 to 1,000 ft (100 to 300 m). The gouge zone is well exposed along the base of the cliff in several places. Note angular blocks of chert, argillite, sandstone, and rare limestone in the fault gouge.

The Nicola above the fault is overturned and facing east as shown by graded beds and cross beds, and is probably part of a large overturned anticlinal fold (cross section B-B', Fig. 3). Minor folds are rare and small faults are mostly about parallel to bedding and are probably partly due to interbed shearing during folding and thrust faulting. Bottom marks on bedding planes are virtually absent in the Nicola, which suggests that lithification took place before folding. In contrast, flute casts and other bottom markings are common in the graded sandstones of the Ashcroft Formation. These beds are folded into a series of near isoclinal folds that are well exposed from 300 to 1,000 ft (100 to 300 m) downhill (east) of the fault contact described above. Fold hinge lines trend due north to north-northwest while axial planes dip steeply west about parallel to the fault that separates Nicola from Ashcroft strata.

Return to the Ashcroft Highway, turn right (south), and continue southeast for about 0.5 mi (0.8 km) to Stop 4.

Stop 4, Ashcroft Highway—cumulative distance: 6.8 mi (11.0 km). The Ashcroft Formation is tightly folded and the outcrops are cut by numerous ductile shear zones that generally lie at low angles to bedding. Individual sandstones can be traced for only a few meters before termination by one of the shear zones. Note that many tightly folded sandstone beds are not fractured much, even in the highly compressed hinges. Axial planes dip west and southwest at various angles. Many hinge lines trend approximately parallel to the highway. Sandstone layers are often terminated by shears but are also seen to end in thin, wispy, pull-apart structures. The shales lack distinct fault planes so the relative movement is distributed throughout the pervasively sheared shale. Taken together, these features suggest that deformation occurred before the sediment was well lithified, that is, shortly after deposition. If so, the deformation is probably of Late Jurassic age. The age of deformation cannot be further constrained, because no Mesozoic rocks younger than Callovian (latest Middle Jurassic) are known from the Cache Creek-Ashcroft area with one exception. On British Columbia 12 (road to Pavilion and Lillooet) there are "chert-pebble conglomerate and sandstone" that contain palynomorphs reported to be of mid-Cretaceous age (Price and others, 1981, p. 307). These are the oldest non-marine strata from this area. They appear to be braided stream deposits and fanglomerates and probably postdated all the ductile deformation described in this text.

Current direction data from flute casts and cross beds in the Ashcroft Formation suggest that the paleo-seafloor slope was to the south and southeast.

Continue south to the Ashcroft Bridge. Note the outcrops in and above the road cuts at the northwest end of the bridge. This is the shaley middle and upper part of the Ashcroft Formation. It is disrupted in the same fashion as seen in Stop 4. Nearly all beds dip west. However, the thin graded sandstones and thick carbonaceous shale are again deformed into nearly isoclinal folds. Numerous changes in facing direction from bed to bed are indicated by bottom marks, graded beds and cross beds. As at Stop 4, axial planes trend north to northwest and generally dip steeply or moderately to the west. A poorly developed slaty cleavage has a similar orientation. On the hillside west of the bridge are intensely disrupted zones of gouge and pervasively sheared shale that may

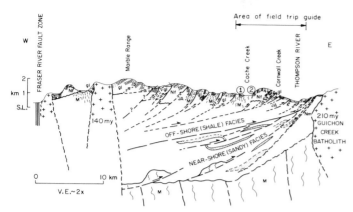

Figure 4. General cross section (see Figure 1 for location). Ashcroft strata are inferred to underlie klippen of Cache Creek and Nicola rocks for reasons discussed in text (modified from Travers, 1978).

be due to major thrust faults, as indicated by the "inferred thrust" fault symbols on Figure 3. Cross the Thompson River on the bridge to Ashcroft. On leaving the bridge, bear left and continue east to the first traffic stop. Turn right (south) and continue through Ashcroft and cross the railroad tracks. Stop 5 is a large dry canyon 0.85 mi (1.3 km) south of the railroad track crossing, on the highway to the Highland Valley copper district.

Stop 5, Four-day Canyon—cumulative distance: 12.2 mi (19.7 km). Cross the fence and go east up the dry wash to see folded and faulted Ashcroft sandstone and shale (cross section C-C', Fig. 3). About 70 ft (20 m) east of the fence approximately 200 ft (60 m) east of highway, note the overturned sandstone on the left (north) bank of the dry creek. The strike of bedding is about due north and the dip is 50°W. Small folds trend approximately N20°W and plunge varies from 10 to 70 degrees. Overturning is shown by load casts, flute casts, and other bottom marks, and by graded beds and cross beds. If the beds are rotated to a horizontal orientation by stereonet, about an axis parallel to the mean hinge line of the folds, a paleo-seafloor slope to the west is indicated. Note also minor reverse faults that, in general, dip moderately to steeply west about parallel with the axial planes of associated drag folds.

Many folds have an overturned east limb. Some folds are of the chevron type with planar, undeformed limbs. Hence, deformation is concentrated in the hinges as shown by thickening of shale and sandstone layers. Other folds have more rounded hinges with chaotic shale cores. Poor to moderately well developed slaty cleavage is sub-parallel to fold axial planes.

Continue up the dry creek bed taking the right (south) fork at 100 ft (30 m) from the fence. At the stream fork, note the overturned syncline that verges east. Just beyond the fork, note the small anticline with a partly sheared core. This structure plunges 22°S, which is approximately the plunge of several more folds seen farther up the canyon. About 230 ft (70 m) from the stream fork, and 360 ft (110 m) from the fence, on the north side of the gulley, is an east verging fold with a sheared core that has become a minor thrust fault, up on the west side, with about 6 ft

(2 m) displacement. "Lizard skin" bottom marks indicate that the east limb is overturned. Numerous flute casts also show facing directions, the majority of which are to the east.

At about 700 ft (200 m) from the fence, a well developed slaty cleavage with a strike of N45°W and a dip of 35°SW is intersected by a locally prominent fracture cleavage to produce a spectacular "pencil slate." The section just traversed contains an interesting collection of minor folds: some are closed and near isoclinal chevron folds, some are open and cylindroidal folds. The axial planes of some of the very tight folds have undergone considerable ductile shear to produce minor reverse faults. Nearly all of the structures, including slaty cleavage, indicate an east or northeast vergence with overturning to the east. Between areas of folding are zones of intensely sheared and chaotic shales. Most of these zones appear to dip steeply west, although rare east-dipping zones are seen. These chaotic zones could be the loci of major thrust faults. However, lack of distinctive stratigraphy and the scarcity of fossils prevent confirmation of this possibility. In general, however, the tight folds and chevron folds are usually located several meters from chaotic zones. Folding is clearly soft-sediment (pre-lithification), and because the orientation of axial planes of these folds is about parallel to the attitude of slaty cleavage, it is assumed that these features are approximately contemporaneous and that they formed fairly soon after deposition. Although no diagnostic fossils have been recovered in Four-day Canyon or nearby, this section was probably deposited some time in the Toarcian to Bathonian interval, so the deformation is likely of Late Jurassic age, possibly Oxfordian. Note that when continuing up the canyon, the intensity of deformation decreases. Near the top of the canyon, sandstone and shale give way to a section of almost entirely very dark carbonaceous shale with a few siltstone and rare very fine grained sandstone beds. These shales may be of Callovian age by correlation to carbonaceous shales along lower Barnes Creek (Fig. 2) and on the east side of Semlin Hill (Frebold and Tipper, 1969). The contact between the lower sandy section and the upper, carbonaceous shale does not outcrop. If the chaotic zones of Four-day Canyon mark major thrust faults, the more sandy section may be thrust over the younger carbonaceous shales which, in turn, may be thrust over older sandy and conglomeratic basal Ashcroft strata, as shown in Figure 3.

Descending back down the canyon, return to the canyon fork mentioned earlier. About 300 ft (100 m) up the north fork of the canyon is another example of pencil slate and small-scale folds.

Return to the Ashcroft-Highland Valley Highway and Ashcroft village and recross the Thompson River. Turn left (south) at the west end of the bridge (the cumulative distance from the start of this Field Guide is now 14.9 mi (24.0 km). For the next 0.8 mi (1.3 km), on the west side of the road, are more outcrops of Ashcroft Formation with numerous folds and ductile shear zone in shales. This land is part of the 105 Mile Post Indian Reservation. As seen at Stop 4, axial planes and slaty cleavage dip west and the eastern limbs of anticlines are often overturned. After another 0.4 mi (0.7 km), the highway turns west and climbs. About 2.5 mi (4.0 km) from the Ashcroft bridge, turn left (south-southwest) onto a poor dirt road. This turn-off is 0.6 mi (0.95 km) from Trans Canada 1. Drive carefully south on the dirt road 0.15 mi (0.25 km) to junk cars near the south edge of Cornwall Creek Canyon. Descend to the dry stream bed on the canyon floor. Walk upstream a few meters to the farthest upstream outcrop.

Stop 6, Cornwall Creek—cumulative distance: 17.5 mi (28.25 km). Various rocks of the Cache Creek Group outcrop at several places in the lower parts of the canyon for a distance of about 0.6 mi (1.0 km) downstream (cross-section C-C′, Fig. 3). Downstream, the southernmost outcrops of Cache Creek Group rock in Cornwall Creek are found just a few meters north of the first tributary gully that drains the nearby high hill, with radio towers (Cheetsum Hill) seen to the southwest.

The Cache Creek Group rocks seen here in Cornwall Creek include pale tan and gray thin bedded chert, dark carbonaceous argillite, and small lensoid blocks of limestone, greenstone, and graywacke which are from 15 to 100 ft (5 to 30 m) thick and from 50 to 200 ft (15 to 60 m) long. The blocks have sheared and faulted margins. The ribbon cherts are folded into a series of small, asymmetric folds. The greenstone and graywacke are folded and faulted, and the argillite is pervasively sheared. Bedding attitudes are highly variable although most dips are steeply west.

Typical Ashcroft Formation strata occur about 50 ft (15 m) west of the southwest end of the Cache Creek block. The Ashcroft strata strike uniformly about N40°W and dip about 25°SW and are mildly deformed near the contact by several small faults(?) and small folds. Ashcroft Formation appears to be in fault contact with the Cache Creek rocks, although the contact is obscured by recent slumping. It appears that the block of Cache Creek rock at Cornwall Creek is more than 300 ft (100 m) above the base of the Ashcroft Formation; however, a determination of the exact relationship is uncertain. Small-scale folding and slumping may obscure a sedimentary contact or faults that may have uplifted and juxtaposed Cache Creek rocks against Ashcroft strata.

Cheetsum Hill (Fig. 3) is composed almost entirely of Ashcroft Formation, except for the far west side; there a few scattered outcrops of Nicola volcanics and sediment outcrop. On the north and east sides of Cheetsum Hill are Ashcroft sandstone and shale that generally dip steeply west and southwest. About one-half of the beds are overturned and face east. Carbonaceous shales are often near chaotically deformed. Several thrust faults of unknown displacement are believed to have disturbed this section. If this is true, Cheetsum Hill (along with Hagan Hill, Stop 3) is a klippe composed mostly of Ashcroft strata and a thin sliver of upper Nicola rocks that have been thrust over Ashcroft and Cache Creek rocks. The Cache Creek rocks are thrust over other parts of the Ashcroft Formation as well. Alternatively, the outcrops of Cache Creek rocks here at Cornwall Creek may be the core of a faulted anticline. If this latter interpretation is correct, the Ash-croft strata could have been deposited directly on Cache Creek mélange. However, in this event, the outcrop on Cornwall Creek would be the only known locality where Ashcroft strata are in depositional contact with Cache Creek rocks. Hence a depositional relationship seems unlikely and certainly cannot be clearly observed here in Cornwall Creek.

If Ashcroft and Cache Creek rocks are in thrust fault contact, as interpreted here (Fig. 4), this has powerful implications for the tectonic history of the western Intermontane Belt, for it constrains the time of suturing of the Cache Creek suspect terrane to Quesnellia, of which the Ashcroft beds are believed to be a part. The preferred interpretation of the rock relations seen at this Field Guide site is that the Cache Creek suspect terrane did not become part of Quesnellia until Late Jurassic time at the earliest.

REFERENCES CITED

Cowan, D. S., 1985, Structural styles in Mesozoic and Cenozoic melanges in the western Cordillera of North America: Geological Society of America Bulletin, v. 96, p. 451–62.

Frebold, H., and Tipper, H. W., 1969, Lower Jurassic rocks and fauna near Ashcroft, British Columbia and their relation to some granitic plutons (92-I): Geological Survey of Canada, Paper 69-23, 20 pp.

Greenly, E., 1919, The geology of Angelsey: Great Britain Geological Survey Memoir, 980 pp.

Hsu, K. J., 1973, Melanges and their distinction from olistostromes: Society of Economic Paleontologists and Mineralogists, Special Publication 19, pp. 321–333.

Karig, D. E., and Sharman, G. F., 1975, Subduction and accretion in trenches: Geological Society of America Bulletin, v. 86, pp. 377–389.

Monger, J.W.H., 1981, Geology of parts of western Ashcroft Map-area (92-I), southwestern British Columbia: in Current Research, Part A, Geological Survey of Canada, Paper 81-1A, p. 185–189.

Monger, J.W.H., and Ross, C. A., 1971, Distribution of Fusulinaceans in the western Canadian Cordillera: Canadian Journal of Earth Sciences, v. 8, p. 259–278.

Orchard, M. J., 1984, Early Permian conodonts from the Harper Ranch beds, Kamloops area, southern British Columbia: Geological Survey of Canada Paper 84-1B, p. 207–215.

Price, R. A., Monger, J.W.H., and Muller, J. E., 1981, Cordilleran cross-section—Calgary to Victoria: Geological Association of Canada, Field Guide to Geology and Mineral Deposits, p. 261–334.

Sada, K., and Danner, W. R., 1974, Early and Middle Pennsylvanian fusulinids from southern British Columbia, Canada and northwestern Washington, U.S.A.: Paleontological Society of Japan Transactions and Proceedings (new series), v. 93, p. 249–265.

Travers, W. B., 1978, Overturned Nicola and Ashcroft strata and their relation to the Cache Creek Group, southwestern Intermontane Belt, British Columbia: Canadian Journal of Earth Sciences, v. 15, p. 99–116.

—— , 1982, Possible large-scale overthrusting near Ashcroft, British Columbia: implications for petroleum prospecting: Bulletin of Canadian Petroleum Geology, v. 30, p. 1–8.

Trettin, H. P., 1961, Geology of the Fraser River Valley between Lillooet and Big Bar Creek: British Columbia Department of Mines and Petroleum Resources, Bulletin 44, 109 pp.

—— , 1980, Permian rocks of the Cache Creek Group in the Marble Range, Clinton area, British Columbia: Geological Survey of Canada, Paper 79-17, 17 pp.

Wood, D. S., 1973, Ophiolites, mélanges, blueschists, and ignimbrites: Early Caledonian subduction in Wales: Society of Economic Paleontologists and Mineralogists, Special Publication 19, pp. 334–344.

Structural relationships on the eastern margin of the Shuswap Metamorphic Complex near Revelstoke, British Columbia, Canada

Larry S. Lane and Richard L. Brown, *Ottawa–Carleton Centre for Geoscience Studies, Carleton University, Ottawa, Ontario K1S 5B6, Canada*

LOCATION AND ACCESS

Revelstoke (population 10,000) lies at the confluence of the Illecillewaet and Columbia Rivers in southern British Columbia (51°N, 118°12′W); and is 435 mi (700 km) east of Vancouver, British Columbia, and 250 mi (400 km) west of Calgary, Alberta (Fig. 1). Principal access is by the Trans-Canada Highway and Canadian Pacific Railway. Provincial highways, secondary, and logging roads provide access to many geologically significant localities throughout the area (Brown and others, 1981). The Big Bend Highway (British Columbia 23) runs northward from the Trans-Canada Highway, past Revelstoke Dam 3 mi (5 km) north, to Mica Dam 96 mi (155 km) north.

SIGNIFICANCE

At Revelstoke Dam and other localities along the Big Bend Highway, there are excellent exposures of the shallowly eastward dipping mylonitic and cataclastic shear zone that defines the tectonic contact between hanging wall rocks of the Selkirk Allochthon and footwall rocks of the Monashee Complex (Fig. 1; Read and Brown, 1981; Brown and Murphy, 1982; Brown and Read, 1983; Lane, 1984a, 1984b; Brown and others, 1986). Through 120 m.y., this zone has been repeatedly active in contractional and extensional tectonic environments, and illustrates many important relationships pertinent to Cordilleran Core Complex models (Crittenden and others, 1980; Armstrong, 1982).

SITE INFORMATION

Glaciated peaks, which rise to the west of Revelstoke above the Columbia River, are carved from early Precambrian core gneisses and mantling metasediments of the Monashee Complex. In the Jurassic these sillimanite grade metamorphic rocks were buried to a depth of at least 15 mi (25 km) and were overthrust from the west by late Proterozoic to Mesozoic strata that comprise the Selkirk Allochthon (Figs. 2, 3). The rocks of the allochthon now stand in high peaks of the Selkirk Mountains, which extend from above the Columbia River Valley eastward to the southern Rocky Mountain Trench. Eastward displacement of the Allochthon relative to the Monashee Complex, now exposed as a tectonic window (Figs. 1, 2), is estimated at 50 to 60 mi (80 to 100 km; Read and Brown, 1981; Brown and others, 1986). The base of the allochthon is exposed in the Columbia River Valley along British Columbia 23 (the Big Bend Highway) and up to 3,300 ft (1,000 m) of mylonite has been developed in the shear zone.

Ductile shearing along the Columbia River segment of the

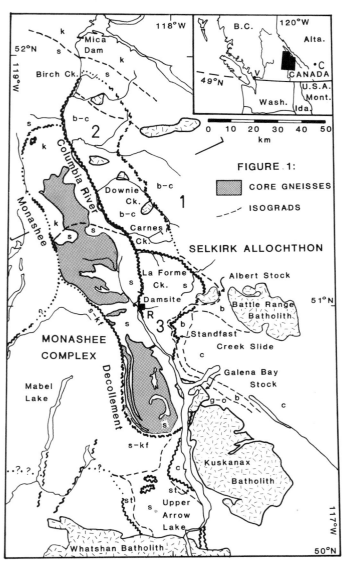

Figure 1. Map showing geologic context, southeastern British Columbia. Key to metamorphic grades: c, chlorite; b, biotite; b-c, biotite-chlorite; g-o, garnet-oligoclase; st, staurolite; k, kyanite; s, sillimanite; s-kf, sillimanite-K-feldspar. Key to cities: C, Calgary; R, Revelstoke; V, Vancouver. Key to numbers: 1, Illecillewaet Slice; 2, Goldstream Slice; 3, Clachnacudainn Slice. Section line refers to Figure 2.

tectonic boundary culminated in the Late Jurassic (Brown and Read, 1983; Lane, 1984b). Uplift and arching of the shear zone occurred from Middle Jurassic to the Paleocene. This period of uplift is interpreted to be a result of crustal scale duplexing that shortened and thickened the crystalline rocks of the Monashee Complex (Fig. 2; Brown and others, 1986).

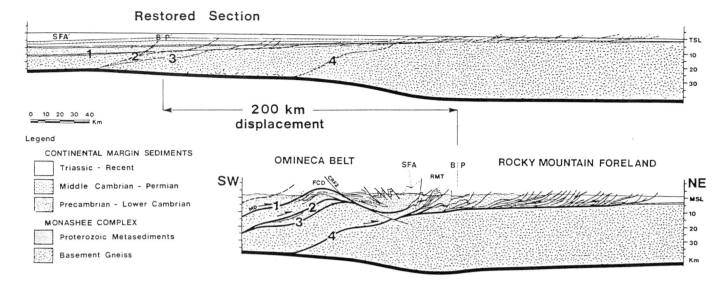

Figure 2. Cross section through the southeastern Canadian Cordillera at the latitude of Downie Creek, British Columbia, together with palinspastic reconstruction. Restored section portrays the inferred geometry of the miogeocline and its attenuated basement prior to shortening along east directed shear zones. Ductile shear zones within the Omineca belt are constructed to balance 124 mi (200 km) of shortening within the Rocky Mountain Foreland and to satisfy existing geometric and petrologic constraints. Area of basement is equal in both sections. Ductile shear zones in the Omineca belt: Monashee décollement (MD): (1), (2), (3), Purcell thrust (4). Other important tectonic elements: Frenchman Cap dome (FCD), Columbia River fault zone (CRFZ), Selkirk fan axis (SFA), Rocky Mountain Trench (RMT), lower Paleozoic platform to basin transition (B/P). (Refer to Brown and others, 1986, for sources and discussion.)

By Eocene time the shear zone at the present level of exposure was no longer ductile, and reactivation related to east-west extension induced cataclasis and development of brittle shears and clay gouge zones. The brittle reactivation zone closely follows the ductile shear zone along much of the Columbia River Valley, but the zones diverge where the Selkirk Allochthon wraps around the northern and southern ends of the footwall culmination. The northern zone of divergence is crossed by British Columbia 23, 59 mi (95 km) north of Revelstoke. These zones of divergence are evidence of a brittle breakaway zone that is bounded to the west by the uplifted Monashee Complex. The wrap around of the easterly-directed ductile shear zone and its westward tracking into the west-rooted Monashee décollement (Fig. 1; Journeay and Brown, 1986) demonstrate that the ductile shear zone developed during easterly thrusting of the Selkirk Allochthon.

LOCALITIES

1. Revelstoke Dam. The Revelstoke Dam area (Fig. 4) on British Columbia 23, 3 mi (5 km) north of the Trans-Canada Highway straddles the shear zone in the Columbia Valley. The west abutment and foundation of the concrete dam comprise pelite, calc-silicate, quartzite, and carbonate of the footwall Monashee Complex. The pelitic rocks contain locally abundant

sillimanite oriented in a southeast trending extension lineation or in radiating clusters lying in the foliation. Relict kyanite is kinked, bent, corroded, and partly altered to andalusite. Large quartzite units are mylonitic; thin quartzite layers in the pelite are incompletely recrystallized. Highly strained carbonate layers are dark grey to black mylonite; the less strained protolith is white coarse-grained marble.

The roadcut for British Columbia 23 at the dam crest exposes garnetiferous amphibolite boudins in sillimanite-bearing pelitic and semi-pelitic metasediments, cut by abundant pegmatite. This unit is one of several tectonic lozenges locally emplaced along the tectonic contact between Selkirk and Monashee terranes. The stratigraphic origin of the unit is uncertain but appears to have been stripped from the footwall and accreted to the hanging wall early in the deformation history. The tectonic contact with recognizable strata of the Monashee Complex lies just below the dam crest, and is gradational in the sense that some graphitic pelitic layers typical of the dam foundation exposures appear to have been tectonically mixed with the lowest layers in the tectonic lozenge.

The Clachnacudainn Slice of the Selkirk Allochthon tectonically overlies the lozenge in the wooded slopes above the roadcut. The contact is not exposed along the road, but white quartz-monzonite of the slice forms high roadcuts 2.5 mi (4 km) north of the dam. Excellent examples of protomylonitic to mylo-

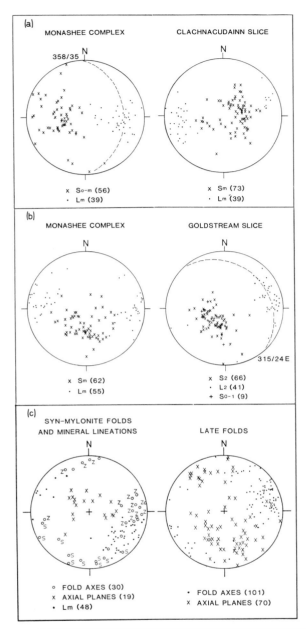

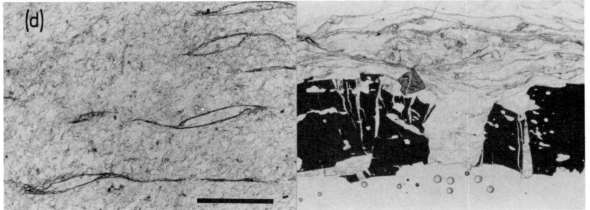

nitic textures can be observed in these exposures. Mesoscopic and microscopic kinematic indicators in mylonitic rocks of the footwall and hanging wall consistently give hanging wall to the east sense of shear (Fig. 3). Late-ductile Z-folds that deform the mylonitic fabrics are consistent with the kinematic indicators. One such fold is visible in the west bank exposure between the powerhouse and spillway.

Timing constraints on the ductile shearing are based on cross-cutting relationships of dated intrusions. At Galena Bay, 28 mi (45 km) south of Revelstoke (Fig. 1), the shear zone is intruded by a 157 Ma granitic stock (Brown and Read, 1983). Significant motion younger than this age in the vicinity of the stock appears unlikely. Elsewhere along the shear zone, ductile activity of Cretaceous or younger age can be documented by minor offset of the contact of a mid-Cretaceous(?) stock.

At the Revelstoke Dam, intense fracturing, clay gouge generation, and hydrothermal alteration have affected a zone 1,312 ft (400 m) thick that includes rocks of all tectonic units. The principal displacement shear, and intense alteration adjacent to it, occur in the rocks of the Monashee Complex, which were temporarily exposed during construction of the dam and other facilities (Lane, 1984a). Alteration consists of chloritization of biotite and garnet, principally adjacent to fractures. Where alteration is most intense, illite, kaolinite, sericite, and carbonate replace plagioclase and mafic minerals. Intense fracturing and minor alteration are exposed in roadcuts of the hanging wall quartz-monzonite pluton, northward from the dam to Coursier Creek (Fig. 4). K-Ar isotopic cooling ages of hornblende and

Figure 3. Summary of the fabric elements and kinematic indicators typical of the Monashee décollement. (a) Fabric attitudes of the Revelstoke Dam area. (b) Fabric attitudes of the Carnes Creek area. (c) Attitudes of syn-mylonite and late-ductile folds from the Revelstoke damsite excavations. Rotation of syn-mylonite structures indicates hanging wall displaced eastward. (d) Asymmetric microstructures indicative of sense of displacement. Right micrograph shows clockwise rotation of tourmaline fragment by 45° (based on axis of maximum absorption) and low angle shear bands. Left micrograph shows muscovite "fish" and c′-fabric tails. Both views look northward in plane light. Scale bar represents 0.08 in (2 mm) in left photo and 0.02 in (0.5 mm) in right photo. (Data from Lane, 1984b, also see Lane and Brown, 1987.)

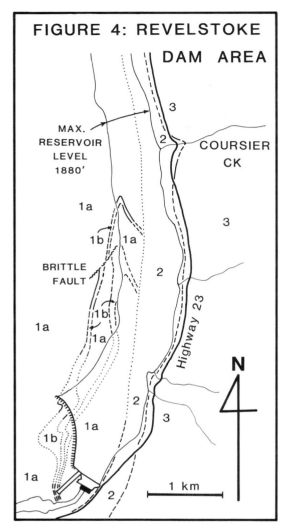

Figure 4. Locality map of the Revelstoke Damsite area. Units: 1: Monashee Complex: 1a, pelite, calc-silicate, minor marble; 1b, quartzite. 2: Tectonic lozenge of sillimanite pelite and semipelite, amphibolite boudins, and pegmatite. 3: Selkirk Allochthon: Quartz-monzonite pluton of the Clachnacudainn Slice, contains protomylonitic to mylonitic microstructures. Tectonic affiliation of unit 2 is uncertain, but is tentatively related to Monashee Complex, based on Rb-Sr isotopic characteristics (Lane, 1984b), and lithologic similarity to amphibolite-bearing units exposed near the northwestern margin of the Monashee Complex (Scammell, 1985). Large quartzite units are defined in the damsite area based on excavations and diamond drilling. North-closing folds in flooded ground north of the dam have east-plunging axes and are syn-mylonitic. The western band of quartzite approximates western limit of mylonitic textures. Contacts between units 1, 2, and 3 are tectonic. Slope between powerhouse and spillway exposes folds that deform mylonitic fabrics.

biotite are consistent with an Eocene age for the brittle displacement (Lane, 1984b). Brittle motion has been estimated to be on the order of 0.6 to 6 mi (1 to 10 km; Lane, 1984a).

2. Sale Creek Area. At the viewpoint 11.8 mi (19 km) north of the Trans-Canada Highway, low roadcuts expose Monashee Complex metapelite, calc-silicate, quartzite, and pegmatite

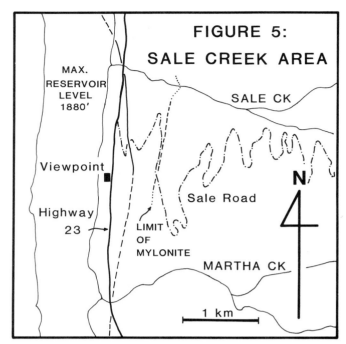

Figure 5. Locality map of Sale Road area. Units: Monashee Complex: pelite, quartzite, carbonate, and pegmatite of the same succession as the Revelstoke Dam. Clachnacudainn Slice of the Selkirk Allochthon: quartz-monzonite pluton. In this area, Monashee Complex metasediments are in direct contact with Selkirk Allochthon. Contacts are dashed where inferred.

of the same unit as at Revelstoke Dam (Fig. 5). The siliceous units have well-developed mylonitic textures. Superimposed on these mylonites are open to closed folds that deform the foliation and extension lineation. These late-ductile folds are well developed in the well-layered quartzite, pelite, and carbonate footwall succession, but are absent in the more uniform metaplutonic rock of the hanging wall (Murphy, 1980). Also, fracturing and chlorite alteration associated with the Eocene brittle faulting is well developed.

A logging road (Sale Road) 1,640 ft (500 m) farther north climbs the ridge to the east (Fig. 5). (This road may be in use by loggers; check with the local Forestry Office in Revelstoke, listed under Government of British Columbia, regarding safety and access.) This road exposes successively higher levels through the quartz-monzonite pluton, showing a progressive decrease in intensity of mylonitization toward the ridge top.

Approximately 1,970 ft (600 m) of mylonitic rocks are exposed on the east bank of the Columbia River, of which 990 ft (300 m) comprise hanging wall granitoid rocks of the Clachnacudainn Slice, which are emplaced directly against the Monashee Complex metasediments in this area. Asymmetric microstructures in the granitoid, together with well-developed C and S fabrics document a hanging wall to the east sense of shear (Murphy, 1980; Brown and Murphy, 1982).

Outside the mylonite zone, the pluton is coarse grained and

contains K-feldspar megacrysts up to 1.2 in (3 cm) long. It has a weak foliation defined by planar alignment of micas, indicative of a previous tectonic history. The transition to mylonitic textures occurs over a distance of about 16.4 ft (5 m), and is well exposed in Sale Creek, as well as in discontinuous exposures near 2,700 ft (823 m) elevation on Sale Road (Brown and Murphy, 1982).

3. Carnes Creek Area. At Carnes Creek, 21.1 mi (34 km) north of the Trans-Canada Highway, sillimanite grade metasediments of the Monashee Complex are juxtaposed against upper greenschist grade rocks of the lower Paleozoic Lardeau Group in the Goldstream slice of the Selkirk Allochthon (Fig. 6). About 1.9 mi (3 km) of high roadcuts leading to the Carnes Creek Bridge illustrate aspects of the ductile mylonitization and the later brittle deformation. The tectonic contact is constrained within 16.4 ft (5 m) near the crest of a small knoll between the parking area and Carnes Creek (Fig. 6).

Hydrothermal alteration associated with the brittle deformation, and recent intense weathering mask the metasedimentary nature of much of the roadcut. Locally preserved calc-silicate and marble, together with sillimanite-bearing pelite in less intensely altered units on the old highway (now flooded), confirm the sedimentary origin of the succession.

Footwall rocks of the Monashee Complex exhibit well developed mylonitic textures, but the hanging wall rocks (calcareous meta-tuff of the Jowett Formation, Lardeau Group) show mylonitic textures only within 98 ft (30 m) of the contact. Units in the footwall are truncated by this contact: near Carnes Creek, units dip shallowly northward. Farther south they dip westward. Thus, from north to south, the tectonic contact crosses successively lower, then higher units in the structural section. Also, units in the hanging wall are truncated at the contact (Fig. 6).

In this area, metamorphic conditions of mylonitization were lower grade than at Revelstoke Dam. Sillimanite occurs here only as porphyroclasts, not the syn-mylonitic masses in strain shadows characteristic of the damsite. This difference along strike in the footwall appears to be coincident with a change in the hanging wall, from the sillimanite grade Clachnacudainn Slice south of La Forme Creek, to greenschist grade Goldstream slice farther north.

Asymmetric microstructures in Monashee Complex rocks near Carnes Creek are consistent with those from other parts of the zone, and together with an east-west extension lineation (Fig. 3), indicate hanging wall east sense of shear (Lane and Brown, 1987).

Brittle fractures in this area represent second-order fractures in a zone dominated by a principal displacement shear. The Carnes Creek fault (Lane, 1984a) may be this principal shear, which is exposed below and east of the south abutment of the bridge. It lies some 430 ft (130 m) below the tectonic contact of the Monashee and Selkirk terranes, and is oriented at 295°/45°N, sub-parallel to the strike of the encompassing Columbia River fault zone as well as the Columbia River Valley in the Carnes Creek area (Fig. 1). Although the local orientation of fractures differs from the damsite, the principal features are the same in

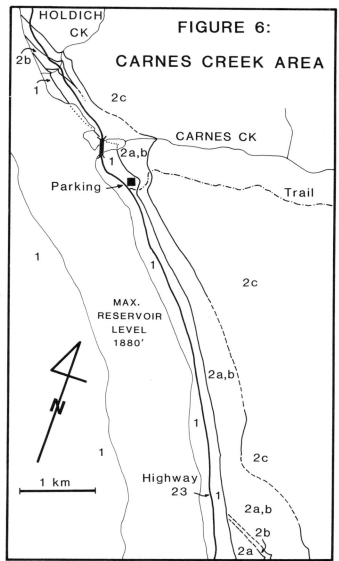

Figure 6. Locality map of Carnes Creek area. Units: 1: Monashee Complex pelite, calc-silicate and minor marble. 2: Goldstream Slice, Selkirk Allochthon: 2a, calcareous tuff, Jowett Formation (late Paleozoic Lardeau Group). 2b, Lensoid dolomitic marble, Jowett Formation. 2c, Siliceous phyllites, locally graphitic, and interbedded psammite, Index Formation (Lardeau Group). Rocks of unit 1 reached sillimanite-muscovite grade of metamorphism, while unit 2 rocks reached middle greenschist (biotite in 2c; actinolite in 2a). Tectonic contact between units 1 and 2 truncates a carbonate unit near Holdich Creek. Eocene brittle faulting and hydrothermal activity have intensely altered many of the exposures along the highway roadcuts. Contacts are dashed where inferred.

both areas: intense fracturing and hydrothermal alteration occur predominantly in the footwall Monashee Complex; the principal gouge/breccia zone lies 430 to 460 ft (130 to 140 m) below the tectonic contact between Monashee and overlying units; and the principal gouge/breccia zone strikes parallel to the Columbia River fault zone and has apparently similar dips. Brittle deformation at Carnes Creek differs from Revelstoke in that fracturing does not reactivate the mylonitic foliation, which is less well developed here than at Revelstoke.

SUMMARY

The shallowly eastward-dipping Monashee decollement defines a terrane boundary in southeast British Columbia, separating a tectonic window of early Proterozoic, sillimanite-kyanite grade metasediments and gneisses (Monashee Complex) from overlying metasediments of upper Proterozoic to Jurassic age, and metamorphic grades varying from chlorite to sillimanite-K-feldspar (Selkirk Allochthon). Major ductile displacements began in the Jurassic, accommodating eastward-directed crustal shortening; but by Eocene time it was the locus of down-to-the-east detachment faulting that produced intense fracturing hydrothermal alteration and generation of abundant illite-rich gouge. Although early motions wrap around the Monashee Complex, the later displacements define a breakaway zone that continues in the Columbia Valley as far as Birch Creek where it dies out.

REFERENCES CITED

Armstrong, R. L., 1982, Cordilleran metamorphic core complexes; From Arizona to southern Canada: Annual Reviews of Earth and Planetary Science, v. 10, p. 129–154.

Brown, R. L., and Murphy, D. C., 1982, Kinematic interpretation of mylonitic rocks in part of the Columbia River fault zone, Shuswap terrane, British Columbia: Canadian Journal of Earth Sciences, v. 19, p. 456–465.

Brown, R. L., and Read, P. B., 1983, Shuswap terrane of British Columbia; A Mesozoic "Core Complex": Geology, v. 11, p. 164–168.

Brown, R. L., Fyles, J. T., Glover, J. K., Höy, T., Okulitch, A. V., Preto, V. A., and Read, P. B., 1981, Southern Cordillera cross-section; Cranbrook to Kamloops, *in* Thompson, R. I., and Cook, D. G., eds., Field guides to geology and mineral deposits: Geological Association of Canada, Mineralogical Association of Canada, Canadian Geophysical Union Joint Annual Meeting, Calgary, Alberta, p. 335–372.

Brown, R. L., Journeay, J. M., Lane, L. S., Murphy, D. C., and Rees, C. J., 1986, Obduction, backfolding and piggyback thrusting in the metamorphic hinterland of the southeastern Canadian Cordillera: Journal of Structural Geology, v. 8, p. 255–268.

Crittenden, M. D., Jr., Coney, P. J., and Davis, G. H., 1980, Cordilleran Metamorphic Core Complexes: Geological Society of America Memoir 153, 490 p.

Journeay, J. M., and Brown, R. L., 1986, Major tectonic boundaries of the Omineca belt in southeastern British Columbia; A progress report, *in* Current Research, Part A: Geological Survey of Canada, Paper 86-1A, p. 81–88.

Lane, L. S., 1984a, Brittle deformation in the Columbia River fault zone, British Columbia: Canadian Journal of Earth Sciences, v. 21, p. 584–598.

—— , 1984b, Deformation history of the Monashee décollement and Columbia River fault zone, British Columbia [Ph.D. thesis]: Ottawa, Ontario, Carleton University, 240 p.

Lane, L. S., and Brown, R. L., 1987, Kinematics of ductile shearing on the Monashee décollement, British Columbia Canada: Geological Society of America Bulletin (in review).

Murphy, D. C., 1980, Mylonite genesis: Columbia River fault zone, British Columbia [M.A. thesis]: Stanford, California, Stanford University, 106 p.

Read, P. B., and Brown, R. L., 1981, Columbia River fault zone; Southeastern margin of the Shuswap and Monashee complexes, southern British Columbia: Canadian Journal of Earth Sciences, v. 18, p. 1127–1145.

Scammell, R. J., 1985, Stratigraphy and structure of the northwestern flank of Frenchman Cap Dome, Monashee Complex, British Columbia; Preliminary results, *in* Current Research, Part A, Geological Survey of Canada, Paper 85-1A, p. 311–316.

Sedimentology of the Upper Cretaceous Nanaimo Basin, British Columbia

Jory A. Pacht, ARCO Oil and Gas Company, 2300 W. Plano Parkway, Plano, Texas 75075

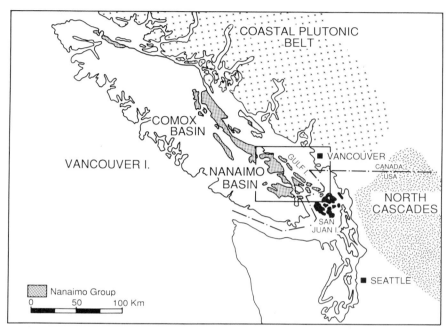

Figure 1. Location of the Nanaimo Basin.

LOCATION AND ACCESSIBILITY

The Nanaimo Group is an assemblage of conglomerate, sandstone, siltstone, shale, and minor coal deposited in the Nanaimo Basin during Santonian through Maastrichtian time (Muller and Jeletzky, 1970; Ward, 1978). It crops out on southeastern Vancouver Island, the Gulf Islands of British Columbia, and the San Juan Islands of Washington (Figs. 1, 2). Nearly continuous outcrops of the Namaimo Group occur along seacliffs on Vancouver Island, and on the Gulf and San Juan islands. Good exposures are also present along roadcuts and major rivers on Vancouver Island.

Ferries from the mainland to Vancouver Island and most of the Gulf and San Juan islands can be boarded at Anacortes, Washington, and Tawassen, British Columbia, and taken to Sidney, British Columbia (approximately 20 mi, 32 km, north of Victoria on Vancouver Island), and to the San Juan and Gulf islands, respectively. Ferries can also be taken from Port Angeles, Washington, to Victoria, British Columbia. In addition, ferries run regularly between islands in both Canada and Washington. Current ferry schedules are necessary for travel in this area.

Tidelands are considered public property in both Canada and the United States; access to the outcrops is available at many public beaches.

SIGNIFICANCE OF THE STUDY AREA

Geologic Setting. The Nanaimo Basin developed within an "orogenic collage." The basin is bordered by a pre-Cretaceous structural high to the north. Its other borders are major tectonic terranes: the Coastal Plutonic Belt to the east, the North Cascades to the southeast, late Paleozoic and early Mesozoic terranes of the San Juan Islands to the south, and Wrangellia to the west. Although some of these terranes were allochthonous, the Nanaimo Basin received sediment from all of them, which suggests that they were assembled relative to one another by Late Cretaceous time (Johnson, 1978; Pacht, 1980, 1984).

The Nanaimo Basin and the Comox Basin to the north form the Georgia Basin (Eisbacher, 1974), which may be an intramassif forearc basin (term from Dickinson and Seely, 1979). It was probably formed by oblique convergence between the Kula(?) and the North American plates during the Late Cretaceous (Johnson, 1984; Pacht, 1980, 1984). Development of this basin records formation of a major dextral transcurrent fault that truncated the pre-Tertiary continental framework of western Washington and southern Vancouver Island during the Late Cretaceous and early Tertiary (Johnson, 1984).

The Nanaimo Basin is a graben, divided into two subbasins by a midbasin horst (Pacht, 1984). This horst trends from Van-

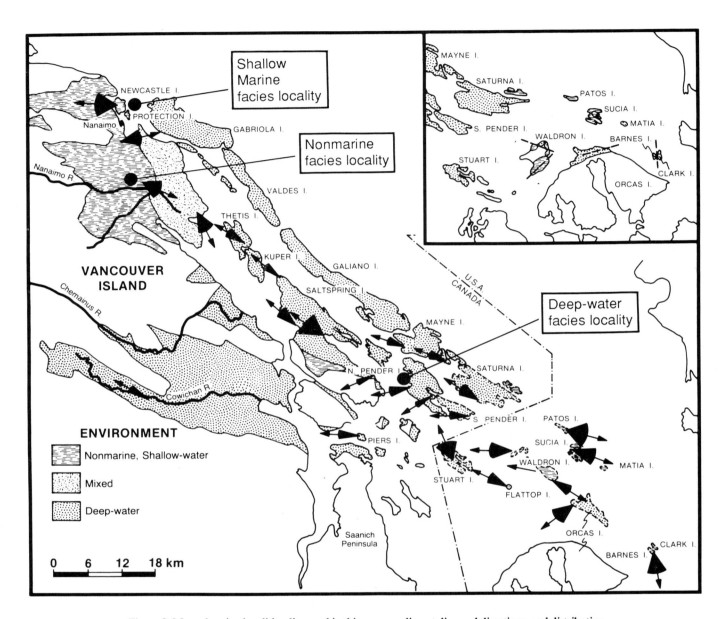

Figure 2. Maps showing localities discussed in this paper, sediment dispersal directions, and distribution of sedimentary environments in the Nanaimo Basin. Pie-shaped wedges refer to 95 percent confidence levels determined by 'T' distributions on measurements. Island outcrops in the southeastern portion of the basin have been rotated to correct for refolding. Detailed discussion of this rotation is presented in Pacht (1980). Inset in upper right corner shows unrotated island outcrops (from Pacht, 1984).

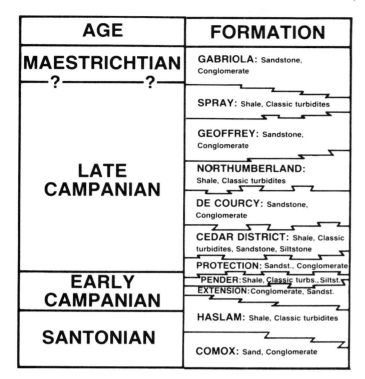

Figure 3. Formational nomenclature of the Nanaimo Group (from Ward, 1978).

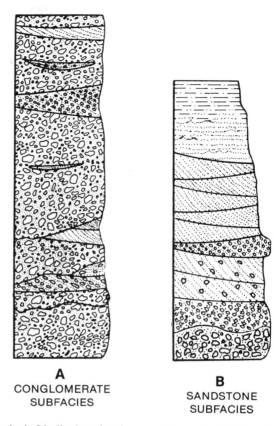

A	B
CONGLOMERATE SUBFACIES	SANDSTONE SUBFACIES

Figure 4. A. Idealized section from conglomerate subfacies of nonmarine facies. B. Idealized section from sandstone subfacies of nonmarine facies. Both sections typify stratification sequences observed along the Nanaimo River.

couver Island, immediately north of the Cowichan River drainage basin onto Saltspring Island (Fig. 2).

Nonmarine and shallow-marine facies occur along the margins of the Nanaimo Basin, whereas deep-marine facies are present in the central part (Fig. 2). Excellent exposures, detailed mapping (Muller and Jeletzky, 1970; Muller, 1977), and good biostratigraphic control (Muller and Jeletzky, 1970; Ward, 1978) (Fig. 3) make it a very good area in which to study distribution of depositional facies in a confined basin (Pacht, 1980; Ward and Stanley, 1982).

DEPOSITIONAL FACIES

Nonmarine Facies. Nonmarine facies are best exposed in the Extension and Pender formations and part of the Protection Formation (Fig. 3) along the Nanaimo River (Fig. 4). Nonmarine rocks can be divided into conglomerate subfacies, sandstone subfacies, and coal, mudstone, and siltstone subfacies, based on dominant rock type. The conglomerate subfacies comprises stratification sequences that contain more than 50% conglomerate. The remainder of the facies is largely sandstone with rare interbeds of siltstone, mudstone, and coal (Fig. 4). Most conglomerate clasts are 0.4 to 2 in (1 to 5 cm) in diameter, although the largest clasts are more than 12 in (30 cm) in size. Disorganized, cross-bedded,

and horizontally-bedded conglomerates are present. Individual beds are very discontinuous and exhibit large variations in thickness along strike. They are commonly arranged in poorly defined fining-upward sequences. The sequences are initiated with conglomerate beds and are overlain by conglomerate sandstone and/or sandstone beds. Sequences are generally truncated by a scour surface, which marks the base of the overlying sequence (Fig. 4A).

The sandstone subfacies (deposits contain greater than 50% sandstone) is dominated by medium-scale cross-bedded sandstone. Most deposits contain less than 20% conglomerate. Fair to good fining-upward sequences are present in this subfacies. The basal part of these sequences commonly consists of conglomerate or cross-bedded pebbly sandstone. The middle part is dominated by medium-scale cross-bedded sandstone. The upper part is characterized by ripple cross-laminated and horizontally-bedded very fine-grained sandstone and siltstone (Fig. 4B).

The coal, mudstone, and siltstone subfacies are characterized by areally extensive coal seams intercalated with beds of siltstone, mudstone, and rare sandstone.

Shallow Marine Facies. Shallow-marine rocks are present

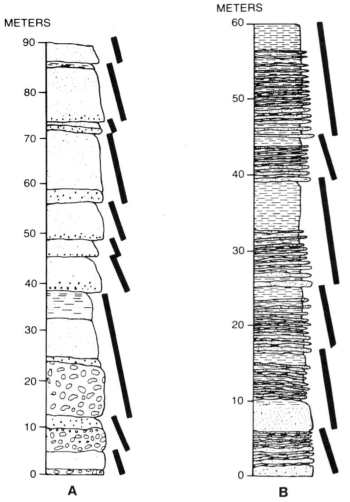

Figure 5. A. Thinning- and fining-upward sequences present in part of the Protection Formation at Mouat Point. B. Thinning-upward sequences observed in classic turbidites in part of the Cedar District Formation at Mouat Point.

bedded units are commonly intercalated with cross-bedded strata.

Siltstone and very fine-grained sandstone beds are also observed in this facies on Newcastle Island. These beds commonly exhibit horizontal bedding and current and oscillatory ripple cross-stratification. An erosional surface normally separates these units from overlying coarser-grained strata.

Deep-water Facies. Most of the Nanaimo Group consists of deep-water strata (Pacht, 1980). Deep-water deposits occur in parts of the Nanaimo River drainage basin throughout the Gulf Islands, in the Cowichan River drainage basin, on Saanich Peninsula and nearby islands, and on some of the northwestern San Juan Islands (Johnson, 1978; Pacht, 1980, 1984; Ward and Stanley, 1982; Fig. 2). Outcrops of the Extension through Cedar District formations along Mouat Point on Pender Island (Fig. 5) are typical of deep-water facies in the Nanaimo Basin.

The Extension and Protection formations (Fig. 3) are characterized by interbedded lenticular units of sandstone, pebbly sandstone, conglomerate, and rare siltstone and mudstone. These beds commonly exhibit well-developed channels that cut into underlying units.

Conglomerates exhibit both disorganized fabric, normal grading, and rare, poorly defined inverse grading. Clasts in some normally-graded beds show poor to fair a(i) a(p) clast orientation. Sandstones at this locality are largely featureless. Some, however, exhibit horizontal-bedding, convolute-bedding (including well-developed flame structures), ripple cross-lamination, and dish structures. Some sandstones contain large mudstone clasts that exhibit a(i) a(p) orientation.

Both the Extension and the Protection formations at this locality exhibit thinning- and fining-upward sequences initiated with conglomerate or pebbly sandstone beds. Erosional surfaces are commonly present at the base of these sequences. Conglomerate beds are overlain by pebbly sandstone and then by sandstone. Some sequences are capped by siltstone, mudstone, and/or classic turbidite beds (Fig. 5A). An 82-ft-thick (25 m) zone of classic turbidites occurs within the Protection Formation.

The Pender Formation (largely covered at Mouat Point), which is developed between the Extension and Protection formations (Ward, 1978; Fig. 3), is largely mudstone. Some thin Tcde Bouma-sequence turbidites are observed.

The Cedar District formation along Mouat Point exhibits Tabcde, Tbcde, and Tcde Bouma sequences in very fine-grained to fine-grained sandstone/mudstone couplets. Some siltstone/mudstone couplets are present near the top of the formation. Turbidites are arranged in thinning-upward sequences generally initiated with massive, featureless sandstone beds or Tabcde turbidites 1.6 to 3.3 ft (0.5 to 1 m) thick. The middle part of the sequences is comprised of Tbcde and subordinate Tcde turbidites. The uppermost parts of the sequences are very thin-bedded (sandstone or siltstone units less than 2 in (5 cm) thick) Tcde turbidites. Some sequences are capped by thick mudstones (Fig. 5B).

DEPOSITIONAL SETTING

Distribution of stratification sequences, and petrographic

along the northwest and the southeast margins of the Nanaimo Basin. They are best developed in the Protection Formation (Fig. 3) on Newcastle Island. This island is public land and can be reached by water-taxi from the town of Nanaimo.

The Protection Formation on Newcastle Island is dominated by well-sorted fine- to medium-grained sandstone, which exhibits a variety of sedimentary structures, including trace fossils (Thallasinoides and possible Ophiomorpha). Some sandstones exhibit low-angle (1 to 10 degrees) inclined stratification. Rare oscillatory and current ripples are superimposed on these beds. The dominant sedimentary structure, however, is trough and planar, medium-scale cross-bedding. Sets range in thickness from 2 in to 3 ft (5 cm to 1 m). Oscillatory and combined-flow ripples are commonly superimposed on cross-bed foresets. Horizontally-

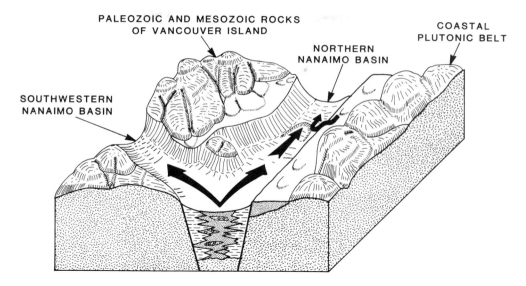

PALEOZOIC AND MESOZOIC ROCKS
OF VANCOUVER ISLAND

NORTHERN
NANAIMO BASIN

COASTAL
PLUTONIC BELT

SOUTHWESTERN
NANAIMO BASIN

Figure 6. Paleogeographic reconstruction of the Nanaimo Basin during Late Cretaceous (from Pacht, 1984).

and paleocurrent data from the Nanaimo Group suggest that deep-water facies were deposited in a confined basin, characterized by sedimentation from several source areas and strong control of sediment distribution by basin topography (Pacht, 1980, 1984; Fig. 6).

Detritus was probably transported to the Nanaimo Basin from surrounding highlands in braided streams. This is suggested by sedimentary structures and stratification sequences in nonmarine facies. Paleocurrent measurements in nonmarine rocks along the Nanaimo river suggest transport to the east toward the central part of the basin. Braided streams probably terminated along the shoreline in small wave-dominated deltas.

Detritus was then transported by longshore currents along shorelines developed at basin margins. Transport occurred to the southeast along the western and eastern sides of the basin and to the west on Newcastle Island, which marks the northern boundary of the Nanaimo Basin (Fig. 2).

Most detritus in the Nanaimo Basin was resedimented into the central part of the basin by sediment gravity flows. Transport and accumulation of sediment in deep water was greatly affected by basin morphology. Sediment was transported parallel to basin margins, to the northwest in the northern arm of the basin (north of the midbasin horst) and to the west in the southern arm (south of the midbasin horst; Pacht, 1984; Figs. 2 and 6).

Accumulation of deep-water strata was greatly controlled by topography. Classic submarine fans did not develop. Nanaimo strata are dominated by thinning-upward sequences, which suggests that deposition probably occurred within a channel complex. The complex was fed from source areas at many points along basin margins (Pacht, 1980, 1984; Fig. 6).

REFERENCES CITED

Dickinson, W. R., and Seely, D. R., 1979, Structure and stratigraphy of forearc regions: American Association of Petroleum Geologists Bulletin, v. 63, p. 2–31.

Eisbacher, J. H., 1974, Evolution of successor basins in the Canadian Cordillera, *in* Dott, R. H., and Shaver, R. H., eds., Modern and ancient geosynclinal sedimentation: Society of Economic Paleontologists and Mineralogists Special Publication 19, p. 274–291.

Johnson, S. Y., 1978, Sedimentology, petrology, and structure of Mesozoic strata in the northwestern San Juan Islands, Washington [M.S. thesis]: Seattle, University of Washington, 105 p.

—— , 1984, Evidence for a margin-truncating fault (pre–Late Eocene) in western Washington: Geology, v. 12, p. 538–541.

Muller, J. E., 1977, Geology of Vancouver Island: Geologic Survey Canada Open-File Map OF-463, scale, 1:250,000.

Muller, J. E., and Jeletzky, J. A., 1970, Geology of the Upper Cretaceous Nanaimo Group, Vancouver Island and Gulf Islands, British Columbia: Geologic Survey of Canada Paper 69-25, 77 p.

Pacht, J. A., 1980, Sedimentology and petrology of the Late Cretaceous Nanaimo Group in the Nanaimo Basin, Washington and British Columbia; Implications for Late Cretaceous tectonics [Ph.D. thesis]: Columbus, Ohio State University, 361 p.

—— , 1984, Petrologic evolution and paleogeography of the Late Cretaceous Nanaimo Basin, Washington and British Columbia; Implications for Cretaceous tectonics: Geological Society of America Bulletin, v. 95, p. 766–778.

Ward, P. D., 1978, Revisions to the stratigraphy and biochronology of the Upper Cretaceous Nanaimo Group, British Columbia and Washington state: Canadian Journal of Earth Science, v. 15, p. 405–423.

Ward, P. D., and Stanley, K. O., 1982, The Haslam Formation; A late Santonian–early Campanian forearc basin deposit in the Insular belt of southwestern British Columbia and adjacent Washington: Journal of Sedimentary Petrology, v. 52, p. 975–990.

Valley of Ten Thousand Smokes, Katmai National Park, Alaska

Wes Hildreth and Judy Fierstein, U.S. Geological Survey, 345 Middlefield Road, Menlo Park, California 94025

LOCATION

The pyroclastic deposits of the Valley of Ten Thousand Smokes (VTTS) lie within Katmai National Park, 450 km southwest of Anchorage and 170 km west of Kodiak (Fig. 1). The only convenient access is by boat or amphibious aircraft from King Salmon to Brooks Camp on Naknek Lake; scheduled commercial service is available during summer months. Brooks Lodge offers a daily commercial shuttle by bus or van to the Overlook Cabin, perched on a scenic knoll near the terminus of the VTTS ash-flow deposit, at the end of a 37-km dirt road from Brooks Camp. Cabins, meals, camping, and National Park Service programs are available at Brooks Camp, but the Overlook Cabin and the Baked Mountain Hut (a day's walk up the VTTS) provide no amenities other than primitive shelter from the normally foul weather. The distal part of the ash-flow sheet can be reached in an hour's walk by rough trail from the Overlook Cabin, but anything more than a brief visit demands a backpack, boots *and* sneakers, high-quality rain gear, a sturdy tent, warm clothing, maps and compass, water bottles, food, an ice axe or staff, goggles for ash storms, a rope for fording streams, and a plan (or philosophy) for dealing with brown bears—which are commonly encountered with little warning. High winds, rain, bears, icy streams, and remoteness make the VTTS a true wilderness, exhilarating but dangerous, occasionally glorious, usually uncomfortable, and never to be trifled with. Information can be obtained by writing to the Superintendent, P.O. Box 7, King Salmon, AK 96613. U.S. Geological Survey 1:62,500 topographic maps, *Mt. Katmai A3, A4, B3, B4,* and *B5,* cover the VTTS and several neighboring stratovolcanoes, and the *Mt. Katmai* 1:250,000 sheet provides a desirable regional perspective.

SIGNIFICANCE

The eruption of 6–9 June 1912 was the most voluminous of the twentieth century, one of the three largest in recorded history, and one of the few historic eruptions to produce welded tuff. Widespread tephra falls, compositionally banded pumice, and the "ten thousand" fumaroles in the ash-flow sheet attracted broad scientific and popular attention, principally owing to the work of the National Geographic Society expeditions of 1915–1919 led by R. F. Griggs (Griggs, 1922). Among historic eruptions, this event was virtually unique in that it generated a large volume of pumiceous pyroclastic flows that came to rest on land. Study of its superbly exposed deposits has strongly influenced our understanding of the eruption, emplacement, degassing, and cooling of silicic ejecta.

INTRODUCTION

A sequence of eruptive pulses (indicated, by the record of discrete ash falls at Kodiak village, to have spanned ~60 hr) produced three principal groups of pyroclastic deposits (Curtis, 1968; Hildreth, 1983): (1) Plinian pumice-fall and surge deposits, dominantly rhyolitic, that preceded and accompanied ash-flow emplacement; (2) a 120-km[2] ash-flow sheet consisting of several flow units and zoned from rhyolite to andesite; and (3) Plinian pumice-fall deposits, dominantly dacitic, that overlie both the ash flow and the stratified rhyolitic tephra. Virtually all of the more than 30 km[3] of pyroclastic material vented within the 2-km-wide Novarupta depression (see Fig. 1, 4), which is thought to be a flaring funnel-shaped structure backfilled by its own ejecta. The small rhyolitic lava dome, named Novarupta, was extruded within the depression after the explosive sequence had ended.

At Mt. Katmai, 10 km east of the 1912 vent, a 3-km-wide caldera (more than 600 m deep) (Fig. 1) collapsed at some unknown time during the eruption, probably in response to withdrawal of magma hydraulically connected with that erupting from the Novarupta flank vent. A small dacite lava dome (the horseshoe island of Griggs, 1922), extruded on the caldera floor and now concealed by more than 250 m of lake water, was the only 1912 eruptive unit confidently attributable to Mt. Katmai itself (Hildreth, 1983).

The five locations described below exhibit some of the best evidence for the sequence and mechanisms of emplacement of the 1912 eruptive products, as well as for subsequent phreatic, fumarolic, and secondary mass-flow processes. The first location requires only half a day from the Overlook Cabin; the second, third, and fourth can be combined into a three-day loop hike; and finally, a visit to the caldera rim of Mt. Katmai would require an additional all-day roped ascent from a camp near Location 3 or 4 (Fig. 1).

LOCATION 1. PYROCLASTIC DEPOSITS NEAR THREE FORKS

Three Forks is the area below the Overlook Cabin where the Ukak River is formed by the confluence of Knife Creek, the River Lethe, and Windy Creek (Figs. 1 and 2). Each of these rushing streams occupies a gorge incised completely through distal part of the 1912 ash-flow sheet into underlying glacial and fluvial deposits and Jurassic siltstone of the Naknek Formation. A 2-km-long trail winds northward from the Overlook Cabin to a footbridge over the Ukak, and an additional 2-km walk upstream along the east bank brings one to Three Forks. Be sure to ask the National Park Service about the status of the footbridge, which spans a dangerous crossing and is sometimes destroyed in periods of heavy runoff.

Deposits present here, 16 km from the Novarupta vent, include (1) the distal tongue of the main ash-flow deposit, here nonwelded, largely rhyolitic, and only 10 to 20 m thick, resting

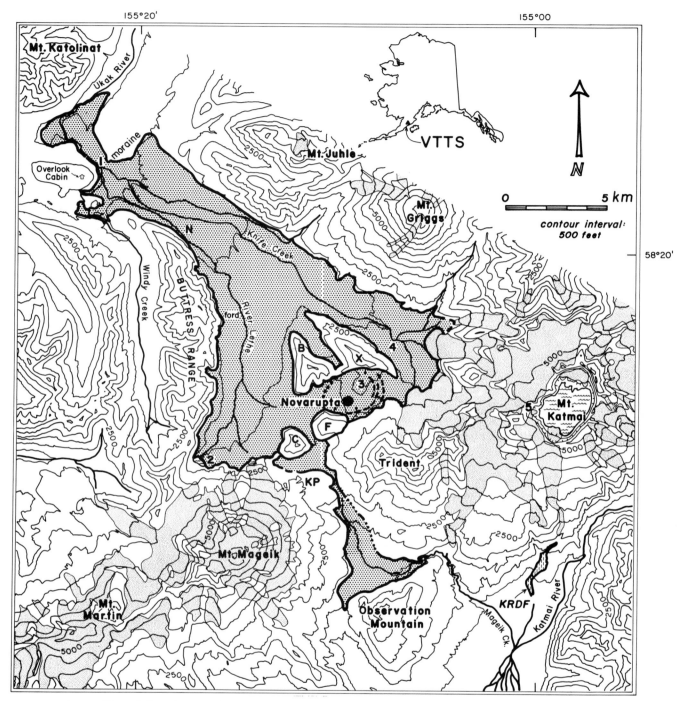

Figure 1. Location map for the Valley of Ten Thousand Smokes (VTTS), Alaska (inset), showing the 120-km^2 extent of the 1912 ash-flow sheet and its source area at Novarupta. One lobe penetrated Katmai Pass (KP) and flowed down the Pacific slope of the volcanic chain. Mts. Griggs, Katmai, Trident, Mageik, and Martin are all andesite-dacite stratovolcanoes, each of which has active fumaroles and has experienced Holocene eruptions. Lighter pattern on stratovolcanoes indicates glaciers. Mt. Cerberus (C) and Falling Mountain (F) are Holocene dacite domes adjacent to the 1912 vent. The Overlook Cabin is the starting point for almost all hikes in the VTTS. Locations 1 through 5 are discussed in the accompanying text. Baked Mountain (B), Broken Mountain (X), the Buttress Range, and most of Mts. Katolinat and Juhle consist of Jurassic sedimentary rocks. The Katmai River debris flow (KRDF) and ash-flow depositional features related to narrowing of the valley near location (N) are discussed in Hildreth (1983). The indicated ford is a rare place where the River Lethe can be waded.

upon charred remnants of willow thickets that grew on the overrun moraine now exposed at creek level; (2) an overlying, dark gray to brown, ash-flow unit, generally only 3 to 5 m thick, consisting of a mixture of andesitic, dacitic, rhyolitic, and banded pumice and occupying a broad shallow swale it scoured into the subjacent flow unit; (3) dacitic airfall strata consisting of four graded couplets of coarse to fine ash, altogether only 40 to 45 cm thick here, but equivalent to the 10-m pumice-fall section (layers C to H of Curtis, 1968) to be visited at Location 4; and (4) a few meters of capping debris-flow deposits, consisting largely of remobilized ash and pumice lumps but also containing blocks of welded tuff from far upvalley and scattered rip-ups of fine-grained airfall strata.

All parts of the ash-flow deposit near Three Forks contain rhyolitic, dacitic, andesitic, and banded pumice lumps in various proportions (Hildreth, 1983). The rhyolite (77% SiO_2) is white and contains only 1–2 wt% phenocrysts, whereas the light-gray dacite (64–66% SiO_2) and black andesite (58–62% SiO_2) both have 30–45 wt% phenocrysts. All 1912 ejecta contain plagioclase, hypersthene, titanomagnetite, ilmenite, apatite, and pyrrhotite; in addition, quartz is present in the rhyolite and augite, and traces of olivine are present in the andesite and dacite.

An ash-flow sheet (ignimbrite) usually consists of multiple flow units, each of which is generally made up of a poorly sorted, unstratified mixture of accidental lithic fragments, juvenile pumice fragments, and crystal-vitric ash produced by explosive and abrasive comminution of the pumice. If a flow travels far enough, however, it commonly develops a measure of crude internal sorting that is best displayed at the base and top of its deposit (Sparks, 1976; Wilson, 1984). Three common manifestations of such sorting can be seen between Three Forks and the footbridge site: (1) fine-grained basal layers, typically 5 to 30 cm thick and gradational into the main deposit, probably produced by grain-dispersive forces that result in the coarser fragments being driven away from the base of each moving flow; (2) density dependent grading of coarse fragments—normal grading for dense lithics but inverse grading for buoyant pumice—sometimes culminating in concentrations of coarse pumice blocks at tops and toes of flow units; (3) a discontinuous ground layer, typically a few centimeters thick, enriched in crystals and lithics but depleted in fine ash relative to the main deposit. Normally in sharp contact beneath the fine-grained basal layer, a ground layer is thought to result from rapid sedimentation from the fluidized head of the moving flow, which may become highly expanded as it ingests air during turbulent encounters with willow thickets, bears, and rough terrain.

In addition to flow units, segregation units a few centimeters to a few meters thick are unusually well developed in the Three Forks area. Defined by alternating enrichments and depletions of pumice lapilli, and ranging from vague grading to abrupt stratification that simulates true flow units, they are thought to result from internal shear promoted by vertical velocity gradients that develop within ash flows as they slow to a halt—rather like a sliding deck of cards. Of the many apparent subunits exposed

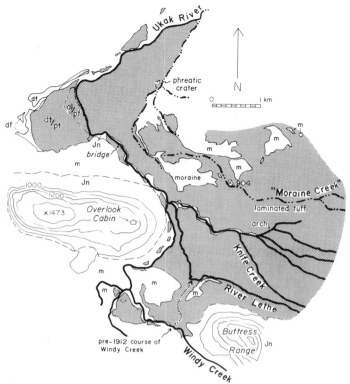

Figure 2. Simplified map of the Three Forks area, Location 1. The pattern indicates areas underlain by the 1912 ash-flow deposit, which is discontinuously mantled by as much as a few meters of alluvial, debris-flow, and airfall deposits. The base of the tuff is generally accessible between the site of the footbridge and the natural arch 4 km up Knife Creek. Check with the National Park Service about the status of the footbridge. Symbols: *al* = alluvium; *df* = debris-flow deposits; *pf* = pyroclastic-flow deposits; *m* = moraine, probably largely Neoglacial in age; *Jn* = Jurassic marine strata of Naknek Formation, largely siltstone. The morainal exposure just south of the sharp bend in the River Lethe includes glaciofluvial beds rich in reworked dacite pumice, probably representing a Holocene eruption of Mt. Mageik.

near Three Forks, only the dark-colored top one and a subjacent light-gray one (prominent along lowermost Knife Creek) independently flowed the length of the VTTS. Most originated within the main rhyolite-rich ash flow not long before it came to rest; a few traveled far enough as coherent pulses, following segregation, that their internal sorting characteristics are as well developed as those of true flow units that originated near the vent. Transition from virtually structureless tuff to extreme laminar segregation can be examined about 2 km east of Three Forks by comparing exposures near the natural arch along a Knife Creek tributary with those in the deep gorge of "Moraine Creek" 700 m northward (Fig. 2).

Surprisingly sluggish emplacement of the 1912 ash flows is indicated by relations between the deposits and the partially engulfed moraine that extends across the VTTS northeast of Three Forks (Fig. 2). In contrast to high-velocity and strongly fluidized

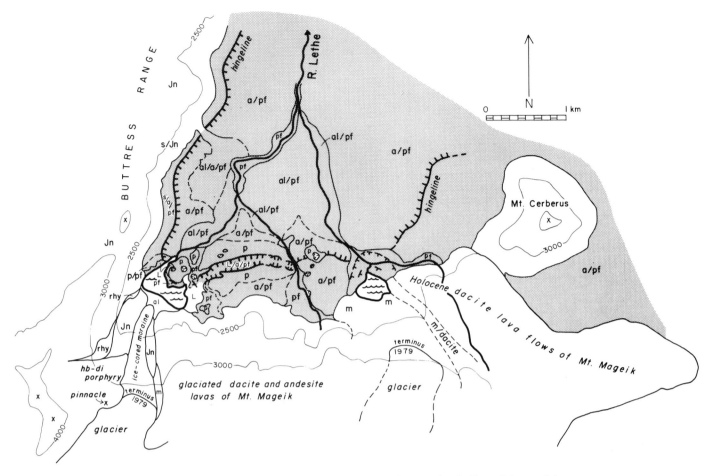

Figure 3. Simplified map of southwestern corner of the VTTS, Location 2. Sites of the two lakes were occupied in 1912 by snouts of glaciers that have since receded. Differential compaction of thick tuff beneath the valley floor produced the marginal bench and its fissured hingeline, which was the locus of many large fumaroles and phreatic explosions. Elongate valley extending northeast from the western lake was excavated along the hingeline by such explosions; called "Fissure Lake" by Griggs (1922), it is now filled with ashy sediments. Dip symbols indicate deformation of the tuff, both by compaction and by marginal downwarping in former ice-contact areas. Symbols: s = pumiceous scree; a = airfall deposits; rhy = rhyolitic sill; p = phreatic explosion deposits; L = lake deposits; others as in Figure 2. The undated "pinnacle porphyry" is hornblende diorite.

ash flows (Wilson, 1984), the 1912 flows were too slow and deflated to surmount this 30-m-high obstacle perpendicular to its path (Hildreth, 1983). One flow passed through saddles along the moraine crest, and some were diverted around its low western end, but much of the 20 to 30-m-thick tuff simply wedges out against the till. Low velocity, poor fluidization, and deflection around obstacles may have helped promote the internal shear thought responsible for the remarkable degree of laminar segregation in this area.

Additional features to note in the Three Forks area include: (1) upright charred trees enclosed in the tuff along Windy Creek and near the cross-valley moraine; (2) compaction faults in the tuff, typically having approximately 1 m displacement; (3)

preferential salmon-pink oxidation of upper and/or coarser grained zones of flow units; (4) multicolored zones of fumarolic alteration controlled by gas-escape pipes, faults, and contraction cracks in indurated tuffs, and wavy irregular conduits in permeable tuff. The ash-flow corridor near the footbridge (Fig. 2) follows a buried pre-1912 channel of the Ukak River; the old course is marked by a compactional swale atop approximately 25 m of infilling tuff that thins drastically toward the present-day riverbed only 300 m westward. A similarly buried pre-1912 course of Windy Creek lies about 3 km southward (Fig. 2).

LOCATION 2. SOUTHWEST CORNER OF THE VTTS

The scenic lake basin at the northwest toe of Mt. Mageik, a

full day's walk from the Overlook Cabin, displays striking evidence for welding, alteration, and deformation of the tuff and for violent tuff-water interactions. A trail southeast of the Overlook Cabin descends to a shallow ford on Windy Creek, beyond which the simplest route hugs the eastern base of the Buttress Range all the way to the southwest corner of the VTTS (Figs. 1 and 3).

The bench ringing the Lethe arm of the upper VTTS, called the "high sand mark" by Griggs (1922), resulted neither from the passage of a highly inflated ash flow nor by axial drainaway of a high-riding flow. Rather, it reflects differential compaction and welding of tuff that may be as thick as 200 m along the axis of the buried glacial valley (Curtis, 1968; Hildreth, 1983). Along the hingeline of the beach, the dip of the tuff toward the valley floor steepens abruptly from approximately 5° to 20°, and numerous faults (many of them antithetic) displace both the ash-flow and overlying air-fall deposits. About 500 m north of the lake, the dacite air-fall strata (here about 80 cm thick) are stepped down about 4 m into a 30-m-wide hingeline graben walled by partially welded and fumarolically indurated ash-flow tuff. Offset of even the youngest air-fall strata along the hinge shows that ash-flow compaction lasted for at least a few days, whereas channeling of late flow units along axial swales on the valley floor suggests that it had been underway immediately after emplacement.

The faults focused such vigorous fumarolic discharge along the hingeline that the top few meters of nonwelded deposits are commonly burnt orange and indurated by vapor-phase crystallization; in slightly deeper exposures the tuff is partially welded and commonly brick red. For comparison, fresher exposures of dark-gray welded tuff rich in andesite and dacite pumice can be examined on the valley floor along the nearby gorge of the upper Lethe.

The chaotically disintegrating tuff bluffs just west of the lake provide a rare upvalley exposure (Hildreth, 1983, Fig. 5) down into the rhyolite-rich flow units that underlie the little-incised andesite-dacite welded tuffs sealing much of the upper VTTS.

The lake basin was occupied in 1912 by a glacier, which had melted back sufficiently to produce the single modern lake only by the 1950s. The ash flows banked against the glacier and probably smothered adjacent snowbanks and ice-cored moraine. Evidence for resulting ash-water interactions includes (1) at least six phreatic craters close to the present north shore of the lake; (2) a fringing apron, 1 to 5 m thick, of pumiceous diamicton and cross-bedded ejecta, containing many blocks that had been welded and fumarolically altered prior to ejection, some of which were still hot enough to develop prismatic joints after redeposition; and (3) peripheral fissuring, slumping, and lakeward tilting of the ash-flow sheet, as a result of the melting of subjacent and adjacent snow and ice. When first visited in 1917, an elongate lake had developed between the ice front and the steaming tuff margin (photo in Griggs, 1922, p. 212). The lake also extended 600 m northeast of the present lake, confined along a narrow cleft that had been explosively excavated by a chain of coalescent phreatic craters along the fissured hingeline of the compactional

bench. Called "Fissure Lake" by Griggs, this extension ultimately silted up, and today more than 10 m of its ashy lacustrine and alluvial fill have been incised by a stream flowing back into the modern lake.

LOCATION 3. NOVARUPTA CALDERA: THE 1912 VENT

From the last stop, the 8-km trek to Novarupta is most informatively made by staying up on the bench below the cliffy andesite lavas of Mt. Mageik, first to another lake (having a history much like that of the lake just visited) and then along the northern bases of Mt. Cerberus and Falling Mountain, both of which are Holocene domes of pyroxene dacite (Figs. 1 and 4).

The 2-km-wide depression separating Falling and Broken mountains is thought to be a funnel-like vent structure nearly filled with its own fallback ejecta (Hildreth, 1983). The sheared-off face of Falling Mountain is one margin and, although relations are now concealed, the Naknek siltstones of Broken Mountain and andesitic lavas of Trident are also truncated by the steep walls of the vent. Lithic fragments of these lithologies are abundant among the ejecta. The oval vent area is outlined in plan (Fig. 4) by arcuate fractures and compaction-related faults in the deposits; the oval is partly occupied by a 250-m-high mound of coarse ejecta (the "Turtle"), in which the small hollow containing the Novarupta dome is eccentrically nested.

The blocky lava dome is 380 m wide, 65 m high, and consists mostly of glassy, phenocryst-poor rhyolite that was extruded after the main pyroclastic outbursts had ceased. Flow banding is defined largely by variable vesicularity, oxidation, and microlite growth in the rhyolite, but approximately 5% of the bands are dacite and andesite, which both contain conspicuous plagioclase and pyroxenes and crop out most abundantly along the dome's southeast margin.

The "Turtle" is a thick pile of coarse fallout, predominantly dacite pumice but including breadcrusted and vitrophyric blocks of dacite, reejected blocks of agglutinate and welded tuff, basement lithics, and a few layers rich in black andesitic scoria. Rhyolite is rare except as tumbled blocks adjacent to the dome. Whether some of the Turtle's elevation is a result of a buried basement high is unknown but is thought unlikely. Mutually cross-cutting sets of faults on its summit (Fig. 4) suggest the possibility of some distensional uplift by intrusion of a crypto-dome. Faulting postdated the final pumice-fall layers and, apparently, formation of the hollow containing Novarupta.

The stratified pyroclastic deposits exposed on the Turtle and in the ejecta ring extending around Novarupta postdate the ash-flow phase of the eruption. Most probably fell on the second and third days when regional fallout continued to rain down on Kodiak, but less dispersed outbursts may have taken place much later. Fine exposures of these proximal andesite-dacite-fall deposits, more than 12 m thick, can be visited in gulches 1 km east and 700 m south of the dome. Along the great cleft just east of Falling

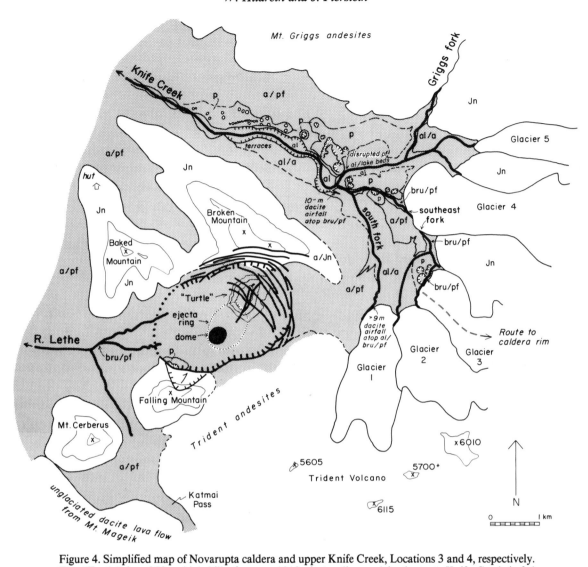

Figure 4. Simplified map of Novarupta caldera and upper Knife Creek, Locations 3 and 4, respectively. The "Turtle" is a 250-m-high hill of coarse 1912 ejecta, discussed in the text. Knife Creek is fed by meltwater from the five Knife Creek Glaciers, and its course across the ash-flow surface is adjoined by numerous phreatic craters and their stratified deposits (p). Block-rich pumice-flow units (bru) overlying the main ash-flow deposit (pf) are discussed in the text. Thick dacite airfall deposits (a) mantle much of the upper VTTS. Other symbols as in Figure 2. Route up Glacier 3 to Location 5 is indicated.

Mountain, inboard-dipping strata drape the sharp rim of the vent funnel.

Along the pair of gullies that converge 2 km west of the dome (Fig. 4) are extensive exposures of two andesite-dacite ash-flow units, each 1 to 3 m thick, intercalated within the 7 to 8 m stack of Plinian dacite fall units. A third local flow unit, extraordinarily rich in dacite pumice blocks, crops out near the confluence of the gullies, only about 1 m above the partially welded surface of the main ash-flow sheet that extends down valley. The three intraplinian ash flows were swale-confined and extended only a few kilometers from source, yet developed

pronounced pumice-block concentrations at their upper and marginal surfaces.

Near-vent exposures of deposits representing the earliest phases of the eruptive sequence are sparse, the most accessible being on the three steep southerly spurs of Baked Mountain (Fig. 4) and along the brink of the Falling Mountain scarp. On Baked Mountain, 0.5 to 2 m of tan to light-gray, rhyolite-rich, cross- to plane-bedded surge and fall deposits are variably preserved beneath 1 to 3 m of brick-red to terra cotta cross-bedded surge and flow deposits rich in lithics and in andesite and dacite pumice. These units are interpreted as products of near-vent blasts and as lag veneers left behind by the zoned sequence of ash flows that moved on to fill the VTTS.

The strongly oxidized surge and flow veneers apparently inspired the naming of Baked Mountain, as neither its gray Naknek siltstones nor the widely mantling light-gray dacite fall units are oxidized, except within the vent area and on the valley floor where fumaroles rooted in the ash-flow sheet permeated the overlying air-fall strata for years.

Truncation of Falling Mountain probably took place on the first day (June 6th), just after the dominantly rhyolitic phases of the eruption had ended, as the brow of its inward-facing scarp is mantled only by the rhyolite-poor surge deposits and the later dacite fall units. The base of the scarp is marked by shallow craters ringed by as much as 4 m of surge-bedded phreatic deposits resting atop the dacite airfall; there is also a litter of rockfall debris made up of Falling Mountain dacite (still falling today), some of it in shattered disintegrating piles as high as 12 m. Another interesting aspect of Falling Mountain (and Cerberus) is its chemical and mineralogical similarity (64% SiO$_2$; plagioclase and two pyroxenes) to the 1912 dacite pumice. Because Falling Mountain dacite erupted at the site of the 1912 vent during Holocene time, either the rhyolitic magma of 1912 evolved within a few thousand years or the rhyolitic reservoir lay elsewhere and was separate from the reservoir yielding the dacite domes.

LOCATION 4. UPPER KNIFE CREEK

From Novarupta, a 3-km eastward hike to the snout of Glacier 1 positions one for a fascinating 4-km walk adjacent to the termini of Glaciers 1, 2, 3, and 4 and on down Knife Creek's southeast fork to the junction with its Griggs fork (Fig. 4). From the junction, plan on a full day's walk back to the Overlook Cabin, either by recrossing the River Lethe and Windy Creek or by following the valley's northeast margin down to the Ukak footbridge. (Be sure to check with the National Park Service about the status of the footbridge before attempting the latter route.)

Where the south fork emerges from its steep slot along the northwest snout of Glacier 1 (Fig. 4), about 9 m of stratified dacite airfall (Curtis' layers C through H) rest upon more than 12 m of pumiceous alluvium that had been immediately reworked by torrential runoff of meltwater from glacial surfaces overridden by the pyroclastic flows. Near the base of the fluvial section here and for 3 km along the southeast fork from Glacier 3 to the junction, are exposures of a partially welded pyroclastic flow, which is 2 to 3 m thick and extremely rich in coarse blocks of dacite pumice. Though intercalated with fluvially reworked ejecta here in proglacial Knife Creek, this late flow unit (*bru* in Fig. 4) is otherwise stratigraphically, compositionally, and lithologically similar to the lowest of the intraplinian units west of Novarupta (see Location 3); it similarly lies only about 1 m above the main welded ash-flow deposit, and was similarly swale-

confined, probably owing to differential compaction of the thick subjacent welded tuff filling a pre-1912 ice-marginal stream channel. Lateral pinch-outs of the block-rich unit are exposed in cross section 150 m below Glacier 4 on the southeast fork and 100 m above the junction on the Griggs fork.

Only the top few meters of the ash-flow deposits have yet been incised in the upper Knife Creek area. Best exposed along the margins of Glaciers 3 and 4 and near the confluence of the south and southeast forks, the tuff is dark gray, partially welded, rich in both andesite and dacite pumice, shows widespread fumarolic alteration, and forms the fluted walls of shallow but treacherous stream gorges. Glaciers 4 and 5 have each advanced several hundred meters since first photographed in 1917, and Glacier 3 (although partly beheaded by the caldera collapse atop Mt. Katmai) has also advanced slightly. As a result, active glacial ice overlies the welded tuff at stream level along the snouts of both Glaciers 3 and 4.

Phreatic craters ringed by aprons of cross-bedded ejecta 1 to 15 m thick are abundant along the southeast fork and along Knife Creek for another 5 km below the junction (Fig. 4). Deposits contain blocks of welded tuff as large as 2 m, some more densely welded than any tuff yet exposed in place. All the craters cut the dacite air-fall layers and spread their ejecta blankets on top of them, indicating that phreatic explosions from within the ash-flow sheet were delayed for at least the two additional days (7–8 June) during which the dacite pumice falls accumulated. Time may have been required for compaction and welding to reduce permeability, enhancing confinement, and to generate fractures providing access for water to the hot interior of the tuff.

The thickest phreatic deposits dip radially away from the lower Griggs fork just above its junction with the southeast fork, an area marked by a cluster of about 20 craters and by chaotic disruption of the welded-tuff sheet. It is thought that all this debris temporarily dammed the Griggs fork, impounding a 1.5-km-long lake in which more than 10 m of mudflow, lacustrine, and fluvial deposits accumulated prior to breaching. Catastrophic breakout (probably in the summer of 1912) was apparently responsible for the debris-flow deposit that mantles much of the lower VTTS (Hildreth, 1983). The flood scoured part of the central VTTS, transported coarse lithics, dacite pumice, and welded-tuff blocks all the way to the lower VTTS and actually outdistanced the ash-flow sheet, dumping debris as thick as 8 m atop its northwest terminus and running on out into the woods where it engulfed upright uncharred trees.

Remnants of the lake deposits not swept away form the lower bluffs along the Griggs fork, whereas the higher bluffs and hills are mostly phreatic deposits atop the air-fall strata. A gulch 200 m south of the junction provides splendid exposures of a 10-m-thick section of dacite fall units C through H, resting atop pumiceous alluvium and the block-rich pyroclastic-flow unit. The same set of air-fall strata was 45 cm thick at Three Forks (Location 1) and 80 cm thick near the western Mageik lake (Location 2).

LOCATION 5. MT. KATMAI CALDERA RIM

From upper Knife Creek, the west notch on the caldera rim of Mt. Katmai can be reached by a roped hike up Glacier 3 (Figs. 1 and 4). The 3–5 hour ascent should not be undertaken lightly, as the weather is treacherous, progress tedious on hummocky ash-covered ice, and the kettles, crevasses, and englacial rivulets hazardous. If you have little climbing experience, charter a scenic flight at Brooks Camp; you're likely to see more from the air, anyway.

Exposed parts of the unstable caldera wall appear to be entirely andesite and dacite, but Naknek siltstone, which crops out as high as the 4,500 ft contour on the cleaver between Glaciers 3 and 4, could conceivably intersect the wall near or beneath lake level. The stratified (and glacially striated) pyroclastic unit forming much of the surface along the west rim is poorly sorted andesite-dacite airfall that ranges from loose scoria to eutaxitic agglutinate. As far as the tooth near the south notch, the southwest wall also appears to be rimmed by pre-1912 agglutinates. The 1912 fallout has largely been blown away or reworked into ashy periglacial diamicton, but pumice lapilli remaining on the caldera rim seem to be entirely ordinary 1912 ejecta from Novarupta. Apart from the small dacite lava on the caldera floor that briefly formed the "horseshoe island" (Griggs, 1922), there is no good evidence that any 1912 eruptive products vented at Mt. Katmai.

Intracaldera glaciers have gradually accumulated atop slump-block benches on the north and south walls, and another ice tongue now flows back into the caldera from the southwest-rim icefield. Calving of icebergs supplements the ample precipitation in sustaining the caldera lake, which is now more than 250 m deep and still adding a meter or two each year. On the lake's blue-green surface, patches of yellow sulfurous froth and sporadic bubbling attest to sublacustrine fumaroles, as does the persistent odor of H_2S. R. J. Motyka measured summer water temperatures of 5°–6°C and pH values of 2.5–3. The lake surface has frozen during some recent winters, at least intermittently.

THE TEN THOUSAND SMOKES

Most fumaroles in the ash-flow sheet had died out by 1930, and those remaining today are odorless wisps of water vapor that issue along faults in the vent region, notably atop the Turtle. Seven years after the eruption, some fumarolic orifice temperatures were still as hot as 645 °C, but recent measurements by T.E.C. Keith indicate none hotter than 90 °C.

Soluble high-temperature fumarolic chlorides, fluorides, and sulfates (Zies, 1929) have been extensively leached, and the vapor-deposited oxide and sulfide minerals generally retrograded as discharge temperature declined. Around Novarupta, low-temperature hydrothermal alteration of the glassy ejecta to clays is actively overprinting the vapor-phase alteration; Keith's work there indicates abundant silica, kaolinite, amorphous clay precursors, and minor smectite, the resulting warm goo being thoroughly mixed with colorful goethite and amorphous iron-hydroxides and locally with gray, amorphous iron sulfides.

Despite retrogression and leaching, many fumarolic deposits retain spectacular zoning in color, mineralogy, and composition. Irregular, pipelike, and funnel-shaped fossil fumaroles are readily accessible in nonwelded tuff near Three Forks and along the lower gorge of the River Lethe, whereas the best exposed fissure fumaroles (controlled by cooling joints in welded tuff) are along the upper River Lethe opposite Baked Mountain. Red coloration of crusts and altered ejecta is caused by hematite or by goethite and amorphous iron hydroxides, which can also yield oranges and yellows—as can sulfur and arsenic sulfides. Black fumarolic coatings are mostly vapor-phase magnetite; white material is predominantly silica and aluminum fluoride hydroxyhydrate, but includes fluorite, kaolinite, alunite, and gypsum.

REFERENCES CITED

Curtis, G. H., 1968, The stratigraphy of the ejecta from the 1912 eruption of Mount Katmai and Novarupta, Alaska, *in* Coats, R. R., Hay, R. L., and Anderson, C. A., eds., Studies in volcanology: Geological Society of America Memoir 116, p. 153–210.

Griggs, R. F., 1922, The Valley of Ten Thousand Smokes: Washington, D.C., National Geographic Society, 340 p.

Hildreth, W., 1983, The compositionally zoned eruption of 1912 in the Valley of Ten Thousand Smokes, Katmai National Park, Alaska: Journal of Volcanology and Geothermal Research, v. 18, p. 1–56.

Sparks, R.S.J., 1976, Grain size variations in ignimbrites and implications for the transport of pyroclastic flows: Sedimentology, v. 23, p. 147–188.

Wilson, C.J.N., 1984, The role of fluidization in the emplacement of pyroclastic flows, 2; Experimental results and their interpretation: Journal of Volcanology and Geothermal Research, v. 20, p. 55–84.

Zies, E.G., 1929, The Valley of Ten Thousand Smokes; I. The fumarolic incrustations and the bearing on ore deposition. II. The acid gases contributed to the sea during volcanic activity: National Geographic Society, Contributed Technical Papers, Katmai Series, v. 4, p. 1–79.

Resurrection Peninsula and Knight Island ophiolites and recent faulting on Montague Island, southern Alaska*

Steven W. Nelson, Marti L. Miller, and J. A. Dumoulin, U.S. Geological Survey, Branch of Alaskan Geology, 4200 University Drive, Anchorage, Alaska 99508-4667

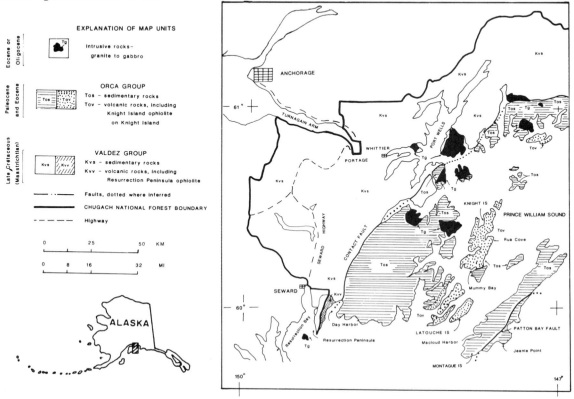

Figure 1. Location map and generalized geologic map of the western half of Chugach National Forest.

LOCATION AND ACCESSIBILITY

The Resurrection Peninsula forms the east side of Resurrection Bay (Fig. 1). The city of Seward is located at the head of the bay and can be reached from Anchorage by highway (127 mi; 204 km). Relief ranges from 1,434 ft (437 m) at the southern end of the peninsula to more than 4,800 ft (1,463 m) 17 mi (28 km) to the north. All rock units composing the informally named Resurrection Peninsula ophiolite are visible and (or) accessible by boat. The eastern half of the peninsula is located within the Chugach National Forest; the western half is mainly state land, but there is some private land with recreational cabins. The Seward A6 and A7 and Blying Sound D6 and D7 maps at 1:63,360 scale (mile-to-the-inch) cover the entire Resurrection Peninsula.

Knight Island is located 53 mi (85 km) east of Seward

(Fig. 1). Numerous fiords indent the 31-mi-long (50 km) by 7.4-mi-wide (12 km) island and offer excellent bedrock exposures. The island is rugged and has a maximum elevation of 3,000 ft (914 m). It has numerous mineral prospects (Tysdal, 1978; Nelson and others, 1984; Jansons and others, 1984; Koski and others, 1985), and several abandoned canneries are located on the island. Knight Island lies entirely within the Chugach National Forest—state and private inholdings constitute less than five percent of its total land area. The Seward A2, A3, B2, B3, and C2, 1:63,360-scale U.S. Geological Survey topographic maps cover the entire island.

Montague Island, 50 mi (80 km) long and up to 11 mi (18 km) wide, lies 10.6 mi (17 km) southeast of Knight Island. It belongs to an island group that forms the southern margin of Prince William Sound (Fig. 1). Montague Island is less rugged and less heavily vegetated than either the Resurrection Peninsula or Knight Island. Rock exposures are excellent along the beaches, and ground disruption due to recent fault movements is clearly

*The sections titled "Resurrection Peninsula" and "Knight Island" are taken directly from Miller (1984), updated by minor additions and deletions. No quotation marks are used for these sections.

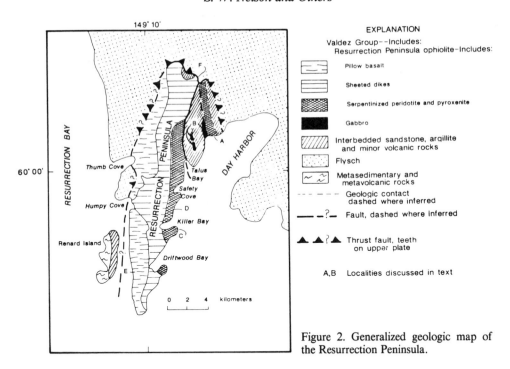

Figure 2. Generalized geologic map of the Resurrection Peninsula.

visible. The Seward A1 and A2 and Blying Sound D1, D2, and D3 maps cover the areas of interest on Montague Island.

In all areas, access is by float-equipped aircraft, helicopter, or boat. Wheel-equipped aircraft can land on the beaches or at several landing strips on Montague Island.

SIGNIFICANCE OF THE SITES

Taken together, the three sites offer a spectacularly well-exposed cross section of ophiolite and accretionary wedge deposits, as well as evidence of recent tectonic activity. These rocks are included in the accretionary wedge deposits of what may be the world's most complete paired belt of mélange and flysch, extending more than 1,240 mi (2,000 km) along the Gulf of Alaska continental margin (Winkler, 1984).

On both the Resurrection Peninsula and Knight Island, we have the opportunity to study young, relatively unmetamorphosed, well-exposed ophiolites of probable Late Cretaceous and lower Tertiary age that represent the most complete ophiolite sequences in Alaska. In addition, these two areas contain volcanogenic massive sulfide deposits that are second only to Kennicott in Alaskan copper production (Jansons and others, 1984).

The accretionary wedge deposits in the Prince William Sound area are divided into two groups: the Valdez Group, of Maastrichtian (Late Cretaceous) age and the Paleocene to middle Eocene Orca Group (Plafker and others, 1984, 1985). Both groups consist largely of graywacke, siltstone, and shale; the finer-grained rocks commonly display a slaty fabric. The Orca Group has traditionally been considered to be somewhat less metamorphosed than the Valdez Group and to be further distinguished from it by the presence of mafic volcanic rocks and local beds of conglomerate. Other mapping, however, has shown that both groups contain similar rock types (Moffit, 1954; Tysdal and Case, 1979; Nelson and others, 1985) including conglomerate and ophiolite sequences that consist of sheeted dikes, pillow basalt flows, layered gabbro, and minor ultramafic rocks.

The contact between the Orca and Valdez groups is designated as the Contact fault (Winkler and Plafker, 1975; and Fig. 1). In the western Prince William Sound area, parallelism in the regional strike and the close lithologic similarities of the two groups make accurate location of the Contact fault problematic (Dumoulin, 1986). Sandstone petrography from the Prince William Sound area is summarized by Dumoulin (1984, 1986), who shows that in the western part of the Prince William Sound area, compositions of Valdez and Orca groups graywackes change systematically from west to east, without an abrupt change at the Contact fault system.

Montague Island contains well-exposed turbidites of the Orca Group deposited in a middle fan setting (Winkler, 1976) as well as hemipelagic rocks that may have formed in a slope basin (Dumoulin and Miller, 1984). The turbidites have locally been offset by faults that had as much as 26 ft (8 m) of vertical movement during the 1964 Alaskan earthquake (Plafker, 1969).

SITE INFORMATION

Resurrection Peninsula. From east to west, the Resurrection Peninsula ophiolite consists of gabbro (in part layered), sheeted mafic dikes, and pillow basalt (Fig. 2); stratigraphic tops face to the west, as indicated by both pillow morphology and the dip of the siltstone interbedded with pillow basalt. Serpentinized peridotite and pyroxenite occur as pods in the gabbro (Fig. 2,

Figure 3. Sheeted dikes on the Resurrection Peninsula. Five dikes are visible oriented parallel to pencil. Lower dike shows chilled margins.

Figure 4. Pillow basalt on the Resurrection Peninsula.

locality A) and as tectonic slices in sedimentary rocks (Fig. 2, locality B). Dikes and small pod-shaped septa of plagiogranite intrude the gabbro unit (Fig. 2, locality C; Fig. 3) to form a mixed zone of gabbro and mafic dikes (Fig. 2, locality D). This sequence of rocks is considered ophiolitic, but lacks the basal tectonized peridote and the cap of deep-water sediment of a complete ophiolite (Coleman, 1977).

The rocks of the Resurrection Bay and Peninsula were first described by Grant and Higgins (1909) and Grant (1915). These workers recognized "ellipsoidal lavas" on the western side of the peninsula and "deeper seated rocks, such as gabbros and periodotites" on the eastern side (Grant, 1915, p. 223). More recently, Tysdal and others (1977) mapped the mafic sequence and believed it to be interbedded with Valdez Group flysch, and therefore Cretaceous in age. They also compared it to a similar mafic sequence in the Orca Group on Knight Island. In both sequences, Tysdal and others (1977) noted a petrographic and chemical similarity to oceanic tholeiites and suggested an ophiolitic affinity. However, because they did not observe layered gabbro, and because there was no basal peridotite, they did not use the term ophiolite. They postulated several possible tectonic settings for formation of the sequence, but favored a "leaky" transform fault adjacent to a continental margin. Nelson and others (1985) showed that indeed there are small local occurrences of siltstone interbedded with pillow basalts (Fig. 2, locality E); however, they did not observe typical Valdez Group flysch interbedded with mafic flow rocks. In fact, at locality F (Fig. 2), it is clear that typical Valdez Group flysch has been thrust over the ophiolite sequence in a west-southwest direction. At this locality a 3 ft-thick (1 m) iron-stained shear zone 2,300 ft (700 m) long marks the fault. Lower plate rocks are interbedded pillow basalt, broken pillow breccia, and minor siltstone. Bedding in the upper plate flysch sequence shows evidence of drag folding near the fault contact. Along the western side of the Resurrection Peninsula, at

Thumb Cove and Humpy Cove (Fig. 2), the contact is interpreted as a fault because of the presence of fractured rock and the difference in metamorphic grade between greenschist facies upper plate and prehnite-pumpellyite facies lower plate rocks.

The igneous rocks of the Resurrection Peninsula consist of an ophiolitic sequence that contains, from base to top, and east to west: (1) gabbro with pods of ultramafic rocks, and dikes and small bodies of plagiogranite, (2) sheeted dikes (Fig. 3), and (3) pillow basalt (Fig. 4).

Ultramafic rocks (mostly serpentinized) occur as pods and small bodies in gabbro and as tectonic blocks in sedimentary rock. In most places, the relict texture and mineralogy indicate that the original rock was clinopyroxenite, dunite, and peridotite (Miller, 1984). The pods in the gabbro range from a few yards (meters) to a few tens of yards (meters) in size. At locality B (Fig. 2), these rocks occur along linear trends and show many fractures, some with slickensides. Heavy vegetation in the area probably obscures more ultramafic rock.

The gabbro occurs as two distinct bodies (referred to as the eastern gabbro and the western gabbro) separated by a fault-bounded, synformal block of interbedded volcanic and sedimentary rocks and more typical Valdez Group flysch. Locally, well-developed, west-dipping magmatic mineral layering occurs in the structurally lowest part of the western gabbro. This mineral layering can be observed in shoreline outcrops between Driftwood Bay and Talus Bay where alternating light- and dark-gray layers indicate different proportions of pyroxene and feldspar. The layered gabbro grades westward into massive gabbro containing an increasing percentage of mafic dikes, and then grades farther westward into the sheeted dike unit. Typical exposures consist of medium- to coarse-grained, subophitic clinopyroxene gabbro with the plagioclase composition ranging from An 60 to 70. Clinopyroxene is always partly altered to actinolite plus chlorite (uralitized); the plagioclase is sericitized; and secondary epi-

Figure 5. Plagiogranite outcrop at locality C on the Resurrection Peninsula.

dote is common. Primary hornblende gabbro (with plagioclase of An 55) crops out stratigraphically higher in the sequence, and pods and dikes of plagiogranite are present. The largest exposure of plagiogranite is near the south entrance to Killer Bay (Fig. 2, locality C; Fig. 5). At this location, a 60-ft-wide (20 m) stockwork zone of light gray hornblende plagiogranite has intruded gabbro. Other small dikes and pods of plagiogranite intrude the gabbro unit in a zone extending at least 3 km to the north. The plagiogranite is fine- to medium-grained and contains about 15 percent green hornblende that is slightly altered to actinolite and chlorite. Other constituents include sodic plagioclase, quartz, and intergranular alteration minerals such as epidote and prehnite. Two chemical analyses of the plagiogranite indicate average values of SiO_2, 75 percent; CaO, 3 percent; Na_2O, 5 percent; and K_2O, $\leqslant$ .05 percent.

Within the western gabbro unit, mafic dikes become more numerous to the west (stratigraphically up). These dikes range from 4 to 20 in (10 to 50 cm) in width, are diabasic in texture, and contain primary hornblende altered to actinolite plus chlorite, plagioclase microlites with a brown cryptocrystalline alteration, and interstitial fine-grained epidote or sphene (Miller, 1984).

The petrographic characteristics suggest that the eastern gabbro was originally part of the same sequence as the western gabbro, but was stratigraphically lower. The eastern gabbro does not show mineral layering in outcrop, but it does contain clinopyroxene cumulate textures. Additionally, more pods of ultramafic rock are found in the eastern gabbro than in the western. The eastern gabbro has probably been faulted up relative to the western gabbro.

Dikes in the gabbro unit become increasingly numerous to the west, and sheeted dikes form the rugged crest of the central part of the peninsula. The dikes generally trend from north-northwest to north-northeast and generally dip steeply to the east, although at the southern end of the peninsula the trend is west-northwest. Locally, crosscutting dikes intrude preexisting dikes at

low angles. Most dikes range in width from 1 to 3 ft (30 cm to 1 m), with their chilled margins indicating sheeted intrusion in a tensional environment. Some dikes may be as large as 6 ft (2 m) across. Aphanitic, porphyritic, and diabasic textures are common. Low-grade metamorphic mineral assemblages include actinolite, chlorite, epidote, and veins of prehnite (Miller, 1984). Westward from the sheeted dike unit, through a zone of pillow basalt screens, volcanic breccia, and dikes, is the pillow basalt unit.

Pillow basalt, subordinate amounts of massive basalt, and broken pillow breccia make up most of the western flank of the Resurrection Peninsula. The pillow basalt forms a west-dipping sequence and contains minor interbedded siltstone. The units strike north and dip 30° to 45° west. Discrete pillows average 0.5 m in diameter; other well-preserved forms include: (1) long thin pillows, (2) tubes, and (3) stubby, budded pillows (Fig. 4). Interpillow spaces are locally filled with red and green chert. In thin section the pillows show axiolitic and glassy textures. Amygdules are composed of chlorite, zeolites, and in some places, epidote; prehnite veins are sparse. The intercalated siltstone shows no visible alteration or metamorphism.

The Resurrection Peninsula ophiolite has been affected by serpentinization and low-grade thermal and hydrothermal metamorphism, both typical of ophiolitic rocks (Coleman, 1977); its lack of penetrative fabric suggests hydrothermal alteration rather than regional metamorphism. The secondary mineral assemblages are those typical of low greenschist facies metamorphism with a later local overprinting by prehnite alteration. Greenschist facies metamorphism is thought to occur in ocean-floor settings of high heat flow and hot circulating sea water (Ernst, 1976; Coleman, 1977). The serpentinization of the ultramafic rocks might have occurred by hydration during the ocean-floor metamorphism or after emplacement.

The upper plate Valdez Group rocks on the peninsula show weak to well-developed penetrative fabric and contain visible to abundant metamorphic biotite, and lesser amounts of muscovite and chlorite. The sedimentary rocks, lying structurally between the east and west gabbro bodies, are less metamorphosed.

Knight Island. Knight Island, 56 mi (90 km) to the east in Prince William Sound (Fig. 6), is composed of mafic rocks similar to those on the Resurrection Peninsula. Pillow basalts occupy the west, north, and portions of the east sides of the island; and a sheeted dike unit (Fig. 7) forms the central part of the island. Small gabbro stocks intrude the sheeted dikes, and ultramafic pods are found as inclusions in the sheeted dikes and as tectonic (structural) blocks in the Port Audry shear zone (Fig. 6). Small pods and dikes of altered plagiogranite, similar to those on the Resurrection Peninsula, are also found in the sheeted dikes. Orca Group flysch crops out at the southern end of the island and may stratigraphically or structurally underlie the igneous rocks.

The lithologic similarities of rocks cropping out on the Resurrection Peninsula and Knight Island are striking. Early workers (Grant and Higgins, 1909; Grant, 1915) considered the two ophiolites to be the same and assigned a Mesozoic age to them. Later workers (Tysdal and others, 1977; Tysdal and Case,

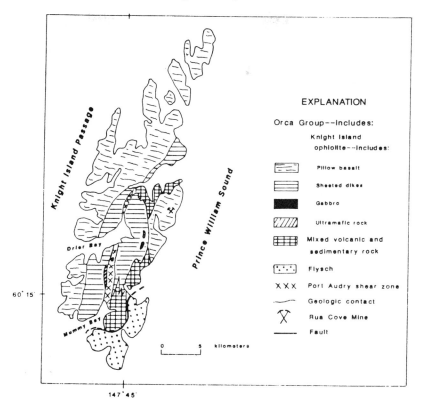

EXPLANATION

Orca Group--Includes:

Knight Island
ophiolite--Includes:

	Pillow basalt
	Sheeted dikes
	Gabbro
	Ultramafic rock
	Mixed volcanic and sedimentary rock
	Flysch
XXX	Port Audry shear zone
	Geologic contact
	Rua Cove Mine
	Fault

Figure 6. Generalized geologic map of Knight Island.

1979) recognized the lithologic and chemical similarities of the two ophiolites, but considered them to have different ages because of their relationships with dissimilar surrounding rocks. The Resurrection Peninsula ophiolite was considered to be of Cretaceous age because of its association with the Valdez Group flysch; the Knight Island ophiolite was considered to be of early Tertiary age because of its association with Orca Group flysch.

Tysdal and others (1977) reported values for major oxides and minor elements of ten samples from the Resurrection Peninsula and nine samples from Knight Island. For both ophiolites, the major oxide values are within the ranges reported for oceanic tholeiites (Cann, 1970; Pearce, 1975). The minor elements, Ti, Y, Zr, Cr, and Ni, indicate an oceanic tholeiite affinity for both the Resurrection Peninsula and Knight Island ophiolites.

Montague Island. Montague Island is the largest island in the Prince William Sound area and shows spectacular offsets along faults as a result of the 1964 earthquake. The island is underlain entirely by sedimentary rocks of the Orca Group, and most shoreline exposures have excellent outcrops of the Tertiary turbidite sequence. Such primary features as flute casts, ripple marks, graded and cross bedding, and rhythmic bedding are usually well preserved. Studies on Hinchinbrook Island just to the northeast by Winkler (1976) suggest that the Orca Group sedimentary rocks in this area were deposited on the middle parts of a complex westward-sloping deep-sea fan. Near Jeanie Point (Fig. 1), on the south side of Montague Island, Dumoulin and

Figure 7. Sheeted dikes on Knight Island.

Miller (1984) have described a sequence of fine-grained hemipelagic rocks including limestone, chert, tuff, mudstone, argillite, and sandstone. These rocks are probably part of the Orca Group and appear to have been deposited in a slope environment in a small, restricted basin (Dumoulin and Miller, 1984).

Among the most interesting features on Montague Island are the physiographic effects of the 1964 earthquake—one of the most violent earthquakes to occur in North America during this century (Hansen and others, 1966). The epicenter of this earthquake is located 79 mi (128 km) north of the southern end of Montague Island, and its magnitude is calculated at 8.4 on the Richter scale (Plafker, 1969). Most of the area north of Montague Island subsided; for instance, the town of Portage (Fig. 1) subsided about 6.6 ft (2 m); however, the southeast side of Montague Island was uplifted 26 ft (8 m) along the Patton Bay fault. This uplift affected shipping lanes and disrupted salmon spawning streams and harbor facilities in parts of Prince William Sound (Hansen and others, 1966). Inferred large-scale uplift of the continental shelf and slope southeast of Montague Island probably generated the seismic sea waves that spread across the Pacific Ocean (Van Dorn, 1964; Plafker, 1965). Several fault scarps (Fig. 8) are still well preserved and display the impressive force of this large earthquake.

Figure 8. Fault scarp, 6.6 ft (2 m) high, on Montague Island. View northeast.

REFERENCES CITED

Cann, J. R., 1970, Rb, Sr, V, Zr, and Nb in some ocean floor basaltic rocks: Earth and Planetary Science Letters, v. 10, p. 7–11.

Coleman, R. G., 1977, Ophiolites: New York, Springer-Verlag, 299 p.

Dumoulin, J. A., 1984, Composition and provenance of sandstones of the Orca and Valdez groups, Prince William Sound, Alaska: Geological Society of America Abstracts with Programs, v. 16, p. 280.

Dumoulin, J. A., 1986, Sandstone composition of the Valdez and Orca groups, Prince William Sound, Alaska: U.S. Geological Survey Bulletin 1774 (in press).

Dumoulin, J. A., and Miller, M. L., 1984, The Jeanie Point complex revisited, *in* Coonrad, W. L., and Elliott, R. L., eds., The United States Geological Survey in Alaska; Accomplishments during 1981: U.S. Geological Survey Circular 868, p. 75–77.

Ernst, W. G., 1976, Petrologic phase equilibria: San Francisco, W. H. Freeman and Company, 333 p.

Grant, U. S., 1915, The southeastern coast of Kenai Peninsula, *in* Martin, G. C., Johnson, B. L., and Grant, U. S., eds., Geology and mineral resources of Kenai Peninsula, Alaska: U.S. Geological Survey Bulletin 587, p. 209–238.

Grant, U. S., and Higgins, D. F., Jr., 1909, Notes on the geology and mineral prospects in the vicinity of Seward, Kenai Peninsula: U.S. Geological Survey Bulletin 379, p. 98–107.

Hansen, W. R., Eckel, E. B., Schaem, W. E., Lyle, R. E., George, W., and Chance, G., 1966, The Alaska earthquake March 27, 1964; Field investigations and reconstruction effort: U.S. Geological Survey Professional Paper 541, 111 p.

Jansons, U., Hoekzema, R. B., Kurtak, J. M., and Fechner, S. A., 1984, Mineral occurrences in the Chugach National Forest, south-central Alaska: U.S. Bureau of Mines Report MLA 5-84, 217 p.

Koski, R. A., Silberman, M. L., Nelson, S. W., and Dumoulin, J. A., 1985, Rua Cove; Anatomy of a volcanogenic Fe-Cu sulfide deposit in ophiolite on Knight Island, Alaska [abs.]: American Association of Petroleum Geologists Bulletin, v. 69, p. 667.

Miller, M. L., 1984, Geology of the Resurrection Peninsula, *in* Winkler, G. L., Miller, M. L., Hoekzema, R. B., and Dumoulin, J. A., Guide to the bedrock geology of a traverse of the Chugach Mountains from Anchorage to Cape Resurrection: Alaska Geological Society Guidebook, p. 25–34.

Moffit, F. H., 1954, Geology of the Prince William Sound region, Alaska: U.S. Geological Survey Bulletin 989-E, p. 225–310.

Nelson, S. W., and 11 others, 1984, Mineral resource potential of the Chugach National Forest, Alaska; Summary report: U.S. Geological Survey Miscellaneous Field Studies Map MF 1645-A, scale 1:250,000, 24 p.

Nelson, S. W., Dumoulin, J. A., and Miller, M. L., 1985, Geologic map of the Chugach National Forest, Alaska: U.S. Geological Survey Miscellaneous Field Studies Map 1645-B, scale 1:250,000, 15 p.

Pearce, J. A., 1975, Basalt geochemistry used to investigate past tectonic environments on Cyprus: Tectonophysics, v. 25, p. 41–67.

Plafker, G., 1965, Tectonic deformation associated with the 1964 Alaskan earthquake: Science, v. 148, no. 3678, p. 1675–1687.

——, 1969, Tectonics of the March 27, 1964, Alaska earthquake: U.S. Geological Survey Professional Paper 453-I, 74 p.

Plafker, G., Keller, G., Barron, J. A., and Blueford, J. R., 1984, Paleontologic data on the age of the Orca Group, Alaska: U.S. Geological Survey Open-File Report 85-429, 24 p.

Plafker, G., Keller, G., Nelson, S. W., Dumoulin, J. A., and Miller, M. L., 1985, Summary of data on the age of the Orca Group, *in* Bartsch-Winkler, S., ed., The United States Geological Survey in Alaska; Accomplishments during 1984: U.S. Geological Survey Circular 967, p. 74–76.

Tysdal, R. G., 1978, Mines, prospects, and occurrences map of the Seward and Blying Sound quadrangles, Alaska: U.S. Geological Survey Miscellaneous Field Studies Map MF 880A, scale 1:250,000.

Tysdal, R. G., and Case, J. E., 1979, Geologic map of the Seward and Blying Sound quadrangles, Alaska: U.S. Geological Survey Miscellaneous Investigation Series Map I-1150, 1 sheet, scale 1:250,000, 12 p.

Tysdal, R. G., Case, J. E., Winkler, G. R., and Clark, S.H.B., 1977, Sheeted dikes, gabbro, and pillow basalt in flysch of coastal southern Alaska: Geology, v. 5, p. 377–383.

Van Dorn, W. G., 1964, Source mechanism of the tsunami of March 28, 1964, in Alaska: Proceedings, Coastal Engineers Conference, 9th, Lisbon, 1964, p. 166–190.

Winkler, G. R., 1976, Deep-sea fan deposition of the lower Tertiary Orca Group, eastern Prince William Sound, Alaska, *in* Miller, T. P., ed., Recent and ancient sedimentary environments in Alaska: Proceedings, Alaska Geological Society Symposium, Anchorage, 1975, p. R1–R20.

——, 1984, Geologic setting, *in* Winkler, G. R., Miller, M. L., Hoekzema, R. B., and Dumoulin, J. A., Guide to the bedrock geology of a traverse of the Chugach Mountains from Anchorage to Cape Resurrection: Alaska Geological Society Guidebook, p. 1–14.

Winkler, G. R., and Plafker, G., 1975, The Landlock fault; Part of a major early Tertiary plate boundary in southern Alaska, *in* Yount, M. E., ed., The United States Geological Survey Alaska Program, 1975: U.S. Geological Survey Circular 722, p. 49.

Chulitna region, south-central Alaska

C. C. Hawley, *Hawley Resource Group, Inc., Anchorage, Alaska 99518*
D. L. Jones, *Department of Geology, University of California, Berkeley, California 94720*
T. E. Smith, *Alaska Division of Geological and Geophysical Surveys, Fairbanks, Alaska 99701*

LOCATION AND ACCESS

The Upper Chulitna region of Alaska is on the south flank of the central Alaska Range on the southeast side of Denali National Park (Fig. 1). It was first recognized as an unusual entity by S. R. Capps, who mapped a belt of Triassic rocks (1919) during a reconnaissance of mineral deposits; he called the area the Upper Chulitna region.

The Chulitna River forms a barrier to access, but once across the Chulitna the country is open and ranges from rolling, easily hikeable country to rugged and difficult. Some of the key sections of the Chulitna region are strikingly exposed in canyon walls, and flybys from either fixed- or rotary-wing aircraft provide excellent overviews of the geology (Figs. 2, 3).

SIGNIFICANCE OF THE REGION

The relatively good exposures and fossiliferous nature of several formations in the Upper Chulitna region were key elements in erecting the hypotheses that much of the Cordilleran region is composed of individual tectonostratigraphic terranes and that some of these terranes have moved northerly for distances perhaps measured in thousands of kilometers.

The Chulitna terrane (Jones and others, 1980) includes the only recognized rocks (limestone) of Early Triassic age in south-central Alaska; further, these rocks have southern affinity and very likely have been transported northerly at least 1,860 mi (3,000 km). The terrane has a base of ophiolite of Late Devonian age and includes clastic red beds of Late Triassic age, which, locally, are in unconformable contact with the Early Triassic limestone. Red beds contain clasts of radiolarian chert derived from ophiolite; they also contain quartzose clasts, indicating that by Late Triassic time the terrane was adjacent to a continental land mass.

The Chulitna terrane is one of three miniterranes that were mashed between the Talkeetna superterrane and the paleo–North American continent (Jones and others, 1980, 1982; Dsejtey and others, 1982).

Finally, the region is strongly mineralized with intrusion-related deposits emplaced in Late Cretaceous and Paleocene time, or about the time of final intense tectonic activity related to accretion of the composite terranes to the North American continent.

GEOLOGY OF THE REGION

History. In 1907, prospector John Coffee discovered placer gold on Bryn Mawr Creek, a tributary to the West Fork of the

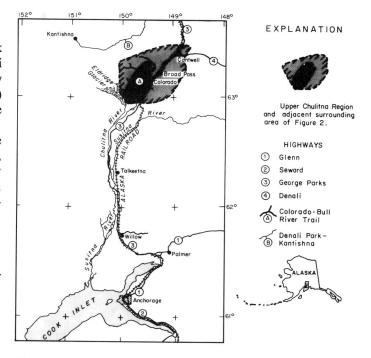

Figure 1. The Upper Chulitna region is on the south flank of the Alaska Range about 140 mi (225 km) north of Anchorage. The area is relatively inaccessible, but is best reached from Colorado Station on the Alaska Railroad or by rotary-wing aircraft. Some key sections in the region are well exposed for viewing from light aircraft, and the Alaska Railroad and Alaska Highway 3 (Parks) give access to adjacent structural terranes.

Chulitna River. By 1909, a series of complex silver-gold deposits had been discovered and staked in the vicinity. In 1917, anticipating that the Alaska Railroad would reach as far as Broad Pass (which could allow development of the deposits), S. R. Capps, of the U.S. Geological Survey, surveyed rocks and mineral deposits (1919). Capps recognized a belt of rocks of Triassic age that included most of the metallic deposits. Interest in mineral deposits along the Alaska Railroad produced a more comprehensive investigation by C. P. Ross in 1931. Ross subdivided Capps' units and, in addition, recognized Devonian(?) and Permian rocks near Triassic units. Both Capps and Ross recognized red beds, tuffs, conglomerates, and the major components of the Chulitna terrane. Paleontologists G. H. Girty and J. B. Reeside, Jr. (Ross, 1933, p. 298–300) noted that although Carboniferous and Triassic species were present in the Chulitna region, they were rather different from other time-equivalent species reported in Alaska.

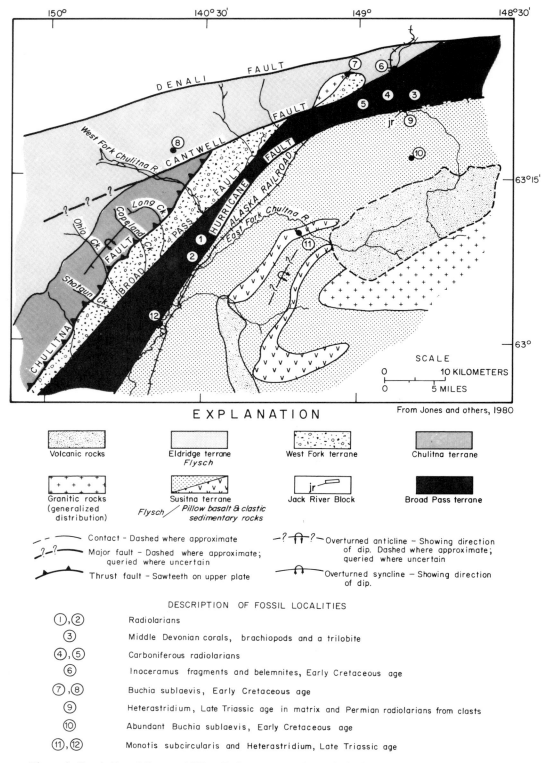

EXPLANATION

From Jones and others, 1980

Volcanic rocks

Eldridge terrane
Flysch

West Fork terrane

Chulitna terrane

Granitic rocks
(generalized
distribution)

Susitna terrane
Flysch / Pillow basalt & clastic
sedimentary rocks

jr

Jack River Block

Broad Pass terrane

— — — Contact – Dashed where approximate

?—?— Major fault – Dashed where approximate;
queried where uncertain

▲▲▲ Thrust fault – Sawteeth on upper plate

—?—⋔—?— Overturned anticline – Showing direction
of dip. Dashed where approximate;
queried where uncertain

—⋔— Overturned syncline – Showing direction
of dip.

DESCRIPTION OF FOSSIL LOCALITIES

①,②   Radiolarians

③   Middle Devonian corals, brachiopods and a trilobite

④,⑤   Carboniferous radiolarians

⑥   Inoceramus fragments and belemnites, Early Cretaceous age

⑦,⑧   Buchia sublaevis, Early Cretaceous age

⑨   Heterastridium, Late Triassic age in matrix and Permian radiolarians from clasts

⑩   Abundant Buchia sublaevis, Early Cretaceous age

⑪,⑫   Monotis subcircularis and Heterastridium, Late Triassic age

Figure 2. Exotic Broad Pass and West Fork terranes and matrix Susitna and Eldridge terranes are exposed along the Alaska Railroad and Denali and Parks highways. The exotic West Fork and Chulitna terranes are inaccessible but are especially well exposed in steep canyon walls of Partin, Shotgun, Ohio, and Copeland creeks, which trend southeasterly across section.

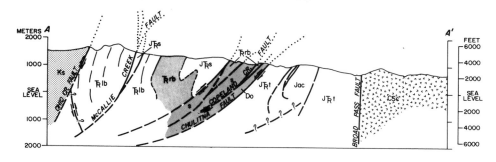

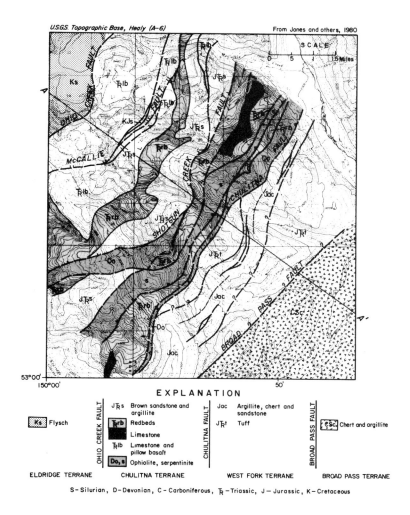

From Jones and others, 1980

U.S.G.S. Topographic Base, Healy (A-6)

EXPLANATION

	JꝚs	Brown sandstone and argillite		Jac	Argillite, chert and sandstone		CSc	Chert and argillite
Ks Flysch	Ꝗrb	Redbeds		JꝚt	Tuff			
	Ꝗlb	Limestone						
		Limestone and pillow basalt						
	Do, s	Ophiolite, serpentinite						

ELDRIDGE TERRANE   CHULITNA TERRANE   WEST FORK TERRANE   BROAD PASS TERRANE

S - Silurian, D - Devonian, C - Carboniferous, Ꝗ - Triassic, J - Jurassic, K - Cretaceous

Figure 3. The Shotgun Creek area has an overturned but relatively unfaulted and complete section of the Chulitna terrane typical of the central and southern part of the Chulitna region. Starting an aircraft flyby at the upper cirque in Shotgun Creek at about 5,000-ft (1,500 m) elevation, the section going downstream is the limestone and pillow basalt unit of Upper Triassic age, red beds of Upper Triassic age, and the brown sandstone and argillite unit of Upper Triassic and Lower Jurassic age. On the south side of the canyon about 2 mi (3.2 km) below the cirque, the Shotgun Creek fault places red beds against brown sandstone and argillite unit. The red-bed unit crops out for about 1,500 ft (460 m) and is succeeded by about 2,000 ft (600 m) of serpentinite of the basal ophiolite unit (Devonian) of the Chulitna terrane. Variants of the same section, but somewhat more faulted, are exposed in Partin and Ohio creeks. Only mountain-qualified pilots should be used to fly these traverses, and aircraft should have adequate power to fly the rough terrain.

Interest in mineral deposits again stimulated activity in the region in 1967 to 1968, when Hawley, Allen Clark, and S.H.B. Clark surveyed the region at a much greater level of detail. During this mapping, units were distinguished lithologically, and faults and fold structures were fairly well laid out (Hawley and Clark, 1974). Additional fossil collections were made; among these was a collection of ammonites of Early Triassic age and unusual affinity (N. J. Silberling in Hawley and Clark, 1974, p. B5–B6; Nichols and Silberling, 1979). Small hypabyssal granitic intrusives were recognized, and a series of gabbro-to-ultramafic rocks was proposed to be related in time to the granitic intrusives. Further work, especially by Sandra H. B. Clark, led to the suggestion that the mafic-ultramafic rocks and accompanying cherts were more likely an older ophiolitic terrane (Clark and others, 1972). The granitic units have affinities with units of the composite Alaska Range batholith of Reed and Lanphere (1973).

The recognition of ophiolite, the existence of Lower Triassic rocks, and the unusual occurrence of Mesozoic red beds led to a comprehensive scientific investigation by a team of geologists, geophysicists, and paleontologists, including David L. Jones, Bela Csejtey, Jr., N. J. Silberling, W. H. Nelson, and Charles D. Blume. This work and other related investigations in the Pacific Northwest (Jones and others, 1982) have led to the recognition that the Upper Chulitna region has several exotic terranes that were mashed together and folded and extensively faulted by earliest Tertiary time.

Commercial investigation of the mineral deposits has continued parallel to scientific study. Two deposits, the Coal Creek tin porphyry and the Golden Zone breccia-pipe/shear zone deposits are drill-indicated mineral resources measured in millions of tons. Geological and geophysical work indicates drill targets at several other deposits.

Geologic Setting. In essential form, the Upper Chulitna terrane is ". . . a mosaic of pre-Cretaceous fragments, representing more than one lithospheric plate, set in a matrix of upper Mesozoic flysch", as recognized by Jones and others (1980, p. A20). The structural grain of rock units, fold axes and faults is northeasterly or parallel to the local trend of the Alaska Range. It can be inferred that the final major juxtaposition of rocks took place either in latest Cretaceous or Paleocene time (Lanphere and Reed, 1985).

The rock strata dip northwesterly, and many strata are overturned. Small hypabyssal igneous bodies of at least two ages form dikes and small stocks or plugs in the stratiform sequence; except for original mineral deposits of ophiolite affinity, complex gold- and silver–bearing sulfide-rich deposits occur with intermediate composition rocks of about 68 Ma age and cassiterite-sulfide deposits with granite of about 58 Ma age.

In later Tertiary time, the complexly folded rocks were locally covered by temperate-environment swamps that probably were originally nearly continuous with those of the Cook Inlet–Susitna Basin. Still later, the region was again uplifted and subjected to both fluviatile and glacial processes that shaped the present topography.

TECTONOSTRATIGRAPHIC COMPLEXES

The region contains three main exotic terranes: Chulitna, West Fork, and Broad Pass. Two matrix terranes are also present: Eldridge and Susitna; these two are dominated by flysch.

Chulitna Terrane. Two main stratigraphic sequences are preserved in the Chulitna terrane; the nomenclature is after Jones and others, 1980. In the northeastern part of the belt, near Long Creek (Fig. 2), an ophiolite unit of late Devonian age is conformably overlain by chert-argillite of Mississippian age. An unconformity separates the Mississippian from volcanic sandstone and conglomerate and chert of Permian age. An Early Triassic limestone unit that crops out in fault slivers near Golden Zone Mine and south of Long Creek contains ammonites characteristic of the *Meekoceras gracilitatus* zone (Fig. 4). This limestone is in turn overlain by Upper Triassic clastic red beds.

The other sequence is well exposed in Partin and Shotgun creeks (Fig. 3). The basal unit consists of interlayered limestone and pillow basalt, with individual limestone units up to several tens of meters thick. The unit is poorly fossiliferous, but is Upper Triassic, strikes northeasterly, and is steeply overturned to the west. The overlying red-bed unit contains dominant clastic red argillite, sandstone, and conglomerate and lesser marine fossiliferous sandstone and limestone. It is also Upper Triassic in age. Red beds are, in turn, overlain by the brown sandstone and argillite unit, which, mainly on the basis of ammonites, ranges from Upper Triassic to Early Jurassic.

The Uppermost unit of the Chulitna terrane is an argillite, sandstone, and chert unit of Upper Jurassic to Early Cretaceous age, which is in fault contact, mainly with the limestone and pillow basalt unit.

The ophiolite complex of Late Devonian–Mississippian age consists of serpentinite, gabbro, pillow basalt, and red radiolarian chert of Late Devonian age, dated on radiolaria by Brian K. Holdsworth and on conodonts by Anita G. Harris and B. R. Wardlaw (Jones and others, 1980). The red chert grades upward into pale green chert of Mississippian age.

Although the red beds are generally separated both by time and intervening strata from ophiolite, nevertheless, red beds contain red radiolarian chert clasts identical to that of the ophiolite.

In addition to the key Chulitna terrane, the Upper Chulitna district contains the transported West Fork and Broad Pass terranes and the matrix Eldridge and Susitna flysch (Jones and others, 1980). In the synthesis of Csejtey and others (1982), the Chulitna, West Fork, and Broad Pass terranes are miniterranes caught between the Talkeetna superterrane and the North America continent, especially represented by the Nixon Fork and Yukon-Tanana terranes.

West Fork and Broad Pass terranes. The West Fork and Broad Pass terranes are, respectively, dominantly Jurassic clastic and Paleozoic chert-argillite–rich sequences that lie between the Chulitna terrane and autochthonous Susitna terrane.

IGNEOUS ROCKS AND ORE DEPOSITS

Although there are a few transported syngenetic deposits of

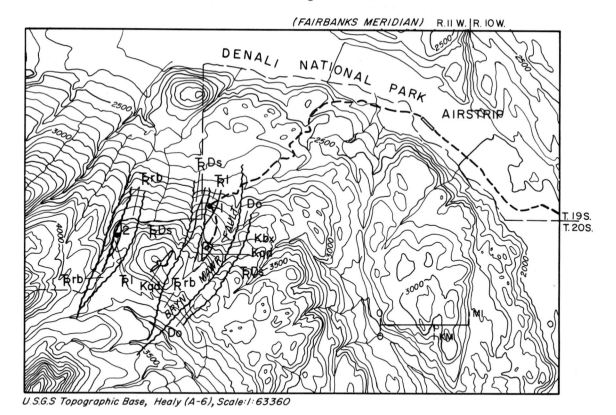

U.S.G.S Topographic Base, Healy (A-6), Scale:1:63360

EXPLANATION

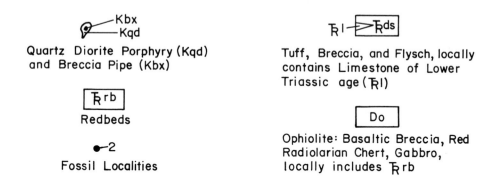

Quartz Diorite Porphyry (Kqd)
and Breccia Pipe (Kbx)

℞rb
Redbeds

●—2
Fossil Localities

Tuff, Breccia, and Flysch, locally
contains Limestone of Lower
Triassic age (℞l)

Do
Ophiolite: Basaltic Breccia, Red
Radiolarian Chert, Gabbro,
locally includes ℞rb

Figure 4. Lower Triassic limestone with the *Meekoceras gracilitatis* Zone is exposed at a switchback in the south branch of the Bull River– Colorado Trail (Locality 1) and above the Golden Zone water supply ditch (NW¼,Sec 4—Locality 2). Ammonites from these two localities are of Smithian age. These locales are the only occurrences of Early Triassic (Smithian) age known along the western border region of North America. The dominant red-bed sequence is entirely bleached and sulfidized from Bryn Mawr to Long Creek. About 0.5 mi (0.8 km) farther south at the end of the trail is a quartz diorite porphyry plug with a central intrusion-related breccia pipe. The Colorado–Bull River Trail is a state right-of-way, but in this area is traverses mining claims. Old buildings near the trail are dangerous and off limits. Permission to visit the breccia pipe should be sought if miners are on the claims. This part of the Colorado–Bull River Trail is presently only accessible by foot or horseback at very low water. An uncontrolled airstrip in S½,Sec.25, suitable for Supercub-type aircraft, gives access to the trail.

chrome and asbestos minerals in the Devonian ophiolite, most of the mineral deposits of the region are associated with hypabyssal to shallow plutonic intrusions, which came in late in the accretionary process. The quartz-diorite host and breccia pipe at Golden Zone have been dated by K-Ar methods, with corrected 1976 IUGS constants, at 69.7–70.2 Ma (Swainbank and others, 1977). The quartz diorite strongly resembles, in chemical composition, units of the Yentna, Hartman, and Mount Estelle intrusives of Reed and Lanphere's (1973) composite Alaska Range batho-

lith. These rocks, with corrected constants, are apparently about 66.0–69.1 Ma.

Empirically, quartz diorites of the Alaska Range correlate with gold- and arsenic-rich intrusive related mineral deposits. Tin-silver deposits of the region, including Coal Creek and Ohio Creek, are probably related to Reed and Lanphere's McKinley-type granite of about 57–60 Ma.

Geologically, the McKinley series rocks appear post-kinematic and were largely intruded into flysch of the Eldridge terrane.

REFERENCES CITED

Capps, S. R., 1919, Mineral resources of the Upper Chulitna region: U.S. Geological Survey Bulletin 692, p. 207–232.

Clark, A. L., Clark, S.H.B., and Hawley, C. C., 1972, Significance of upper Paleozoic oceanic crust in the Upper Chulitna district, west-central Alaska Range: U.S. Geological Survey Professional Paper 800-C, p. 95–101.

Csejtey, B., Jr., Cox, D. P., and Evarts, R. C., 1982, The Cenozoic Denali fault system and the Cretaceous accretionary development of southern Alaska: Journal of Geophysical Research, v. 87, p. 3741–3754.

Hawley, C. C., and Clark, A. L., 1974, Geology and Mineral deposits of the Upper Chulitna district, Alaska: U.S. Geological Survey Professional Paper 758-B, 47 p.

Jones, D. L., and others, 1980, Age and structural significance of ophiolite and adjoining rocks in the Upper Chulitna district, south-central Alaska: U.S. Geological Survey Professional Paper 1121-A, 21 p.

Jones, D. L., Silberling, N. J., Gilbert, W., and Coney, P. J., 1982, Character, distribution, and tectonic significance of accretionary terranes in the central

Alaska Range: Journal of Geophysical Research, v. 87, p. 3709–3717.

Lanphere, M. A., and Reed, B. L., 1985, The McKinley sequence of granitic rocks; A key element in the accretionary history of southern Alaska: Journal of Geophysical Research, v. 90, p. 11413–11430.

Nichols, K. M., and Silberling, N. J., 1979, Early Triassic (Smithian) ammonites of paleoequatorial affinity from the Chulitna terrane, south-central Alaska: U.S. Geological Survey Professional Paper 1121-B, 5 p.

Reed, B. L., and Lanphere, M. A., 1973, Alaska-Aleutian Range, batholith; Geochronology, chemistry, and relation to circum-Pacific plutonism: Geological Society of America Bulletin, v. 84, p. 2583–2610.

Ross, C. P., 1933, Mineral deposits near the West Fork of the Chulitna River, Alaska: U.S. Geological Survey Bulletin 849-E, p. 289–333.

Swainbank, R. C., Smith, T. E., and Turner, D. L., 1977, Geology and K-Ar age of mineralized intrusive rocks from the Chulitna mining district, central Alaska, *in* Short notes on Alaska geology, 1977: State of Alaska Division of Geological and Geophysical Surveys, p. 23–28.

The Cenozoic section at Suntrana, Alaska

Clyde Wahrhaftig, U.S. Geological Survey, 345 Middlefield Road, Menlo Park, California 94025

LOCATION AND ACCESSIBILITY

The Cenozoic section at Suntrana in the Nenana coal field is located at Lat.63°52′N, Long.148°51′W, Healy D-4 Quadrangle, 3 to 3.5 mi (5 to 6 km) east of the Nenana River, on the north bank of Healy Creek and on Suntrana Creek, its 1.2-mi-long (2 km) tributary at this point (Figs. 1 and 2). It is approximately 6.7 mi (10.8 km) east, by automobile road, of mile 248.8 (km 400.4) on the George Parks Highway. Mile 248.8 is 7 mi (11.5 km) north on the highway from the entrance to Denali National Park. Suntrana Creek is narrow and shallow and is easily crossed dryshod most of the year. Its upper 1 mi (1.5 km) is usually dry. Vehicles should not be parked close to the crossing of Suntrana Creek, and one should not be in its canyon whenever storms are imminent, for during heavy rains masses of mud and water as much as 25 ft (8 m) deep have come raging down the creek.

SIGNIFICANCE OF THE LOCALITY

During much of Miocene (and perhaps Oligocene) time, central Alaska had moderate relief, and rivers flowed south across the site of the Alaska Range (which did not then exist) toward the Gulf of Alaska. Large areas of slow subsidence accumulated alluvial, lacustrine, and swamp deposits. Peat and woody vegetation of the swamps compacted into coal beds. Starting in late Miocene time the Alaska Range rose athwart this drainage, damming it and creating one or more lakes that ultimately spilled northwestward to form the present Tanana River system.

The record of this geologic history is contained in the Tertiary rocks of the Nenana coal field on the north flank of the Alaska Range, which form the herein-named Usibelli Group (formerly the informally designated coal-bearing formation or group) and the overlying Nenana Gravel. The five formations of the Usibelli Group, totaling 1,940 ft (590 m) in thickness at Suntrana, record the period of south-flowing drainage. The overlying Nenana Gravel represents an alluvial apron that developed on the north side of the Alaska Range as it rose in late Miocene and Pliocene time.

The Usibelli Group was once almost continuously exposed in magnificent cliffed badlands and cutbanks along Healy and Suntrana creeks at Suntrana (Fig. 4). Coal mining, road building, and channel diversions have destroyed much of the exposure or caused it to become overgrown, but enough remains to make this one of the best places to see the characteristic features of each formation. More than 3,000 ft (1,000 m) of the Nenana Gravel are continuously exposed in the upper basin of Suntrana Creek.

These Tertiary rocks are poorly consolidated, and would normally be quickly reduced to vegetation-covered slopes. Their magnificent badland outcrops are kept free of debris and vegeta-

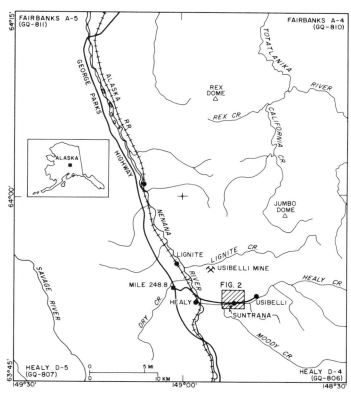

Figure 1. Map of the area of Healy D-4 and D-5 and Fairbanks A-4 and A-5 quadrangles, showing locations of Suntrana, Healy, Usibelli, the George Parks Highway, the Alaska Railroad, Figure 2, places mentioned in text, and numbers of U.S. Geological Survey Geologic quadrangle (GQ) maps covering these quadrangles.

tion by an extremely rapid rate of erosion, amounting to more than 1 cm per year for the drainage basin of Suntrana Creek. The erosion process involves frost-riving in winter and debris flows in early summer.

The coal of the Nenana coal field is a major fuel resource of Alaska and has contributed greatly to the economic development of the state.

Coal was first mined in the Nenana coal field when the Alaska Railroad reached Lignite in 1918. The Suntrana Mine (Fig. 2) operated from 1922 to the mid-1950s and consisted of a crosscut intersecting the coal beds about 30 to 50 ft (10 to 15 m) above river level and haulage tunnels along the coal beds east and west from the crosscut. Coal above the haulage-tunnel level was mined by the room-and-pillar method. At present, all coal mined from the coal field, by the Usibelli Coal Mine, Incorporated, is mined by open-pit methods.

Underground coal fires have been a natural component of the erosion process of the coal field since Tertiary times, because

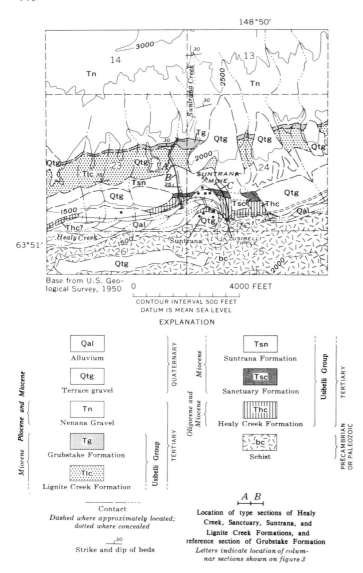

Figure 2. Geologic map of Suntrana and vicinity (part of T.12S.,R.7W., Healy D-4 Quadrangle, Alaska) showing the type section of the Usibelli Group. Modified from Wahrhaftig and others, 1969, Figure 2.

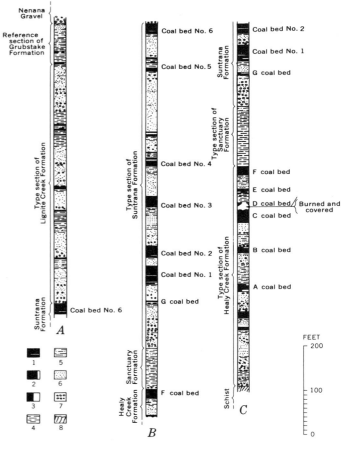

Figure 3. Stratigraphic column of the Usibelli Group at Suntrana. Section A, measured along Suntrana Creek from coal bed No. 6 to the base of the Nenana Gravel; section B, measured along Suntrana Creek from near the railroad bridge to coal bed No. 6; section C, measured on the north bank of Healy Creek from the base of the Usibelli Group to the top of the badland exposure east of the Suntrana Mine. 1, coal (showing bone or clay parting); 2, bony coal; 3, bone (mixture of clay and organic matter); 4, claystone and shale; 5, siltstone; 6, sandstone, in part cross-bedded; 7, pebbles and conglomerate; 8, schist (unconformity at top). Slightly modified from Wahrhaftig and others, 1969, Figure 3.

clinker and baked shale from burning of coal beds occur as pebbles in the Nenana Gravel. Bright yellow, orange, and red outcrops of baked sandstone and shale, as well as brown to purple patches of lavalike clinker, decorate the exposures of the Usibelli Group. Extreme care should be exercised in approaching fumaroles produced by the burning coal, for the ground around them is subject to caving, possibly into abandoned mine workings, and the fumaroles emit noxious gases such as carbon monoxide.

SITE INFORMATION

The geology of the site and its environs is shown in Figure 2 (see also Wahrhaftig, 1970a). A columnar section of the Usibelli

Group as exposed in 1944 on Suntrana Creek and at Suntrana is shown in Figure 3. Figure 4 is an aerial view of the exposures. Strata of the Usibelli Group dip 30° to 35° N and strike due E to N75°E.

Usibelli Group. The rocks informally termed the coal-bearing group by Wahrhaftig and others, 1969, are here formally named the Usibelli Group. The Usibelli Group is named for the settlement of Usibelli, located on Healy Creek approximately 2 mi (3 km) east of the group's type section, shown on the U.S. Geological Survey topographic map of the Healy D-4 Quadrangle, 1950 edition, revised 1976 (see also Orth, 1967, p. 1014).

The type section of the Usibelli Group is defined as consisting of the exposures on the north bank of Healy Creek and along

Figure 4. Aerial view of Suntrana and Suntrana Creek from the southwest. Base of the Usibelli Group on the lower right. Suntrana Creek on the left. Exposures of the Nenana Gravel occur in the upper left, north of the forks of Suntrana Creek. Photograph by Bradford Washburn, 12:50 pm, 18 September, 1938.

the lower part of Suntrana Creek, as shown in Figures 2, 3, and 4. It consists of the type sections of the Healy Creek, Sanctuary, Suntrana, and Lignite Creek formations, and the reference section of the Grubstake Formation at this locality. Lithologic descriptions of each of the component formations at their respective type or reference sections or type localities are given in Figure 3; they are summarized in the following paragraphs and are given in more detail in Wahrhaftig and others, 1969. Lithologic variations in the formations away from their respective type or reference sections or type localities are given in the same reference.

The Usibelli Group is here considered to comprise all the coal-bearing Tertiary rocks on the north flank of the Alaska Range between the Toklat River on the west and Jarvis Creek on the east that have been variously assigned to these formations, to

the coal-bearing group, or, before 1969, to the coal-bearing formation of Tertiary age (e.g., Wahrhaftig and Hickcox, 1955).

Healy Creek Formation. The Healy Creek Formation is the lowest formation that is here assigned to the Usibelli Group. It is exposed along the northeast wall of the canyon of Healy Creek at Suntrana, from its basal contact with the unconformably underlying quartz-mica schists and associated metamorphic rocks formerly assigned to the now-abandoned Birch Creek Schist, about 750 ft (230 m) south of the road to Usibelli, northward for about 1,000 ft (300 m) to the top of the F coal bed (the coal bed immediately beneath a thick brown-weathering shale). It consists at this locality of 500 ft (150 m) of interbedded poorly sorted and poorly consolidated sandstone, conglomerate, claystone, and subbituminous coal. Beds in this formation tend to be lenticular, and the sandstone and conglomerate have much interstitial silt and clay. According to Triplehorn (1976), the clay fraction is 50 to 75 percent kaolinite-chlorite, 20 to 45 percent illite, and less than 15 percent montmorillonite. The sandstone and conglomerate characteristically contain abundant materials derived from nearby basement rocks. Quartz, gray quartzite, and black to gray chert predominate, and the sandstone tends to be micaceous, all suggesting a provenance including the underlying metamorphic rocks. The formation thickens abruptly and irregularly, suggesting that it covers an irregular surface eroded on the metamorphic rocks. The basal gravel in this section is interpreted by Buffler and Triplehorn (1976) to be a gravelly braided-stream deposit and the remaining clastic strata to be point-bar deposits. The exposures of the Healy Creek Formation at Suntrana were previously assigned by Wolfe and Leopold (in Wahrhaftig and others, 1969) to the early Miocene, but have been more recently reassigned by Wolfe and Tanai (1980) to the early and middle Miocene (lower part of the Seldovian Stage). However, in the Rex Creek area (Fig. 1), beds assigned to the Healy Creek Formation are as old as late Oligocene (Wolfe and Tanai, 1980).

Sanctuary Formation. The Sanctuary Formation is here assigned to the Usibelli Group; it conformably overlies the Healy Creek Formation and consists of 130 ft (40 m) of gray shale that weathers to a characteristic chocolate brown or yellowish brown. It is exposed as a brown band on the cliff east of the old Suntrana Mine adit, and also crops out beside the road at the crossing of Suntrana Creek. It has alternating pale-weathering and dark-weathering laminae a fraction of a cm to 3 cm thick. Triplehorn (1976) reports the shale to consist 50 percent of kaolinite-chlorite, about 35 percent of illite, and about 15 percent of montmorillonite. The Sanctuary apparently accumulated in a large shallow lake. It was previously assigned by Wolfe and Leopold (in Wahrhaftig and others, 1969) to the early and(or) middle Miocene and more recently reassigned by Wolfe and Tanai (1980) to the middle Miocene.

Suntrana Formation. The Suntrana Formation, which is here assigned to the Usibelli Group, rests conformably on the Sanctuary Formation and forms the upper part of the bluff north and northeast of the old Suntrana Mine adit and the canyon walls of the downstream 1,150 ft (350 m) of the canyon of Suntrana Creek. It consists here of eight fining-upward cycles of sandstone and pebbly sandstone, overlain by claystone, overlain by one or more coal beds 1.5 to 40 ft (0.5 to 12 m) thick. The eight cycles total 720 ft (220 m) in thickness. The Suntrana Formation contains the major coal resources of the Nenana coal field.

The sandstone of the Suntrana Formation is typically clean, well sorted, cross-bedded, and consists predominantly of quartz. Cross-bedding directions are variable, but suggest current directions toward the south or southwest. Pebbles in the sandstone consist predominantly of quartz, schist, chert, and quartzite, and suggest a thoroughly weathered provenance region. Clay below the No. 1 coal bed (Fig. 3) is nearly pure kaolinite; above that bed, it is 35 to 60 percent montmorillonite, 25 to 50 percent kaolinite-chlorite, and about 15 percent illite (Triplehorn, 1976).

The conglomeratic basal 80 ft (25 m) of the formation at Suntrana is interpreted by Buffler and Triplehorn (1976) as a gravelly braided-stream deposit and the remaining clastic strata as sandy braided-stream deposits. The coal beds formerly exposed in these cliffs have either, in part, been mined out or consumed by fire.

Wolfe and Leopold (*in* Wahrhaftig and others, 1969) and Wolfe and Tanai (1980) assign the Suntrana Formation to the middle Miocene.

Lignite Creek Formation. The overlying Lignite Creek Formation, which is here assigned to the Usibelli Group, is 540 ft (165 m) thick and is exposed in the walls of Suntrana Creek Canyon, upstream from the exposures of the Suntrana Formation. It consists of five repeated sequences of sandstone, pebbly at the base, overlain by claystone and siltstone, with 1 to 3 coal beds between 1.5 and 6 ft (0.5 and 2 m) thick at the top. The sandstone is clean and cross-bedded as in the Suntrana Formation, and cross-bedding indicates consistent current directions toward the south. However, unlike the Suntrana Formation, the sandstone in the Lignite Creek Formation contains numerous varicolored mineral grains, including 15 to 25 percent feldspar, and it has an overall buff color. The pebbles consist of a variety of rock types, including volcanic and plutonic rocks. Montmorillonite predominates in the clay, suggesting a component of volcanic ash (Triplehorn, 1976). Buffler and Triplehorn (1976) regard the clastic strata as predominantly point-bar deposits. The thin coal beds in the Lignite Creek (as well as the topmost meter or so of the No. 5 and No. 6 coal beds of the Suntrana Formation) appear to consist of mats of roots, stems, and branches, unlike the blocky-fracturing, thoroughly macerated coals lower in the coal-bearing section. Wolfe and Leopold (in Wahrhaftig and others, 1969) previously assigned the Lignite Creek Formation to the middle Miocene (late Seldovian Stage), whereas Wolfe and Tanai (1980) have more recently reassigned the Lignite Creek to the late middle to early late Miocene Homerian Stage.

Grubstake Formation. The Grubstake Formation is the highest formation here assigned to the Usibelli Group; it is exposed at the forks of Suntrana Creek, where it rests conformably on the Lignite Creek Formation. The lowermost 30 ft (9 m) consist largely of thinly laminated shale with a thin coal bed near the base, the middle 33 ft (10 m) consist of alternating micaceous siltstone and clay shale layers, and the upper 33 ft (10 m) consist largely of siltstone with thin beds of fine sandstone. A hard layer about 30 ft (9 m) above the base of the formation contains fine white markings perpendicular to the bedding and abundant shards of rhyolitic glass.

The type locality of the Grubstake Formation is located on Tatlanika Creek near the mouth of Grubstake Creek in Fairbanks A-3 Quadrangle, about 25 mi (40 km) northeast of the Suntrana section. At its type locality, the Grubstake is 1,000 to 1,500 ft (300 to 460 m) thick and consists of dark silty claystone interbedded with dark sandstone containing clasts apparently derived from the Cantwell Formation (see below). In its lower part are two thick beds of white vitric ash whose glass has an index of refraction identical to that in the bed 30 ft (9 m) above the base of the

Grubstake on Suntrana Creek. A radiometric K-Ar age on glass from the Tatlanika Creek site, by M. A. Lanphere (in Wahrhaftig and others, 1969) is 8.3 ± 0.4 Ma, using the new decay constants, which is considered late Miocene in age. The Suntrana Creek locality has not been independently dated. Plant megafossils buried by the ashfall are assigned a late Homerian (late Miocene) age (Wolfe and Tanai, 1980).

The laminated character and fine grain-size of the Grubstake Formation suggests an origin in a broad shallow lake. The apparent southern provenance of the interbedded sandstone, together with its lithologic similarity to interstitial sand and sandstone lenses in the overlying Nenana Gravel, suggest a source to the south in the Alaska Range; therefore the Grubstake Formation represents the blocking of southward drainage by the beginning of growth of the Alaska Range.

Nenana Gravel. Approximately 3,400 ft (1,040 m) of the Nenana Gravel are exposed in the upper basin of Suntrana Creek, dipping 30° to the north and resting with apparent conformity on a 10-ft-thick (3 m) sandstone bed at the top of the Grubstake. An additional 600 ft (180 m) of stratigraphically higher beds of the Nenana Gravel lie north of the basin of Suntrana Creek.

The Nenana Gravel on Suntrana Creek is massive to thick-bedded, poorly consolidated, generally well-sorted conglomerate. Near the base, the average clast size is 1 to 3 cm and the maximum clast size is 7 to 12 cm; this increases to an average clast size of 6 to 8 cm and a maximum clast size of 30 cm at the top of the formation. Interstitial material is coarse to very coarse dark sand, apparently derived largely from the Cantwell Formation and other dark rocks in the Alaska Range to the south. This sand also occurs throughout the Nenana Gravel as lenses 3 to 10 ft (1 to 3 m) thick, 50 to 100 ft (15 to 30 m) long, and spaced 30 to 50 ft (10 to 15 m) apart. Pebble imbrication and cross-bedding indicate deposition by north-flowing streams (Wahrhaftig, 1958, and unpublished data; Wahrhaftig and others, 1951).

The most prominent lithologies in the larger clasts of the Nenana Gravel on Suntrana Creek are dark sandstone and conglomerate derived from the Cantwell Formation; these make up 30 to 50 percent of clasts in the lower 2,000 ft (600 m) of the formation and 15 to 30 percent in the upper part. Light blue dacite of unknown provenance with prominent phenocrysts of quartz and feldspar makes up 5 to 10 percent of the lower 900 ft (275 m) and 1 to 5 percent of the overlying 800 ft (240 m), but it is absent above. Cobbles and boulders of dark green, medium- to coarse-grained, saussuritized ophitic diorite and gabbro, derived mainly from large pre-Jurassic sills in the Alaska Range between the Hines Creek and McKinley strands of the Denali fault system, make their first appearance about 1,600 ft (485 m) above the base, and constitute 30 percent or more of all clasts over 2,400 ft (730 m) above the base. Other abundant clast lithologies include quartz, chert, and quartzite, generally as pebbles, and granitic rocks and rhyolite.

The lowermost 2,600 ft (800 m) of the Nenana Gravel is brown to tan; above its lowermost 2,600 ft (800 m) the Nenana is generally reddish brown and contains several thin beds of purplish clay that are commonly overlain by a thin layer of fine-grained white gravel, interpreted to be reworked material from the Usibelli Group. Pebbles and sand grains throughout the Nenana Gravel are commonly coated with a thin layer of iron oxide, and the clasts are generally slightly weathered. Weathering that occurred shortly after deposition and before burial by overlying beds within the Nenana Gravel has given sandstone and conglomerate cobbles weathering rinds as much as 2.5 cm thick and has reduced some granite and rhyolite clasts to grus (Wahrhaftig, 1958).

The Nenana Gravel is composed of the detritus shed north from the Alaska Range during the orogeny that brought that range into existence (Wahrhaftig, 1970b). It was deposited as a series of coalescing alluvial fans along the site of the present-day northern foothills of the Alaska Range into a basin whose floor subsided as much as 3,000 ft (900 m), by streams that coincided approximately with the present-day Nenana, Wood, and Teklanika rivers. The orogeny began in the southern part of the range and spread northward, ultimately involving the Nenana Gravel. The depth of erosion in the source areas was roughly comparable to thickness of the accumulation in the basin of deposition. The uppermost part of the Nenana Gravel was derived partly from young basement uplifts within the area of deposition.

The Nenana Gravel has been deformed into irregular box folds with amplitudes of 2,000 to 6,000 ft (600 to 1800 m), wavelengths of 5 to 10 mi (8 to 16 km), and steep and locally faulted limbs (Wahrhaftig, 1970b). The orogeny appears to have begun between 8 and 10 Ma, and to have ended well before the emplacement of Jumbo Dome (Fig. 1) at 2.8 Ma (Lanphere, in Wahrhaftig, 1970a, corrected for new decay constants).

Erosion of the outcrops. The poorly consolidated Tertiary rocks are typically exposed in intricately dissected badland gullies. Badlands total 3 to 6 percent of the area in the basins of Healy and Lignite creeks underlain by Tertiary rocks, and individual badlands cover from a few square meters to as much as 0.4 mi^2 (1.0 km^2). Many badlands are being eroded at an average rate exceeding 1 cm per year, based on measurements of volume and frequency of debris flows and on repeated photography over periods of 10 to 70 years (Wahrhaftig and Birman, 1954).

The badlands are initiated (1) as meander scars on the banks of major streams, such as Healy and Lignite creeks, (2) as the headwalls of landslides, and (3) as nickpoints of rejuvenation in minor tributaries.

Badland erosion has an annual cycle. It begins with ravelling of coarse clasts and detachment of slabs of sandstone, due to accumulation of ice veins behind the clasts and slabs during freezing spells, and thawing during warm sunny days in late fall and winter. The loosened clasts tumble to the base of the steep badland slopes, knocking others loose on their way, and accumulate in a single winter as talus cones several meters thick and high, completely covering the floors of minor badland gullies (Fig. 5a). Early summer cloudbursts mobilize the winter's talus accumulation, producing catastrophic debris flows that choke minor tribu-

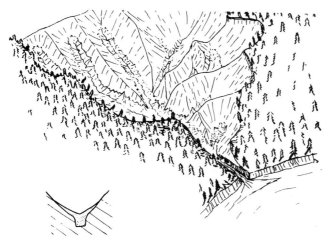

a. May: Talus from winter frost-riving fills gully bottoms.

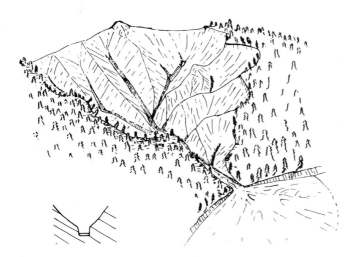

b. July: Debris flows flush gullies and build alluvial fans.

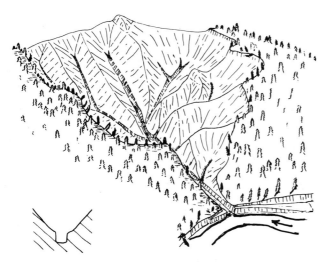

o. October: Major streams remove alluvial fans.

tary channels and are deposited as alluvial fans across trunk streams such as Healy Creek (Fig. 5b). Finally, the tributary channels are cleared and alluvial-fan debris is removed during high-water stages of trunk streams, typically during cyclonic storms in late summer and early autumn (Figure 5c). Records kept by the Alaska Railroad of track repair at the bridge across the mouth of Suntrana Creek, showed that for the 23 years preceding 1948, a mudflow buried the tracks an average of once every two years.

Actively eroding badland slopes, which are steeper than the angle of repose, tend to migrate upslope parallel to themselves. Badlands that start as meander scars migrate up the face of the riverbank, leaving a talus-mantled slope at the angle of repose below them, and disappear at the top of the bank. Badlands, such as that on Suntrana Creek, that are an intricate maze of gullies, reach stability and are revegetated when the gully floors grow so wide that debris flows cannot remove all the talus. The closely spaced, straight-sided, sharp-crested ridges on the north side of Healy Creek appear to have originated through badland erosion.

REFERENCES CITED

Buffler, R. T., and Triplehorn, D. M., 1976, Depositional environments of the Tertiary coal-bearing group, central Alaska, *in* Miller, T. P., ed., Recent and ancient sedimentary environments in Alaska, Proceedings of Symposium, April 2–4, Anchorage: Anchorage, Alaska Geological Society, p. H1–H10.

Orth, D. J., 1967, Dictionary of Alaska place names: U.S. Geological Survey Professional Paper 567, 1084 p.

Triplehorn, D. M., 1976, Clay mineralogy and petrology of the coal-bearing group near Healy, Alaska: Alaska Division of Geological and Geophysical Surveys Geologic Report 52, 14 p.

Wahrhaftig, C., 1958, Quaternary geology of the Nenana River valley and adjacent parts of the Alaska Range: U.S. Geological Survey Professional Paper 293, p. 1–68.

——— , 1970a, Geologic map of the Healy D-4 Quadrangle, Alaska: U.S. Geological Survey Map GQ-806, scale 1:63,360.

——— , 1970b, Late Cenozoic orogeny in the Alaska Range: Geological Society of America Abstracts with Programs, v. 2, no. 7, p. 713–714.

Wahrhaftig, C., and Birman, J. H., 1954, Stripping coal deposits on Lower Lignite Creek, Nenana Coal Field, Alaska: U.S. Geological Survey Circular 310, 11 p.

Wahrhaftig, C., and Hickcox, C. A., 1955, Geology and coal deposits, Jarvis Creek Coal Field, Alaska: U.S. Geological Survey Bulletin 989-G, p. 352–367.

Wahrhaftig, C., Hickcox, C. A., and Freedman, J., 1951, Coal deposits on Healy and Lignite creeks, Nenana Coal Field, Alaska, with a section on clay deposits on Healy Creek by E. H. Cobb: U.S. Geological Survey Bulletin 963-E, p. 141–168.

Wahrhaftig, C., Wolfe, J. A., Leopold, E. B., and Lanphere, M. A., 1969, The coal-bearing group in the Nenana Coal Field, Alaska: U.S. Geological Survey Bulletin 1274-D, p. D1–D30.

Wolfe, J. A., and Tanai, T., 1980, The Miocene Seldovia Point flora from the Kenai Group, Alaska: U.S. Geological Survey Professional Paper 1105, 52 p.

Figure 5. Three stages in the annual cycle of erosion of a typical badlands area in the Nenana coal field.

Delta River area, Alaska Range, Alaska

Troy L. Péwé, Department of Geology, Arizona State University, Tempe, Arizona 85287

LOCATION

The Alaska Range is a glacially sculptured, arcuate mountain wall extending west and southwest 600 mi (1,000 km) in central Alaska from the Canadian border to the Aleutian Range. It is extremely rugged and includes Mt. McKinley (20,443 ft; 6,195 m elevation), the highest mountain on the North American continent. The Delta River, 75 mi (125 km) long, originates on the south side of the range in the Tangle Lakes and flows northward through the range to the Tanana River. The Richardson Highway (Fig. 1) crosses the range through the Delta River valley, which is 0.6 to 12 mi (1 to 20 km) wide and 60 mi (100 km) long. Because of the lack of villages, cities, and towns, references to geographical points in the valley have been referenced to mile posts on the Richardson Highway since the early part of this century. This practice will be continued in this report. Mile 0 on the highway is at Valdez, a town on the Gulf of Alaska and the terminus of the Trans-Alaska Pipeline System; the highway ends at mile Fairbanks 363.8.

SIGNIFICANCE

The Delta River valley is one of the most striking glaciated valleys in Alaska; it is easily accessible and exhibits a great variety of classic geologic features. This U-shaped valley was occupied by glacial ice as recently as 10,000 years ago. It displays moraines and ice-contact features in the southern area; ice-carved valley walls, roches moutonnées, and other glacial features in the central part; and massive terminal moraines in the north. Glacier-sculptured features, such as cirques, aretes, horns, and hanging valleys are exhibited in the central part of the range. Today, four 4–30-mi (6–50-km) long valley glaciers extend almost to the Richardson Highway and are accessible on foot. One, Black Rapids Glacier, surged rapidly in 1937 and may surge again soon. At least one active rock glacier is also accessible from the highway. Loess is actively being deposited in this valley and has been for the past 7,000 years.

The valley and its deposits clearly exhibit evidence of past and present permafrost and its effect on man. Large-scale patterned ground (ice-wedge-cast polygons) and ice-wedge casts of late Wisconsinan age are well displayed in outwash deposits at the northern part of the valley near Donnelly Dome. They indicate that permafrost was formerly more extensive. The entire valley lies in the discontinuous zone of the perennially frozen ground of Alaska; type and distribution of vegetation reflect the underlying frozen ground. The 480-mi- (800-km-) long Trans-Alaska Pipeline System extends through the Delta River valley and clearly exhibits various construction modes needed to accommodate problems due to permafrost and other geologic hazards. The longest active crustal break in Alaska, the Denali

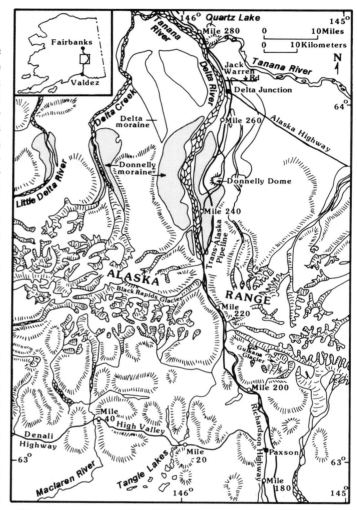

Figure 1. Index map of the Delta River area, Alaska Range, central Alaska.

fault, crosses the valley, highway, and pipeline, and is topographically well expressed; its crossing required highly original pipeline construction to accommodate future movements along the fault.

For the past 35 years the Delta River Valley has been used as a showcase for geological features for national and international field trips and for student trips from Alaskan and other universities. The reader is especially referred to the latest guidebook by Péwé and Reger (1983) published for the Fourth International Conference on Permafrost by the Alaskan Division of Geological and Geophysical Surveys, and the colored two-sheet geological map of the Mt. Hayes D-4 Quadrangle (Péwé and Holmes, 1964) published by the U.S. Geological Survey (MGI-394).

Figure 2. Oblique aerial view (to the southwest) of the outwash plains and moraines of Donnelly age near Donnelly Dome; the Delta River and Alaska Range are in the distance. Photograph 210RT-55RT-M864-55SRW-MM58: U.S. Air Force, August 29, 1949.

Figure 3. Oblique aerial view (to the north) of clouds of silt transported by wind from the Delta River flood plain near Donnelly Dome, central Alaska (from Péwé, 1951a, pl. 1-A).

QUATERNARY GLACIATIONS

At least four Quaternary glaciations, each apparently less extensive than the previous, are recorded in the Delta River area (Péwé and others, 1953; Péwé and Holmes, 1964). Glaciers pushed south and north from the crest of the range, and some of the ice on the south side exited north through the Delta River valley. On the north side, the glaciers largely remained in mountain valleys, spreading terminal bulbs onto the lowland of the Tanana River valley. On the south side, glaciers coalesced to form large piedmont ice sheets that covered the lowlands and pushed south into Copper River Basin (Péwé, 1965).

The earliest recognized glacial advance is the Darling Creek Glaciation of early Quaternary age. It is identified from patches of drift 2,000 to 3,000 ft (610 to 920 m) above the floor of the Delta River valley and from isolated erratics up to 15 ft (4.6 m) in diameter in the mountains on the south side of the Alaska Range. The succeeding glacial advance was the Delta Glaciation, recognized by fairly well-preserved breached moraines on the north side of the range (Fig. 1).

The next two glacial advances occurred in late Quaternary (Wisconsin) time and are closely related in extent and age. On the north side of the Alaska Range, they have been grouped into one broad glaciation, termed the Donnelly Glaciation (Fig. 1); on the south side they have been named the Denali I and Denali II Glaciations. On both sides of the range, these deposits are characterized by fresh knob-and-kettle topography.

At Mile 252.7, the subdued topography of the moraine is

well exposed and the till is well exhibited in the road cut. The fresh hummocky topography of the Donnelly end moraine can be seen at Mile 245.2 (Fig. 2). The fresh till here is much less weathered than the Delta till, contains more schist clasts, and only about 5% of the granitic clasts are significantly weathered. The weathered zone of the Delta till is 5 to 6 ft (1.5 to 2 m) thick, but that of the Donnelly till is 1.5 to 2 ft (0.5 to 0.7 m) thick. Boulder counts are higher on Donnelly moraine, and clast surfaces are less affected by granular disintegration than on Delta moraines (Péwé and Reger, 1983, Fig. 34).

The age of the Delta Glaciation remains unsettled. Ten Brink and Waythomas (1982) believed it to be early Wisconsinan. However, because soil profiles are much deeper on Delta moraines than on Donnelly moraines, because many Delta moraines are considerably modified by slope processes, and because of significant differences between physiographic and sedimentologic parameters of Delta- and Donnelly-age moraines in the type area (see Péwé and Reger, 1983, p. 59, for details), Péwé believes the Delta Glaciation is pre-Wisconsin in age.

EOLIAN PHENOMENENA

The lower part of the Delta River valley has a high frequency of high winds compared to other places in interior Alaska. It is one of the best places to see great clouds of dust being blown from the mile-wide, braided, vegetation-free Delta River

Figure 4. Oblique aerial view of large-scale patterned ground reflecting ice wedge casts in outwash gravel in the Donnelly Dome area, Alaska. Vegetation cover is mixed evergreen-deciduous scrub and shrub. (Photograph 2021 by T. L. Péwé, July 14, 1961.)

floodplain, and to observe the deposition of modern loess (Fig. 3) (Péwé, 1951a). Clouds of silt are blown as high as 7,000 ft (1,200 m) and over hundreds of square kilometers. Holocene loess has been deposited on the bluffs adjacent to the Delta River at least for the past 7,000 years at a rate of about 0.2 to 2 mm per year, based on radiocarbon dates of a series of superimposed white spruce stumps enclosed in the loess, if little or no erosion on the almost flat, heavily vegetated areas during the measured interval is assumed (Péwé, 1968).

The sand blown by strong winds from the unvegetated Tanana River and Delta River floodplains and outwash fans in Pleistocene time cut, grooved, and polished boulders and cobbles to form ventifacts along the southern flank of the Delta moraine (Mile 251.3) and on the north side of the Tanana River (Fig. 1; see Péwé and Holmes, 1964, and Péwé and Reger, 1983, for location of easily accessible ventifacts).

POLYGONAL GROUND AND ICE WEDGE CASTS

Large-scale polygons on the surface of the outwash plain of the Donnelly Glaciation (Fig. 2) are outlined by a network of intersecting, trenchlike depressions 1 to 3 ft (0.3 to 0.9 m) deep and 3 to 6 ft (0.9 to 1.8 m) wide on both sides of the highway at Mile 251.2 (Fig. 4). The difference in vegetation types between the polygon centers and troughs delineates the polygonal pattern. The polygons are 80 to 130 ft (24 to 40 m) in diameter and are three- to six-sided. Wedge-shaped masses of fine-grained sediments underlie the slight surface depressions that mark polygon boundaries and cross cut the massive outwash gravel.

Polygons do not form in the seasonally frozen ground in the Donnelly Dome area today. Apparently, when contraction cracks formed, the climate was colder. With greater cooling of the ground, permafrost was more widespread, and conventional ice wedges formed by contraction cracking of former permafrost. These wedges formed during the Donnelly glaciation and subsequently melted. The resulting voids were filled with sediment from the adjacent gravel and overlying loess.

The climate in the Donnelly Dome area at the time the large-scale ice-wedge polygons formed was colder and more rigorous than today. Tree line was 1,500 to 1,800 ft (460 to 550 m) lower in the adjacent Yukon-Tanana Upland, and snowline was 1,500 ft (460 m) lower, based on a study of cirque floors. The mean annual air temperature was at least 3°C colder, about –5.8°C, in contrast to the modern mean annual air temperature of

Figure 5. Aerial view (to the northeast) of the Delta River and terminus of the Black Rapids Glacier. Pre-1937 moraines can be seen near the river. Photograph 627 RT-55RT-M864-55SRW-9M58: U.S. Air Force, August 29, 1949.

−2.8°C. (Refer to Péwé and others [1969] for a discussion of the origin and paleoclimatic significance of large polygonal ground in the Donnelly Dome area.)

MODERN VALLEY GLACIERS

The Alaska Range has spectacular valley glaciers 0.5 to 40 mi (1 to 65 km) long (Fig. 1). They are largest and most numerous on the south side of the range, where they are nourished chiefly by air masses moving north to northeast from the northern Pacific Ocean (Péwé, 1975). On the south-center side of the range, modern snowline is about 5,500 ft (1,670 m) elevation; on the north side, it is 6,530 ft (1,980 m) elevation (Péwé and Reger, 1972).

The termini of Black Rapids, Castner, Canwell, and Gulkana glaciers are within a few meters to a few kilometers from the Richardson Highway and accessible by foot. (The Delta River must be crossed by boat, footbridge, or cable.) All four have well-developed moraines of late Holocene advances that have been studied, and since 1960 Gulkana Glacier has been the site of detailed glaciology studies by scientists of the University of Alaska and the U.S. Geological Survey.

Black Rapids Glacier gained worldwide attention in 1937 when it advanced at spectacular rates—up to 200 ft (61 m) per day (Geist and Péwé, 1957; Hance, 1937). It did not cross the Delta River; however, earlier Holocene advances did cross the river (Péwé, 1951b; Fig. 5). (U.S. Geological Survey topographic

maps [1/63,360] Mt. Hayes B-4 and C-4 clearly illustrate geologic features in the middle Delta River valley.)

At Mile 226.6 near the mouth of Falls Creek, the Richardson Highway cuts through bouldery till of the oldest recognized Holocene advance of Black Rapids Glacier. Two thin, discontinuous layers of turf on one block were dated at 3,120 ±120 Ma (I-12, 109) and 4,350 ±140 Ma (I-12, 108). This indicates that the earliest Holocene advance of Black Rapids Glacier occurred sometime after 3,100 Ma.

Two radiocarbon dates of felted peat and a branch recovered from between thin distal gravel of the Falls Creek fan and the underlying oldest Holocene till are dated at less than 190 Ma (I-12, 110) and less than 230 Ma (I-12, 111), respectively. They provide a minimum age for the oldest Holocene advance.

Attempts to date this moraine and the other prehistoric Holocene moraines of Black Rapids, Castner, and Canwell glaciers first utilized dendrochronology and radiocarbon dating. Péwé's dendrochronologic studies from 1951 through 1957 convinced him that the multiple Holocene moraines of Black Rapids, Castner, and Canwell glaciers probably were built by concurrent advances during the past 400 years. His observations of spruce colonizing the 1937 terminal moraine of Black Rapids Glacier demonstrate that near-tree-line conditions, including cool summer temperatures, severe winds, and shifting substrates due to melting ice masses, delayed tree growth on moraines in this part of the central Alaska Range for at least 15 to 20 years following construction of the moraines. This delay factor must be added to ring counts of trees to estimate the ages of Holocene moraines (Péwé and Reger, 1983, Table 4).

The oldest Holocene terminal moraine of Black Rapids Glacier is compound and adjoins a forest of spruce that is significantly older than the trees growing on the moraine. West of the Delta River, a dense forest with thick turf and many older fallen and decayed trees (estimated to be 350 to 500 years old) is just beyond the limit of Holocene glaciation. The oldest solid-center tree sampled in this forest in 1951 was 228 years old. A similar forest exists beyond the oldest Holocene moraine east of the Delta River and south of Gunnysack Creek. In contrast, trees estimated to be as old as 133 years—and older trees with rotten centers— were growing on the oldest Holocene moraine of Black Rapids Glacier in 1951. Thus, spruce and poplar trees were apparently present in this part of the Delta River valley before the earliest Holocene advance of Black Rapids Glacier. On the basis of tree-ring evidence and the development of an incipient soil on this moraine (which indicates it was built more than 200 years ago), Péwé estimates the age of the oldest Holocene moraine of Black Rapids Glacier at 330 Ma and correlates it with similar moraines at nearby Castner and Canwell glaciers.

An arcuate terminal moraine 1 mi (1.6 km) in front of the 1937 terminal moraine appears fresh, has no turf cover, and locally contains a small ice core. In 1951, 80-yr-old trees in a first-generation forest on this moraine were cored. According to Péwé, the moraine formed about 130 Ma. Mendenhall (1900) visited the terminus of Black Rapids Glacier in 1898 and stated that the glacier had evidently advanced a few years earlier, leaving a fresh-looking moraine.

To support the 400-Ma age for the earliest Holocene advance of the valley glaciers entering the middle Delta River valley (Fig. 1), Péwé collected a spruce log in 1953 in the perennially frozen till of the oldest Holocene moraine of Canwell Glacier. It was dated at less than 200 Ma by the radiocarbon method (W-268); however, Meyer Rubin (1964, written communication, 1964) said the log could be older than 200 Ma. Another piece of the log was dated in 1985 and was determined to be 310 ±75 years old (GX-10982).

The terminal moraine of the younger Holocene advance of Canwell Glacier is well preserved near the highway. In 1951, the oldest tree encountered on this moraine was 102 years old (Péwé, 1957). A log collected by Péwé in 1953 in the till of this moraine was dated in 1985 at 200 ±75 yr B.P. (GX-10093).

An attempt to date the Holocene advances of these four glaciers was made by Reger in 1963 (Reger, 1964, 1968; Reger and Péwé, 1969; Péwé and Reger, 1983, Fig. 52) using lichenometry; i.e., thalli diameters of *Rhizocarbon geographicum.* If the lichens studied are first generation and if published lichen-growth curves for the St. Elias Mountains (Denton and Karlén, 1973, 1977) are used, the oldest Holocene advances of Black Rapids, Castner, and Canwell glaciers should be 1,265 to 1,450 years old (Péwé and Reger, 1983) and the younger about 390 years old. Early advances of Gulkana Glacier may be as old as 5,700 years. At the present time, the dating controversy has not been resolved, but it is possible that the lichen used in dating may not have been first generation specimens.

A wonderful view of Gulkana Glacier is available from Mile 197.7 where a gravel road branches off and extends to within 2 mi (3 km) of the glacier. The meterology, mass budget, surface flow, subglacial topography, and ice strucure of Gulkana Glacier have been studied in detail from 1960 to 1966 (Moores, 1962; Sellmann, 1962; Mayo and Péwé, 1963; Reger, 1968; Rutter, 1965; Ostenso and others, 1965; Péwé, 1965; Meier and others, 1971; Péwé and Reger, 1983). The glacier has been retreating rapidly since about 1900.

TRANS-ALASKA PIPELINE SYSTEM

The Trans-Alaska Pipeline System (TAPS) was the largest privately funded construction effort in history. It was built between 1974 and 1977 and transports about 1.2 million barrels of crude oil daily to Valdez from arctic Alaska. About 386 mi (643 km) of the 771-mi- (1,285-km)-long line extends south from Fairbanks (Fig. 1) and traverses the entire 60 mi (100 km) of the Delta River valley.

TAPS cost about $8 billion to build. Probably about $1 billion of that amount was necessary to learn about, combat, accommodate, and otherwise work with the perennially and seasonally frozen ground.

The 4 ft (1.2-m-) diameter pipeline was designed for burial in permafrost along most of the route, and the oil temperature (at

Figure 6. Oblique aerial view (to the northeast) of the Richardson Highway, the Trans-Alaska Pipeline System route, location of the Denali Fault, and the termini of Canwell and Castner glaciers in the central Alaska Range. Dashed line around glacial moraines indicates terminus of younger Holocene advances Photograph 620RI-55RT-M864-55SRW-9M58: U.S. Air Force, August 29, 1949.

full production) was estimated at 70–80°C. Such an installation could have thawed the surrounding permafrost. Thawing of the widespread ice-rich permafrost by a buried warm-oil line could cause liquefaction, loss of bearing strength, and soil flow. Differential settlement of the line could occur, and thawed mud could flow down slopes. The greatest differential settlement could occur in areas of ice wedges, where troughs could form and deflect surface water into the trenches, causing erosion and more thawing. About half the pipeline (369 mi; 615 km) is built above ground because of the presence of ice-rich permafrost.

The Denali fault is one of the longest crustal breaks in Alaska. This right-lateral strike-slip fault is topographically expressed as an arcuate trough that can be traced without interruption from the southwestern Alaska Range through the crest of the Alaska Range into the Shakwak Trench, Yukon Territory, Canada, and perhaps into southeastern Alaska. Some of the largest glaciers in the Alaska Range, many of which surge, occupy segments of the Denali fault-line valley, including Chistochina, Gakona, Canwell, Black Rapids, Susitna, and Muldrow glaciers. Geologic evidence indicates that average rates of displacement along the Denali fault, measured over 10,000–65,000,000-year periods, vary from 0.1 to 3.5 cm per year. Offsets of glacial deposits of Donnelly age, Holocene alluvial fans, and at least 20 drainages along the McKinley strand west of the Richardson Highway demonstrate that 15 to 200 ft (5 to 60 m) of right-lateral movement and 20 to 30 ft (6 to 10 m) of vertical movement have occurred during the past 10,000 years (Stout and others, 1973). Although numerous microseisims and a few small earthquakes have been detected along the eastern Denali fault system, historic offset is not well documented.

The TAPS route crosses the McKinley strand of the Denali fault at Mile 216 on the Richardson Highway (Fig. 6). The pipeline is so constructed here that it can slide laterally as much as 20 ft (6 m) on Teflon "shoes" to accommodate lateral displacement along this right-lateral strike-slip fault. It can also accommodate 6 ft (1.8 m) vertical displacement (Péwé and Reger, 1983). Just south, on Rainbow Mountain (Fig. 7), rock glaciers extend almost to the highway.

REFERENCES CITED

Denton, G. H., and Karlén, W., 1973, Lichenometry; Its application to Holocene moraine studies in southern Alaska and Swedish Lapland: Arctic and Alpine Research, v. 5, no. 4, p. 347–372.
—— , 1977, Holocene glacial and tree-line variations in the White River valley and Skolai Pass, Alaska, and Yukon Territory: Quaternary Research, v. 7, no. 1, p. 63–111.
Geist, O. W., and Péwé, T. L., 1957, Quantitative measurements of the 1937 advance of the Black Rapids Glacier, Alaska [abs.]: Alaska Science Conference, 5th, Anchorage, 1954, Proceedings, p. 51–52.
Hance, J. H., 1937, The recent advance of Black Rapids Glacier, Alaska: Journal of Geology, v. 45, p. 775–783.
Hanson, L. G., 1963, Bedrock geology of the Rainbow Mountain area, Alaska Range, Alaska: Alaska Division of Mines and Minerals Geologic Report 2, 82 p.
Mayo, L. R., and Péwé, T. L., 1963, Ablation and net total radiation, Gulkana Glacier, *in* Kingery, W. D., ed., Ice and snow; Properties, processes, and applications: Cambridge, Massachusetts Institute of Technology Press, p. 633–643.
Meier, M. F., Tangborn, W. V., Mayo, L. R., and Post, A., 1971, Combined ice and water balances of Gulkana and Wolverine Glaciers, Alaska, and South Cascade Glacier, Washington, 1965 and 1966 hydrologic years: U.S. Geological Survey Professional Paper 715-A, 23 p.
Mendenhall, W. C., 1900, A reconnaissance from Resurrection Bay to the Tanana River, Alaska, in 1898: U.S. Geological Survey 20th Annual Report, pt. 7, p. 265–340.
Moores, E. A., 1962, Configuration of the surface velocity profile of Gulkana Glacier, central Alaska Range, Alaska [M.S. thesis]: Fairbanks, University of Alaska, 47 p.
Ostenso, N. A., Sellmann, P. V., and Péwé, T. L., 1965, The bottom topography of Gulkana Glacier, Alaska Range: Journal of Glaciology, v. 5, p. 651–660.
Péwé, T. L., 1951a, An observation on wind-blown silt: Journal of Geology, v. 59, no. 4, p. 399–401.
—— , 1951b, Recent history of Black Rapids Glacier, Alaska [abs.]: Geological Society of America Bulletin, v. 62, no. 12, p. 1558.
—— , 1957, Recent history of Canwell and Castner glaciers, Alaska [abs.]: Geological Society of America Bulletin, v. 68, no. 12, pt. 2, p. 1779.
—— , 1965, Delta River area, Alaska Range, *in* Péwé, T. L., Ferrians, O. J., Jr., Nichols, D. R., and Karlstrom, T.N.V., Guidebook for field conference F, central and south-central Alaska, International Association for Quaternary Research, 7th Congress, Fairbanks, 1965: Lincoln, Nebraska Academy of Sciences, p. 55–93 (reprinted 1977, College, Alaska Division of Geological and Geophysical Surveys).
—— , 1968, Loess deposits of Alaska: International Geological Congress, 23rd Session, Prague, 1968, Proceedings, v. 8, p. 297–309.

Figure 7. Oblique aerial view (to the east) of Rainbow Mountain area near the crest of the Alaska Range, Mt. Hayes B-4 Quadrangle, Alaska. Photograph by U.S. Army Air Corps, 1943.

—— , 1975, Quaternary geology of Alaska: U.S. Geological Survey Professional Paper 835, 145 p.
Péwé, T. L., and Holmes, G. W., 1964, Geology of the Mt. Hayes (D-4) Quadrangle, Alaska: U.S. Geological Survey Miscellaneous Geologic Investigations Map I-394, scale 1:63:360, 2 sheets.
Péwé, T. L., and Reger, R. D., 1972, Modern and Wisconsinan snowlines in Alaska: International Geological Congress, 24th Session, Montreal, 1972, Proceedings, p. 187–197.
—— , 1983, eds., Guidebook to permafrost and Quaternary geology along the Richardson and Glenn highways between Fairbanks and Anchorage, Alaska: Guidebook 1, Alaska Division of Geological and Geophysical Surveys, 263 p.
Péwé, T. L., and others, 1953, Tentative correlation of glaciations in Alaska, *in* Péwé, T. L., and others, eds., Multiple glaciation in Alaska: U.S. Geological Survey Circular 289, p. 12–13.
Péwé, T. L., Church, R. E., Andresen, M. J., 1969, Origin and paleoclimatic significance of large-scale polygons in the Donnelly Dome area, Alaska: Geological Society of America Special Paper 103, 87 p.
Reger, R. D., 1964, Recent glacial history of Gulkana and College Glaciers, central Alaska Range, Alaska [M.S. thesis]: Fairbanks, University of Alaska, 75 p.
—— , 1968, Recent history of Gulkana and College Glaciers, central Alaska Range, Alaska: Journal of Geology, v. 76, no. 1, p. 2–16.
Reger, R. D., and Péwé, T. L., 1969, Lichenometric dating in the central Alaska Range, *in* Péwé, T. L., ed., The periglacial environment, past and present: Montreal, McGill-Queens University Press, p. 223–247.
Rutter, N. W., 1965, Foliation pattern of Gulkana Glacier, Alaska Range, Alaska: Journal of Glaciology, v. 5, p. 711–718.
Sellmann, P. V., 1962, Flow and ablation of Gulkana Glacier, Alaska [M.S. thesis]: Fairbanks, University of Alaska, 36 p.
Stout, J. H., Brady, J. B., Weber, F. R., and Page, R. A., 1973, Evidence for Quaternary movement on the McKinley strand of the Denali fault in the Delta River area, Alaska: Geological Society of America Bulletin, v. 84, no. 3, p. 939–948.
Ten Brink, N. W., and Waythomas, C. F., 1982, Late Wisconsin glacial chronology of the north-central Alaska Range; A regional synthesis and its implications for early human settlements: Unpublished report, 27 p.

The Alaska Tatonduk River section; An exceptional Proterozoic into Permian continental margin record

Carol Wagner Allison, Marine Science Institute, University of California, Santa Barbara, California 93106

LOCATION

The Tatonduk River section is located in the Nation Arch, a wedge-shaped area in eastern Alaska bounded by the Alaska–Yukon Territory border and the Dawson, Tintina, and Glenn Creek fault complexes (Fig. 1). The Yukon River flows north through the western portion of this wedge. The Tatonduk River flows west from the Yukon Territory, joining the Yukon River 13 mi (21 km) north of the village of Eagle.

Scheduled commercial air service is available to Eagle from Fairbanks, and there is road access to Eagle from May through September. Access to the Tatonduk River section is via riverboat or helicopter. There are no established trails into or within the Taonduk River valley. The high gradient and boulder bed of the river require a very shallow-draft boat with sufficient power to maintain control in the strong current. Shoaling near the mouth and multiple channels upstream demand skill in identification of the main channel.

This region falls partly within the boundaries of Yukon-Charley National Rivers Preserve, supervised by National Park Service personnel based in Eagle, and by the Doyon Regional Corporation, based in Fairbanks. Current land management policies allow access to the Tatonduk River valley; however, written permission is advisable for fossil collection, and helicopter usage in specific areas and periods is restricted.

SIGNIFICANCE OF THE SITE

The Nation Arch preserves the most complete contiguous record of unmetamorphosed Proterozoic into Cretaceous continental margin sedimentation known in a relatively small area in North America. The term Tatonduk Terrane has been proposed by Churkin and others (1982) for the Alaska portion of the arch. It includes the westernmost exposures unequivocally related to the western Cordilleran pre-Cretaceous sedimentary and volcanic sequence and is depositionally unrelated to the Yukon Crystalline Terrane juxtaposed along the Tintina Fault. Nation Arch pre-Mesozoic units are reported in the subsurface of the Kandik Basin west of the Glenn Creek Fault (Louisiana Land and Exploration Company, unpublished well data). Thus, the unusual completeness of the Nation Arch record enhances the importance of the region as the foundation to which remaining Alaska terranes have been sutured and with which they can be compared. The great variety of fossils present in Nation Arch rocks, and their stratigraphically, as well as geographically, widespread occurrence in the region, further magnifies usefulness of this record in interpretation of northwestern North American tectonic, sedimentary, paleobiogeographic, and paleoenvironmental history.

Examination of various areas within the Nation Arch is

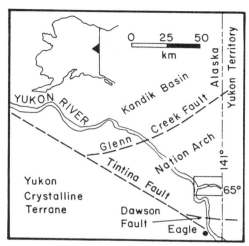

Figure 1. Nation Arch region, east-central Alaska. Tatonduk River area indicated by box in lower right.

necessary to consider the entire regional history; however, the outstanding continuous section exposed in the valley of the Tatonduk River offers a nearly complete record of Late Proterozoic into Permian accumulation along a persistent continental margin. A boat trip down the lower 7.5 mi (12 km) of this swift-flowing river provides a rare opportunity to traverse some 400 m.y. of geologic history, expressed in varied and distinctive units, in a visually beautiful setting.

SITE INFORMATION

Topography. Elevations in the Tatonduk River valley range from aroune 900 ft (270 m) at the river to around 2,000 ft (600 m) on the adjacent slopes and more than 5,000 ft (1,500 m) in the northeastern peaks. Lower elevations are heavily vegetated, with exposure confined to river and stream banks and to ridges and bluffs formed by the more resistant units.

Structure. The portion of the Tatonduk River section described here is a homoclinal sequence of generally steeply dipping beds trending north-south at the river (Fig. 2). It continues into a west-plunging anticline to the north and a southwest-plunging anticline to the south. The axis of the northern anticline shows dextral offset along the Hard Luck Fault, interpreted as an east-directed thrust by some workers. An unnamed fault at the east end of the described section marks the contact of Late Proterozoic fine clastics with massive Early Cambrian carbonates. Additional small faults, not shown on Figure 2, are present in the area.

Stratigraphy. Rocks exposed in the Tatonduk River section are referred to 14 units (Figs. 2 and 3). First measured and described in detail by Mertie (1933), additional units were recog-

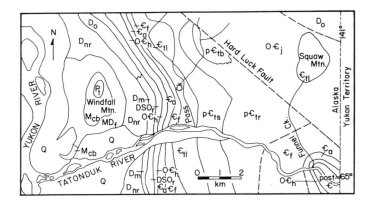

Figure 2. Geologic map of the Tatonduk River valley. Map symbols as in Figure 3. Compiled from Brabb and Churkin, 1969; Young, 1982.

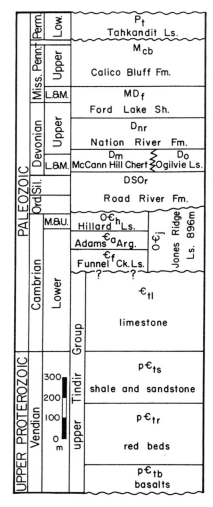

Figure 3. Stratigraphic column for the Tatonduk River area. Compiled from Mertie, 1933; Brabb and Churkin, 1969; Young, 1982; Clough and Blodgett, 1984.

nized and age assignments refined by Brabb and Churkin (1969). Discussion of regional history is presented in Payne and Allison (1981). Fossils were reported from several units in Mertie (1933) and in separately published works resulting from studies in conjunction with geologic mapping of Brabb and Churkin. Many of these are cited below in discussion of the individual units. Subsequent investigations have expanded knowledge of the depositional history, biotas, particularly microfossils, and age of several units. In some cases, fossils from Tatonduk section rocks exposed in areas beyond the river valley are mentioned. Only a few of the Tatonduk River exposures have been exhaustively studied; thus, it is likely that further investigation will add to the biotas known from the river section.

SITE DESCRIPTION

Description of the river-level exposures begins with Proterozoic rocks exposed 7.6 mi (12. 3 km) east of the mouth of the river and continues downstream 6 mi (10 km) to Early Permian outcrops on Windfall Mountain (Fig. 2). The north bank of the river and slope above it, which offer the best exposures of most of the units, is described here. Lithologies and paleontologic information, with sources for more detailed coverage, are given for each unit, followed by an interpretation of local Proterozoic and Paleozoic depositional history. Thicknesses indicated on Figure 3 are for the local section; regional thicknesses of many of the units are greater than those cited for the Tatonduk River outcrops.

Tindir Group. Four lithologically distinct upper Tindir Group units, which have not been formally named, are represented in the river section. A bed-by-bed description of these outcrops is included in Mertie (1933), and they are discussed in detail and interpreted in Young (1982). Dark gray amygdaloidal and pillow basalts with associated tuff, visible in upper Pass Creek (Fig. 2), lie at the base of the upper Tindir. The succeeding unit is characterized by maroon finely laminated mudstones with lesser siltstones. Prominent maroon diamictites less than one to more than 20 in (50 cm) thick (Fig. 4) are intermittently present in the red beds, which are especially well exposed in a low bluff

1.4 mi (2.3 km) downstream from their first appearance. Dolomitic and siliceous beds are present as well as rare high-grade hematitic cherts and copper-rich green horizons. Sporadic altered tuffs confirm volcanic activity during red bed deposition, with concurrent glacial influence indicated by striated clasts and apparent dropstones. Thick, unstratified maroon conglomerates overlying the red beds north of the river are interpreted by Young (1982) as a separate unit, rapidly dumped downslope from the east near the end of red bed accumulation. Rare black cherty beds within the red mudstones have yielded a few cyanobacterial coccoids and branched tubes of apparent fungal origin, and a bright red chert near the top of the unit at the river level contains abundant iron bacteria (Allison, 1987).

Gray shales that mark the beginning of deposition of the overlying unit appear in the section just above the red chert and intergrade with maroon mudstones for several meters along the river bank in the contact zone between the two units. Quartz- or

Figure 4. Diamictite bed in maroon mudstones of the Tindir red beds, north bank of the Tatonduk River. (Source: Bruce I. Clardy.)

Figure 5. Nodular horizon in the Tindir shale and sandstone unit, north bank of the Tatonduk River.

dolomite-rich sandstones, many of which are cross laminated and sole marked, are interbedded with the gray shales, as are a few tan-weathering conglomerates (Fig. 5). This unit contains minor volcanic breccias and is compositionally similar to the red beds, but it lacks the high iron content and striking red coloration. No fossils have been identified in the shale and sandstone unit.

The shift from dominantly dolomite to limestone carbonate deposition recorded in many similar age sedimentary sections worldwide near the Proterozoic-Paleozoic boundary is represented in the Tatonduk River section where tan-weathering Tindir gray shales and sandstones are gradationally overlain by recessive thin-bedded, increasingly dark limestones and rusty weathering shales. This contact, and the uppermost Tindir limestone unit, are better exposed in upper Pass Creek than at the river. Pyrite-rich, fetid-smelling beds typical of the limestone unit are, however, intermittently exposed on the banks of lower Pass Creek. Oolitic cherty horizons to 230 ft (70 m) below the top of the limestone unit in Pass Creek contain cyanobacterial filament sheaths, in masses and as the coiled *Obruchevella*; rare cyanobacterial coccoids and enigmatic annularly ribbed tubular microfossils are also present (Kline, 1977; Allison, 1987). An exceptionally varied and well-preserved biota of bacteria, cyanobacteria, acritarchs, chrysophytes, protistans, fungi, and poriferan spicules occurs in cherts in the upper 276 ft (84 m) of the limestone unit 14 mi (22 km) northeast of the Tatonduk River (Awramik and Allison, 1980; Allison and Hilbert, 1986).

Funnel Creek Limestone. About 0.7 mi (1.2 km) downstream from the mouth of Pass Creek, the resistant Funnel Creek Limestone forms a prominent rib in the north bank of the river, concordantly overlying the recessive Tindir limestone (Fig. 6). The Funnel Creek is also well exposed in pinnacles and ridges in lower-dipping beds immediately upriver from the Tindir red beds (Fig. 2). This pale gray, fine- to medium-grained, laminated limestone is pervasively silicified and contains many oolitic lay-

ers. Fossils have not been located in the river section of the Funnel Creek; however, archeocyathids occur 500 ft (150 m) above the base of the laterally equivalent lower Jones Ridge Limestone 7 mi (11.5 km) northeast (Brabb, 1967). Outcrops of the Funnel Creek Limestone 14 mi (22 km) northeast of the Tatonduk River contain cyanobacterial, chrysophyte, and acritarch microfossils (Allison and Hilgert, 1986).

Adams Argillite. The Funnel Creek Limestone is succeeded downstream on the Tatonduk River by the recessive weathering Adams Argillite, a light olive-gray, fine-grained argillite unit with silty and quartzitic phases (Brabb, 1967). Sandy, oolitic limestone lenses, exposed part way up on the north slope of the river and elsewhere in the region, contain very well preserved archeocyathids in biohermal structures that include also the calcareous green algae *Renalcis* and *Epiphyton,* poriferan spicules, the coelenterate *Tabulaconus,* coeloscleritophore chancelloriids, and trilobite fragments. This biota indicates a late Early Cambrian age and, in contradistinction to similar age archeocyathid biotas further south in the Cordillera, has Siberian affinities (R. A. Gangloff, unpublished data). South and northeast of the Tatonduk River, trilobites of middle Early Cambrian age occur in the Adams Argillite (Palmer, 1968), which is apparently laterally equivalent to the middle portion of the Jones Ridge Limestone.

Hillard Limestone. A second cliff-forming limestone at the top of the Adams Argillite also forms prominent ribs in the north bank of the Tatonduk River (Fig. 7). The lower beds of this dominantly carbonate unit, the Hillard Limestone, include two boulder conglomerates with clasts of oolitic limestone and with trilobites of late Early Cambrian age and Siberian affinities (Palmer, 1968). Succeeding sandy and gritty limestones, dolomites, and dolomitic limestones contain Middle and early Late Cambrian trilobites, and Early Ordovician shelly fossils occur in the uppermost Hillard on the north slope above the river. Echi-

Figure 6. Panoramic view of the Tatonduk River section looking northwest from just below the mouth of Pass Creek to the Yukon River. Map symbols as in Figure 3.

Figure 7. Massive lower portion of the Hillard Limestone overlying the recessive weathering Adams Argillite, north bank of the Tatonduk River. Archeocyathid-rich limestone lens in the Adams Argillite is exposed in the treeless area on the slope above the river.

noderm fragments are abundant in some thin beds in the river section. The darker beds in this unit are very rich in organic debris. Pyrite and phosphatic material are common in some horizons, including one thin bed made up primarily of phosphate chips (Brabb, 1967). The Hillard Limestone is laterally equivalent to the upper portion of the lower Jones Ridge Limestone, which contains shelly biotas of Middle and Late Cambrian and Early Ordovician ages. The upper Jones Ridge Limestone has yielded fossils of Middle or Late Ordovician age, younger than the uppermost fossils known in the Hillard on the Tatonduk River.

Road River Formation. The Hillard Limestone is in fault contact at the river level with the strongly contrasting dark, thin-bedded, recessive weathering Road River Formation, which elsewhere lies unconformably on the Hillard. Black shales with lesser cherty, sandy, and conglomeratic beds characterize the Road River, a unit also recognized to the north in Alaska and to the east in the Yukon Territory. Graptolites, abundant in several horizons in the river level section (Fig. 8), include biotas of Middle Ordovician (Llandeilo to middle Caradoc) and Early to Late Silurian (Llandovery to Ludlow) age (Churkin and Brabb, 1965). A chert-rich bed near the top of the Road River in this section contains gastropods, nautiloids, and plant fragments. Elsewhere in the region conodonts, crustaceans, and tentaculitids are locally abundant (Churkin and Brabb, 1965). Thin sections reveal the presence also of fungi, radiolarians?, and spores.

McCann Hill Chert and Ogilvie Limestone. Covered intervals in the upper Road River Formation at the river level tend to mask the concordant contact with the overlying Early, Middle, and Late Devonian McCann Hill Chert, composed also of dark, thin, fine-grained beds with minor cherty grits. These beds do not contain graptolites but do contain spores, fungal hyphae, and rare conodonts and mollusks, and locally abundant radiolarians, sponge spicules, and plant fragments. South and north of the

Tatonduk River outcrops, dark limestones in the basal part of the formation contain a rich and varied fauna including corals, bryozoans, brachiopods, tentaculitids, pelecypods, conodonts, ostracods, and fish remains (Churkin and Brabb, 1965, and papers cited therein). In more recent work by Clough and Blodgett (1984), these shelly limestones are referred to the Ogilvie Formation, exposed north of Squaw Mountain (Fig. 2) and in adjacent Yukon Territory.

Nation River Formation. Appearance of grits in the upper McCann Hill foreshadows a change in depositional regime at the contact with the typically tan-weathering gray sandstones and conglomerates of the Late Devonian Nation River Formation. Mudstones in the Nation River locally contain abundant plant fragments and poorly preserved spores (Scott and Doher, 1967). Impressions of sphenopsids and lycopsids in gray sandstone in the Tatonduk River section, and psilophytes, lycopods, and progymnosperms reported in rare pockets 11 to 18.6 mi (18 to 30 km) to the northwest, indicate the nature of the adjacent land flora (Tiffney, 1976). A black shale referred to the Nation River Formation and exposed on the north side of the bluff at Eagle contains abundant poorly preserved radiolarians.

Ford Lake Shale. At least portions of all the units described thus far in the Tatonduk River section can be seen at the river level. The remaining three, with the exception of an isolated small outcrop of the Calico Bluff Formation, can only be examined on or at the base of the adjacent slopes in the lower portion of the river where the valley widens markedly.

Grayish black laminated beds of the Ford Lake Shale gradationally overlie the Nation River Formation. The dominantly siliceous and fine-grained Ford Lake commonly has a yellow sulfur coating on weathered surfaces and includes minor siltstone, quartzite, and limestone and, especially near the top, phosphate nodules. Fossils include common plant stems and wood fragments, and rare conodonts and brachiopods as well as locally abundant radiolarians that confirm a Late Devonian into Early and Middle? Mississippian age for the Ford Lake (Holdsworth and others, 1978).

Calico Bluff Formation. Thin, pale-weathering dark limestones near the top of the Ford Lake grade rhythmically into interbedded bioclastic limestones and dark shales of the Calico Bluff Formation, exposed on the slopes of Windfall Mountain and in isolated outcrop on the lower Tatonduk River (Fig. 2). The strikingly banded Calico Bluff Formation reflects deposition by turbidity currents on a submarine slope and rise system and locally shows good examples of early postdepositional slumping (Moore, 1979). It is a richly fossiliferous unit estimated by Mertie (1933), based on collection of exposures at and near Calico Bluff on the Yukon River south of the Tatonduk, to contain more than 250 species of corals, bryozoans, brachiopods, pelecypods, gastropods, cephalopods, and crinoids. Foraminifera indicate that the Late Mississippian Calico Bluff may range into the Pennsylvanian (Armstrong, 1975). Outcrops on the Tatonduk River and on Windfall Mountain contain sphenopsid and lycopsid plant remains as well as shelly fossils, including species additional to those at Calico Bluff.

Tahkandit Limestone. Pale-weathering, mostly fine-grained beds of the Tahkandit Limestone cap Windfall Mountain. This is the youngest pre-Quaternary unit in the Tatonduk River valley. As elsewhere, the Windfall Mountain Tahkandit exposures contain abundant brachiopods with less common gastropods, pelecypods, corals and ostracods. More complete exposures of this unit to the north contain abundant fossils including also algae, bryozoa, crinoids, and foraminifers that have Arctic affinities. A glauconitic sandstone underlying the prominent shelly limestone to the north also contains spores, pollen, and hystrichosphaerids (Brabb and Grant, 1971). The Step Conglomerate, a lateral equivalent of the Tahkandit Limestone, is reported both north (Brabb and Churkin, 1969) and south (Foster, 1976) of the Tatonduk River, but has not been identified in the latter area.

Continental Margin Evolution. The basal upper Tindir basalts, probably as young as 600 Ma or as old as 720 Ma, lie apparently unconformably, although typically concordantly, on oolitic shallow marine dolomites at the top of the 5,900-ft-thick (1,800 m) lower Tindir Group, a precursor continental margin sequence. The pillowed structure and tholeiitic composition of the basal upper Tindir basalts suggest a rifting event widely re-

Figure 8. Slumped graptolitic shales of the Road River Formation, north bank of the Tatonduk River. (Source: Earl E. Brabb.)

corded at this time along the western Cordillera (Stewart, 1972). Association of continuing volcanic activity and of glacial activity during ensuing red bed deposition is also widely recorded in similar-age rocks of the western Cordillera, China, Australia, and elsewhere. The Tatonduk area record of early postrifting reflects a quiet subaqueous setting with recurrent influx of coarser material from an easterly source via turbidity currents driven by intermittent nearby fault activity (Young, 1982). Abundant fan complex features are present in the red beds; it is postulated that the interaction of hydrothermal leaching of volcanic rocks with tidewater-derived cold water was the mechanism responsible for the chert hematite beds (Young, 1982).

Cessation of iron accumulation marks the shift to deposition of the gradationally overlying gray shales, sandstones, and conglomerates, which record sporadic volcanic activity and possibly glacial activity in lobe, channel, and levee deposits in a deepening basin (Young, 1982). Generally deep marine conditions prevailed in the Tatonduk River area throughout accumulation of much of the gradationally overlying Tindir limestone unit, although presence of oolites and photosynthetic cyanobacteria well below the top of the limestone unit in Pass Creek and obvious shallowing upward features in the upper 276 ft (84 m) to the northeast clearly foreshadow the appearance of shallow-water carbonates in the concordantly overlying Funnel Creek and Jones Ridge limestones.

Sponge spicules in the Tindir limestone unit northeast of the Tatonduk River are the oldest-known metazoan fossils in the Nation Arch. By middle Early Cambrian time, shelly metazoans had become continuing residents in the area; they continued to be through early Late Cretaceous time, when marine conditions ceased to exist in the region and terrestrial sediments began to accumulate.

The carbonate platform that was established in the Alaska-Yukon border area northeast of the Tatonduk River at the onset of the Funnel Creek and Jones Ridge limestone deposition occupies the same location as shallow-water stromatolitic carbonates intermittently present in the lower Tindir Group. Oscillation of the margin between these carbonates and deeper water shales to the south and west is marked by contraction of the platform during middle Early Cambrian Adams Argillite accumulation and with onset of Road River Formation deposition in the Early Ordovician. A 4-ft-thick (1.2 m) shale and siltstone on top of a 16-ft (5-m) covered interval at the top of the Jones Ridge Limestone 7 mi (11 km) north of the Tatonduk River that contains graptolites of late Early Silurian (late Llandovery) age indicates maximum easterly transgression documented in the region. By Early Devonian Emsian time, carbonate platform deposits were again forming in the border region with appearance of the Ogilvie Limestone, which shifts from platform to open-shelf deposits in the latest Emsian, followed by transgression and deposition of the McCann Hill Chert during the Middle to early Late Devonian (Clough and Blodgett, 1984).

The appearance of coarse clastics in late McCann Hill time probably signaled the beginning of uplift in the northwestern Cordillera and the westward shedding of sandstones and conglomerates in deep-sea fans during accumulation of the Late Devonian Nation River Formation (Nilsen and others, 1976). The wide presence of fragmented land plants in the Nation River Formation, as well as the overlying Ford Lake Shale, suggest continuing erosion of a well-vegetated source area on the adjacent landmass.

Return to a quiet, deeper-water sedimentary regime is reflected during deposition of the Late Devonian to early Late Mississippian Ford Lake Shale, in which open ocean conditions are indicated by the local presence of radiolarian-rich deposits (Holdsworth and others, 1978). Shelly, shelf to deep-marine metazoans return in abundance to the local fossil record during accumulation of the conformably overlying Late Mississippian and earliest Pennsylvanian Calico Bluff Formation. Regionwide emergence took place sometime between Early Pennsylvanian and Early Permian time. The northernmost record of the Calico Bluff Formation is at Windfall Mountain, north of which shallow-water Early Permian Tahkandit beds lie directly on the Nation River Formation. These relationships suggest uplift to the northwest or north during Pennsylvanian time, as indicated by southerly paleocurrent directions in the Calico Bluff Formation (Moore, 1979).

The remnant Tahkandit Limestone capping Windfall Mountain marks the return of shallow-water carbonate deposition in the Tatonduk River area during Leonardian and Guadalupian time. Regionwide emergence took place again in the Nation Arch between Late Permian and Middle Triassic time, when deep-water, fine clastics of the thick Middle Triassic to Early Cretaceous Glenn Shale began to accumulate in both the Nation Arch and Kandik Basin.

REFERENCES CITED

Allison, C. W., 1987, Paleontology of Late Proterozoic and Early Cambrian rocks of east-central Alaska: U.S. Geological Survey Professional Paper 1449 (in press).

Allison, C. W., and Hilgert, J.W., 1986, Scale microfossils from the Early Cambrian of northwest Canada: Journal of Paleontology, v. 60, p. 973–1015.

Armstrong, A. K., 1975, Carboniferous corals of Alaska, a preliminary report: U.S. Geological Survey Professional Paper 823-C, p. C45–C56.

Awramik, S. M., and Allison, C. W., 1980, Earliest Cambrian blue-green algal communities from the Yukon Territory of Canada: Abstracts 26th International Geological Congress, Paris, v. 1, p. 198.

Brabb, E. E., 1967, Stratigraphy of the Cambrian and Ordovician rocks of east-central Alaska: U.S. Geological Survey Professional Paper 559-A, 30 p.

Brabb, E. E., and Churkin, M., Jr., 1969, Geologic map of the Charley River Quadrangle, east-central Alaska: U.S. Geological Survey Miscellaneous Investigation Map I-573, scale 1:250,000.

Brabb, E. E., and Grant, R. E., 1971, Stratigraphy and paleontology of the revised type section for the Tahkandit Limestone (Permian) in east-central Alaska: U.S. Geological Survey Professional Paper 703, 26 p.

Churkin, M., Jr., and Brabb, E. E., 1965, Ordovician, Silurian and Devonian biostratigraphy of east-central Alaska: American Association of Petroleum Geologists Bulletin, v. 49, p. 172–185.

Churkin, M., Jr., Foster, H. L., Chapman, R. H., and Weber, F. R., 1982, Terranes and suture zones in east-central Alaska: Journal of Geophysical Research, v. 87, p. 3718–3730.

Clough, J. G., and Blodgett, R. B., 1984, Lower Devonian basin to shelf carbonates in outcrop from the western Ogilvie Mountains, Alaska and Yukon Territory, in Eliuk, L. L., Kaldi, J., Watts, N., and Harrison, G., eds., Carbonates in subsurface and outcrop: Canadian Society of Petroleum Geologists, p. 57–81.

Foster, H. L., 1976, Geologic map of the Eagle Quadrangle, east-central Alaska: U.S. Geological Survey Miscellaneous Investigation Series Map I-922, scale 1:250,000.

Holdsworth, B. K., Jones, D. L., and Allison, C. W., 1978, Upper Devonian radiolarians separated from chert of the Ford Lake Shale, Alaska: U.S. Geological Survey Journal of Research, v. 6, p. 775–788.

Kline, G. L., 1977, Earliest Cambrian (Tommotian) age of the upper Tindir Group, east-central Alaska: Geological Society of America Abstracts with Programs, v. 9, p. 448.

Mertie, J. B., Jr., 1933, The Tatonduk-Nation district, Alaska: U.S. Geological Survey Bulletin 836-E, p. 345–454.

Moore, M. G., 1979, Stratigraphy and depositional environments of the Carboniferous Calico Bluff Formation, east-central Alaska [M.S. thesis]: Fairbanks, University of Alaska, 102 p.

Nilsen, T. H., Brabb, E. E., and Simoni, T. R., Jr., 1976, Stratigraphy and sedimentology of the Nation River Formation, a Devonian deep-sea fan deposit in east-central Alaska: Proceedings of the Alaska Geological Society Symposium held April 2–4, 1975, at Anchorage, Alaska, p. E1–E20.

Palmer, A. R., 1968, Cambrian trilobites of east-central Alaska: U.S. Geological Survey Professional Paper 559-B, 115 p.

Payne, M. W., and Allison, C. W., 1981, Paleozoic continental-margin sedimentation in east-central Alaska: Geology, v. 9, p. 274–279.

Scott, R. A., and Doher, L. I., 1967, Palynological evidence for Devonian age of the National River Formation, east-central Alaska: U.S. Geological Survey Professional Paper 575-B, B45–B49.

Stewart, J. H., 1972, Initial deposits in the Cordilleran Geosyncline; Evidence of a Late Precambrian (<850 m.y.) continental margin separation: Geological Society of America Bulletin, v. 83, p. 1345–1360.

Tiffney, B. H., 1976, A survey of paleobotanical sites within the proposed Yukon-Charley National Rivers area: Hannover, New Hampshire, The Center for Northern Studies, 32 p.

Young, G. M., 1982, The late Proterozoic Tindir Group, east-central Alaska; Evolution of a continental margin: Geological Society of America Bulletin, v. 93, p. 759–783.

The Ambler sequence at Arctic Ridge, Ambler District, Alaska

Jeanine M. Schmidt, *U.S. Geological Survey, Branch of Alaskan Geology, 4200 University Drive, Anchorage, Alaska 99508-4667*

LOCATION AND ACCESS

Arctic Ridge (67°09'18"N, 156°23'18"W to 67°12'42"N, 156°21'12"W) is located in the Ambler mineral district on the southern flank of the Brooks Range, northwest Alaska (Fig. 1), in the Ambler River A-1 Quadrangle. The ridgeline is 1.5 mi (2.5 km) due east of VABM Riley, between the Kogoluktuk and Shungnak rivers (Fig. 2).

Access to the area is easiest by helicopter. Arctic Ridge lies 270 mi (430 km) northwest of Fairbanks, 170 mi (270 km) east of Kotzebue, and 310 mi (500 km) northeast of Nome. Commercial helicopters are usually available for charter in Fairbanks and Nome. The traverse of the Ambler sequence begins on the highest peak (locally known as Arctic peak, elevation 3,560 ft; 1,085 m, Fig. 2) of the southern ridgeline, southeast of the buildings in Subarctic Creek valley (Fig. 3). Permission to cross private land and pate 'ed claims should be obtained prior to a visit from the NANA regional corporation (Northwest Alaska Native Association, based in Kotzebue) and Kennecott Mining Company (Salt Lake City, Utah), respectively. An alternative is to charter a small single-engine plane to Arctic airstrip (Fig. 2). This is a private airstrip, and permission for landing must be obtained from Kennecott Mining Company prior to any visit. From the airstrip, walk approximately 7.5 mi (12 km) northeast along a dirt road to Arctic camp, and from there up to Arctic peak.

The ridge is often covered by patchy snow into early July, but most outcrops are accessible by early June. The area usually remains free of snow cover until mid-September; however flurries can occur in any month. The ridgeline is steep, craggy, and locally difficult to traverse. Allow 5 to 10 hours in addition to the time required to arrive in the area. Interesting sidehill outcrops are 500 to 1,500 ft (150 to 450 m) in elevation below the ridgeline and often on steep slopes; allow extra time or use a helicopter to reach these.

SIGNIFICANCE OF SITE

Arctic Ridge exposes the thickest known (>5,900 ft; 1,800 m) stratigraphic section of the Ambler sequence (Hitzman and others, 1982), a succession of interlayered calcareous, pelitic, and graphitic metasedimentary rocks and bimodal basaltic and rhyolitic metavolcanic and volcaniclastic rocks. The Ambler sequence is in part Late Devonian in age, and is interpreted to have formed during a discrete, short-lived period of ensialic rifting along the North American continental margin (Hitzman and others, 1986). These rift-related volcanic rocks host at least six major occurrences of volcanogenic massive sulfide mineralization along the 80 mi (130 km) strike length of the Ambler District. The Arctic Ridge section includes the largest of these deposits yet discovered, at the Arctic prospect, where published reserves are 35 million short tons of 4.0% Cu, 5.5% Zn, 1.0% Pb, 1.5 oz/ton Ag,

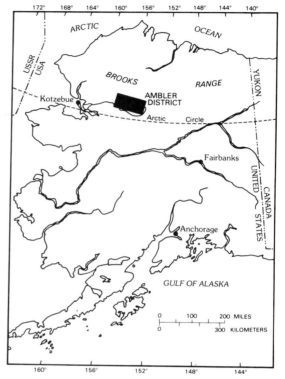

Figure 1. Location map of the Ambler district, northwest Alaska.

and 0.019 oz/ton Au (Sichermann and others, 1976), in an open-pit minable configuration.

SITE INFORMATION

The geology of the Arctic Ridge area comprises a single, south-southwest dipping (10° to 45°) exposure through the Ambler sequence (Fig. 4; Schmidt, 1986). The Ambler sequence lies with apparent conformity within the Anirak Schist (Hitzman and others, 1982), a 6,900 to 20,000 ft (2,100 to 6,000 m) thick Paleozoic section of fine-grained metapelite, with minor marble and metabasalt. The Ambler sequence reaches its greatest known thickness (6,070 ft; 1,850 m) on Arctic Ridge, where it is composed of approximately 65% metavolcanic and volcaniclastic rocks and 35% metasedimentary rocks. Along its exposed length, the percentage of metasedimentary rocks in the Ambler sequence varies from 20–55%, and is commonly 35–50%. These rocks consist of carbonaceous and pelitic metasedimentary rocks, calcareous and dolomitic marbles and calcareous schists, and minor calc-silicate gneiss (informally referred to as "gnurgle gneiss" by Hitzman and others, 1982). Along the Arctic Ridge traverse, carbonate rocks are most abundant in the lower one-third, and comprise about 15% of the total stratigraphic column. Pelitic metasedimentary rocks are relatively rare (6%), and are intercalated

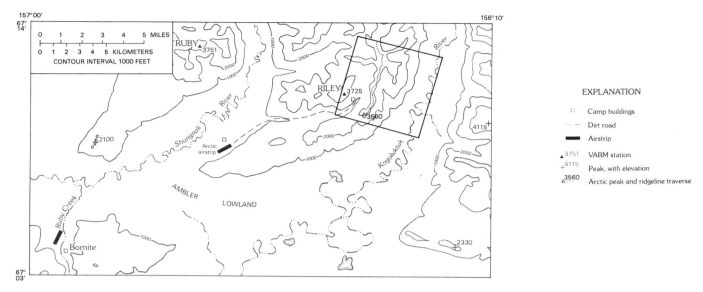

Figure 2. Location map of the central Ambler district, showing Arctic Ridge and main physiographic features. Outline shows the location of Figure 3.

mainly with impure calcareous metasedimentary rocks. Carbonaceous metasedimentary rocks are slightly more common (9%) and are intercalated with all varieties of metarhyolite and metabasalts, and most other metasedimentary rocks. At Arctic, metarhyolitic rocks are approximately three times as abundant as metabasalts, but proportions across the district vary from 1:1 to 10:1.

Metarhyolitic rocks of the Ambler sequence have been divided into three major types. These are (1) coarsely porphyritic metarhyolites (crystal and crystal-lithic ash-flow tuffs); (2) aphanitic metarhyolites (lava flows, domes, and dikes); and (3) a variable assemblage of schistose metarhyolites, derived from tuffs, lapilli tuffs, and mixed epiclastic and volcaniclastic material.

Metamafic rocks include locally pillowed and brecciated flows, numerous plugs and sill-like bodies. Metabasites have been divided into extrusive rocks (flows, flow-breccias, and tuffs) and intrusive rocks (plugs, sills, and dikes), based on field relations and textures.

The upper part of the sequence at Arctic Ridge is dated as latest Devonian and Early Mississippian(?) by a Pb-U zircon date of 360± 10 Ma (Dillon and others, 1980). Macrofossils from carbonate rocks near the base of the Ambler sequence farther to the east yield Middle(?) Devonian to Early Mississippian(?) ages (Hitzman and others, 1982; Smith and others, 1978). A reconstructed stratigraphic column through the area (Fig. 5) assumes that intrafolial folds produced during the Jurassic(?) to mid-Cretaceous Brooks Range orogeny have not substantially thickened the Ambler sequence. All thicknesses are estimates, since millimeter- to meter-scale folds are moderate to common in rocks of intermediate to low competence (most metasedimentary rocks, schistose rhyolites, and phyllosilicate alteration associated with the volcanogenic sulfide deposit).

Fold deformation of the area was accompanied by high-pressure, moderate-temperature greenschist to glaucophane schist facies metamorphism. Biotite does not occur in any pelitic rocks and garnet is only locally present in metabasalts and calc-silicate rocks. Blue amphibole distribution in the district is irregular and compositionally controlled. No blue amphibole has been identified from Arctic Ridge. Primary igneous and sedimentary textures and mineralogy are poorly to very well preserved, depending on the competence and original grain size of the rocks. Primary igneous and sedimentary protolith terms are applied wherever possible throughout this guide; where origin is uncertain, metamorphic terminology has been retained. Metamorphism was essentially isochemical and did not obscure strong chemical gradients associated with sulfide deposition, or the premetamorphic compositions of igneous rocks (Schmidt, 1983). The age of metamorphism at Arctic Ridge is suggested as Early Cretaceous or older based on K-Ar cooling ages of 109 to 163 Ma on muscovite and 105 to 116 Ma on biotite from a variety of schists (R. Wilson, written communication, 1985; ages recalculated with new constants from data reported by Turner and others, 1978).

TRAVERSE NOTES

At Arctic peak, the traverse begins within the part of the Anirak Schist that overlies the Ambler sequence (Figs. 4A, 4B, 5). The Anirak Schist is a brownish gray, variably graphitic, medium-grained quartz–white mica metapelite with minor amounts of chlorite, albite, epidote, and Fe-sulfides. It has a characteristic widely spaced (1-10 mm), wavy, discontinuous foliation, defined by white mica and quartz lenses and segregations (1 cm to 1 m in longest dimension). Metamorphic grade decreases gradually to the south, and approximately 2.5 mi (4 km) southwest of Arctic Ridge, these metapelitic rocks are phyllitic in texture. The Anirak Schist is interpreted to have formed from a fine-grained sandstone, siltstone, or mudstone.

From Arctic peak and high points farther north along the ridge, the glacial ice cap of Mount Oyokuk is visible in clear weather, 37 mi (60 km) to the northeast. To the east, the Kogoluktuk River develops southward from a braided stream into a broader river with classic meanders and oxbow lakes. Polygonal ground characteristic of permafrost occurs locally in the river valley, and solifluction lobes occur on some steeper tundra slopes. West of the ridgeline, the U-shaped cross section of Subarctic Creek valley is one record of Pleistocene glaciation, as are the small cirque lakes west of Riley Ridge.

Stop A. The uppermost major unit of the Ambler sequence seen on the traverse is a locally pillowed metabasalt, the only mafic extrusive rock type noted in the Arctic area (A; Fig. 4A). This subaqueous flow is interlayered with, and has locally baked underlying carbonaceous mudstones. Its upper parts contain thin marble lenses and pods, in many cases between pillows. This metabasalt is thickest in the Subarctic Creek area, suggesting a trough or vent area there. It is variably overlain by a graphitic schist north of VABM Riley and a thick calcareous schist on Arctic Ridge, suggesting a significant difference in either water depth (paleotopography) or sedimentation source/rate between the two areas.

The metabasalt flow on Arctic Ridge is granoblastic to semischistose, medium to dark grayish green, with weathering rinds on pillows commonly a few centimeters thick, and more calcareous, pitted, and schistose than interiors. Its mineralogy is actinolite-chlorite-albite-sphene-epidote, with minor calcite, white mica, quartz, biotite, and Fe-sulfides. Neither garnet nor glaucophane occur within this flow on Arctic Ridge; however, the same flow contains both minerals on Riley Ridge.

Stop B. Below the metabasalt is a 250 to 750 ft (75 to 230 m) thick unit of thinly interlayered aphanitic and schistose metarhyolite, with a distinctly porphyroblastic unit at its base (B, Fig. 4A). This interval at Arctic Ridge has been informally referred to as the Margarita unit (Schmidt, 1983). It is composed of tan, gray, yellow and light-green weathering rhyolite with poor foliation in the aphanitic rocks and moderate foliation developed in the coarser-grained schistose and porphyroblastic varieties. Aphanitic rhyolites within the Margarita unit are light gray, sugary in appearance, generally lack flow banding, and unlike most Ambler sequence aphanites, locally contain sparse 1–2% small (<5mm) phenocrysts of quartz or less common K-feldspar. Their matrix is a fine-grained (<5–25 μm) intergrowth of quartz, albite, and white mica with some K-feldspar, 0–2% black or dark brown biotite, and 0–5% Fe-sulfides.

The interlayered schistose and feldspathic rhyolites are grayish green to tan in color, and are composed of quartz, white mica, and albite, with minor sphene, chlorite, and calcite, and trace biotite, zircon, and epidote. Albitic feldspar commonly occurs as large, poikiloblastic grains coalescing in the matrix, and incorporating up to 35% large inclusions. Where these albite agglomerates are large and relatively free of inclusions, the rock is

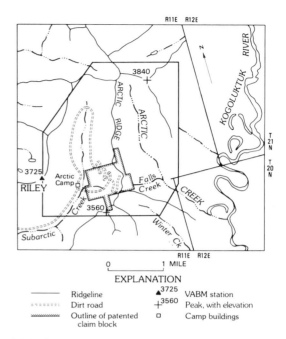

Figure 3. Physiographic map of Arctic Ridge area showing natural and man-made features. Outline shows the approximate area of the geologic map (Fig. 4A).

visibly porphyroblastic. Rare lapilli-sized lenses of rhyolite within a similar matrix are the only primary textures recognized in the schistose/porphyroblastic rhyolites. The aphanitic rhyolites were probably deposited as flows, but the schistose varieties were probably tuffaceous. Some of the least feldspathic quartz–white mica schists may have been derived by sedimentary reworking of rhyolitic tuffs between episodes of deposition. A similar interval of mixed rhyolites occurs near the top of the Ambler sequence over a wide area (19 mi; 30 km to the east, 2+ mi; 3+ km to the west), and because it hosts massive sulfide mineralization elsewhere, it is considered to be subaqueous.

The Margarita metarhyolite unit overlies the uppermost (mrp3) of the three major metarhyolite porphyry units at Arctic. The rhyolite porphyries contain up to 45% (20–25% common) phenocrysts of blue quartz (1–10 mm), gray K-feldspar (2 mm to 5 cm) and minor reddish brown biotite (<2 mm). The poorly foliated, very fine-grained (<10 μm) groundmass contains wide mica, quartz, and feldspar derived from a devitrified glass, and weathers recessively to give a distinctive "button" rhyolite appearance. Bipyramidal quartz is sparse, broken phenocrysts are common and resorption channels occur in both quartz and K-feldspar. Some parts of the metarhyolite porphyry contain pumice lapilli, which show little elongation (2:1 to 4:1 length:thickness ratios); a few areas have a eutaxitic flattened fabric suggesting partial welding of molten material.

Textures and mineralogy of all of the metarhyolite porphyries vary on a scale of centimeters to meters; mappable subunits have not been distinguished over most of the district. More primary textures in these rocks can be seen on this traverse when

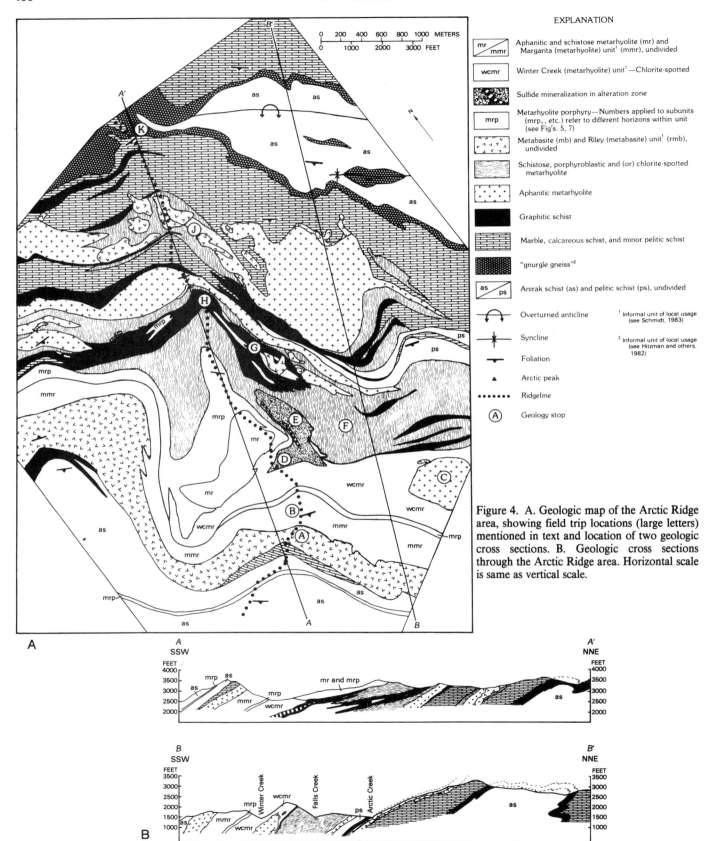

EXPLANATION

mr / mmr	Aphanitic and schistose metarhyolite (mr) and Margarita (metarhyolite) unit[1] (mmr), undivided
wcmr	Winter Creek (metarhyolite) unit[1]—Chlorite-spotted
	Sulfide mineralization in alteration zone
mrp	Metarhyolite porphyry—Numbers applied to subunits (mrp₁, etc.) refer to different horizons within unit (see Fig's. 5, 7)
	Metabasite (mb) and Riley (metabasite) unit[1] (rmb), undivided
	Schistose, porphyroblastic and (or) chlorite-spotted metarhyolite
	Aphanitic metarhyolite
	Graphitic schist
	Marble, calcareous schist, and minor pelitic schist
	"gnurgle gneiss"[2]
as / ps	Anirak schist (as) and pelitic schist (ps), undivided

↷ Overturned anticline

✴ Syncline

— Foliation

▲ Arctic peak

•••• Ridgeline

Ⓐ Geology stop

[1] Informal unit of local usage (see Schmidt, 1983)

[2] Informal unit of local usage (see Hitzman and others, 1982)

Figure 4. A. Geologic map of the Arctic Ridge area, showing field trip locations (large letters) mentioned in text and location of two geologic cross sections. B. Geologic cross sections through the Arctic Ridge area. Horizontal scale is same as vertical scale.

Sections are generalized in areas

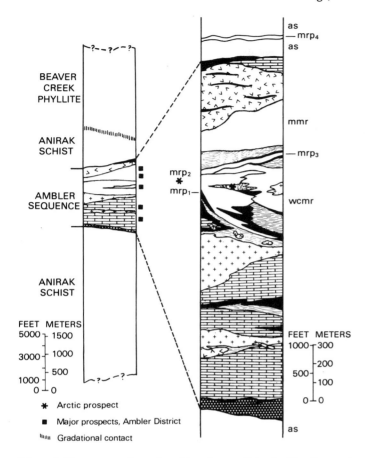

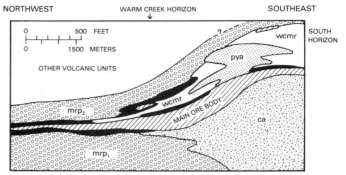

Figure 6. Idealized geologic cross section through the Arctic prospect. Sulfide mineralization shown by hachured lines. Black areas are graphitic schist. mrp, metarhyolite porphyry; ca, chloritic footwall alteration; pya, hanging wall pyritic alteration; wcmr, Winter Creek volcanoclastic metarhyolite.

Figure 5. Reconstructed stratigraphic column of the Ambler Group at Arctic Ridge. All thicknesses shown are estimates. See Figure 4 for explanation.

crossing mrp2 farther north along the ridge; however, the best preserved textures are seen in unweathered drill core from the various prospects.

The phenocryst and groundmass mineralogy and modes, textures, and the interlayering of the porphyries with minor carbonaceous mud (graphitic schist) and mud-matrix autobreccias, indicate that these rhyolites were deposited as subaqueous, crystal-rich ash-flow tuffs. Arctic Ridge preserves the thickest and most numerous metarhyolite porphyries in the district, suggesting a depositional trough in this area. Subsurface drill information on the percentage of rhyolite breccia fragments and carbonaceous mud matrix in mrp1 suggests emplacement from south to north.

The 490 to 1,380 ft (150 to 420 m) of section immediately beneath mrp3 on Arctic Ridge is lithologically different than most of the Ambler district, and includes the largest known volcanogenic massive sulfide prospect. An enigmatic chlorite-spotted feldspathic schistose metarhyolite immediately underlies mrp3 and thickens rapidly eastward toward Arctic Creek (to 600 ft; 0 to 185 m thick). This rhyolite, informally called the Winter Creek unit (Schmidt, 1983), contains isolated blocks and pods of calc-schist up to 4 in (10 cm) in size, biotite or brown clay-rich schist, and mafic material (2 in to tens of yards; 5 cm to tens of meters).

The pods are usually rounded with sharp contacts, isolated, and comprise 0.5–20% of the metarhyolite locally and <3% overall.

The calc-schist blocks are interpreted as wall-rock inclusions within the rhyolite. Several of the mafic pods and blocks are surrounded by reaction rims 1 in to 6 ft (2 cm to 2m) wide, and these rims, and local mafic mineral segregations suggest either that the mafic blocks are partially resorbed wall-rock fragments, or were molten and result from magma mixing. Metamafic and biotite-rich rocks are apparently most common in the upper part of the Winter Creek unit; these may have been derived from a xenolith-rich horizon, or alternatively, from hydrothermally altered or weathered mafic ash deposited between episodes of rhyolitic volcanism.

Stop C. An aphanitic rhyolite dome (C; Fig. 4A) between Winter and Falls creeks, preserves millimeter- to centimeter-scale flow banding around its margins. Its position within the thickest part of the Winter Creek unit suggests that it may have been a feeder dome for that tuffaceous rhyolite.

Interfingered to the north and west of, and below, the schistose rhyolite of the Winter Creek unit are two discrete metarhyolite porphyry units with graphitic schist and semimassive sulfide mineralization between them (mrp1 below, and mrp2 above the sulfides). Complex interfingering at the edge of the Winter Creek unit, and the hydrothermal alteration associated with sulfides have obscured primary lithologies and contacts, but mrp2 and mrp3 may represent time-equivalent ash-flow units, which occupied different topographic levels. Both are overlain by mixed aphanitic and schistose rhyolite and both are thin; mrp2 is 50 to 250 ft (15 to 75m), and mrp3 is <100 ft (<30 m) thick. In contrast, mrp1 below mineralization is at least 425 ft (130 m) thick, and contains numerous intervals of graphitic schist as well as abundant monomict breccias of porphyry fragments in a black schist matrix.

Stop D. The Arctic volcanogenic sulfide prospect (Fig. 6) consists of a number of lenses of semimassive sulfides, interfingered with graphitic schists, underlain by chloritic alteration and overlain by pyritic alteration (D; Fig. 4A). This overlying laminated pyrite-calcite-quartz-barite rock produces the orange color anomaly noted near the deposit. Surface exposures of sulfide minerals are limited to several small trenches along the eastern side of the ridge, and several tens of meters below the ridgeline (E; Fig. 4A).

The sulfides and alteration minerals occur along the northwest (down-dropped) side of an apparent syndepositional fault (<820 ft; 250 m offset) controlling fluid flow. To the southeast of this paleofracture, the stratigraphic section is composed of thick chlorite-spotted and porphyroblastic metarhyolite tuffs, which locally contain pumice lapilli (F; Fig. 4A), and cannot easily be correlated with the mrp1/mrp2 sequence. Several garnet-bearing, coarse-grained metabasite plugs (G; Fig. 4A) crop out along the surface projection of this paleo-fracture toward Arctic Creek.

Stop H. Northward from the mineral deposit, metarhyolite porphyry 1 becomes increasingly interlayered with graphitic schist (carbonaceous mud) and monomict autobreccias. At its outcrop in a wide black saddle along the ridgeline, it is composed of more than 90% graphitic schist (H; Fig. 4A). The black schist saddle is underlain northward by a metarhyolitic tuff horizon (100 ft; 30 m thick) and an aphanitic rhyolite flow (150 to 980 ft; 45 to 300 m thick). These metarhyolites are similar in both mineralogy and textures to individual units within the Margarita unit seen above.

Beneath these rhyolites, which mark the end of significant carbonate deposition, lies a 690 to 1,410 ft (210 to 430 m) thick interval of mixed carbonate, pelitic, and carbonaceous metasedimentary rocks, which record input of terrigenous and volcaniclastic detritus into the basin as well as carbonate deposition. The pelitic metasedimentary rocks are gray to brown in color, well foliated, variably graphitic, and often slightly calcareous. They interfinger with, and grade into more pure calc-schists. The calc-schists are light to medium grayish green in color, granoblastic to moderately well foliated, and are composed of calcite, quartz, and white mica with minor sphene. They weather gray to brown, with a pockmarked, salt-and-pepper appearance due to the preferential dissolution of calcite, and usually preserve a structural lineation but no obvious bedding.

Below the mixed metasedimentary rocks is the lowest volcanic horizon in the Ambler sequence on Arctic ridge—an aphanitic metarhyolite flow. This metarhyolite is 150 to 500 ft (45 to 155 m) thick, locally very well flow banded, and is associated with pyritic exhalative(?) quartz and minor graphitic schist (J; Fig. 4A). This lowest volcanic horizon is one along which later gabbroic intrusion was common.

Beneath the lowest metarhyolite, the lowermost part of the Ambler sequence consists of 690 to 1,000 ft (210 to 305 m) of calcareous sedimentary rocks and limestone, now metamorphosed to calc-schist and marble. Several rugose(?) corals of indeterminate age (Hitzman and others, 1982) have been found at Arctic Ridge; the same stratigraphic interval of carbonate rocks farther east yielded a variety of Middle Devonian to Early Mississippian(?) macrofossils (Smith and others, 1978). The presence of crinoids and corals in the calcite marbles suggests fairly shallow water (<330 ft; 100 m) deposition with little terrigenous/volcaniclastic sediment.

Stop K. The lowermost unit of the Ambler sequence along Arctic Ridge is a banded calc-silicate gneiss informally referred to as "gnurgle gneiss" (K; Fig. 4A). This gneiss is composed of thinly laminated (0.2 to 4 in; 0.5 to 10 cm) mono- and bi-mineralic bands of actinolite, garnet, calcite, quartz, and epidote, with minor magnetite, sphene, and clinozoisite. It also contains sporadic disseminated chalcopyrite, pyrite, and sphalerite in irregular pods up to 12 in (30 cm) in diameter. The "gnurgle gneiss" was originally a metal-rich chemical sediment similar to a ferruginous chert or dolomitic marl, and may have formed by exhalative processes related to incipient rifting and preceding active volcanism. The gneiss thins rapidly away from Arctic Ridge, and 3.7 mi (6 km) to the east, angular fragments of gneiss in a deformed matrix of calc-schist are the only evidence of this horizon.

The "gnurgle gneiss" preserves the only examples of cross-folds (egg-crate folds) known from this area. Intrafolial folds, often rootless and with stretched-out limbs, are best preserved in the interlayered marble and calc-schist lithologies that dominate in the lower third of the Ambler sequence. Most of the rhyolitic lithologies behaved as rigid entities during deformation, show little evidence of folding, and only minor brittle fracture.

REFERENCES CITED

Dillon, J. T., Pessel, G. H., Chen, J. H., and Veach, N. C., 1980, Middle Paleozoic magmatism and orogenesis in the Brooks Range, Alaska: Geology, v. 8, p. 338–343.

Hitzman, M. W., Smith, T. E., and Proffett, J. M., 1982, Bedrock geology of the Ambler District, southwestern Brooks Range, Alaska: Alaska State Division Geologic and Geophysical Surveys, Geologic Report no. 75, 2 plates, scale 1:125,000.

Hitzman, M. W., Proffett, J. M., Schmidt, J. J., and Smith, T. E., 1986, Geology and mineralization of the Ambler District, northwestern Alaska: Economic Geology, v. 81, p. 1592–1618.

Schmidt, J. M., 1983, Geology and geochemistry of the Arctic prospect, Ambler District, Alaska: [Ph.D. thesis]: Stanford University, 253 p.

Schmidt, J. M., 1986, Stratigraphic setting and mineralogy of the Arctic volcanogenic massive sulfide prospect, Ambler District, Alaska: Economic Geology,

v. 81, p. 1619–1643.

Sichermann, H. A., Russell, R., and Fikkan, P., 1976, The geology and mineralization of the Ambler District, Alaska: Spokane, unpublished Bear Creek Mining Company manuscript, 22 p.

Smith, T. E., Webster, G. D., Heatwole, D. A., Proffett, J. M., Kelsey, G., and Glavinovich, P. S., 1978, Evidence for mid-Paleozoic depositional age of volcanogenic base-metal massive sulfide occurrences and enclosing strata, Ambler district, northwest Alaska: Geological Society of America Abstracts with Programs, v. 10, p. 148.

Turner, D. L., Forbes, R. B., and Mayfield, C. F., 1978, K-Ar geochronology of the Survey Pass, Ambler River, and eastern Baird Mountains quadrangles, southwestern Brooks Range, Alaska: U.S. Geological Survey Open-File Report 78-254, 41 p.

The Doonerak fenster, central Brooks Range, Alaska

C. G. Mull and J. T. Dillon, *Alaska Division of Geological and Geophysical Surveys, 794 University Avenue, Fairbanks, Alaska 99709*
Karen E. Adams, *Department of Geological Sciences, The University of Alaska, Fairbanks, Alaska 99701*

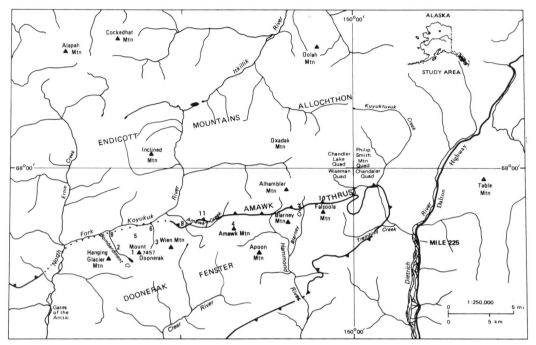

Figure 1. Index map of eastern end of Doonerak fenster showing location of major geographic features and significant geologic localities.

LOCATION AND ACCESSIBILITY

The Doonerak fenster is located in the central Brooks Range about 35 mi (55 km) south of the Endicott Mountains front. It can be traced for at least 70 mi (110 km) southwest from the Dalton Highway (the Trans-Alaska Pipeline haul road) about 50 mi (80 km) north of the village of Coldfoot. The area is located on the Wiseman and Chandalar 1:250,000 quadrangle maps and the Wiseman D1 and D2 and Chandalar D6 1:63,360 quadrangle maps.

The most significant stratigraphic and structural relationships in the fenster and at the base of the overlying Endicott Mountains allochthon are best exposed near Mount Doonerak and in the canyon of Amawk Creek 15 to 22 mi (24 to 35 km) west of the Dalton Highway. These areas are most easily reached by helicopter, or by bush planes, which can be landed on a gravel bar near the junction of Bombardment Creek and the North Fork of the Koyukuk River. The area is within the Gates of the Arctic National Park, and permission for helicopter access must be obtained in advance from the Superintendent, Gates of the Arctic National Park, P.O. Box 74680, Fairbanks, Alaska 99707.

The eastern end of the fenster is accessible by foot from the Dalton Highway from between miles 225 and 226. By wading the

Dietrich River and hiking 5 mi (8 km) to ridges northwest of Kuyuktuvuk Creek and Trembley Creek (an elevation gain of 2,000 to 3,000 ft [600 to 900 m]), a visitor gains access to some of the major stratigraphic units exposed in the fenster and at the base of the allochthon. However, in this area, exposures are not as good, and the structure is much more complicated than in the area of Mount Doonerak and Amawk Creek to the west.

Only the base of the Endicott Mountains allochthon is present in the area of the Doonerak fenster. The upper part of the Upper Devonian clastic sequence and the Mississippian to Triassic rocks on the allochthon are best exposed from the Brooks Range crest at Atigun Pass north to the mountain front, and are most easily reached from the Dalton Highway in the northern Endicott Mountains.

SIGNIFICANCE OF LOCALITY

The Doonerak fenster (Brosge and Reiser, 1971; Dillon and others, 1986) is the most significant locality for understanding the regional structural style and tectonic history of the central Brooks Range. The fenster is a major northeast-southwest–

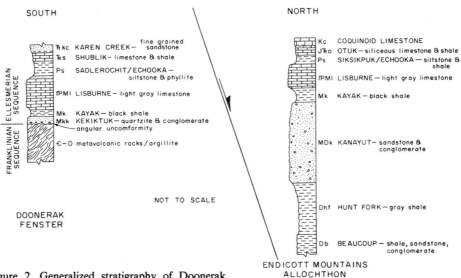

Figure 2. Generalized stratigraphy of Doonerak fenster and Endicott Mountains allochthon.

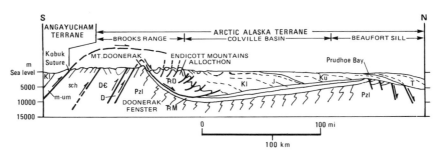

Figure 3. Diagrammatic cross section of Brooks Range and Arctic Slope showing generalized relationship of Doonerak fenster and Endicott Mountains allochthon. T, Tertiary; Ku, Upper Cretaceous; Kl, Lower Cretaceous; J, Jurassic; ℞M, Mississippian to Triassic; ℞D, Devonian to Triassic on Endicott Mountains allochthon; D, Middle and lower Upper Devonian on Endicott Mountains allochthon; Pzl, lower Paleozoic; DЄ, Devonian to Cambrian; sch, schist; m-um, mafic-ultramafic rocks.

trending antiform bounded by the Endicott Mountains allochthon, which to the north forms the northern Endicott Mountains of the central Brooks Range (Fig. 1). The core of the fenster contains a distinctive Lower Mississippian to Upper Triassic section that unconformably overlies lower Paleozoic volcanic and metamorphic rocks (Fig. 2). A similar, coeval stratigraphic sequence is found 150 mi (240 km) to the north and northeast in the Arctic Slope subsurface and in outcrop in the northeastern Brooks Range. Rocks in the fenster are overlain by the Amawk thrust, the sole fault of the Endicott Mountains allochthon (Fig. 3).

Distinctive features that distinguish the stratigraphic sequence on the allochthon from the sequence in the fenster include the following: (1) a thick sequence of Upper Devonian deltaic deposits on the allochthon that conformably underlie Mississippian to Jurassic sediments, in contrast to lower Paleozoic argillite and metavolcanic rocks in the fenster that unconformably underlie Mississippian to Triassic sediments; (2) the presence of siliceous Permian and Triassic strata on the allochthon in contrast to

nonsiliceous coeval silica in the fenster; and (3) the presence of maroon and green shale with associated siderite and barite nodules in the Permian section on the allochthon, in contrast to correlative dark gray to black shale in the fenster.

Restoration of the Endicott Mountains allochthon to south of the Doonerak fenster results in a logical reconstruction of the Mississippian through Triassic depositional basin. A minimum of 55 mi (88 km) of tectonic overlap of the allochthon over the Doonerak fenster is indicated by regional relationships; the actual shortening is at least an order of magnitude greater. Thrust emplacement of the allochthon probably occurred during the Early Cretaceous part of the Brookian orogeny and was followed by Late Cretaceous folding, which formed the Doonerak antiform.

GENERAL STRATIGRAPHY

Two major depositional sequences are present in the Doonerak fenster. The Lower Paleozoic Franklinian sequence consists of low-rank metasedimentary and metavolcanic rocks, overlain by the Ellesmerian sequence of Mississippian to Triassic age shelf-

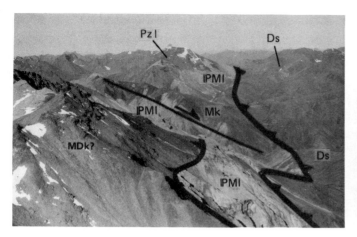

Figure 4. View to the west from Falsoola Mountain along north flank of Doonerak fenster toward Mount Doonerak (center skyline). North-dipping Lisburne Group limestone (ℙM1) and Kayak Shale and Kekiktuk Conglomerate (Mk) are overlain by the Amawk thrust, the sole fault of the Endicott Mountains allochthon. Base of allochthon consists of Beaucoup Formation and Hunt Fork Shale undifferentiated (Ds), and possible Kanayut Conglomerate (MDk?).

Figure 5. View to southwest, of Mount Doonerak, showing high-angle fault separating lower Paleozoic metavolcanic rocks and argillite(Pzl) from Lisburne Group limestone (ℙM1). Throw is more than 1,000 ft (300 m).

carbonate and shallow-marine clastic rocks. A regional unconformity at the base of the Mississippian separates the Ellesmerian sequence from the Franklinian sequence (Armstrong and others 1976). Details of the stratigraphy in the fenster are given by Dutro and others (1976) and by Mull and others (1987).

In contrast to the rocks of the fenster, the Endicott Mountains allochthon consists of a thick Upper Devonian deltaic sequence conformably overlain by Mississippian to Jurassic shelf carbonate and clastic rocks.

STRUCTURE

The generalized structural relationship between the Doonerak fenster and the Endicott Mountains allochthon is illustrated in Figure 3. The Doonerak fenster is defined by major thrust faults that dip away from the axis of the antiform. The Amawk thrust (Fig. 4), which forms the base of the Endicott Mountains allochthon, can be readily traced from the north flank of the eastern Doonerak fenster around the eastern plunge to the south flank of the antiform. On the north flank, northwest of Mount Doonerak, the Ellesmerian stratigraphic sequence is a complete section with relatively uniform regional north dip, broken only by minor faulting. However, to the east, deformation within the Ellesmerian rocks becomes progressively more complex, and is characterized by thrust imbrication and isoclinal folding. Thrust faults have also been identified in the Franklinian sequence; however, these faults may have been pre-Mississippian faults reactivated during the Cretaceous orogeny. In addition, a series of east-west–trending high-angle faults cut the Franklinian and Ellesmerian sequences; some of these faults also cut the overlying Amawk thrust and the lower beds of the Endicott Mountains allochthon.

SIGNIFICANT GEOLOGIC LOCALITIES

The locations of some of the most significant structural and stratigraphic features in the Doonerak fenster are plotted on Figure 1 and are discussed below.

Franklinian sequence. Pre-Mississippian argillite and Cambrian-Ordovician metavolcanic rocks form the main mass of Mount Doonerak (locality 1; Fig. 5), and are also well exposed in the canyon of Bombardment Creek (locality 2) and near the head of an unnamed canyon between Mount Doonerak and Wien Mountain (locality 3).

Kekiktuk Conglomerate and Kayak Shale. Kekiktuk Conglomerate and Kayak Shale are generally buried beneath limestone talus from the overlying Lisburne. However, Kekiktuk unconformably overlying the Franklinian sequence is present in several localities near the head of the unnamed canyon between Mount Doonerak and Wien Mountain (locality 3). Kayak Shale and Kekiktuk Conglomerate are both well exposed on large north-dipping flatirons on the north flank of Amawk Mountain near the head of Amawk Creek (locality 4).

Lisburne Group. Complete sections of the limestone of the Lisburne Group are present in cliff exposures in Bombardment Creek canyon and in the two canyons to the east north of Mount Doonerak (localities 2, 5, and 6; Fig. 5). The Lisburne is also well exposed on the north side of Amawk Creek canyon (locality 7, Fig. 6).

Sadlerochit Group. Fine-grained calcareous sandstone of the Echooka Formation of the Sadlerochit Group is present on a number of north-dipping flatirons north of Mount Doonerak. One of the best exposures of the Echooka and an overlying black phyllitic shale interval is on the north side of Amawk Creek canyon (locality 7; Fig. 6). The trace fossil *Zoophycos,* which is characteristic of the Echooka in northern Alaska, and scattered brachiopods are present at this locality. The Sadlerochit is also

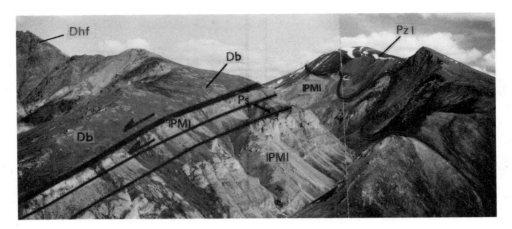

Figure 6. View to the east up Amawk Creek canyon showing good exposure of Sadlerochit Group (Ps) beneath imbricate thrust slice of Lisburne Group limestone (ℙM1). Devonian Hunt Fork Shale (Dhf) and Beaucorp Formation (Db) on lower part of Endicott Mountains allochthon are separated from the Lisburne Group by the Amawk thrust. North-dipping flatirons of Lisburne; Kayak Shale and Kekiktuk Congloemrate depositionally overlie lower Paleozoic rocks (Pzl) on Amawk Mountain (right center skyline.)

well exposed in a narrow gorge near the mouth of Bombardment Creek (locality 8). The upper part of the section, and the contact with the overlying Shublik Formation is readily accessible in the canyon; however, the lower part of the section is difficult to reach in the very narrow upper part of the gorge. The depositional contact of the Echooka with the underlying Lisburne Group carbonates can be examined just above a waterfall at the head of the gorge. This locality may be reached by climbing above the west side of the gorge.

Shublik Formation and Karen Creek Sandstone. The youngest strata in the Ellesmerian sequence in the Doonerak fenster consist of the Shublik Formation and the Karen Creek Sandstone. These units are exposed at only one locality—near the mouth of the canyon of Bombardment Creek (locality 8). The thin-bedded black earthy limestone and shale of the Shublik is exposed on both sides of the canyon, but the Karen Creek, 6 ft (2 m) thick, is exposed only on the east side of the downstream end of the canyon. Phosphatic nodules and the flat pelecypods *Halobia sp.* and *Monotis sp.* are common in the limestone beds of the Shublik.

Amawk thrust. The Amawk thrust is generally not exposed. On the north wall of the Amawk Creek canyon (locality 7; Fig. 6) it is in a covered interval between the gray limestone cliffs of an imbricate thrust sheet of Lisburne and the overlying argillite of the Beaucoup and Hunt Fork formations near the base of the Endicott Mountains allochthon. On a low flatiron just west of the mouth of the Amawk Creek Canyon (locality 9), the Amawk thrust is in a narrow covered interval between the Lisburne Group limestone and a schistose quartzite, which forms the base of the allochthon. The thrust is also present in some places in the valley of Trembley Creek near the east end of the fenster.

High-angle faults. A high-angle fault that cuts the Franklinian sequence and the Lisburne Group is well exposed in the canyons north of Mount Doonerak (localities 3 and 5; Fig. 5). North of Falsoola Mountain, high-angle faults also cut the Amawk thrust (locality 10).

Basal beds on the Endicott Mountains allochthon. Intensely deformed phyllite and schist with abundant quartz veining is present in several river bluff exposures along the North Fork of the Koyukuk River both upstream and downstream from the mouth of Bombardment Creek (locality 8). Thick purple and green argillite of the Beaucoup Formation and dark gray phyllitic shale of the basal Hunt Fork are exposed on the mountain slopes north of the Amawk Creek canyon (locality 11; Fig. 6). Isolated blocks of intensely fractured and quartz-veined schistose quartzite and conglomerate are also present in the lower parts of these slopes. The quartzite and conglomerate may be part of the Beaucoup Formation or may be tectonic blocks derived from an unknown source.

REFERENCES CITED

Armstrong, A. K., Mamet, B. L., Brosgé, W. P., and Reiser, H. N., 1976, Carboniferous section and unconformity at Mount Doonerak, Brooks Range, northern Alaska: American Association of Petroleum Geologists Bulletin, v. 60, no. 6, p. 962–972.

Brosgé, W. P., and Reiser, H. N., 1971, Preliminary bedrock geologic map, Wiseman and eastern Survey Pass quadrangles, Alaska: U.S. Geological Survey Open-File Map OF-71-56, scale 1:250,000, 2 sheets.

Dillon, J. T., Brosgé, W. P., and Dutro, J. T., Jr., 1986, Generalized geologic map of the Wiseman Quadrangle, Alaska: U.S. Geological Survey Open-File Map OF—86-219, scale 1:250,000, 1 sheet.

Dutro, J. T., Jr., Brosgé, W. P., Lanphere, M. A., and Reiser, H. N., 1976, Geologic significance of the Doonerak structural high, central Brooks Range, Alaska: American Association of Petroleum Geologists Bulletin, v. 60, no. 6, p. 952–961.

Mull, C. G., Adams, K. E., and Dillon, J. T., 1987, Stratigraphy and structure of the Doonerak fenster and Endicott Mountains allochthon, central Brooks Range, Alaska, in Tailleur, I. L., and Weimer, P., eds., Alaskan North Slope geology: Society of Economic Paleontologists and Mineralogists, Pacific Section (in press).

Geologic features of Ignek Valley and adjacent mountains, northeastern Alaska

C. M. Molenaar, U.S. Geological Survey, Box 25046, MS 940, Denver Federal Center, Denver, Colorado 80225
C. G. Mull, Alaska Division of Geological and Geophysical Surveys, Pouch 7028, Anchorage, Alaska 99510
D. A. Swauger, Arco Alaska, Inc., P.O. Box 100360, Anchorage, Alaska 99510

LOCATION AND ACCESSIBILITY

Ignek Valley lies along the northernmost salient of the Brooks Range front in the Arctic National Wildlife Refuge in northeastern Alaska (Figs. 1 and 2). Lying between the Shublik Mountains on the south and the Sadlerochit Mountains on the north, Ignek Valley includes the entire drainage of Ignek Creek and the headwaters of the Katakturuk River, which drains northward through the Sadlerochit Mountains as an antecedent river. The entire area is in the Mt. Michelson 1:250,000-scale Quadrangle or on the Mt. Michelson C-3 and C-4 1:63,360-scale quadrangles.

Access to the area is easiest by helicopter, which can be chartered at Deadhorse (Prudhoe Bay), 85 mi (137 km) to the northwest. In addition, bush planes have landed on nearby gravel bars and terraces, or at Schrader Lake 15 mi (24 km) southeast of Ignek mesa (informal name). Charter air service is available at Deadhorse, or Kaktovik, 60 mi (96 km) to the northeast. Because Ignek Valley is within the Arctic National Wildlife Refuge, permission to use helicopters or to conduct geological investigations within the refuge must be obtained by writing to the Refuge Manager, Arctic National Wildlife Refuge, Box 20, 101 12th Avenue, Fairbanks, Alaska 99701.

SIGNIFICANCE OF SITE

Ignek Valley and the adjacent Sadlerochit and Shublik Mountains is a prime area for stratigraphic, sedimentologic, and structural studies (Fig. 1). This locality has some of the best exposures of the Proterozoic through Cretaceous part of the stratigraphic section that underlies the North Slope including the formations that are the producing reservoirs at the giant Prudhoe Bay oil field (Fig. 2). Lithologies and depositional environments range from nonmarine rocks to shelf sandstones, shales, and carbonates, to deep-water turbidites and distal, condensed sections of euxinic organic-rich shale and bentonite. In addition, compressionally folded and faulted structures are well exposed in the mountains flanking the valley.

GENERAL STRATIGRAPHY

The stratigraphic section in northeastern Alaska consists of three depositional sequences; the Franklinian sequence of Precambrian to Middle Devonian age, the Ellesmerian sequence of Mississippian to Early Cretaceous (Barremian) age, and the Brookian sequence of Early Cretaceous (Aptian?–Albian) to

Figure 1. Aerial view to west of Ignek Valley. Resistant bed in valley bottom is Kemik Sandstone, which unconformably overlies Kingak Shale (covered). Light-toned outcrops in valley are bentonitic shales of Hue Shale. Sadlerochit Mountains are on right and Shublik Mountains are on left.

Quaternary age. Parts of all three sequences are exposed in Ignek Valley and the adjacent mountains. Figure 3 shows the stratigraphic section exposed in this area.

Throughout most of the North Slope and Brooks Range, the Franklinian sequence consists of low-rank metamorphic rocks, but in the Sadlerochit and Shublik Mountains, a 13,000-ft-(4,000-m-) thick section of carbonate rocks is present. The Ellesmerian sequence, which unconformably overlies the Franklinian sequence, consists dominantly of shelf carbonate rocks and shallow-marine, mineralogically mature sandstone and shale. Deposition occurred along a slowly subsiding continental margin in which the land area was to the north and the seaway was to the south. Uplift and erosion of the rifted continental margin of northernmost Alaska in Early Cretaceous (Neocomian) time was followed by subsidence and northward transgression of the sea during which the Kemik Sandstone and pebble shale unit were deposited. This ended the northern-source Ellesmerian depositional cycle. The producing reservoirs in the Prudhoe Bay area oil fields are in Ellesmerian rocks, primarily sandstones of the Ivishak Formation and lesser amounts from carbonates of the Lisburne Group, the Kekiktuk Conglomerate, the Karen Creek Sandstone (Sag River Sandstone), and equivalents of the Kemik Sandstone.

Following subsidence of the northern land area, all subsequent deposition, the Brookian sequence, was derived from the ancestral Brooks Range orogenic belt to the south and southwest. Regionally, the Brookian sequence consists of more than 12,000

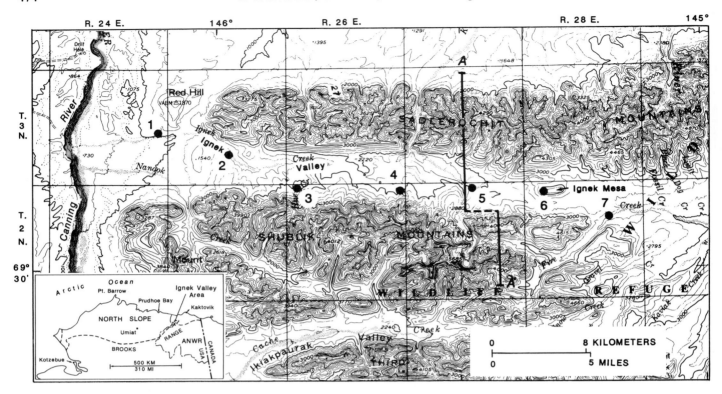

Figure 2. Topographic map of Ignek Valley area (reduced from U.S. Geological Survey Mt. Michelson Quadrangle map, 1:250,000 scale, and inset index map. Contour interval is 200 ft. Numbers refer to localities of some of the geologic features discussed.

ft (3,660 m) of northeasterly and northerly prograding deep-marine to nonmarine diachronous facies (Molenaar and others, 1987). Only the lower 1,800 ft (550 m) of the Brookian sequence is preserved in Ignek Valley (Fig. 3).

For additional stratigraphic details of pre-Mississippian rocks, see Dutro (1970), and Blodgett and others (1986); for Mississippian and Pennsylvanian rocks, see Brosgé and others (1962), and Armstrong and Mamet (1974, 1975); for Permian through Jurassic rocks, see Detterman and others (1975); for Cretaceous and Tertiary rocks, see Molenaar (1983), and Molenaar and others (1987). Mull (1987) discussed the Kemik Sandstone (or Formation) and Bird and Molenaar (1987) discussed the entire stratigrahic section. For geologic maps of the area, see Reiser and others (1970, 1971) and Bader and Bird (1986).

STRUCTURE

Ignek Valley is a synclinal valley between two doubly plunging, north-verging, faulted anticlinal mountain blocks, the Shublik Mountains on the south and the Sadlerochit Mountains on the north (Figs. 1 and 2). The structures are the result of compressional tectonics involving thrusting (Fig. 4). Leiggi and Russell (1985) suggested that deformation occurred between the mid-Eocene and the present with possibly two deformational events.

The structural detachments or sole thrusts are thought to be in the Kingak Shale, along the pre-Mississippian unconformity (or Kayak Shale), and within the pre-Mississippian section (Kelley and Molenaar, 1985). Field relations along the north side of the Sadlerochit Mountains suggest that the Sadlerochit and Shublik Mountains are probably late Tertiary structures that have less horizontal displacement than earlier thrust slices (Kelley and Foland, 1987).

GEOLOGIC FEATURES AND SIGNIFICANT LOCALITIES

Significant exposures of stratigraphic units and structural features are discussed starting with more obvious features in the valley; some localities are numbered from west to east for orientation purposes on Figure 2. Figure 5 shows photographs of some of the features.

Hue Shale (Localities 1 and 3). The conspicuous outcrops that make Ignek Valley so colorful are red-weathering tuff beds in the Hue Shale (Fig. 3). Ignek is the native word for "fire," and Ignek Creek was named either for these red-weathering beds, or for a former occurrence of naturally burning organic-rich shale at the western end of the valley. The natives reported that these beds were still smoking when their ancestors came into the country

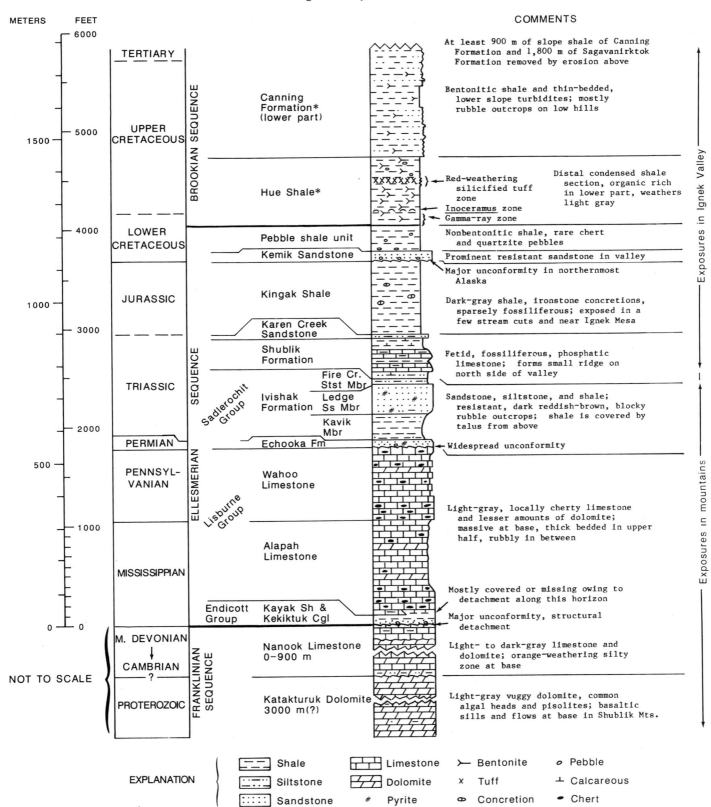

Figure 3. Columnar section of rocks exposed in Ignek Valley and adjacent mountains. *Hue Shale and Canning Formation terminology was recently proposed by Molenaar and others (1987). These units were formerly referred to as the Colville Group.

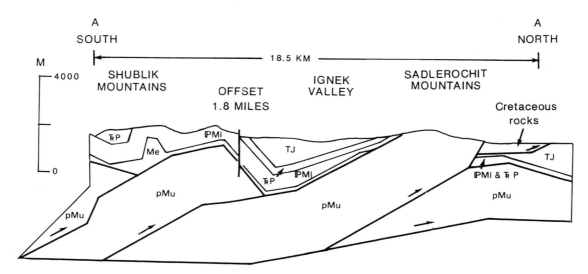

Figure 4. Structural cross section across Ignek Valley. Structural shortening is approximately 27%. pMu = pre-Mississippian rocks (Nanook Limestone, Katakturuk Dolomite, and Neruokpuk Quartzite); Me = Endicott Group (mainly Kayak Shale); ℙMl = Lisburne Group; ℞P = Triassic and Permian rocks (Shublik Formation and Sadlerochit Group); TJ = Tertiary to Jurassic rocks. Modified from J. S. Namson and W. K. Wallace, ARCO Oil and Gas Company, unpublished report. Location of section is shown in Figure 2.

(Leffingwell, 1919). Red and black clinkers in the lower part of the Hue Shale along the north bank of Ignek Creek (locality 1 near the center of Sec.24,T3N,R24E) are probably the result of this burn.

The red-weathering zone is 65 to 100 ft (20 to 30 m) thick and consists of interbedded black shale, bentonite, and silicified tuff. The tuff weathers bright red in subdued rubble outcrops on hills. The light tones of other subdued outcrops of Hue Shale in Ignek Valley are due to its bentonitic character. The most complete section of Hue Shale is exposed at its type section on the west side of Hue Creek (locality 3—NE¼Sec.6,T2N,R26E) (Fig. 5A).

Kemik Sandstone (Localities 3, 5, and 6). The Kemik Sandstone is a conspicuous resistant unit that can be traced as a nearly continuous rubble-covered ridge the entire length of Ignek Valley. It also caps Ignek mesa at the east end of the valley. Its outcrop pattern delineates two or three folds in the eastern part of the valley (Fig. 1). In some areas, the Kemik appears as two separate sandstone units, but most, if not all, of these duplications are due to repetition by small thrust faults. Complete sections of the Kemik, including the regional Lower Cretaceous unconformity at the base, are exposed along Ignek Creek east of locality 1, along Hue Creek (locality 3), and near the head of the Katakturuk River at locality 5 near the center of N½Sec.3,T2N,R27E.

Pebble Shale Unit (Localities 2, 3, and 5). The pebble shale unit, which overlies the Kemik Sandstone, contains frosted-quartz grains and rare pebbles and cobbles throughout the unit. It is well exposed on the west side of Hue Creek (locality 3) (Fig. 5A), along the southwest side of Ignek Creek (locality 2 in the center of W½Sec.27,T3N,R25E), and in a small gully on the

southwest side of the Katakuruk River at locality 5. A concentration of pebbles also occurs at the base of the shale unit.

Canning Formation (Localities 3 and 4). Turbidites of the Canning Formation generally form rubble outcrops near the synclinal parts of the valley, but in-place outcrops are present on the west side of Hue Creek (locality 3) and along the creek at locality 4 in the NE¼Sec.1,T2N,R26E. The youngest bedrock exposure in the valley is possibly Paleocene in age. It is a thin-bedded turbidite sequence located on the east side of a rib separating a tributary from Hue Creek just east of locality 3 in NW¼Sec.5,T2N,R26E. This outcrop also shows the asymmetric Ignek Valley synclinal axis.

Kingak Shale (Localities 3, 5, and 6). Exposures of Kingak Shale are rare in Ignek Valley. However, fair exposures of parts of the formation are found in bare slopes below the northwest and southwest side of Ignek mesa (locality 6). Scattered ammonites occur in red-weathering ironstone concretions at these localities. In addition, stream-cut exposures are located in the SE¼Sec.3 and NE¼Sec.11,T2N,R27E (south and southeast of locality 5) and along Hue Creek (locality 3).

North Flank of Shublik Mountains. A thrust fault along the north side of the west half of the Shublik Mountains juxtaposes south-dipping Proterozoic Katakturuk Dolomite against steeply dipping to overturned Jurassic and Cretaceous rocks (Fig. 5A), a stratigraphic separation of about 10,000 ft (3,000 m). Fault slivers of Shublik Formation and Kingak Shale occur in a few places along the fault zone. In the SE¼Sec.1,T2N,R26E, the Katakturuk Dolomite is in thrust contact with either the pebble shale unit or the lower part of the Hue Shale. To the east, the thrust fault cuts into the Shublik Mountains, and the bounding mountain front is a steeply dipping, north-verging, fault-bend fold

Figure 5. Photographs of outcrops: A: Aerial view to southwest of Shublik Mountains and Hue Creek. South-dipping Katakturuk Dolomite of Shublik Mountains is thrust against Mesozoic rocks. A complete Cretaceous section is exposed on west side of Hue Creek in center of photo. Pebble shale unit (overturned) is dark outcrop, and Hue Shale is light-toned outcrop. Tight syncline in lower part of photo is in Paleocene(?) turbidites of Canning Formation. B: Aerial view to north of Katakturuk River canyon through Sadlerochit Mountains. All strata are dipping toward viewer and range from the red-weathering tuff of the Hue Shale at bottom to Katakturuk Dolomite in canyon. C: Aerial view to southeast over Nanook Creek and west end of Shublik Mountains. Dark-colored rocks are gabbroic and basaltic sills or flows interlayered in lower part of Katakturuk Dolomite. Distant glacier-clad peak is Mt. Chamberlin, the highest peak in the Brooks Range (elev. 9,020 ft; 2,750 m). D: View to east in western part of Sadlerochit Mountains of angular unconformity or structural detachment surface between Lisburne Group above and Katakturuk Dolomite (Secs. 11, 12, T3N, R25E). Missing Endicott Group may have been removed by faulting or was never deposited. E: View northeast down Fire Creek showing section from upper part of Ledge Sandstone Member of Ivishak Formation to lower part of Kingak Shale. South flank of eastern Sadlerochit Mountains is in background.

made up of the Lisburne Group and overlying Sadlerochit Group (Fig. 4).

South Flank of Sadlerochit Mountains. The north side of Ignek Valley is bounded by the 20–25-degree dip slope on the top of the resistant Sadlerochit Group, which forms the south flank of the Sadlerochit Mountains. The Shublik Formation crops out discontinuously along small ridges and spurs at the base of the dip slope. Good exposures of the Shublik are rare, but part of the formation is well exposed along a small stream west of Ignek Mesa in the NW¼Sec.1,T2N,R27E (between localities 5 and 6). The Shublik is abundantly fossiliferous; the flat pelecypods *Halobia* and *Monotis* are common.

High on the south flank of the Sadlerochit Mountains, the contact of the Sadlerochit and the Lisburne Groups is conspicuously marked by the sharp contrast between the light-gray limestone of the Lisburne and the overlying dark-reddish-brown–weathering clastics of the Sadlerochit Group; the color is due to oxidation of pyrite (Fig. 1). This obvious contact marks a widespread unconformity in which, in this area, Upper Pennsylvanian and Lower Permian rocks are missing. Locally, the smooth contact is interrupted by a channel cut, such on the north side of a tributary of the Katakturuk River in the SE¼Sec.25,T3N,R27E (northeast of locality 5), where a channel is filled with a chert breccia. In that same area, nearly complete sections of Lisburne and Sadlerochit Groups are present.

Pre-Mississippian Unconformity and Detachment Surface. The angular unconformity or detachment plane that separates the Franklinian sequence from the overlying Ellesmerian sequence is visible on peaks and ridges of the Sadlerochit Mountains and south of the frontal ridge of the Shublik Mountains. Because both the Franklinian and Ellesmerian rocks strike east-west, the angular relations are better displayed on north-south exposures. This discordance can be seen from a distance in many areas near the crest of the Sadlerochit Mountains (Fig. 5D) and in the western Shublik Mountains. In these areas, a thin Endicott Group (mostly Kayak Shale) or carbonates of the Lisburne Group overlie the unconformity. However, there has been structural detachment of unknown displacement along the unconformity surface (Kelley and Molenaar, 1985; Leigge and Russell, 1985). The detachment may account for the missing or thin Endicott Group in many areas.

Fire Creek Section (Locality 7). A final location that should be noted includes excellent exposures along the gorge of Fire Creek on the eastern plunge of the Shublik Mountains near the intersections of Secs.11,14,15,T2N,R28E (locality 7), 3 mi (5 km) southeast of Ignek mesa. The exposed interval includes the upper part of the Lisburne Group, the Sadlerochit Group, the Shublik Formation, the Karen Creek Sandstone, and the lower part of the Kingak Shale (Fig. 5E). Parts of the gorge through the Sadlerochit are so narrow that they can not be traversed.

REFERENCES CITED

Armstrong, A. K., and Mamet, B. L., 1974, Carboniferous biostratigraphy, Prudhoe Bay State 1 to northeastern Brook Range, Arctic Alaska: American Association of Petroleum Geologists Bulletin, v. 58, no. 4, p. 646–660.

—— , 1975, Carboniferous biostratigraphy, northeastern Brooks Range, Arctic Alaska: U.S. Geological Survey Professional Paper 884, 29 p.

Bader, J. W., and Bird, K. J., 1986, Geologic map of the Demarcation Point, Mt. Michelson, Flaxman Island, and Barter Island quadrangles, Alaska: U.S. Geological Survey Miscellaneous Field Investigations Map I-1791, scale 1:250,000.

Bird, K. J., and Molenaar, C. M., 1987, Stratigraphy, in Bird, K. J., and Magoon, L. B., eds., Petroleum geology of the northern part of the Arctic National Wildlife Refuge, northeastern Alaska: U.S. Geological Survey Bulletin 1778 (in press).

Blodgett, R. B., Clough, J. G., Dutro, J. T., Jr., Ormiston, A. R., Palmer, A. R., and Taylor, M. E., 1986, Age revisions of the Nanook Limestone and Katakturuk Dolomite, northeastern Brooks Range, Alaska, in Bartsch-Winkler, S., and Reed, K. M., eds., Geologic studies in Alaska by the U.S. Geological Survey during 1985: U.S. Geological Survey Circular 978, p. 5–10.

Brosgé, W. P., Dutro, J. T., Jr., Mangus, M. D., and Reiser, H. N., 1962, Paleozoic sequence in eastern Brooks Range, Alaska: American Association of Petroleum Geologists Bulletin, v. 46, no. 12, p. 2174–2198.

Detterman, R. L., Reiser, H. N., Brosgé, W. P., and Dutro, J. T., Jr., 1975, Post-Carboniferous stratigraphy, northeastern Alaska: U.S. Geological Survey Professional Paper 886, 46 p.

Dutro, J. T., Jr., 1970, Pre-Carboniferous carbonate rocks, northeastern Alaska, in Adkison, W. L., and Brosgé, M. M., eds., Proceedings of the geological seminar on the North Slope of Alaska: Los Angeles, California, American Association of Petroleum Geologists, Pacific Section, p. M1–M8.

Kelley, J. S., and Foland, R. L., 1987, Structural interpretation of surface studies and multichannel seismic profiles, in Bird, K. J., and Magoon, L. B., eds.,

Petroleum geology of the northern part of the Arctic National Wildlife Refuge, northeastern Alaska: U.S. Geological Survey Bulletin 1778 (in press).

Kelley, J. S., and Molenaar, C. M., 1985, Detachment tectonics in Sadlerochit and Shublik Mountains and applications for exploration beneath coastal plain, Arctic National Wildlife Range, Alaska [abs.]: American Association of Petroleum Geologists Bulletin, v. 69, no. 4, p. 667.

Leffingwell, E. de K., 1919, The Canning River region, northern Alaska: U.S. Geological Survey Professional Paper 109, 251 p.

Leiggi, P. A., and Russell, B. J., 1985, Style and age of tectonism of Sadlerochit Mountains to Franklin Mountains, Arctic National Wildlife Refuge (ANWR), Alaska [abs.]: American Association of Petroleum Geologists Bulletin, v. 69, no. 4, p. 668.

Molenaar, C. M., 1983, Depositional relations of Cretaceous and Lower Tertiary rocks, northeastern Alaska: American Association of Petroleum Geologists Bulletin, v. 67, no. 7, p. 1066–1080.

Molenaar, C. M., Bird, K. J., and Kirk, A. R., 1987, Cretaceous and Tertiary stratigraphy of northeastern Alaska, in Tailleur, I. L., and Weimer, P., eds., Alaskan North Slope geology: Pacific Section, Society of Economic Paleontologists and Mineralogists and Alaska Geological Society, v. 50 (in press).

Mull, C. G., 1987, Kemik Formation, Arctic National Wildlife Refuge, northeastern Alaska, in Tailleur, I. L., and Weimer, P., eds., Alaskan North Slope geology: Pacific Section, Society of Economic Paleontologists and Mineralogists and Alaska Geological Society, v. 50 (in press).

Reiser, H. N., Dutro, J. T., Jr., Brosgé, W. P., Armstrong, A. K., and Detterman, R. L., 1970, Progress map, geology of the Sadlerochit and Shublik Mountains, Mt. Michelson C-1, C-2, C-3, and C-4 quadrangles, Alaska: U.S. Geological Survey Open-File Report 70-273, scale 1:63,360.

Reiser, H. N., Brosgé, W. P., Dutro, J. T., Jr., and Detterman, R. L., 1971, Preliminary geologic map, Mt. Michelson quadrangle, Alaska: U.S. Geological Survey Open-File Report 71-237, scale 1:200,000.

Index

(Italic page numbers indicate major references)

Typeset by WESType Publishing Services, Inc., Boulder, Colorado
Printed in U.S.A. by Malloy Lithographing, Inc., Ann Arbor, Michigan

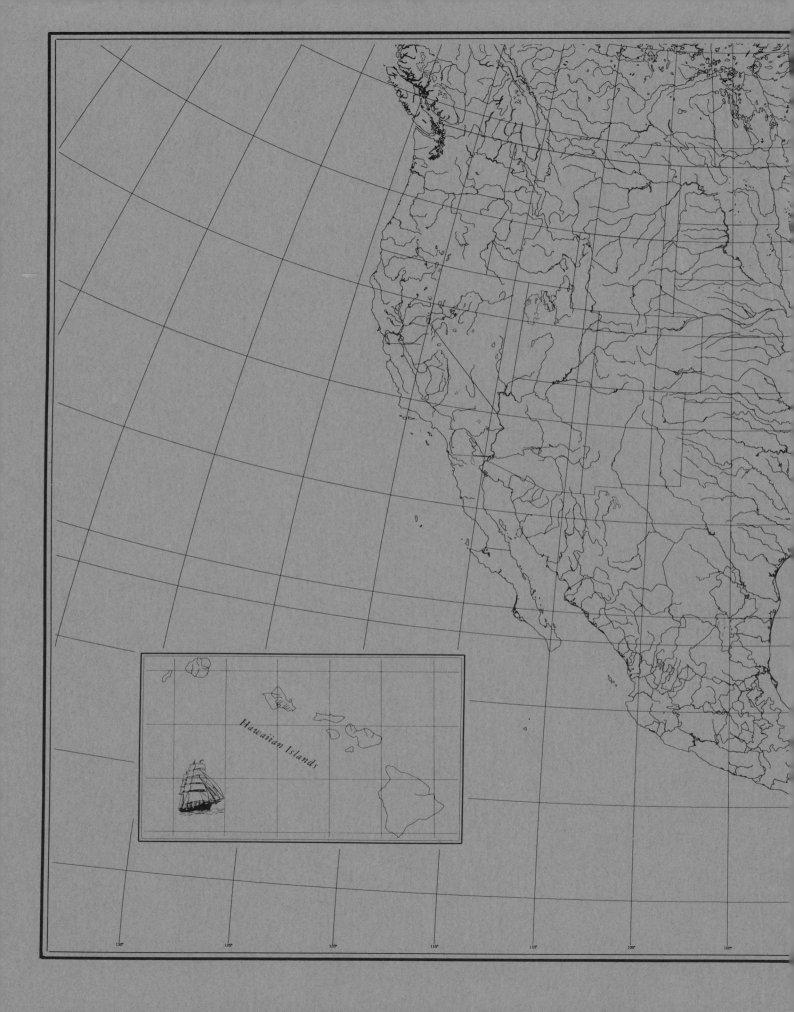

Hawaiian Islands